T0344759

MICROBIAL BIOFILMS

SECOND EDITION

MICROBIAL BIOFILMS

SECOND EDITION

Editors

Mahmoud Ghannoum
Matthew Parsek
Marvin Whiteley
Pranab K. Mukherjee

Washington, DC

Library of Congress Cataloging-in-Publication Data

Microbial biofilms / editors, Mahmoud Ghannoum, Case Western University, Cleveland, OH; Matthew Parsek, University of Washington, Seattle, WA; Marvin Whiteley, University of Texas at Austin, Austin, TX; Pranab K. Mukherjee, Case Western University, Cleveland, OH. -- 2nd edition.
pages cm
Includes bibliographical references and index.
ISBN 978-1-55581-745-9 (alk. paper) -- ISBN 978-1-55581-746-6 (alk. paper) 1. Biofilms. I. Ghannoum, Mahmoud A. (Mahmoud Afif), editor. II. Parsek, Matthew R., editor. III. Whiteley, Marvin, editor. IV. Mukherjee, Pranab K., editor.
QR100.8.B55M5252 2015
579'.17--dc23
2015030735

ISBN 978-1-55581-745-9
e-ISBN 978-1-55581-746-6
doi:10.1128/9781555817466

Printed in the United States of America

10 9 8 7 6 5 4 3 2 1

Address editorial correspondence to: ASM Press, 1752 N St., N.W., Washington, DC 20036-2904, USA.

Send orders to: ASM Press, P.O. Box 605, Herndon, VA 20172, USA.

Phone: 800-546-2416; 703-661-1593. Fax: 703-661-1501.

E-mail: books@asmusa.org

Online: http://estore.asm.org.

Contents

Contributors

Tatsuya Akiyama
Center for Biofilm Engineering
Department of Microbiology and Immunology
Montana State University-Bozeman
Bozeman, MT 59717

David R. Andes
Department of Medicine and Department of Medical Microbiology and Immunology
University of Wisconsin
Madison, WI 53706

Nicolas Barraud
Centre for Marine Bio-Innovation and School of Biotechnology and Biomolecular Sciences
The University of New South Wales
Sydney, NSW 2052, Australia

Anne Beauvais
Unité des Aspergillus
Institut Pasteur
75015 Paris, France

Cécile Berne
Department of Biology
Jordan Hall JH142
Indiana University
Bloomington, IN 47405

Brian Bothner
Center for Biofilm Engineering
Department of Chemistry and Biochemistry
Montana State University-Bozeman
Bozeman, MT 59717

Yves V. Brun
Department of Biology
Jordan Hall JH142
Indiana University
Bloomington, IN 47405

Arturo Casadevall
Department of Molecular Microbiology and Immunology
Johns Hopkins Bloomberg School of Public Health
Baltimore, MD 21205

Jyotsna Chandra
Center for Medical Mycology and Mycology Reference Laboratory
Department of Dermatology
University Hospitals of Cleveland and Case Western Reserve University
Cleveland, OH 44106

Connie Chang
Center for Biofilm Engineering
Department of Chemical and Biological Engineering
Montana State University-Bozeman
Bozeman, MT 59717

Matthew R. Chapman
Department of Molecular, Cellular, and Developmental Biology
University of Michigan
Ann Arbor, MI 48109

William H. DePas
Department of Microbiology and Immunology
University of Michigan
Ann Arbor, MI 48109

Jigar V. Desai
Department of Biological Sciences
Carnegie Mellon University
Pittsburgh, PA 15213

Adrien Ducret
Department of Biology
Jordan Hall JH142
Indiana University
Bloomington, IN 47405

Jiunn N. C. Fong
Department of Microbiology and Environmental Toxicology
University of California, Santa Cruz
Santa Cruz, CA 95064

Michael J. Franklin
Center for Biofilm Engineering
Department of Microbiology and Immunology
Montana State University-Bozeman
Bozeman, MT 59717

Jean-Marc Ghigo
Institut Pasteur
Unité de Génétique des Biofilms
Département de Microbiologie
F-75015 Paris, France

Michael Givskov
Singapore Centre on Environmental Life Sciences Engineering
Nanyang Technological University
Singapore
Costerton Biofilm Center
Department of International Health, Immunology, and Microbiology
University of Copenhagen
2200 København N, Denmark

Dae-Gon Ha
Departments of Microbiology and Immunology
Geisel School of Medicine at Dartmouth
Hanover, NH 03755

Gail G. Hardy
Department of Biology
Jordan Hall JH142
Indiana University
Bloomington, IN 47405

David A. Hufnagel
Department of Molecular, Cellular, and Developmental Biology
University of Michigan
Ann Arbor, MI 48109

Christopher J. Jones
Department of Microbiology and Environmental Toxicology
University of California, Santa Cruz
Santa Cruz, CA 95064

Aspasia Katragkou
Infectious Diseases Unit
3rd Department of Pediatrics
Faculty of Medicine
Aristotle University School of Health Sciences
Hippokration Hospital
54642 Thessaloniki, Greece
Transplantation-Oncology Infectious Diseases Program
Weill Cornell Medical Center of Cornell University
New York, NY 14850

Staffan Kjelleberg
Centre for Marine Bio-Innovation and School of Biotechnology and Biomolecular Sciences
The University of New South Wales
Sydney, NSW 2052, Australia
Singapore Centre on Environmental Life Sciences Engineering, and the School of Biological Sciences
Nanyang Technological University
Singapore 639798

Roberto Kolter
Department of Microbiology and Immunobiology
Harvard Medical School
Boston, MA 02115

Jean-Paul Latgé
Unité des Aspergillus
Institut Pasteur
75015 Paris, France

Dominique H. Limoli
Department of Microbial Infection and Immunity
Ohio State University
Columbus, OH 43210

José L. López-Ribot
Department of Biology
South Texas Center for Emerging Infectious Diseases
The University of Texas at San Antonio
San Antonio, TX 78249

Luis R. Martinez
Department of Biomedical Sciences
College of Osteopathic Medicine
New York Institute of Technology
Old Westbury, NY 11568

Aaron P. Mitchell
Department of Biological Sciences
Carnegie Mellon University
Pittsburgh, PA 15213

Pranab K. Mukherjee
Center for Medical Mycology and Mycology Reference Laboratory
Department of Dermatology
University Hospitals of Cleveland and Case Western Reserve University
Cleveland, OH 44106

Jeniel E. Nett
Department of Medicine and Department of Medical Microbiology and Immunology
University of Wisconsin
Madison, WI 53706

George A. O'Toole
Departments of Microbiology and Immunology
Geisel School of Medicine at Dartmouth
Hanover, NH 03755

Christopher G. Pierce
Department of Biology
South Texas Center for Emerging Infectious Diseases
The University of Texas at San Antonio
San Antonio, TX 78249

Anand K. Ramasubramanian
Department of Biomedical Engineering
South Texas Center for Emerging Infectious Diseases
The University of Texas at San Antonio
San Antonio, TX 78249

Olaya Rendueles
Institute for Integrative Biology
ETH Zürich
8092 Zürich, Switzerland

Scott A. Rice
Centre for Marine Bio-Innovation and School of Biotechnology and Biomolecular Sciences
The University of New South Wales
Sydney, NSW 2052, Australia
Singapore Centre on Environmental Life Sciences Engineering, and the School of Biological Sciences
Nanyang Technological University
Singapore 639798

Emmanuel Roilides
Infectious Diseases Unit
3rd Department of Pediatrics
Faculty of Medicine
Aristotle University School of Health Sciences
Hippokration Hospital
54642 Thessaloniki, Greece

Maria Simitsopoulou
Infectious Diseases Unit
3rd Department of Pediatrics
Faculty of Medicine
Aristotle University School of Health Sciences
Hippokration Hospital
54642 Thessaloniki, Greece

Anand Srinivasan
Department of Biomedical Engineering
South Texas Center for Emerging Infectious Diseases
The University of Texas at San Antonio
San Antonio, TX 78249

Philip S. Stewart
Center for Biofilm Engineering
Montana State University-Bozeman
Bozeman, MT 59717

Tim Tolker-Nielsen
Department of Immunology and Microbiology
Faculty of Health and Medical Sciences
University of Copenhagen
DK 2000 Copenhagen, Denmark

Jordi van Gestel
Department of Microbiology and Immunobiology
Harvard Medical School
Boston, MA 02115

Hera Vlamakis
Department of Microbiology and Immunobiology
Harvard Medical School
Boston, MA 02115

Thomas J. Walsh
Transplantation-Oncology Infectious Diseases Program
Weill Cornell Medical Center of Cornell University
New York, NY 14850

Daniel J. Wozniak
Department of Microbial Infection and Immunity
Ohio State University
Columbus, OH 43210

Liang Yang
Singapore Centre on Environmental Life Sciences Engineering
Nanyang Technological University
Singapore
School of Biological Sciences
Nanyang Technological University
Singapore 639798

Fitnat H. Yildiz
Department of Microbiology and Environmental Toxicology
University of California, Santa Cruz
Santa Cruz, CA 95064

Preface

It is hard to imagine that 10 years have already passed since the first edition of *Microbial Biofilms* was published. The success of the first edition of this book prompted the American Society for Microbiology to commission this second edition. Unlike the first edition, when the field was finding its way and gathering strength to become an area of intense interest that spurred a decade of breathtaking research, now we are in a period in which the fruits of this work are being harvested. While in the first monograph the majority of the research was discovery-based, which is the case in most research areas at their inception, the findings of the last decade have resulted in an in-depth understanding of the molecular mechanisms underlying the biology and antimicrobial resistance of biofilms and their pathogenesis. Furthermore, proteomic and genomic approaches identified targets that can be exploited in our discovery efforts of antibiofilm therapeutic approaches, as well as tools that will aid these efforts. We also witnessed efforts at expanding the findings from basic research into translational applications, aimed at preventing and treating biofilm-associated diseases.

In spite of tremendous scientific advances, significant hurdles remain. More research is needed to study mixed-species biofilms and link the complex environments of biofilms and the microbial (bacterial, fungal, and viral) biome. Although reductive analyses based on single genes and proteins, as well as systems biology approaches, have identified several key modulators of biofilm growth, a unified mechanistic model that encompasses the role of these modulators and explains the distinct biology and pathogenesis of biofilms is still lacking. Cross-network analyses of these findings should go a long way toward creating such a model. Moreover, systematic analyses of these interactions may allow us to exploit the novel features of microbial biofilms to discover potent antibiofilm agents. A prerequisite for attaining this goal is the development of standardized methods to evaluate biofilm growth and antibiofilm activity as well as a more thorough understanding of *in vivo* biofilms. Standardized methods will pave the way for the development of new therapeutic strategies that can be tested in relevant *in vivo* biofilm models. Our efforts to discover therapeutic approaches by translating the scientific findings of the last decade should be accelerated, together with discovery and development of devices that prevent or inhibit biofilm formation.

In this edition, some chapters are updated by established investigators based on recent findings, while others are by new contributors, providing unique and fresh insights into the field. This book will take the reader on an exciting journey of bacterial and fungal biofilms, ranging all the way from basic molecular interactions to innovative therapies, with stops along the way to examine the division of labor in biofilms, new approaches to combat the threat of microbial biofilms, and how biofilms evade the host defense. We would like to extend our thanks to all contributors, without whom completion of this exciting monograph would not have been possible. We thank them for their patience during the publication process and for giving up their valuable time to meet our deadlines. It is our hope that this volume will be of interest to basic science researchers and clinicians, and that it may provide inspiration for graduates and postgraduates to be attracted to the field of microbial biofilms.

Mahmoud Ghannoum
Matthew Parsek
Marvin Whiteley
Pranab K. Mukherjee

Acknowledgments

First off, we would like to acknowledge the authors. Under the best of circumstances, writing a book chapter is a lot of work. The authors have produced a collection of chapters of great breadth and depth. We thank them for supporting this publication and for their patience and efforts bringing it to fruition.

Since the publication of the first ASM biofilms book in 2004, the field has lost two of its great leaders, Bill Costerton and Peter Gilbert. We would like to pay special tribute to these individuals. Both were founding figures in biofilm microbiology, carrying out innovative, pioneering research. Both laid the early groundwork that shaped the field and trained a generation of scientists. Both conducted research that brought together diverse scientific disciplines. Both were also colorful characters that are sorely missed.

J. William Costerton lost his battle with cancer in 2012. He carried out much of his ground-breaking research at the University of Calgary. In 1995, he took on the leadership of the Center for Biofilm Engineering in Bozeman, Montana. He is well known for his key recognition of the prevalence and importance of biofilms in the environment and disease. He will also be remembered as a powerful advocate for biofilm research at a time when microbiology was somewhat reluctant to embrace the importance of biofilms. Another key aspect of Bill's legacy was the extraordinary efforts he made to promote young researchers.

Peter Gilbert passed away in 2008. He was recognized for his contributions toward understanding biofilm antimicrobial tolerance. He was also known for his provocative ideas that inspired new research directions. During the course of his 30-year career at Manchester University, he authored over 250 research papers and reviews. Underneath his famous gruff exterior were a quick wit and a sharp sense of humor. A very telling insight into Peter was how beloved he was by his former trainees.

New Technologies for Studying Biofilms

1

MICHAEL J. FRANKLIN,[1,2] CONNIE CHANG,[1,3] TATSUYA AKIYAMA,[1,2] and BRIAN BOTHNER[1,4]

INTRODUCTION

The results of recent biofilm characterizations have helped reveal the complexities of these surface-associated communities of microorganisms. The activities of the cells and the structure of the extracellular matrix material demonstrate that biofilm bacteria engage in a variety of physiological behaviors that are distinct from planktonic cells (1–3). For example, bacteria in biofilms are adapted to growth on surfaces, and most produce adhesins and extracellular polymers that allow the cells to firmly adhere to the surfaces or to neighboring cells (4–6). The extracellular material of biofilms contains polysaccharides, proteins, and DNA that form a glue-like substance for adhesion to the surface and for the three-dimensional (3D) biofilm architecture (4). The matrix material, although produced by the individual cells, forms structures that provide benefits for the entire community, including protection of the cells from various environmental stresses (7–9). Biofilm cells form a community and engage in intercellular signaling activities (10–19). Diffusible signaling molecules and metabolites provide cues for expression of genes that

[1]Center for Biofilm Engineering; [2]Department of Microbiology and Immunology; [3]Department of Chemical and Biological Engineering; [4]Department of Chemistry and Biochemistry, Montana State University-Bozeman, Bozeman, MT 59717.

Microbial Biofilms 2nd Edition
Edited by Mahmoud Ghannoum, Matthew Parsek, Marvin Whiteley, and Pranab K. Mukherjee

doi:10.1128/microbiolspec.MB-0016-2014

may benefit the entire community, such as genes for production of extracellular enzymes that allow the biofilm bacteria to utilize complex nutrient sources (18, 20–22). Biofilm cells are not static. Many microorganisms have adapted to surface-associated motility, such as twitching and swarming motility (23–28). Cellular activities, including matrix production, intercellular signaling, and surface-associated swarming motility suggest that biofilms engage in communal activities. As a result, biofilms have been compared to multicellular organs where cells differentiate with specialized functions (2, 29). However, bacteria do not always cooperate with each other. Biofilms are also sites of intense competition. The bacteria within biofilms compete for nutrients and space by producing toxic chemicals to inhibit or kill neighboring cells or inject toxins directly into neighboring cells through type VI secretion (30–33). Therefore, biofilm cells exhibit both communal and competitive activities.

The complexity of biofilms (and the complexity of the technologies required to study biofilm activities) is compounded by the fact that biofilms are inherently heterogeneous (34). In natural biofilm communities, the biofilm structure may be stratified as different organisms migrate to their optimal position for access to light, oxygen, nutrients, secondary metabolites, and signaling compounds (35–37). Even in biofilms composed of one species, subpopulations of cells show heterogeneous activities. In a recent review (34), three general factors that contribute to biofilm heterogeneity were described: (i) physiological heterogeneity, where the bacteria adapt to their local environmental conditions. As oxygen or nutrients diffuse into the biofilms from their sources and are utilized by the bacteria, chemical concentration gradients develop. The chemical gradients may intersect and overlap with gradients of waste products or signaling compounds, forming many unique microenvironments within biofilms that are not mimicked by growth of planktonic bacteria. The bacteria respond to their local environmental conditions, and therefore the physiology of individual cells may differ from other cells that are in close proximity (38–40). (ii) Genetic variability, where mutations may occur in initially clonal populations of cells. Cells within the community may develop mutations causing cellular differentiation (41, 42). This genetic variability may account for the identification of biofilm subpopulations that differ from the rest of the community, such as the rugos or mucoid strains that arise during biofilm growth (43–47). (iii) Stochastic gene expression events, where subsets of cells express the same genes at different levels, even when cells experience very similar environmental conditions (48, 49). Stochastic events promote division of labor, increasing the functional and morphological complexity of biofilms (50). Stochastic events may help account for the formation of some subpopulations of cells that differ from the rest of the community, such as the persister cells, which are cells that have enhanced tolerance to antibiotics compared to other cells in the community (51–54). Since chemical and physiological heterogeneities are tightly linked, variation within biofilm communities is an important factor to consider in designing experiments for biofilm studies.

Many discoveries concerning the physiology, genetics, and ecology of microorganisms growing in biofilms have been made since the first edition of this book. These discoveries have been made possible in part due to the rapid technological advances that have occurred in biological research. Included in these advances is the adaptation of new molecular, analytical, and imaging strategies. Technological developments include the application of “omics” technologies in biofilm research. For example, high-throughput DNA sequencing technologies (next-generation sequencing) have been used for genomics and metagenomics studies and have provided insight into the genetic coding potential of biofilm organisms and into biofilm community structures (55–58). Transcriptomics approaches, including RNAseq, microarrays, and reverse transcription quantitative PCR (RT-qPCR), have advanced our understanding of global and localized gene expression processes that occur within biofilms (59–62). Mass spectrometry has also facilitated our understanding of biofilm proteomics and will be useful for metabolic profiling of biofilms (metabolomics) (63, 64).

Biofilm imaging has also provided a greater appreciation for the complexity and dynamics of biofilms. Advances in imaging technology have led to the ability to obtain 3D images of hydrated biofilms in real time. Fluorescent proteins, which are available in a variety of colors, allow imaging and differentiation of the bacterial cells (36, 65–68). When fused to promoter sequences or other proteins, the fluorescent proteins also enable imaging of localization in microbial gene expression (5, 69). Fluorescent staining combined with confocal scanning laser microscopy (CSLM) and high-speed computing provides information on cell localization and heterogeneity. A few fluorescent stains are now available for certain components of extracellular materials, providing 3D structural analysis of the biofilm extracellular material (70–73). Fluorescent *in situ* hybridization (FISH) has provided information on the community structure of biofilms (74–78). Other imaging approaches allow characterization of biofilm chemistry, including nuclear magnetic resonance imaging that provides information on water dynamics within biofilms (79, 80), while Fourier transform infrared spectroscopy (FT-IR) analysis and Raman imaging of biofilms allow characterization of cellular and extracellular compositions.

The use of new technology for the study of biofilms is dependent upon the scales of interest in characterizing biofilms. These scales can be at the systems level (using omics or imaging approaches), the cellular level (using approaches such as microfluidics and laser microdissection), or the gene/enzyme level (using mutagenesis, enzyme activity, and gene fusion approaches). By combining these different strategies, it may be possible to obtain a comprehensive, systems-level analysis of the structure, function, and dynamics of microbial biofilms. In this article, we review some of these new technologies to study biofilms and provide information on some emerging technologies that will likely be applied to biofilm research.

TRADITIONAL METHODS FOR BIOFILM GROWTH UNDER LABORATORY CONDITIONS

Biofilms are studied in their natural environment and in laboratory scale bioreactors, which are designed to provide models of natural biofilms. The bioreactors can be considered in two general formats: operating under static conditions or continuous flow conditions (Table 1). Using static conditions, biofilm bacteria may undergo growth phases, (lag, exponential, and stationary phases) as in typical planktonic laboratory cultures. Continuous flow reactors are usually operated under "wash-out" conditions. In this case, the residence time of the bioreactor chamber is shorter than the doubling time of the bacteria. This results in most of the planktonic cells being washed out of the reactor, while only those bacteria that are able to adhere to a surface remain within the reactor. The surfaces containing the biofilms may be coupons that can be removed from the reactor for analyses, or even the walls of the reactors, which can be imaged directly.

The static biofilms have been particularly useful for a number of biofilm studies (Fig. 1). In particular, O'Toole and coworkers utilized a relatively simple method to perform high-throughput screening of transposon mutants that were impaired in biofilm formation (26, 81). In this screening method, strains of bacteria, each with a random transposon insertion, are incubated in the wells of microtiter plates. Following incubation, the planktonic cells are removed by washing, and the biofilm cells that remain associated with the wells of

TABLE 1 Methods for *in vitro* cultivation of biofilms

Format/technique	Experimental features	Applications and examples	References
Static biofilms	• Low or no shear • No replacement of medium • No cell washout		
Microtiter plate	• High throughput • Limited biomass	• Phenotypic screening of mutant libraries • Attachment and early biofilm development studies • Biomass quantification with staining	26, 81
Calgary device (MBEC)	• High throughput • Peg material may be modified • Biomass may be recovered from pegs • Limited amount of biomass	• Phenotypic screening of mutant libraries • Antibiotic susceptibility studies • Microscopy with fluorescent probes • Biomass quantification with staining	82
Colony biofilm	• Large biomass in short amount of time • Inexpensive laboratory materials • Low throughput	• Antibiotic susceptibility and penetration studies • Chemical gradient measurements using microelectrodes • Heterogeneity studies using microscopy and fluorescent probes • Cryosectioning studies for gene expression heterogeneity	38, 40, 61, 62, 85, 87
Continuous flow biofilms	• Continuous supply of fresh medium • Adjustable shear force • Low to medium throughput		
CDC reactor	• Special surface materials may be used • Multiple biofilms are formed simultaneously • Suitable for time-course study • May be used for anaerobic cultures	• Antibiotic susceptibility/viability studies • Microscopy studies with fluorescent probes • Applicable for omics studies	88
Drip flow reactor	• Special surface materials may be used • High gas transfer • Heterogeneous biofilm • Large biomass in short time	• Antibiotic susceptibility/viability studies • Chemical gradient measurements using microelectrodes • Heterogeneity studies using microscopy fluorescent probes • Cryosection and laser capture microdissection followed by transcriptomic analysis • Biofilm-immune cell interaction	38, 61, 89
Imaging flow cells	• Real-time detection • Surfaces can be modified • Appropriate for short-time experiments	• Real-time imaging • Monitoring attachment, development, and detachment phases • Microscopy with fluorescent tags • Attenuated total reflection Fourier transform infrared spectrometry • Hydrodynamics in biofilm by nuclear magnetic resonance	40, 79, 80, 91

(Continued on next page)

TABLE 1 Methods for *in vitro* cultivation of biofilms *(Continued)*

Format/technique	Experimental features	Applications and examples	References
Microfluidics	• Miniaturization • Custom platform • Adjustable shear force		
Microfluidics flow cells	• Precise control of hydrodynamics • Low to medium throughput	• Microscopy with fluorescent tags • Spatial heterogeneity • Quorum sensing • Effect of hydrodynamics on biofilm growth and detachment • Formation of streamer phenotype • Measurements of adhesion forces	123, 124, 125, 126, 127, 131, 133
Micropatterned polymer stencils	• High throughput • Both static and flow conditions • Surface materials can be changed or coated • Topographically patterned surfaces can be used	• Microscopy with fluorescent tags • Chemical heterogeneity and quorum sensing • Attachment and early biofilm development • Spatial heterogeneity	103, 121

the plate are stained with a colorimetric indicator such as crystal violet. The amount of crystal violet taken up by the cells is then assayed, providing a quantitative indicator of the cellular amount of the biofilm, and the mutant strains are compared to the wild-type strains. This approach has become a well-accepted model for biofilm formation and is often used as a first step in determining if a mutation in a particular gene affects the ability of the bacteria to adhere to a surface and develop into a biofilm. Once the putative biofilm-impairment mutation is identified by sequencing the site of transposon insertion, the mutant strains can be compared to the wild type using lower-throughput approaches (such as imaging or molecular approaches). A modification of this microtiter plate biofilm assay uses the Calgary device (82). In the Calgary device, rather than assaying the biofilm that grows on the wells of a microtiter plate, the biofilms grow on pegs that are immersed in the wells of the microtiter plate. The advantage of this approach is that the bacteria must adhere to the pegs to be assayed. This reduces the interference of planktonic cells, which may remain at the bottom of the microtiter plate wells following washing but are not actually biofilm bacteria.

Biofilms may also be cultivated under static conditions on coupons placed in the wells of microtiter plates or as pellicles at the air-water interface (83). Depending on the organism cultivated (and the native environment of the isolate) certain cells may prefer growth under static conditions. For example, Chimileski et al. (84) showed that the *Archaeon Haloferax volcanii* forms thick biofilms when cultured under these static conditions but did not form substantial biofilms when cultured under continuous flow. This type of growth may reflect the preference for static conditions where these *Archaea* are commonly found in their natural environment.

Another type of static biofilm is the "colony biofilm" (85). Colony biofilms are cultivated on filters that are placed on the surface of agar petri plates. The filters are then transferred to fresh medium at regular intervals, giving the biofilms a semi-continuous supply of fresh nutrients. The greatest advantage of this approach is that these biofilms are easy to grow, using inexpensive laboratory material. The biofilms formed using this approach are generally thick and are therefore useful for a variety of experiments, such as cryoembedding and thin-sectioning to obtain vertical cross-sections of the biofilms (86). By using differential staining on colony biofilms, information on the heterogeneity of biofilms has been obtained

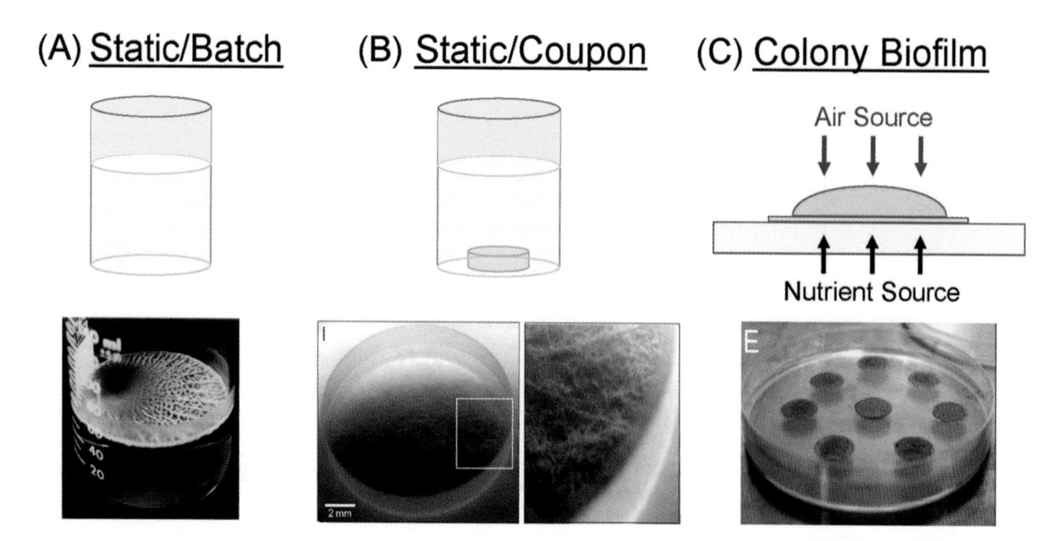

FIGURE 1 Examples of methods for biofilm cultivation under static conditions. (A) Biofilm cultured at the air-water interface, forming a pellicle. Published with permission from reference 83. (B) Biofilm cultured on a glass coupon under static conditions. Published with permission from reference 84. (C) Example of biofilm growth as a colony biofilm. Published with permission from reference 84. doi:10.1128/microbiolspec.MB-0016-2014.f1

(38). In addition, colony biofilms have been useful for determining gene expression heterogeneity within biofilms (61, 62). Colony biofilms have also been used to assay the rate of diffusion of antibiotics through model biofilms (87). Of course, the disadvantage of these biofilms is that since there is no continuous flow of medium, the bacteria are not forced to adhere to a surface or to the matrix material. Since there is no wash-out, planktonic cells may interfere with the biofilm assays.

Continuous-flow reactors have advanced biofilm studies by providing better mimics of natural biofilms. Reactors commonly used for continuous flow include the CDC reactor (88), the drip-flow reactor (89), and flow reactors designed for CSLM imaging (90–92) (Fig. 2). The CDC reactor is a fairly large-scale reactor that provides medium inlet and outlet ports. The reactor chamber contains rods with removable coupons that can be made of a variety of different materials, such as glass, hydroxyapatite, or metals. The biofilms form on the surface of the coupons, which may then be removed aseptically over time. Experiments performed on coupon biofilms include direct imaging of the samples using CSLM or other imaging approaches, cell viability studies (colony forming unit and direct counts of bacteria), or omics studies. Since the CDC reactors are usually run under wash-out conditions with fluid shear, the biofilms formed on the coupons are often thick and mimic natural biofilms. The disadvantage of this approach is that the reactors often have large medium reservoirs, making this approach low-throughput.

A similar type of continuous flow reactor is the drip-flow reactor (89). In these reactors, medium is dripped onto the surface of a coupon, and the medium flows across the coupon. Biofilms cultured using this format are often thick and provide a good mimic for some natural biofilms, where there is fluid flow across a surface with an air interface. Biofilms cultured using the drip flow format often show heterogeneity across the slide from the initial site of the medium input to the outlet. In some cases the biofilms develop streamers, as observed in many environments such as streams with rapid fluid flow.

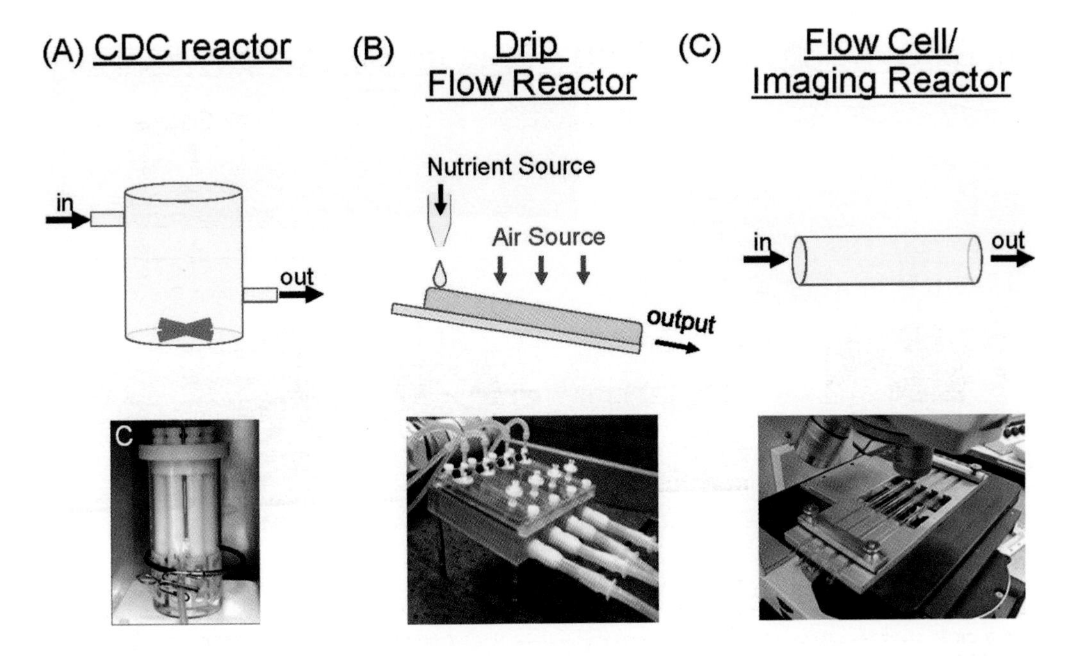

FIGURE 2 Examples of continuous-flow reactors for biofilm cultivation. (A) CDC reactor with medium inlet and outlet ports. Biofilms form on coupons arranged on removable Teflon rods. Published with permission from reference 88. (B) Drip-flow reactor with medium inlet and outlet ports and air exchange ports. Biofilms form on removable slides. Published with permission from reference 89. (C) Capillary flow cell for imaging biofilms. Published with permission from http://centerforgenomicsciences.org/research/biofilm_flow.html. doi:10.1128/microbiolspec.MB-0016-2014.f2

Biofilms cultured under drip-flow conditions are amenable to cryoembedding and thin-sectioning, allowing studies of biofilm vertical heterogeneity (61, 93). Drip-flow cultured biofilms have also been used in combination with microelectrode sensors to study chemical gradients within biofilms (39, 94).

Biofilms may also be cultured under continuous flow conditions using imaging flow cells (90–92). These flow cells usually contain a window, composed of a microscope cover slip, that allows observation of the biofilms by microscopy. Imaging of these biofilms by CSLM provides information on the 3D architecture of the biofilms. A modification of this approach is the capillary flow cell biofilm (95–97), in which biofilms are cultured in small square capillary tubes. Biofilms that form on the walls of the capillary tubes can then be directly imaged. A disadvantage of this approach is that because the tubes are so small, they clog easily and therefore are most appropriate for short-term experiments.

MINIATURIZATION APPROACHES TO BIOFILM CULTIVATION AND CHARACTERIZATION

In many cases, high-throughput genetic and microbial physiology studies have relied on miniaturization of bioreactors (Fig. 3). An important development in the study of biofilms at the cellular level has been the use of microfluidics. Microfluidics is the precise manipulation and control of fluids in micro-scale channels that are typically less than 100 μm. The channel dimensions define the volumes that are handled, which range from nanoliters to picoliters. Microfluidics is a field that has grown since its inception in the 1990s, following the advent of technology

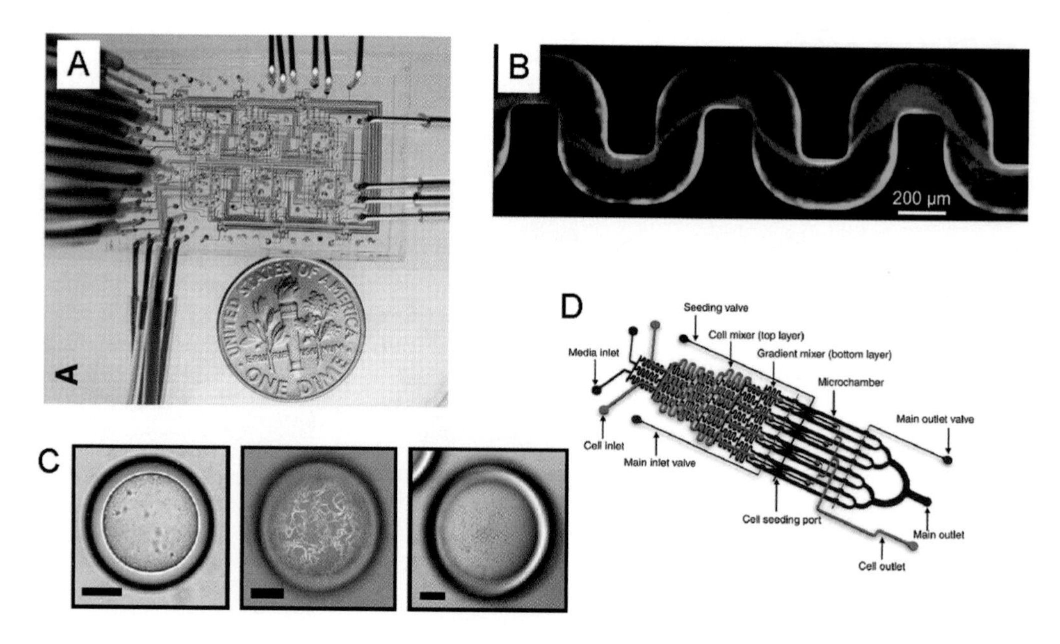

FIGURE 3 Examples of microfluidics approaches applied to biofilm research. (A) Example of a microfluidics device for precise control of fluids. Published with permission from reference 110. (B) Biofilm streamers forming within a microfluidics flow channel. Published with permission from reference 127. Microdroplet biofilm reactor showing phenotypic switching of cells and simultaneous change in expression from cyan fluorescent protein to the yellow fluorescent protein Published with permission from reference 259. (D) Schematic representation of microfluidics flow cell. Published with permission from reference 133. doi:10.1128/microbiolspec.MB-0016-2014.f3

from the semiconductor industry and the field of microelectromechanical systems. Microfluidic devices, commonly called "lab-on-a-chip" or "miniaturized total analysis system (μTAS)" devices, allow rapid sample assaying by a reduction in sample volumes (98).

The field of microfluidics is characterized by the study of fluid flow at microscale lengths, which greatly differs from macroscale flow. At this length scale, the flow of fluids is laminar and is described by low Reynolds numbers (*Re*). *Re* is a dimensionless quantity defined by the ratio of inertial to viscous forces in a fluid. In microfluidic systems, *Re* lies in the laminar regime, where mixing due to turbulent flow is nonexistent and molecular transport across streamlines occurs through diffusion, instead of convective mixing. The modeling and computational analysis of fluid flow in microfluidic channels is simplified under these conditions. At submillimeter-length scales, forces that become significant include interfacial tension and capillary forces. These can be tuned to manipulate the wetting properties, and therefore flow or movement, of fluids in microfluidic devices. Due to the tunability of the surrounding fluid flow and chemical composition, microfluidics allows a remarkable degree of control over the physical and chemical environment of microorganisms, whose sizes are within the same order of magnitude as that of microfluidic channels. Since these small microenvironments can be easily manipulated using microfluidics, this technique is useful for studying samples from individual cells to small populations. In addition, microfluidic devices are often made from materials that are compatible with light microscopy, such as optically trans parent polydimethylsiloxane (PDMS) bonded to glass, which is optimal for viewing single cells and fluorescent particles or dyes.

There has been a wealth of information obtained from the study of single bacterial cells and small populations of cells in microfluidic devices and in other microfabricated environments (99–104). Microfluidics has been used as a new approach for cultivating biofilms, for studying the dynamics of biofilm formation and detachment, for high-throughput assays, and for biofilm sensing. These studies may be important in understanding the early stages of biofilm growth. Examples include the effect of antibiotics upon individually isolated bacterium (105), the evolution of bacterial antibiotic resistance in connected microenvironments (106), cell variability in bacterial persistence (107, 108), quorum sensing in small numbers and aggregates of bacteria (109), and growth rates and lineages of single cells or small populations in microscale channels or chemostats (99, 110–112). Microfluidics has also been applied to multigene analysis of single cells using microfluidic digital polymerase chain reaction (113), the sorting of individual cells in droplet microfluidics (114), isolation of bacteria in fluids (115–117), and bacterial chemotaxis (118, 119). There are alternative methods for studying small numbers of bacteria in confined spaces that are not microfluidics-based but offer the same advantages as microfluidics, including controlled microenvironments and microscopically viewable substrates (104, 120).

Since small volumes are easily handled using microfluidics, large-parameter spaces can be systematically explored at fast rates, making high-throughput sampling possible. Thus, these lab-on-a-chip technologies become complementary to traditional experimental methods in the laboratory. In one study, arrays of biofilms were created using micropatterned polymer stencils (121). Various bacterial and fungal biofilms were grown upon hundreds of circular stencil arrays that were 100 μm in diameter. These high-throughput arrays of biofilm substrates were interfaced within a microfluidic device to study biofilm growth in both static and flow conditions. In another study, biofilms were exposed to eight different concentrations of signals in a device containing eight microchambers with a single gradient diffusive mixer to perform high-throughput testing of chemical compounds (122, 123). Another chip was able to generate gradients of dissolved oxygen (124).

Microfluidics has been used to study the effect of hydrodynamics on the growth and detachment of biofilms. By changing the flow velocity and fluid shear in locations within the devices, biofilms can be grown to various thicknesses within the channels. For example, *Staphylococcus epidermidis* was found to grow as a monolayer in regions of high flow velocity and multilayered in regions of low flow velocity (125). Other studies have examined suspended filamentous biofilms or streamers of *Pseudomonas aeruginosa* that are formed in curved microfluidic channels (126). Those studies showed that secondary flows within the device cause the formation of streamers. These biofilm streamers are a major component of biofilms in natural, medical, and industrial environments (127).

Biofilms are comprised of rigid bacteria, soft viscoelastic extracellular matrix, and small molecules that may provide signaling or generate forces for spreading and survival (128, 129). The physical properties of biofilms may be characterized using microfluidics and can provide a fundamental understanding of the physical properties of biofilms at the microscale. For example, a combination of CSLM and a flexible microfluidic device was used to characterize the viscoelastic properties of *S. epidermidis* and *Klebsiella pneumoniae* biofilms (130). In addition, precise control of fluid forces by microfluidics has been used to measure adhesion forces of pili in cellular attachment to surfaces (131).

Since biofilms frequently exist as a heterogeneous community of cells (34), microfluidic techniques have been used to organize the spatial and temporal location of cells and to simplify the complexity of these heterogeneous populations. One study used a combination of molecular genetics, microfluidics, and microscopy to show a fitness trade-off between two strains of *Vibrio cholerae* biofilms, in which one strain that was able to produce extracellular polymeric substance (EPS) competed with another that was deficient in its ability to produce EPS (132). That study showed that a single phenotype, extracellular polymeric substance secretion, governs a fitness tradeoff between colonization of biovolume and dispersal to new locations. In another study, a quorum sensing circuit was used to create a dual-species biofilm where a microfluidic device was used to control the consortia of bacteria (133).

OMICS APPROACHES TO BIOFILM STUDIES

Genomics and Metagenomics Approaches for Biofilm Studies

The advent of high-throughput sequencing technologies (such as Illumina, Roche 454, SOLiD, and Pacific Biosciences [PacBio]) (134) has allowed investigators to obtain genome-scale information on biofilms. Some of these technologies also provide information on genome modification (epigenetics) (135), which may be one mechanism that affects gene expression and physiology of biofilm cells. These new sequencing technologies have been used to obtain complete genome sequences of thousands of microorganisms, and the cost of obtaining genome sequences continues to drop, making complete genome sequences of many more strains feasible. The next-generation sequencing technologies combined with functional genomics studies (such as transposon mutant libraries and transcriptomics studies) have provided information on cellular activities within biofilms, the role of essential genes in biofilm formation, as well as the community structure of natural biofilms.

Next-generation sequencing technologies can provide billions of nucleotides of sequence for an individual sample. Therefore, this technology can be used to obtain sequence information on entire microbial communities (metagenomes). Analyzing data of conserved genes from the community (such as 16S ribosomal RNA and multilocus sequence typing) allows the characterization of the microbial community structure and, to some extent, the abundance of individual microorganisms. Since metagenome data are useful for obtaining information on all genes (not just conserved genes), this information can be used to obtain the genetic potential of microbial communities.

Metagenome analysis has been applied to low-diversity biofilms, including the microbial community associated with the later stages of lung degeneration of cystic fibrosis patients, which is composed of very few species, particularly *P. aeruginosa* and *Burkholderia cepacia* (136). Another low-diversity environment analyzed by metagenomics is a deep subsurface sample, where the complete genome of an essentially monospecies community was assembled (137). Biofilms may have an intermediate level of diversity, particularly when environmental conditions impinge on a community, allowing growth of only certain types of microorganisms that are adapted for that environment. An example of biofilm communities that have been analyzed by metagenomics is the biofilms associated with acid mine drainage (138–140). Biofilms associated with thermal features of hot-spring communities have also been characterized by metagenome analysis (141–143). Next-generation sequencing technology is amenable to very complex microbial communities and has been used to characterize community structures of the human microbiome, including the oral biofilm microbiome (58, 144, 145) and the gut microbiome (146–148).

While obtaining genomic and metagenomic data is now relatively inexpensive and fast, analyzing the sequence information can be complicated and time-consuming. With most next-generation sequencing technologies, the sequence reads are short, and therefore the data must be assembled or else the reads must be binned onto the sequence of known microorganisms. New sequencing methods include long reads with PacBio sequencing and paired-end reads (149). Bioinformatics tools are being developed that allow the assembly and analysis of genomic and metagenomic DNA sequence data (e.g., 150–154). While these and other bioinformatics approaches have provided vast amounts of information on the structure and coding potential of a variety of different natural biofilms, complete assembly of genomes is still difficult. Repetitive and similar sequences in metagenome data are difficult for computer programs to handle and can lead to errors such as chimeric sequences and gaps in genome sequences.

Metagenome assemblies are complicated by the diversity of the biofilm community, which in natural environments can range from low diversity to very complex community structures. Several approaches have been developed to generate metagenome information from short reads of next-generation sequence information from biofilm samples. McLean and Kakirde (155) described two general approaches. One approach uses the isolation of DNA directly from the environment, fragmenting the DNA and performing direct sequencing using next-generation sequencing. This approach has the advantage of providing high coverage of DNA sequence from the environmental sample. One disadvantage of this approach lies in the difficulty in determining the source of the DNA (for example, whether the DNA is derived from cellular or extracellular material or if it is from live or dead cells). Assembling the reads into genome information is also complicated, since assembly may generate chimeric sequences or incomplete data sets. Binning the data based on known marker genes or genome sequences is useful for analyzing the number of reads associated with a particular organism. Binning is dependent on whether genome sequence information is available for the organisms isolated from that particular environment. McLean and Kakirde (155) also discussed the use of metagenomic sequencing for biofilms by first using large DNA fragment libraries cloned into fosmid, cosmid, or bacterial artificial chromosomes (BACs). The advantage of this approach is that the linkage of genes can be obtained, since the sequence reads are obtained from adjacent large DNA fragments. This helps reduce misassembly and chimeric assembly. A disadvantage of this approach is that an additional DNA cloning step is required, which may discriminate against certain DNA sequences and therefore reduce the diversity of DNA sequences available in the metagenome library.

An emerging approach for obtaining DNA sequences directly from environmental samples without first cultivating the bacteria is the use of single-cell genomics (156, 157). This approach was recently applied to a biofilm from a hospital sink, where a new phylum of uncultivated bacteria was identified (158). With this approach, single cells are first isolated by fluorescent activated cell sorting, laser tweezers, or microfluidics. The genomic DNA of the isolated cells is then amplified by using multiple strand displacement amplification to obtain enough DNA for sequencing (159). The advantage of this approach is that chimeric sequences are reduced, since the sequence information is known to be derived from individual cells. Therefore, this approach provides not only information on the type of organism in the environment, but also information on the organisms' metabolic capabilities, as was shown for the novel hospital sink isolate. With the decrease in cost and time for sequencing as well as improvement in sequencing approaches, single-cell genomics has the potential to become a major technique for metagenomic analysis of environmental biofilms.

Transcriptomics Approaches for Biofilm Studies

For many years investigators have asked the question "What makes biofilm cells different from planktonic cells?" Questions about the physiology of biofilm bacteria include: Why do infectious biofilms have enhanced resistance to antibiotics compared to planktonic cells? Why do they show enhanced resistance to host defensive processes? How do environmental biofilm cells differ in metabolic activities compared to the same strains growing planktonically? One approach to address these questions is to analyze the global expression profile of the bacteria when cultured under biofilm conditions and under planktonic conditions. The RNA expressed under the different conditions is isolated, and all RNA transcripts are quantified. These transcriptomics studies have been applied to a variety of biofilms and have provided important information on the gene expression patterns that are unique to biofilm cells (e.g., 60, 160–164). Until recently, transcriptomics studies have relied on microarrays and RT-qPCR. With the advent of next-generation sequencing technologies, it is now possible to analyze the entire transcriptome of biofilms by RNAseq, the sequencing of cDNA libraries generated from pools of RNA. With this RNAseq approach, the number of RNA transcripts for each gene can be quantified for cells growing under different conditions such as in biofilm and in planktonic cells.

For microarray studies, the total RNA from cells cultured under the different conditions is isolated. The RNA is reverse transcribed to cDNA using reverse transcriptase and hybridized to oligonucleotide probes arrayed on a surface. The fluorescently labeled cDNA is then quantified, using either one-color arrays (Affymetrix-style arrays) or two-color technology, where the cDNAs from the different samples are labeled with different fluorescent dyes. The array data are normalized so that a direct comparison between samples may be made. The data from these biofilm studies as well as other transcriptome studies are deposited in the Gene Expression Omnibus database and are freely available for other investigators to examine for new patterns in gene expression (165, 166). While microarrays have been very informative for transcriptome studies, standard microarrays are made to detect known and predicted genes and thus cannot detect small RNAs located at intergenic regions. Several studies used tiling array, a subtype of microarray with customized chips carrying partially or nonpartially overlapping oligonucleotide probes to cover the whole-genome sequence. With tiling array, an abundance of small RNAs as well as posttranscriptionally modified RNAs can be detected (167, 168).

With the advent of next-generation sequencing, it is now possible to obtain a quantitative description of the bacterial transcriptome using sequencing (RNAseq) (169). RNAseq has several advantages over microarray studies for transcriptome analysis of biofilms. The greatest advantage of RNAseq is that an array platform does not need to be available for every organism of interest, making this approach amenable to any species of *Bacteria*, *Archaea*, or *Eukarya*. RNAseq is amenable to metatranscriptomics studies, since the genome sequences of the organisms of interest do not need to be known *a priori* (170, 171). It has been applied to several environmental biofilm samples, including characterizing the transcriptome of the human oral microbiome (172) and hot-spring microbial mats (173, 174). Another advantage is that RNAseq provides information that may not be available with the microarray approach, such as the presence of small noncoding RNAs expressed from intergenic regions and information on the sites of promoter sequences and operon structures (175–178).

While the global transcriptomics approaches provide an average value for gene expression over the entire biofilm population, they do not provide information on local heterogeneity of the cells (34). Several approaches, including microfluidics, have been developed to address the question of localized transcriptional processes within biofilm. Another approach to studying localized biofilm processes is the combination of laser capture microdissection and transcriptomics (61, 62) (Fig. 4). Using these approaches, subpopulations of bacteria from different regions of biofilms are isolated and captured using laser capture microdissection. RNA and DNA are extracted from the captured cells and analyzed directly by using RT-qPCR and qPCR for individual gene analyses (61, 179). Alternatively, the RNA or DNA may be amplified using multiple strand displacement for transcriptomics analysis (62). This approach has the advantage of excellent sensitivity and large dynamic range and therefore can be performed for quantitative analysis of gene expression for a few cells from defined regions within the biofilms.

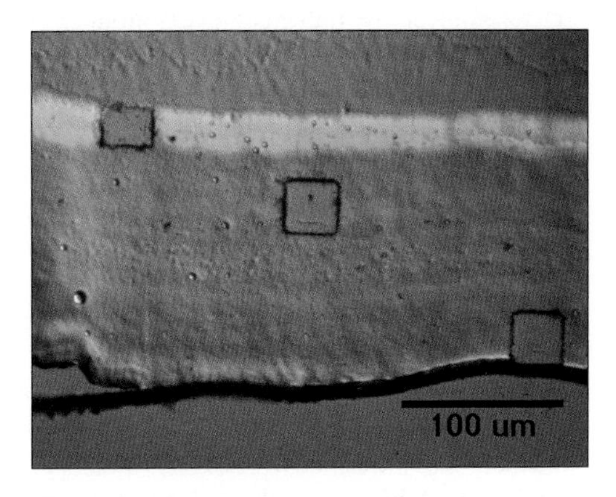

FIGURE 4 Example of a biofilm vertical transect, showing GFP–gene expression heterogeneity. Areas were cut from different biofilm strata and captured for transcriptomics analyses. Published with permission from reference 61. doi:10.1128/microbiolspec.MB-0016-2014.f4

Proteomics and Metabolomics Approaches for Biofilm Studies

The technological advances and a general realization of the power of omics approaches to the study of complex biological systems have advanced the application of these studies to biofilm biology. Two rapidly expanding approaches within this field are proteomics and metabolomics (180, 181). From an information standpoint, proteomics and metabolomics are a step closer to biofilm physiology than genomics and transcriptomics, since the proteins and metabolites are biomolecules that comprise most of the enzymatic activity of the cells and the metabolic substrates and products of those enzymes. However, neither proteomics nor metabolomics provide data sets that are as complete as methods focused on nucleic acids (180, 181). It is now routine to obtain deep coverage of transcribed regions of a genome using transcriptomics approaches, but there has yet to be a truly comprehensive proteomics or metabolomics investigation. This is largely due to a combination of technical and practical challenges, one of which is that amplification of the target is not possible in proteomics. This puts a technical limit on detection in that protein concentration can span five orders of magnitude in a single cell. This obstacle can be overcome through fractionation, but this has practical limitations based on time, cost, and value. Fortunately, the balance between value-added data and expense is now strongly in favor of collecting proteomics and metabolomics data from biological samples.

Key to both proteomics and metabolomics are advances in chromatography and mass analysis. Liquid chromatography (LC) coupled with electrospray ionization and a high resolution mass analyzer (LCMS) has evolved into a straightforward and very sensitive method for the analysis of a wide array of biomolecules. Quantitative information can be compiled from thousands of proteins from bacterial systems, and posttranslational modifications can readily be tracked.

There are two primary approaches used in proteomics: two-dimensional differential gel electrophoresis (2D DIGE) and shotgun proteomics analysis. Variants of each approach exist. 2D DIGE was developed first and has been used in several studies of biofilm proteomes. For example, Sauer et al. (182) used 2D DIGE to characterize the changes in proteome patterns of *P. aeruginosa* over time, demonstrating that biofilm formation of this organism is a developmental process. 2D DIGE has also been used to characterize strain variations in biofilm proteomes and the effect of environmental conditions on biofilm intercellular and extracellular proteomes (183–191). With 2D DIGE, proteins are identified by first cutting the protein spots from the gel. The protein in the dissected piece of gel is analyzed using in-gel proteolysis, which amounts to adding trypsin and then analyzing the peptides that are released using liquid chromatography/mass spectrometry. The peptide mixture is separated over a reverse-phase column, and then the mass of the intact peptide and the fragmentation pattern are collected. Peptide fragmentation is highly predictable, allowing sequence tags to be generated. The experimental data is compared with an *in silico* pattern generated from the predicted protein coding regions of the organism of interest. This process can be automated, and femtomolar sensitivity is now routine.

Traditionally, the 2D DIGE approach has required staining of the proteins with stains such as silver or coomassie blue. The proteomes of cells cultured under separate conditions are run on separate gels, and then the intensities of individual protein spots are compared using image analysis software. A more recent advance allows relative quantification between samples by labeling each protein sample with different colors of fluorescent tags (192, 193). Samples are then mixed together and separated by charge (pI) and molecular weight. After the gel has been run, the intensity of protein spots from each sample is determined using high-resolution imaging. Over a thousand protein spots can be visualized on a single 24-cm gel, and the specific sample of origin for a given spot can be determined by which color tag it has. Often an internal standard is added to the gel to facilitate analysis across gels and experiments. Sophisticated image analysis software that can warp images and adjust for variations in background and protein loading has been developed (194). 2D DIGE is well suited for rapid screening of samples for global changes. In addition, because posttranslational modifications of proteins usually lead to changes in protein pI or molecular weight, 2D DIGE is extremely sensitive to global and specific changes in proteins such as phosphorylation (195, 196). The capability to investigate posttranslational modification in an untargeted way is a strength of 2D DIGE.

Shotgun proteomics has now been applied to biofilm samples for metaproteomics studies of cellular and extracellular proteins (197, 198) and for characterizing the effects of biofilm growth, environmental conditions, and intercellular signaling of individual strains of biofilm organisms (e.g., 199–203). Shotgun proteomics uses direct digestion of a complex protein sample with trypsin and LCMS analysis of the ensuing peptide mixture (204). This step produces a highly complex mixture of peptides, which from a bacterial cell may contain >50,000 distinct peptides. To handle this complex solution, multidimensional chromatography is normally used. The standard approach separates peptides into fractions using strong cation exchange, with each fraction then being analyzed using reverse-phase LCMS as above. The advantage of the shotgun approach is that 2,000 to 4,000 proteins can routinely be identified, providing a deeper view into the proteome. The method can also be entirely automated, allowing for high-throughput work. Recent advances using long reverse-phase gradients allow direct analysis of total proteome digests with 5,000 to 10,000 protein IDs from a sample.

Comparisons between biofilm samples can be made directly using label-free approaches or through the addition of isobaric tags for relative and absolute quantitation (iTRAQ) (205–210). While iTRAQ adds additional steps to sample preparation, it facilitates multiplexing of up to eight samples in a single LCMS run, dramatically reducing the amount of instrument time that is required.

Metabolomics is the most recent addition to the omics quiver that has been applied to biofilms (211–216). Metabolomics has a distinct advantage over other omics approaches, such as transcriptomics and proteomics, since the latter two only report on biological potential and not the activity, whereas the presence and changes in abundance of metabolites are direct readouts of biological activity. As with proteomics, the sensitivity and speed of leading-edge LCMS instruments facilitate detection of thousands of compounds in cellular extracts and extracellular solutions.

Untargeted metabolomics primarily relies on LCMS because it enables the analysis of the widest variety of compounds with excellent quantification, reproducibility, and sensitivity (217, 218). In addition, the high accuracy of modern mass analyzers allows chemical formulas to be generated for a majority of the detected molecular features. The downside to using LCMS is that chemical formulas are highly redundant, and fragmentation patterns cannot be readily predicted as with peptides. Fortunately, data analysis tools such as XCMS, metaXCMS, and MZmine are freely available (219). These tools link to databases such as METLIN (220, 221) and the Human Metabolome (222), which contain >80,000 compounds. More specific analyses, such as those focusing on central carbon metabolism and lipids, are well served by gas chromatography–mass spectrometry, for which spectral libraries based on fragmentation patterns are available, facilitating direct compound identification. The specific technique to be used again depends on the question being asked. Metabolomics experiments can be designed to be untargeted surveys, which are a powerful approach for the identification of novel compounds and unbiased analysis of microbial states (64). On the other hand, highly targeted approaches can make quantitative analyses of hundreds of compounds in under ten minutes if large numbers of samples must be processed or high time resolutions of a biological response are needed.

BIOFILM IMAGING

One of the most important advances in the study of biofilm structure, function, and dynamics has been the ability to visualize hydrated living biofilms in three dimensions, over time, using CSLM. The first 3D images of biofilms were obtained by CSLM approximately 25 years ago (90). Now, CSLM is available in many labs and core facilities, and therefore this technology is now used in almost all biofilm studies (Table 2). Important advances in CSLM for the analysis of microbial biofilms have been the optics and lasers, image analysis software, and high-speed computing power which allows image analysis of large data files and also allows time-lapse imaging of biofilm developmental processes. Chemical and molecular biological advances have also improved the ability to image biofilms in three dimensions. In particular, the use of fluorescent proteins and fluorescent probes allows imaging by multiple fluorochromes for assays of the individual biofilm components simultaneously. Another new development in imaging technology is super-resolution microscopy such as photo-activated localization microscopy (PLAM), fluorescence photoactivation localization microscopy (FPLAM), and stochastic optical reconstruction microscopy (STORM) (223–225), imagining techniques that give resolution down to tens of nanometers, below the diffraction limit of light. These techniques use photoactivatable fluorescent proteins or probes to perform imaging with much greater

TABLE 2 Imaging strategies in biofilm studies

Technique	Experimental features	Types and modifications	Application and examples	References
Fluorescent imaging				
Fluorescent protein	• Wide range of colors • Multiple functions • Genetically encoded • No need for fixation • Real-time detection • Molecular techniques need to be developed for organisms • Cell physiology may be altered	• Photoactivatable • Photoconvertable • Photoswitchable • FP timer • FRET/BRET • Oxygen-independent FP • Unstable GFP	• Cell differentiation • Cell tracking • Expression profile • Protein localization • Activity indicator • High-resolution imaging	36, 40, 48, 62, 65, 68, 91, 132, 227, 228, 229, 231
Fluorescent *in situ* hybridization (FISH)	• Wide range of colors • High target specificity • No genetic modification necessary • Exogenous addition • Fixation of sample is necessary	• Oligonucleotide probe • Peptide nucleic acid probe • CLASI-FISH • CARD-FISH • MAR-FISH	• Differentiation of multiple organisms • Cell localization • mRNA expression profile • High-resolution imaging	78, 234, 235, 236, 237, 238, 239, 242, 260, 261
Fluorescent stain	• No genetic modification necessary • Exogenous addition • Can be specific for biomolecules and cellular components	• SYBR-green • FM1-43 • CTC • BrdU-fluorescent antibody • Calcofluor • TOTO-1	• Matrix material and cellular component visualization • Activity indicator • Viability indicator	38, 73, 226, 240, 241, 242, 246
Analytical imaging				
Fourier transform infrared spectrometry (FT-IR)	• Nondestructive • Real-time detection	• ATR/FT-IR • SR-FTIR	• Chemical composition of EPS • Quantification of biomass • Spatial distribution of biomolecules	91, 247, 248, 249, 250
Raman spectrometry	• Nondestructive • Real-time detection	• Raman microscopy • SERS	• Chemical composition of EPS • Quantification of biomass • Spatial distribution of biomolecules	249, 251, 252, 253, 254, 255, 256, 257, 258
Nuclear magnetic resonance microscopy	• Nondestructive • Real-time detection	• Use of isotope-labeled substrates	• Water flow and water dynamics of EPS • Dynamics of metabolites	79, 80, 262, 263

[a]Abbreviations: CLASI, combinatorial labeling and spectral imaging; ATR/FT-IR, attenuated total reflection Fourier transform infrared spectrometry; BrdU, bromodeoxyuridine; BRET, bioluminescence resonance energy transfer; CARD-FISH, catalyzed reporter deposition-FISH; CTC, cyanoditolyl tetrazolium chloride; FP, fluorescent protein; FRET, Förster resonance energy transfer; MAR-FISH, microautoradiography-FISH; SR-FTIR, synchrotron-radiation-based FTIR; SERS, surface-enhanced Raman scattering.

resolution than traditional light or CSLM microscopy. It is likely that high-resolution approaches will soon find an application in biofilm research.

CSLM uses a combination of laser excitation of fluorescent molecules and fluorescence emission through a pinhole to filter out light that is not in the optimal focal plane. As a result, with CSLM there is no out-of-focus haze, which is typical for 3D specimens imaged using conventional light or epifluorescence microscopy. With CSLM, a series of in-focus 2D images in the X and Y planes are obtained at submicrometer intervals in the Z-direction. The 2D stacks are converted into a 3D image using image analysis software. CSLM is ideal for imaging biofilms, since most biofilms have extensive 3D structure and can be stained with fluorescent probes or fluorescent proteins. Neu and Lawrence (226) published a comprehensive review of CSLM applied to biofilm research, including many published examples of CSLM techniques and applications. Here, we briefly describe a few of the approaches that are most commonly used to obtain 3D images of microbial biofilms.

The most widely used approach of CSLM imaging in biofilm research is to culture biofilms of bacteria that contain fluorescent proteins (such as the green fluorescent protein [GFP])in flow cells and to image the biofilms that form on the walls of the flow cells. The greatest advantage of using GFP-labeled cells is that the 3D architecture of the biofilm cells can be visualized during biofilm development, without the need to add external fluorescent dyes. Generally, the GFP is expressed from a constitutive promoter, allowing comparison of the biofilm formation properties of a wild-type strain and a mutant derivative that is impaired in biofilm formation (e.g., 91, 227). Since fluorescent proteins are now available in many colors (68), it is possible to differentially label multiple strains, species, or mutants with different colors of fluorescent proteins. These strains can then be cultured separately or in multispecies biofilms and imaged by CSLM (36, 65). These experiments may be used to assay the competitive advantage of a particular strain or the effect of a gene mutation on the competitive index of an organism (132, 228). While GFP-labeling of cells has provided excellent images of a variety of biofilms, there are some disadvantages to this approach. First, a genetic system must be available for the organism of interest to introduce and express the *gfp* gene in the cell. Second, the *gfp* expression may affect the growth of the cells, since energy is required to express this nonnative gene, often from a multi–copy number of plasmids. Third, although the GFP-labeling provides excellent images of the cellular material in the biofilms, it does not provide information on the extracellular component of the biofilm matrix material, which must be stained separately to provide a comprehensive view of the biofilm architecture.

A modification of the fluorescent protein labeling approach is to use the GFP (or another fluorescent protein) as a reporter gene (Fig. 5). Fluorescent reporter genes are used to assay the expression of a particular gene within the biofilms. Reporter gene fusions can act as transcriptional reporters (if the *gfp* is fused to a gene's promoter region) or translation reporters (if the *gfp* is fused in frame with another gene, generating a chimeric protein). Fluorescent reporter gene fusions combined with CSLM have allowed investigators to examine gene expression and protein production in biofilms and the effect of environmental conditions on gene expression. They have also been used to serve as a measure of metabolic activity within regions of the biofilms. Translational reporter fusions may be used to determine the localization of tagged proteins within cells (229–232). Combined with CSLM, fluorescent reporter genes are particularly useful for examining gene expression heterogeneity in biofilms and for studying stochastic gene expression events (48, 233).

Biofilms may also be imaged by labeling the cells with fluorescent oligonucleotide probes.

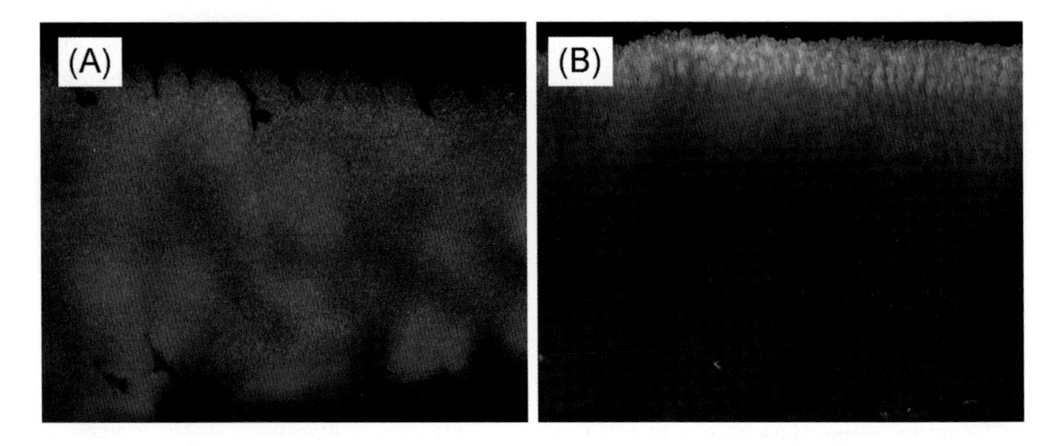

FIGURE 5 Gene expression heterogeneity, demonstrated by translational fusions of target proteins to the yellow fluorescent protein (YFP). (A) Translational fusion of the IbpA protein to YFP, showing uniform distribution of IbpA throughout the biofilm. Cells were counterstained with mCherry fluorescent protein (mCFP). (B) Translational fusion of Rmf protein to YFP, showing that most Rmf production occurs in cells at the top of the biofilm. Cells counterstained with mCFP (M.J. Franklin, unpublished data). doi:10.1128/microbiolspec.MB-0016-2014.f5

Fluorescent *in situ* hybridization (FISH) has traditionally been used to detect the abundance of multiple bacteria species in environmental samples (74) based on hybridization to the 16S rRNA. FISH has been used on a variety of biofilm samples. For example, FISH was used to determine spatial distribution of ammonia-oxidizing and nitrite-oxidizing bacteria in nitrifying aggregates or biofilms (78, 234) and the methanogen and sulfate-reducer components of anaerobic biofilms (75). Traditional and modified FISH have also been used to detect localized abundance of specific mRNAs in biofilm or termite gut microflora (235, 236). The microbial community in natural biofilm can be highly diverse. Recent development of combinatorial labeling and spectral imaging FISH (CLASI-FISH) has the potential to detect hundreds of microbial taxa in single microscopy imaging (237–239). The disadvantage of FISH approaches is that the samples have to be fixed with chemicals prior to the probe hybridization, which may disturb the structure of the biofilm and makes time-course studies difficult.

Fluorescent stains are also used in combination with CSLM for imaging biofilm cellular and extracellular matrix material. Neu and Lawrence (226) provided tables with comprehensive lists of fluorescent stains commonly used in biofilm research. The fluorescent stains are generally designed to bind a specific cellular component, such as DNA (e.g., propidium iodide, SYBR-green, and ToTo-1) or protein (e.g., Sypro-Ruby). The FM1-43 stain is interesting in that it intercalates into the cell membrane and provides definition of the bacterial cell wall when imaged using CSLM (240). In addition to staining cellular materials, fluorescent stains are available for studying metabolic activities within biofilms. For example, tetrazolium salts precipitate when reduced by the biofilm cells, forming a zone of fluorescence around the active cells (241, 242). In another example, bromodeoxyuridine (BrdU) is incorporated into the DNA of actively growing cells. Following incorporation, the active cells can be identified by staining the BrdU with fluorescent anti-BrdU antibodies. This approach was used to study spatial heterogeneity in *S. epidermidis* biofilms (38). A third approach which will likely find an application in biofilm research is the "click-chemistry" technology (243–245). In this approach, amino acid analogs are incorporated into the protein (or peptidoglycan) of actively growing cells. The analogs are then labeled with a fluorescent probe to identify the cells that are actively involved in protein or cell wall biosynthesis.

Most fluorescent dyes used in biofilm studies stain cellular components. Characterizing the extracellular matrix material by CSLM has been more challenging. In fact, the extracellular matrix of biofilms has been referred to as the "dark matter of biofilms" since it is difficult to image, having approximately the same refractive index as water (4). The biofilm matrix material is difficult to stain because it is complex. The matrix material is composed of a combination of polysaccharides, proteins, and extracellular DNA. The composition, particularly the polysaccharide components, and the relative amounts of the components vary for different species, and even for different strains, of bacteria. Therefore, it is unlikely that there will ever be a universally effective biofilm matrix stain. A few fluorescent stains are becoming available for CSLM studies of biofilm matrix components (Fig. 6). In particular, fluorescent lectins bind to specific sugars and can be used to stain certain extracellular polysaccharides. Since lectins are large molecules, they do not penetrate biofilms well. Ideally, small molecule stains will become available for staining the polysaccharide component of biofilms, such as Calcofluor, which is used to stain biofilms of strains that produce cellulose (246). Another component of the biofilm extracellular matrix material is extracellular DNA. Various stains are available that bind DNA, with the TOTO-1 iodide stain providing excellent contrast between the biofilm eDNA component and the biofilm cells (73).

While the extracellular matrix material of biofilms is still difficult to image, it can be characterized by spectroscopic techniques. Attenuated total reflection Fourier transform infrared spectrometry (ATR/FT-IR) is a nondestructive technique that has been used to determine the chemical composition of biofilms by analysis of absorbance of infrared light by particular chemical groups (91, 247–249). For example, ATR/FT-IR was used to quantify changes in the biomass and chemical composition of extracellular matrix material in *P. aeruginosa* and alginate O acetylation–deficient mutants of *P. aeruginosa* (91). Synchrotron-radiation-based FT-IR with a microfluidic system was used to determine the responses of biofilm to antibiotic exposure and assaying uptake of antibiotics, cell lysis, and stability of biofilm by

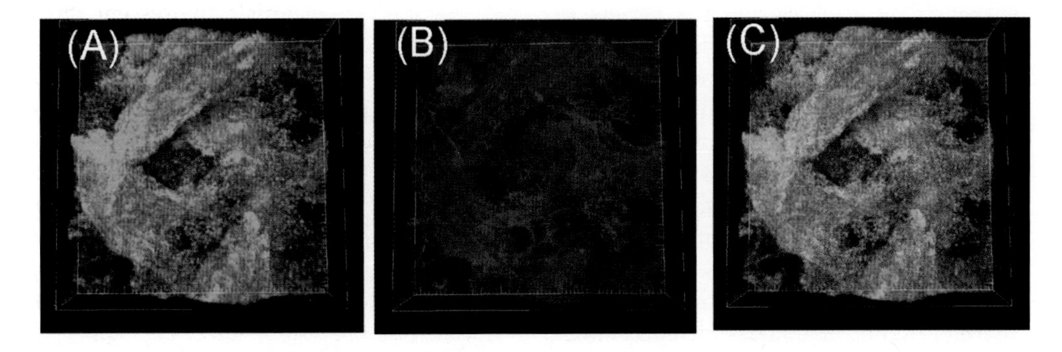

FIGURE 6 (A) Images of *P. aeruginosa* biofilms, where the cells constitutively express the GFP. (B) Extracellular matrix material stained with Bodipy 630/650-X NHS from Life Technologies. (C) Combined image showing GFP-fluorescent bacteria and Bodipy-stained matrix (M.J. Franklin, unpublished data). doi:10.1128/microbiolspec.MB-0016-2014.f6

detecting changes in chemical composition (250). Raman spectrometry has also been used to define the chemistry of matrix materials in biofilms (251, 252). Raman spectrometry is based on the light scattering patterns detected after irradiation of a sample with monochromatic light. The frequency of scattered light differs among substrates and can be used to study the chemical composition of biofilm. Raman spectrometry with CSLM was used to determine spatial distribution biomass and water as well as chemical composition of wild-type and small colony variant *P. aeruginosa* biofilms (253, 254). Modified Raman spectroscopy–based analysis such as surface-enhanced Raman scattering has increased sensitivity and may detect components in the EPS that are not detectable by Raman microscopy (255–258). Finally, nuclear magnetic resonance imaging has been used to characterize water dynamics within biofilms, as well as molecular dynamics and diffusion of biomolecules in biofilms (79, 80). These analytical techniques are nondestructive methods for real-time detection of spatial heterogeneity of the chemical environment within biofilms.

SUMMARY

Most microbial life grows in association with surfaces. Microorganisms that adapt to growth on surfaces adopt a variety of complex behaviors that include adhesion to the surface, expression of genes, and enzymatic activities that allow adaptation to a sessile lifestyle. The activities of microorganisms in biofilms are not uniform across the entire biofilm, and cells adapt to their local environments. Many complementary technologies are required to study the metabolic activities and the physical and chemical environments of biofilm-associated communities. These technologies include molecular approaches, such as gene expression and metabolomics studies to characterize the activity of the microbial cells, analytical approaches to understand the chemistry of biofilm organisms and their matrix materials, imaging approaches to characterize structure-function relationships of the microbial cells, and miniaturization studies to characterize the physical and heterogeneous properties of biofilms. By combining these and other new techniques, it is now possible to gain insight into these complex microbial communities from the gene to the systems level.

ACKNOWLEDGMENTS

This review is dedicated to the late Dr. David C. White, who spent much of his productive career developing innovative approaches designed to characterize microorganisms and their metabolic activities in natural environments. The authors thank Dr. Phil Stewart, Ms. Kerry Williamson, and Ms. Betsey Pitts for their helpful discussions of this review. This work was supported in part by NIH/NIAID award AI113330 (M.J.F.).

Conflicts of interest: We disclose no conflicts.

CITATION

Franklin MJ, Chang C, Akiyama T, Bothner B. 2015. New technologies for studying biofilms. Microbiol Spectrum 3(4):MB-0016-2014.

REFERENCES

1. **Costerton JW, Cheng KJ, Geesey GG, Ladd TI, Nickel NC, Dosgupton M, Marine IJ.** 1987. Bacterial biofilms in nature and disease. *Annu Rev Microbiol* **41:**435–461.
2. **Costerton JW, Lewandovski Z, Caldwell DE, Korber DR, Lappin-Scott HM.** 1995. Microbial biofilms. *Annu Rev Microbiol* **49:**711–745.
3. **Costerton JW, Stewart PS, Greenberg EP.** 1999. Bacterial biofilms: a common cause of persistent infections. *Science* **284:**1318–1322.
4. **Flemming HC, Wingender J.** 2010. The biofilm matrix. *Nat Rev* **8:**623–633.
5. **Hung C, Zhou Y, Pinkner JS, Dodson KW, Crowley JR, Heuser J, Chapman MR, Hadjifrangiskou M, Henderson JP, Hultgren**

SJ. 2013. *Escherichia coli* biofilms have an organized and complex extracellular matrix structure. *MBio* **4:**e00645-00613. doi:10.1128/mBio.00645-13.

6. **Soto GE, Hultgren SJ.** 1999. Bacterial adhesins: common themes and variations in architecture and assembly. *J Bacteriol* **181:** 1059–1071.
7. **Fux CA, Costerton JW, Stewart PS, Stoodley P.** 2005. Survival strategies of infectious biofilms. *Trends Microbiol* **13:**34–40.
8. **Singh R, Ray P, Das A, Sharma M.** 2009. Role of persisters and small-colony variants in antibiotic resistance of planktonic and biofilm-associated *Staphylococcus aureus*: an *in vitro* study. *J Med Microbiol* **58:**1067–1073.
9. **van de Mortel M, Halverson LJ.** 2004. Cell envelope components contributing to biofilm growth and survival of *Pseudomonas putida* in low-water-content habitats. *Mol Microbiol* **52:**735–750.
10. **Davies DG, Parsek MR, Pearson JP, Iglewski BH, Costerton JW, Greenberg EP.** 1998. The involvement of cell-to-cell signals in the development of a bacterial biofilm. *Science* **280:**295–298.
11. **De Kievit TR, Iglewski BH.** 1999. Quorum sensing, gene expression, and *Pseudomonas* biofilms. *Methods Enzymol* **310:**117–128.
12. **Fuqua WC, Winans SC, Greenberg EP.** 1994. Quorum sensing in bacteria: the LuxR-LuxI family of cell density-responsive transcriptional regulators. *J Bacteriol* **176:**269–275.
13. **Parsek MR, Greenberg EP.** 1999. Quorum sensing signals in development of *Pseudomonas aeruginosa* biofilms. *Methods Enzymol* **310:**43–55.
14. **Parsek MR, Greenberg EP.** 2000. Acyl-homoserine lactone quorum sensing in Gram-negative bacteria: a signaling mechanism involved in associations with higher organisms. *Proc Natl Acad Sci USA* **97:**8789–8793.
15. **Passador L, Iglewski BH.** 1995. Quorum sensing and virulence gene regulation in *Pseudomonas aeruginosa*, p 65–78. *In* Roth JA et al. (ed), *Virulence Mechanisms of Bacterial Pathogens.* American Society for Microbiology, Washington D.C.
16. **Pesci EC, Pearson JP, Seed PC, Iglewski BH.** 1997. Regulation of *las* and *rhl* quorum sensing in *Pseudomonas aeruginosa. J Bacteriol* **179:**3127–3132.
17. **Ueda A, Wood TK.** 2009. Connecting quorum sensing, c-di-GMP, pel polysaccharide, and biofilm formation in *Pseudomonas aeruginosa* through tyrosine phosphatase TpbA (PA3885). *PLoS Pathog* **5:**e1000483. doi:10.1371/journal.ppat.1000483.
18. **Whiteley M, Lee KM, Greenberg EP.** 1999. Identification of genes controlled by quorum sensing in *Pseudomonas aeruginosa. Proc Natl Acad Sci USA* **96:**13904–13909.
19. **Zhu J, Mekalanos JJ.** 2003. Quorum sensing-dependent biofilms enhance colonization in *Vibrio cholerae. Dev Cell* **5:**647–656.
20. **Brint JM, Ohman DE.** 1995. Synthesis of multiple exoproducts in *Pseudomonas aeruginosa* is under the control of RhlR-RhlI, another set of regulators in strain PAO1 with homology to the autoinducer-responsive LuxR-LuxI family. *J Bacteriol* **177:**7155–7163.
21. **Pearson JP, Pesci EC, Iglewski BH.** 1997. Roles of *Pseudomonas aeruginosa* las and rhl quorum-sensing systems in control of elastase and rhamnolipid biosynthesis genes. *J Bacteriol* **179:**5756–5767.
22. **Tielen P, Rosenau F, Wilhelm S, Jaeger KE, Flemming HC, Wingender J.** 2010. Extracellular enzymes affect biofilm formation of mucoid *Pseudomonas aeruginosa. Microbiology* **156:**2239–2252.
23. **Glessner A, Smith RS, Iglewski BH, Robinson JA.** 1999. Roles of *Pseudomonas aeruginosa las* and *rhl* quorum-sensing systems in control of twitching motility. *J Bacteriol* **181:**1623–1629.
24. **Heydorn A, Ersboll B, Kato J, Hentzer M, Parsek MR, Tolker-Nielsen T, Givskov M, Molin S.** 2002. Statistical analysis of *Pseudomonas aeruginosa* biofilm development: impact of mutations in genes involved in twitching motility, cell-to-cell signaling, and stationary-phase sigma factor expression. *Appl Environ Microbiol* **68:**2008–2017.
25. **Heurlier K, Williams F, Heeb S, Dormond C, Pessi G, Singer D, Camara M, Williams P, Haas D.** 2004. Positive control of swarming, rhamnolipid synthesis, and lipase production by the posttranscriptional RsmA/RsmZ system in *Pseudomonas aeruginosa* PAO1. *J Bacteriol* **186:**2936–2945.
26. **O'Toole GA, Kolter R.** 1998. Flagellar and twitching motility are necessary for *Pseudomonas aeruginosa* biofilm development. *Mol Microbiol* **30:**295–304.
27. **Shrout JD, Chopp DL, Just CL, Hentzer M, Givskov M, Parsek MR.** 2006. The impact of quorum sensing and swarming motility on *Pseudomonas aeruginosa* biofilm formation is nutritionally conditional. *Mol Microbiol* **62:**1264–1277.
28. **Wang Q, Frye JG, McClelland M, Harshey RM.** 2004. Gene expression patterns during swarming in *Salmonella typhimurium*: genes specific to surface growth and putative new motility and pathogenicity genes. *Mol Microbiol* **52:**169–187.

29. **Aguilar C, Vlamakis H, Losick R, Kolter R.** 2007. Thinking about *Bacillus subtilis* as a multicellular organism. *Curr Opin Microbiol* **10:**638–643.
30. **Basler M, Ho BT, Mekalanos JJ.** 2013. Tit-for-tat: type VI secretion system counterattack during bacterial cell-cell interactions. *Cell* **152:**884–894.
31. **Basler M, Pilhofer M, Henderson GP, Jensen GJ, Mekalanos JJ.** 2012. Type VI secretion requires a dynamic contractile phage tail-like structure. *Nature* **483:**182–186.
32. **Gibbs KA, Urbanowski ML, Greenberg EP.** 2008. Genetic determinants of self identity and social recognition in bacteria. *Science* **321:** 256–259.
33. **Moscoso JA, Mikkelsen H, Heeb S, Williams P, Filloux A.** 2011. The *Pseudomonas aeruginosa* sensor RetS switches type III and type VI secretion via c-di-GMP signalling. *Environ Microbiol* **13:**3128–3138.
34. **Stewart PS, Franklin MJ.** 2008. Physiological heterogeneity in biofilms. *Nat Rev* **6:**199–210.
35. **Becraft ED, Cohan FM, Kuhl M, Jensen SI, Ward DM.** 2011. Fine-scale distribution patterns of *Synechococcus* ecological diversity in microbial mats of Mushroom Spring, Yellowstone National Park. *Appl Environ Microbiol* **77:**7689–7697.
36. **Kim W, Racimo F, Schluter J, Levy SB, Foster KR.** 2014. Importance of positioning for microbial evolution. *Proc Natl Acad Sci USA* **111:**E1639–E1647.
37. **Ramsing NB, Ferris MJ, Ward DM.** 2000. Highly ordered vertical structure of *Synechococcus* populations within the one-millimeter-thick photic zone of a hot spring cyanobacterial mat. *Appl Environ Microbiol* **66:**1038–1049.
38. **Rani SA, Pitts B, Beyenal H, Veluchamy RA, Lewandowski Z, Davison WM, Buckingham-Meyer K, Stewart PS.** 2007. Spatial patterns of DNA replication, protein synthesis, and oxygen concentration within bacterial biofilms reveal diverse physiological states. *J Bacteriol* **189:**4223–4233.
39. **Rasmussen K, Lewandowski Z.** 1998. Microelectrode measurements of local mass transport rates in heterogeneous biofilms. *Biotechnol Bioeng* **59:**302–309.
40. **Werner E, Roe F, Bugnicourt A, Franklin MJ, Heydorn A, Molin S, Pitts B, Stewart PS.** 2004. Stratified growth in *Pseudomonas aeruginosa* biofilms. *Appl Environ Microbiol* **70:**6188–6196.
41. **Boles BR, Thoendel M, Singh PK.** 2004. Self-generated diversity produces "insurance effects" in biofilm communities. *Proc Natl Acad Sci USA* **101:**16630–16635.
42. **Hansen SK, Rainey PB, Haagensen JA, Molin S.** 2007. Evolution of species interactions in a biofilm community. *Nature* **445:**533–536.
43. **Allegrucci M, Sauer K.** 2007. Characterization of colony morphology variants isolated from *Streptococcus pneumoniae* biofilms. *J Bacteriol* **189:**2030–2038.
44. **Hansen SK, Haagensen JA, Gjermansen M, Jorgensen TM, Tolker-Nielsen T, Molin S.** 2007. Characterization of a *Pseudomonas putida* rough variant evolved in a mixed-species biofilm with *Acinetobacter* sp. strain C6. *J Bacteriol* **189:**4932–4943.
45. **Kirisits MJ, Prost L, Starkey M, Parsek MR.** 2005. Characterization of colony morphology variants isolated from *Pseudomonas aeruginosa* biofilms. *Appl Environ Microbiol* **71:**4809–4821.
46. **McEllistrem MC, Ransford JV, Khan SA.** 2007. Characterization of *in vitro* biofilm-associated pneumococcal phase variants of a clinically relevant serotype 3 clone. *J Clin Microbiol* **45:**97–101.
47. **Valle J, Vergara-Irigaray M, Merino N, Penades JR, Lasa I.** 2007. sigmaB regulates IS256-mediated *Staphylococcus aureus* biofilm phenotypic variation. *J Bacteriol* **189:**2886–2896.
48. **Baty AM 3rd, Eastburn CC, Diwu Z, Techkarnjanaruk S, Goodman AE, Geesey GG.** 2000. Differentiation of chitinase-active and non-chitinase-active subpopulations of a marine bacterium during chitin degradation. *Appl Environ Microbiol* **66:**3566–3573.
49. **Baty AM 3rd, Eastburn CC, Techkarnjanaruk S, Goodman AE, Geesey GG.** 2000. Spatial and temporal variations in chitinolytic gene expression and bacterial biomass production during chitin degradation. *Appl Environ Microbiol* **66:**3574–3585.
50. **Vlamakis H, Aguilar C, Losick R, Kolter R.** 2008. Control of cell fate by the formation of an architecturally complex bacterial community. *Genes Dev* **22:**945–953.
51. **Gelens L, Hill L, Vandervelde A, Danckaert J, Loris R.** 2013. A general model for toxin-antitoxin module dynamics can explain persister cell formation in *E. coli*. *PLoS Comput Biol* **9:** e1003190. doi:10.1371/journal.pcbi.1003190.
52. **Koh RS, Dunlop MJ.** 2012. Modeling suggests that gene circuit architecture controls phenotypic variability in a bacterial persistence network. *BMC Syst Biol* **6:**47.
53. **Lewis K.** 2007. Persister cells, dormancy and infectious disease. *Nat Rev* **5:**48–56.
54. **Mulcahy LR, Burns JL, Lory S, Lewis K.** 2010. Emergence of *Pseudomonas aeruginosa*

strains producing high levels of persister cells in patients with cystic fibrosis. *J Bacteriol* **192:**6191–6199.

55. **Denef VJ, Mueller RS, Banfield JF.** 2010. AMD biofilms: using model communities to study microbial evolution and ecological complexity in nature. *ISME J* **4:**599–610.
56. **Grzymski JJ, Murray AE, Campbell BJ, Kaplarevic M, Gao GR, Lee C, Daniel R, Ghadiri A, Feldman RA, Cary SC.** 2008. Metagenome analysis of an extreme microbial symbiosis reveals eurythermal adaptation and metabolic flexibility. *Proc Natl Acad Sci USA* **105:**17516–17521.
57. **Inskeep WP, Rusch DB, Jay ZJ, Herrgard MJ, Kozubal MA, Richardson TH, Macur RE, Hamamura N, Jennings R, Fouke BW, Reysenbach AL, Roberto F, Young M, Schwartz A, Boyd ES, Badger JH, Mathur EJ, Ortmann AC, Bateson M, Geesey G, Frazier M.** 2010. Metagenomes from high-temperature chemotrophic systems reveal geochemical controls on microbial community structure and function. *PLoS One* **5:**e9773. doi:10.1371/journal.pone.0009773.
58. **Xu P, Gunsolley J.** 2014. Application of metagenomics in understanding oral health and disease. *Virulence* **5:**424–432.
59. **Ishii S, Suzuki S, Norden-Krichmar TM, Tenney A, Chain PS, Scholz MB, Nealson KH, Bretschger O.** 2013. A novel metatranscriptomic approach to identify gene expression dynamics during extracellular electron transfer. *Nat Commun* **4:**1601.
60. **Folsom JP, Richards L, Pitts B, Roe F, Ehrlich GD, Parker A, Mazurie A, Stewart PS.** 2010. Physiology of *Pseudomonas aeruginosa* in biofilms as revealed by transcriptome analysis. *BMC Microbiol* **10:**294.
61. **Lenz AP, Williamson KS, Pitts B, Stewart PS, Franklin MJ.** 2008. Localized gene expression in *Pseudomonas aeruginosa* biofilms. *Appl Environ Microbiol* **74:**4463–4471.
62. **Williamson KS, Richards LA, Perez-Osorio AC, Pitts B, McInnerney K, Stewart PS, Franklin MJ.** 2012. Heterogeneity in *Pseudomonas aeruginosa* biofilms includes expression of ribosome hibernation factors in the antibiotic-tolerant subpopulation and hypoxia-induced stress response in the metabolically active population. *J Bacteriol* **194:**2062–2073.
63. **Beale DJ, Barratt R, Marlow DR, Dunn MS, Palombo EA, Morrison PD, Key C.** 2013. Application of metabolomics to understanding biofilms in water distribution systems: a pilot study. *Biofouling* **29:**283–294.
64. **Secor PR, Jennings LK, James GA, Kirker KR, Pulcini ED, McInnerney K, Gerlach R, Livinghouse T, Hilmer JK, Bothner B, Fleckman P, Olerud JE, Stewart PS.** 2013. Phevalin (aureusimine B) production by *Staphylococcus aureus* biofilm and impacts on human keratinocyte gene expression. *PLoS One* **7:** e40973. doi:10.1371/journal.pone.0040973.
65. **Klayman BJ, Klapper I, Stewart PS, Camper AK.** 2008. Measurements of accumulation and displacement at the single cell cluster level in *Pseudomonas aeruginosa* biofilms. *Environ Microbiol* **10:**2344–2354.
66. **Lam AJ, St-Pierre F, Gong Y, Marshall JD, Cranfill PJ, Baird MA, McKeown MR, Wiedenmann J, Davidson MW, Schnitzer MJ, Tsien RY, Lin MZ.** 2012. Improving FRET dynamic range with bright green and red fluorescent proteins. *Nat Methods* **9:**1005–1012.
67. **Shaner NC, Lin MZ, McKeown MR, Steinbach PA, Hazelwood KL, Davidson MW, Tsien RY.** 2008. Improving the photostability of bright monomeric orange and red fluorescent proteins. *Nat Methods* **5:**545–551.
68. **Shaner NC, Steinbach PA, Tsien RY.** 2005. A guide to choosing fluorescent proteins. *Nat Methods* **2:**905–909.
69. **McLoon AL, Kolodkin-Gal I, Rubinstein SM, Kolter R, Losick R.** 2011. Spatial regulation of histidine kinases governing biofilm formation in *Bacillus subtilis*. *J Bacteriol* **193:**679–685.
70. **Allesen-Holm M, Barken KB, Yang L, Klausen M, Webb JS, Kjelleberg S, Molin S, Givskov M, Tolker-Nielsen T.** 2006. A characterization of DNA release in *Pseudomonas aeruginosa* cultures and biofilms. *Mol Microbiol* **59:**1114–1128.
71. **Baird FJ, Wadsworth MP, Hill JE.** 2012. Evaluation and optimization of multiple fluorophore analysis of a *Pseudomonas aeruginosa* biofilm. *J Microbiol Methods* **90:**192–196.
72. **Chen MY, Lee DJ, Tay JH, Show KY.** 2007. Staining of extracellular polymeric substances and cells in bioaggregates. *Appl Microbiol Biotechnol* **75:**467–474.
73. **Gloag ES, Turnbull L, Huang A, Vallotton P, Wang H, Nolan LM, Mililli L, Hunt C, Lu J, Osvath SR, Monahan LG, Cavaliere R, Charles IG, Wand MP, Gee ML, Prabhakar R, Whitchurch CB.** 2013. Self-organization of bacterial biofilms is facilitated by extracellular DNA. *Proc Natl Acad Sci USA* **110:**11541–11546.
74. **Amann R, Ludwig W.** 2000. Ribosomal RNA-targeted nucleic acid probes for studies in microbial ecology. *FEMS Microbiol Rev* **24:**555–565.

75. **Brileya KA, Camilleri LB, Fields MW.** 2014. 3D-fluorescence *in situ* hybridization of intact, anaerobic biofilm. *Methods Mol Biol* **1151:**189–197.
76. **DeLong EF, Wickham GS, Pace NR.** 1989. Phylogenetic stains: ribosomal RNA-based probes for the identification of single cells. *Science* **243:**1360–1363.
77. **Moller S, Pedersen AR, Poulsen LK, Arvin E, Molin S.** 1996. Activity and three-dimensional distribution of toluene-degrading *Pseudomonas putida* in a multispecies biofilm assessed by quantitative *in situ* hybridization and scanning confocal laser microscopy. *Appl Environ Microbiol* **62:**4632–4640.
78. **Schramm A, De Beer D, Wagner M, Amann R.** 1998. Identification and activities *in situ* of *Nitrosospira* and *Nitrospira* spp. as dominant populations in a nitrifying fluidized bed reactor. *Appl Environ Microbiol* **64:**3480–3485.
79. **Hornemann JA, Codd SL, Fell RJ, Stewart PS, Seymour JD.** 2009. Secondary flow mixing due to biofilm growth in capillaries of varying dimensions. *Biotechnol Bioeng* **103:**353–360.
80. **Hornemann JA, Lysova AA, Codd SL, Seymour JD, Busse SC, Stewart PS, Brown JR.** 2008. Biopolymer and water dynamics in microbial biofilm extracellular polymeric substance. *Biomacromolecules* **9:**2322–2328.
81. **O'Toole GA, Kolter R.** 1998. Initiation of biofilm formation in *Pseudomonas fluorescens* WCS365 proceeds via multiple, convergent signalling pathways: a genetic analysis. *Mol Microbiol* **28:**449–461.
82. **Ceri H, Olson ME, Stremick C, Read RR, Morck D, Buret A.** 1999. The Calgary biofilm device: new technology for rapid determination of antibiotic susceptibilities of bacterial biofilms. *J Clin Microbiol* **37:**1771–1776.
83. **Branda SS, González-Pastor JE, Ben-Yehuda S, Losick R, Kolter R.** 2001. Fruiting body formation by *Bacillus subtilis*. *Proc Natl Acad Sci USA* **98:**11621–11626.
84. **Chimileski S, Franklin MJ, Papke RT.** 2014. Biofilms formed by the archaeon *Haloferax volcanii* exhibit cellular differentiation and social motility, and facilitate horizontal gene transfer. *BMC Biol* **12:**65.
85. **Wentland E, Stewart PS, Huang C-T, McFeters GA.** 1996. Spatial variations in growth rate within *Klebsiella pneumoniae* colonies and biofilm. *Biotechnol Prog* **12:**316–321.
86. **Huang C, McFeters G, Stewart P.** 1996. Evaluation of physiological staining, cryoembedding, and autofluorescence quenching techniques on fouling biofilms. *Biofouling* **9:**269–277.
87. **Anderl JN, Franklin MJ, Stewart PS.** 2000. Role of antibiotic penetration limitation in *Klebsiella pneumoniae* biofilm resistance to ampicillin and ciprofloxacin. *Antimicrob Agents Chemother* **44:**1818–1824.
88. **Goeres DM, Loetterle LR, Hamilton MA, Murga R, Kirby DW, Donlan RM.** 2005. Statistical assessment of a laboratory method for growing biofilms. *Microbiology* **151:**757–762.
89. **Goeres DM, Hamilton MA, Beck NA, Buckingham-Meyer K, Hilyard JD, Loetterle LR, Lorenz LA, Walker DK, Stewart PS.** 2009. A method for growing a biofilm under low shear at the air-liquid interface using the drip flow biofilm reactor. *Nat Protoc* **4:**783–788.
90. **Lawrence JR, Korber DR, Hoyle BD, Costerton JW, Caldwell DE.** 1991. Optical sectioning of microbial biofilms. *J Bacteriol* **173:**6558–6567.
91. **Nivens DE, Ohman DE, Williams J, Franklin MJ.** 2001. Role of alginate and its O acetylation in formation of *Pseudomonas aeruginosa* microcolonies and biofilms. *J Bacteriol* **183:**1047–1057.
92. **Stapper AP, Narasimhan G, Ohman DE, Barakat J, Hentzer M, Molin S, Kharazmi A, Hoiby N, Mathee K.** 2004. Alginate production affects *Pseudomonas aeruginosa* biofilm development and architecture, but is not essential for biofilm formation. *J Med Microbiol* **53:**679–690.
93. **Xu KD, Franklin MJ, Park CH, McFeters GA, Stewart PS.** 2001. Gene expression and protein levels of the stationary phase sigma factor, RpoS, in continuously-fed *Pseudomonas aeruginosa* biofilms. *FEMS Microbiol Lett* **199:**67–71.
94. **Lewandowski Z, Beyenal H.** 2001. Limiting-current-type microelectrodes for quantifying mass transport dynamics in biofilms. *Methods Enzymol* **337:**339–359.
95. **Dunsmore BC, Jacobsen A, Hall-Stoodley L, Bass CJ, Lappin-Scott HM, Stoodley P.** 2002. The influence of fluid shear on the structure and material properties of sulphate-reducing bacterial biofilms. *J Ind Microbiol Biotechnol* **29:**347–353.
96. **Rani SA, Pitts B, Stewart PS.** 2005. Rapid diffusion of fluorescent tracers into *Staphylococcus epidermidis* biofilms visualized by time lapse microscopy. *Antimicrob Agents Chemother* **49:**728–732.
97. **Xi C, Marks D, Schlachter S, Luo W, Boppart SA.** 2006. High-resolution three-dimensional imaging of biofilm development using optical coherence tomography. *J Biomed Opt* **11:**34001.

98. **Sackmann EK, Fulton AL, Beebe DJ.** 2014. The present and future role of microfluidics in biomedical research. *Nature* **507:**181–189.
99. **Groisman A, Lobo C, Cho H, Campbell JK, Dufour YS, Stevens AM, Levchenko A.** 2005. A microfluidic chemostat for experiments with bacterial and yeast cells. *Nature Methods* **2:**685–689.
100. **Leung K, Zahn H, Leaver T, Konwar KM, Hanson NW, Pagé AP, Lo C-C, Chain PS, Hallam SJ, Hansen CL.** 2012. A programmable droplet-based microfluidic device applied to multiparameter analysis of single microbes and microbial communities. *Proc Natl Acad Sci USA* **109:**7665–7670.
101. **Locke JC, Elowitz MB.** 2009. Using movies to analyse gene circuit dynamics in single cells. *Nat Rev Microbiol* **7:**383–392.
102. **Nichols D, Cahoon N, Trakhtenberg E, Pham L, Mehta A, Belanger A, Kanigan T, Lewis K, Epstein S.** 2010. Use of ichip for high-throughput *in situ* cultivation of "uncultivable" microbial species. *Appl Environ Microbiol* **76:**2445–2450.
103. **Weibel DB, DiLuzio WR, Whitesides GM.** 2007. Microfabrication meets microbiology. *Nat Rev Microbiol* **5:**209–218.
104. **Wessel AK, Hmelo L, Parsek MR, Whiteley M.** 2013. Going local: technologies for exploring bacterial microenvironments. *Nat Rev Microbiol* **11:**337–348.
105. **Boedicker JQ, Li L, Kline TR, Ismagilov RF.** 2008. Detecting bacteria and determining their susceptibility to antibiotics by stochastic confinement in nanoliter droplets using plug-based microfluidics. *Lab Chip* **8:**1265–1272.
106. **Zhang Q, Lambert G, Liao D, Kim H, Robin K, Tung C-K, Pourmand N, Austin RH.** 2011. Acceleration of emergence of bacterial antibiotic resistance in connected microenvironments. *Science* **333:**1764–1767.
107. **Balaban NQ, Merrin J, Chait R, Kowalik L, Leibler S.** 2004. Bacterial persistence as a phenotypic switch. *Science* **305:**1622–1625.
108. **Gefen O, Gabay C, Mumcuoglu M, Engel G, Balaban NQ.** 2008. Single-cell protein induction dynamics reveals a period of vulnerability to antibiotics in persister bacteria. *Proc Natl Acad Sci USA* **105:**6145–6149.
109. **Boedicker JQ, Vincent ME, Ismagilov RF.** 2009. Microfluidic confinement of single cells of bacteria in small volumes initiates high-density behavior of quorum sensing and growth and reveals Its variability. *Angew Chem Int Ed Engl* **48:**5908–5911.
110. **Balagaddé FK, You L, Hansen CL, Arnold FH, Quake SR.** 2005. Long-term monitoring of bacteria undergoing programmed population control in a microchemostat. *Science* **309:**137–140.
111. **Rowat AC, Bird JC, Agresti JJ, Rando OJ, Weitz DA.** 2009. Tracking lineages of single cells in lines using a microfluidic device. *Proc Natl Acad Sci USA* **106:**18149–18154.
112. **Wang P, Robert L, Pelletier J, Dang WL, Taddei F, Wright A, Jun S.** 2010. Robust growth of *Escherichia coli*. *Curr Biol* **20:**1099–1103.
113. **Ottesen EA, Hong JW, Quake SR, Leadbetter JR.** 2006. Microfluidic digital PCR enables multigene analysis of individual environmental bacteria. *Science* **314:**1464–1467.
114. **Baret J-C, Miller OJ, Taly V, Ryckelynck M, El-Harrak A, Frenz L, Rick C, Samuels ML, Hutchison JB, Agresti JJ.** 2009. Fluorescence-activated droplet sorting (FADS): efficient microfluidic cell sorting based on enzymatic activity. *Lab Chip* **9:**1850–1858.
115. **Gomez-Sjoberg R, Morisette DT, Bashir R.** 2005. Impedance microbiology-on-a-chip: microfluidic bioprocessor for rapid detection of bacterial metabolism. *J Microelectromechanic Syst* **14:**829–838.
116. **Mach AJ, Di Carlo D.** 2010. Continuous scalable blood filtration device using inertial microfluidics. *Biotechnol Bioeng* **107:**302–311.
117. **Wu Z, Willing B, Bjerketorp J, Jansson JK, Hjort K.** 2009. Soft inertial microfluidics for high throughput separation of bacteria from human blood cells. *Lab Chip* **9:**1193–1199.
118. **Mao H, Cremer PS, Manson MD.** 2003. A sensitive, versatile microfluidic assay for bacterial chemotaxis. *Proc Natl Acad Sci USA* **100:**5449–5454.
119. **Stocker R, Seymour JR, Samadani A, Hunt DE, Polz MF.** 2008. Rapid chemotactic response enables marine bacteria to exploit ephemeral microscale nutrient patches. *Proc Natl Acad Sci USA* **105:**4209–4214.
120. **Connell JL, Wessel AK, Parsek MR, Ellington AD, Whiteley M, Shear JB.** 2010. Probing prokaryotic social behaviors with bacterial "lobster traps." *MBio* **1:**e00202-00210. doi:10.1128/mBio.00202-10.
121. **Eun Y-J, Weibel DB.** 2009. Fabrication of microbial biofilm arrays by geometric control of cell adhesion. *Langmuir* **25:**4643–4654.
122. **Kim KP, Kim Y-G, Choi C-H, Kim H-E, Lee S-H, Chang W-S, Lee C-S.** 2010. *In situ* monitoring of antibiotic susceptibility of bacterial biofilms in a microfluidic device. *Lab Chip* **10:**3296–3299.
123. **Kim J, Hegde M, Kim SH, Wood TK, Jayaraman A.** 2012. A microfluidic device for

high throughput bacterial biofilm studies. *Lab Chip* **12:**1157–1163.

124. **Skolimowski M, Nielsen MW, Emnéus J, Molin S, Taboryski R, Sternberg C, Dufva M, Geschke O.** 2010. Microfluidic dissolved oxygen gradient generator biochip as a useful tool in bacterial biofilm studies. *Lab Chip* **10:**2162–2169.
125. **Lee J-H, Kaplan J, Lee W.** 2008. Microfluidic devices for studying growth and detachment of *Staphylococcus epidermidis* biofilms. *Biomed Microdevices* **10:**489–498.
126. **Rusconi R, Lecuyer S, Guglielmini L, Stone HA.** 2010. Laminar flow around corners triggers the formation of biofilm streamers. *J R Soc Interface* **7:**1293–1299.
127. **Drescher K, Shen Y, Bassler BL, Stone HA.** 2013. Biofilm streamers cause catastrophic disruption of flow with consequences for environmental and medical systems. *Proc Natl Acad Sci USA* **110:**4345–4350.
128. **Seminara A, Angelini TE, Wilking JN, Vlamakis H, Ebrahim S, Kolter R, Weitz DA, Brenner MP.** 2011. Osmotic spreading of *Bacillus subtilis* biofilms driven by an extracellular matrix. *Proc Natl Acad Sci USA* **109:**1116–1121.
129. **Wilking JN, Angelini TE, Seminara A, Brenner MP, Weitz DA.** 2011. Biofilms as complex fluids. *MRS Bull* **36:**385–391.
130. **Hohne DN, Younger JG, Solomon MJ.** 2009. Flexible microfluidic device for mechanical property characterization of soft viscoelastic solids such as bacterial biofilms. *Langmuir* **25:**7743–7751.
131. **De La Fuente L, Montanes E, Meng Y, Li Y, Burr TJ, Hoch H, Wu M.** 2007. Assessing adhesion forces of type I and type IV pili of *Xylella fastidiosa* bacteria by use of a microfluidic flow chamber. *Appl Environ Microbiol* **73:**2690–2696.
132. **Nadell CD, Bassler BL.** 2011. A fitness trade-off between local competition and dispersal in *Vibrio cholerae* biofilms. *Proc Natl Acad Sci USA* **108:**14181–14185.
133. **Hong SH, Hegde M, Kim J, Wang X, Jayaraman A, Wood TK.** 2012. Synthetic quorum-sensing circuit to control consortial biofilm formation and dispersal in a microfluidic device. *Nat Commun* **3:**613.
134. **Metzker ML.** 2010. Sequencing technologies: the next generation. *Nat Rev Genet* **11:**31–46.
135. **Flusberg BA, Webster DR, Lee JH, Travers KJ, Olivares EC, Clark TA, Korlach J, Turner SW.** 2010. Direct detection of DNA methylation during single-molecule, real-time sequencing. *Nat Methods* **7:**461–465.
136. **Goddard AF, Staudinger BJ, Dowd SE, Joshi-Datar A, Wolcott RD, Aitken ML, Fligner CL, Singh PK.** 2012. Direct sampling of cystic fibrosis lungs indicates that DNA-based analyses of upper-airway specimens can misrepresent lung microbiota. *Proc Natl Acad Sci USA* **109:**13769–13774.
137. **Chivian D, Brodie EL, Alm EJ, Culley DE, Dehal PS, DeSantis TZ, Gihring TM, Lapidus A, Lin LH, Lowry SR, Moser DP, Richardson PM, Southam G, Wanger G, Pratt LM, Andersen GL, Hazen TC, Brockman FJ, Arkin AP, Onstott TC.** 2008. Environmental genomics reveals a single-species ecosystem deep within Earth. *Science* **322:**275–278.
138. **Dick GJ, Andersson AF, Baker BJ, Simmons SL, Thomas BC, Yelton AP, Banfield JF.** 2009. Community-wide analysis of microbial genome sequence signatures. *Genome Biol* **10:** R85.
139. **Goltsman DS, Dasari M, Thomas BC, Shah MB, VerBerkmoes NC, Hettich RL, Banfield JF.** 2013. New group in the *Leptospirillum* clade: cultivation-independent community genomics, proteomics, and transcriptomics of the new species "*Leptospirillum* group IV UBA BS." *Appl Environ Microbiol* **79:**5384–5393.
140. **Yelton AP, Comolli LR, Justice NB, Castelle C, Denef VJ, Thomas BC, Banfield JF.** 2013. Comparative genomics in acid mine drainage biofilm communities reveals metabolic and structural differentiation of co-occurring archaea. *BMC Genomics* **14:**485.
141. **Inskeep WP, Jay ZJ, Herrgard MJ, Kozubal MA, Rusch DB, Tringe SG, Macur RE, Jennings R, Boyd ES, Spear JR, Roberto FF.** 2013. Phylogenetic and functional analysis of metagenome sequence from high-temperature archaeal habitats demonstrate linkages between metabolic potential and geochemistry. *Front Microbiol* **4:**95.
142. **Klatt CG, Wood JM, Rusch DB, Bateson MM, Hamamura N, Heidelberg JF, Grossman AR, Bhaya D, Cohan FM, Kuhl M, Bryant DA, Ward DM.** 2011. Community ecology of hot spring cyanobacterial mats: predominant populations and their functional potential. *ISME J* **5:**1262–1278.
143. **Klatt CG, Inskeep WP, Herrgard MJ, Jay ZJ, Rusch DB, Tringe SG, Niki Parenteau M, Ward DM, Boomer SM, Bryant DA, Miller SR.** 2013. Community structure and function of high-temperature chlorophototrophic microbial mats inhabiting diverse geothermal environments. *Front Microbiol* **4:**106.
144. **Hasan NA, Young BA, Minard-Smith AT, Saeed K, Li H, Heizer EM, McMillan NJ,**

Isom R, Abdullah AS, Bornman DM, Faith SA, Choi SY, Dickens ML, Cebula TA, Colwell RR. 2014. Microbial community profiling of human saliva using shotgun metagenomic sequencing. *PLoS One* **9:**e97699. doi:10.1371/journal.pone.0097699.

145. **Wang J, Qi J, Zhao H, He S, Zhang Y, Wei S, Zhao F.** 2013. Metagenomic sequencing reveals microbiota and its functional potential associated with periodontal disease. *Sci Rep* **3:**1843.
146. **Lozupone CA, Stombaugh J, Gonzalez A, Ackermann G, Wendel D, Vazquez-Baeza Y, Jansson JK, Gordon JI, Knight R.** 2013. Meta-analyses of studies of the human microbiota. *Genome Res* **23:**1704–1714.
147. **Turnbaugh PJ, Ley RE, Hamady M, Fraser-Liggett CM, Knight R, Gordon JI.** 2007. The human microbiome project. *Nature* **449:**804–810.
148. **Turnbaugh PJ, Hamady M, Yatsunenko T, Cantarel BL, Duncan A, Ley RE, Sogin ML, Jones WJ, Roe BA, Affourtit JP, Egholm M, Henrissat B, Heath AC, Knight R, Gordon JI.** 2009. A core gut microbiome in obese and lean twins. *Nature* **457:**480–484.
149. **Fullwood MJ, Wei CL, Liu ET, Ruan Y.** 2009. Next-generation DNA sequencing of paired-end tags (PET) for transcriptome and genome analyses. *Genome Res* **19:**521–532.
150. **Bankevich A, Nurk S, Antipov D, Gurevich AA, Dvorkin M, Kulikov AS, Lesin VM, Nikolenko SI, Pham S, Prjibelski AD, Pyshkin AV, Sirotkin AV, Vyahhi N, Tesler G, Alekseyev MA, Pevzner PA.** 2012. SPAdes: a new genome assembly algorithm and its applications to single-cell sequencing. *J Comput Biol* **19:**455–477.
151. **Huson DH, Auch AF, Qi J, Schuster SC.** 2007. MEGAN analysis of metagenomic data. *Genome Res* **17:**377–386.
152. **Huson DH, Richter DC, Mitra S, Auch AF, Schuster SC.** 2009. Methods for comparative metagenomics. *BMC Bioinformatics* **10**(Suppl 1):S12.
153. **Namiki T, Hachiya T, Tanaka H, Sakakibara Y.** 2012. MetaVelvet: an extension of Velvet assembler to *de novo* metagenome assembly from short sequence reads. *Nucleic Acids Res* **40:**e155.
154. **Zerbino DR, Birney E.** 2008. Velvet: algorithms for *de novo* short read assembly using de Bruijn graphs. *Genome Res* **18:**821–829.
155. **McLean RJ, Kakirde KS.** 2013. Enhancing metagenomics investigations of microbial interactions with biofilm technology. *Int J Mol Sci* **14:**22246–22257.
156. **Blainey PC.** 2013. The future is now: single-cell genomics of bacteria and archaea. *FEMS Microbiol Rev* **37:**407–427.
157. **Landry ZC, Giovanonni SJ, Quake SR, Blainey PC.** 2013. Optofluidic cell selection from complex microbial communities for single-genome analysis. *Methods Enzymol* **531:**61–90.
158. **McLean JS, Lombardo MJ, Badger JH, Edlund A, Novotny M, Yee-Greenbaum J, Vyahhi N, Hall AP, Yang Y, Dupont CL, Ziegler MG, Chitsaz H, Allen AE, Yooseph S, Tesler G, Pevzner PA, Friedman RM, Nealson KH, Venter JC, Lasken RS.** 2013. Candidate phylum TM6 genome recovered from a hospital sink biofilm provides genomic insights into this uncultivated phylum. *Proc Natl Acad Sci USA* **110:**E2390–E2399.
159. **Dean FB, Hosono S, Fang L, Wu X, Faruqi AF, Bray-Ward P, Sun Z, Zong Q, Du Y, Du J, Driscoll M, Song W, Kingsmore SF, Egholm M, Lasken RS.** 2002. Comprehensive human genome amplification using multiple displacement amplification. *Proc Natl Acad Sci USA* **99:**5261–5266.
160. **Hentzer M, Eberl L, Givskov M.** 2005. Transcriptome analysis of *Pseudomonas aeruginosa* biofilm development: anaerobic respiration and iron limitation. *Biofilms* **2:**37–62.
161. **Mikkelsen H, Duck Z, Lilley KS, Welch M.** 2007. Interrelationships between colonies, biofilms, and planktonic cells of *Pseudomonas aeruginosa*. *J Bacteriol* **189:**2411–2416.
162. **Schoolnik GK, Voskuil MI, Schnappinger D, Yildiz FH, Meibom K, Dolganov NA, Wilson MA, Chong KH.** 2001. Whole genome DNA microarray expression analysis of biofilm development by *Vibrio cholerae* O1 E1 Tor. *Methods Enzymol* **336:**3–18.
163. **Waite RD, Paccanaro A, Papakonstantinopoulou A, Hurst JM, Saqi M, Littler E, Curtis MA.** 2006. Clustering of *Pseudomonas aeruginosa* transcriptomes from planktonic cultures, developing and mature biofilms reveals distinct expression profiles. *BMC Genomics* **7:**162.
164. **Whiteley M, Bangera MG, Bumgarner RE, Parsek MR, Teitzel GM, Lory S, Greenberg EP.** 2001. Gene expression in *Pseudomonas aeruginosa* biofilms. *Nature* **413:**860–864.
165. **Barrett T, Wilhite SE, Ledoux P, Evangelista C, Kim IF, Tomashevsky M, Marshall KA, Phillippy KH, Sherman PM, Holko M, Yefanov A, Lee H, Zhang N, Robertson CL, Serova N, Davis S, Soboleva A.** 2013. NCBI GEO: archive for functional genomics data sets: update. *Nucleic Acids Res* **41:**D991–D995.

166. **Edgar R, Domrachev M, Lash AE.** 2002. Gene Expression Omnibus: NCBI gene expression and hybridization array data repository. *Nucleic Acids Res* **30:**207–210.
167. **Stead MB, Marshburn S, Mohanty BK, Mitra J, Pena Castillo L, Ray D, van Bakel H, Hughes TR, Kushner SR.** 2011. Analysis of *Escherichia coli* RNase E and RNase III activity *in vivo* using tiling microarrays. *Nucleic Acids Res* **39:**3188–3203.
168. **Toledo-Arana A, Dussurget O, Nikitas G, Sesto N, Guet-Revillet H, Balestrino D, Loh E, Gripenland J, Tiensuu T, Vaitkevicius K, Barthelemy M, Vergassola M, Nahori MA, Soubigou G, Regnault B, Coppee JY, Lecuit M, Johansson J, Cossart P.** 2009. The *Listeria* transcriptional landscape from saprophytism to virulence. *Nature* **459:**950–956.
169. **Wang Z, Gerstein M, Snyder M.** 2009. RNA-Seq: a revolutionary tool for transcriptomics. *Nat Rev Genet* **10:**57–63.
170. **Gifford S, Satinsky B, Moran MA.** 2014. Quantitative microbial metatranscriptomics. *Methods Mol Biol* **1096:**213–229.
171. **Moran MA, Satinsky B, Gifford SM, Luo H, Rivers A, Chan LK, Meng J, Durham BP, Shen C, Varaljay VA, Smith CB, Yager PL, Hopkinson BM.** 2013. Sizing up metatranscriptomics. *ISME J* **7:**237–243.
172. **Jorth P, Turner KH, Gumus P, Nizam N, Buduneli N, Whiteley M.** 2014. Metatranscriptomics of the human oral microbiome during health and disease. *MBio* **5:**e01012-01014. doi:10.1128/mBio.01012-14.
173. **Klatt CG, Liu Z, Ludwig M, Kuhl M, Jensen SI, Bryant DA, Ward DM.** 2013. Temporal metatranscriptomic patterning in phototrophic *Chloroflexi* inhabiting a microbial mat in a geothermal spring. *ISME J* **7:**1775–1789.
174. **Liu Z, Klatt CG, Wood JM, Rusch DB, Ludwig M, Wittekindt N, Tomsho LP, Schuster SC, Ward DM, Bryant DA.** 2011. Metatranscriptomic analyses of chlorophototrophs of a hot-spring microbial mat. *ISME J* **5:**1279–1290.
175. **Cattoir V, Narasimhan G, Skurnik D, Aschard H, Roux D, Ramphal R, Jyot J, Lory S.** 2013. Transcriptional response of mucoid *Pseudomonas aeruginosa* to human respiratory mucus. *MBio* **3:**e00410-00412. doi:10.1128/mBio.00410-12.
176. **Dugar G, Herbig A, Forstner KU, Heidrich N, Reinhardt R, Nieselt K, Sharma CM.** 2013. High-resolution transcriptome maps reveal strain-specific regulatory features of multiple *Campylobacter jejuni* isolates. *PLoS Genet* **9:** e1003495. doi:10.1371/journal.pgen.1003495.
177. **Mandlik A, Livny J, Robins WP, Ritchie JM, Mekalanos JJ, Waldor MK.** 2011. RNA-Seq-based monitoring of infection-linked changes in *Vibrio cholerae* gene expression. *Cell Host Microbe* **10:**165–174.
178. **Taveirne ME, Theriot CM, Livny J, DiRita VJ.** 2013. The complete *Campylobacter jejuni* transcriptome during colonization of a natural host determined by RNAseq. *PLoS One* **8:** e73586. doi:10.1371/journal.pone.0073586.
179. **Pérez-Osorio AC, Williamson KS, Franklin MJ.** 2010. Heterogeneous *rpoS* and *rhlR* mRNA levels and 16S rRNA/rDNA (rRNA gene) ratios within *Pseudomonas aeruginosa* biofilms, sampled by laser capture microdissection. *J Bacteriol* **192:**2991–3000.
180. **Van Oudenhove L, Devreese B.** 2013. A review on recent developments in mass spectrometry instrumentation and quantitative tools advancing bacterial proteomics. *Appl Microbiol Biotechnol* **97:**4749–4762.
181. **Vaudel M, Sickmann A, Martens L.** 2013. Introduction to opportunities and pitfalls in functional mass spectrometry based proteomics. *Biochim Biophys Acta* **1844:**12–20.
182. **Sauer K, Camper AK, Ehrlich GD, Costerton JW, Davies DG.** 2002. *Pseudomonas aeruginosa* displays multiple phenotypes during development as a biofilm. *J Bacteriol* **184:**1140–1154.
183. **Boes N, Schreiber K, Hartig E, Jaensch L, Schobert M.** 2006. The *Pseudomonas aeruginosa* universal stress protein PA4352 is essential for surviving anaerobic energy stress. *J Bacteriol* **188:**6529–6538.
184. **Fong JC, Karplus K, Schoolnik GK, Yildiz FH.** 2006. Identification and characterization of RbmA, a novel protein required for the development of rugose colony morphology and biofilm structure in *Vibrio cholerae*. *J Bacteriol* **188:**1049–1059.
185. **Kalmokoff M, Lanthier P, Tremblay TL, Foss M, Lau PC, Sanders G, Austin J, Kelly J, Szymanski CM.** 2006. Proteomic analysis of *Campylobacter jejuni* 11168 biofilms reveals a role for the motility complex in biofilm formation. *J Bacteriol* **188:**4312–4320.
186. **Patrauchan MA, Sarkisova SA, Franklin MJ.** 2007. Strain-specific proteome responses of *Pseudomonas aeruginosa* to biofilm-associated growth and to calcium. *Microbiology* **153:**3838–3851.
187. **Sarkisova S, Patrauchan MA, Berglund D, Nivens DE, Franklin MJ.** 2005. Calcium-induced virulence factors associated with the extracellular matrix of mucoid *Pseudomonas aeruginosa* biofilms. *J Bacteriol* **187:**4327–4337.

188. **Rao J, Damron FH, Basler M, Digiandomenico A, Sherman NE, Fox JW, Mekalanos JJ, Goldberg JB.** 2011. Comparisons of two proteomic analyses of non-mucoid and mucoid *Pseudomonas aeruginosa* clinical isolates from a cystic fibrosis patient. *Front Microbiol* **2:**162.
189. **Sauer K, Camper AK.** 2001. Characterization of phenotypic changes in *Pseudomonas putida* in response to surface-associated growth. *J Bacteriol* **183:**6579–6589.
190. **Schreiber K, Boes N, Eschbach M, Jaensch L, Wehland J, Bjarnsholt T, Givskov M, Hentzer M, Schobert M.** 2006. Anaerobic survival of *Pseudomonas aeruginosa* by pyruvate fermentation requires an Usp-type stress protein. *J Bacteriol* **188:**659–668.
191. **Wen ZT, Baker HV, Burne RA.** 2006. Influence of BrpA on critical virulence attributes of *Streptococcus mutans. J Bacteriol* **188:**2983–2992.
192. **Unlu M, Morgan ME, Minden JS.** 1997. Difference gel electrophoresis: a single gel method for detecting changes in protein extracts. *Electrophoresis* **18:**2071–2077.
193. **Tonge R, Shaw J, Middleton B, Rowlinson R, Rayner S, Young J, Pognan F, Hawkins E, Currie I, Davison M.** 2001. Validation and development of fluorescence two-dimensional differential gel electrophoresis proteomics technology. *Proteomics* **1:**377–396.
194. **Marouga R, David S, Hawkins E.** 2005. The development of the DIGE system: 2D fluorescence difference gel analysis technology. *Anal Bioanal Chem* **382:**669–678.
195. **Maaty WS, Wiedenheft B, Tarlykov P, Schaff N, Heinemann J, Robison-Cox J, Valenzuela J, Dougherty A, Blum P, Lawrence CM, Douglas T, Young MJ, Bothner B.** 2009. Something old, something new, something borrowed; how the thermoacidophilic archaeon *Sulfolobus solfataricus* responds to oxidative stress. *PLoS One* **4:**e6964. doi:10.1371/journal.pone.0006964.
196. **Petrova OE, Sauer K.** 2012. Dispersion by *Pseudomonas aeruginosa* requires an unusual posttranslational modification of BdlA. *Proc Natl Acad Sci USA* **109:**16690–16695.
197. **Denef VJ, Shah MB, Verberkmoes NC, Hettich RL, Banfield JF.** 2007. Implications of strain- and species-level sequence divergence for community and isolate shotgun proteomic analysis. *J Proteome Res* **6:**3152–3161.
198. **Jiao Y, D'Haeseleer P, Dill BD, Shah M, Verberkmoes NC, Hettich RL, Banfield JF, Thelen MP.** 2011. Identification of biofilm matrix-associated proteins from an acid mine drainage microbial community. *Appl Environ Microbiol* **77:**5230–5237.
199. **Pessi G, Braunwalder R, Grunau A, Omasits U, Ahrens CH, Eberl L.** 2013. Response of *Burkholderia cenocepacia* H111 to micro-oxia. *PLoS One* **8:**e72939. doi:10.1371/journal.pone.0072939.
200. **Santi L, Beys-da-Silva WO, Berger M, Calzolari D, Guimaraes JA, Moresco JJ, Yates JR 3rd.** 2014. Proteomic profile of *Cryptococcus neoformans* biofilm reveals changes in metabolic processes. *J Proteome Res* **13:**1545–1559.
201. **Schmid N, Pessi G, Deng Y, Aguilar C, Carlier AL, Grunau A, Omasits U, Zhang LH, Ahrens CH, Eberl L.** 2012. The AHL- and BDSF-dependent quorum sensing systems control specific and overlapping sets of genes in *Burkholderia cenocepacia* H111. *PLoS One* **7:**e49966. doi:10.1371/journal.pone.0049966.
202. **Wu WL, Liao JH, Lin GH, Lin MH, Chang YC, Liang SY, Yang FL, Khoo KH, Wu SH.** 2013. Phosphoproteomic analysis reveals the effects of PilF phosphorylation on type IV pilus and biofilm formation in *Thermus thermophilus* HB27. *Mol Cell Proteomics* **12:**2701–2713.
203. **Vera M, Krok B, Bellenberg S, Sand W, Poetsch A.** 2013. Shotgun proteomics study of early biofilm formation process of *Acidithiobacillus ferrooxidans* ATCC 23270 on pyrite. *Proteomics* **13:**1133–1144.
204. **Wolters DA, Washburn MP, Yates JR 3rd.** 2001. An automated multidimensional protein identification technology for shotgun proteomics. *Anal Chem* **73:**5683–5690.
205. **Cabral MP, Soares NC, Aranda J, Parreira JR, Rumbo C, Poza M, Valle J, Calamia V, Lasa I, Bou G.** 2011. Proteomic and functional analyses reveal a unique lifestyle for *Acinetobacter baumannii* biofilms and a key role for histidine metabolism. *J Proteome Res* **10:**3399–3417.
206. **Chopra S, Ramkissoon K, Anderson DC.** 2013. A systematic quantitative proteomic examination of multidrug resistance in *Acinetobacter baumannii. J Proteomics* **84:**17–39.
207. **Evans C, Noirel J, Ow SY, Salim M, Pereira-Medrano AG, Couto N, Pandhal J, Smith D, Pham TK, Karunakaran E, Zou X, Biggs CA, Wright PC.** 2012. An insight into iTRAQ: where do we stand now? *Anal Bioanal Chem* **404:**1011–1027.
208. **Huynh TT, McDougald D, Klebensberger J, Al Qarni B, Barraud N, Rice SA, Kjelleberg S, Schleheck D.** 2012. Glucose starvation-

induced dispersal of *Pseudomonas aeruginosa* biofilms is cAMP and energy dependent. *PLoS One* **7:**e42874. doi:10.1371/journal.pone.0042874.

209. **Pham TK, Roy S, Noirel J, Douglas I, Wright PC, Stafford GP.** 2010. A quantitative proteomic analysis of biofilm adaptation by the periodontal pathogen *Tannerella forsythia*. *Proteomics* **10:**3130–3141.
210. **Solis N, Parker BL, Kwong SM, Robinson G, Firth N, Cordwell SJ.** 2014. *Staphylococcus aureus* surface proteins involved in adaptation to oxacillin identified using a novel cell shaving approach. *J Proteome Res* **13:**2954–2972.
211. **Booth SC, Workentine ML, Wen J, Shaykhutdinov R, Vogel HJ, Ceri H, Turner RJ, Weljie AM.** 2011. Differences in metabolism between the biofilm and planktonic response to metal stress. *J Proteome Res* **10:**3190–3199.
212. **Kouremenos KA, Beale DJ, Antti H, Palombo EA.** 2014. Liquid chromatography time of flight mass spectrometry based environmental metabolomics for the analysis of *Pseudomonas putida* bacteria in potable water. *J Chromatogr B Analyt Technol Biomed Life Sci* **966:**179–86.
213. **Martinez P, Galvez S, Ohtsuka N, Budinich M, Cortes MP, Serpell C, Nakahigashi K, Hirayama A, Tomita M, Soga T, Martinez S, Maass A, Parada P.** 2013. Metabolomic study of Chilean biomining bacteria *Acidithiobacillus ferrooxidans* strain Wenelen and *Acidithiobacillus thiooxidans* strain Licanantay. *Metabolomics* **9:**247–257.
214. **Mosier AC, Justice NB, Bowen BP, Baran R, Thomas BC, Northen TR, Banfield JF.** 2013. Metabolites associated with adaptation of microorganisms to an acidophilic, metal-rich environment identified by stable-isotope-enabled metabolomics. *MBio* **4:**e00484-00412. doi:10.1128/mBio.00484-12.
215. **White AP, Weljie AM, Apel D, Zhang P, Shaykhutdinov R, Vogel HJ, Surette MG.** 2010. A global metabolic shift is linked to *Salmonella* multicellular development. *PLoS One* **5:**e11814. doi:10.1371/journal.pone.0011814.
216. **Workentine ML, Harrison JJ, Weljie AM, Tran VA, Stenroos PU, Tremaroli V, Vogel HJ, Ceri H, Turner RJ.** 2010. Phenotypic and metabolic profiling of colony morphology variants evolved from *Pseudomonas fluorescens* biofilms. *Environ Microbiol* **12:**1565–1577.
217. **Want EJ, O'Maille G, Smith CA, Brandon TR, Uritboonthai W, Qin C, Trauger SA, Siuzdak G.** 2006. Solvent-dependent metabolite distribution, clustering, and protein extraction for serum profiling with mass spectrometry. *Anal Chem* **78:**743–752.
218. **Want EJ, Cravatt BF, Siuzdak G.** 2005. The expanding role of mass spectrometry in metabolite profiling and characterization. *Chembiochem* **6:**1941–1951.
219. **Tautenhahn R, Patti GJ, Kalisiak E, Miyamoto T, Schmidt M, Lo FY, McBee J, Baliga NS, Siuzdak G.** 2010. metaXCMS: second-order analysis of untargeted metabolomics data. *Anal Chem* **83:**696–700.
220. **Smith CA, O'Maille G, Want EJ, Qin C, Trauger SA, Brandon TR, Custodio DE, Abagyan R, Siuzdak G.** 2005. METLIN: a metabolite mass spectral database. *Ther Drug Monit* **27:**747–751.
221. **Tautenhahn R, Cho K, Uritboonthai W, Zhu Z, Patti GJ, Siuzdak G.** 2012. An accelerated workflow for untargeted metabolomics using the METLIN database. *Nat Biotechnol* **30:**826–828.
222. **Wishart DS, Knox C, Guo AC, Eisner R, Young N, Gautam B, Hau DD, Psychogios N, Dong E, Bouatra S, Mandal R, Sinelnikov I, Xia J, Jia L, Cruz JA, Lim E, Sobsey CA, Shrivastava S, Huang P, Liu P, Fang L, Peng J, Fradette R, Cheng D, Tzur D, Clements M, Lewis A, De Souza A, Zuniga A, Dawe M, Xiong Y, Clive D, Greiner R, Nazyrova A, Shaykhutdinov R, Li L, Vogel HJ, Forsythe I.** 2009. HMDB: a knowledgebase for the human metabolome. *Nucleic Acids Res* **37:**D603–D610.
223. **Betzig E, Patterson GH, Sougrat R, Lindwasser OW, Olenych S, Bonifacino JS, Davidson MW, Lippincott-Schwartz J, Hess HF.** 2006. Imaging intracellular fluorescent proteins at nanometer resolution. *Science* **313:**1642–1645.
224. **Hess ST, Girirajan TP, Mason MD.** 2006. Ultra-high resolution imaging by fluorescence photoactivation localization microscopy. *Biophys J* **91:**4258–4272.
225. **Rust MJ, Bates M, Zhuang X.** 2006. Sub-diffraction-limit imaging by stochastic optical reconstruction microscopy (STORM). *Nat Methods* **3:**793–795.
226. **Neu TR, Lawrence JR.** 2014. Investigation of microbial biofilm structure by laser scanning microscopy. *Adv Biochem Eng Biotechnol* **146:**1–51.
227. **Jackson KD, Starkey M, Kremer S, Parsek MR, Wozniak DJ.** 2004. Identification of *psl*, a locus encoding a potential exopolysaccharide that is essential for *Pseudomonas aeruginosa* PAO1 biofilm formation. *J Bacteriol* **186:**4466–4475.
228. **Barken KB, Pamp SJ, Yang L, Gjermansen M, Bertrand JJ, Klausen M, Givskov M,**

Whitchurch CB, Engel JN, Tolker-Nielsen T. 2008. Roles of type IV pili, flagellum-mediated motility and extracellular DNA in the formation of mature multicellular structures in *Pseudomonas aeruginosa* biofilms. *Environ Microbiol* **10:**2331–2343.

229. **Christen B, Fero MJ, Hillson NJ, Bowman G, Hong SH, Shapiro L, McAdams HH.** 2010. High-throughput identification of protein localization dependency networks. *Proc Natl Acad Sci USA* **107:**4681–4686.

230. **Lew MD, Lee SF, Ptacin JL, Lee MK, Twieg RJ, Shapiro L, Moerner WE.** 2011. Three-dimensional superresolution colocalization of intracellular protein superstructures and the cell surface in live *Caulobacter crescentus*. *Proc Natl Acad Sci USA* **108:**E1102–E1110.

231. **Li G, Young KD.** 2012. Isolation and identification of new inner membrane-associated proteins that localize to cell poles in *Escherichia coli*. *Mol Microbiol* **84:**276–295.

232. **Lindner AB, Madden R, Demarez A, Stewart EJ, Taddei F.** 2008. Asymmetric segregation of protein aggregates is associated with cellular aging and rejuvenation. *Proc Natl Acad Sci USA* **105:**3076–3081.

233. **Chai Y, Chu F, Kolter R, Losick R.** 2008. Bistability and biofilm formation in *Bacillus subtilis*. *Mol Microbiol* **67:**254–263.

234. **Lydmark P, Lind M, Sørensson F, Hermansson M.** 2006. Vertical distribution of nitrifying populations in bacterial biofilms from a full-scale nitrifying trickling filter. *Environ Microbiol* **8:**2036–2049.

235. **Kofoed MV, Nielsen DA, Revsbech NP, Schramm A.** 2012. Fluorescence *in situ* hybridization (FISH) detection of nitrite reductase transcripts (nirS mRNA) in *Pseudomonas stutzeri* biofilms relative to a microscale oxygen gradient. *Syst Appl Microbiol* **35:**513–517.

236. **Rosenthal AZ, Zhang X, Lucey KS, Ottesen EA, Trivedi V, Choi HM, Pierce NA, Leadbetter JR.** 2013. Localizing transcripts to single cells suggests an important role of uncultured deltaproteobacteria in the termite gut hydrogen economy. *Proc Natl Acad Sci USA* **110:**16163–16168.

237. **Behnam F, Vilcinskas A, Wagner M, Stoecker K.** 2012. A straightforward DOPE (double labeling of oligonucleotide probes)-FISH (fluorescence *in situ* hybridization) method for simultaneous multicolor detection of six microbial populations. *Appl Environ Microbiol* **78:**5138–5142.

238. **Valm AM, Mark Welch JL, Rieken CW, Hasegawa Y, Sogin ML, Oldenbourg R, Dewhirst FE, Borisy GG.** 2011. Systems-level analysis of microbial community organization through combinatorial labeling and spectral imaging. *Proc Natl Acad Sci USA* **108:**4152–4157.

239. **Valm AM, Mark Welch JL, Borisy GG.** 2012. CLASI-FISH: principles of combinatorial labeling and spectral imaging. *Syst Appl Microbiol* **35:**496–502.

240. **Schneider D, Fuhrmann E, Scholz I, Hess WR, Graumann PL.** 2007. Fluorescence staining of live cyanobacterial cells suggest non-stringent chromosome segregation and absence of a connection between cytoplasmic and thylakoid membranes. *BMC Cell Biol* **8:**39.

241. **Créach V, Baudoux AC, Bertru G, Rouzic BL.** 2003. Direct estimate of active bacteria: CTC use and limitations. *J Microbiol Methods* **52:**19–28.

242. **Nielsen JL, Aquino de Muro M, Nielsen PH.** 2003. Evaluation of the redox dye 5-cyano-2,3-tolyl-tetrazolium chloride for activity studies by simultaneous use of microautoradiography and fluorescence *in situ* hybridization. *Appl Environ Microbiol* **69:**641–643.

243. **Breinbauer R, Kohn M.** 2003. Azide-alkyne coupling: a powerful reaction for bioconjugate chemistry. *Chembiochem* **4:**1147–1149.

244. **Liechti GW, Kuru E, Hall E, Kalinda A, Brun YV, VanNieuwenhze M, Maurelli AT.** 2014. A new metabolic cell-wall labelling method reveals peptidoglycan in *Chlamydia trachomatis*. *Nature* **506:**507–510.

245. **Siegrist MS, Whiteside S, Jewett JC, Aditham A, Cava F, Bertozzi CR.** 2013. (D)-amino acid chemical reporters reveal peptidoglycan dynamics of an intracellular pathogen. *ACS Chem Biol* **8:**500–505.

246. **Spiers AJ, Bohannon J, Gehrig SM, Rainey PB.** 2003. Biofilm formation at the air-liquid interface by the *Pseudomonas fluorescens* SBW25 wrinkly spreader requires an acetylated form of cellulose. *Mol Microbiol* **50:**15–27.

247. **Elzinga EJ, Huang JH, Chorover J, Kretzschmar R.** 2012. ATR-FTIR spectroscopy study of the influence of pH and contact time on the adhesion of *Shewanella putrefaciens* bacterial cells to the surface of hematite. *Environ Sci Technol* **46:**12848–12855.

248. **Quilès F, Humbert F.** 2014. On the production of glycogen by *Pseudomonas fluorescens* during biofilm development: an *in situ* study by attenuated total reflection-infrared with chemometrics. *Biofouling* **30:**709–718.

249. **Suci PA, Geesey GG, Tyler BJ.** 2001. Integration of Raman microscopy, differential interference contrast microscopy, and attenuated total reflection Fourier transform infrared

spectroscopy to investigate chlorhexidine spatial and temporal distribution in *Candida albicans* biofilms. *J Microbiol Methods* **46:**193–208.

250. **Holman HY, Miles R, Hao Z, Wozei E, Anderson LM, Yang H.** 2009. Real-time chemical imaging of bacterial activity in biofilms using open-channel microfluidics and synchrotron FTIR spectromicroscopy. *Anal Chem* **81:**8564–8570.
251. **Pätzold R, Keuntje M, Theophile K, Muller J, Mielcarek E, Ngezahayo A, Anders-von Ahlften A.** 2008. *In situ* mapping of nitrifiers and anammox bacteria in microbial aggregates by means of confocal resonance Raman microscopy. *J Microbiol Methods* **72:**241–248.
252. **Pätzold R, Keuntje M, Anders-von Ahlften A.** 2006. A new approach to non-destructive analysis of biofilms by confocal Raman microscopy. *Anal Bioanal Chem* **386:**286–292.
253. **Sandt C, Smith-Palmer T, Pink J, Brennan L, Pink D.** 2007. Confocal Raman microspectroscopy as a tool for studying the chemical heterogeneities of biofilms *in situ*. *J Appl Microbiol* **103:**1808–1820.
254. **Sandt C, Smith-Palmer T, Comeau J, Pink D.** 2009. Quantification of water and biomass in small colony variant PAO1 biofilms by confocal Raman microspectroscopy. *Appl Microbiol Biotechnol* **83:**1171–1182.
255. **Chao Y, Zhang T.** 2012. Surface-enhanced Raman scattering (SERS) revealing chemical variation during biofilm formation: from initial attachment to mature biofilm. *Anal Bioanal Chem* **404:**1465–1475.
256. **Ivleva NP, Wagner M, Horn H, Niessner R, Haisch C.** 2008. *In situ* surface-enhanced Raman scattering analysis of biofilm. *Anal Chem* **80:**8538–8544.
257. **Ivleva NP, Wagner M, Szkola A, Horn H, Niessner R, Haisch C.** 2010. Label-free *in situ* SERS imaging of biofilms. *J Phys Chem B* **114:**10184–10194.
258. **Wagner M, Ivleva NP, Haisch C, Niessner R, Horn H.** 2009. Combined use of confocal laser scanning microscopy (CLSM) and Raman microscopy (RM): investigations on EPS-matrix. *Water Res* **43:**63–76.
259. **Chang CB, Wilking JN, Kim S-H, Shum HC, Weitz DA.** 2015. Monodisperse emulsion drop microenvironments for bacterial biofilm growth. *Small* doi:10.1002/smll.201403125.
260. **Alonso C.** 2012. Tips and tricks for high quality MAR-FISH preparations: focus on bacterioplankton analysis. *Syst Appl Microbiol* **35:**503–512.
261. **Tischer K, Zeder M, Klug R, Pernthaler J, Schattenhofer M, Harms H, Wendeberg A.** 2012. Fluorescence in situ hybridization (CARD-FISH) of microorganisms in hydrocarbon contaminated aquifer sediment samples. *Syst Appl Microbiol* **35:**526–532.
262. **Majors PD, McLean JS, Pinchuk GE, Fredrickson JK, Gorby YA, Minard KR, Wind RA.** 2005. NMR methods for in situ biofilm metabolism studies. *J Microbiol Methods* **62:**337–344.
263. **Renslow RS, Majors PD, McLean JS, Fredrickson JK, Ahmed B, Beyenal H.** 2010. In situ effective diffusion coefficient profiles in live biofilms using pulsed-field gradient nuclear magnetic resonance. *Biotechnol Bioeng* **106:**928–937.

Fungal Biofilms: *In Vivo* Models for Discovery of Anti-Biofilm Drugs

2

JENIEL E. NETT[1] and DAVID R. ANDES[1]

SIGNIFICANCE OF FUNGAL BIOFILMS IN INFECTION

Many fungal and bacterial pathogens build biofilms, establishing resilient communities on a variety of clinical surfaces (1, 2). Biofilm formation has become increasingly appreciated as one of the most common modes of growth. Medically, these adherent conglomerates of cells pose a serious obstacle for the treatment of infection. Compared to nonbiofilm, planktonic cells, they are extraordinarily tolerant to anti-infective therapies and resist killing by host defenses (3). Biofilm formation has been well described for *Candida albicans*, the most common fungal pathogen in the hospital setting (4, 5). More recently, the majority of clinically encountered fungi have been shown to produce biofilms. This group includes filamentous fungi (*Aspergillus, Fusarium*, zygomycetes), *Pneumocystis*, and yeasts (*Blastoschizomyces, Saccharomyces, Malassezia, Trichosporon, Cryptococcus*, and numerous *Candida* spp.) (Table 1) (6–16).

One of the distinguishing traits of biofilm communities is their ability to adhere to a surface. In the medical setting, indwelling devices such as catheters provide an ideal niche for biofilm formation (1, 17). As technology

[1]Department of Medicine and Department of Medical Microbiology and Immunology, University of Wisconsin, Madison WI 53706.

Microbial Biofilms 2nd Edition
Edited by Mahmoud Ghannoum, Matthew Parsek, Marvin Whiteley, and Pranab K. Mukherjee

doi:10.1128/microbiolspec.MB-0008-2014

TABLE 1 Medically important fungi forming biofilms

Aspergillus
Blastoschizomyces
Candida
Cryptococcus
Fusarium
Malassezia
Pneumocystis
Trichosporon
Zygomycetes

advances, the use of medical devices continues to escalate. At least 35 million devices are implanted yearly in the United States alone, and the majority of hospital-acquired infections are device-associated (18). Many types of devices are at risk for biofilm infection, including catheters, dentures, implants, pacemakers, artificial heart valves, and central nervous system shunts (1, 18). Biofilm infections may be catastrophic, resulting in device malfunction or life-threatening, systemic infection (1). Candidiasis in the hospital setting usually involves biofilm infection of a medical device. *Candida* spp. are the fourth most common nosocomial bloodstream pathogens and the third most common urinary tract pathogens (19–22). While biofilm infection was initially described for *C. albicans*, the majority of *Candida* spp., including *Candida dubliniensis, Candida glabrata, Candida krusei, Candida tropicalis*, and *Candida parapsilosis*, have now been shown to cause biofilm infections (23).

Mucosal candidiasis is widespread. Vaginal candidiasis effects approximately 30 to 50% of women, and many suffer recurrent infections (24). The clinical relevance of biofilms on biotic surfaces has become increasingly evident, and mucosal biofilms have been described for *Candida*. In an animal model of oropharyngeal candidiasis, *Candida* forms a biofilm of yeast, hyphae, and commensal bacteria within the epithelial surface (25). *Candida* biofilms growing on the vaginal epithelial lining similarly demonstrate a typical biofilm architecture with adherent cells embedded in an extracellular matrix (26). Mucosal biofilms appear to have many similarities to biofilms growing on abiotic surfaces, including sessile growth, protection from environmental factors, and variable access to nutrients (27). However, biofilms on a mucosal surface participate in a dynamic interaction with the adjacent epithelial lining. The host epithelial lining may deliver immune component, nutrients, and antifungal components. Therefore, the basal aspect of a mucosal biofilm would be expected to be exposed to a vastly different environment when compared to a biofilm on an abiotic device surface.

Like *Candida*, other fungal pathogens, including *Cryptococcus* and *Aspergillus*, have been found in biofilms adherent to abiotic device surfaces (15, 28). However, for *Aspergillus*, the biofilm mode of growth is usually observed in the absence of a foreign body. Even in the absence of a device, *Aspergillus* has been shown to proliferate as a cellular aggregate encased in extracellular matrix with properties similar to device-associated biofilms (28). Clinical examples of these non-device associated biofilms include fungal sinusitis, pulmonary aspergillosis, and aspergilloma (7). Because the biofilm lifestyle of *Aspergillus* is a relatively new field of investigation, further investigations are needed to explore this growth mode.

FUNGAL BIOFILM TRAITS

Structure

Fungal biofilms are comprised of adherent cells covered by an extracellular polymeric matrix. The process of fungal biofilm formation *in vitro* was initially described for *C. albicans* in three main stages (29). First, during early biofilm formation, *Candida* cells adhere to the biofilm substrate. For *C. albicans*, germ tube formation may be elicited. During the intermediate biofilm formation stage, the extracellular matrix begins to appear and covers proliferating fungal colonies. Finally, mature biofilm formation is marked by extracellular matrix encasing the adherent yeast and developing hyphae. It is now recognized that dispersion is a key component of the dynamic nature of biofilms (30). Release of cells from biofilms is a regulated process by which organisms can disseminate throughout the host and establish new sites of infection. This developmental process appears to hold true for *in vivo Candida* biofilms (31, 32).

For *Aspergillus*, aerial, static conditions promote biofilm formation (33). Under these conditions, *Aspergillus fumigatus* grows as a cohesive mycelial structure. The cells become progressively encased in a hydrophobic extracellular material over time. Ultrastructure analysis reveals air channels embedded within the conglomerate (33). These multicellular communities, surrounded by acellular matrix material, also develop *in vivo* in a murine model of pulmonary aspergilloma and invasive aspergillosis (7).

Tolerance to Antifungals

Biofilms are notoriously difficult to treat in the clinical setting, and their physical removal is often required for eradication of infection (34). Available antifungal medications are seldom effective, given the high tolerance of biofilms to these commonly used anti-infective therapies. *Candida* biofilms proliferate in the face of antifungal concentrations up to 1,000-fold higher than those needed to inhibit nonbiofilm, planktonic cells (29, 35–37).

The biofilm lifestyle of *C. albicans* is associated with resistance to available drug classes compared to activity against planktonic cells (29, 35, 38–42). Resistance to the azoles (fluconazole, itraconazole, voriconazole, posaconazole) is particularly pronounced, while the echinocandins (caspofungin, micafungin, anidulafungin) and liposomal amphotericin are more effective. Various mechanisms have been shown to contribute to this resistant phenotype, including the production of an extracellular matrix, an increase in efflux pump activity, alteration of sterols, production of resistant "persister cells," activation of stress responses, and an increase in cell density (40, 43–50). Biofilms formed by *Aspergillus* and *Cryptococcus* also display this multidrug resistance (8, 33, 51, 52). For *Aspergillus* biofilms, antifungal tolerance has been linked to the presence of an extracellular matrix, as well as an increase in efflux pump activity (52–54). Few investigations have examined the mechanisms underlying the antifungal resistance of *Cryptococcus* biofilms, but this phenotype has been correlated with the production of melanin (51).

Immune Resistance

In addition to the well-described drug resistance, biofilm growth also appears to afford protection from the host immune response (55–58). Compared to planktonic cells, both neutrophils and mononuclear cells are less effective in killing *Candida* biofilm cells. Mononuclear cells become entrapped in biofilms but do not efficiently activate or phagocytize fungal cells (57, 59). Neutrophils have impaired function against both *C. albicans* and *C. parapsilosis* biofilms (55, 56, 58). Although most studies involving the leukocyte response to fungal biofilms have been limited to *in vitro* studies, the phenotype appears clinically relevant. An intact immune response is not sufficient to clear *Candida* biofilms. When coupled with antifungal therapy, improved clinical outcomes are observed when *Candida*-infected medical devices are removed (34, 60). Studies examining the innate immune response to *Candida neoformans* biofilms also demonstrated a resistant phenotype. *In vitro*, these biofilms are more tolerant of defensins and oxidative stress than are planktonic cells (61).

HOST FACTORS INFLUENCING FUNGAL BIOFILMS

Fungal biofilms form in a variety of clinical niches. These sites of infection can vary quite significantly with regard to available nutrients, flow conditions, immune components, pH, and the substrate for cell adhesion and initiation of biofilm growth (Fig. 1). Each of these factors is likely to influence the biofilm properties and structure, as discussed below. Models most closely mimicking the clinical niche are necessary to best reproduce the host environmental conditions and ultimately a clinical biofilm infection.

Flow Conditions

One of the greatest environmental differences among the sites of common fungal biofilms is the flow conditions. For example, *Candida* can form a biofilm in the face of a low rate of salivary flow (denture stomatitis), a rapid current of blood (endocarditis), or an intermittent flow (vascular and urinary catheter infection) (18). The Douglas group investigated the influence of flow conditions on *C. albicans* biofilm architecture using *in vitro* models (62, 63). Compared to biofilms grown in static conditions, those propagated in a continuous flow device were encased in a higher concentration of extracellular matrix. The continuous-flow biofilms also exhibited increased resistance to antifungals, including amphotericin B and fluconazole. Investigation of continuous flow by an independent laboratory confirmed the drug resistance phenotype, as well as the altered biofilm structure (64). Biofilms formed in the flow environment were more dense and compact. One might expect this architectural change to greatly impact other aspects of biofilm physiology, such as the availability of oxygen and nutrients.

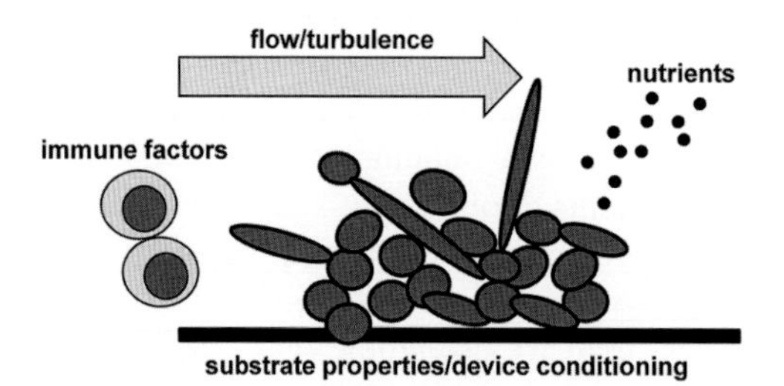

FIGURE 1 Host factors influence fungal biofilm formation. This schematic depicts host conditions that may impact fungal biofilm formation and architecture. *In vivo* animal models can closely mimic biofilm infection niches, incorporating many of these host and environmental conditions. doi:10.1128/microbiolspec.MB-0008-2014.f1

As described above, *Aspergillus* has been shown to form biofilms during invasive pulmonary aspergillosis and pulmonary aspergilloma (7). In the lung, these biofilms proliferate in a static, aerial environment. Unlike *Candida* biofilms, the extracellular matrix production is mostly supported by these low flow conditions (33). The variability in key properties, including matrix production and drug resistance, of biofilms formed under a variety of flow conditions points to the importance of using models that best fit a particular physiologic niche.

Substrates and Conditioning

One of the most influential factors for biofilm initiation is the substrate for adherence (4, 65). Although many materials have been shown to support biofilm formation, the topography and hydrophobicity of the substrate may greatly impact fungal adherence. Indwelling medical devices are often designed to resist microbial adherence. Compared to other plastics, these materials, such as silicone, may even require a preconditioning, or protein coating, for robust biofilm formation *in vitro* (66, 67). However, *in vivo*, medical devices are rapidly conditioned with host factors from the surrounding fluids, such as blood, saliva, urine, or other fluids (68–72). Several of the factors that may coat various devices, influencing adherence and biofilm formation, include fibrinogen, fibronectin, vitronectin, thrombospondin-1, albumin, and von Willebrand factor (68–72). Because the concentrations of these substances differ among clinical niche sites, *in vivo* models best account for the conditioning of medical devices prior to biofilm initiation. When considering mucosal biofilms, substrate representation becomes even more complex. Here, the epithelial layer of cells provides a surface for microbial adhesion which also involves receptor-ligand interactions (27).

Nutrient Composition

Biofilm niche sites vary greatly with respect to the availability of nutrients. For example, blood is a fairly nutrient-rich environment, while urine has a lower abundance of sugars and proteins. In addition, the conditions may be altered by both illness and diet. Untreated diabetes mellitus raises the glucose content throughout the host, while a diet high in sugar primarily raises the glucose content in the oral cavity. The carbon source (galactose or glucose) and abundance can greatly influence biofilm integrity for both *C. albicans* and *C. neoformans* (65, 73). Another factor shown to influence *Candida* biofilm formation is the concentration of metal ions (Ni^{2+}, Fe^{3+}, Cr^{3+}) (74). These variables can impact the rate of biofilm growth, the production of extracellular matrix, and the strength of the biofilm. Animal models of biofilm infection which utilize an equivalent anatomical site provide the ideal composition of nutrients and minerals to mimic patient biofilm infections.

Host Immune Components

Mounting evidence suggests a complex interaction between host immune cells and fungal biofilms. For example, leukocytes, important for controlling fungal infections, have also been shown to promote biofilm growth (59). *In vitro*, *C. albicans* biofilms were observed to proliferate in response to a soluble factor released by mononuclear peripheral blood cells. Ultimately, mononuclear cells became entangled within the basal level of a *C. albicans* biofilm and were not able to phagocytize the biofilm cells. When examining oral mucosal *Candida* biofilms, Dongari-Bagtzoglou et al. observed the migration of neutrophils throughout the biofilm (25). Host immune cells appear to incorporate into biofilm, and even augment biofilm growth, but are usually unable to contain the infection. To understand how the immune system impacts the biofilm lifestyle, models encompassing immune components at the site of infection are optimal.

IN VIVO MODELS OF FUNGAL BIOFILMS AND DRUG DISCOVERY

Animal models best integrate the influence of host factors on the formation of biofilms and acquisition of their phenotypic traits. Utilization of animal models incorporates not only the influence of the immune system, but also niche-specific factors, such as the flow conditions, nutrients in the environment, and the substrate or surface of adherence. Because *Candida* has served as a model organism for fungal biofilm infection in this arena, models involving this pathogen will be much of the focus of discussion in this article (Table 2 and Fig. 2).

Vascular Catheter Model

Perhaps the most commonly used animal model for *in vivo* biofilm study is the venous catheter model. This model has been adapted for use in a rat, a rabbit, and a mouse and has been instrumental for examining the efficacy of antifungals against biofilms formed *in vivo*

TABLE 2 Animal models of *Candida* biofilm infection

Central venous catheter	Rat (31)
	Rabbit (75)
	Mouse (76)
Urinary catheter	Mouse (98)
Subcutaneous implant	Mouse (93)
	Rat (94)
Denture stomatitis	Rat (32, 87)
Oral mucosal	Mouse (25)
Vaginal mucosal	Mouse (26)

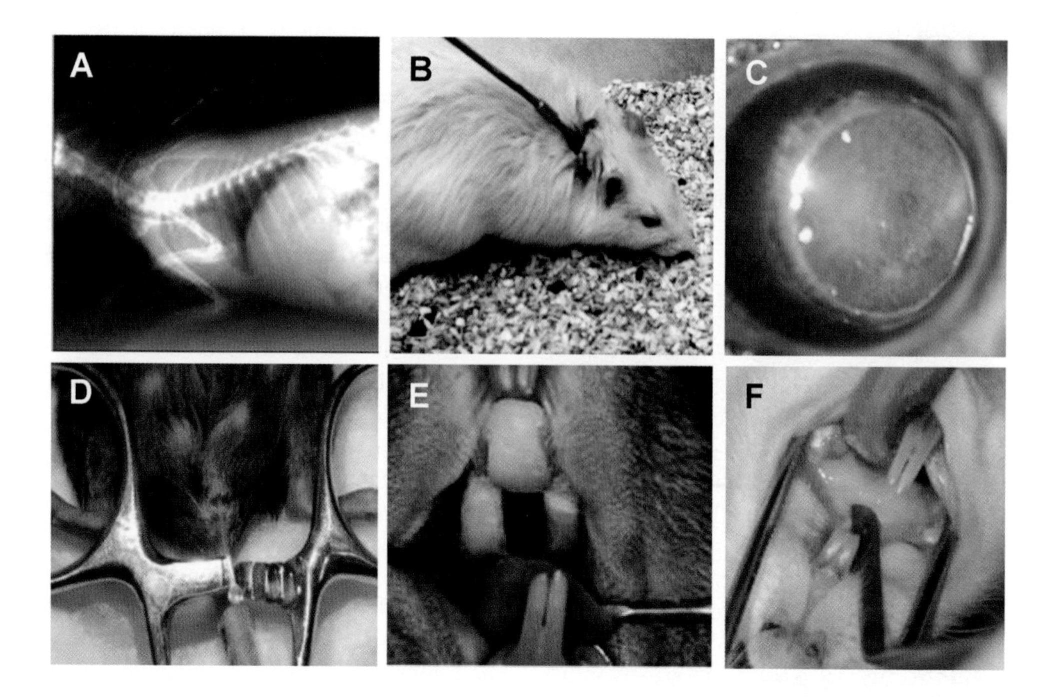

FIGURE 2 Animal models of fungal biofilm infection. (A) Rabbit venous catheter–associated *Candida* biofilm infection (75). (B) Rat venous catheter–associated *Candida* biofilm infection (31). (C) Mouse contact lens–associated *Fusarium* biofilm model (102). (D) Mouse urinary catheter–associated *Candida* biofilm infection (98). (E) Rat denture-associated *Candida* biofilm infection (removable intraoral device) (87). (F) Rat denture-associated *Candida* biofilm infection (32). Images adapted from prior publications (32, 75, 87, 98, 102, 114). doi:10.1128/microbiolspec.MB-0008-2014.f2

(31, 75, 76). As a close mimic of one of the most common clinical biofilm infections, biofilms on the luminal catheter surface are exposed to host conditions, flow, serum proteins, and immune components. The model involves surgical vascular catheter insertion (often jugular vein) followed by subcutaneous tunneling and securing with a protective device. Performing the procedure prior to luminal inoculation allows for a period for host protein conditioning of the device surface. The influence of anti-infectives on biofilm growth can be assessed following systemic administration of drug or by instilling in the lumen as a lock therapy. Common techniques include microscopy for evaluation of biofilm extent and architecture or viable burden determination (Fig. 3). Results of these studies have corroborated the multi-drug-resistant phenotype of *in vitro Candida* biofilms and the need for discovery of new strategies and drugs to treat these infections (31).

One approach to circumvent the drug tolerance of biofilms is to directly administer antifungal in the form of lock therapy (75, 77–79) (Table 3). By avoiding the majority of systemic toxicities, significantly higher drug doses may be safely delivered. Vascular catheter animal models have been valuable for analyzing the efficacy of various antifungal lock therapies against *Candida* biofilms *in vivo*. When instilled in the lumen of *C. albicans*–infected catheters, several lock therapies were found to successfully treat the biofilm infections. Of the clinically available antifungals, the more efficacious solutions have included liposomal amphotericin B

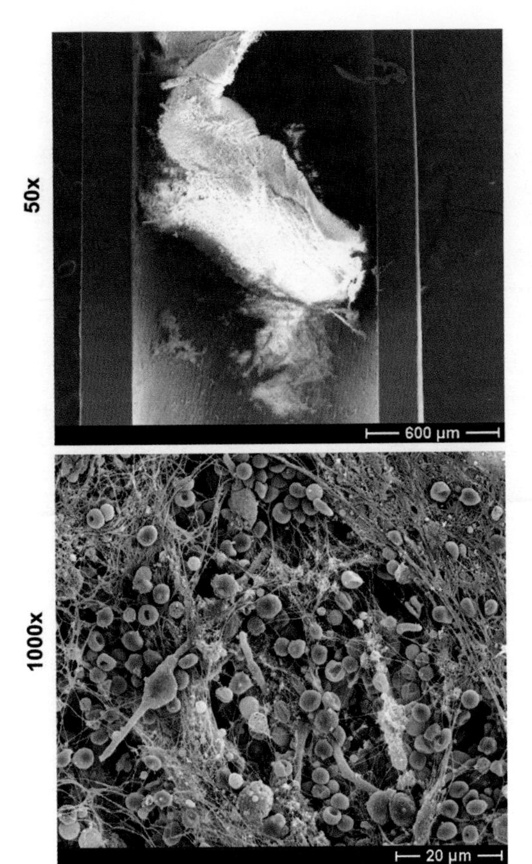

FIGURE 3 *C. albicans* biofilm infection of rat jugular venous catheter. *C. albicans* was instilled in the lumen of a subcutaneously tunneled jugular venous catheter and allowed to dwell for 6 hours. After a growth period of 24 hours, the catheter was harvested, fixed, and dehydrated. Catheter segments were imaged by scanning electron microscopy on a JEOL JSM-6100 at 10 kV (50x and 1,000x). The biofilm is composed of both yeast and hyphae encased in an extracellular matrix. Host components, including red blood cells, appear to associate with the biofilm. doi:10.1128/microbiolspec.MB-0008-2014.f3

(10 mg/ml), amphotericin B lipid complex (5 mg/ml), and caspofungin (6.67 mg/ml) (75, 77–79). As might be predicted by *in vitro* biofilm susceptibility studies, the azole drugs and lower doses of echinocandin drugs have significantly less activity against vascular catheter biofilms and are not ideal for catheter lock therapy (75, 76). One concern regarding the use of available antifungals for lock therapy is the potential for fostering a resistance-promoting environment. As a method to avoid this possibility, studies have also explored the use of alternative agents such as biocides. Encouraging results have been observed *in vitro* for ethanol, ethylenediaminetetraacetic acid (EDTA), and high-dose minocycline (3 mg/ml) lock solutions (80–82).

A second tactic to overcome the profound antifungal tolerance of biofilms is combination drug therapy. The rat vascular catheter biofilm infection model has successfully been used to evaluate the *in vivo* efficacy of several combination therapy lock solutions (48, 49). These studies have explored the impact of adding agents targeting cellular stress responses on azole drug resistance. Uppuluri et al. demonstrated the efficacy of combining calcineurin inhibitors and fluconazole for lock therapy treatment of *C. albicans* biofilms (48). An agent inhibiting the calcineurin pathway (tacrolimus) was found to augment the activity of fluconazole against *C. albicans* catheter biofilms. Subsequent investigations suggest that calcineurin inhibitors similarly potentiate the activity of agents in other drug classes, including the echinocandins and amphotericin B (83). Robbins et al. also used a rat venous catheter model to test the efficacy of combination lock therapy (49). Combining an inhibitor of the Hsp90 pathway (17-AAG) with fluconazole improved the activity against *C. albicans* biofilms. The mechanism of this action is thought to involve a decrease in biofilm extracellular matrix, limiting the capacity of the matrix to sequester antifungal.

Denture Model

Denture stomatitis involves biofilm formation on a denture surface and inflammation of the adjacent oral mucosal surface (84, 85). These infections are common, occurring in up to 70% of denture wearers, and are often

TABLE 3 Strategies for treatment of *Candida* biofilms with demonstrated efficacy in animal models

Lock therapies	Liposomal amphotericin B (10 mg/ml) (75)
	Amphotericin B lipid complex (5 mg/ml) (77)
	Caspofungin (6.67 mg/ml) (78)
Combination therapies	Calcineurin inhibitor (tacrolimus) and fluconazole (48)
	Hsp90 inhibitor (17-AAG) and fluconazole (49)
	Diclofenac (NSAID) and caspofungin (95)

quite painful, even impairing the ability to eat. Biofilms are frequently polymicrobial, with *Candida* spp. playing a key role. Several models have been developed to explore the pathogenesis and treatment of denture stomatitis (32, 86, 87). Early models included primarily Macaca irus monkeys with custom-fitted acrylic plates and Wistar rats fitted with prefabricated acrylic devices (88–90). The focus of these investigations was examination of the mucosal inflammatory process associated with the infected device. *Candida*-infected animals with oral devices were observed to develop mucosal lesions similar to those seen in patients with denture stomatitis (88–90). Although both models were useful for describing the host response to denture biofilms, the rat model was more suited for drug efficacy studies, primarily related to cost. In these investigations, the incorporation of either chlorhexidine or miconazole to the denture acrylic material prevented the development of mucosal lesions of palatal candidiasis (88, 89). However, the chlorhexidine product was poorly tolerated, with rats undergoing weight loss from poor dietary intake.

With the discovery of the role of biofilms in device-associated infections, there has been renewed interest in animal models to mimic denture stomatitis (32, 87, 91). Two models have been developed to replicate this clinical scenario in rats. In the first model, a Sprague-Dawley rat undergoes placement of an acrylic dental device over the hard palate, which is secured in place by orthodontic wire (32). Because the device is fitted to the individual rat, there is close approximation of the device with the oral mucosa, and this space can be inoculated with *Candida* to produce a biofilm device infection and associated mucosal inflammation over the course of 24 to 72 hours (Fig. 4). This model represents an acute infection in the setting of immunosuppression, because rats are treated with a single dose of cortisone prior to infection. In a second rat model, Wistar rats are custom fitted to palatal acrylic devices retained by orthodontic wires (87, 91). However, a portion of the device is secured by embedded magnets and is easily removable throughout the experimental course. Following inoculation of *Candida*, biofilm develops on the device surface over weeks. In addition, mucosal biofilm infection and inflammation ensue, mimicking clinical infection. Because the devices can remain in place for an extended time (8 weeks), this model offers the opportunity to longitudinally follow the course of an individual animal with a chronic infection.

Studies have only just begun to explore the antifungal treatment in the rodent denture models. As might be predicted from clinical scenarios and other infection models, the *C. albicans* biofilms on the denture surface were found to exhibit high tolerance of both fluconazole and micafungin upon either topical or systemic administration (32). The model has also been helpful for exploring the role of gene products on denture biofilm infection *in vivo*. Chen et al. described the importance of the calcineurin pathway for *C. dubliniensis* in the processes of both filamentation and biofilm formation in a rat denture biofilm (92). This suggests that calcineurin inhibitors may be a viable option for treatment of *C. dubliniensis*. Because drugs in this class exert synergistic activity with azole and echinocandin drugs, combination therapy is an attractive possibility for treatment of *C. dubliniensis* biofilm infections (92).

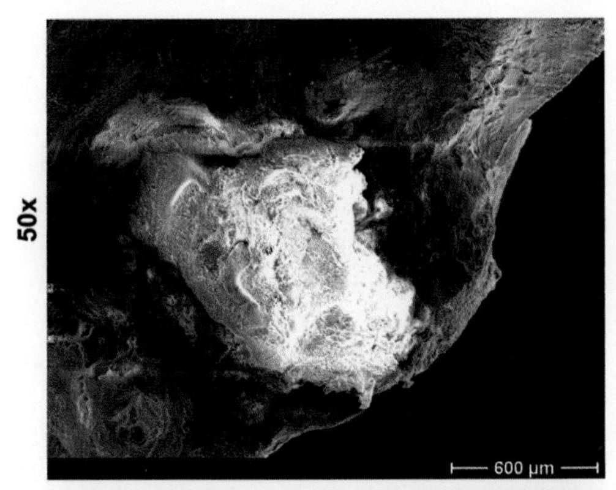

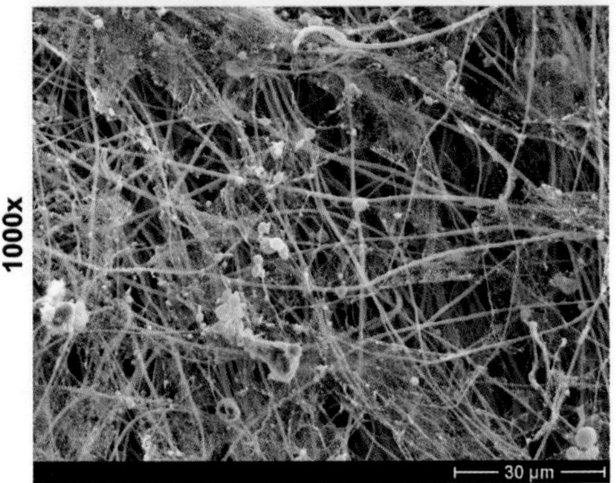

FIGURE 4 *C. albicans* biofilm infection in a rat denture model. *C. albicans* was inoculated between the hard palate and an acrylic device secured with orthodontic wire. After a growth period of 48 hours, the denture was harvested, fixed, and dehydrated. Oral devices were imaged by scanning electron microscopy on a JEOL JSM-6100 at 10 kV (50x and 1,000x). The biofilm is composed of both yeast and hyphae encased in an extracellular matrix. Larger host cells were observed as well. Microbiologic evaluation identified a polymicrobial infection consisting of *C. albicans* and various bacteria. doi:10.1128/microbiolspec.MB-0008-2014.f4

Subcutaneous Implant Model

To study the activity of a novel antifungal formulation against biofilms, Zumbuehl et al. developed a murine model of subcutaneous *Candida* biofilm infection (93). Disks containing amphogel, a dextran-based hydrogel loaded with amphotericin B, were inoculated with *C. albicans* and surgically implanted in the subcutaneous flank tissue of BALB/c mice. After 3 days, *Candida* had been cleared from the surface of the disks containing amphogel. In contrast, control disks with hydrogel only were coated with *C. albicans* biofilm and host cells. The amphogel was well tolerated, eliciting only a minimal or mild inflammatory response. Because this antifungal hydrogel maintains efficacy for over 50 days, it is ideally suited for the prevention of device-associated infection.

As an alternative, simpler model for the study of *Candida* biofilm infections, Ricicova et al. fabricated a subcutaneous catheter implant model. In this model, a polyurethane catheter segment is inoculated with *Candida* and implanted under the skin of a rat (94). Over the course of 1 to 6 days, a multilayer biofilm consisting of both yeast and hyphae forms on the catheter surface. Compared to the vascular catheter model, this procedure is less invasive and requires a shorter period of anesthesia. In terms of mimicking patient infection, the model has similarities to both vascular catheter infections and wound infections. The implanted catheter material is a close mimic of the vascular catheter material used in patients. The model is avascular, so biofilm cells are not subjected to blood flow conditions and are not exposed to the same concentrations of serum protein and blood cells. However, the devices can be treated with serum prior to implantation to partially mimic this exposure. The anatomical location of the implantation is most similar to a biofilm wound infection. The model allows for the interaction between host immune components and *Candida* biofilms in the subcutaneous tissue. One clear advantage is that the procedure allows for placement of multiple segments of catheter for potential comparison of several *Candida* strains or conditions within the same animal.

The subcutaneous implant model has also proven to be efficacious for evaluation of antibiofilm treatments. Bink et al. used this model to test the efficacy of combining a nonsteroidal anti-inflammatory drug (NSAID) and an echinocandin for treatment of *C. albicans* biofilms *in vivo* (95). NSAIDs impair prostaglandin synthesis by targeting mammalian cyclooxygenases. Agents in this class are available for the treatment of pain and inflammation. However, the activity of this drug class is not limited to mammalian systems, because they have also been shown to disrupt filamentation and biofilm formation in *C. albicans*, likely through inhibition of prostaglandin E2 synthesis (96, 97). To examine the impact of disrupting this pathway *in vivo*, rats received diclofenac treatment in the setting of subcutaneous catheter implant infection (95). In rats that had been treated with diclofenac prior to development of *C. albicans* biofilm infection, the antibiofilm activity of caspofungin was enhanced. It is of great interest to identify an available drug class able to potentiate the activity of echinocandins. However, the animals in the study received diclofenac prior to infection, and whether or not it would be a helpful adjuvant later in the course of infection is not known.

Urinary Catheter Model

To investigate catheter-associated urinary tract infections and candiduria, Wang et al. developed a murine model representing this clinical scenario (98). In this model, a guide wire is inserted through the urethra of a female mouse, and a catheter segment is threaded over the guide wire and into the bladder. The segment is secured by suture through the bladder wall. After 5 to 7 days, the animal is infected with *C. albicans* intravesicularly. Candiduria is detectable quickly after infection and persists for 28 days. A dense biofilm of adherent yeast and hyphae forms on both the luminal and extraluminal surfaces. To increase the susceptibility to *Candida* infection, mice lacking lysozyme M, an important effector for mucosal innate immunity, can be used. The model closely mimics patient *Candida* biofilm formation with regard to the use of biofilm substrate (catheter) and anatomic location (bladder). Because only a segment of catheter is in place, the flow conditions are likely less than would be observed for a patient catheter functioning to drain the bladder. The model incorporates the mammalian immune system with the option of using wild type or immunocompromised animals. To our knowledge, this model has not yet been used for investigating the activity of antibiofilm therapies.

Mucosal Candidiasis Models

Biofilms have frequently been described in association with medical devices and abiotic surfaces. However, there is mounting evidence that *Candida* spp. exhibit similar characteristics when growing on mucosal surfaces (25, 26). Murine models of both oropharyngeal and vaginal candidiasis demonstrate that *Candida* produces conglomerates of yeast, hyphae, and extracellular material associated with mucosal surfaces. In an oropharyngeal candidiasis model, the biofilms appear to be complex, involving commensal bacteria, neutrophils, and keratin (25). A murine model of vaginal candidiasis shows that *C. albicans* regulators of biofilm formation on abiotic surfaces are similar to those required for the development of vaginal biofilms. Although mucosal biofilms share many characteristics with device-associated biofilms, it is not clear that they exhibit the same degree of drug resistance. Clinically, mucosal biofilms are usually responsive to antifungal therapies, including azoles (99, 100).

Fusarium keratitis

Outbreaks of *Fusarium* keratitis have prompted interest in models to investigate this contact lens–associated biofilm infection (101). Two murine models have been developed to investigate the pathogenesis of this process (102, 103). Sun et al. demonstrated the ability of *Fusarium oxysporum* to form a biofilm on the surface of a contact lens and for this to induce keratitis on an injured mouse cornea (102). The model has successfully been implemented for the study of the host aspects of *Fusarium* keratitis. Zhang et al. developed a similar model of murine fungal keratitis employing *Fusarium solani* (103). Although this model also involves inoculation of an injured cornea, the organisms are directly seeded and no contact lens is involved. The model utilizes fluorescently labeled fungi for visualization of the infective process.

FUTURE DIRECTIONS

Recognition of the importance of animal models for the discovery of antibiofilm drugs has only recently emerged. Most investigations have focused on *C. albicans* as a model pathogen, and the vascular catheter models of biofilm infection have been the most popular. It will be interesting to see how biofilms formed under the conditions of other clinically relevant niches respond to antifungal therapies. The murine urinary catheter model, the rat subcutaneous model, and rat denture models will be of great value for these investigations of *Candida* (32, 87, 94, 98). The ocular lens model should be similarly useful for identification of preventative and therapeutic compounds for fungal keratitis. The models allow for testing of antibiofilm compounds under physiologic conditions very similar to those encountered clinically. In addition to these animal models of device-associated infections, models of mucosal *Candida* biofilms will surely be helpful for the study of these common infections (25, 26).

As it is becoming increasingly clear that infections caused by diverse fungal pathogens involve biofilm communities, animal models of these infections will be beneficial for pathogenesis and drug discovery studies. Models are underway for several of the fungal pathogens. One example, discussed earlier, is *Fusarium* keratitis in the setting of contact lens biofilm infection. Murine models have been developed to study both host and pathogen aspects of this process and could be utilized to evaluate novel antibiofilm therapies in this unique clinical niche (102, 103). Also, *Aspergillus* spp., including *A. fumigatus*, have been shown to produce fungal aggregates during pulmonary infection (7). Models mimicking aspergilloma or invasive aspergillosis will be helpful for exploring the impact of antifungal treatment on this mode of growth. Although *C. albicans* has been the model pathogen for many *in vivo* biofilm investigations, the *in vivo* biofilm models can likely be adapted to biofilm infections caused by a variety of non-*albicans Candida* spp. and other yeasts, such as *Cryptococcus*. Of note, the rat vascular catheter model has been successfully used for the study of *C. parapsilosis* and *C. glabrata*, while a rat denture model has been employed for the investigation of *C. dubliniensis* (92, 104).

There are many approaches to the discovery of new anti-infectives. One strategy is to screen large libraries of compounds. Using *in vitro* models, the mining of pharmaceutical and natural product libraries has identified novel compounds with antibiofilm activity (105–107). An alternative approach is to determine a mechanism leading to drug resistance and identify or develop an anti-infective that disrupts the process. For *C. albicans*, the biofilm property most closely linked to resistance is the extracellular matrix. Enzymatic degradation of key matrix components, such as extracellular DNA and β-1,3 glucan, has been shown to enhance antifungal activity, suggesting these as potential drug targets (67, 108). A therapy directed at extracellular DNA degradation has been shown to be beneficial for patients with cystic fibrosis. It is thought that dornase alfa (Pulmozyme), a clinically available inhaled enzymatic treatment, works by degrading extracellular DNA of bacterial biofilms (109). Regardless of the path of drug discovery, animal models will be beneficial for testing the efficacy of compounds against clinical biofilms and establishing safety.

One of the unique aspects of exploring the antibiofilm activity of drugs in animal models is the opportunity to vary the mode of antifungal delivery. For example, compounds may be systemically administered, topically administered, coated on a device, or embedded in a device. Another interesting delivery method is direct administration of a gel with prolonged elution of high antifungal concentrations, such as was developed for amphotericin B (110). Systemic administration is feasible to test in all models, while direct, topical administration of a compound is easily achievable in either the denture models via topical therapy or the venous catheter models via lock therapy. The subcutaneous tissue model may be ideal for exploring the utility of embedding or coating a device with an antibiofilm compound, because numerous devices can be tested in a single animal. Another potential application is the investigation of vaccine efficacy, such as the NDV-3 vaccine in clinical trials, vaccines found to be efficacious in nonbiofilm models of infections, or future vaccines designed specifically to inhibit the biofilm mode of growth (111–113).

ACKNOWLEDGMENTS

J.N. is supported by the Burroughs Wellcome Fund 1012299 and NIH K08 AI108727. D.A. is supported by NIH R01 AI073289. The authors acknowledge the use of instrumentation supported by the UW MRSEC (DMR-1121288) and the UW NSEC (DMR-0832760).

Conflicts of interest: We declare no conflicts.

CITATION

Nett JE, Andes DR. 2015. Fungal biofilms: *in vivo* models for discovery of anti-biofilm drugs. Microbiol Spectrum 3(4):MB-0008-2014.

REFERENCES

1. **Donlan RM.** 2001. Biofilms and device-associated infections. *Emerg Infect Dis* **7:** 277–281.
2. **Hoyle BD, Jass J, Costerton JW.** 1990. The biofilm glycocalyx as a resistance factor. *J Antimicrob Chemother* **26:**1–5.
3. **Hawser SP, Baillie GS, Douglas LJ.** 1998. Production of extracellular matrix by *Candida albicans* biofilms. *J Med Microbiol* **47:**253–256.
4. **Douglas LJ.** 2002. Medical importance of biofilms in *Candida infections. Rev Iberoam Micol* **19:**139–143.
5. **Donlan RM.** 2001. Biofilm formation: a clinically relevant microbiological process. *Clin Infect Dis* **33:**1387–1392.
6. **Ramage G, Rajendran R, Sherry L, Williams C.** 2012. Fungal biofilm resistance. *Int J Microbiol* **2012:**528521.
7. **Loussert C, Schmitt C, Prevost MC, Balloy V, Fadel E, Philippe B, Kauffmann-Lacroix C, Latge JP, Beauvais A.** 2010. *In vivo* biofilm composition of *Aspergillus fumigatus. Cell Microbiol* **12:**405–410.
8. **Seidler MJ, Salvenmoser S, Muller FM.** 2008. *Aspergillus fumigatus* forms biofilms with reduced antifungal drug susceptibility on bronchial epithelial cells. *Antimicrob Agents Chemother* **52:**4130–4136.
9. **Davis LE, Cook G, Costerton JW.** 2002. Biofilm on ventriculo-peritoneal shunt tubing as a cause of treatment failure in coccidioidal meningitis. *Emerg Infect Dis* **8:**376–379.
10. **Singh R, Shivaprakash MR, Chakrabarti A.** 2011. Biofilm formation by zygomycetes: quantification, structure and matrix composition. *Microbiology* **157:**2611–2618.
11. **D'Antonio D, Parruti G, Pontieri E, Di Bonaventura G, Manzoli L, Sferra R, Vetuschi A, Piccolomini R, Romano F, Staniscia T.** 2004. Slime production by clinical isolates of *Blastoschizomyces capitatus* from patients with hematological malignancies and catheter-related fungemia. *Eur J Clin Microbiol Infect Dis* **23:**787–789.
12. **Reynolds TB, Fink GR.** 2001. Bakers' yeast, a model for fungal biofilm formation. *Science* **291:**878–881.
13. **Cannizzo FT, Eraso E, Ezkurra PA, Villar-Vidal M, Bollo E, Castella G, Cabanes FJ, Vidotto V, Quindos G.** 2007. Biofilm development by clinical isolates of *Malassezia pachydermatis. Med Mycol* **45:**357–361.
14. **Di Bonaventura G, Pompilio A, Picciani C, Iezzi M, D'Antonio D, Piccolomini R.** 2006. Biofilm formation by the emerging fungal pathogen *Trichosporon asahii:* development, architecture, and antifungal resistance. *Antimicrob Agents Chemother* **50:**3269–3276.
15. **Walsh TJ, Schlegel R, Moody MM, Costerton JW, Salcman M.** 1986. Ventriculoatrial shunt

infection due to *Cryptococcus neoformans*: an ultrastructural and quantitative microbiological study. *Neurosurgery* **18:**373–375.

16. **Dyavaiah M, Ramani R, Chu DS, Ritterband DC, Shah MK, Samsonoff WA, Chaturvedi S, Chaturvedi V.** 2007. Molecular characterization, biofilm analysis and experimental biofouling study of *Fusarium* isolates from recent cases of fungal keratitis in New York State. *BMC Ophthalmol* **7:**1.
17. **Passerini L, Lam K, Costerton JW, King EG.** 1992. Biofilms on indwelling vascular catheters. *Crit Care Med* **20:**665–673.
18. **Kojic EM, Darouiche RO.** 2004. *Candida* infections of medical devices. *Clin Microbiol Rev* **17:**255–267.
19. **Groeger JS, Lucas AB, Thaler HT, Friedlander-Klar H, Brown AE, Kiehn TE, Armstrong D.** 1993. Infectious morbidity associated with long-term use of venous access devices in patients with cancer. *Ann Intern Med* **119:**1168–1174.
20. **Richards MJ, Edwards JR, Culver DH, Gaynes RP.** 1999. Nosocomial infections in medical intensive care units in the United States. National Nosocomial Infections Surveillance System. *Crit Care Med* **27:**887–892.
21. **Edmond MB, Wallace SE, McClish DK, Pfaller MA, Jones RN, Wenzel RP.** 1999. Nosocomial bloodstream infections in United States hospitals: a three-year analysis. *Clin Infect Dis* **29:**239–244.
22. **Pfaller MA, Diekema DJ.** 2007. Epidemiology of invasive candidiasis: a persistent public health problem. *Clin Microbiol Rev* **20:**133–163.
23. **Shin JH, Kee SJ, Shin MG, Kim SH, Shin DH, Lee SK, Suh SP, Ryang DW.** 2002. Biofilm production by isolates of *Candida* species recovered from nonneutropenic patients: comparison of bloodstream isolates with isolates from other sources. *J Clin Microbiol* **40:**1244–1248.
24. **Foxman B, Muraglia R, Dietz JP, Sobel JD, Wagner J.** 2013. Prevalence of recurrent vulvovaginal candidiasis in 5 European countries and the United States: results from an Internet panel survey. *J Low Genit Tract Dis* **17:**340–345.
25. **Dongari-Bagtzoglou A, Kashleva H, Dwivedi P, Diaz P, Vasilakos J.** 2009. Characterization of mucosal *Candida albicans* biofilms. *PLoS One* **4:**e7967. doi:10.1371/journal.pone.0007967.
26. **Harriott MM, Lilly EA, Rodriguez TE, Fidel PL Jr, Noverr MC.** 2010. *Candida albicans* forms biofilms on the vaginal mucosa. *Microbiology* **156:**3635–3644.
27. **Dongari-Bagtzoglou A.** 2008. Mucosal biofilms: challenges and future directions. *Expert Rev Anti Infect Ther* **6:**141–144.
28. **Ramage G, Rajendran R, Gutierrez-Correa M, Jones B, Williams C.** 2011. *Aspergillus* biofilms: clinical and industrial significance. *FEMS Microbiol Lett* **324:**89–97.
29. **Chandra J, Kuhn DM, Mukherjee PK, Hoyer LL, McCormick T, Ghannoum MA.** 2001. Biofilm formation by the fungal pathogen *Candida albicans*: development, architecture, and drug resistance. *J Bacteriol* **183:**5385–5394.
30. **Uppuluri P, Chaturvedi AK, Srinivasan A, Banerjee M, Ramasubramaniam AK, Kohler JR, Kadosh D, Lopez-Ribot JL.** 2010. Dispersion as an important step in the *Candida albicans* biofilm developmental cycle. *PLoS Pathog* **6:**e1000828. doi:10.1371/journal.ppat.1000828.
31. **Andes D, Nett J, Oschel P, Albrecht R, Marchillo K, Pitula A.** 2004. Development and characterization of an *in vivo* central venous catheter *Candida albicans* biofilm model. *Infect Immun* **72:**6023–6031.
32. **Nett JE, Marchillo K, Spiegel CA, Andes DR.** 2010. Development and validation of an *in vivo Candida albicans* biofilm denture model. *Infect Immun* **78:**3650–3659.
33. **Beauvais A, Schmidt C, Guadagnini S, Roux P, Perret E, Henry C, Paris S, Mallet A, Prevost MC, Latge JP.** 2007. An extracellular matrix glues together the aerial-grown hyphae of *Aspergillus fumigatus*. *Cell Microbiol* **9:**1588–1600.
34. **Pappas PG, Kauffman CA, Andes D, Benjamin DK Jr, Calandra TF, Edwards JE Jr, Filler SG, Fisher JF, Kullberg BJ, Ostrosky-Zeichner L, Reboli AC, Rex JH, Walsh TJ, Sobel JD.** 2009. Clinical practice guidelines for the management of candidiasis: 2009 update by the Infectious Diseases Society of America. *Clin Infect Dis* **48:**503–535.
35. **Hawser SP, Douglas LJ.** 1994. Biofilm formation by *Candida* species on the surface of catheter materials *in vitro*. *Infect Immun* **62:**915–921.
36. **Mah TF, Pitts B, Pellock B, Walker GC, Stewart PS, O'Toole GA.** 2003. A genetic basis for *Pseudomonas aeruginosa* biofilm antibiotic resistance. *Nature* **426:**306–310.
37. **O'Toole GA.** 2003. To build a biofilm. *J Bacteriol* **185:**2687–2689.
38. **Baillie GS, Douglas LJ.** 1998. Effect of growth rate on resistance of *Candida albicans* biofilms to antifungal agents. *Antimicrob Agents Chemother* **42:**1900–1905.
39. **Lewis RE, Kontoyiannis DP, Darouiche RO, Raad II, Prince RA.** 2002. Antifungal activity of amphotericin B, fluconazole, and

voriconazole in an *in vitro* model of *Candida* catheter-related bloodstream infection. *Antimicrob Agents Chemother* **46:**3499–3505.

40. **Mukherjee PK, Chandra J, Kuhn DM, Ghannoum MA.** 2003. Mechanism of fluconazole resistance in *Candida albicans* biofilms: phase-specific role of efflux pumps and membrane sterols. *Infect Immun* **71:**4333–4340.
41. **Ramage G, VandeWalle K, Bachmann SP, Wickes BL, Lopez-Ribot JL.** 2002. *In vitro* pharmacodynamic properties of three antifungal agents against preformed *Candida albicans* biofilms determined by time-kill studies. *Antimicrob Agents Chemother* **46:**3634–3636.
42. **Kuhn DM, George T, Chandra J, Mukherjee PK, Ghannoum MA.** 2002. Antifungal susceptibility of *Candida* biofilms: unique efficacy of amphotericin B lipid formulations and echinocandins. *Antimicrob Agents Chemother* **46:**1773–1780.
43. **Ramage G, Bachmann S, Patterson TF, Wickes BL, Lopez-Ribot JL.** 2002. Investigation of multidrug efflux pumps in relation to fluconazole resistance in *Candida albicans* biofilms. *J Antimicrob Chemother* **49:**973–980.
44. **Kumamoto CA.** 2005. A contact-activated kinase signals *Candida albicans* invasive growth and biofilm development. *Proc Natl Acad Sci USA* **102:**5576–5581.
45. **Khot PD, Suci PA, Miller RL, Nelson RD, Tyler BJ.** 2006. A small subpopulation of blastospores in *Candida albicans* biofilms exhibit resistance to amphotericin B associated with differential regulation of ergosterol and beta-1,6-glucan pathway genes. *Antimicrob Agents Chemother* **50:**3708–3716.
46. **LaFleur MD, Kumamoto CA, Lewis K.** 2006. *Candida albicans* biofilms produce antifungal-tolerant persister cells. *Antimicrob Agents Chemother* **50:**3839–3846.
47. **Perumal P, Mekala S, Chaffin WL.** 2007. Role for cell density in antifungal drug resistance in *Candida albicans* biofilms. *Antimicrob Agents Chemother* **51:**2454–2463.
48. **Uppuluri P, Nett J, Heitman J, Andes D.** 2008. Synergistic effect of calcineurin inhibitors and fluconazole against *Candida albicans* biofilms. *Antimicrob Agents Chemother* **52:** 1127–1132.
49. **Robbins N, Uppuluri P, Nett J, Rajendran R, Ramage G, Lopez-Ribot JL, Andes D, Cowen LE.** 2011. Hsp90 governs dispersion and drug resistance of fungal biofilms. *PLoS Pathog* **7:**e1002257. doi:10.1371/journal.ppat.1002257.
50. **Mukherjee PK, Zhou G, Munyon R, Ghannoum MA.** 2005. *Candida* biofilm: a well-designed protected environment. *Med Mycol* **43:**191–208.
51. **Martinez LR, Casadevall A.** 2006. Susceptibility of *Cryptococcus neoformans* biofilms to antifungal agents *in vitro*. *Antimicrob Agents Chemother* **50:**1021–1033.
52. **Mowat E, Lang S, Williams C, McCulloch E, Jones B, Ramage G.** 2008. Phase-dependent antifungal activity against *Aspergillus fumigatus* developing multicellular filamentous biofilms. *J Antimicrob Chemother* **62:**1281–1284.
53. **Rajendran R, Williams C, Lappin DF, Millington O, Martins M, Ramage G.** 2013. Extracellular DNA release acts as an antifungal resistance mechanism in mature *Aspergillus fumigatus* biofilms. *Eukaryot Cell* **12:**420–429.
54. **Bugli F, Posteraro B, Papi M, Torelli R, Maiorana A, Paroni Sterbini F, Posteraro P, Sanguinetti M, De Spirito M.** 2013. *In vitro* interaction between alginate lyase and amphotericin B against *Aspergillus fumigatus* biofilm determined by different methods. *Antimicrob Agents Chemother* **57:**1275–1282.
55. **Katragkou A, Chatzimoschou A, Simitsopoulou M, Georgiadou E, Roilides E.** 2011. Additive antifungal activity of anidulafungin and human neutrophils against *Candida parapsilosis* biofilms. *J Antimicrob Chemother* **66:**588–591.
56. **Katragkou A, Simitsopoulou M, Chatzimoschou A, Georgiadou E, Walsh TJ, Roilides E.** 2011. Effects of interferon-gamma and granulocyte colony-stimulating factor on antifungal activity of human polymorphonuclear neutrophils against *Candida albicans* grown as biofilms or planktonic cells. *Cytokine* **55:**330–334.
57. **Katragkou A, Kruhlak MJ, Simitsopoulou M, Chatzimoschou A, Taparkou A, Cotten CJ, Paliogianni F, Diza-Mataftsi E, Tsantali C, Walsh TJ, Roilides E.** 2010. Interactions between human phagocytes and *Candida albicans* biofilms alone and in combination with antifungal agents. *J Infect Dis* **201:**1941–1949.
58. **Xie Z, Thompson A, Sobue T, Kashleva H, Xu H, Vasilakos J, Dongari-Bagtzoglou A.** 2012. *Candida albicans* biofilms do not trigger reactive oxygen species and evade neutrophil killing. *J Infect Dis* **206:**1936–1945.
59. **Chandra J, McCormick TS, Imamura Y, Mukherjee PK, Ghannoum MA.** 2007. Interaction of *Candida albicans* with adherent human peripheral blood mononuclear cells increases *C. albicans* biofilm formation and results in differential expression of pro- and anti-inflammatory cytokines. *Infect Immun* **75:** 2612–2620.

60. **Andes DR, Safdar N, Baddley JW, Playford G, Reboli AC, Rex JH, Sobel JD, Pappas PG, Kullberg BJ.** 2012. Impact of treatment strategy on outcomes in patients with candidemia and other forms of invasive candidiasis: a patient-level quantitative review of randomized trials. *Clin Infect Dis* **54:**1110–1122.
61. **Martinez LR, Casadevall A.** 2006. *Cryptococcus neoformans* cells in biofilms are less susceptible than planktonic cells to antimicrobial molecules produced by the innate immune system. *Infect Immun* **74:**6118–6123.
62. **Al-Fattani MA, Douglas LJ.** 2004. Penetration of *Candida* biofilms by antifungal agents. *Antimicrob Agents Chemother* **48:**3291–3297.
63. **Al-Fattani MA, Douglas LJ.** 2006. Biofilm matrix of *Candida albicans* and *Candida tropicalis*: chemical composition and role in drug resistance. *J Med Microbiol* **55:**999–1008.
64. **Uppuluri P, Chaturvedi AK, Lopez-Ribot JL.** 2009. Design of a simple model of *Candida albicans* biofilms formed under conditions of flow: development, architecture, and drug resistance. *Mycopathologia* **168:**101–109.
65. **Martinez LR, Casadevall A.** 2007. *Cryptococcus neoformans* biofilm formation depends on surface support and carbon source and reduces fungal cell susceptibility to heat, cold, and UV light. *Appl Environ Microbiol* **73:**4592–4601.
66. **Nobile CJ, Mitchell AP.** 2005. Regulation of cell-surface genes and biofilm formation by the *C. albicans* transcription factor Bcr1p. *Curr Biol* **15:**1150–1155.
67. **Nett J, Lincoln L, Marchillo K, Massey R, Holoyda K, Hoff B, VanHandel M, Andes D.** 2007. Putative role of beta-1,3 glucans in *Candida albicans* biofilm resistance. *Antimicrob Agents Chemother* **51:**510–520.
68. **Francois P, Schrenzel J, Stoerman-Chopard C, Favre H, Herrmann M, Foster TJ, Lew DP, Vaudaux P.** 2000. Identification of plasma proteins adsorbed on hemodialysis tubing that promote *Staphylococcus aureus* adhesion. *J Lab Clin Med* **135:**32–42.
69. **Proctor RA.** 2000. Toward an understanding of biomaterial infections: a complex interplay between the host and bacteria. *J Lab Clin Med* **135:**14–15.
70. **Jenney CR, Anderson JM.** 2000. Adsorbed serum proteins responsible for surface dependent human macrophage behavior. *J Biomed Mater Res* **49:**435–447.
71. **Brash JL, Ten Hove P.** 1993. Protein adsorption studies on 'standard' polymeric materials. *J Biomater Sci Polym Ed* **4:**591–599.
72. **Yanagisawa N, Li DQ, Ljungh A.** 2004. Protein adsorption on *ex vivo* catheters and polymers exposed to peritoneal dialysis effluent. *Perit Dial Int* **24:**264–273.
73. **Jin Y, Samaranayake LP, Samaranayake Y, Yip HK.** 2004. Biofilm formation of *Candida albicans* is variably affected by saliva and dietary sugars. *Arch Oral Biol* **49:**789–798.
74. **Ronsani MM, Mores Rymovicz AU, Meira TM, Trindade Gregio AM, Guariza Filho O, Tanaka OM, Ribeiro Rosa EA.** 2011. Virulence modulation of *Candida albicans* biofilms by metal ions commonly released from orthodontic devices. *Microb Pathog* **51:**421–425.
75. **Schinabeck MK, Long LA, Hossain MA, Chandra J, Mukherjee PK, Mohamed S, Ghannoum MA.** 2004. Rabbit model of *Candida albicans* biofilm infection: liposomal amphotericin B antifungal lock therapy. *Antimicrob Agents Chemother* **48:**1727–1732.
76. **Lazzell AL, Chaturvedi AK, Pierce CG, Prasad D, Uppuluri P, Lopez-Ribot JL.** 2009. Treatment and prevention of *Candida albicans* biofilms with caspofungin in a novel central venous catheter murine model of candidiasis. *J Antimicrob Chemother* **64:**567–570.
77. **Mukherjee PK, Long L, Kim HG, Ghannoum MA.** 2009. Amphotericin B lipid complex is efficacious in the treatment of *Candida albicans* biofilms using a model of catheter-associated *Candida* biofilms. *Int J Antimicrob Agents* **33:** 149–153.
78. **Shuford JA, Rouse MS, Piper KE, Steckelberg JM, Patel R.** 2006. Evaluation of caspofungin and amphotericin B deoxycholate against *Candida albicans* biofilms in an experimental intravascular catheter infection model. *J Infect Dis* **194:**710–713.
79. **Walraven CJ, Lee SA.** 2013. Antifungal lock therapy. *Antimicrob Agents Chemother* **57:**1–8.
80. **Raad I, Hanna H, Dvorak T, Chaiban G, Hachem R.** 2007. Optimal antimicrobial catheter lock solution, using different combinations of minocycline, EDTA, and 25-percent ethanol, rapidly eradicates organisms embedded in biofilm. *Antimicrob Agents Chemother* **51:**78–83.
81. **Sherertz RJ, Boger MS, Collins CA, Mason L, Raad II.** 2006. Comparative *in vitro* efficacies of various catheter lock solutions. *Antimicrob Agents Chemother* **50:**1865–1868.
82. **Raad I, Chatzinikolaou I, Chaiban G, Hanna H, Hachem R, Dvorak T, Cook G, Costerton W.** 2003. *In vitro* and *ex vivo* activities of minocycline and EDTA against microorganisms embedded in biofilm on catheter surfaces. *Antimicrob Agents Chemother* **47:**3580–3585.

83. **Shinde RB, Chauhan NM, Raut JS, Karuppayil SM.** 2012. Sensitization of *Candida albicans* biofilms to various antifungal drugs by cyclosporine A. *Ann Clin Microbiol Antimicrob* **11:**27.
84. **Webb BC, Thomas CJ, Willcox MD, Harty DW, Knox KW.** 1998. *Candida*-associated denture stomatitis. Aetiology and management: a review. Part 1. Factors influencing distribution of *Candida* species in the oral cavity. *Aust Dent J* **43:**45–50.
85. **Webb BC, Thomas CJ, Willcox MD, Harty DW, Knox KW.** 1998. *Candida*-associated denture stomatitis. Aetiology and management: a review. Part 2. Oral diseases caused by *Candida* species. *Aust Dent J* **43:**160–166.
86. **Samaranayake YH, Samaranayake LP.** 2001. Experimental oral candidiasis in animal models. *Clin Microbiol Rev* **14:**398–429.
87. **Johnson CC, Yu A, Lee H, Fidel PL Jr, Noverr MC.** 2012. Development of a contemporary animal model of *Candida albicans*-associated denture stomatitis using a novel intraoral denture system. *Infect Immun* **80:** 1736–1743.
88. **Norris MM, Lamb DJ, Craig GT, Martin MV.** 1985. The effect of miconazole on palatal candidosis induced in the Wistar rat. *J Dent* **13:**288–294.
89. **Lamb DJ, Martin MV.** 1983. An *in vitro* and *in vivo* study of the effect of incorporation of chlorhexidine into autopolymerizing acrylic resin plates upon the growth of *Candida albicans*. *Biomaterials* **4:**205–209.
90. **Budtz-Jorgensen E.** 1971. Denture stomatitis. IV. An experimental model in monkeys. *Acta Odontol Scand* **29:**513–526.
91. **Lee H, Yu A, Johnson CC, Lilly EA, Noverr MC, Fidel PL Jr.** 2011. Fabrication of a multi-applicable removable intraoral denture system for rodent research. *J Oral Rehabil* **38:**686–690.
92. **Chen YL, Brand A, Morrison EL, Silao FG, Bigol UG, Malbas FF Jr, Nett JE, Andes DR, Solis NV, Filler SG, Averette A, Heitman J.** 2011. Calcineurin controls drug tolerance, hyphal growth, and virulence in *Candida dubliniensis*. *Eukaryot Cell* **10:**803–819.
93. **Zumbuehl A, Ferreira L, Kuhn D, Astashkina A, Long L, Yeo Y, Iaconis T, Ghannoum M, Fink GR, Langer R, Kohane DS.** 2007. Antifungal hydrogels. *Proc Natl Acad Sci USA* **104:**12994–12998.
94. **Ricicova M, Kucharikova S, Tournu H, Hendrix J, Bujdakova H, Van Eldere J, Lagrou K, Van Dijck P.** 2010. *Candida albicans* biofilm formation in a new *in vivo* rat model. *Microbiology* **156:**909–919.
95. **Bink A, Kucharikova S, Neirinck B, Vleugels J, Van Dijck P, Cammue BP, Thevissen K.** 2012. The nonsteroidal antiinflammatory drug diclofenac potentiates the *in vivo* activity of caspofungin against *Candida albicans* biofilms. *J Infect Dis* **206:**1790–1797.
96. **Alem MA, Douglas LJ.** 2005. Prostaglandin production during growth of *Candida albicans* biofilms. *J Med Microbiol* **54:**1001–1005.
97. **Ghalehnoo ZR, Rashki A, Najimi M, Dominguez A.** 2010. The role of diclofenac sodium in the dimorphic transition in *Candida albicans*. *Microb Pathog* **48:**110–115.
98. **Wang X, Fries BC.** 2011. A murine model for catheter-associated candiduria. *J Med Microbiol* **60:**1523–1529.
99. **Sobel JD, Faro S, Force RW, Foxman B, Ledger WJ, Nyirjesy PR, Reed BD, Summers PR.** 1998. Vulvovaginal candidiasis: epidemiologic, diagnostic, and therapeutic considerations. *Am J Obstet Gynecol* **178:**203–211.
100. **Graybill JR, Vazquez J, Darouiche RO, Morhart R, Greenspan D, Tuazon C, Wheat LJ, Carey J, Leviton I, Hewitt RG, MacGregor RR, Valenti W, Restrepo M, Moskovitz BL.** 1998. Randomized trial of itraconazole oral solution for oropharyngeal candidiasis in HIV/AIDS patients. *Am J Med* **104:**33–39.
101. **Chang DC, Grant GB, O'Donnell K, Wannemuehler KA, Noble-Wang J, Rao CY, Jacobson LM, Crowell CS, Sneed RS, Lewis FM, Schaffzin JK, Kainer MA, Genese CA, Alfonso EC, Jones DB, Srinivasan A, Fridkin SK, Park BJ, Fusarium Keratitis Investigation Team.** 2006. Multistate outbreak of *Fusarium* keratitis associated with use of a contact lens solution. *JAMA* **296:**953–963.
102. **Sun Y, Chandra J, Mukherjee P, Szczotka-Flynn L, Ghannoum MA, Pearlman E.** 2010. A murine model of contact lens-associated *Fusarium* keratitis. *Invest Ophthalmol Vis Sci* **51:**1511–1516.
103. **Zhang H, Wang L, Li Z, Liu S, Xie Y, He S, Deng X, Yang B, Liu H, Chen G, Zhao H, Zhang J.** 2013. A novel murine model of *Fusarium solani* keratitis utilizing fluorescent labeled fungi. *Exp Eye Res* **110:**107–112.
104. **Nett J, Lincoln L, Marchillo K, Andes D.** 2007. Beta -1,3 glucan as a test for central venous catheter biofilm infection. *J Infect Dis* **195:**1705–1712.
105. **Lafleur MD, Sun L, Lister I, Keating J, Nantel A, Long L, Ghannoum M, North J, Lee RE, Coleman K, Dahl T, Lewis K.** 2013. Potentiation of azole antifungals by 2-adamantanamine. *Antimicrob Agents Chemother* **57:**3585–3592.

106. **Sherry L, Jose A, Murray C, Williams C, Jones B, Millington O, Bagg J, Ramage G.** 2012. Carbohydrate derived fulvic acid: an *in vitro* investigation of a novel membrane active antiseptic agent against *Candida albicans* biofilms. *Front Microbiol* **3:**116.

107. **Coleman JJ, Okoli I, Tegos GP, Holson EB, Wagner FF, Hamblin MR, Mylonakis E.** 2010. Characterization of plant-derived saponin natural products against *Candida albicans*. *ACS Chem Biol* **5:**321–332.

108. **Martins M, Henriques M, Lopez-Ribot JL, Oliveira R.** 2012. Addition of DNase improves the *in vitro* activity of antifungal drugs against *Candida albicans* biofilms. *Mycoses* **55:** 80–85.

109. **Frederiksen B, Pressler T, Hansen A, Koch C, Hoiby N.** 2006. Effect of aerosolized rhDNase (Pulmozyme) on pulmonary colonization in patients with cystic fibrosis. *Acta Paediatr* **95:**1070–1074.

110. **Hudson SP, Langer R, Fink GR, Kohane DS.** 2010. Injectable *in situ* cross-linking hydrogels for local antifungal therapy. *Biomaterials* **31:** 1444–1452.

111. **Schmidt CS, White CJ, Ibrahim AS, Filler SG, Fu Y, Yeaman MR, Edwards JE Jr, Hennessey JP Jr.** 2012. NDV-3, a recombinant alum-adjuvanted vaccine for *Candida* and *Staphylococcus aureus*, is safe and immunogenic in healthy adults. *Vaccine* **30:**7594–7600.

112. **Luo G, Ibrahim AS, Spellberg B, Nobile CJ, Mitchell AP, Fu Y.** 2010. *Candida albicans* Hyr1p confers resistance to neutrophil killing and is a potential vaccine target. *J Infect Dis* **201:**1718–1728.

113. **Cassone A.** 2013. Development of vaccines for *Candida albicans*: fighting a skilled transformer. *Nat Rev Microbiol* **11:**884–891.

114. **Nett JE, Marchillo K, Andes DR.** 2012. Modeling of fungal biofilms using a rat central vein catheter. *Methods Mol Biol* **845:**547–556.

Biofilm Development

3

TIM TOLKER-NIELSEN[1]

INTRODUCTION

Experimental approaches primarily focused on genetic and microscopic techniques have laid the foundation for our current models of bacterial biofilm formation. This work has enabled researchers to define biofilm formation as a process that consists of specific stages. The biofilm developmental cycle is believed to include (i) initial attachment of microbes to a surface or each other, (ii) formation of microcolonies, (iii) maturation of the biofilm, and (iv) dispersal of the biofilm (e.g., reference 1). The different biofilm stages include bacterial physiology and phenotypic responses suggestive of the existence of a unique biofilm biology which is not found for planktonic bacteria.

The switch from the solitary planktonic bacterial lifestyle to the communal biofilm lifestyle involves a change in the bacteria so that they initiate the production of adhesins and extracellular matrix compounds which interconnect them in the biofilm. The extracellular biofilm matrix serves as a scaffold that has an essential cell-to-cell connecting and structural function in biofilms and plays a role in a number of processes including cell attachment, cell-to-cell interactions, and antimicrobial tolerance (2–7). The biofilm matrix that is produced by the bacteria contains mainly polysaccharides, proteins, and extracellular DNA (8).

[1]Department of Immunology and Microbiology, Faculty of Health and Medical Sciences, University of Copenhagen, DK 2000 Copenhagen, Denmark.

Microbial Biofilms 2nd Edition
Edited by Mahmoud Ghannoum, Matthew Parsek, Marvin Whiteley, and Pranab K. Mukherjee

doi:10.1128/microbiolspec.MB-0001-2014

This article focuses on several issues of biofilm formation, including initiation of biofilm formation, development of biofilm structure, formation of biofilms through multiple pathways, bacterial motility during biofilm formation, roles of quorum-sensing in biofilm formation, subpopulation development and interactions, biofilm formation governed by adaptive responses, and termination of biofilm formation in response to environmental cues.

BIOFILM FORMATION INITIATES IN RESPONSE TO SPECIFIC ENVIRONMENTAL CUES

Recent work in a number of bacterial species, including *Pseudomonas aeruginosa, Pseudomonas fluorescens, Pseudomonas putida, Vibrio cholerae, Yersinia pestis, Escherichia coli, Salmonella enterica, Burkholderia cenocepacia, Bacillus subtilis,* and *Clostridium difficile,* indicates that initiation of biofilm formation occurs in response to an increase in the level of the intracellular second messenger c-di-GMP (9–24). Synthesis and degradation of c-di-GMP in the bacteria is accomplished by two distinct classes of proteins with opposing enzymatic activities (see reference 25 for a review). Diguanylate cyclases harboring GGDEF domains synthesize c-di-GMP from two GTP molecules, whereas phosphodiesterases harboring EAL or HD-GYP domains degrade c-di-GMP. Sensory domains are frequently associated with the GGDEF and EAL/HD-GYP domain proteins, translating diverse environmental cues into c-di-GMP levels. The bacteria usually produce several different c-di-GMP diguanylate cyclases and phosphodiesterases, and evidence is accruing that they work in separate c-di-GMP circuits (26). Our current knowledge suggests that a number of different environmental cues and transducer mechanisms can lead to increases in local pools of c-di-GMP, which in turn can activate the production of adhesins and extracellular matrix products (e.g., references 27, 28). For example, in *P. aeruginosa* the WspA protein was found to be a membrane-bound receptor that detects a signal associated with contact to a surface (27, 29, 30). The signal is mediated to the histidine kinase WspE, which catalyzes phosphotransfer to the response regulator-diguanylate cyclase WspR, which in turn produces c-di-GMP. Synthesis of biofilm matrix products such as the CdrA adhesin and the Psl, Pel, and alginate exopolysaccharides in turn is positively regulated by c-di-GMP in *P. aeruginosa* (18, 31).

In addition to c-di-GMP-mediated regulation, biofilm formation is also regulated via small regulatory RNAs (sRNA) in many bacterial species (32). For example, work with *P. aeruginosa* suggests that synthesis of the exopolysaccharides Psl and Pel is induced in response to environmental signals sensed by the sensor kinases LadS and RetS and the sensor kinase/response regulator pair GacS/GacA (33–38). Phosphorylation of GacA by GacS is antagonized by the RetS sensor kinase and stimulated by the LadS sensor kinase. After phosphorylation of GacA by GacS, GacA induces transcription of the RsmY and RsmZ sRNAs. The RsmY and RsmZ sRNAs bind to and reduce the activity of the RsmA protein, which otherwise inhibits expression of a number of genes, including those encoding Psl and Pel polysaccharide production. In addition, sensor kinases and response regulators encoded by the *bfiSR, bfmSR,* and *mifSR* genes are evidently involved in the regulation of *P. aeruginosa* biofilm formation (39). Evidence was provided that the BfiSR system regulates biofilm formation by affecting synthesis of the CafA ribonuclease, which controls the level of the RsmZ sRNA (40). Moreover, the protein SagS was shown to contribute to the motile-to-sessile switch and act in concert with BfiSR to initiate *P. aeruginosa* biofilm formation (41).

Moscoso et al. (42) provided evidence that a *P. aeruginosa retS* mutant contains increased levels of c-di-GMP, indicating a link between sRNA-mediated and c-di-GMP-mediated biofilm regulation in *P. aeruginosa.*

THE ARCHITECTURE AND ORGANIZATION OF BIOFILMS ARE SPECIES DEPENDENT

Although microcolonies are the basic unit in most biofilms, the structure of the microcolonies can vary greatly depending on the biofilm-forming bacterial species. For example, it has been demonstrated that under identical conditions in a flow chamber *P. putida* forms loose protruding microcolonies (Fig. 1A), whereas *Pseudomonas knackmussii* (formerly termed *Pseudomonas* sp. B13) forms spherical microcolonies (Fig. 1B) (43). Moreover, when the two *Pseudomonas* species were grown together in dual-species biofilms, they still formed their characteristic microcolony structures (Fig. 1C), apparently without affecting each other (43). The architecture and organization of the three different biofilms are therefore dependent on the biofilm-forming bacterial species. Multiple factors are involved in the formation of particular structures in biofilms, and currently the mechanisms underlying the difference in biofilm structure displayed by *P. putida* and *P. knackmussii* is not known. However, in the case of *P. putida*, biofilm formation in flow chambers is mainly governed by the large adhesive protein LapA (9, 10, 44), whereas for other pseudomonads, such as *P. aeruginosa*, biofilm formation in flow chambers is mainly dependent on the exopolysaccharides Psl and Pel (7, 45, 46). Differences between the extracellular matrix components that interconnect bacteria in biofilms may give rise to different structures of the microcolonies.

STRUCTURE DEVELOPMENT IN BIOFILMS IS DEPENDENT ON NUTRITIONAL CONDITIONS

As described above, different bacterial species may form different biofilm structures under identical conditions. In addition, the same bacterial species may form different biofilm structures under different environmental conditions. For example, Klausen et al. (47) demonstrated that *P. aeruginosa* forms mushroom-shaped microcolonies when it grows in flow chambers that are irrigated with glucose medium, whereas it forms flat biofilms when it grows in flow

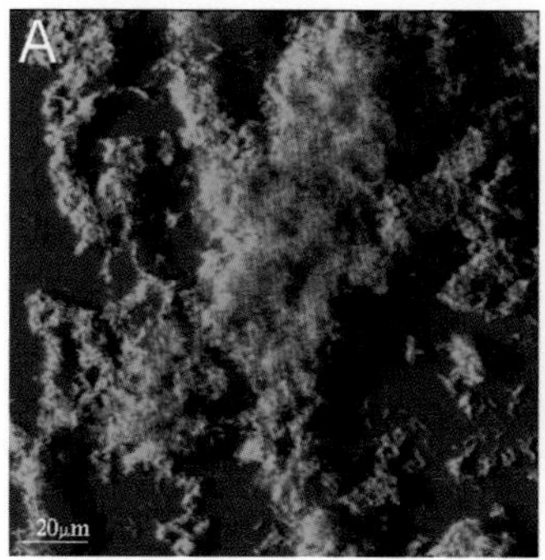

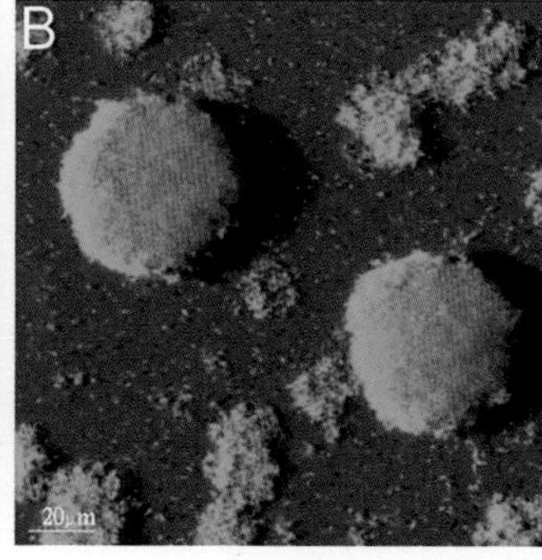

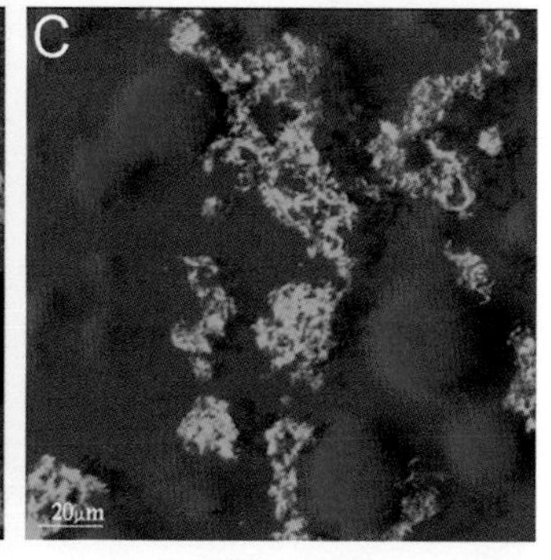

FIGURE 1 Confocal laser scanning microscopy (CLSM) images showing spatial structures in flow-chamber-grown 5-day-old biofilms formed by (A) Gfp-tagged (green fluorescent) *P. putida*, (B) Gfp-tagged *P. knackmussii*, and (C) a mixture of Gfp-tagged *P. putida* and DsRed-tagged (red fluorescent) *P. knackmussii*. Bars, 20 µm. Adapted from reference 43 with permission from the American Society for Microbiology. doi:10.1128/microbiolspec.MB-0001-2014.f1

chambers that are irrigated with citrate medium (Fig. 2). Moreover, the structure of an established biofilm can change in response to a change in nutritional conditions. Nielsen et al. (48) studied biofilm formation in flow chambers of a mixture consisting of *P. knackmussii* and *Burkholderia xenovorans* (formerly termed *Burkholderia* sp. LB400). These bacteria have the potential to interact metabolically because *P. knackmussii* can metabolize chlorobenzoate produced by *B. xenovorans* when grown on chlorobiphenyl. When the dual-species biofilm was fed with medium containing chlorobiphenyl, mixed-species microcolonies consisting of associated *P. knackmussii* and *B. xenovorans* bacteria were formed. In contrast, when the mixture was fed citrate, which can be metabolized by both species, the two species formed separate microcolonies. After a shift in carbon source from a citrate medium to a chlorobiphenyl medium, movement of the *P. knackmussii* bacteria led to a change in the spatial structure of the biofilm from the separate microcolonies toward the mixed-species microcolonies.

Similar observations were made by Wolfaardt et al. (49), who studied a microbial mixture capable of degrading the herbicide diclofop. When this mixture was grown in flow chambers irrigated with diclofob, a highly structured biofilm with specific patterns of intergeneric cellular coaggregation was formed. But when the mixture was grown on tryptic soy broth (TSB), a biofilm lacking variation in thickness and structure was formed. After a shift in carbon source from TSB to diclofob, it took TSB-grown biofilms only two days to acquire the typical structure of diclofob-grown biofilms. Although the organisms and the nature of the metabolic interactions in the study (49) were unknown, this study provided substantial evidence that structure development in biofilms is dependent on nutritional conditions and that the structure of an established biofilm can change in response to a change in nutritional conditions.

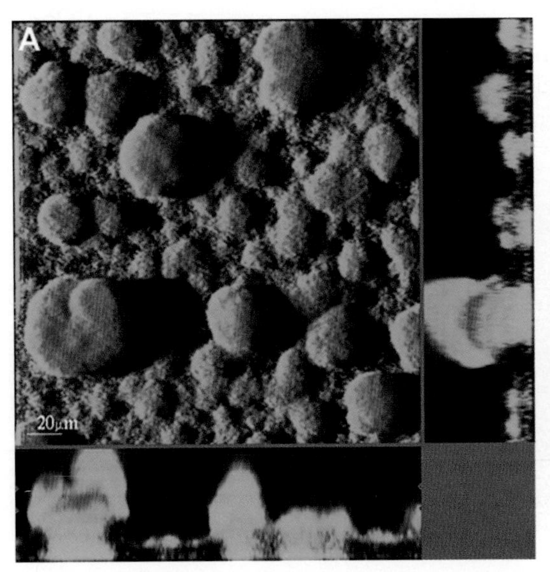

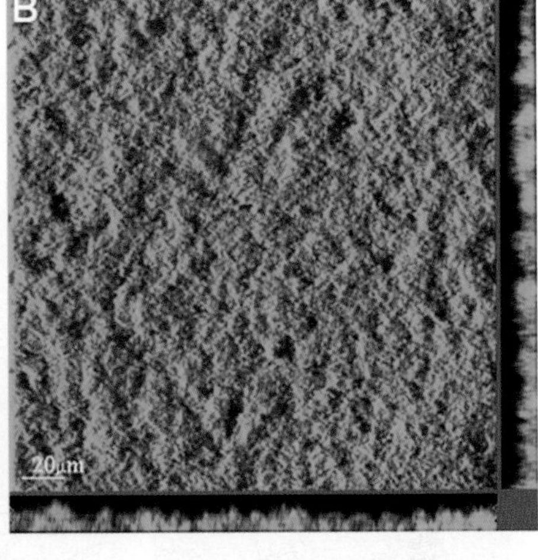

FIGURE 2 CLSM micrographs acquired in 5-day-old *P. aeruginosa* PAO1 biofilms grown on (A) glucose minimal medium and (B) citrate minimal medium. The central pictures show-top down fluorescence projections, and the flanking pictures show vertical sections. Bars, 20 µm. Adapted from reference 47 with permission from Wiley-Blackwell publishing. doi:10.1128/microbiolspec.MB-0001-2014.f2

BACTERIAL MOTILITY MAY BE AN INTEGRATED PART OF BIOFILM FORMATION

The studies described above suggest that bacterial migration is involved in the structural changes that can occur in biofilms in response to a change in nutritional conditions. However, it appears that bacterial migration can be important for the structural development of biofilms also under stable environmental conditions. The available evidence suggests that *P. aeruginosa* forms flat biofilm in flow chambers irrigated with citrate medium because the bacteria migrate extensively during the initial phase of biofilm formation and do not settle to form microcolonies (47). *P. aeruginosa pilA* mutants (deficient in biogenesis of type IV pili) were shown to form protruding microcolonies in flow chambers irrigated with citrate medium, indicating that the motility that occurs in citrate-grown *P. aeruginosa* wild type biofilms is driven by type IV pili (47). In flow chambers irrigated with glucose medium *P. aeruginosa* forms two distinct subpopulations in the initial phase of biofilm formation: a nonmotile subpopulation that forms initial microcolonies and a motile subpopulation that initially migrates on the substratum (50). Formation of the mushroom-shaped structures in the glucose-grown biofilms evidently involves colonization of the initial microcolonies by bacteria from the migrating subpopulation that subsequently form mushroom caps on top of the initial microcolonies, which then correspond to the stalk of the mushroom-shaped structures (50). In glucose-grown *P. aeruginosa* biofilms containing a mixture of Cfp-tagged *pilA* mutants and Yfp-tagged wild type bacteria, mushroom-shaped structures that contain *pilA* mutants in the stalk and wild type in the cap are formed (50) (Fig. 3). In addition, mushroom-shaped structures that contain wild type bacteria in both the stalk and cap can also be found (Fig. 3). These findings provide evidence that bacterial migration may be an integrated part of biofilm formation.

BIOFILM FORMATION CAN OCCUR THROUGH MULTIPLE PATHWAYS

As described above, the formation of biofilm structures is both species-specific and dependent on environmental conditions, suggesting that biofilm formation can occur through multiple pathways. Under similar conditions the mechanisms underlying biofilm formation for closely related species such as *P. aeruginosa* and *P. putida* may be very different. As indicated above, the formation of *P. putida* biofilm in flow chambers is mainly dependent on the large

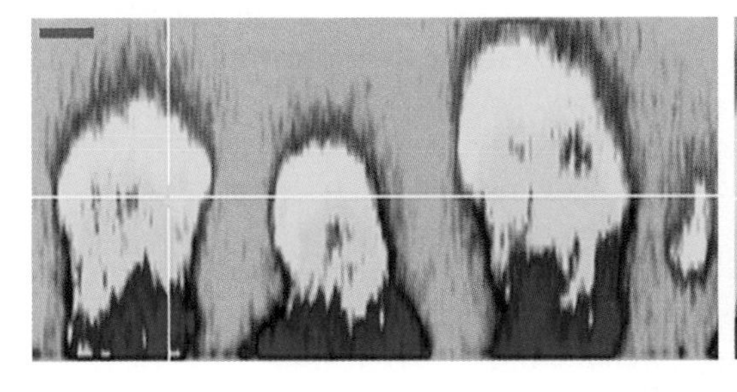
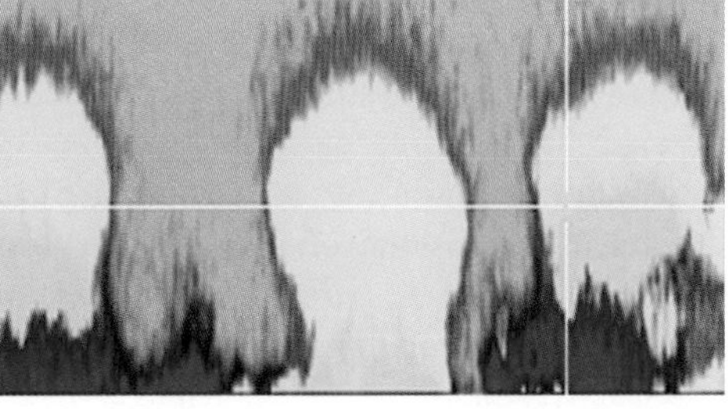

FIGURE 3 CLSM vertical sections acquired in a color-coded *P. aeruginosa* biofilm grown in a flow chamber on glucose minimal medium. The CLSM micrographs were acquired in a 4-day-old biofilm which was initiated with a 1:1 mixture of Yfp-tagged (yellow fluorescent) *P. aeruginosa* PAO1 wild type and Cfp-tagged (cyan fluorescent) *P. aeruginosa pilA* mutants. Bars, 20 μm. Adapted from reference 50 with permission from Wiley-Blackwell publishing. doi:10.1128/microbiolspec.MB-0001-2014.f3

adhesive surface protein LapA (9, 10, 44), whereas the formation of *P. aeruginosa* biofilm in flow chambers usually depends on exopolysaccharides such as Psl and Pel (7, 45, 46). However, it is also conceivable that a specific bacterial species may form biofilm through different pathways under different conditions. *P. aeruginosa*, for example, can produce a number of biofilm matrix products, including the Psl, Pel, and alginate polysaccharides; extracellular DNA; and the type IV pili, Cup, CdrA, LecA, LecB, and Fap protein components (2, 31, 45, 51–57). It may be expected that some *P. aeruginosa* strains selected for under specific conditions will produce some of these biofilm matrix components in higher quantities than other strains and that the biofilm pathways will vary accordingly.

Whitchurch et al. (2) reported that the presence of DNase in the medium prevented biofilm formation by the *P. aeruginosa* PAO1 laboratory strain in microtiter trays and flow chambers, suggesting that extracellular DNA is an important matrix component in these biofilms. Addition of DNase to young flow-chamber-grown *P. aeruginosa* PAO1 biofilms led to dispersal but did not disperse mature biofilms, probably due to production of increasing amounts of other biofilm matrix components during biofilm maturation. In contrast, Nemoto et al. (58) found that mature biofilms formed by four independent clinical *P. aeruginosa* isolates could be dispersed by DNase treatment, suggesting that extracellular DNA is the primary cell-to-cell interconnecting compound in mature biofilms formed by these *P. aeruginosa* strains. In addition, Murakawa (59, 60) investigated the chemical composition of the biofilm matrix from 20 clinical *P. aeruginosa* isolates and found that the biofilm matrix from 18 strains consisted primarily of DNA, while 2 strains with a mucoid phenotype produced slimes composed primarily of alginate.

Differences in biofilm structure may give rise to different phenotypic responses, e.g., to antibiotic treatment. Many types of biofilms display a remarkable increased tolerance to antimicrobial treatment compared to planktonic bacteria (61). The mechanisms that contribute to biofilm-associated antimicrobial tolerance include restricted antimicrobial diffusion, differential physiological activity, induction of specific tolerance mechanisms, and persister cell formation (62). However, similar phenotypic responses to antibiotic treatment may have different underlying conditions even for biofilms formed by the same species and displaying similar structure. For example, tolerance to tobramycin in *P. aeruginosa* PA14 biofilms can be caused by periplasmic glucans that bind tobramycin and prevent cell death most likely by sequestering the antibiotic (63). Synthesis of the periplasmic glucans requires the *ndvB* gene in *P. aeruginosa* PA14, and biofilms formed by a *P. aeruginosa ndvB* mutant were found to be much more sensitive to tobramycin than wild type biofilms. In contrast, the *ndvB* mutant and wild type showed no difference in tobramycin sensitivity when grown in planktonic culture. Reverse transcriptase PCR provided evidence that the *ndvB* gene was expressed specifically in *P. aeruginosa* PA14 biofilms and not in planktonic cells (63). However, microarray analysis has provided evidence that *ndvB* is expressed at the same low level in biofilm and planktonic cells of *P. aeruginosa* PAO1 (64), and therefore the *ndvB*-mediated mechanism appears to be restricted to specific *P. aeruginosa* strains. However, *P. aeruginosa* PAO1 biofilms also display tolerance to tobramycin treatment, and in this case extracellular DNA seems to be of primary importance because it binds tobramycin and shields the antibiotic (65).

QUORUM-SENSING CAN PLAY IMPORTANT ROLES IN BIOFILM FORMATION

In agreement with the fact that bacteria are closely associated in biofilms, quorum-sensing has been shown to play a role in biofilm formation for various bacterial species, e.g., *P. aeruginosa* and *B. cenocepacia*. Davies et al. (66) showed that a quorum-sensing-defective mutant strain of *P. aeruginosa* formed flat and undifferentiated biofilms in flow-chambers under conditions where the wild type strain formed biofilms with large mushroom-shaped structures. In addition, Hentzer et al. (67) demonstrated that acylated homoserine lactone analogues, which inhibit quorum-sensing, affected *P. aeruginosa* biofilm formation in flow chambers. Subsequent studies provided further evidence for the importance of quorum-sensing signaling in the structural development of *P. aeruginosa* biofilms (e.g., references 4, 68). Quorum-sensing plays a role in the generation of extracellular DNA in *P. aeruginosa* biofilms (69). In addition to a small amount of extracellular DNA present in the initial phase in *P. aeruginosa* biofilms, release of a large amount of extracellular DNA occurs at a later stage of *P. aeruginosa* biofilm formation, regulated via the *Pseudomonas* quinolone signal (PQS)-based quorum-sensing system (69). Evidence has been provided that the release of DNA in *P. aeruginosa* biofilm occurs as a consequence of lysis of a small subpopulation of the bacteria (69). In agreement with a role for PQS-based quorum-sensing in the generation of extracellular DNA that functions as a biofilm matrix component, *P. aeruginosa pqs* mutants were found to have defects in biofilm formation, and the thin biofilms that were formed by the *pqs* mutants contained little extracellular DNA (4, 68–70). The mechanism involved in PQS-mediated DNA-release is unknown, but evidence has been provided that PQS can act as a cell-sensitizing pro-oxidant (71), which might cause lysis of a subpopulation of the *P. aeruginosa* bacteria during biofilm formation.

In addition to extracellular DNA, quorum-sensing signaling controls the production of the biosurfactant rhamnolipid, which was shown to be important for biofilm formation by *P. aeruginosa* (72–74). Furthermore, the production of the *P. aeruginosa* LecA and LecB lectins was shown to be regulated by quorum-sensing (75), and the lectins were shown to play a role in *P. aeruginosa* biofilm formation (56, 57). Quorum-sensing signaling in *P. aeruginosa* also controls the production of siderophores such as pyoverdine and pyochelin, which are also of importance for biofilm formation (76).

Although quorum-sensing evidently controls the production of many factors which are important for *P. aeruginosa* biofilm formation, the role of quorum-sensing appears to be conditional, because quorum-sensing mutant strains evidently form biofilms indistinguishable from wild type biofilms under some conditions (77). A more definitive requirement for quorum-sensing in biofilm formation seems to be the case for *B. cenocepacia*. Inhülsen et al. (78) and Schmid et al. (79) identified three main factors which potentially link quorum-sensing signaling to *B. cenocepacia* biofilm formation: type-1 fimbriae, the BclACB lectins, and the large surface protein BapA. These factors are all under quorum-sensing regulation in *B. cenocepacia* and were investigated in detail for their roles in biofilm formation. Deletion of neither the type-1 fimbriae genes nor the adjacent chaperone-usher transport system genes affected biofilm formation by *B. cenocepacia* in microtiter trays or flow chambers, indicating that type-1 fimbriae are not essential for biofilm formation by *B. cenocepacia* on abiotic surfaces. Deletion of the *bclACB* operon did not result in any significant difference in biofilm formation in microtiter trays compared to the wild type strain. In contrast, the deletion of *bapA*, as well as the adjacent genes encoding a putative ABC transporter system responsible for the secretion of BapA, resulted in a significant reduction in biofilm formation in microtiter trays and flow chambers. Downregulation of BapA expression through a rhamnose-inducible promoter in the *bclACB* mutant resulted in a defect in biofilm formation compared to the corresponding wild type strain, with reduced *bapA* expression, suggesting that lectins are important in surface colonization when BapA is limiting and that BapA is the major factor for biofilm formation.

In addition to the fimbriae, lectins, and BapA, quorum-sensing also seems to regulate the production of extracellular DNA in *B. cenocepacia* biofilms. McCarthy et al. (80) provided evidence that four-day-old *B. cenocepacia* wild type biofilms grown in flow chambers contained substantial amounts of extracellular DNA as a part of the biofilm matrix. In contrast, biofilms formed by a quorum-sensing-defective *B. cenocepacia* mutant never passed the initial microcolony stage, and the thin biofilms showed little staining by propidium iodide, indicating that extracellular DNA was not generated in large quantities (80).

BACTERIAL SUBPOPULATIONS DEVELOP AND INTERACT DURING BIOFILM FORMATION

As described above, the different bacteria in mixed-species biofilms can engage in metabolic interactions, and this affects biofilm formation and the structure of the biofilm. However, it also appears that subpopulations develop and interact in monospecies biofilms and that these interactions affect biofilm formation and the structure of the biofilms. For example, the motile and nonmotile subpopulations that can develop during *P. aeruginosa* biofilm formation appear to interact in a number of ways. Yang et al. (4) provided evidence that the PQS quorum-sensing system and genes necessary for pyoverdine synthesis are expressed only in the stalk part of the mushroom-shaped microcolonies that form in glucose-grown *P. aeruginosa* biofilms. PQS quorum-sensing controls the production of extracellular DNA in *P. aeruginosa* biofilms (69), and evidence was provided that release of extracellular DNA by the bacteria in the initial microcolonies (which subsequently constitute the stalk part of the mature mushroom-shaped microcolonies) is necessary for development of the mushroom-shaped microcolonies (4). Furthermore, evidence was presented that pyoverdine production in the initial microcolonies is necessary for iron uptake in the cap-forming subpopulation and for development of the mushroom-shaped microcolonies (4).

The propositions made by Yang et al. (4) were mainly based on genetic evidence. Experiments involving *P. aeruginosa* strains with *pqsA::gfp* and *pvdA::gfp* fluorescent reporters provided evidence that the genes required for PQS quorum-sensing and pyoverdine synthesis were expressed specifically in the stalk part of the mushroom-shaped microcolonies in *P. aeruginosa* biofilms. The three mutant strains *P. aeruginosa pilA*, *P. aeruginosa pvdA* (deficient in pyoverdine production), and *P. aeruginosa pqsA* (deficient in PQS quorum-sensing) were shown to be unable to form mushroom-shaped structures individually in single-strain biofilms. *P. aeruginosa pilA* mutants could only form the initial microcolonies because type IV pili are necessary for cap formation, and it appeared that the *P. aeruginosa pvdA* and *P. aeruginosa pqsA* mutant strains also had a defect in cap formation. However, mushroom-shaped structures with *pilA* mutants in the stalk and *pqsA* mutants in the cap were formed in *pilA/pqsA* mixed-strain biofilms. Likewise, mushroom-shaped structures with *pilA* mutants in the stalk and *pvdA* mutants in the cap were formed in *pilA/pvdA* mixed-strain biofilms. It appears that the *pilA*, *pvdA*, and *pqsA* mutants cannot form mushroom-shaped microcolonies individually in single-strain biofilms, but in *pilA/pqsA* mixed-strain biofilms and *pilA/pvdA* mixed-strain biofilms the subpopulations interact with each other, and together they form mushroom-shaped microcolonies.

BIOFILM FORMATION IS PROGRAMMED IN THE SENSE THAT REGULATED SYNTHESIS OF EXTRACELLULAR MATRIX COMPONENTS IS INVOLVED, BUT IT IS ALSO GOVERNED BY ADAPTIVE RESPONSES

Observations of the mushroom-shaped microcolonies that can form in flow-chamber-grown *P. aeruginosa* biofilms have led to comparisons between *P. aeruginosa* biofilm microcolonies and the fruiting bodies that are formed by *Myxococcus* bacteria (66), and it has led to speculation about biofilm formation as a highly programmed developmental process with hierarchically ordered genetic pathways controlling attachment, microcolony formation, and microcolony maturation (81). The developmental model implies that the structure of biofilms has evolved to provide a specific function. For example, it has been proposed that the ability of *P. aeruginosa* to form mushroom-shaped microcolonies in biofilms has developed as a result of group-level evolution, because these structures give efficient nutrient supply to the bacteria and efficient removal of waste products (82). However, the available evidence suggests that the mushroom-shaped microcolonies are formed in the flow-chamber-grown *P. aeruginosa* biofilms mainly as a consequence of the special conditions in the flow chamber and the existence of motile and nonmotile subpopulations. Because of bacterial nutrient consumption, there is a decreasing nutrient gradient from the top to the bottom of the biofilm in the flow chamber (83). The motile bacteria that accumulate on the top of the microcolonies therefore have a growth advantage in comparison to the bacteria in the lower part of the biofilm and can proliferate to form the mushroom caps (84). As described above, extracellular DNA released by the nonmotile bacteria appears to be important for settling of the motile bacteria in the cap portion of the mushroom-shaped microcolonies (4). In addition, exopolysaccharide production, rhamnolipid surfactant production, quorum-sensing, and siderophores are also known to be involved in formation of the mushroom-shaped microcolonies (85). It is presently not understood why the *P. aeruginosa* population differentiates into nonmotile and motile subpopulations in the initial stage of biofilm formation in glucose-fed flow chambers and not in citrate-fed flow chambers.

If biofilm formation is a highly programmed process, it would be expected that a core set of "biofilm genes" would be expressed in all biofilms of a given bacterium. In the case of *P. aeruginosa* a number of microarray analyses have been performed to monitor genes that are expressed during biofilm formation (e.g., 64, 86–88). However, the transcriptomic analyses performed by various research groups have failed to consistently identify specific biofilm regulons. The failure to detect common themes in the gene expression profile of *P. aeruginosa* biofilm cells suggests that biofilm formation is mainly governed by adaptive responses. This is in agreement with the fact that the structural development of *P. aeruginosa* biofilms and different mixed-species biofilms is dependent on the nutritional conditions and can change in response to changing nutritional conditions, as described above. However, formation of biofilms does require expression of biofilm matrix products. In the case of *P. aeruginosa* these biofilm matrix products include Psl, Pel, alginate, extracellular DNA, type IV pili, Cup, CdrA, LecA, LecB, and Fap (2, 6, 31, 45, 52–57). The synthesis of many of these products is positively regulated by the intracellular signaling molecule c-di-GMP (e.g., see reference 25 for a review). As mentioned previously, synthesis and degradation of c-di-GMP in the bacteria is accomplished by diguanylate cyclases and phosphodiesterases that contain sensory domains, enabling translation of diverse environmental cues into synthesis or degradation of c-di-GMP, which binds to downstream effector molecules and modulates their function, resulting in regulation of the production of different adhesins and biofilm matrix products. It therefore appears that biofilm formation is programmed in the sense that regulated synthesis of extracellular matrix components is involved. However, although researchers have defined a number of biofilm stages, there are currently no data that couple the initial c-di-GMP-regulated matrix production to specific stages occurring later during biofilm formation. In addition, there is currently no evidence that the different c-di-GMP signaling circuits are hierarchically ordered to control transitions through specific stages in biofilm formation. Biofilm-specific pathways may be limited to the initial regulation of adhesin and matrix production, whereas the successive stages of biofilm formation may be governed by adaptive responses.

BIOFILM FORMATION IS LARGELY GOVERNED BY ADAPTIVE RESPONSES OF INDIVIDUAL BACTERIA, BUT GROUP-LEVEL ACTIVITIES ARE ALSO INVOLVED

As mentioned above, biofilm formation can be viewed as a developmental process or as a process governed by adaptive responses of individual bacteria. Biofilm formation as a developmental process involves biofilm-specific genes that are part of hierarchically ordered pathways dedicated to controlling the transition through specific biofilm stages. Biofilm formation governed by adaptive responses involves the ability of the individual bacteria to regulate cellular adhesiveness and motility in response to micro-environmental cues. The developmental and adaptive response hypotheses can be distinguished in evolutionary terms, because the former involves selection of a given trait because of its benefit to the group, whereas the latter involves selection of a given trait because of its benefit to the individual bacterium. The apparently cooperative traits in biofilms can in many cases alternatively be explained by the fitness advantages of this behavior for the individual bacterium. For example, the formation of regular mushroom-shaped structures in *P. aeruginosa* biofilms via a pathway that involves type IV pili-dependent cellular migration may be a coordinated social process that creates biofilms with architectures which allow optimal circulation and nutrient supply to the population, or it may, because nutrient gradients develop in biofilms, be the result of chemotaxis of individual bacteria moving to a favorable nutrient-containing micro-environment.

The production of the cell-to-cell interconnecting components in biofilms may be a cost each bacterium pays to contribute to the social activity of creating a protective biofilm domicile, or it may simply increase the adhesiveness of single bacterial cells, allowing them to persist in the specific environment. Quorum-sensing-dependent production of extracellular DNA and lectins in *P. aeruginosa* biofilms (69, 75), as well as quorum-sensing-dependent production of surface protein, fimbriae, and lectins in *B. cenocepacia* biofilms (78), may be interpreted as group-level activities. However, instead of a means to regulate production of specific factors at the group level, quorum-sensing may be diffusion-sensing that enables the individual bacterium to determine whether secreted molecules rapidly move away from the cell (89). Yet the observed involvement of quorum-sensing in DNA release from a lysing subpopulation in *P. aeruginosa* biofilms, resulting in biofilm stabilization, may be interpreted as social behavior because the lysing cells that provide the DNA obviously are not themselves benefited directly.

BIOFILM FORMATION TERMINATES IN RESPONSE TO SPECIFIC ENVIRONMENTAL CUES

In agreement with the suggestion that biofilm formation is mainly governed by adaptive responses, and does not involve stable or meta-stable developmental changes, the bacteria in biofilms can at any stage sense environmental cues and terminate biofilm formation in response to specific signals. In the case of *P. putida*, for example, it was shown that established flow-chamber-grown biofilms can disperse completely within a few minutes in response to a shift from a medium with a carbon source to a medium without a carbon source or in response to a stoppage in the flow of growth medium (10). Biofilm formation by *P. putida* in flow chambers is mainly governed by the large adhesive protein LapA (9, 44). Evidence has been provided that c-di-GMP signaling regulates biofilm formation in *P. putida* by controlling the presence of LapA on the cell surface (9). The presence of LapA on the cell surface is controlled by the proteins LapD and LapG in response to the intracellular level of c-di-GMP. The LapG protein is a periplasmic proteinase, and it is able to cleave LapA off the cell surface when it is not repressed. The LapD protein spans the cytoplasmic membrane and contains degenerate GGDEF and EAL domains which can bind c-di-GMP, and it regulates the activity of the LapG proteinase by repressing it when the intracellular level of c-di-GMP is high and derepressing it when the intracellular level of c-di-GMP is low. A c-di-GMP phosphodisterase gene in *P. putida*, homologous to the *P. aeruginosa bifA* gene, has recently been identified as being essential for starvation-induced dispersal of *P. putida* biofilms along with the *lapG* gene, suggesting a link between *P. putida* BifA phosphodisterase activity and LapG-mediated dispersal (90).

The work described above, together with structural and biochemical work done in *P. fluorescens* (91, 92), suggests that *P. putida* can rapidly terminate biofilm formation in response to starvation through the following mechanisms: (i) nutrient limitation leads to activation of the BifA phosphodiesterase via an unknown pathway; (ii) activation of the BifA phosphodiesterase leads to a decrease in the c-di-GMP level in the vicinity of the transmembrane LapD protein; (iii) the reduction in the c-di-GMP level causes dissociation of c-di-GMP from the LapD protein; (iv) dissociation of c-di-GMP from LapD causes derepression of the periplasmic LapG protease; (v) the derepressed LapG protease cleaves off the cell-associated LapA protein; (v) cleavage of the LapA protein leads to the release of the cells from the biofilm.

CONCLUDING REMARKS

During the past decade we have gained much knowledge about the molecular mechanisms that are involved in the initiation and termination of biofilm formation. In many bacteria, these processes appear to occur in response to specific environmental cues and result in, respectively, induction or termination of biofilm matrix production via the second messenger molecule c-di-GMP. In between initiation and termination of biofilm formation we have defined specific biofilm stages, but the currently available evidence suggests that these transitions are mainly governed by adaptive responses, and not by specific genetic programs. It appears that biofilm formation can occur through multiple pathways and that the spatial structure of the biofilms is species dependent as well as dependent on the environmental conditions. Bacterial subpopulations, e.g., motile and nonmotile subpopulations, can develop and interact during biofilm formation, and these interactions can affect the structure of the biofilm. The available evidence suggests that biofilm formation is programmed in the sense that regulated synthesis of extracellular matrix components is involved. Furthermore, our current knowledge suggests that biofilm formation is mainly governed by adaptive responses of individual bacteria, although group-level activities are also involved.

The realization that a number of separate c-di-GMP signaling circuits regulate the production of different biofilm matrix components in bacteria in response to specific environmental cues has been a major step forward in biofilm research. However, although we have defined a number of biofilm stages, there is currently no data that couple the initial c-di-GMP-regulated matrix production to specific stages occurring later during biofilm formation, and there is currently no evidence that the various c-di-GMP signaling circuits are hierarchically ordered to control the transition through specific stages in biofilm formation. Future research will show whether biofilm-specific pathways are limited to the initial and terminal c-di-GMP-mediated regulation of adhesin and matrix production, with the in-between stages of biofilm formation governed by adaptive responses, or whether the different c-di-GMP signaling circuits are hierarchically ordered to control the transition through specific stages in biofilm formation.

ACKNOWLEDGMENTS

This work was supported by grants from the Danish Council for Independent Research and the Lundbeck foundation.

Conflicts of interest: I declare no conflicts.

CITATION

Tolker-Nielsen T. 2015. Biofilm development. Microbiol Spectrum 3(2):MB-0001-2014.

REFERENCES

1. **Sauer K, Camper AK, Ehrlich GD, Costerton JW, Davies DG.** 2002. *Pseudomonas aeruginosa* displays multiple phenotypes during development as a biofilm. *J Bacteriol* **184:**1140–1154.
2. **Whitchurch CB, Tolker-Nielsen T, Ragas PC, Mattick JS.** 2002. Extracellular DNA required for bacterial biofilm formation. *Science* **295:** 1487.
3. **Ma L, Jackson KD, Landry RM, Parsek MR, Wozniak DJ.** 2006. Analysis of *Pseudomonas aeruginosa* conditional psl variants reveals roles for the psl polysaccharide in adhesion and maintaining biofilm structure post-attachment. *J Bacteriol* **188:**8213–8221.
4. **Yang L, Nilsson M, Gjermansen M, Givskov M, Tolker-Nielsen T.** 2009. Pyoverdine and PQS mediated subpopulation interactions involved in *Pseudomonas aeruginosa* biofilm formation. *Mol Microbiol* **74:**1380–1392.
5. **Molin S, Tolker-Nielsen T.** 2003. Gene transfer occurs with enhanced efficiency in biofilms and induces enhanced stabilisation of the biofilm structure. *Curr Opin Biotechnol* **14:** 255–261.
6. **Friedman L, Kolter R.** 2003. Genes involved in matrix formation in *Pseudomonas aeruginosa* PA14 biofilms. *Mol Microbiol* **51:**675–690.
7. **Jackson KD, Starkey M, Kremer S, Parsek MR, Wozniak DJ.** 2004. Identification of psl, a locus encoding a potential exopolysaccharide that is essential for *Pseudomonas aeruginosa* PAO1 biofilm formation. *J Bacteriol* **186:**4466–4475.
8. **Pamp SJ, Gjermansen M, Tolker-Nielsen T.** 2007. The biofilm matrix: a sticky framework, p 37–69. *In* Kjelleberg S, Givskov M (ed), *The Biofilm Mode of Life: Mechanisms and Adaptations.* Horizon Bioscience, Norfolk, UK.
9. **Gjermansen M, Nilsson M, Yang L, Tolker-Nielsen T.** 2010. Characterization of starvation-induced dispersion in *Pseudomonas putida* biofilms: genetic elements and molecular mechanisms. *Mol Microbiol* **75:**815–826.
10. **Gjermansen M, Ragas P, Sternberg C, Molin S, Tolker-Nielsen T.** 2005. Characterization of starvation-induced dispersion in *Pseudomonas putida* biofilms. *Environ Microbiol* **7:**894–906.
11. **Gjermansen M, Ragas P, Tolker-Nielsen T.** 2006. Proteins with GGDEF and EAL domains regulate *Pseudomonas putida* biofilm formation and dispersal. *FEMS Microbiol Lett* **265:**215–224.
12. **Ryan RP, Lucey J, O'Donovan K, McCarthy Y, Yang L, Tolker-Nielsen T, Dow JM.** 2009. HD-GYP domain proteins regulate biofilm formation and virulence in *Pseudomonas aeruginosa. Environ Microbiol* **11:**1126–1136.

13. **Simm R, Morr M, Kader A, Nimtz M, Romling U.** 2004. GGDEF and EAL domains inversely regulate cyclic di-GMP levels and transition from sessility to motility. *Mol Microbiol* **53:**1123–1134.
14. **Ross P, Weinhouse H, Aloni Y, Michaeli D, Weinberger-Ohana P, Mayer R, Braun S, de Vroom E, van der Marel GA, van Boom JH, Benziman M.** 1987. Regulation of cellulose synthesis in *Acetobacter xylinum* by cyclic diguanylic acid. *Nature* **325:**279–281.
15. **Garcia B, Latasa C, Solano C, Garcia-del Portillo F, Gamazo C, Lasa I.** 2004. Role of the GGDEF protein family in *Salmonella* cellulose biosynthesis and biofilm formation. *Mol Microbiol* **54:**264–277.
16. **Tischler AD, Camilli A.** 2004. Cyclic diguanylate (c-di-GMP) regulates *Vibrio cholerae* biofilm formation. *Mol Microbiol* **53:**857–869.
17. **Kirillina O, Fetherston JD, Bobrov AG, Abney J, Perry RD.** 2004. HmsP, a putative phosphodiesterase, and HmsT, a putative diguanylate cyclase, control Hms-dependent biofilm formation in *Yersinia pestis. Mol Microbiol* **54:**75–88.
18. **Hickman JW, Tifrea DF, Harwood CS.** 2005. A chemosensory system that regulates biofilm formation through modulation of cyclic diguanylate levels. *Proc Natl Acad Sci USA* **102:** 14422–14427.
19. **Kulasakara H, Lee V, Brencic A, Liberati N, Urbach J, Miyata S, Lee DG, Neely AN, Hyodo M, Hayakawa Y, Ausubel FM, Lory S.** 2006. Analysis of *Pseudomonas aeruginosa* diguanylate cyclases and phosphodiesterases reveals a role for bis-(3′-5′)-cyclic-GMP in virulence. *Proc Natl Acad Sci USA* **103:**2839–2844.
20. **Goymer P, Kahn SG, Malone JG, Gehrig SM, Spiers AJ, Rainey PB.** 2006. Adaptive divergence in experimental populations of *Pseudomonas fluorescens*. II. Role of the GGDEF regulator WspR in evolution and development of the wrinkly spreader phenotype. *Genetics* **173:**515–526.
21. **Lim B, Beyhan S, Meir J, Yildiz FH.** 2006. Cyclic-diGMP signal transduction systems in *Vibrio cholerae*: modulation of rugosity and biofilm formation. *Mol Microbiol* **60:**331–348.
22. **Fazli M, O'Connell A, Nilsson M, Niehaus K, Dow JM, Givskov M, Ryan RP, Tolker-Nielsen T.** 2011. The CRP/FNR family protein Bcam1349 is a c-di-GMP effector that regulates biofilm formation in the respiratory pathogen *Burkholderia cenocepacia. Mol Microbiol* **82:**327–341.
23. **Chen Y, Chai Y, Guo JH, Losick R.** 2012. Evidence for cyclic Di-GMP-mediated signaling in *Bacillus subtilis. J Bacteriol* **194:**5080–5090.
24. **Purcell EB, McKee RW, McBride SM, Waters CM, Tamayo R.** 2012. Cyclic diguanylate inversely regulates motility and aggregation in *Clostridium difficile. J Bacteriol* **194:** 3307–3316.
25. **Hengge R.** 2009. Principles of c-di-GMP signalling in bacteria. *Nat Rev Microbiol* **7:** 263–273.
26. **Massie JP, Reynolds EL, Koestler BJ, Cong JP, Agostoni M, Waters CM.** 2012. Quantification of high-specificity cyclic diguanylate signaling. *Proc Natl Acad Sci USA* **109:**12746–12751.
27. **O'Connor JR, Kuwada NJ, Huangyutitham V, Wiggins PA, Harwood CS.** 2012. Surface sensing and lateral subcellular localization of WspA, the receptor in a chemosensory-like system leading to c-di-GMP production. *Mol Microbiol* **86:**720–729.
28. **Monds RD, Newell PD, Gross RH, O'Toole GA.** 2007. Phosphate-dependent modulation of c-di-GMP levels regulates *Pseudomonas fluorescens* Pf0-1 biofilm formation by controlling secretion of the adhesin LapA. *Mol Microbiol* **63:**656–679.
29. **Guvener ZT, Harwood CS.** 2007. Subcellular location characteristics of the *Pseudomonas aeruginosa* GGDEF protein, WspR, indicate that it produces cyclic-di-GMP in response to growth on surfaces. *Mol Microbiol* **66:**1459–1473.
30. **Guvener ZT, Tifrea DF, Harwood CS.** 2006. Two different *Pseudomonas aeruginosa* chemosensory signal transduction complexes localize to cell poles and form and remould in stationary phase. *Mol Microbiol* **61:**106–118.
31. **Borlee BR, Goldman AD, Murakami K, Samudrala R, Wozniak DJ, Parsek MR.** 2010. *Pseudomonas aeruginosa* uses a cyclic-di-GMP-regulated adhesin to reinforce the biofilm extracellular matrix. *Mol Microbiol* **75:**827–842.
32. **Chambers JR, Sauer K.** 2013. Small RNAs and their role in biofilm formation. *Trends Microbiol* **21:**39–49.
33. **Goodman AL, Kulasekara B, Rietsch A, Boyd D, Smith RS, Lory S.** 2004. A signaling network reciprocally regulates genes associated with acute infection and chronic persistence in *Pseudomonas aeruginosa. Dev Cell* **7:**745–754.
34. **Goodman AL, Merighi M, Hyodo M, Ventre I, Filloux A, Lory S.** 2009. Direct interaction between sensor kinase proteins mediates acute

and chronic disease phenotypes in a bacterial pathogen. *Genes Dev* **23:**249–259.

35. **Ventre I, Goodman AL, Vallet-Gely I, Vasseur P, Soscia C, Molin S, Bleves S, Lazdunski A, Lory S, Filloux A.** 2006. Multiple sensors control reciprocal expression of *Pseudomonas aeruginosa* regulatory RNA and virulence genes. *Proc Natl Acad Sci USA* **103:**171–176.
36. **Brencic A, Lory S.** 2009. Determination of the regulon and identification of novel mRNA targets of *Pseudomonas aeruginosa* RsmA. *Mol Microbiol* **72:**612–632.
37. **Brencic A, McFarland KA, McManus HR, Castang S, Mogno I, Dove SL, Lory S.** 2009. The GacS/GacA signal transduction system of *Pseudomonas aeruginosa* acts exclusively through its control over the transcription of the RsmY and RsmZ regulatory small RNAs. *Mol Microbiol* **73:**434–445.
38. **Irie Y, Starkey M, Edwards AN, Wozniak DJ, Romeo T, Parsek MR.** 2010. *Pseudomonas aeruginosa* biofilm matrix polysaccharide Psl is regulated transcriptionally by RpoS and post-transcriptionally by RsmA. *Mol Microbiol* **78:**158–172.
39. **Petrova OE, Sauer K.** 2009. A novel signaling network essential for regulating *Pseudomonas aeruginosa* biofilm development. *PLoS Pathog* **5:**e1000668. doi:10.1371/journal.ppat.1000668.
40. **Petrova OE, Sauer K.** 2010. The novel two-component regulatory system BfiSR regulates biofilm development by controlling the small RNA rsmZ through CafA. *J Bacteriol* **192:**5275–5288.
41. **Petrova OE, Sauer K.** 2011. SagS contributes to the motile-sessile switch and acts in concert with BfiSR to enable *Pseudomonas aeruginosa* biofilm formation. *J Bacteriol* **193:**6614–6628.
42. **Moscoso JA, Mikkelsen H, Heeb S, Williams P, Filloux A.** 2011. The *Pseudomonas aeruginosa* sensor RetS switches type III and type VI secretion via c-di-GMP signalling. *Environ Microbiol* **13:**3128–3138.
43. **Tolker-Nielsen T, Brinch UC, Ragas PC, Andersen JB, Jacobsen CS, Molin S.** 2000. Development and dynamics of *Pseudomonas* sp. biofilms. *J Bacteriol* **182:**6482–6489.
44. **Nilsson M, Chiang WC, Fazli M, Gjermansen M, Givskov M, Tolker-Nielsen T.** 2011. Influence of putative exopolysaccharide genes on *Pseudomonas putida* KT2440 biofilm stability. *Environ Microbiol* **13:**1357–1369.
45. **Matsukawa M, Greenberg EP.** 2004. Putative exopolysaccharide synthesis genes influence *Pseudomonas aeruginosa* biofilm development. *J Bacteriol* **186:**4449–4456.
46. **Wozniak DJ, Wyckoff TJ, Starkey M, Keyser R, Azadi P, O'Toole GA, Parsek MR.** 2003. Alginate is not a significant component of the extracellular polysaccharide matrix of PA14 and PAO1 *Pseudomonas aeruginosa* biofilms. *Proc Natl Acad Sci USA* **100:**7907–7912.
47. **Klausen M, Heydorn A, Ragas P, Lambertsen L, Aaes-Jorgensen A, Molin S, Tolker-Nielsen T.** 2003. Biofilm formation by *Pseudomonas aeruginosa* wild type, flagella and type IV pili mutants. *Mol Microbiol* **48:**1511–1524.
48. **Nielsen AT, Tolker-Nielsen T, Barken KB, Molin S.** 2000. Role of commensal relationships on the spatial structure of a surface-attached microbial consortium. *Environ Microbiol* **2:**59–68.
49. **Wolfaardt GM, Lawrence JR, Robarts RD, Caldwell SJ, Caldwell DE.** 1994. Multicellular organization in a degradative biofilm community. *Appl Environ Microbiol* **60:**434–446.
50. **Klausen M, Aaes-Jorgensen A, Molin S, Tolker-Nielsen T.** 2003. Involvement of bacterial migration in the development of complex multicellular structures in *Pseudomonas aeruginosa* biofilms. *Mol Microbiol* **50:**61–68.
51. **Friedman L, Kolter R.** 2004. Two genetic loci produce distinct carbohydrate-rich structural components of the *Pseudomonas aeruginosa* biofilm matrix. *J Bacteriol* **186:**4457–4465.
52. **Hoiby N.** 1974. *Pseudomonas aeruginosa* infection in cystic fibrosis. Relationship between mucoid strains of *Pseudomonas aeruginosa* and the humoral immune response. *Acta Pathol Microbiol Scand B Microbiol Immunol* **82:**551–558.
53. **O'Toole GA, Kolter R.** 1998. Flagellar and twitching motility are necessary for *Pseudomonas aeruginosa* biofilm development. *Mol Microbiol* **30:**295–304.
54. **Vallet I, Olson JW, Lory S, Lazdunski A, Filloux A.** 2001. The chaperone/usher pathways of *Pseudomonas aeruginosa*: identification of fimbrial gene clusters (cup) and their involvement in biofilm formation. *Proc Natl Acad Sci USA* **98:**6911–6916.
55. **Dueholm MS, Sondergaard MT, Nilsson M, Christiansen G, Stensballe A, Overgaard MT, Givskov M, Tolker-Nielsen T, Otzen DE, Nielsen PH.** 2013. Expression of Fap amyloids in *Pseudomonas aeruginosa, P. fluorescens,* and *P. putida* results in aggregation and increased biofilm formation. *Microbiologyopen* **2:**365–382.
56. **Tielker D, Hacker S, Loris R, Strathmann M, Wingender J, Wilhelm S, Rosenau F, Jaeger KE.** 2005. *Pseudomonas aeruginosa* lectin

LecB is located in the outer membrane and is involved in biofilm formation. *Microbiology* **151:**1313–1323.

57. **Diggle SP, Stacey RE, Dodd C, Camara M, Williams P, Winzer K.** 2006. The galactophilic lectin, LecA, contributes to biofilm development in *Pseudomonas aeruginosa*. *Environ Microbiol* **8:**1095–1104.
58. **Nemoto K, Hirota K, Murakami K, Taniguti K, Murata H, Viducic D, Miyake Y.** 2003. Effect of Varidase (streptodornase) on biofilm formed by *Pseudomonas aeruginosa*. *Chemotherapy* **49:**121–125.
59. **Murakawa T.** 1973. Slime production by *Pseudomonas aeruginosa*. IV. Chemical analysis of two varieties of slime produced by *Pseudomonas aeruginosa*. *Jpn J Microbiol* **17:**513–520.
60. **Murakawa T.** 1973. Slime production by *Pseudomonas aeruginosa*. III. Purification of slime and its physicochemical properties. *Jpn J Microbiol* **17:**273–281.
61. **Costerton JW, Stewart PS, Greenberg EP.** 1999. Bacterial biofilms: a common cause of persistent infections. *Science* **284:**1318–1322.
62. **Ciofu O, Tolker-Nielsen T.** 2010. Antibiotic tolerance and resistance in biofilms, p 215–230. *In* Høiby N, Jensen PØ, Moser C (ed), *Biofilm Infections*. Springer, New York.
63. **Mah TF, Pitts B, Pellock B, Walker GC, Stewart PS, O'Toole GA.** 2003. A genetic basis for *Pseudomonas aeruginosa* biofilm antibiotic resistance. *Nature* **426:**306–310.
64. **Hentzer M, Eberl L, Givskov M.** 2005. Transcriptome analysis of *Pseudomonas aeruginosa* biofilm development: anaerobic respiration and iron limitation. *Biofilms* **2:**37–61.
65. **Chiang WC, Nilsson M, Jensen PO, Hoiby N, Nielsen TE, Givskov M, Tolker-Nielsen T.** 2013. Extracellular DNA shields against aminoglycosides in *Pseudomonas aeruginosa* biofilms. *Antimicrob Agents Chemother* **57:**2352–2361.
66. **Davies DG, Parsek MR, Pearson JP, Iglewski BH, Costerton JW, Greenberg EP.** 1998. The involvement of cell-to-cell signals in the development of a bacterial biofilm. *Science* **280:**295–298.
67. **Hentzer M, Riedel K, Rasmussen TB, Heydorn A, Andersen JB, Parsek MR, Rice SA, Eberl L, Molin S, Hoiby N, Kjelleberg S, Givskov M.** 2002. Inhibition of quorum sensing in *Pseudomonas aeruginosa* biofilm bacteria by a halogenated furanone compound. *Microbiology* **148:**87–102.
68. **Yang L, Barken KB, Skindersoe ME, Christensen AB, Givskov M, Tolker-Nielsen T.** 2007. Effects of iron on DNA release and biofilm development by *Pseudomonas aeruginosa*. *Microbiology* **153:**1318–1328.
69. **Allesen-Holm M, Barken KB, Yang L, Klausen M, Webb JS, Kjelleberg S, Molin S, Givskov M, Tolker-Nielsen T.** 2006. A characterization of DNA release in *Pseudomonas aeruginosa* cultures and biofilms. *Mol Microbiol* **59:**1114–1128.
70. **Diggle SP, Winzer K, Chhabra SR, Worrall KE, Camara M, Williams P.** 2003. The *Pseudomonas aeruginosa* quinolone signal molecule overcomes the cell density-dependency of the quorum sensing hierarchy, regulates rhl-dependent genes at the onset of stationary phase and can be produced in the absence of LasR. *Mol Microbiol* **50:**29–43.
71. **Haussler S, Becker T.** 2008. The pseudomonas quinolone signal (PQS) balances life and death in *Pseudomonas aeruginosa* populations. *PLoS Pathog* **4:**e1000166. doi:10.1371/journal.ppat.1000166.
72. **Davey ME, Caiazza NC, O'Toole GA.** 2003. Rhamnolipid surfactant production affects biofilm architecture in *Pseudomonas aeruginosa* PAO1. *J Bacteriol* **185:**1027–1036.
73. **Boles BR, Thoendel M, Singh PK.** 2005. Rhamnolipids mediate detachment of *Pseudomonas aeruginosa* from biofilms. *Mol Microbiol* **57:**1210–1223.
74. **Pamp SJ, Tolker-Nielsen T.** 2007. Multiple roles of biosurfactants in structural biofilm development by *Pseudomonas aeruginosa*. *J Bacteriol* **189:**2531–2539.
75. **Winzer K, Falconer C, Garber NC, Diggle SP, Camara M, Williams P.** 2000. The *Pseudomonas aeruginosa* lectins PA-IL and PA-IIL are controlled by quorum sensing and by RpoS. *J Bacteriol* **182:**6401–6411.
76. **Banin E, Vasil ML, Greenberg EP.** 2005. Iron and *Pseudomonas aeruginosa* biofilm formation. *Proc Natl Acad Sci USA* **102:**11076–11081.
77. **Heydorn A, Ersboll B, Kato J, Hentzer M, Parsek MR, Tolker-Nielsen T, Givskov M, Molin S.** 2002. Statistical analysis of *Pseudomonas aeruginosa* biofilm development: impact of mutations in genes involved in twitching motility, cell-to-cell signaling, and stationary-phase sigma factor expression. *Appl Environ Microbiol* **68:**2008–2017.
78. **Inhülsen S, Aguilar C, Schmid N, Suppiger A, Riedel K, Eberl L.** 2012. Identification of functions linking quorum sensing with biofilm formation in *Burkholderia cenocepacia* H111. *Microbiologyopen* **1:**225–242.
79. **Schmid N, Pessi G, Deng Y, Aguilar C, Carlier AL, Grunau A, Omasits U, Zhang**

LH, Ahrens CH, Eberl L. 2012. The AHL- and BDSF-dependent quorum sensing systems control specific and overlapping sets of genes in *Burkholderia cenocepacia* H111. *PLoS One* **7:** e49966. doi:10.1371/journal.pone.0049966.

80. **McCarthy Y, Yang L, Twomey KB, Sass A, Tolker-Nielsen T, Mahenthiralingam E, Dow JM, Ryan RP.** 2010. A sensor kinase recognizing the cell-cell signal BDSF (cis-2-dodecenoic acid) regulates virulence in *Burkholderia cenocepacia*. *Mol Microbiol* **77:**1220–1236.
81. **O'Toole G, Kaplan HB, Kolter R.** 2000. Biofilm formation as microbial development. *Annu Rev Microbiol* **54:**49–79.
82. **Costerton JW, Lewandowski Z, Caldwell DE, Korber DR, Lappin-Scott HM.** 1995. Microbial biofilms. *Annu Rev Microbiol* **49:**711–745.
83. **Pamp SJ, Gjermansen M, Johansen HK, Tolker-Nielsen T.** 2008. Tolerance to the antimicrobial peptide colistin in *Pseudomonas aeruginosa* biofilms is linked to metabolically active cells, and depends on the *pmr* and *mexAB-oprM* genes. *Mol Microbiol* **68:**223–240.
84. **Barken KB, Pamp SJ, Yang L, Gjermansen M, Bertrand JJ, Klausen M, Givskov M, Whitchurch CB, Engel JN, Tolker-Nielsen T.** 2008. Roles of type IV pili, flagellum-mediated motility and extracellular DNA in the formation of mature multicellular structures in *Pseudomonas aeruginosa* biofilms. *Environ Microbiol* **10:**2331–2343.
85. **Harmsen M, Yang L, Pamp SJ, Tolker-Nielsen T.** 2010. An update on *Pseudomonas aeruginosa* biofilm formation, tolerance, and dispersal. *FEMS Immunol Med Microbiol* **59:**253–268.
86. **Whiteley M, Bangera MG, Bumgarner RE, Parsek MR, Teitzel GM, Lory S, Greenberg EP.** 2001. Gene expression in *Pseudomonas aeruginosa* biofilms. *Nature* **413:**860–864.
87. **Waite RD, Papakonstantinopoulou A, Littler E, Curtis MA.** 2005. Transcriptome analysis of *Pseudomonas aeruginosa* growth: comparison of gene expression in planktonic cultures and developing and mature biofilms. *J Bacteriol* **187:**6571–6576.
88. **Folsom JP, Richards L, Pitts B, Roe F, Ehrlich GD, Parker A, Mazurie A, Stewart PS.** 2010. Physiology of *Pseudomonas aeruginosa* in biofilms as revealed by transcriptome analysis. *BMC Microbiol* **10:**294.
89. **Redfield RJ.** 2002. Is quorum sensing a side effect of diffusion sensing? *Trends Microbiol* **10:**365–370.
90. **Lopez-Sanchez A, Jimenez-Fernandez A, Calero P, Gallego LD, Govantes F.** 2013. New methods for the isolation and characterization of biofilm-persistent mutants in *Pseudomonas putida*. *Environ Microbiol Rep* **5:**679–685.
91. **Navarro MV, Newell PD, Krasteva PV, Chatterjee D, Madden DR, O'Toole GA, Sondermann H.** 2011. Structural basis for c-di-GMP-mediated inside-out signaling controlling periplasmic proteolysis. *PLoS Biol* **9:**e1000588. doi:10.1371/journal.pbio.1000588.
92. **Newell PD, Boyd CD, Sondermann H, O'Toole GA.** 2011. A c-di-GMP effector system controls cell adhesion by inside-out signaling and surface protein cleavage. *PLoS Biol* **9:** e1000587. doi:10.1371/journal.pbio.1000587.

Division of Labor in Biofilms: the Ecology of Cell Differentiation

4

JORDI VAN GESTEL,[1] HERA VLAMAKIS,[1] and ROBERTO KOLTER[1]

INTRODUCTION

One of the most remarkable features of the evolutionary process is its capacity to construct. In billions of years a primordial soup of organic compounds evolved to the theater of life extant today. This ability to construct is best illustrated by a number of transitions that have occurred during the natural history of our planet, such as the evolution of the first prebiotic cells, eukaryotes, multicellularity, and eusociality (1, 2). These transitions all bear a number of striking similarities (2). First, construction evolves through cooperation (3–6). That is, new organizational layers come about through the cooperative interaction of biological units that previously functioned independently. For example, organelles evolved from microbes that engaged in mutualistic interactions through endosymbiosis, and multicellularity evolved from cells that cooperate by sticking together, either via incomplete cell division or through aggregation (7). In addition to cooperation, a second aspect characterizes major evolutionary transitions: the division of labor (8, 9). A precise definition of the division of labor will be given below, but one can loosely speak of division of labor when individuals—during their cooperative interactions—specialize in

[1]Department of Microbiology and Immunobiology, Harvard Medical School, Boston, MA 02115.

Microbial Biofilms 2nd Edition
Edited by Mahmoud Ghannoum, Matthew Parsek, Marvin Whiteley, and Pranab K. Mukherjee

doi:10.1128/microbiolspec.MB-0002-2014

performing different "tasks." Perhaps the most striking example comes from multicellular development. Multicellular organisms consist of many specialized cell types (e.g., muscle cells, neurons, epithelia, etc.). Despite being genetically identical, these cells have differentiated and thereby organized themselves in different physiological and morphological structures (e.g., organs) that together make up the individual.

In this article we focus on the division of labor within bacterial biofilms. In contrast to multicellular organisms, the division of labor among bacterial cells in biofilms is less self-evident and thus the subject of some debate. This is partly because many classical evolutionary concepts, such as individuality, are mainly inspired by metazoan life and are therefore not readily applicable to microorganisms (1, 10, 11). For example, views differ as to whether bacterial biofilms are a primordial form of multicellular development or, rather, an aggregate of individuals (12–14).

To get a common understanding of the concepts that we use throughout this article, we first discuss the theoretical and conceptual basis of cooperation, phenotypic heterogeneity, and the division of labor. This is particularly important because the conceptual grounds of the division of labor are strongly embedded in both evolutionary and ecological theory. This first part ends with a few well-examined case studies of multicellularity and the division of labor in microbes. In the second part of the article, we discuss the division of labor in biofilms with a particular emphasis on *Myxococcus xanthus*, *Bacillus subtilis*, and *Pseudomonas aeruginosa*.

COOPERATION, SPECIALIZATION, AND THE DIVISION OF LABOR

Cooperation: an Alignment of Fitness Interests

The division of labor requires a cooperative interaction between specialized individuals. Cooperation is defined as a phenotypic behavior that is costly to perform for an individual but benefits its interaction partner. In a well-mixed population that consists of cooperative and noncooperative individuals, one expects that the latter have a selective benefit because they receive the benefits of cooperation without paying the costs (3). For example, imagine that there is a population of bacterial cells that produce a siderophore to scavenge iron (15, 16). If it is costly to produce the siderophore, a mutant that stops producing it and still benefits from that produced by others is expected to have a selective advantage (Fig. 1A). It is therefore challenging to explain why cooperation is not exploited. Since the division of labor cannot evolve or be maintained in the presence of such exploitation, we first have to explain the premise of cooperation. There are multiple mechanisms that can explain the evolution of cooperation, which all result in the emergence of assortative interactions: cooperators are more likely to interact with each other than defectors are to interact with cooperators (3, 17).

Perhaps the simplest mechanism to explain the evolution of cooperation is spatial segregation (18–21). When cooperative genotypes grow separately from noncooperative genotypes, exploitation is impossible. Under those conditions cooperation will evolve because groups of cooperative individuals perform better than groups of noncooperative individuals. In multicellular organisms that go through a single-cell bottleneck (e.g., metazoans), cells are genetically identical and cooperation can easily evolve because there is no risk of exploitation (5). By the same token, we expect that cooperation more readily evolves in monoclonal biofilms (or clonal pockets inside nonclonal biofilms), because there are fewer genetic variants that could exploit cooperation. In this article, we limit our discussion to monoclonal biofilms. This does not imply that cooperation and, hence, the division of labor cannot occur between genetically distinct individuals or

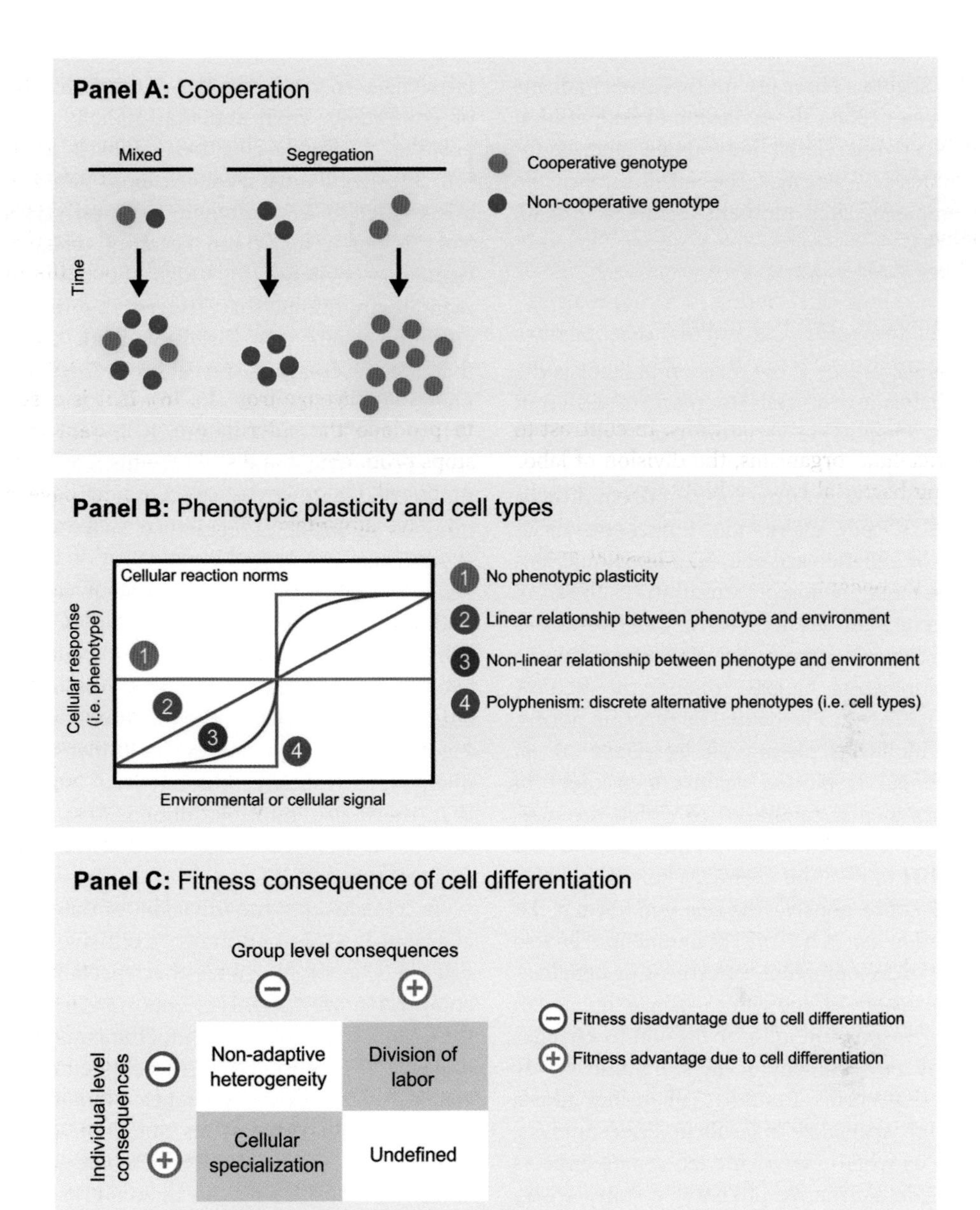

FIGURE 1 Conceptual and theoretical basis for the division of labor. (A) Growth of cooperative and noncooperative cells when mixed (left) or segregated (right). When mixed, the noncooperative genotype performs better than the cooperative phenotype; it benefits from cooperation without paying the costs. When segregated, the cooperative genotype performs better. (B) Reaction norms. Different colored lines and associated numbers show different types of reaction norms as indicated on the right. (C) Fitness consequences of cell differentiation at the individual level (i.e., cell) and group level (i.e., colony or part of the colony). When cell differentiation is not beneficial at either level, phenotypic heterogeneity is nonadaptive. When it is only beneficial at the cell level, there is cellular specialization. When it is only beneficial at the colony level, there is division of labor. When it is beneficial at both levels, one cannot directly determine the function of cell differentiation. doi:10.1128/microbiolspec.MB-0002-2014.f1

even species. There are multiple mechanisms that can explain the evolution of cooperation between nonrelated individuals, and endosymbiosis is perhaps the most remarkable example of such mutualistic interaction (3, 22–24).

Phenotypic Heterogeneity

A crucial aspect of the division of labor is the specialization of cells to perform different tasks. Phenotypic variation can result from various proximate mechanisms: plastic responses to local environmental conditions, noise in gene expression, epigenetic variation, or genetic variation. In monoclonal biofilms, most phenotypic variation results from nongenetic differences. Here, we refer to this variation as phenotypic heterogeneity. In developmental biology, phenotypic heterogeneity is typically studied by reaction norms, in which the phenotypic response of an individual is plotted against a gradient of environmental conditions to which the individual is exposed (25, 26). Figure 1B shows a number of reaction norms. When an individual is nonresponsive, the reaction norm is flat (reaction norm 1); this phenomenon is also called developmental robustness or canalization. When an individual is plastic, it can either respond in a linear fashion to changes in the environment or in a nonlinear way (reaction norms 2 and 3). When individuals strongly specialize to perform different tasks, such as when they divide labor, one expects multiple alternative phenotypic states (reaction norm 4). The presence of discrete phenotypic states is also known as polyphenism (as opposed to polymorphism, in which alternative phenotypic states result from genetic variation).

A number of features characterize reaction norms underlying the division of labor. First, as mentioned above, one expects that cells express a small number of relatively discrete phenotypic states that represent the alternative cell types (27). Cell types can develop in response to discrete environmental signals or via a regulatory amplification of continuous environmental signals using positive or double-negative feedback loops (28, 29). Regulatory feedback loops can also affect a cell's commitment to a cell type. That is, such loops can result in bistable regulatory switches in which the environmental conditions that trigger the differentiation event are different from those that are necessary to revert to the original phenotypic state (e.g., hysteresis [30, 31]). When a cell's commitment is irreversible, we speak of terminal differentiation. In the second half of the article we discuss a number of bistable regulatory switches in *B. subtilis*.

A second property of reaction norms underlying the division of labor is that different phenotypic states are mutually exclusive: a cell expresses either one of the alternative cell types. In the reaction norm of Figure 1B we only show a one-dimensional phenotypic response, but one can imagine that there are multiple dimensions; each dimension would correspond to an alternative cell type.

In conclusion, the division of labor is characterized by mutually exclusive and discrete phenotypic states that specialize in complementary tasks. It is important to note that various regulatory mechanisms can underlie the presence of these phenotypic states. Moreover, the presence of mutually exclusive phenotypic states does not necessarily imply that cells divide labor. There are many lifestyle switches that can result in the same reaction norms. For example, for certain bacteria the onset of biofilm formation is defined when some cells switch from a motile planktonic lifestyle to a surface-attached aggregative lifestyle. This switch is accompanied by two mutually exclusive and discrete phenotypic states: motile cells and matrix-producing cells (32). Thus, like cooperation, phenotypic specialization is a requirement for the division of labor but by itself is not sufficient to prove that there is division of labor.

Division of Labor

It is important to make a distinction between phenotypic specialization on the one hand and the division of labor on the other hand. Phenotypic specialization is a cell-level property, and the division of labor is a colony-level property. Even though the division of labor by definition requires the presence of different cell types, the presence of different cell types does not imply that cells divide labor. In a biofilm, cell specialization might simply be an adaptive response to the local environmental conditions to which cells are exposed and thereby does not involve cooperative interactions between different cell types. The only way to disentangle phenotypic specialization and the division of labor is by examining the fitness consequences of differentiation at both the cellular and colony levels (Fig. 1C) (by colony level we do not strictly mean the whole colony, but rather the level at which cells interact as a group). When cells divide labor, a colony that consists of multiple cell types performs better than a colony that consists solely of any one of them. It is important to note that this advantage is not a mere consequence of cells adapting to local environmental conditions, but rather an emergent property from the interaction between cells. For example, the division of labor allows cells to carry out specific tasks, thereby avoiding the regulatory or metabolic burden of switching between different tasks or expressing them simultaneously. This is, of course, only possible when cells specialize in complementary tasks and share the associated benefits. Thus, one can recognize the division of labor from the emergent fitness benefits that occur at the colony level due to the cooperative interaction of specialized cell types, in which the various cell types are somehow interdependent (33).

An alternative way, albeit indirect, to recognize the division of labor is to examine the cell-level consequences of cell differentiation. When cell differentiation reduces the fitness of a cell, its evolutionary origin can only be explained by colony-level fitness benefits that result from the division of labor (Fig. 1C) (when excluding alternatives like nonadaptive phenotypic heterogeneity or bet-hedging, see discussion below). Perhaps the most compelling example of this can be found in metazoans. Some cells in metazoans are destined to become gametes, whereas others terminally differentiate into somatic cells. Terminal differentiation can never be viewed as a cell-level adaptation because cells that become somatic do not contribute to reproduction and therefore have a relative fitness of zero. This is particularly apparent for cells that undergo so-called programmed cell death (34–37). The type of division of labor in which only a fraction of cells contribute to reproduction is called reproductive division of labor. Terminal differentiation is also present in some microbes (38). We discuss several examples of this below.

Although terminal differentiation excludes alternative hypotheses that can explain cell differentiation, it does not explain the benefits that are associated with the division of labor. These benefits are nevertheless responsible for the remarkable evolutionary success of the division of labor. For many biological systems a detailed understanding of the emergent properties that result in the interaction between cell types is lacking. In the second part of the article we argue that such understanding can often be acquired through ecology. In the following paragraphs we discuss some case studies of the division of labor in microbes.

BACTERIAL MULTICELLULARITY AND THE DIVISION OF LABOR

Although multicellular eukaryotes are well known for their remarkable organismic adaptations, multicellularity evolved about two billion years earlier in bacteria (39, 40). Here we focus on two instances of bacterial multicellularity: filamentous multicellularity in cyanobacteria and aerial hyphae in actinobacteria (5). These clear examples of division of labor are used as a stepping stone toward discussing different cell fates within biofilms.

Cyanobacteria present a beautiful example of bacterial differentiation. Some cyanobacteria form filaments that can express up to four different cell types (41–43): photosynthetic cells, heterocysts, akinetes, and hormogonia. The first two cell types are known to divide labor. Heterocysts fix nitrogen using the enzyme nitrogenase. Since nitrogenase is sensitive to oxygen, nitrogen fixation is incompatible with photosynthesis. Consequently, cells cannot fix nitrogen and carbon at the same time. Because cells need both carbon and nitrogen, the strong phenotypic trade-off resulted in the evolution of two specialized cell types—heterocysts and photosynthetic cells—that share their resources, as opposed to a generalist that inefficiently fixes both nitrogen and carbon (see Fig. 2 for the role of phenotypic trade-offs on the division of labor). In cyanobacteria that divide labor, such as *Anabaena*, large filaments of photosynthetic cells are typically interspersed by a smaller number of heterocysts (44, 45). Heterocyst development is triggered by nitrogen deprivation (46); a positive regulatory feedback loop subsequently ensures a cell's developmental

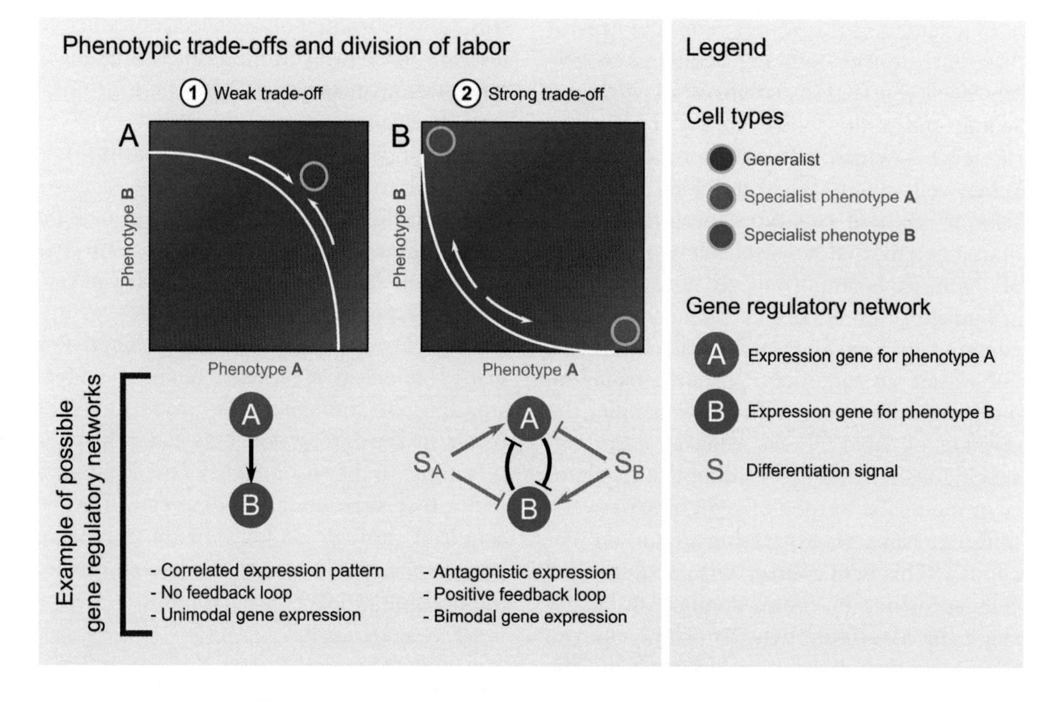

FIGURE 2 Phenotypic trade-offs and the division of labor. Trade-off between two tasks: phenotype A and B. The trade-off constrains a cell such that expressing phenotype A (e.g., photosynthesis) is at the expense of phenotype B (e.g., nitrogen fixation). The trade-off can be weak (concave shape) or strong (convex shape). (A) Expected evolutionary outcome when the trade-off between phenotypes A and B is weak: phenotypic generalist. The regulatory network that controls the expression of phenotypes A and B should result in coexpression. (B) Expected evolutionary outcome when the trade-off between phenotypes A and B is strong: cell specialization and the division of labor. In this case, the regulatory network that controls the expression of phenotypes A and B should result in antagonistic expression and commit cells to a given cell type (positive feedback loops). Consequently, each cell expresses only phenotype A or B. doi:10.1128/microbiolspec.MB-0002-2014.f2

commitment (47). At the same time, lateral inhibition—via a signaling peptide (PatS)—prevents neighboring cells from differentiating into heterocysts (Fig. 3A), resulting in a semi-regular spacing of heterocysts along the filament (48–51). Heterocysts are terminally differentiated and cannot divide, so in addition to metabolic cooperation there is also reproductive division of labor (52, 53). In contrast to *Anabaena* species, the filamentous cyanobacteria *Plectonema boryanum* has evolved an alternative strategy: rather than separating nitrogen and carbon fixation in space, it separates the processes in time by switching back and forth between nitrogen and carbon fixation (54). The advantage of temporal, instead of spatial, differentiation is that it does not require cooperation between cells. It is therefore plausible that the regulatory mechanisms for temporal cell differentiation evolved first and were later coopted in some species during the evolution of spatial division of labor, which presumably is more efficient (55). Phenotypic trade-offs, like the one described here, also underlie the division of labor in other species (56, 57).

The developmental pattern of many actinobacteria is another example of bacterial multicellularity where division of labor is clear. Most cells are part of a vegetative mycelium, consisting of branching hyphae, and others form aerial hyphae that produce spores. Although many details of the regulation of cell differentiation in actinomycetes are yet to be discovered, there are some good indications for the division of labor, especially, during aerial hyphae formation in *Streptomyces coelicolor* (Fig. 3B). Upon starvation, aerial hyphae develop from the vegetative mycelium and grow into the air by locally breaking the water tension (58, 59). The aerial hyphae go through a tightly regulated developmental cascade that results in an apical sporogenic cell consisting of prespore

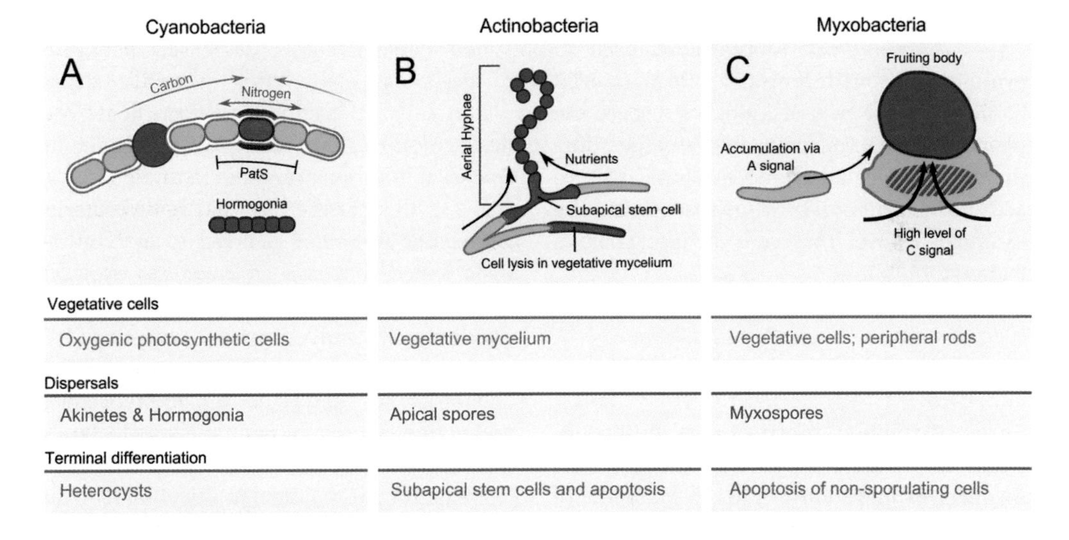

FIGURE 3 Bacterial multicellularity. For each form of multicellularity we show a number of different cell types: green is vegetative cells, blue is spores, and red is terminally differentiated cells, including cells that undergo lysis. (A) Filamentous multicellularity and cell differentiation in cyanobacteria. PatS is a signaling peptide that blocks heterocyst formation in the neighboring cells in the cyanobacteria filaments. (B) Filamentous multicellularity and aerial hyphae formation in actinobacteria. (C) Colonial multicellularity and fruiting body formation in myxobacteria. A-signal-dependent aggregation is illustrated by an arrow, and the level of C-signal is highest in the base and center of a fruiting body. doi:10.1128/microbiolspec.MB-0002-2014.f3

compartments and a subapical stem cell. In each of the prespore compartments a spore matures (59). In the case of aerial hyphae formation, there are no apparent phenotypic trade-offs that could explain the interaction between specialist phenotypes as described above for cyanobacteria. However, the different cell types do cooperate. For example, the vegetative mycelium secretes proteases that presumably help break down the substrate mycelium, thereby providing nutrients for sporulation (60, 61). In addition, a significant fraction of the vegetative mycelium undergoes cell lysis (61–64). The nutrients that are liberated from these dead cells are thought to benefit aerial hyphae formation (61, 64). Since cell lysis cannot be a local adaptation, this phenomenon can only be explained by division of labor that results in a colony-level advantage (assuming that cell lysis is an evolutionarily selected trait). In conclusion, these examples of bacterial multicellularity exhibit cell differentiation and cooperative interactions between cell types, indicating that division of labor does exist. These examples can therefore be used to evaluate cell differentiation in bacterial biofilms. In the next section, we discuss the phenotypic heterogeneity that emerges during biofilm formation and evaluate if there are indications, like those seen in the examples above, that cells divide labor in bacterial biofilms.

DIVISION OF LABOR IN BIOFILMS

Surface-attached biofilms are heterogeneous by nature. The accumulation of cells on surfaces inevitably results in gradients of nutrient sources, electron acceptors, waste products or any other products that are generated by cells (65–67). Not surprisingly, cells respond to these gradients by physiological adaptation (67). One would therefore expect that biofilm formation invariably results in the phenotypic heterogeneity of its constituent cells. At the same time, the environmental gradients also afford an organizing potential (68). The chemical gradients—akin to morphogen gradients in eukaryotic development—confer positional information to cells, which allows them to spatially organize themselves via cell differentiation (69, 70). This could subsequently facilitate cooperative interactions between cell types and, hence, the division of labor. Here, we discuss whether cell differentiation processes in biofilms might indeed be explained by division of labor, as opposed to being simply the result of local physiological adaptation. In particular, we focus on biofilm formation in *M. xanthus*, *B. subtilis*, and *P. aeruginosa*.

M. xanthus Multicellularity

Myxobacteria are social bacteria that survive in groups during both nutrient-rich growth and starvation. These bacteria are predatory and secrete antibacterial compounds. Cells then grow on nutrients obtained as they break down macromolecules that are released after prey bacteria are killed (71). When nutrient levels decrease, thousands of cells aggregate into mounds within which cells differentiate to form heat- and desiccation-resistant spores, often producing remarkable aerial structures (fruiting bodies) (72, 73). Despite the fact that myxobacterial aggregates were not referred to as biofilms historically, they are an excellent example of robust bacterial biofilms during both nutrient-rich and -replete conditions (14, 74–76). In contrast to the filamentous multicellularity we described for cyanobacteria and actinobacteria, biofilm and fruiting body formation in myxobacteria result from cell aggregation (i.e., colonial multicellularity; see Fig. 3C). As a consequence, there is no unicellular bottleneck, and multiple genotypes could partake in its development. As we describe below, this has allowed for extensive studies analyzing fruiting body development in "chimeric" populations where strains harboring different mutations are mixed.

When starved, *M. xanthus* cells undergo an elaborate developmental program that culminates in at least three different cell fates: spores, peripheral rods, and cells that will lyse (72, 73). Sporulation only occurs within fruiting bodies, and cells that will sporulate differentially express certain genes required for sporulation (77) and show different protein profiles than nonsporulating cells, termed "peripheral rods" (73, 78). Peripheral rods are a discrete subpopulation of cells that remain outside of the fruiting bodies. This subpopulation is proposed to function as persister cells which do not undergo cell division but are likely ready to respond to any sudden increase in nutrients (73, 79). In addition to exhibiting a different protein profile than spores and presporulating cells, peripheral rods also lack the presence of lipid bodies that are found in cells that are destined to become spores (73, 80). As cells undergo the morphological changes required to become spores, the lipid bodies are consumed. Therefore, it has been proposed that the lipid bodies provide the energy required for cells to sporulate (80). There is also a significant portion of the cells that lyse during fruiting body formation. However, the overall number of lysed cells, the timing of lysis, and the proposed mechanism of lysis varies depending on the strain studied and conditions that are used (73). Historically, cell lysis was proposed to provide nutrients that allow spore differentiation. In addition, more recent results suggest that cell lysis may also play a role in the aggregation of cells (73, 78).

Since *M. xanthus* cells must coordinate their behavior during vegetative motility, predation, and fruiting body formation, cells must be able to communicate with each other. Indeed, an *M. xanthus* cell interacts with its neighbors using both contact-dependent mechanisms and secreted signals (76). Amazingly, *M. xanthus* cells that come in contact with each other exchange outer membrane proteins and lipids in a regulated manner that results in phenotypic changes in cells (76). When nutrients are sparse, a mixture of extracellular amino acids and peptides (termed the A-signal) is important for the onset of fruiting body formation by ensuring that cellular aggregation only starts when there is a critical mass of starving cells (81, 82). Subsequent to A-signaling, a second contact-dependent signal becomes important. The so-called C-signal is a processed form of the CsgA protein (p17), and C-signaling functions in a threshold-dependent manner to regulate different behaviors (72, 83). C-signaling is required for aggregation, and aggregation stimulates C-signaling. Therefore, there is a positive feedback loop that ensures the continuation of fruiting body formation once it has started (84). The level of C-signal that a cell senses may be an important determinant of its fate as a spore, peripheral rod, or lysed cell (72, 73, 76). In other myxobacterial species, such as *Chondromyces apiculatus* and *Stigmatella aurantiaca*, nonsporulating cells in the fruiting body can function as stalks that presumably aid in the process of spore dispersal (85). For convenience, below we refer to the nonsporulating cells inside an *M. xanthus* fruiting body as "stalk" cells as well.

Using Chimeric Fruiting Bodies to Determine the Role of Different *M. xanthus* Cell Types

Since cell lysis in stalk cells is under regulatory control (78, 86), the behavior of stalk cells cannot be explained by local adaptation. Instead, cell lysis is expected to yield benefits at the colony level, much like the lysis of vegetative mycelium in *Streptomyces* discussed above. However, in contrast to *Streptomyces*, the fitness consequences and interaction of different cell types can be examined directly by studying chimeras: fruiting bodies formed by multiple genotypes (87). Numerous studies of *M. xanthus* have examined chimeras consisting of genetically engineered mutants, wild isolates, and laboratory-evolved genotypes (87–94). This plethora of experiments resulted in some unique insights. For example, fruiting body formation depends on cooperation. When A-signal or C-signal mutants were mixed with a wild-type strain, the total spore production of a fruiting body dropped and the mutants produced a disproportionately high share of spores in comparison to the wild type (90). In other words, the mutants behave like developmental cheaters: by avoiding the costs of signal production they were able to increase the production of spores. Similar results were obtained when mixing a wild-type strain with its evolved descendants. Velicer and colleagues (95) evolved *M. xanthus* for approximately 1,000 generations under "asocial" conditions. The evolved genotypes were highly aberrant in social behaviors such as fruiting body formation. However, when mixed with their "social" ancestor, they could partake in fruiting body formation and had a competitive advantage by producing a disproportionately high fraction of spores at the expense of the overall spore production (90). These studies show that stalk cells express cooperative traits that contribute to the total spore production of a fruiting body and can be exploited by other genotypes. Consequently, fruiting body formation can be seen as a developmental process in which the division of labor between spores and stalk cells results in an effective dispersal organ.

B. subtilis Differentiation

As explained above, when cells divide labor, phenotypic differentiation is characterized by discrete phenotypic states (i.e., cell types) that are mutually exclusive. These phenotypic states can be recognized from the multimodal distribution of gene expression, concerning the genes that encode for the respective phenotypes. For *B. subtilis*, bimodal distributions in gene expression have been associated with a number of (not necessarily mutually exclusive) phenotypes that appear during biofilm formation (96–98): motility, surfactin production, matrix production, protease production, and sporulation. These phenotypes are typically referred to as cell types. While matrix-producing cells are the only cells that are essential for the formation of biofilms, all of these cell types can be found within a biofilm. Motile cells have an upregulated expression of the *fla/che* operon, which is required for the biosynthesis of flagella (99). The expression of this operon does not necessarily mean that cells within the biofilm are actually motile, because once cells begin to express the *epsA-O* operon (which encodes enzymes that produce the exopolysaccharide component of matrix), there is a feedback where the EpsE glycosyltransferase protein physically binds to and inhibits FliG, a component of the flagellar motor. This interaction inhibits flagella rotation, thereby inhibiting motility in cells that have begun to produce extracellular matrix (100).

Surfactin-producing cells produce the surfactant surfactin (101–103), which also functions as a communicative signal that triggers matrix production (104, 105) and an antimicrobial (106, 107). Matrix-producing cells express the *epsA-O* and *tapA-sipW-tasA* operons, which results in the production of, respectively, extracellular polysaccharides (EPS) and the structural protein TasA (108–110). TasA assembles into amyloid-like fibers that attach to cell walls via an accessory protein, TapA (111, 112). Another protein that contributes to the extracellular matrix is BslA, which is important for surface hydrophobicity (113–116). However, unlike the *epsA-O* and *tapA* operons, *bslA* expression occurs in all of the cells within the population (116). In addition, matrix-producing cells also produce antimicrobial toxins (Skf and Sdp) that kill other species or those sibling cells which do not express the immunity genes (i.e., cannibalism) (117–120). Protease-producing cells secrete bacillopeptidase and subtilisin, two proteolytic enzymes, which are encoded by the *bpr* and *aprE* genes (98, 121, 122). Finally, spores are stress-resistant dormant cells that are formed during the developmental process of endosporulation (84, 123–125).

Regulation of Differentiation in *B. subtilis*

In general, cell differentiation in *B. subtilis* is triggered in response to some environmental signals that—via sensor kinases—initiate a phosphorylation cascade. The phosphorylation cascade integrates the environmental information by funneling the regulatory input toward the phosphorylation of one (or a few) downstream regulatory protein(s). This key regulatory protein, which is typically subject to a positive or double-negative regulatory feedback loop, converts the continuous environmental information into a discrete Boolean switch (i.e., bistable or binary switch) that controls if a phenotype is either expressed or not. In the case of *B. subtilis,* a few cell types have been shown to be controlled by bistable regulatory switches (30, 126). Figure 4A shows a simplified scheme of a part of the regulatory pathways that underlie these bistable switches (for details see references 127, 128, and references therein).

One of the key regulatory proteins controlling biofilm formation is Spo0A, which is a transcriptional regulator involved in the regulation of motility, matrix production, protease production, and sporulation (97, 128, 129). There is a graded response to the level of phosphorylated Spo0A~P (130, 131). When levels of Spo0A~P are low, cells are motile (132). In response to intermediate levels of Spo0A~P, cells produce matrix (i.e., EPS and TasA) and—dependent on the phosphorylation of another regulatory gene (DegU)—secrete proteases (96, 122, 130). At high levels of Spo0A~P, cells initiate sporulation.

The phosphorylation state of Spo0A can be modulated by five histidine kinases, four of which appear to be important for biofilm formation (133, 134). These kinases sense a variety of environmental signals, including self-generated products like surfactin and matrix (105, 135–138). Once phosphorylated, Spo0A~P indirectly represses the expression of *sinR* (139–141). SinR represses the *epsA-O* and *tapA-sipW-tasA* operon by competing for the binding sites of an activating protein, RemA (142). Spo0A~P also represses AbrB, which, like SinR, is a repressor of the *epsA-O* and *tapA-sipW-tasA* operon as well as of *bslA* (113, 143, 144). Both SinR and AbrB are part of a double-negative feedback loop. The repression of SinR derepresses the expression of *slrR* (145–147). SlrR subsequently sequesters SinR by forming a SinR-SlrR complex, which further relieves SinR-mediated repression of the matrix genes, and *slrR* and also represses motility genes. The sequestering of SinR by SlrR therefore results in a bistable switch. Consequently, matrix production and motility become two mutually exclusive cell types (128, 148). At high levels of Spo0A~P the repression of SinR weakens (not shown in Fig. 4A), which downregulates matrix production and simultaneously triggers the sporulation process (126, 149).

Besides translating environmental signals to binary phenotypic responses (29), regulatory feedback loops can also amplify stochastic fluctuations (i.e., noise) in the levels or activities of regulatory components (28, 96, 150, 151). As a consequence, cells can differentiate into different cell types, despite being exposed to the same environmental conditions (152). Matrix production, protease production, and sporulation are subject to probabilistic cell differentiation (122, 148, 153), which is generally viewed as a product of evolutionary adaptation (28, 151). The probability of differentiation can be manipulated by changing the level of noise or the regulatory circuit that underlies differentiation (154–157). The aforementioned switch toward matrix production, via the double-negative feedback loop between SinR and SlrR, is a well-studied example of probabilistic cell differentiation. Norman and colleagues (148) showed—using a highly controlled microfluidics device—that the differentiation toward matrix production is,

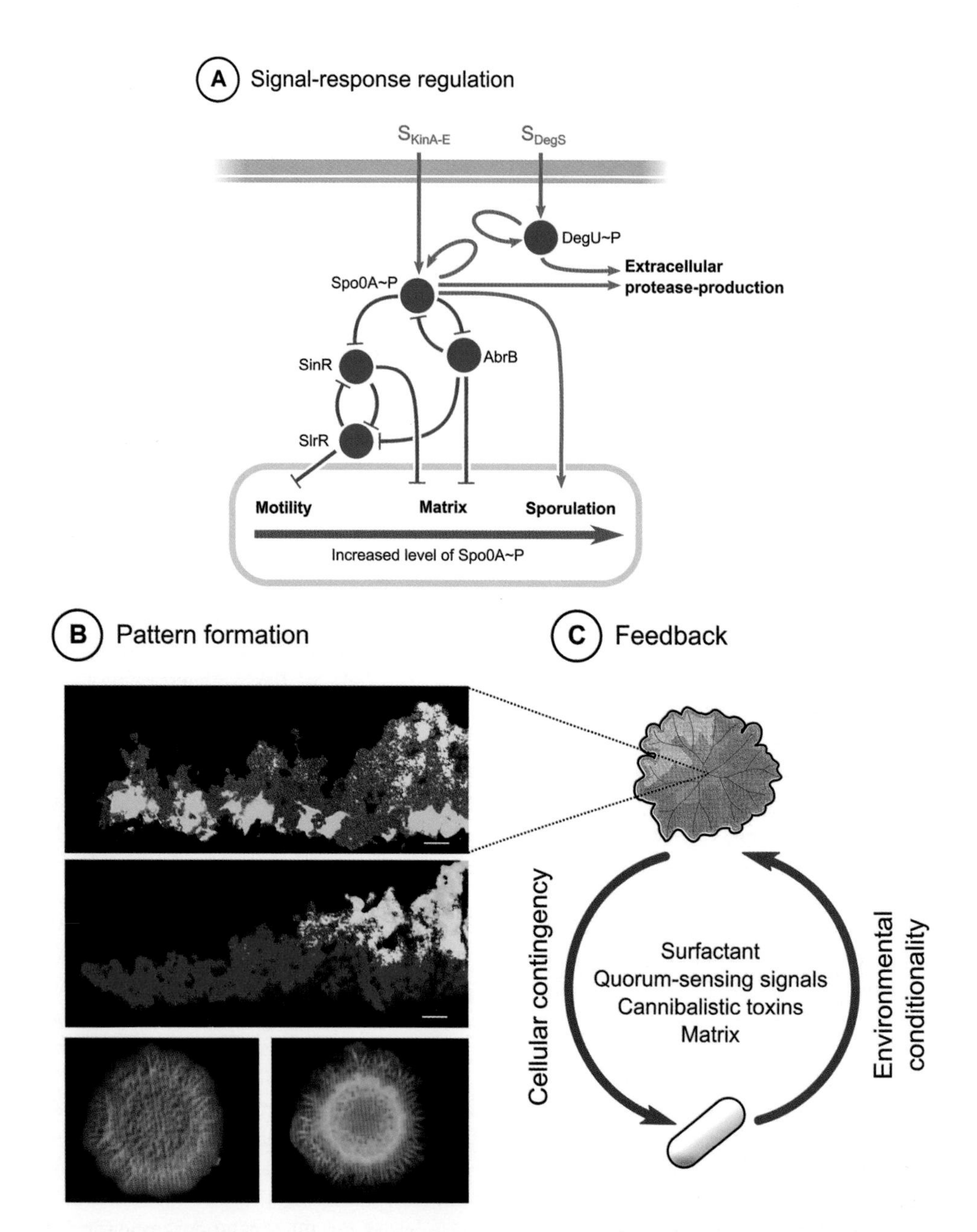

FIGURE 4 **Cell differentiation and pattern formation in *B. subtilis* biofilms. (A) Simplified scheme of the regulatory circuit that controls cell differentiation. Regulatory repression (red T-bars) or stimulation (green arrows) can involve both transcriptional regulation and (de)phosphorylation. The gray box shows the expected developmental transition in time throughout biofilm formation: motile cells differentiate to matrix-producing cells, which later sporulate. S_{KinA-E} and S_{DegS} are environmental signals that affect the sensory kinases KinA-E and DegS. (B) Pattern formation in cross-sections and top view of *B. subtilis* colony biofilms. Cell types shown in cross-sections are sporulating cells (artificially colored yellow or green), motile cells (blue), and matrix-producing cells (red). In the top view, sporulating cells are shown in green and colocalize with the biofilm wrinkles. (C) Feedback between cellular contingency and environmental conditionality. Images are adapted from references 144 and 169. doi:10.1128/microbiolspec.MB-0002-2014.f4**

at least in part, stochastic. However, once differentiated, the time spent as a matrix producer is tightly controlled, such that cells are committed to matrix production for a number of generations (148, 158, 159). As suggested by Norman and colleagues, this commitment may allow for the cooperation between the progeny of a differentiated cell (148).

B. subtilis Division of Labor

Knowing the regulatory mechanisms and their consequences, we are left with the question of what such probabilistic cell differentiation tells us about the division of labor in biofilms. If cells indeed respond differently to the same environmental conditions, cell differentiation cannot possibly be explained by local physiological adaptation, because supposedly there is only one "optimal" phenotype. Assuming that the phenotypic heterogeneity is adaptive, there are only two alternative explanations for the presence of probabilistic differentiation: bet-hedging and the division of labor. Bet-hedging is an adaptive strategy to cope with unpredictable environmental fluctuations (160). When there are relatively infrequent, unpredictable, and strong environmental changes, a cell can get a fitness advantage by producing a mixture of progeny that is phenotypically diverse (96, 150, 161, 162). In this way, a cell ensures that at least a fraction of its progeny is adapted to the unforeseen environmental changes. A commonly used example of bet-hedging is bacterial persistence in which a small fraction (10^{-5} to 10^{-6}) of cells differentiate into a slow-growing state and become resistant against environmental stressors such as antibiotics (163). Since only a small fraction of cells differentiate into persister cells, a bet-hedging genotype hardly pays a cost for producing them, while, at the same time, it does ensure its survival in the event of a sudden antibiotic influx. Bet-hedging would only evolve when environmental conditions change in an unpredictable way and do not allow for a direct phenotypic response (164–167). Although it is unknown how predictable the environmental changes are during biofilm formation, we would argue that they are relatively gradual and therefore predictable (see discussion below). Thus, probabilistic cell differentiation during biofilm formation might be better explained by the division of labor.

Although direct evidence for cooperative interactions between different cell types in *B. subtilis* is lacking, some properties of probabilistic cell differentiation in *B. subtilis* make the division of labor plausible. Contingent on the actual environmental conditions, the rates at which differentiation occurs are relatively high in comparison to persistence (148, 154, 155, 168). In addition, even though the onset of cell differentiation is sensitive to noise, the regulation thereafter is more deterministic, which hints at some form of coordination. Finally, many cell types secrete products into the environment, which allows for a direct interaction between differentiated and nondifferentiated cells. For example, matrix and protease producers all secrete products that are in principle available to their nondifferentiated siblings.

Even though the bistable switches are sensitive to internal regulatory noise, cell differentiation is largely regulated by external factors because the kinases that phosphorylate Spo0A respond to specific signals. The conditionality of cell differentiation on the local environmental conditions leads to spatial pattern formation, in which certain cell types preferentially occur in specific regions of the biofilm (134, 169). Vlamakis and colleagues (169) showed that motile cells mainly occur on the edges and lower parts of *B. subtilis* biofilms. Matrix producers occur more in the center, and sporulating cells more on the top of biofilms (Fig. 4B). Other studies furthermore showed that spores are largely localized in the wrinkles of a biofilm and in structures that resemble fruiting bodies (103, 144). Cells not only respond to their environment, but also strongly shape their environment. For example, cells consume resources, such as nutrients and oxygen, and secrete surfactin, matrix (e.g., EPS, TasA, and BslA), communicative signals (e.g., surfactin), antimicrobial toxins (Skf and Sdp; see discussion below), and proteases. These products significantly affect the structure of a biofilm, which becomes immediately apparent from studying mutants (108, 115, 144). In addition, some of the products (e.g., the communicative signals) directly affect cell differentiation by triggering one of the sensor kinases (105, 136). As a consequence, there is feedback between the contingency of cell differentiation on environmental conditions and the subsequent influence of these cell types on their environment (i.e., environmental conditionality) (Fig. 4C). Although this feedback results in pattern formation, this by itself does not necessarily mean that different cell types are interacting in a cooperative manner (170).

Despite the detailed knowledge of cell differentiation in *B. subtilis*, relatively little is known about how the different cell types interact and what the fitness consequences are of their interaction. It is, however, plausible that they do cooperate, because many of the cell types presumably pay a cost for producing products that are secreted in the environment and benefit other cells. A recent study, for example, showed that EPS is costly to produce, while it facilitates colony spreading (262). EPS-producing cells could be exploited by EPS-deficient mutants, thereby showing that non-matrix-producing cells benefit from EPS produced by others. Furthermore, similar to developmental chimeras in *M. xanthus* (88), matrix-deficient mutants, *eps* and *tasA* or *eps tasA* and *bslA* (formerly *yuaB*), can complement each other when they are mixed and thereby form a biofilm that is indistinguishable from that of the wild type (109, 171). A recent study showed that *eps* and *tasA* mutants can also complement each other during plant root colonization: despite being unable to colonize the root by themselves, together they can (172). These chimera studies confirm that cells can interact by sharing products they secrete in the environment. Since all these studies are based on interactions between mutants, it is still unknown if similar interactions also occur between cell types in wild-type biofilms. However, it is likely that interactions do occur, given the prevalence of probabilistic cell differentiation (as discussed above).

Another interesting aspect of cell differentiation with respect to the division of labor is the production of antimicrobial toxins. Only matrix-producing cells produce toxins, which can kill sibling cells that have a low level of Spo0A~P and therefore do not express the necessary immunity genes (118). The nutrients that become available through cell lysis are consumed by the matrix-producing cells, which consequently delay sporulation (117, 119). Although antimicrobial toxins are more effective against other soil-dwelling bacteria than sibling cells (120), it is surprising that not all biofilm-inhabiting cells express the necessary immunity genes. Toxin-induced cell lysis is therefore often compared to programmed cell death and viewed as an altruistic trait that benefits matrix-producing cells (37, 119). Cells within *B. subtilis* biofilms also lyse in a manner that is independent of the Skf and Sdp toxins. Localized patterned cell death coupled with the production of extracellular matrix results in the complex wrinkling pattern observed in *B. subtilis* biofilms (173). Cell lysis is commonly observed during biofilm formation in other organisms. For example, in *Pseudomonas aeruginosa* cell lysis has been associated with biofilm dispersal (174, 175). In the next section we evaluate biofilm formation in *P. aeruginosa* and, in particular, the interaction between various subpopulations that appear during biofilm growth.

P. aeruginosa Microcolony Division of Labor

P. aeruginosa is an opportunistic pathogen with a broad host range and, like *B. subtilis* and *M. xanthus*, a common inhabitant of the soil (176). Biofilm formation in *P. aeruginosa* is typically studied in flow chambers. The biofilms that are formed in flow chambers are much smaller than the colony biofilms or pellicles studied in *B. subtilis* and are referred to as microcolonies (175, 177). Microcolonies are at most a few hundred micrometers thick and can easily be examined using scanning confocal laser microscopy, which allows for a detailed examination of the three-dimensional structure. This showed that the shape of microcolonies depends on the nutrient conditions (178). When cells are grown on citrate as the sole carbon source, colonies are flat. In contrast, when grown on glucose, colonies have a mushroom-like shape: there is a relatively narrow stalk at the bottom that is topped by a wider cap (Fig. 5).

Given their interesting morphology, mushroom-shaped microcolonies have been intensely studied. To understand their development, many studies have examined biofilms composed of two different strains (176, 178–184). Like in the case of *M. xanthus*, multiple predefined mutants were mixed to examine how these mixtures affect microcolony development. Although many of these studies were not intended to examine cooperation, which was the case for chimera studies in *M. xanthus*, they did provide a number of interesting insights. For example, Klausen and colleagues (178) showed that the mushroom-shaped structures resulted from the interaction between two subpopulations: motile and nonmotile cells (185). This was shown by studying chimeras of motile wild-type cells with twitching motility-deficient mutants (*pilA*), each labeled with a distinct fluorescent protein (Fig. 5A). Initially, the nonmotile cells formed small "stalk" colonies by localized clonal growth. After approxi-

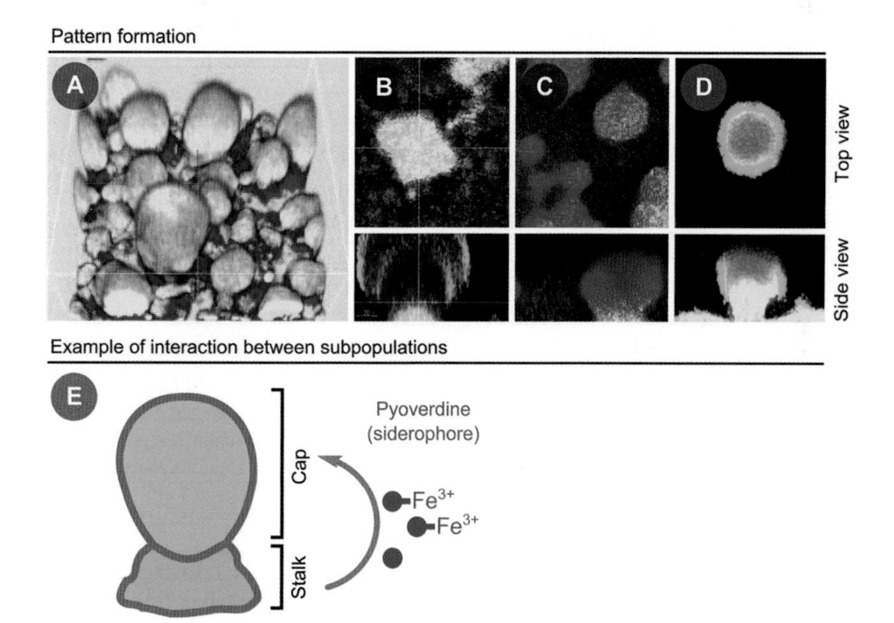

FIGURE 5 Pattern formation in *P. aeruginosa* microcolonies. (A) Fruiting bodies consisting of nonmotile stalk cells (blue) and motile cap cells (yellow). (B) Localization of eDNA in microcolonies (red). (C) Localization of rhamnolipid production (yellow). (D) Live (green) and dead (yellow/red) cells after EDTA treatment in a 4-day-old microcolony. (E) Schematic overview of mushroom-shaped microcolonies and the interaction between the stalk and cap cells through the production of the iron-scavenging siderophore pyoverdine. Images adapted from references 178, 190, 192, and 196. doi:10.1128/microbiolspec.MB-0002-2014.f5

mately 4 days, the motile cells moved on top of these stalk colonies via type IV pili-mediated twitching motility. This migration results in the formation of caps and, hence, the mushroom-shaped microcolonies (178). Perhaps through chemotaxis (186), motile cells might climb on top of the stalk cells to access more nutrients (182, 187). In citrate minimal medium the absence of mushroom-shaped microcolonies can be explained by the lack of nonmotile cells, and therefore no stalks can form (179). Finally, it is important to note that even though Klausen and colleagues (178) examined genetic chimeras, the same phenotypic subpopulations are present in the absence of genetic variation (i.e., phenotypic heterogeneity) (185).

Signaling During Microcolony Formation

There are a number of phenotypic differences between the stalk and cap subpopulations. On average, the stalk subpopulation has a higher cell density than the cap subpopulation and is less metabolically active (184, 188). In addition, stalk cells produce quorum-sensing signals, surfactants, and siderophores, while cap cells do not (183, 189–192). There are three important quorum-sensing systems that regulate microcolony formation: Las, Rhl, and Pqs. The Las and Rhl systems involve two homoserine lactone signals, and the Pqs system involves a quinolone. All three quorum-sensing signals are predominantly expressed in the stalk of the microcolony, which might be explained by the high cell density (183, 189, 192). For some growth conditions the Las system is essential for microcolony development (193), whereas for others it is not (194–196). These differences might be explained by the nutrient-dependent influence of quorum sensing on motility in *P. aeruginosa* biofilms (197). The Rhl system is required for rhamnolipid production (198). Given that the *rhl* genes are expressed more in the stalk cells, it is perhaps not surprising that rhamnolipids are also predominantly produced in the stalk (Fig. 5C) (190). Rhamnolipid surfactants are important for a number of biofilm-related properties. They affect the spacing between the microcolonies (199). They are necessary for early stages of microcolony formation and cap formation (181). Finally, depending on the culturing conditions, rhamnolipids play a role in the dispersal of cells at the end of microcolony formation (180, 200).

The quorum-sensing systems are also involved in the production of extracellular DNA (eDNA) during microcolony formation (196). eDNA is a major component of the extracellular matrix and is essential for both the establishment of microcolonies and their early development (201, 202). Treatment with DNAse prevents colony establishment and triggers dispersal in young microcolonies (201). The eDNA consists of chromosomal DNA and originates from a small subpopulation of cells that undergoes cell lysis (196). The degree of cell lysis and, hence, the production of eDNA depends on the Pqs system (196, 203): high levels of quinolone signal result in high levels of cell lysis, while low levels of quinolone reduce the degree of cell lysis. Interestingly, eDNA production is regulated in time and space. eDNA is produced before cap formation by the stalk subpopulation and is primarily localized at the outermost edge of the stalk colony (Fig. 5B). It has been suggested that eDNA facilitates cap cells' migration on top of the stalk subpopulation (182, 183). Treatment of stalk colonies with DNAse prior to migration inhibits cap formation, as does lack of the Pqs system (182). In agreement with the localization of eDNA (Fig. 5B), Yang and colleagues (192) showed that *pqsA*, a gene that encodes for an essential component of the Pqs system (203), is expressed before cap formation in the outermost edge of the stalk. In addition, they showed that both the expression of *pqsA* and the production of eDNA depend on the iron concentration in the medium. At high iron concentrations *pqsA* expression was repressed and, consequently, mushroom-shaped microcolonies were not formed. The lack of structure made biofilms more susceptible to antibiotic treatment (192). Other studies have also shown that mushroom-shaped microcolonies are more resilient against antimicrobial and EDTA treatments (Fig. 5D) because of the different physiological states of stalk and cap cells (185, 188, 204, 205).

Sharing of Common Goods within the Biofilm

Given that cell lysis is part of the regulatory circuit of microcolony formation, it could be viewed as a cooperative trait. That is, it may be argued that a fraction of stalk cells sacrifice themselves to help cap cells migrating on top of them. By the same token, it may be argued that the production of quorum-sensing signals, siderophores, surfactants, and polysaccharides are cooperative traits. Many of these public goods are solely produced by the stalk cells. The question therefore arises as to whether they are somehow shared with the cap cells. That is, is there an interaction between the stalk and cap subpopulation? Yang and colleagues (183) showed in a number of elegant experiments that these subpopulations indeed interact. They focused on two particular common goods: Pqs signaling and pyoverdine production (siderophore). Both common goods are necessary for the formation of mushroom-shaped microcolonies (206). For each experiment, Yang and colleagues made a chimera consisting of two genotypes: one motile and one nonmotile (*pilA*). As described previously, the *pilA* genotype always occurs in the stalk subpopulation (178). In addition, either the motile or the nonmotile genotype is given an additional mutation that abolishes common good production: either a *pqsA* mutation that prevents quorum sensing or a *pvdA* mutation that prevents pyoverdine production. If the cap subpopulation depends on the common good produced by the stalk cells, the chimeras with defective stalk cells should be aberrant in normal microcolony formation, while this should not be the case for those of defective cap cells. The *pilA/pqsA* and *pilA/pvdA* chimeras both produce normal mushroom-shaped microcolonies, in which nonmotile *pilA* cells form the stalk and either *pqsA* or *pvdA* cells form the cap. This is not surprising because also in wild-type microcolonies, cap cells do not express *pqsA* or *pvdA*, so a mutation in the cap cells should not affect their phenotype (191, 192). In contrast, both the *pilApqsA*/wild-type and *pilApvdA*/wild-type chimeras are defective in cap formation. This shows that cap-formation indeed depends on the Pqs system and pyoverdine production of the stalk cells.

In the case of the Pqs system, the results are in agreement with the studies discussed above. Pqs signaling results in localized cell lysis and the release of DNA. This eDNA is necessary for the cap cells to migrate on the stalk cells. This was furthermore confirmed by the fact that *pilApqsA*/wild-type chimeras did produce caps in the presence of exogenously added DNA (183). In the case of pyoverdine production, the cap subpopulation is dependent on the pyoverdine produced by stalk subpopulation (Fig. 5E). In *pilA/pvdA* chimeras' cap cells express *fpvA*, a gene that encodes the ferric-pyoverdine uptake system, indicating that cap cells take up the pyoverdine produced by the stalk cells. The interaction between the stalk and cap cells was further confirmed by the fact that *pilA/fpvA* chimeras were also defective in cap formation. In that case, stalk cells do produce pyoverdines, but *fpvA*-mutant cap cells cannot access them because they lack the uptake system.

All in all, Yang and colleagues (183) demonstrated that *P. aeruginosa* microcolony formation comes about through the interaction of heterogeneous subpopulations. Mutants defective in type-IV pili formation, Pqs signaling, or pyoverdine production cannot produce normal microcolonies alone. However, when mixed, they produce mushroom-shaped microcolonies indistinguishable from wild type. Although it is yet unclear if there is an ecological advantage associated with cap formation and if common good production is costly (although in the case of Pqs-mediated cell lysis there obviously is a fitness cost), this study shows that heterogeneous subpopulations inside a biofilm can engage in a cooperative interaction.

Dispersal of *P. aeruginosa* Microcolonies

Another interesting stage in *P. aeruginosa* biofilm formation occurs after approximately 7 days when cells start to disperse (194). Like the onset of biofilm formation, this dispersal stage is subject to regulation and therefore differs from other dispersal events (e.g., biofilm sloughing) that primarily result from shear stress (184, 207, 208). Since mostly single cells disperse from the microcolonies during this phase, this type of dispersal is also called seeding dispersal. Seeding dispersal depends on a number of environmental factors (209, 210) such as nutrients (211), rhamnolipids (180), quorum-sensing signals (200, 212), nitric oxide (213), and oxygen (214). The phenotype of dispersing cells is more similar to those of motile cells that initiate biofilm formation than it is to the cells inhabiting a mature microcolony (194). In general, the dispersal stage is characterized by a phenomenon called "'hollowing"; cells localized in the center of the microcolony disperse, while those on the edges remain statically attached to the surface (174, 194, 215). Webb and colleagues (174) discovered that the occurrence of hollowing is consistently associated with localized cell death and lysis. Only a subpopulation of cells in the center of the microcolony remains viable, becomes motile, and, subsequently, disperses. Surprisingly, given the consistent timing of hollowing, cell lysis was largely mediated by the hyperinfectivity and lytic effects of an otherwise nonlytic filamentous prophage. The induction of the prophage is more common in the biofilm when compared to planktonic cells and can, under some conditions, be affected by quorum-sensing signaling (174, 200, 203, 212, 216).

Hollowing at the end of microcolony maturation is often hypothesized to be an adaptive trait that facilitates dispersal (175, 215, 217). In agreement with this hypothesis, Rice and colleagues (218) showed that the virulence of a wild-type strain was significantly higher than that of a prophage-deficient mutant when infecting mice. In other words, phage-mediated hollowing seems to facilitate *P. aeruginosa* virulence. In addition, it has been shown that phage-mediated hollowing occurs in natural isolates from cystic fibrosis patients (219). Not only *P. aeruginosa* shows the hollowing phenotype; in *Pseudoalteromonas tunicata* a similar phenomenon occurs (220–222). *P. tunicata* is a common inhabitant of the marine environment, where it colonizes surfaces of eukaryotic organisms. Similar to *P. aeruginosa*, during hollowing in *P. tunicata* the majority of cells in the center of the microcolony lyse (220). The remaining viable cells become motile and disperse. Dispersal is highly reproducible and always occurs after the same period of biofilm development. In contrast to *P. aeruginosa*, cell lysis in *P. tunicata* is not induced by a prophage, but through the expression of an autotoxic protein called AlpP (220, 222). In the absence of this autotoxin, hollowing did not occur, and the number of dispersing cells was greatly reduced. Besides increasing the number of dispersing cells, the dispersing cells that result from AlpP-mediated hollowing have a higher metabolic activity and express a wider range of phenotypes, which might be beneficial for subsequent colonization (221). Thus, colony hollowing in *P. aeruginosa* and *P. tunicata* illustrates that a nonadaptive cellular behavior (i.e., cell lysis) can facilitate a colony-level function (i.e., dispersal). In consideration of the division of labor, the interaction between cell types should therefore be evaluated with respect to the ecological functionality that emerges at the colony level (or at a lower organizational level; e.g., clonal pockets in multispecies biofilms).

Although one might be inclined, based on the previous examples, to assume that cell lysis evolved to facilitate dispersal, there is insufficient evidence for this assumption. In fact, similar to Skf and Sdp produced by *B. subtilis* (120), AlpP is effective in direct competition with other species, and its autotoxicity might just be a side effect (223–226). Likewise, the prophage-induced cell lysis in *P. aeruginosa* might simply result from the classic conflict between a phage and its host (13). In general, one should be cautious in giving an evolutionary interpretation without considering the ecological context under which traits have evolved (13, 226, 227).

Dispersal Organs: Microcolonies and Fruiting Bodies

To interpret the patterns observed in *P. aeruginosa* microcolonies, microcolony formation is often loosely compared to the developmental process of fruiting body formation in *M. xanthus* (14, 74). There are a number of similarities. In short, both species initiate colony formation in response to nutrient depletion (72, 84, 228). Initially they form a monolayer of cells that aggregate into colonies via pilus-mediated movement and cell division (74, 229). Colony formation is, at least in part, regulated by a number of quorum-sensing signals and involves the secretion of extracellular matrix as well as localized cell lysis (82, 86, 189, 196). Eventually, both species form colonies with a mushroom-shaped structure, from which only a fraction of cells eventually disperse. Therefore, both microcolonies and fruiting bodies are often hypothesized to facilitate dispersal (84, 175).

BENEFITS OF DIFFERENTIATION AND DIVISION OF LABOR

In the sections above, we described three organisms which have independently evolved to display phenotypic heterogeneity in biofilms. Furthermore, we showed that many of the evolved cell types interact inside these biofilms. Yet the question often remains: Why do cells differentiate? As pointed out by previous reviews (14, 187, 207, 230), any form of multicellular pattern formation—with or without developmental regulation—results from the feedback between cellular responsiveness and environmental conditionality (Fig. 4C) (69, 70, 170). To understand the significance of pattern formation, one should ask not only how cells respond, but also why they do so (13, 14, 231). As described in the beginning of this article, when cells differentiate to divide labor they are expected to cooperate with each another. This cooperative interaction should result in an emergent benefit (i.e., ecological functionality) at the colony level. One therefore needs to determine the fitness costs and consequences that are associated with the expression of a cell type: Is cell differentiation costly? Who benefits from cell differentiation? What is the emergent benefit from a cooperative interaction between different cell types? Studies in which chimeras of different genotypes are examined, such as those discussed above for *M. xanthus*, help answer these questions. In this way, a number of phenotypes that are associated with biofilm formation have been shown to be cooperative traits (16, 232–236). For example, the stalk cells in microcolonies of *P. aeruginosa* produce pyoverdine. A number of studies showed that pyoverdine production is costly for a cell (i.e., reduced cell division rate) and that, in unstructured environments, pyoverdine-deficient mutants can exploit pyoverdine-producing wild-type cells (15, 232, 237, 238). Aggregation, however, constrains pyoverdine diffusion. This might explain why pyoverdine-producing cells are not exploited in natural settings (239). Knowing that some cell types express cooperative traits is only the first step in understanding the interaction between different cell types. The second step, which is perhaps the most challenging, is to understand why different cell types cooperate, such as the apparent cooperation between stalk and cap cells in microcolonies.

To characterize the potential ecological advantages that are associated with biofilm formation, one typically compares mature biofilms with planktonic cells. In this way, two major ecological advantages have been discovered (67, 228, 240, 241): (i) protection against environmental stress and (ii) metabolic cooperation. These benefits are based on a rather static comparison between biofilms and planktonic cells. We believe that the ecological functionality of cell types may be better understood when we appreciate the dynamic process of biofilm formation. Biofilms are a transient stage in the life cycle of bacteria. Considering this complete life cycle can help us in determining the potential functions and benefits that might emerge from the cooperative interaction between different cell types.

LIFE CYCLE BIOLOGY: THE ECOLOGY OF CELL DIFFERENTIATION

Most biofilms form a transient stage in the bacterium's life cycle, as shown in *M. xanthus*, *B. subtilis*, and *P. aeruginosa* (208, 242). Consequently, the life cycle of bacteria can roughly be divided into two life phases which alternate over time: a unicellular life phase and a multicellular (biofilm) life phase. This biphasic life cycle is analogous to life cycles seen in multicellular organisms, in which the multicellular life phase is typically associated with resource acquisition, while the unicellular life phase is required for sexual reproduction or dispersal (170, 208). If we greatly simplify the life cycle of a biofilm-forming bacterium, there are two overarching environmental conditions that cells encounter: those that result in a transition toward biofilm formation and those that result in motility. In the first transition cells have to colonize a surface and build a biofilm. In the second transition the biofilm breaks down and cells disperse (Fig. 6).

When cells encounter a new surface, they benefit if they adhere to this surface before their competitors do, spread quickly, and prevent their competitors from colonizing the surface as well. The advantage derived from colonizing a surface first is also known as the "founder effect" and can simply result from the fact that colonization is only possible when cells can directly access the substrate. An and colleagues (243), for example, showed that the outcome of the competitive interaction between *P. aeruginosa* and *Agrobacterium tumefaciens* depends on the order in which these species colonize the surface. Once adhered to the surface, a colony can spread passively by division or actively by motility and surfactant production (244). Both *B. subtilis* and *P. aeruginosa* produce large amounts of surfactants during a process called swarming, in which hyperflagellated cells spread over the surface (245–249). In the case of swarming in *B. subtilis*, the majority of surfactin is produced in the center of the swarm, from which it spreads and covers the whole colony (250, 251).

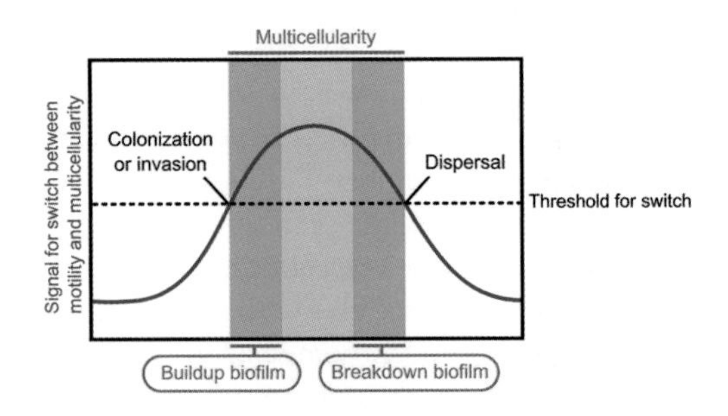

FIGURE 6 Simplified schematic view of the life cycle of biofilm formation. The life cycle is divided into two life phases: the multicellular phase and the dispersal phase. Various environmental conditions influence the switch toward aggregation, typically mediated by a second messenger. When the second messenger passes a certain threshold, aggregation is initiated and, the other way around, when it drops below a certain threshold, cells revert to the dispersal phase. doi:10.1128/microbiolspec.MB-0002-2014.f6

Finally, cells are expected to make the surface uninhabitable for close competitors. In both *B. subtilis* and *P. aeruginosa*, the regulatory pathways of antimicrobial production are closely entangled with those of biofilm formation, because matrix and antimicrobial genes are coexpressed (252–258). In the case of *B. subtilis*, matrix production can be triggered by antimicrobials produced by competitors (104, 106, 107, 137). In other words, *B. subtilis* cells likely cooperate to compete against other soil-dwelling organisms. All in all, one can imagine that the division of labor can enhance the effectiveness of colonization by having cells specializing in the different tasks.

The breakdown of biofilms and subsequent cell dispersal also involve a number of tasks. To efficiently disperse, cells should anticipate adverse conditions, such that dispersal units (e.g., spores) develop in a timely fashion but not too early (259). For biofilm dispersal, cells should break down the biofilm matrix, develop dispersal units, and facilitate these units to actually disperse. Biofilm degradation provides nutrients to cells, which can increase the total number of dispersal units that eventually develop. However, having many dispersal units is only effective when they detach from the surface. As seen in the dispersal organs of *M. xanthus* and *S. coelicolor*, dispersal can be facilitated if cells break their surface tension with water and rise from the colony. Furthermore, the mere breakdown of matrix already facilitates dispersal by weakening the adherence that cells have to each other. As for colonization, the division of labor might greatly enhance the effectiveness of dispersal (260, 261). For example, imagine that a *B. subtilis* biofilm is confronted with a sudden drop in nutrient availability. There might be too few nutrients for all cells to sporulate on time. One can imagine that some cells specialize in producing proteases to break down the biofilm. Other cells might produce antimicrobial toxins that primarily kill cells that are unlikely to sporulate anyway. The lysed cells subsequently provide nutrients for the sporulating cells, which therefore can finish sporulation. In the end, this task specialization might maximize the total number of spores that disperse. To examine if cells indeed divide labor, competition experiments are required, and one should study the effect of cell differentiation on total spore production. One could, for example, examine if total production is lower in the absence of cell lysis (i.e., when all cells are immune against Skf and Sdp).

CONCLUSIONS

Biofilms are remarkable examples of biological construction. In many bacteria, cells that can live independently decide otherwise by aggregating into biofilms. These biofilms are characterized by heterogeneity. The mere presence of a surface results in environmental gradients that trigger differentiation into various cell types. In this article we discussed cell types that are not simply adapted to the local environment, but can engage in cooperative interactions that allow for the division of labor. This division of labor requires both cooperation and coordination between cell types, much like that seen in primitive multicellular organisms. Despite this similarity, it is futile to make a very strict comparison between biofilms and multicellular individuals. In the end, "organismality" and "individuality" cannot be defined as an all-or-none concept. Clear examples of intermediate levels of functional integration, in which different degrees of division of labor may be at play, have been studied. With this in mind, biofilms might be the ideal system for acquiring a more general understanding of the evolution of biological organization.

ACKNOWLEDGMENTS

Conflicts of interest: We declare no conflicts.

CITATION

van Gestel J, Vlamakis H, Kolter R. 2015. Division of labor in biofilms: the ecology of cell differentiation. Microbiol Spectrum 3(2): MB-0002-2014.

REFERENCES

1. **Buss LW.** 1987. *The Evolution of Individuality.* Princeton University Press, Princeton, NJ.
2. **Maynard Smith J, Szathmary E.** 1997. *The Major Transitions in Evolution.* Oxford University Press, Oxford, UK.
3. **Nowak MA.** 2006. Five rules for the evolution of cooperation. *Science* **314:**1560–1563.
4. **Michod RE, Herron MD.** 2006. Cooperation and conflict during evolutionary transitions in individuality. *J Evol Biol* **19:**1406–1409.
5. **Grosberg RK, Strathmann RR.** 2007. The evolution of multicellularity: a minor major transition? *Annu Rev Ecol Evol Syst* **38:**621–654.
6. **Nowak MA, Tarnita CE, Antal T.** 2010. Evolutionary dynamics in structured populations. *Philos Trans R Soc B Biol Sci* **365:**19–30.
7. **Tarnita CE, Taubes CH, Nowak MA.** 2013. Evolutionary construction by staying together and coming together. *J Theor Biol* **320:**10–22.
8. **Kirk DL.** 2005. A twelve-step program for evolving multicellularity and a division of labor. *Bioessays* **27:**299–310.
9. **Michod RE.** 2007. Evolution of individuality during the transition from unicellular to multicellular life. *Proc Natl Acad Sci USA* **104:**8613–8618.
10. **Pepper JW, Herron MD.** 2008. Does biology need an organism concept? *Biol Rev Camb Philos Soc* **83:**621–627.
11. **Herron MD, Rashidi A, Shelton DE, Driscoll WW.** 2013. Cellular differentiation and individuality in the "minor" multicellular taxa. *Biol Rev* **88:**844–861.
12. **Watnick P, Kolter R.** 2000. Biofilm, city of microbes. *J Bacteriol* **182:**2675–2679.
13. **Nadell CD, Xavier JB, Foster KR.** 2009. The sociobiology of biofilms. *FEMS Microbiol Rev* **33:**206–224.
14. **Monds RD, O'Toole GA.** 2009. The developmental model of microbial biofilms: ten years of a paradigm up for review. *Trends Microbiol* **17:**73–87.
15. **Griffin AS, West SA, Buckling A.** 2004. Cooperation and competition in pathogenic bacteria. *Nature* **430:**1024–1027.
16. **West SA, Griffin AS, Gardner A, Diggle SP.** 2006. Social evolution theory for microorganisms. *Nat Rev Microbiol* **4:**597–607.
17. **Fletcher JA, Doebeli M.** 2009. A simple and general explanation for the evolution of altruism. *Proc R Soc B Biol Sci* **276:**13–19.
18. **Nowak MA, Sigmund K.** 1992. Tit for tat in heterogeneous populations. *Nature* **355:**250–253.
19. **Nowak MA, Bonhoeffer S, May RM.** 1994. Spatial games and the maintenance of cooperation. *Proc Natl Acad Sci USA* **91:**4877–4881.
20. **Kreft JU.** 2004. Biofilms promote altruism. *Microbiology* **150:**2751–2760.
21. **Xavier JB, Foster KR.** 2007. Cooperation and conflict in microbial biofilms. *Proc Natl Acad Sci USA* **104:**876–881.
22. **Timmis JN, Ayliffe MA, Huang CY, Martin W.** 2004. Endosymbiotic gene transfer: organelle genomes forge eukaryotic chromosomes. *Nat Rev Genet* **5:**123–135.
23. **Wernegreen JJ.** 2004. Endosymbiosis: lessons in conflict resolution. *PLoS Biol* **2:**e68. doi:10.1371/journal.pbio.0020068.
24. **Husnik F, Nikoh N, Koga R, Ross L, Duncan RP, Fujie M, Tanaka M, Satoh N, Bachtrog D, Wilson ACC, von Dohlen CD, Fukatsu T, McCutcheon JP.** 2013. Horizontal gene transfer from diverse bacteria to an insect genome enables a tripartite nested mealybug symbiosis. *Cell* **153:**1567–1578.
25. **Schlichting CD, Pigliucci M.** 1998. *Phenotypic Evolution: A Reaction Norm Perspective.* Sinauer, Sunderland, MA.
26. **Beldade P, Mateus ARA, Keller RA.** 2011. Evolution and molecular mechanisms of adaptive developmental plasticity. *Mol Ecol* **20:**1347–1363.
27. **Schlichting CD.** 2003. Origins of differentiation via phenotypic plasticity. *Evol Dev* **5:**98–105.
28. **Smits WK, Kuipers OP, Veening JW.** 2006. Phenotypic variation in bacteria: the role of feedback regulation. *Nat Rev Microbiol* **4:**259–271.
29. **Alon U.** 2007. Network motifs: theory and experimental approaches. *Nat Rev Genet* **8:**450–461.
30. **Dubnau D, Losick R.** 2006. Bistability in bacteria. *Mol Microbiol* **61:**564–572.
31. **Mitrophanov AY, Groisman EA.** 2008. Positive feedback in cellular control systems. *BioEssays* **30:**542–555.
32. **Guttenplan SB, Kearns DB.** 2013. Regulation of flagellar motility during biofilm formation. *FEMS Microbiol Rev* **37:**849–871.
33. **Duarte A, Weissing FJ, Pen I, Keller L.** 2011. An evolutionary perspective on self-organized

division of labor in social insects. *Annu Rev Ecol Evol Syst* **42:**91–110.

34. **Meier P, Finch A, Evan G.** 2000. Apoptosis in development. *Nature* **407:**796–801.
35. **Ameisen JC.** 2002. On the origin, evolution, and nature of programmed cell death: a timeline of four billion years. *Cell Death Differ* **9:**367–393.
36. **Golstein P, Aubry L, Levraud JP.** 2003. Cell-death alternative model organisms: why and which? *Nat Rev Mol Cell Biol* **4:** 798–807.
37. **Engelberg-Kulka H, Amitai S, Kolodkin-Gal I, Hazan R.** 2006. Bacterial programmed cell death and multicellular behavior in bacteria. *PLoS Genet* **2:**e135. doi:10.1371/journal.pgen.0020135.
38. **Claverys JP, Havarstein LS.** 2007. Cannibalism and fratricide: mechanisms and raisons d'etre. *Nat Rev Microbiol* **5:**219–229.
39. **Schopf JW.** 1993. Microfossils of the Early Archean Apex Chert: new evidence of the antiquity of life. *Science* **260:**640–646.
40. **Knoll AH.** 2011. The multiple origins of complex multicellularity. *Annu Rev Earth Planet Sci* **39:**217–239.
41. **Adams DG, Duggan PS.** 1999. Heterocyst and akinete differentiation in cyanobacteria. *New Phytol* **144:**3–33.
42. **Flores E, Herrero A.** 2010. Compartmentalized function through cell differentiation in filamentous cyanobacteria. *Nat Rev Microbiol* **8:**39–50.
43. **Schirrmeister BE, de Vos JM, Antonelli A, Bagheri HC.** 2013. Evolution of multicellularity coincided with increased diversification of cyanobacteria and the Great Oxidation Event. *Proc Natl Acad Sci USA* **110:**1791–1796.
44. **Wolk CP.** 1968. Movement of carbon from vegetative cells to heterocysts in *Anabaena cylindrica. J Bacteriol* **96:**2138–2143.
45. **Wolk CP, Thomas J, Shaffer PW, Austin SM, Galonsky A.** 1976. Pathway of nitrogen metabolism after fixation of 13N-labeled nitrogen gas by the cyanobacterium, *Anabaena cylindrica. J Biol Chem* **251:**5027–5034.
46. **Frías JE, Flores E, Herrero A.** 1994. Requirement of the regulatory protein NtcA for the expression of nitrogen assimilation and heterocyst development genes in the cyanobacterium *Anabaena* sp. PCC7120. *Mol Microbiol* **14:**823–832.
47. **Black TA, Cai Y, Wolk CP.** 1993. Spatial expression and autoregulation of *hetR*, a gene involved in the control of heterocyst development in *Anabaena. Mol Microbiol* **9:**77–84.
48. **Yoon HS, Golden JW.** 1998. Heterocyst pattern formation controlled by a diffusible peptide. *Science* **282:**935–938.
49. **Callahan SM, Buikema WJ.** 2001. The role of HetN in maintenance of the heterocyst pattern in *Anabaena* sp. PCC 7120. *Mol Microbiol* **40:** 941–950.
50. **Yoon HS, Golden JW.** 2001. PatS and products of nitrogen fixation control heterocyst pattern. *J Bacteriol* **183:**2605–2613.
51. **Zhang CC, Laurent S, Sakr S, Peng L, Bédu S.** 2006. Heterocyst differentiation and pattern formation in cyanobacteria: a chorus of signals. *Mol Microbiol* **59:**367–375.
52. **Rossetti V, Schirrmeister BE, Bernasconi MV, Bagheri HC.** 2010. The evolutionary path to terminal differentiation and division of labor in cyanobacteria. *J Theor Biol* **262:** 23–34.
53. **Matias Rodrigues JF, Rankin DJ, Rossetti V, Wagner A, Bagheri HC.** 2012. Differences in cell division rates drive the evolution of terminal differentiation in microbes. *PLoS Comput Biol* **8:**e1002468. doi:10.1371/journal.pcbi.1002468.
54. **Misra HS, Tuli R.** 2000. Differential expression of photosynthesis and nitrogen fixation genes in the cyanobacterium *Plectonema boryanum. Plant Physiol* **122:**731–736.
55. **Tomitani A, Knoll AH, Cavanaugh CM, Ohno T.** 2006. The evolutionary diversification of cyanobacteria: molecular–phylogenetic and paleontological perspectives. *Proc Natl Acad Sci USA* **103:**5442–5447.
56. **Michod RE, Viossat Y, Solari CA, Hurand M, Nedelcu AM.** 2006. Life-history evolution and the origin of multicellularity. *J Theor Biol* **239:**257–272.
57. **Michod RE.** 2006. The group covariance effect and fitness trade-offs during evolutionary transitions in individuality. *Proc Natl Acad Sci USA* **103:**9113–9117.
58. **Flärdh K, Buttner MJ.** 2009. *Streptomyces* morphogenetics: dissecting differentiation in a filamentous bacterium. *Nat Rev Microbiol* **7:**36–49.
59. **McCormick JR, Flärdh K.** 2012. Signals and regulators that govern *Streptomyces* development. *FEMS Microbiol Rev* **36:**206–231.
60. **Kang SG, Lee KJ.** 1997. Kinetic analysis of morphological differentiation and protease production in *Streptomyces albidoflavus* SMF301. *Microbiology* **143:**2709–2714.
61. **Chater KF, Biró S, Lee KJ, Palmer T, Schrempf H.** 2010. The complex extracellular biology of *Streptomyces. FEMS Microbiol Rev* **34:**171–198.

62. **Wildermuth H.** 1970. Development and organization of the aerial mycelium in *Streptomyces coelicolor*. *J Gen Microbiol* **60:**43–50.
63. **Miguélez EM, Hardisson C, Manzanal MB.** 1999. Hyphal death during colony development in *Streptomyces antibioticus*: morphological evidence for the existence of a process of cell deletion in a multicellular prokaryote. *J Cell Biol* **145:**515–525.
64. **Manteca A, Claessen D, Lopez-Iglesias C, Sanchez J.** 2007. Aerial hyphae in surface cultures of *Streptomyces lividans* and *Streptomyces coelicolor* originate from viable segments surviving an early programmed cell death event. *FEMS Microbiol Lett* **274:**118–125.
65. **Stewart PS.** 2003. Diffusion in biofilms. *J Bacteriol* **185:**1485–1491.
66. **Rani SA, Pitts B, Beyenal H, Veluchamy RA, Lewandowski Z, Davison WM, Buckingham-Meyer K, Stewart PS.** 2007. Spatial patterns of DNA replication, protein synthesis, and oxygen concentration within bacterial biofilms reveal diverse physiological states. *J Bacteriol* **189:**4223–4233.
67. **Stewart PS, Franklin MJ.** 2008. Physiological heterogeneity in biofilms. *Nat Rev Microbiol* **6:**199–210.
68. **Kolter R, Greenberg EP.** 2006. Microbial sciences: the superficial life of microbes. *Nature* **441:**300–302.
69. **Wolpert L.** 1969. Positional information and the spatial pattern of cellular differentiation. *J Theor Biol* **25:**1–47.
70. **Wolpert L, Beddington R, Jessell T, Lawrence P, Meyerowitz E, Smith J.** 2002. *Principles of Development*. Oxford University Press, New York.
71. **Berleman J, Auer M.** 2013. The role of bacterial outer membrane vesicles for intra- and interspecies delivery. *Environ Microbiol* **15:**347–354.
72. **Konovalova A, Petters T, Søgaard-Andersen L.** 2010. Extracellular biology of *Myxococcus xanthus*. *FEMS Microbiol Rev* **34:**89–106.
73. **Higgs PI, Hartzell PL, Holkenbrink C, Hoiczyk E.** 2014. *Myxococcus xanthus* vegetative and developmental cell heterogeneity. *In* Yang Z, Higgs PI (ed), *Myxobacteria: Genomics and Molecular Biology*. Horizon Scientific Press, Norfolk, UK.
74. **O'Toole G, Kaplan HB, Kolter R.** 2000. Biofilm formation as microbial development. *Annu Rev Microbiol* **54:**49–79.
75. **Kaiser D.** 2001. Building a multicellular organism. *Annu Rev Genet* **35:**103–123.
76. **Pathak DT, Wei X, Wall D.** 2012. Myxobacterial tools for social interactions. *Res Microbiol* **163:**579–591.
77. **Julien B, Kaiser AD, Garza A.** 2000. Spatial control of cell differentiation in *Myxococcus xanthus*. *Proc Natl Acad Sci USA* **97:**9098–9103.
78. **Lee B, Holkenbrink C, Treuner-Lange A, Higgs PI.** 2012. *Myxococcus xanthus* developmental cell fate production: heterogeneous accumulation of developmental regulatory proteins and reexamination of the role of MazF in developmental lysis. *J Bacteriol* **194:**3058–3068.
79. **O'Connor KA, Zusman DR.** 1991. Development in *Myxococcus xanthus* involves differentiation into two cell types, peripheral rods and spores. *J Bacteriol* **173:**3318–3333.
80. **Hoiczyk E, Ring MW, McHugh CA, Schwär G, Bode E, Krug D, Altmeyer MO, Lu JZ, Bode HB.** 2009. Lipid body formation plays a central role in cell fate determination during developmental differentiation of *Myxococcus xanthus*. *Mol Microbiol* **74:**497–517.
81. **Kuspa A, Plamann L, Kaiser D.** 1992. A-signalling and the cell density requirement for *Myxococcus xanthus* development. *J Bacteriol* **174:**7360–7369.
82. **Kaiser D.** 2004. Signaling in myxobacteria. *Annu Rev Microbiol* **58:**75–98.
83. **Lobedanz S, Søgaard-Andersen L.** 2003. Identification of the C-signal, a contact-dependent morphogen coordinating multiple developmental responses in *Myxococcus xanthus*. *Genes Dev* **17:**2151–2161.
84. **Kroos L.** 2007. The *Bacillus* and *Myxococcus* developmental networks and their transcriptional regulators. *Annu Rev Genet* **41:**13–39.
85. **Spröer C, Reichenbach H, Stackebrandt E.** 1999. The correlation between morphological and phylogenetic classification of myxobacteria. *Int J Syst Bacteriol* **49:**1255–1262.
86. **Nariya H, Inouye M.** 2008. MazF, an mRNA interferase, mediates programmed cell death during multicellular *Myxococcus* development. *Cell* **132:**55–66.
87. **Velicer GJ, Vos M.** 2009. Sociobiology of the *Myxobacteria*. *Annu Rev Microbiol* **63:**599–623.
88. **Hagen DC, Bretscher AP, Kaiser D.** 1978. Synergism between morphogenetic mutants of *Myxococcus xanthus*. *Dev Biol* **64:**284–296.
89. **Kroos L, Kaiser D.** 1987. Expression of many developmentally regulated genes in *Myxococcus* depends on a sequence of cell interactions. *Genes Dev* **1:**840–854.
90. **Velicer GJ, Kroos L, Lenski RE.** 2000. Developmental cheating in the social bacterium *Myxococcus xanthus*. *Nature* **404:**598–601.
91. **Fiegna F, Velicer GJ.** 2003. Competitive fates of bacterial social parasites: persistence and

self-induced extinction of *Myxococcus xanthus* cheaters. *Proc R Soc Lond Ser B Biol Sci* **270:** 1527–1534.

92. **Fiegna F, Velicer GJ.** 2005. Exploitative and hierarchical antagonism in a cooperative bacterium. *PloS Biol* **3:**1980–1987.
93. **Fiegna F, Yu YTN, Kadam SV, Velicer GJ.** 2006. Evolution of an obligate social cheater to a superior cooperator. *Nature* **441:**310–314.
94. **Vos M, Velicer GJ.** 2009. Social conflict in centimeter and global-scale populations of the bacterium *Myxococcus xanthus*. *Curr Biol* **19:**1763–1767.
95. **Velicer GJ, Kroos L, Lenski RE.** 1998. Loss of social behaviors by *Myxococcus xanthus* during evolution in an unstructured habitat. *Proc Natl Acad Sci USA* **95:**12376–12380.
96. **Veening JW, Smits WK, Kuipers OP.** 2008. Bistability, epigenetics, and bet-hedging in bacteria. *Annu Rev Microbiol* **62:**193–210.
97. **Lopez D, Vlamakis H, Kolter R.** 2009. Generation of multiple cell types in *Bacillus subtilis*. *FEMS Microbiol Rev* **33:**152–163.
98. **Marlow VL, Cianfanelli FR, Porter M, Cairns LS, Dale JK, Stanley-Wall NR.** 2014. The prevalence and origin of exoprotease-producing cells in the *Bacillus subtilis* biofilm. *Microbiology* **160:**56–66.
99. **Kearns DB, Losick R.** 2005. Cell population heterogeneity during growth of *Bacillus subtilis*. *Genes Dev* **19:**3083–3094.
100. **Blair KM, Turner L, Winkelman JT, Berg HC, Kearns DB.** 2008. A molecular clutch disables flagella in the *Bacillus subtilis* biofilm. *Science* **320:**1636–1638.
101. **Nakano MM, Magnuson R, Myers A, Curry J, Grossman AD, Zuber P.** 1991. *srfA* is an operon required for surfactin production, competence development, and efficient sporulation in *Bacillus subtilis*. *J Bacteriol* **173:**1770–1778.
102. **Nakano MM, Xia LA, Zuber P.** 1991. Transcription initiation of the *srfA* operon, which is controlled by the ComP-ComA signal transduction system in *Bacillus subtilis*. *J Bacteriol* **173:**5487–5493.
103. **Branda SS, González-Pastor JE, Ben-Yehuda S, Losick R, Kolter R.** 2001. Fruiting body formation by *Bacillus subtilis*. *Proc Natl Acad Sci USA* **98:**11621–11626.
104. **Lopez D, Fischbach MA, Chu F, Losick R, Kolter R.** 2009. Structurally diverse natural products that cause potassium leakage trigger multicellularity in *Bacillus subtilis*. *Proc Natl Acad Sci USA* **106:**280–285.
105. **Lopez D, Vlamakis H, Losick R, Kolter R.** 2009. Paracrine signaling in a bacterium. *Genes Dev* **23:**1631–1638.
106. **Bais HP, Fall R, Vivanco JM.** 2004. Biocontrol of *Bacillus subtilis* against infection of *Arabidopsis* roots by *Pseudomonas syringae* is facilitated by biofilm formation and surfactin production. *Plant Physiol* **134:**307–319.
107. **Gonzalez DJ, Haste NM, Hollands A, Fleming TC, Hamby M, Pogliano K, Nizet V, Dorrestein PC.** 2011. Microbial competition between *Bacillus subtilis* and *Staphylococcus aureus* monitored by imaging mass spectrometry. *Microbiology* **157:**2485–2492.
108. **Branda SS, González-Pastor JE, Dervyn E, Ehrlich SD, Losick R, Kolter R.** 2004. Genes involved in formation of structured multicellular communities by *Bacillus subtilis*. *J Bacteriol* **186:**3970–3979.
109. **Branda SS, Chu F, Kearns DB, Losick R, Kolter R.** 2006. A major protein component of the *Bacillus subtilis* biofilm matrix. *Mol Microbiol* **59:**1229–1238.
110. **Marvasi M, Visscher PT, Casillas Martinez L.** 2010. Exopolymeric substances (EPS) from *Bacillus subtilis*: polymers and genes encoding their synthesis. *FEMS Microbiol Lett* **313:**1–9.
111. **Romero D, Aguilar C, Losick R, Kolter R.** 2010. Amyloid fibers provide structural integrity to *Bacillus subtilis* biofilms. *Proc Natl Acad Sci USA* **107:**2230–2234.
112. **Romero D, Vlamakis H, Losick R, Kolter R.** 2011. An accessory protein required for anchoring and assembly of amyloid fibres in *B. subtilis* biofilms. *Mol Microbiol* **80:**1155–1168.
113. **Verhamme DT, Murray EJ, Stanley-Wall NR.** 2009. DegU and Spo0A jointly control transcription of two loci required for complex colony development by *Bacillus subtilis*. *J Bacteriol* **191:**100–108.
114. **Kovács ÁT, van Gestel J, Kuipers OP.** 2012. The protective layer of biofilm: a repellent function for a new class of amphiphilic proteins. *Mol Microbiol* **85:**8–11.
115. **Kobayashi K, Iwano M.** 2012. BslA (YuaB) forms a hydrophobic layer on the surface of *Bacillus subtilis* biofilms. *Mol Microbiol* **85:**51–66.
116. **Hobley L, Ostrowski A, Rao FV, Bromley KM, Porter M, Prescott AR, MacPhee CE, van Aalten DMF, Stanley-Wall NR.** 2013. BslA is a self-assembling bacterial hydrophobin that coats the *Bacillus subtilis* biofilm. *Proc Natl Acad Sci USA* **110:**13600–13605.
117. **Gonzalez-Pastor JE, Hobbs EC, Losick R.** 2003. Cannibalism by sporulating bacteria. *Science* **301:**510–513.
118. **Ellermeier CD, Hobbs EC, Gonzalez-Pastor JE, Losick R.** 2006. A three-protein signaling pathway governing immunity to a bacterial cannibalism toxin. *Cell* **124:**549–559.

119. **Lopez D, Vlamakis H, Losick R, Kolter R.** 2009. Cannibalism enhances biofilm development in *Bacillus subtilis. Mol Microbiol* **74:**609–618.
120. **Nandy SK, Bapat PM, Venkatesh KV.** 2007. Sporulating bacteria prefers predation to cannibalism in mixed cultures. *FEBS Lett* **581:**151–156.
121. **Msadek T.** 1999. When the going gets tough: survival strategies and environmental signaling networks in *Bacillus subtilis. Trends Microbiol* **7:**201–207.
122. **Veening JW, Igoshin OA, Eijlander RT, Nijland R, Hamoen LW, Kuipers OP.** 2008. Transient heterogeneity in extracellular protease production by *Bacillus subtilis. Mol Syst Biol* **4:**184.
123. **Eichenberger P, Fujita M, Jensen ST, Conlon EM, Rudner DZ, Wang ST, Ferguson C, Haga K, Sato T, Liu JS, Losick R.** 2004. The program of gene transcription for a single differentiating cell type during sporulation in *Bacillus subtilis. PLoS Biol* **2:**1664–1683.
124. **Piggot PJ, Hilbert DW.** 2004. Sporulation of *Bacillus subtilis. Curr Opin Microbiol* **7:**579–586.
125. **Dworkin J, Losick R.** 2005. Developmental commitment in a bacterium. *Cell* **121:**401–409.
126. **Chai YR, Chu F, Kolter R, Losick R.** 2008. Bistability and biofilm formation in *Bacillus subtilis. Mol Microbiol* **67:**254–263.
127. **Murray EJ, Kiley TB, Stanley-Wall NR.** 2009. A pivotal role for the response regulator DegU in controlling multicellular behaviour. *Microbiology* **155:**1–8.
128. **Vlamakis H, Chai Y, Beauregard P, Losick R, Kolter R.** 2013. Sticking together: building a biofilm the *Bacillus subtilis* way. *Nat Rev Microbiol* **11:**157–168.
129. **Hamon MA, Lazazzera BA.** 2001. The sporulation transcription factor Spo0A is required for biofilm development in *Bacillus subtilis. Mol Microbiol* **42:**1199–1209.
130. **Fujita M, Gonzalez-Pastor JE, Losick R.** 2005. High- and low-threshold genes in the Spo0A regulon of *Bacillus subtilis. J Bacteriol* **187:**1357–1368.
131. **Molle V, Fujita M, Jensen ST, Eichenberger P, González-Pastor JE, Liu JS, Losick R.** 2003. The Spo0A regulon of *Bacillus subtilis. Mol Microbiol* **50:**1683–1701.
132. **Verhamme DT, Kiley TB, Stanley-Wall NR.** 2007. DegU co-ordinates multicellular behaviour exhibited by *Bacillus subtilis. Mol Microbiol* **65:**554–568.
133. **Jiang M, Shao WL, Perego M, Hoch JA.** 2000. Multiple histidine kinases regulate entry into stationary phase and sporulation in *Bacillus subtilis. Mol Microbiol* **38:**535–542.
134. **McLoon AL, Kolodkin-Gal I, Rubinstein SM, Kolter R, Losick R.** 2011. Spatial regulation of histidine kinases governing biofilm formation in *Bacillus subtilis. J Bacteriol* **193:**679–685.
135. **Aguilar C, Vlamakis H, Guzman A, Losick R, Kolter R.** 2010. KinD is a checkpoint protein linking spore formation to extracellular-matrix production in *Bacillus subtilis* biofilms. *MBio* **1:**1–7. doi:10.1128/mBio.00035-10.
136. **Lopez D, Kolter R.** 2010. Extracellular signals that define distinct and coexisting cell fates in *Bacillus subtilis. FEMS Microbiol Rev* **34:**134–149.
137. **Shank EA, Klepac-Ceraj V, Collado-Torres L, Powers GE, Losick R, Kolter R.** 2011. Interspecies interactions that result in *Bacillus subtilis* forming biofilms are mediated mainly by members of its own genus. *Proc Natl Acad Sci USA* **108:**E1236–E1243.
138. **Kolodkin-Gal I, Elsholz AKW, Muth C, Girguis PR, Kolter R, Losick R.** 2013. Respiration control of multicellularity in *Bacillus subtilis* by a complex of the cytochrome chain with a membrane-embedded histidine kinase. *Genes Dev* **27:**887–899.
139. **Bai U, Mandic-Mulec I, Smith I.** 1993. SinI modulates the activity of SinR, a developmental switch protein of *Bacillus subtilis*, by protein-protein interaction. *Genes Dev* **7:**139–148.
140. **Lewis RJ, Brannigana JA, Smith I, Wilkinson AJ.** 1996. Crystallisation of the *Bacillus subtilis* sporulation inhibitor SinR, complexed with its antagonist, Sinl. *FEBS Lett* **378:**98–100.
141. **Kearns DB, Chu F, Branda SS, Kolter R, Losick R.** 2004. A master regulator for biofilm formation by *Bacillus subtilis. Mol Microbiol* **55:**739–749.
142. **Winkelman JT, Bree AC, Bate AR, Eichenberger P, Gourse RL, Kearns DB.** 2013. RemA is a DNA-binding protein that activates biofilm matrix gene expression in *Bacillus subtilis. Mol Microbiol* **88:**984–997.
143. **Hamon MA, Stanley NR, Britton RA, Grossman AD, Lazazzera BA.** 2004. Identification of AbrB-regulated genes involved in biofilm formation by *Bacillus subtilis. Mol Microbiol* **52:**847–860.
144. **Veening JW, Kuipers OP, Brul S, Hellingwerf KJ, Kort R.** 2006. Effects of phosphorelay perturbations on architecture, sporulation, and spore resistance in biofilms of *Bacillus subtilis. J Bacteriol* **188:**3099–3109.
145. **Chu F, Kearns DB, McLoon A, Chai Y, Kolter R, Losick R.** 2008. A novel regulatory protein

governing biofilm formation in *Bacillus subtilis. Mol Microbiol* **68:**1117–1127.

146. **Kobayashi K.** 2008. SlrR/SlrA controls the initiation of biofilm formation in *Bacillus subtilis. Mol Microbiol* **69:**1399–1410.
147. **Chai Y, Kolter R, Losick R.** 2009. Paralogous antirepressors acting on the master regulator for biofilm formation in *Bacillus subtilis. Mol Microbiol* **74:**876–887.
148. **Norman TM, Lord ND, Paulsson J, Losick R.** 2013. Memory and modularity in cell-fate decision making. *Nature* **503:**481–486.
149. **Chai Y, Norman T, Kolter R, Losick R.** 2011. Evidence that metabolism and chromosome copy number control mutually exclusive cell fates in *Bacillus subtilis. EMBO J* **30:**1402–1413.
150. **Davidson CJ, Surette MG.** 2008. Individuality in bacteria. *Annu Rev Genet* **42:**253–268.
151. **Eldar A, Elowitz MB.** 2010. Functional roles for noise in genetic circuits. *Nature* **467:**167–173.
152. **Losick R, Desplan C.** 2008. Stochasticity and cell fate. *Science* **320:**65–68.
153. **Veening JW, Stewart EJ, Berngruber TW, Taddei F, Kuipers OP, Hamoen LW.** 2008. Bet-hedging and epigenetic inheritance in bacterial cell development. *Proc Natl Acad Sci USA* **105:**4393–4398.
154. **Süel GM, Garcia-Ojalvo J, Liberman LM, Elowitz MB.** 2006. An excitable gene regulatory circuit induces transient cellular differentiation. *Nature* **440:**545–550.
155. **Maamar H, Raj A, Dubnau D.** 2007. Noise in gene expression determines cell fate in *Bacillus subtilis. Science* **317:**526–529.
156. **Süel GM, Kulkarni RP, Dworkin J, Garcia-Ojalvo J, Elowitz MB.** 2007. Tunability and noise dependence in differentiation dynamics. *Science* **315:**1716–1719.
157. **Çağatay T, Turcotte M, Elowitz MB, Garcia-Ojalvo J, Süel GM.** 2009. Architecture-dependent noise discriminates functionally analogous differentiation circuits. *Cell* **139:** 512–522.
158. **Chai Y, Norman T, Kolter R, Losick R.** 2010. An epigenetic switch governing daughter cell separation in *Bacillus subtilis. Genes Dev* **24:** 754–765.
159. **Chai Y, Kolter R, Losick R.** 2010. Reversal of an epigenetic switch governing cell chaining in *Bacillus subtilis* by protein instability. *Mol Microbiol* **78:**218–229.
160. **Seger J.** 1987. What is bet-hedging? *Oxf Surv Evol Biol* **4:**182–211.
161. **Kussell E, Leibler S.** 2005. Phenotypic diversity, population growth, and information in fluctuating environments. *Science* **309:**2075–2078.
162. **De Jong IG, Haccou P, Kuipers OP.** 2011. Bet hedging or not? A guide to proper classification of microbial survival strategies. *Bioessays* **33:**215–223.
163. **Balaban NQ, Merrin J, Chait R, Kowalik L, Leibler S.** 2004. Bacterial persistence as a phenotypic switch. *Science* **305:**1622–1625.
164. **Thattai M, van Oudenaarden A.** 2004. Stochastic gene expression in fluctuating environments. *Genetics* **167:**523–530.
165. **Donaldson-Matasci MC, Lachmann M, Bergstrom CT.** 2008. Phenotypic diversity as an adaptation to environmental uncertainty. *Evol Ecol Res* **10:**493–515.
166. **Frank SA.** 2011. Natural selection. I. Variable environments and uncertain returns on investment. *J Evol Biol* **24:**2299–2309.
167. **Starrfelt J, Kokko H.** 2012. Bet-hedging: a triple trade-off between means, variances and correlations. *Biol Rev* **87:**742–755.
168. **Maamar H, Dubnau D.** 2005. Bistability in the *Bacillus subtilis* K-state (competence) system requires a positive feedback loop. *Mol Microbiol* **56:**615–624.
169. **Vlamakis H, Aguilar C, Losick R, Kolter R.** 2008. Control of cell fate by the formation of an architecturally complex bacterial community. *Genes Dev* **22:**945–953.
170. **Bonner JT.** 2001. *First Signals: the Evolution of Multicellular Development.* Princeton University Press, Princeton, NJ.
171. **Ostrowski A, Mehert A, Prescott A, Kiley TB, Stanley-Wall NR.** 2011. YuaB functions synergistically with the exopolysaccharide and TasA amyloid fibers to allow biofilm formation by *Bacillus subtilis. J Bacteriol* **193:**4821–4831.
172. **Beauregard PB, Chai Y, Vlamakis H, Losick R, Kolter R.** 2013. *Bacillus subtilis* biofilm induction by plant polysaccharides. *Proc Natl Acad Sci USA* **110:**E1621–E1630.
173. **Asally M, Kittisopikul M, Rué P, Du Y, Hu Z, Cagatay T, Robinson AB, Lu H, Garcia-Ojalvo J, Süel GM.** 2012. Localized cell death focuses mechanical forces during 3D patterning in a biofilm. *Proc Natl Acad Sci USA* **109:**18891–18896.
174. **Webb JS, Thompson LS, James S, Charlton T, Tolker-Nielsen T, Koch B, Givskov M, Kjelleberg S.** 2003. Cell death in *Pseudomonas aeruginosa* biofilm development. *J Bacteriol* **185:**4585–4592.
175. **Webb JS, Givskov M, Kjelleberg S.** 2003. Bacterial biofilms: prokaryotic adventures in multicellularity. *Curr Opin Microbiol* **6:**578–585.
176. **Mikkelsen H, Sivaneson M, Filloux A.** 2011. Key two-component regulatory systems that

control biofilm formation in *Pseudomonas aeruginosa*. *Environ Microbiol* **13:**1666–1681.

177. **Aguilar C, Vlamakis H, Losick R, Kolter R.** 2007. Thinking about *Bacillus subtilis* as a multicellular organism. *Curr Opin Microbiol* **10:**638–643.
178. **Klausen M, Aaes-Jørgensen A, Molin S, Tolker-Nielsen T.** 2003. Involvement of bacterial migration in the development of complex multicellular structures in *Pseudomonas aeruginosa* biofilms. *Mol Microbiol* **50:**61–68.
179. **Klausen M, Heydorn A, Ragas P, Lambertsen L, Aaes-Jørgensen A, Molin S, Tolker-Nielsen T.** 2003. Biofilm formation by *Pseudomonas aeruginosa* wild type, flagella and type IV pili mutants. *Mol Microbiol* **48:**1511–1524.
180. **Boles BR, Thoendel M, Singh PK.** 2005. Rhamnolipids mediate detachment of *Pseudomonas aeruginosa* from biofilms. *Mol Microbiol* **57:**1210–1223.
181. **Pamp SJ, Tolker-Nielsen T.** 2007. Multiple roles of biosurfactants in structural biofilm development by *Pseudomonas aeruginosa*. *J Bacteriol* **189:**2531–2539.
182. **Barken KB, Pamp SJ, Yang L, Gjermansen M, Bertrand JJ, Klausen M, Givskov M, Whitchurch CB, Engel JN, Tolker-Nielsen T.** 2008. Roles of type IV pili, flagellum-mediated motility and extracellular DNA in the formation of mature multicellular structures in *Pseudomonas aeruginosa* biofilms. *Environ Microbiol* **10:**2331–2343.
183. **Yang L, Nilsson M, Gjermansen M, Givskov M, Tolker-Nielsen T.** 2009. Pyoverdine and PQS mediated subpopulation interactions involved in *Pseudomonas aeruginosa* biofilm formation. *Mol Microbiol* **74:**1380–1392.
184. **Harmsen M, Yang L, Pamp SJ, Tolker-Nielsen T.** 2010. An update on *Pseudomonas aeruginosa* biofilm formation, tolerance, and dispersal. *FEMS Immunol Med Microbiol* **59:** 253–268.
185. **Haagensen JAJ, Klausen M, Ernst RK, Miller SI, Folkesson A, Tolker-Nielsen T, Molin S.** 2007. Differentiation and distribution of colistin- and sodium dodecyl sulfate-tolerant cells in *Pseudomonas aeruginosa* biofilms. *J Bacteriol* **189:**28–37.
186. **Beatson SA, Whitchurch CB, Sargent JL, Levesque RC, Mattick JS.** 2002. Differential regulation of twitching motility and elastase production by Vfr in *Pseudomonas aeruginosa*. *J Bacteriol* **184:**3605–3613.
187. **Tolker-Nielsen T, Brinch UC, Ragas PC, Andersen JB, Jacobsen CS, Molin S.** 2000. Development and dynamics of *Pseudomonas* sp. biofilms. *J Bacteriol* **182:**6482–6489.
188. **Pamp SJ, Gjermansen M, Johansen HK, Tolker-Nielsen T.** 2008. Tolerance to the antimicrobial peptide colistin in *Pseudomonas aeruginosa* biofilms is linked to metabolically active cells, and depends on the *pmr* and *mexAB-oprM* genes. *Mol Microbiol* **68:**223–240.
189. **Kievit TRD, Gillis R, Marx S, Brown C, Iglewski BH.** 2001. Quorum-sensing genes in *Pseudomonas aeruginosa* biofilms: their role and expression patterns. *Appl Environ Microbiol* **67:**1865–1873.
190. **Lequette Y, Greenberg EP.** 2005. Timing and localization of rhamnolipid synthesis gene expression in *Pseudomonas aeruginosa* biofilms. *J Bacteriol* **187:**37–44.
191. **Kaneko Y, Thoendel M, Olakanmi O, Britigan BE, Singh PK.** 2007. The transition metal gallium disrupts *Pseudomonas aeruginosa* iron metabolism and has antimicrobial and antibiofilm activity. *J Clin Invest* **117:**877–888.
192. **Yang L, Barken KB, Skindersoe ME, Christensen AB, Givskov M, Tolker-Nielsen T.** 2007. Effects of iron on DNA release and biofilm development by *Pseudomonas aeruginosa*. *Microbiology* **153:**1318–1328.
193. **Davies DG, Parsek MR, Pearson JP, Iglewski BH, Costerton JW, Greenberg EP.** 1998. The involvement of cell-to-cell signals in the development of a bacterial biofilm. *Science* **280:**295–298.
194. **Sauer K, Camper AK, Ehrlich GD, Costerton JW, Davies DG.** 2002. *Pseudomonas aeruginosa* displays multiple phenotypes during development as a biofilm. *J Bacteriol* **184:**1140–1154.
195. **Heydorn A, Ersbøll B, Kato J, Hentzer M, Parsek MR, Tolker-Nielsen T, Givskov M, Molin S.** 2002. Statistical analysis of *Pseudomonas aeruginosa* biofilm development: impact of mutations in genes involved in twitching motility, cell-to-cell signaling, and stationary-phase sigma factor expression. *Appl Environ Microbiol* **68:**2008–2017.
196. **Allesen-Holm M, Barken KB, Yang L, Klausen M, Webb JS, Kjelleberg S, Molin S, Givskov M, Tolker-Nielsen T.** 2006. A characterization of DNA release in *Pseudomonas aeruginosa* cultures and biofilms. *Mol Microbiol* **59:**1114–1128.
197. **Shrout JD, Chopp DL, Just CL, Hentzer M, Givskov M, Parsek MR.** 2006. The impact of quorum sensing and swarming motility on *Pseudomonas aeruginosa* biofilm formation is nutritionally conditional. *Mol Microbiol* **62:** 1264–1277.
198. **Ochsner UA, Reiser J.** 1995. Autoinducer-mediated regulation of rhamnolipid biosurfac-

tant synthesis in *Pseudomonas aeruginosa*. *Proc Natl Acad Sci USA* **92:**6424–6428.

199. **Davey ME, Caiazza NC, O'Toole GA.** 2003. Rhamnolipid surfactant production affects biofilm architecture in *Pseudomonas aeruginosa* PAO1. *J Bacteriol* **185:**1027–1036.
200. **Purevdorj-Gage B, Costerton WJ, Stoodley P.** 2005. Phenotypic differentiation and seeding dispersal in non-mucoid and mucoid *Pseudomonas aeruginosa* biofilms. *Microbiology* **151:**1569–1576.
201. **Whitchurch CB, Tolker-Nielsen T, Ragas PC, Mattick JS.** 2002. Extracellular DNA required for bacterial biofilm formation. *Science* **295:** 1487.
202. **Matsukawa M, Greenberg EP.** 2004. Putative exopolysaccharide synthesis genes influence *Pseudomonas aeruginosa* biofilm development. *J Bacteriol* **186:**4449–4456.
203. **D'Argenio DA, Calfee MW, Rainey PB, Pesci EC.** 2002. Autolysis and autoaggregation in *Pseudomonas aeruginosa* colony morphology mutants. *J Bacteriol* **184:**6481–6489.
204. **Bjarnsholt T, Jensen PØ, Burmølle M, Hentzer M, Haagensen JAJ, Hougen HP, Calum H, Madsen KG, Moser C, Molin S, Høiby N, Givskov M.** 2005. *Pseudomonas aeruginosa* tolerance to tobramycin, hydrogen peroxide and polymorphonuclear leukocytes is quorum-sensing dependent. *Microbiology* **151:**373–383.
205. **Banin E, Brady KM, Greenberg EP.** 2006. Chelator-induced dispersal and killing of *Pseudomonas aeruginosa* cells in a biofilm. *Appl Environ Microbiol* **72:**2064–2069.
206. **Banin E, Vasil ML, Greenberg EP.** 2005. Iron and *Pseudomonas aeruginosa* biofilm formation. *Proc Natl Acad Sci USA* **102:**11076–11081.
207. **Klausen M, Gjermansen M, Kreft JU, Tolker-Nielsen T.** 2006. Dynamics of development and dispersal in sessile microbial communities: examples from *Pseudomonas aeruginosa* and *Pseudomonas putida* model biofilms. *FEMS Microbiol Lett* **261:**1–11.
208. **McDougald D, Rice SA, Barraud N, Steinberg PD, Kjelleberg S.** 2012. Should we stay or should we go: mechanisms and ecological consequences for biofilm dispersal. *Nat Rev Microbiol* **10:**39–50.
209. **Morgan R, Kohn S, Hwang S-H, Hassett DJ, Sauer K.** 2006. BdlA, a chemotaxis regulator essential for biofilm dispersion in *Pseudomonas aeruginosa*. *J Bacteriol* **188:**7335–7343.
210. **Petrova OE, Sauer K.** 2012. Dispersion by *Pseudomonas aeruginosa* requires an unusual posttranslational modification of BdlA. *Proc Natl Acad Sci USA* **109:**16690–16695.
211. **Sauer K, Cullen MC, Rickard AH, Zeef LA, Davies DG, Gilbert P.** 2004. Characterization of nutrient-induced dispersion in *Pseudomonas aeruginosa* PAO1 biofilm. *J Bacteriol* **186:** 7312–7326.
212. **Wagner VE, Bushnell D, Passador L, Brooks AI, Iglewski BH.** 2003. Microarray analysis of *Pseudomonas aeruginosa* quorum-sensing regulons: effects of growth phase and environment. *J Bacteriol* **185:**2080–2095.
213. **Barraud N, Hassett DJ, Hwang S-H, Rice SA, Kjelleberg S, Webb JS.** 2006. Involvement of nitric oxide in biofilm dispersal of *Pseudomonas aeruginosa*. *J Bacteriol* **188:**7344–7353.
214. **An S, Wu J, Zhang LH.** 2010. Modulation of *Pseudomonas aeruginosa* biofilm dispersal by a cyclic-di-GMP phosphodiesterase with a putative hypoxia-sensing domain. *Appl Environ Microbiol* **76:**8160–8173.
215. **Ma L, Conover M, Lu H, Parsek MR, Bayles K, Wozniak DJ.** 2009. Assembly and development of the *Pseudomonas aeruginosa* biofilm matrix. *PLoS Pathog* **5:**e1000354. doi:10.1371/journal.ppat.1000354.
216. **Whiteley M, Bangera MG, Bumgarner RE, Parsek MR, Teitzel GM, Lory S, Greenberg EP.** 2001. Gene expression in *Pseudomonas aeruginosa* biofilms. *Nature* **413:**860–864.
217. **Webb JS, Lau M, Kjelleberg S.** 2004. Bacteriophage and phenotypic variation in *Pseudomonas aeruginosa* biofilm development. *J Bacteriol* **186:**8066–8073.
218. **Rice SA, Tan CH, Mikkelsen PJ, Kung V, Woo J, Tay M, Hauser A, McDougald D, Webb JS, Kjelleberg S.** 2008. The biofilm life cycle and virulence of *Pseudomonas aeruginosa* are dependent on a filamentous prophage. *ISME J* **3:**271–282.
219. **Kirov SM, Webb JS, O'May CY, Reid DW, Woo JKK, Rice SA, Kjelleberg S.** 2007. Biofilm differentiation and dispersal in mucoid *Pseudomonas aeruginosa* isolates from patients with cystic fibrosis. *Microbiology* **153:**3264–3274.
220. **Mai-Prochnow A, Evans F, Dalisay-Saludes D, Stelzer S, Egan S, James S, Webb JS, Kjelleberg S.** 2004. Biofilm development and cell death in the marine bacterium *Pseudoalteromonas tunicata*. *Appl Environ Microbiol* **70:**3232–3238.
221. **Mai-Prochnow A, Webb JS, Ferrari BC, Kjelleberg S.** 2006. Ecological advantages of autolysis during the development and dispersal of *Pseudoalteromonas tunicata* biofilms. *Appl Environ Microbiol* **72:**5414–5420.
222. **Mai-Prochnow A, Lucas-Elio P, Egan S, Thomas T, Webb JS, Sanchez-Amat A,**

Kjelleberg S. 2008. Hydrogen peroxide linked to lysine oxidase activity facilitates biofilm differentiation and dispersal in several Gram-negative bacteria. *J Bacteriol* **190:**5493–5501.

223. **James SG, Holmström C, Kjelleberg S.** 1996. Purification and characterization of a novel antibacterial protein from the marine bacterium D2. *Appl Environ Microbiol* **62:**2783–2788.
224. **Rao D, Webb JS, Kjelleberg S.** 2005. Competitive interactions in mixed-species biofilms containing the marine bacterium *Pseudoalteromonas tunicata*. *Appl Environ Microbiol* **71:**1729–1736.
225. **Rao D, Skovhus T, Tujula N, Holmström C, Dahllöf I, Webb JS, Kjelleberg S.** 2010. Ability of *Pseudoalteromonas tunicata* to colonize natural biofilms and its effect on microbial community structure. *FEMS Microbiol Ecol* **73:**450–457.
226. **Nedelcu AM, Driscoll WW, Durand PM, Herron MD, Rashidi A.** 2011. On the paradigm of altruistic suicide in the unicellular world. *Evolution* **65:**3–20.
227. **Gould SJ, Lewontin RC.** 1979. The spandrels of San Marco and the Panglossian paradigm: a critique of the adaptationist programme. *Proc R Soc Lond B Biol Sci* **205:**581–598.
228. **Costerton JW, Lewandowski Z, Caldwell DE, Korber DR, Lappin-Scott HM.** 1995. Microbial biofilms. *Annu Rev Microbiol* **49:**711–745.
229. **O'Toole GA, Kolter R.** 1998. Flagellar and twitching motility are necessary for *Pseudomonas aeruginosa* biofilm development. *Mol Microbiol* **30:**295–304.
230. **Nadell CD, Xavier JB, Levin SA, Foster KR.** 2008. The evolution of quorum sensing in bacterial biofilms. *PLoS Biol* **6:**e14. doi:10.1371/journal.pbio.0060014.
231. **Nadell CD, Bucci V, Drescher K, Levin SA, Bassler BL, Xavier JB.** 2013. Cutting through the complexity of cell collectives. *Proc R Soc B Biol Sci* **280:**1–11.
232. **West SA, Diggle SP, Buckling A, Gardner A, Griffin AS.** 2007. The social lives of microbes. *Annu Rev Ecol Evol Syst* **38:**53–77.
233. **Diggle SP, Griffin AS, Campbell GS, West SA.** 2007. Cooperation and conflict in quorum-sensing bacterial populations. *Nature* **450:**411–414.
234. **Nadell CD, Bassler BL.** 2011. A fitness trade-off between local competition and dispersal in *Vibrio cholerae* biofilms. *Proc Natl Acad Sci USA* **108:**14181–14185.
235. **Rainey PB, Rainey K.** 2003. Evolution of cooperation and conflict in experimental bacterial populations. *Nature* **425:**72–74.
236. **Xavier JB, Kim W, Foster KR.** 2011. A molecular mechanism that stabilizes cooperative secretions in *Pseudomonas aeruginosa*. *Mol Microbiol* **79:**166–179.
237. **Buckling A, Harrison F, Vos M, Brockhurst MA, Gardner A, West SA, Griffin A.** 2007. Siderophore-mediated cooperation and virulence in *Pseudomonas aeruginosa*. *FEMS Microbiol Ecol* **62:**135–141.
238. **Kümmerli R, Griffin AS, West SA, Buckling A, Harrison F.** 2009. Viscous medium promotes cooperation in the pathogenic bacterium *Pseudomonas aeruginosa*. *Proc Biol Sci* **276:**3531–3538.
239. **Julou T, Mora T, Guillon L, Croquette V, Schalk IJ, Bensimon D, Desprat N.** 2013. Cell–cell contacts confine public goods diffusion inside *Pseudomonas aeruginosa* clonal microcolonies. *Proc Natl Acad Sci USA* **110:**12577–12582.
240. **Davey ME, O'Toole GA.** 2000. Microbial biofilms: from ecology to molecular genetics. *Microbiol Mol Biol Rev* **64:**847–867.
241. **Stewart PS.** 2002. Mechanisms of antibiotic resistance in bacterial biofilms. *Int J Med Microbiol* **292:**107–113.
242. **Boutte CC, Crosson S.** 2013. Bacterial lifestyle shapes stringent response activation. *Trends Microbiol* **21:**174–180.
243. **An D, Danhorn T, Fuqua C, Parsek MR.** 2006. Quorum sensing and motility mediate interactions between *Pseudomonas aeruginosa* and *Agrobacterium tumefaciens* in biofilm cocultures. *Proc Natl Acad Sci USA* **103:**3828–3833.
244. **Henrichsen J.** 1972. Bacterial surface translocation: a survey and a classification. *Bacteriol Rev* **36:**478–503.
245. **Köhler T, Curty LK, Barja F, van Delden C, Pechère JC.** 2000. Swarming of *Pseudomonas aeruginosa* is dependent on cell-to-cell signaling and requires flagella and pili. *J Bacteriol* **182:**5990–5996.
246. **Déziel E, Lépine F, Milot S, Villemur R.** 2003. *rhlA* is required for the production of a novel biosurfactant promoting swarming motility in *Pseudomonas aeruginosa*: 3-(3-hydroxyalkanoyloxy)alkanoic acids (HAAs), the precursors of rhamnolipids. *Microbiology* **149:**2005–2013.
247. **Kearns DB, Losick R.** 2003. Swarming motility in undomesticated *Bacillus subtilis*. *Mol Microbiol* **49:**581–590.
248. **Kearns DB, Chu F, Rudner R, Losick R.** 2004. Genes governing swarming in *Bacillus subtilis*

and evidence for a phase variation mechanism controlling surface motility. *Mol Microbiol* **52:**357–369.

249. **Kearns DB.** 2010. A field guide to bacterial swarming motility. *Nat Rev Microbiol* **8:**634–644.

250. **Debois D, Hamze K, Guérineau V, Le Caër J-P, Holland IB, Lopes P, Ouazzani J, Séror SJ, Brunelle A, Laprévote O.** 2008. *In situ* localisation and quantification of surfactins in a *Bacillus subtilis* swarming community by imaging mass spectrometry. *Proteomics* **8:**3682–3691.

251. **Hamze K, Autret S, Hinc K, Laalami S, Julkowska D, Briandet R, Renault M, Absalon C, Holland IB, Putzer H, Séror SJ.** 2011. Single-cell analysis *in situ* in a *Bacillus subtilis* swarming community identifies distinct spatially separated subpopulations differentially expressing *hag* (flagellin), including specialized swarmers. *Microbiology* **157:**2456–2469.

252. **Fuqua C, Winans SC, Greenberg EP.** 1996. Census and consensus in bacterial ecosystems: the LuxR-LuxI family of quorum-sensing transcriptional regulators. *Annu Rev Microbiol* **50:**727–751.

253. **Van Delden C, Iglewski BH.** 1998. Cell-to-cell signaling and *Pseudomonas aeruginosa* infections. *Emerg Infect Dis* **4:**551–560.

254. **Zheng G, Yan LZ, Vederas JC, Zuber P.** 1999. Genes of the *sbo-alb* locus of *Bacillus subtilis* are required for production of the antilisterial bacteriocin subtilosin. *J Bacteriol* **181:**7346–7355.

255. **Hamoen LW, Venema G, Kuipers OP.** 2003. Controlling competence in *Bacillus subtilis*: shared use of regulators. *Microbiology* **149:**9–17.

256. **Inaoka T, Takahashi K, Ohnishi-Kameyama M, Yoshida M, Ochi K.** 2003. Guanine nucleotides guanosine 5′-diphosphate 3′-diphosphate and GTP co-operatively regulate the production of an antibiotic bacilysin in *Bacillus subtilis*. *J Biol Chem* **278:**2169–2176.

257. **Yazgan A, Cetin S, Ozcengiz G.** 2003. The effects of insertional mutations in *comQ, comP, srfA, spo0H, spo0A* and *abrB* genes on bacilysin biosynthesis in *Bacillus subtilis*. *Biochim Biophys Acta* **1626:**51–56.

258. **Matz C, Bergfeld T, Rice SA, Kjelleberg S.** 2004. Microcolonies, quorum sensing and cytotoxicity determine the survival of *Pseudomonas aeruginosa* biofilms exposed to protozoan grazing. *Environ Microbiol* **6:**218–226.

259. **van Gestel J, Nowak MA, Tarnita CE.** 2012. The evolution of cell-to-cell communication in a sporulating bacterium. *PLoS Comput Biol* **8:**e1002818. doi:10.1371/journal.pcbi.1002818.

260. **Baty AM, Eastburn CC, Diwu Z, Techkarnjanaruk S, Goodman AE, Geesey GG.** 2000. Differentiation of chitinase-active and non-chitinase-active subpopulations of a marine bacterium during chitin degradation. *Appl Environ Microbiol* **66:**3566–3573.

261. **Baty AM, Eastburn CC, Techkarnjanaruk S, Goodman AE, Geesey GG.** 2000. Spatial and temporal variations in chitinolytic gene expression and bacterial biomass production during chitin degradation. *Appl Environ Microbiol* **66:**3574–3585.

262. **van Gestel J, Weissing FJ, Kuipers OP, Kovacs AT.** 2014. Density of founder cells affects spatial pattern formation and cooperation in *Bacillus subtilis* biofilms. *ISME J* **10:**2069–2079.

Candida albicans Biofilm Development and Its Genetic Control

5

JIGAR V. DESAI[1] and AARON P. MITCHELL[1]

The human commensal *Candida albicans* is the leading fungal colonizer of implanted medical devices and a frequent cause of nosocomial infections (1, 2). Several *Candida* species, including *C. albicans*, are part of the mucosal flora of most healthy individuals and reside in the gastrointestinal and genitourinary tracts. These organisms are thus poised to cause infection when a suitable niche becomes available. The use of broad-spectrum antibiotics is an additional risk factor for *Candida* infections, probably because bacterial competitors that are eliminated would otherwise keep fungal populations in check. The extreme resistance of biofilm cells to antifungal therapy is a further complication, and often the infected device has to be removed and replaced to prevent recurrent infection (1). Here, we focus mainly on biofilm formation by *C. albicans*, the most intensively studied of the *Candida* species.

BIOFILM STRUCTURE AND DEVELOPMENT

The first published image of a *Candida* biofilm on an implanted catheter came from the pioneering studies of Marrie and Costerton (3). This and many subsequent reports of *Candida* biofilms on devices prompted Hawser

[1]Department of Biological Sciences, Carnegie Mellon University, Pittsburgh, PA 15213.

Microbial Biofilms 2nd Edition
Edited by Mahmoud Ghannoum, Matthew Parsek, Marvin Whiteley, and Pranab K. Mukherjee

doi:10.1128/microbiolspec.MB-0005-2014

and Douglas to develop an *in vitro* system to study *Candida* biofilm development on catheter material discs (4). Their scanning electron micrographs provided the first glimpse of *C. albicans* biofilm architecture, which has since been studied by confocal imaging as well (see Fig. 1). *C. albicans* can grow either as individual oval cells (called yeast cells or blastospores) or as long filamentous cells attached end-to-end (called pseudohyphae or hyphae, distinguished by specifics of cell structure) (5). Biofilms grown *in vitro* under a variety of conditions have a basal substrate-bound layer of yeast cells (Fig. 1A,B) that ranges from 20 to 100 microns in depth under many conditions. Filamentous cells project from the basal layer and can extend for several hundred microns (Fig. 1A,C). Yeast cells are often found to be produced by the filamentous cells, especially in the apical regions of the biofilm (Fig. 1A,C). Amorphous extracellular matrix material is found throughout the biofilm (Fig. 1A,B,C), which can appear aggregated (shown here) or dispersed (6), depending on staining and fixation. A three-dimensional reconstruction (Fig. 1D) reveals a very dense basal region beneath loosely packed filamentous cells. The loose packing of the upper region may facilitate solvent access to the basal region.

Fungi are nonmotile, and biofilm structure thus reflects the sequence of cell division events that occur during biofilm development. Chandra et al. analyzed time courses of *C. albicans* biofilm development on two different substrates and proposed that biofilm development occurs in stages (7). They used a yeast cell inoculum because yeast cells are more likely than long filamentous cells to be

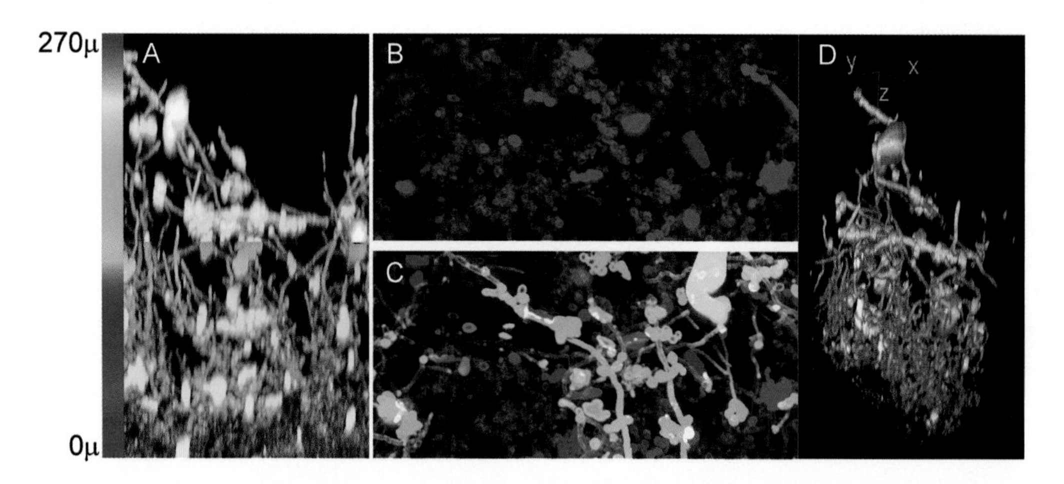

FIGURE 1 Confocal micrographic images of a *C. albicans* biofilm. These images present a biofilm grown *in vitro* in yeast extract-peptone-dextrose medium at 37°C. The sample was prepared by embedding and staining with Alexafluor 594-conjugated Concanavalin A, using a procedure modified from reference 83. (A) Side projection view. Hyphae are clearly visible in the upper portion of the biofilm, as are aggregates of brightly stained extracellular material. A color scale bar represents the 270-micron depth and indicates the pseudocolor scale used for apical projections. (B) Apical projection of basal (substrate-proximal) 50-micron region. A yeast cell layer is evident from the substrate level (red) to 50 microns above the substrate (blue). A few hyphae or pseudohyphae are visible as well. Some amorphous extracellular material is apparent. (C) Apical projection of the entire biofilm. Hyphae are visible above the basal layer, extending from ~150 microns (green) to 270 microns (red) above the substrate. Yeast cells are seen in clusters at the ends of hyphae. (D) Three-dimensional reconstruction of the biofilm sample. Hyphae at the top of the biofilm are readily visible above the dense basal region. doi:10.1128/microbiolspec.MB-0005-2014.f1

able to disseminate to new sites. In the early stage, individual yeast cells adhered to the substrate. They then proliferated as yeast to produce microcolonies, and coalescence of microcolonies yielded the basal layer of the biofilm. Biofilm development then entered an intermediate stage of high metabolic activity along with the emergence of hyphae and production of extracellular matrix material. In the final maturation stage, there was extensive accumulation of extracellular matrix material. The images did not show the presence of apical yeast cells, and they may have been obscured by intensely stained matrix. The authors also found that greatly reduced susceptibility to fluconazole, amphotericin B, nystatin, and chlorhexidine was acquired at the time of transition to the intermediate stage, concomitant with the increase in metabolic activity and accumulation of matrix material. This finding is in keeping with more recent studies that reveal that drug binding by extracellular matrix is a major source of biofilm drug resistance (see below).

The final step in biofilm formation can be considered to be the release of cells, permitting colonization of new sites and, unfortunately, disseminated infection (8). Uppuluri et al. (9) found that cell dispersion occurs throughout biofilm development and does not represent a temporally distinct stage. Cells released from biofilms were mainly yeast cells, not filaments. Remarkably, the released cells were phenotypically distinct from cells grown planktonically for the same amount of time in the same medium. The released cells displayed higher levels of adhesion to plastic or endothelial cells, probably due to their increased propensity to produce hyphae. In addition, the released cells were more virulent than planktonic cells in a disseminated infection model. Thus, biofilm dispersion yields a unique class of yeast cell with increased ability to create new biofilms and cause infection.

Do biofilms follow the same developmental steps described above during a true catheter infection? One cannot reason from first principles to reach a conclusion about how representative an *in vitro* model may be. We believe that the simplest approach to validate *in vitro* observations is to use an animal model of biofilm-based infection. There are animal models (10) for venous catheter infection (11, 12), urinary catheter infection (13), and denture stomatitis infection (14). (For review, see reference 105.) There is also a subcutaneous catheter model that cultures biofilm cells in a host environment, though it may not resemble in detail a device currently in use (10). Finally, there are animal models for both oral and vaginal mucosal infections, which are in essence biofilms that form on mucosal tissue (reviewed in reference 15). No investigation to our knowledge has validated the detailed observations regarding early, intermediate, and mature *in vitro* biofilms with these models. In addition, the detailed architecture of *in vitro* biofilms is generally not recapitulated in the *in vivo* models. Our perspective is that *in vitro* analysis allows clear documentation and characterization of biofilm alterations caused by genetic, physiological, environmental, or pharmacological perturbations. The selection of *in vivo* models then allows one to validate the key findings from *in vitro* studies, even if some of the details in each model may be different.

CELL MORPHOLOGY AND BIOFILM FORMATION

Under most conditions, both yeast and filamentous cells are required for *C. albicans* biofilm formation. Initial support for this conclusion came from mutants that were locked in either yeast or filamentous growth states (16), though the genetic basis for the mutant phenotypes was uncertain. Each mutant produced an altered biofilm with reduced biomass or cell density. A random insertion mutant screen further substantiated a role of hyphal morphogenesis in biofilm development (17). Mutants with insertions in the genes *NUP85*, *MDS3*, *SUV3*, and *KEM1* were identified as biofilm-defective, and there was no known molecular or functional connection among them. However, they were all defective in hyphal formation in several media. In addition, in mixed biofilms formed with wild-type cells, each mutant produced only yeast cells. Therefore, the mutations caused defects in filamentation in the context of a biofilm, arguably the most relevant situation to assay. Ramage et al. found that two well-established hyphal-defective mutant strains, *efg1Δ/Δ* and *efg1Δ/Δ cph1Δ/Δ*, were defective in forming biofilms (18). These mutants yielded only sparse substrate-attached cells, not a true basal layer. Remarkably, though, the substrate-attached mutants displayed no susceptibility to fluconazole and only moderate susceptibility to amphotericin B. These findings indicated that surface-bound growth is sufficient to induce the antifungal resistance of biofilm cells and were consistent with the finding from Chandra et al. (7) that resistance increases substantially before a biofilm fully matures.

Why are filamentous growth forms so important for biofilm formation? Insight into the answer came from the transcription factor Bcr1 (19), identified in the first systematic screen of *C. albicans* transcription factor mutants. The *bcr1Δ/Δ* mutant was defective in biofilm formation and also failed to form hyphae under some conditions. Importantly, though, in mixed biofilms formed with wild-type cells, mutant cells yielded abundant hyphae. Also, the nonadherent cells produced by the mutant under biofilm-inducing conditions included hyphae. These results suggested that the mutant produces hyphae that are defective in a function required for biofilm formation. Transcript profiling and functional analysis pointed to the same conclusion: Bcr1 is required for expression of genes for cell surface adherence proteins (called adhesins), such as *ALS1*, *ALS3*, and *HWP1*. Many of these genes, including *ALS3* and *HWP1*, are induced strongly during hyphal growth. It is important to note that overexpression of adhesin genes *ALS1*, *ALS3*, or *HWP1* in a *bcr1Δ/Δ* mutant background restored biofilm formation ability, both *in vitro* and in a catheter infection model (20). This study was the first to provide evidence that hyphae are required for biofilm formation because of their cell surface adhesins.

BIOFILM-ASSOCIATED GENE EXPRESSION

If biofilm cells have unique phenotypic properties, one might expect that biofilm cells express a set of genes that are different from planktonic cells. Several studies have characterized the biofilm transcriptome (21–25). Although many different growth conditions and comparison conditions were utilized, there is good overall agreement, especially among many of the most highly induced genes in biofilm formation (22). Most importantly, these transcriptome studies have provided leads for functional analysis. For example, in the first such study, Garcia-Sanchez et al. found that amino acid biosynthetic genes were consistently upregulated in biofilms grown under diverse conditions (21). That observation led them to assay biofilm formation by a *gcn4Δ/Δ* mutant, which is defective in the general control of amino acid biosynthetic genes. The *gcn4Δ/Δ* mutant produced a biofilm, but its overall biomass and metabolic activity was substantially reduced compared to the wild type. These assays were conducted in a rich medium in which planktonic growth of the wild-type and mutant strains were equivalent. Hence the mutant may be defective in retention of cells within the biofilm. Such a mutant phenotype would be difficult to detect in a large *in vitro* screen; the profiling data clearly pointed in a unique direction for functional analysis. In addition, these findings fit well with the observation made repeatedly that ribosome biogenesis genes are upregulated in biofilm cells compared to planktonic cells. A simple hypothesis is that both amino acid synthesis genes and ribosomal biogenesis genes allow increased protein synthesis in biofilm cells, or perhaps a subset of biofilm cells, that contributes to biofilm stability and cohesion. Given the *gcn4Δ/Δ* mutant phenotype, those protein products may be adhesins or extracellular matrix components that mediate cell-cell adherence.

Broader surveys of mutants defective in biofilm-induced genes have not always yielded many genes that clearly function in biofilm formation, based on mutant phenotypes (26). One reason for the limited correlation may be functional redundancy of biofilm-associated genes, for which examples are well known (20, 27, 28). A second reason may be the limited spectrum of biofilm properties that have often been assayed. Desai et al. used a panel of assays to explore defects in mutants with insertions in biofilm-induced genes, including adherence, azole tolerance, overall biofilm integrity, and quorum-sensing responses (22). They found that the majority of mutants had a significant phenotypic alteration in at least one assay, though many were not obviously deficient in overall biofilm formation ability. Because many processes contribute to the overall structure of a biofilm, it seems reasonable that functional understanding of many biofilm-induced genes may require assays of several phenotypic parameters.

Because a biofilm is a complex and heterogeneous environment, one might expect that some biofilm-induced genes may be part of response pathways that have little impact on biofilm phenotypes *per se*. Thus, many investigations have sought to prioritize biofilm-induced genes for functional analysis. Perhaps the most elegant prioritization approach was undertaken by Nobile et al. (29), who extended the transcription factor mutant screen (19) to identify six biofilm regulators. They combined genome-wide expression profiling of the transcription factor mutants with chromatin immunoprecipitation assays to define the transcription factors' direct targets. There were over ~1,000 target genes in the overall biofilm network, but only 23 genes were bound by all six regulators. These shared targets may be highly enriched for biofilm-related functions. A second prioritization approach is to focus on genes that are biofilm-induced under diverse conditions, as the d'Enfert group did with a panel of growth conditions (21, 26) or as Desai et al. (22) did by employing two different *C. albicans* isolates to define common biofilm-induced genes.

THE CELL SURFACE AND ADHERENCE

The cell wall is the cellular structure that interacts most directly with the substratum or another cell. The *C. albicans* cell wall is primarily made of carbohydrates and glycoproteins (30). Carbohydrates such as β-glucan and chitin form an inner core of cell wall, responsible for its mechanical strength, and mannoproteins that include adhesins form an outer fibrillar layer (30). Adhesins are defined by their ability to mediate adherence directly or by their structural similarity to proteins that do so (31). Other cell wall or cell surface proteins may affect adhesin levels, processing, or exposure at the cell surface and thus affect adherence indirectly.

Many adhesins of *C. albicans* have a C-terminal sequence that is used for covalent attachment of a glycophosphatidylinositol (GPI) anchor (32). This GPI anchor initially tethers the protein on the outer face of the plasma membrane. The GPI anchor is then cleaved; the protein and anchor remnant are transferred to β-1,6-glucan and remain attached to the cell wall (32). Adhesins of this class include members of the Als (agglutinin like sequence) family (33), Eap1 (enhanced adherence to polystyrene 1) (34), Hwp1 (hyphal wall protein 1) (35, 36), and Rbt1 (repressed by TUP1) (37), all of which are expressed at much higher levels in hyphal cells than in yeast cells. There is an adhesin-like protein expressed at the highest levels in yeast cells, Ywp1 (yeast wall protein 1), but it seems to function as an antiadhesin (38). There are also proteins that may function as adhesins but lack a GPI anchor, including Mp65 (mannoprotein of 65kDa) (39), Csh1 (cell surface hydrophobicity) (40), and Pra1 (pH regulated antigen) (32).

Early approaches to identify adhesins involved analysis of cell wall components that adhered to a surface after the adherent cells were washed away (41, 42). However, the first studies to define *C. albicans* adhesins functionally relied upon heterologous expression in *Saccharomyces cerevisiae* (43, 44). Als1 was identified in a screen of a *C. albicans* expression library in *S. cerevisiae* for clones that improved *S. cerevisiae* adherence to epithelial and endothelial cells (44). Als5 was identified through a similar approach: its expression in *S. cerevisiae* improved adherence to beads coated with fibronectin, laminin, and collagen (43). Adhesins from this *ALS* gene family have since been studied in detail (31, 33). They are organized into four major regions: (i) an N-terminal immunoglobulin-like domain, (ii) a threonine-rich region, (iii) a series of 36 amino acid tandem repeats, and (iv) a highly glycosylated stalk region (31, 45). (All Als proteins have N-terminal signal sequences as well, allowing their entry into the secretion pathway.) Initial adherence has been proposed to be mediated by the N-terminal module, which is capable of ligand binding (46–48). These ligands include a broad range of denatured peptides, reflecting the broad specificity of Als proteins (48). The threonine-rich region and the tandem repeat region are required for cell-cell adherence, as demonstrated through heterologous expression of domain deletion mutants in *S. cerevisiae* (49). The eight Als proteins seem to have redundant functions in biofilm formation for the most part, because high-level expression of any *ALS* gene in a biofilm-defective *als1Δ/Δ als3Δ/Δ* mutant restores biofilm formation *in vitro* and *in vivo* in the rat venous catheter model (50). Thus, our current understanding is that the Als proteins function as a set of interchangeable adhesins to promote biofilm formation.

Recent studies have addressed a long-standing mystery about the Als proteins and other adhesins: how can proteins with such weak affinities for their ligands mediate stable binding? The answer lies in the ability of the threonine-rich region to form multiprotein aggregates, or amyloids (45). When amyloid formation is initiated (by tugging an Als in an atomic force microscope, for example), it spreads across the cell surface to create a nanodomain. The Als aggregate becomes in essence a multivalent adhesin. Thus, even weakly bound ligands are rebound rapidly after they are released (45). Such amyloid-forming regions are found in many other cell surface adhesins, so amyloid formation may be a common mechanism to stabilize ligand-binding interactions.

Several other GPI-linked cell wall proteins function as biofilm adhesins, including Eap1, Hwp1, and Rbt1. Eap1 was identified as a *C. albicans* library clone that enabled adherence to plastic by an otherwise nonadherent *S. cerevisiae* strain (34). Like the Als adhesins, Eap1 has an N-terminal ligand-binding domain followed by serine- and threonine-rich repeats that permit the N-terminal domain to project beyond the cell wall glucan (51). Eap1 is required for biofilm formation, because an *eap1Δ/Δ* mutant is defective in biofilm formation *in vitro* and *in vivo* in the rat venous catheter model (52).

Hwp1 is structurally distinct from the Als proteins and Eap1. It is in essence a set of short peptide repeats followed by a GPI anchor addition site. Its role in host cell binding is remarkable: it is a substrate for host transglutaminases, which link it covalently to epithelial cell surfaces (35). Although it may also serve as a transglutaminase substrate during biofilm formation *in vivo*, it must function differently in biofilms formed *in vitro* because *C. albicans* does not make its own transglutaminases (35). An *hwp1Δ/Δ* mutant has a moderate-to-severe biofilm defect *in vitro* and *in vivo* (36). Two observations argue that Hwp1 has a distinct and complementary role to that of the Als adhesins in biofilm formation (50). First, overexpression of *HWP1* does not allow biofilm formation by the *als1Δ/Δ als3Δ/Δ* mutant, in contrast to overexpression of any *ALS* gene. Second, a mixture of biofilm-defective *als1Δ/Δ als3Δ/Δ* cells and biofilm-defective *hwp1Δ/Δ* cells is able to form a biofilm. The mechanism seems likely to be that Hwp1 and Als1/Als3 can interact on cell surfaces to mediate cell-cell binding. This inference comes from the fact that heterologous expression of *HWP1* in *S. cerevisiae* improves its adherence in wild-type *C. albicans* cells, and not to *als1Δ/Δ als3Δ/Δ* mutant cells (50). Hwp1 and Als1/Als3 may thus function analogously to mating agglutinins of *S. cerevisiae* that permit binding of *MAT***a** and *MAT*α cells (31).

Rbt1 is in the same adhesin family as Hwp1 (37, 53). An *rbt1Δ/Δ* mutant has a mild biofilm defect *in vitro* but shows additive effects with mutations in family members *HWP1* and *HWP2* (53). Its N-terminal region promotes surface hydrophobicity and mediates adherence to polystyrene (37). A central domain is predicted to have high aggregation potential, and amyloid-inhibitor experiments similar to those carried out with Als5 support such a function (37). Although Rbt1 is normally expressed only on hyphal cells, Monniot et al. could create a constitutive *RBT1* allele through fusion to the *TEF1* promoter. This constitutively expressed protein could be recognized by antiepitope antibodies only on hyphal cell surfaces. Recognition on yeast cell surfaces required mild digestion of the cell wall with zymolyase (37). These observations suggest that there is a fundamental structural difference between yeast and hyphal cell walls that affects the exposure of Rbt1 and, potentially, many other adhesins.

One interesting GPI-anchor-containing protein, Ywp1, functions to reduce adherence (38, 54). *YWP1* is expressed at much higher levels in yeast cells than in hyphae, so it is possible that Ywp1 is critical for dispersion of yeast cells from a biofilm. It is yet not known how Ywp1 exerts its antiadhesive effects; it may interact with specific adhesins, or it may alter the cell surface to deny access to adhesins. In that context, it would be interesting to see if Ywp1 is required for the inhibition of Rbt1 epitope access on yeast cells observed by Monniot et al. (37).

How is adherence regulated? As mentioned above, many of the major known adhesins are expressed at the highest levels on hyphal cells. Their expression is regulated by transcription factors that also govern hyphal development (29, 55). In addition, the adherence of yeast cells, which is thought to be the initial step in biofilm formation, appears to be under complex control. Finkel et al. screened for transcription factor mutants with altered adherence to silicone (56) and uncovered 30 transcription factors that are required for adherence. Expression of all known and predicted cell wall protein genes was assayed in the mutants, which allowed provisional assignment of both regulators and cell wall protein genes to pathways. The value of this approach was supported by positive overexpression-rescue tests of several new pathway relationships. For example, the findings indicated that Snf5 and Ace2 lie in a pathway that governs adherence, biofilm formation, and cell wall integrity (56). In addition, the findings argued that the protein kinase Cbk1 and transcription factor Bcr1 act in the same pathway, and contemporaneous studies revealed that Cbk1 phosphorylates Bcr1 (57). A simple interpretation is that a large number of transcriptional regulatory pathways govern adherence, but they ultimately impact a small number of response mechanisms. Interestingly, several of the transcription factors were not required for biofilm formation in an *in vitro* system but were required in the rat catheter *in vivo* model (56). This finding emphasizes the limitations of *in vitro* biofilm models and the potential that our reliance on *in vitro* models may cause us to overlook critical functions that act *in vivo* during infection.

Several upstream regulators that govern adhesin expression have also been identified, thus paving the way to define the actual molecular or physiological signals that govern biofilm formation. As mentioned above, the protein kinase Cbk1 phosphorylates and activates Bcr1, perhaps ensuring that hyphal adhesins are only expressed when Cbk1-dependent cell polarity functions are active (57). In addition, the Tor1 kinase, a central regulator of ribosome biogenesis and starvation responses, is a negative regulator of adhesin genes *ALS1*, *ALS3*, and *HWP1* (58). This relationship may reflect a role for starvation in promoting adherence and biofilm formation. Recent studies have revealed that the stress-responsive MAP kinase Hog1 mediates this effect of Tor1 and that transcription factor Brg1 may be the direct target of this pathway (59). Because Hog1 is activated by high osmolarity as well as oxidative stress (60), these signals may also influence the ability to adhere and form a biofilm. Finally, we note that the cyclic AMP-dependent protein kinase catalytic subunit Tpk1 functions as a negative regulator of adherence and *ALS1* expression (61), perhaps through effects on the cyclic AMP pathway target transcription factor Efg1 (60). This pathway governs hyphal morphogenesis, so it seems possible that the response can modulate the adhesin levels on hyphae in a biofilm. Clearly, these novel pathway relationships will whet our appetites for dissection of signals and responses in biofilm formation for some time to come.

Many genes that have broad effects on cell wall biogenesis or integrity also affect adherence or biofilm formation. For example, *GAL102* and the *PMT* (protein mannosyl transferase) gene family govern protein mannosylation (62, 63). The impact of mutations on biofilm formation may result from altered adhesin glycosylation. Other cell wall proteins that govern adherence but may not be adhesins are Sun41 and Pga1, both of which have roles in cell wall integrity (64–67). However, the fact that a cell wall protein affects cell wall integrity does not rule out the possibility that it is an adhesin. The Als adhesins, in particular, are famous as multifunctional proteins. Als3 is the best example, with roles in adherence to numerous substrates, host receptor binding, host cell invasion, and iron acquisition (68). Als2 is a possible bridge between cell wall integrity and adhesin function: it seems to be essential for viability, and changes in *ALS2* gene dosage have profound effects on cell wall depth and sensitivity to cell wall perturbing agents (61, 69). Thus, a known adhesin seems to have a role in overall cell wall architecture and integrity.

Might the cell wall have a sensory function? The transcription of many genes (including adhesin genes) is induced rapidly after the initial adherence step (24). Perhaps surface binding generates a signal that switches the cell growth program from planktonic to biofilm. In fact, several groups have studied contact sensing phenomena and their regulation (70–73). The transmembrane protein Dfi1, through calmodulin binding, regulates the activity of the MAP kinase Cek1. The MAP kinase Mkc1 is also activated after cells interact with semisolid surfaces (71, 72, 74). Both Cek1 and Mkc1 have roles in biofilm formation (71, 75). Thus, while the evidence now is fragmentary, a fascinating possibility is that physical changes in the cell wall occur upon substrate binding that activate Cek1 and Mkc1 to promote biofilm formation.

EXTRACELLULAR MATRIX MATERIAL

A mature biofilm shows complex architecture with heterogeneous cell types enmeshed in extracellular matrix. Biofilm matrix was first characterized by the Douglas group (76). They found the presence of carbohydrate, protein, hexosamine, phosphorus, and uronic acid. Additionally, they observed that treatment with enzymes such as β-1,3-glucanase, proteinase K, DNase I, chitinase, and β-N-acetylglucosaminidase compromised biofilm cohesion (76). A good portion of the glucose initially detected by the Douglas group is found in soluble β-glucan (77), which Nett and colleagues have shown to be a key matrix determinant of antifungal drug resistance (see below). Thus, the *C. albicans* biofilm matrix functions in both biofilm integrity and drug resistance.

Matrix production can vary considerably with growth conditions. For example, there is less matrix production when biofilms are grown statically than with shaking (78). Also, matrix production is greater in RPMI medium than in Spider medium (79), both of which are commonly used by many investigators. A further complication is that matrix composition has not been dissected under these varied growth conditions. Given the broad functional roles of matrix components, it may be useful to develop some standardized procedures for analysis of biofilm properties.

The most well understood role of a matrix component is the function of β-1,3 glucan in biofilm azole resistance. Nett et al. manipulated the essential *FKS1* gene, which is responsible for cell wall β-1,3 glucan synthesis (80). They showed that decreased or increased *FKS1* expression or activity results in a corresponding change in the amount of biofilm matrix (soluble) β-1,3 glucan. Hence, matrix β-1,3 glucan follows the same biosynthetic pathway as cell wall β-1,3 glucan. Remarkably, the strains with reduced *FKS1* activity produced biofilms *in vitro* and *in vivo* that were exquisitely sensitive to fluconazole, while during planktonic growth there was no change in fluconazole sensitivity (80). These observations showed that β-1,3 glucan synthesis is required for a biofilm-specific drug resistance mechanism. In fact, addition of isolated biofilm matrix to planktonic cells conferred fluconazole resistance. Direct binding assays were used to show that drug sequestration is the mechanism by which β-1,3 glucan confers biofilm fluconazole resistance (80). To understand the biogenesis of matrix β-1,3 glucan, Taff et al. created null mutant strains in candidate glucan modification genes that were upregulated *in vivo* during biofilm development (28). They found three genes (two that encode glucan transferases Bgl2 and Phr1 and one that encodes exoglucanase Xog1) to affect matrix β-1,3 glucan production and fluconazole susceptibility. Because the glucan modification pathway is extracellular, it seems like an excellent target for antibiofilm therapeutics.

Proteins and DNA also constitute an integral part of the matrix material. The protein component has been characterized through a proteomic approach by Lopez-Ribot and colleagues (81). Many of the most abundant proteins found in matrix were similar to the proteins found in supernatants of planktonic cultures. In addition, a large proportion of the matrix proteins are annotated as cytoplasmic. DNA is also a functional matrix component, as indicated by the finding that DNase I treatment compromises biofilm integrity (76). Moreover, addition of DNA improves biofilm formation as indicated by increased biomass (82). It seems possible that cell lysis may be a major source of the cytoplasmic proteins and DNA in the biofilm matrix.

There have been several approaches to identify the regulators of biofilm matrix production. An unusual biofilm morphology led Nobile et al. to identify the zinc acquisition regulator Zap1 as a negative regulator of matrix β-1,3 glucan (27). Transcriptomic and ChIP assays followed by functional analysis revealed the key Zap1 targets to include two glucoamylases (Gca1 and Gca2) and three alcohol dehydrogenases (Csh1, Ifd6, and Adh5) (27). Although Gca1 and Gca2 may act directly on matrix polysaccharides, it seems likely that Csh1, Ifd6, and Adh5 act indirectly, perhaps through effects on quorum sensing molecule production (83). Recently, the Soll lab identified a role for Bcr1 in regulating the impenetrability of MTL-heterozygous biofilms to dyes and polymorphonuclear leukocytes, which are likely to be matrix-associated traits. Through a series of Bcr1 target gene overexpression assays, they found that the extracellular CFEM (common in several fungal extracellular membrane) proteins promote this matrix function (79). The CFEM proteins were shown previously to be required for biofilm formation, but their role in matrix properties was not anticipated (84). It remains to be determined whether the CFEM proteins are themselves matrix components or if they act more indirectly through effects on signaling or nutrient acquisition (85).

The studies of Zap1 and Bcr1 seem to have defined pathways that do not affect *FKS1* regulation (see reference 28, in particular). However, a candidate gene approach based on *S. cerevisiae* ortholog function identified Smi1 as a regulator that acts upstream of *FKS1* (86). Specifically, a *smi1Δ/Δ* mutant had decreased biofilm fluconazole resistance, β-glucan production, and *FKS1* RNA accumulation. Moreover, increased expression of *FKS1* caused increased fluconazole resistance in the *smi1Δ/Δ* mutant. Current evidence indicates that Smi1 acts through the transcription factor Rlm1 to govern *FKS1* expression (86). In addition, the chaperone Hsp90 is required for matrix β-glucan production (87). This role for Hsp90 is independent of its regulatory interactions with the known client proteins calcineurin and Mkc1. Hsp90 may affect *FKS1* expression or activity, perhaps through the Smi1-Rlm1 pathway.

BIOFILM METABOLISM

A central theme that has emerged from transcriptome studies is that the mature *C. albicans* biofilm presents a hypoxic environment. The first general indication of biofilm hypoxia came from the observation that glycolytic genes are upregulated in biofilms (21, 26). This response might be expected if energy from hexoses in biofilms derives from fermentative reactions, which are much less efficient than respiration. Indeed, the Butler group set out to compare gene expression during biofilm growth and during hypoxia with the species *C. parapsilosis* (88). A set of 60 genes was common to the two responses, representing mainly genes involved in glycolysis or in synthesis of fatty acids and ergosterol. In addition, a recent metabolomic comparison of biofilm and planktonic cells revealed that biofilms accumulate lower levels of succinate, fumarate, citrate, and malate (89). This outcome probably reflects diminished flux through the tricarboxylic cycle, as expected if respiration rates are lower in biofilm cells than in planktonic cells. The overall hypoxic metabolism of biofilm cells is functionally significant, based on properties of the transcription factor Tye7. This transcription factor is an activator of glycolytic genes, and its function is critical for growth when respiration is blocked (90). Bonhomme et al. found that a *tye7Δ/Δ* mutant had greatly reduced ability to form a biofilm, in keeping with the hypothesis that the biofilm environment is hypoxic. In addition, the mutant biofilm contained an excess of filamentous cells, and observations with metabolic inhibitors argued that hyperfilamentation was a result of decreased glycolytic flux and ATP synthesis (26). This study leads to two interesting conclusions. First, hypoxic or fermentative carbon metabolism is critical for biofilm formation. Second, it is generally appreciated that biofilm growth leads to abundant hyphal formation in media that induce planktonic hyphae poorly (see references 7 and 17, for example); it seems possible that hypoxia may be the signal that induces hyphal formation during biofilm growth.

The transcription factor Efg1, a central regulator of biofilm formation and hyphal formation (1, 60), may have a pivotal role in coordinating hyphal formation and hypoxic metabolism. Stichternoth and Ernst explored this connection through examination of Efg1-responsive genes under hypoxic conditions (91). Many of the same genes that were activated rapidly by Efg1 corresponded to metabolic genes activated during biofilm formation. In fact, the *TYE7* gene is a direct target of Efg1 (29). Therefore, the metabolic genes that respond to Efg1 during hypoxia may do so through their activation by Tye7.

If fermentation is necessary for biofilm physiology, one might expect biofilms to accumulate increased levels of fermentation products such as ethanol compared to planktonic cells. However, ethanol is not more abundant in biofilms (89, 92), and inhibition of ethanol production leads to increased biofilm formation (92). These observations can be reconciled with the metabolic inferences discussed above if *C. albicans* uses alternative electron acceptors, thus yielding reduced products other than ethanol. For example, hypoxic growth induces the genes involved in sulfur assimilation and methionine and cysteine biosynthesis (91). These genes were also found to be upregulated in biofilms (21, 23–25, 91). It is possible that, when oxygen is scarce, such as in biofilms, the sulfur assimilation pathway, with its multiple reduction-involving steps, provides additional means to balance the reducing equivalents arising from glycolysis.

Many metabolic products have an impact on *C. albicans* cell properties that affect the structure or integrity of the biofilm. The most intensively studied example is the quorum sensing molecule farnesol, which functions as an inhibitor of hyphal morphogenesis and of biofilm formation (93, 94) through its action on the Ras1-cyclic AMP pathway (95). Additionally, farnesol has recently been shown to block Nrg1 degradation (96), and Nrg1 can promote cell dispersion from biofilms (96). Although the biofilm environment may modify responses to quorum sensing molecules (83), the simplest generally accepted model at this time is that farnesol and other quorum sensing molecules promote release of yeast cells from mature biofilms.

One metabolite with enigmatic biological impact is glycerol. It is familiar to most yeast biologists as a major osmoprotectant and net output of the HOG pathway (97). Glycerol levels are considerably elevated in biofilm cells compared to planktonic cells (22, 89), and the glycerol biosynthetic genes are upregulated in biofilms (21, 22, 24, 25). Deletion of the glycerol biosynthetic gene *RHR2* causes a severe biofilm defect *in vitro* and *in vivo* in a rat catheter model (22, 26). Unexpectedly, the reduced glycerol levels cause decreased expression of biofilm adhesin genes (including *ALS1*, *ALS3*, and *HWP1*), and expression of any of these adhesins at elevated levels restores biofilm formation by the *rhr2Δ/Δ* mutant *in vitro* and *in vivo* (22). It is not clear why glycerol and biofilm formation should be so intimately coupled; fermentation of a hexose to glycerol does not allow ATP production, though it could be used to consume reducing equivalents generated under hypoxic conditions. The glycerol-adhesin regulatory relationship may reflect the role of glycerol in synthesis of GPI anchors, or perhaps a coupling of biofilm formation and turgor-requiring tissue invasion in natural contexts. This example illustrates that biofilm metabolites may have an impact that extends far beyond metabolism.

BIOFILM DRUG RESISTANCE

C. albicans biofilm cells are much more resistant than planktonic cells to a spectrum of antifungal drugs. As described above, drug sequestration by matrix β-1,3 glucan is one major resistance mechanism (98). However, the extracellular DNA of biofilm matrix contributes to resistance to amphotericin B, as DNase treatment increases antifungal susceptibility of biofilm cells (99). Several additional processes further contribute to drug resistance. For example, the drug efflux pump genes *CDR1*, *CDR2*, and *MDR1* are upregulated in biofilms and contribute to fluconazole resistance of early, though not mature, biofilms (100, 101). A decrease in ergosterol levels is observed in intermediate and mature biofilms, so there is potentially less target available for amphotericin B (100). Additionally, persister cells have been observed for *C. albicans* biofilms as they have for bacterial biofilms (102). LaFleur et al. identified these phenotypic variants from biofilms as survivors after amphotericin B treatment (102). There has been exciting progress recently in defining the genetic determinants of persister cell formation: the Thevissen group has shown that reactive oxygen species generated by miconazole treatment induce expression of the superoxide dismutase (SOD) gene family. They linked this response to generation of persisters by showing that chemical superoxide dismutase inhibition, or a genetic deletion affecting the major cell surface family members Sod4 and Sod5, causes a severe reduction in the level of persisters (103). The authors note that this mechanism may be specific to miconazole. The challenge in analysis of persisters reflects in part a broader knowledge gap: at this time we do not understand the extent of heterogeneity among fungal biofilm cells (104), nor have we developed the tools to dissect subpopulations. In any case, it is clear that biofilm drug resistance is a multifactorial phenomenon. The most effective therapies may prevent biofilms from forming, rather than trying to eliminate them once they are present.

ACKNOWLEDGMENTS

We are grateful to Drs. Frederick Lanni and Saranna Fanning for many delightful discussions about biofilms.
Biofilm research in our laboratory has been supported by NIH grant R01 AI067703 (APM) and a Stupakoff fellowship (JVD).

Conflicts of interest: We disclose no conflicts.

CITATION

Desai JV, Mitchell AP. 2015. *Candida albicans* biofilm development and its genetic control. Microbiol Spectrum 3(3):MB-0005-2014.

REFERENCES

1. **Finkel JS, Mitchell AP.** 2011. Genetic control of *Candida albicans* biofilm development. *Nat Rev Microbiol* **9:**109–118.
2. **Kojic EM, Darouiche RO.** 2004. *Candida* infections of medical devices. *Clin Microbiol Rev* **17:**255–267.
3. **Marrie TJ, Costerton JW.** 1984. Scanning and transmission electron microscopy of *in situ* bacterial colonization of intravenous and intraarterial catheters. *J Clin Microbiol* **19:**687–693.
4. **Hawser SP, Douglas LJ.** 1994. Biofilm formation by *Candida* species on the surface of catheter materials *in vitro*. *Infect Immun* **62:**915–921.
5. **Berman J, Sudbery PE.** 2002. *Candida albicans*: a molecular revolution built on lessons from budding yeast. *Nat Rev Genet* **3:**918–930.
6. **Daniels KJ, Park YN, Srikantha T, Pujol C, Soll DR.** 2013. Impact of environmental conditions on the form and function of *Candida albicans* biofilms. *Eukaryot Cell* **12:**1389–1402.
7. **Chandra J, Kuhn DM, Mukherjee PK, Hoyer LL, McCormick T, Ghannoum MA.** 2001. Biofilm formation by the fungal pathogen *Candida albicans*: development, architecture, and drug resistance. *J Bacteriol* **183:**5385–5394.
8. **Blankenship JR, Mitchell AP.** 2006. How to build a biofilm: a fungal perspective. *Curr Opin Microbiol* **9:**588–594.
9. **Uppuluri P, Chaturvedi AK, Srinivasan A, Banerjee M, Ramasubramaniam AK, Kohler JR, Kadosh D, Lopez-Ribot JL.** 2010. Dispersion as an important step in the *Candida albicans* biofilm developmental cycle. *PLoS Pathog* **6:**e1000828. doi:10.1371/journal.ppat.1000828.
10. **Tournu H, Van Dijck P.** 2012. *Candida* biofilms and the host: models and new concepts for eradication. *Int J Microbiol* **2012:**845352.
11. **Chandra J, Long L, Ghannoum MA, Mukherjee PK.** 2011. A rabbit model for evaluation of catheter-associated fungal biofilms. *Virulence* **2:**466–474.

12. **Andes D, Nett J, Oschel P, Albrecht R, Marchillo K, Pitula A.** 2004. Development and characterization of an *in vivo* central venous catheter *Candida albicans* biofilm model. *Infect Immun* **72:**6023–6031.
13. **Wang X, Fries BC.** 2011. A murine model for catheter-associated candiduria. *J Med Microbiol* **60:**1523–1529.
14. **Nett JE, Marchillo K, Spiegel CA, Andes DR.** 2010. Development and validation of an *in vivo Candida albicans* biofilm denture model. *Infect Immun* **78:**3650–3659.
15. **Ganguly S, Mitchell AP.** 2011. Mucosal biofilms of *Candida albicans. Curr Opin Microbiol* **14:**380–385.
16. **Baillie GS, Douglas LJ.** 1999. Role of dimorphism in the development of *Candida albicans* biofilms. *J Med Microbiol* **48:**671–679.
17. **Richard ML, Nobile CJ, Bruno VM, Mitchell AP.** 2005. *Candida albicans* biofilm-defective mutants. *Eukaryot Cell* **4:**1493–1502.
18. **Ramage G, VandeWalle K, Lopez-Ribot JL, Wickes BL.** 2002. The filamentation pathway controlled by the Efg1 regulator protein is required for normal biofilm formation and development in *Candida albicans. FEMS Microbiol Lett* **214:**95–100.
19. **Nobile CJ, Mitchell AP.** 2005. Regulation of cell-surface genes and biofilm formation by the *C. albicans* transcription factor Bcr1p. *Curr Biol* **15:**1150–1155.
20. **Nobile CJ, Andes DR, Nett JE, Smith FJ, Yue F, Phan QT, Edwards JE, Filler SG, Mitchell AP.** 2006. Critical role of Bcr1-dependent adhesins in *C. albicans* biofilm formation *in vitro* and *in vivo. PLoS Pathog* **2:**e63.
21. **Garcia-Sanchez S, Aubert S, Iraqui I, Janbon G, Ghigo JM, d'Enfert C.** 2004. *Candida albicans* biofilms: a developmental state associated with specific and stable gene expression patterns. *Eukaryot Cell* **3:**536–545.
22. **Desai JV, Bruno VM, Ganguly S, Stamper RJ, Mitchell KF, Solis N, Hill EM, Xu W, Filler SG, Andes DR, Fanning S, Lanni F, Mitchell AP.** 2013. Regulatory role of glycerol in *Candida albicans* biofilm formation. *MBio* **4:**e00637-00612. doi:10.1128/mBio.00637-12.
23. **Murillo LA, Newport G, Lan CY, Habelitz S, Dungan J, Agabian NM.** 2005. Genome-wide transcription profiling of the early phase of biofilm formation by *Candida albicans. Eukaryot Cell* **4:**1562–1573.
24. **Yeater KM, Chandra J, Cheng G, Mukherjee PK, Zhao X, Rodriguez-Zas SL, Kwast KE, Ghannoum MA, Hoyer LL.** 2007. Temporal analysis of *Candida albicans* gene expression during biofilm development. *Microbiology* **153:**2373–2385.
25. **Nett JE, Lepak AJ, Marchillo K, Andes DR.** 2009. Time course global gene expression analysis of an *in vivo Candida* biofilm. *J Infect Dis* **200:**307–313.
26. **Bonhomme J, Chauvel M, Goyard S, Roux P, Rossignol T, d'Enfert C.** 2011. Contribution of the glycolytic flux and hypoxia adaptation to efficient biofilm formation by *Candida albicans. Mol Microbiol* **80:**995–1013.
27. **Nobile CJ, Nett JE, Hernday AD, Homann OR, Deneault JS, Nantel A, Andes DR, Johnson AD, Mitchell AP.** 2009. Biofilm matrix regulation by *Candida albicans* Zap1. *PLoS Biol* **7:**e1000133. doi:10.1371/journal.pbio.1000133.
28. **Taff HT, Nett JE, Zarnowski R, Ross KM, Sanchez H, Cain MT, Hamaker J, Mitchell AP, Andes DR.** 2012. A *Candida* biofilm-induced pathway for matrix glucan delivery: implications for drug resistance. *PLoS Pathog* **8:**e1002848. doi:10.1371/journal.ppat.1002848.
29. **Nobile CJ, Fox EP, Nett JE, Sorrells TR, Mitrovich QM, Hernday AD, Tuch BB, Andes DR, Johnson AD.** 2012. A recently evolved transcriptional network controls biofilm development in *Candida albicans. Cell* **148:**126–138.
30. **Gow NA, Hube B.** 2012. Importance of the *Candida albicans* cell wall during commensalism and infection. *Curr Opin Microbiol* **15:**406–412.
31. **Dranginis AM, Rauceo JM, Coronado JE, Lipke PN.** 2007. A biochemical guide to yeast adhesins: glycoproteins for social and antisocial occasions. *Microbiol Mol Biol Rev* **71:**282–294.
32. **Chaffin WL.** 2008. *Candida albicans* cell wall proteins. *Microbiol Mol Biol Rev* **72:**495–544.
33. **Hoyer LL, Green CB, Oh SH, Zhao X.** 2008. Discovering the secrets of the *Candida albicans* agglutinin-like sequence (ALS) gene family: a sticky pursuit. *Med Mycol* **46:**1–15.
34. **Li F, Palecek SP.** 2003. EAP1, a *Candida albicans* gene involved in binding human epithelial cells. *Eukaryot Cell* **2:**1266–1273.
35. **Staab JF, Bradway SD, Fidel PL, Sundstrom P.** 1999. Adhesive and mammalian transglutaminase substrate properties of *Candida albicans* Hwp1. *Science* **283:**1535–1538.
36. **Nobile CJ, Nett JE, Andes DR, Mitchell AP.** 2006. Function of *Candida albicans* adhesin Hwp1 in biofilm formation. *Eukaryot Cell* **5:**1604–1610.
37. **Monniot C, Boisrame A, Da Costa G, Chauvel M, Sautour M, Bougnoux ME, Bellon-Fontaine MN, Dalle F, d'Enfert C, Richard ML.** 2013. Rbt1 protein domains analysis in *Candida albicans* brings insights into hyphal surface

modifications and Rbt1 potential role during adhesion and biofilm formation. *PLoS One* **8:** e82395. doi:10.1371/journal.pone.0082395.

38. **Granger BL, Flenniken ML, Davis DA, Mitchell AP, Cutler JE.** 2005. Yeast wall protein 1 of *Candida albicans. Microbiology* **151:** 1631–1644.
39. **Sandini S, Stringaro A, Arancia S, Colone M, Mondello F, Murtas S, Girolamo A, Mastrangelo N, De Bernardis F.** 2011. The MP65 gene is required for cell wall integrity, adherence to epithelial cells and biofilm formation in *Candida albicans. BMC Microbiol* **11:**106.
40. **Singleton DR, Masuoka J, Hazen KC.** 2001. Cloning and analysis of a *Candida albicans* gene that affects cell surface hydrophobicity. *J Bacteriol* **183:**3582–3588.
41. **Chaffin WL, Lopez-Ribot JL, Casanova M, Gozalbo D, Martinez JP.** 1998. Cell wall and secreted proteins of *Candida albicans*: identification, function, and expression. *Microbiol Mol Biol Rev* **62:**130–180.
42. **Tronchin G, Bouchara JP, Robert R.** 1989. Dynamic changes of the cell wall surface of *Candida albicans* associated with germination and adherence. *Eur J Cell Biol* **50:**285–290.
43. **Gaur NK, Klotz SA, Henderson RL.** 1999. Overexpression of the *Candida albicans* ALA1 gene in *Saccharomyces cerevisiae* results in aggregation following attachment of yeast cells to extracellular matrix proteins, adherence properties similar to those of *Candida albicans. Infect Immun* **67:**6040–6047.
44. **Fu Y, Rieg G, Fonzi WA, Belanger PH, Edwards JE Jr, Filler SG.** 1998. Expression of the *Candida albicans* gene ALS1 in *Saccharomyces cerevisiae* induces adherence to endothelial and epithelial cells. *Infect Immun* **66:**1783–1786.
45. **Lipke PN, Garcia MC, Alsteens D, Ramsook CB, Klotz SA, Dufrene YF.** 2012. Strengthening relationships: amyloids create adhesion nanodomains in yeasts. *Trends Microbiol* **20:** 59–65.
46. **Hoyer LL, Hecht JE.** 2001. The ALS5 gene of *Candida albicans* and analysis of the Als5p N-terminal domain. *Yeast* **18:**49–60.
47. **Salgado PS, Yan R, Taylor JD, Burchell L, Jones R, Hoyer LL, Matthews SJ, Simpson PJ, Cota E.** 2011. Structural basis for the broad specificity to host-cell ligands by the pathogenic fungus *Candida albicans. Proc Natl Acad Sci USA* **108:**15775–15779.
48. **Klotz SA, Gaur NK, Lake DF, Chan V, Rauceo J, Lipke PN.** 2004. Degenerate peptide recognition by *Candida albicans* adhesins Als5p and Als1p. *Infect Immun* **72:**2029–2034.
49. **Rauceo JM, De Armond R, Otoo H, Kahn PC, Klotz SA, Gaur NK, Lipke PN.** 2006. Threonine-rich repeats increase fibronectin binding in the *Candida albicans* adhesin Als5p. *Eukaryot Cell* **5:**1664–1673.
50. **Nobile CJ, Schneider HA, Nett JE, Sheppard DC, Filler SG, Andes DR, Mitchell AP.** 2008. Complementary adhesin function in *C. albicans* biofilm formation. *Curr Biol* **18:**1017–1024.
51. **Li F, Palecek SP.** 2008. Distinct domains of the *Candida albicans* adhesin Eap1p mediate cell-cell and cell-substrate interactions. *Microbiology* **154:**1193–1203.
52. **Li F, Svarovsky MJ, Karlsson AJ, Wagner JP, Marchillo K, Oshel P, Andes D, Palecek SP.** 2007. Eap1p, an adhesin that mediates *Candida albicans* biofilm formation *in vitro* and *in vivo. Eukaryot Cell* **6:**931–939.
53. **Ene IV, Bennett RJ.** 2009. Hwp1 and related adhesins contribute to both mating and biofilm formation in *Candida albicans. Eukaryot Cell* **8:**1909–1913.
54. **Granger BL.** 2012. Insight into the anti-adhesive effect of yeast wall protein 1 of *Candida albicans. Eukaryot Cell* **11:**795–805.
55. **Biswas S, Van Dijck P, Datta A.** 2007. Environmental sensing and signal transduction pathways regulating morphopathogenic determinants of *Candida albicans. Microbiol Mol Biol Rev* **71:**348–376.
56. **Finkel JS, Xu W, Huang D, Hill EM, Desai JV, Woolford CA, Nett JE, Taff H, Norice CT, Andes DR, Lanni F, Mitchell AP.** 2012. Portrait of *Candida albicans* adherence regulators. *PLoS Pathog* **8:**e1002525. doi:10.1371/journal.ppat.1002525.
57. **Gutierrez-Escribano P, Zeidler U, Suarez MB, Bachellier-Bassi S, Clemente-Blanco A, Bonhomme J, Vazquez de Aldana CR, d'Enfert C, Correa-Bordes J.** 2012. The NDR/LATS kinase Cbk1 controls the activity of the transcriptional regulator Bcr1 during biofilm formation in *Candida albicans. PLoS Pathog* **8:** e1002683. doi:10.1371/journal.ppat.1002683.
58. **Bastidas RJ, Heitman J, Cardenas ME.** 2009. The protein kinase Tor1 regulates adhesin gene expression in *Candida albicans. PLoS Pathog* **5:** e1000294. doi:10.1371/journal.ppat.1000294.
59. **Su C, Lu Y, Liu H.** 2013. Reduced TOR signaling sustains hyphal development in *Candida albicans* by lowering Hog1 basal activity. *Mol Biol Cell* **24:**385–397.
60. **Shapiro RS, Robbins N, Cowen LE.** 2011. Regulatory circuitry governing fungal development, drug resistance, and disease. *Microbiol Mol Biol Rev* **75:**213–267.

61. **Fanning S, Xu W, Beaurepaire C, Suhan JP, Nantel A, Mitchell AP.** 2012. Functional control of the *Candida albicans* cell wall by catalytic protein kinase A subunit Tpk1. *Mol Microbiol* **86:**284–302.
62. **Sen M, Shah B, Rakshit S, Singh V, Padmanabhan B, Ponnusamy M, Pari K, Vishwakarma R, Nandi D, Sadhale PP.** 2011. UDP-glucose 4, 6-dehydratase activity plays an important role in maintaining cell wall integrity and virulence of *Candida albicans*. *PLoS Pathog* **7:**e1002384. doi:10.1371/journal.ppat.1002384.
63. **Peltroche-Llacsahuanga H, Goyard S, d'Enfert C, Prill SK, Ernst JF.** 2006. Protein O-mannosyltransferase isoforms regulate biofilm formation in *Candida albicans*. *Antimicrob Agents Chemother* **50:**3488–3491.
64. **Hiller E, Heine S, Brunner H, Rupp S.** 2007. *Candida albicans* Sun41p, a putative glycosidase, is involved in morphogenesis, cell wall biogenesis, and biofilm formation. *Eukaryot Cell* **6:**2056–2065.
65. **Norice CT, Smith FJ Jr, Solis N, Filler SG, Mitchell AP.** 2007. Requirement for *Candida albicans* Sun41 in biofilm formation and virulence. *Eukaryot Cell* **6:**2046–2055.
66. **Hashash R, Younes S, Bahnan W, El Koussa J, Maalouf K, Dimassi HI, Khalaf RA.** 2011. Characterisation of Pga1, a putative *Candida albicans* cell wall protein necessary for proper adhesion and biofilm formation. *Mycoses* **54:**491–500.
67. **Finkel JS, Yudanin N, Nett JE, Andes DR, Mitchell AP.** 2011. Application of the systematic "DAmP" approach to create a partially defective *C. albicans* mutant. *Fungal Genet Biol* **48:**1056–1061.
68. **Liu Y, Filler SG.** 2011. *Candida albicans* Als3, a multifunctional adhesin and invasin. *Eukaryot Cell* **10:**168–173.
69. **Zhao X, Oh SH, Yeater KM, Hoyer LL.** 2005. Analysis of the *Candida albicans* Als2p and Als4p adhesins suggests the potential for compensatory function within the Als family. *Microbiology* **151:**1619–1630.
70. **Brand A, Lee K, Veses V, Gow NA.** 2009. Calcium homeostasis is required for contact-dependent helical and sinusoidal tip growth in *Candida albicans* hyphae. *Mol Microbiol* **71:**1155–1164.
71. **Kumamoto CA.** 2005. A contact-activated kinase signals *Candida albicans* invasive growth and biofilm development. *Proc Natl Acad Sci USA* **102:**5576–5581.
72. **Zucchi PC, Davis TR, Kumamoto CA.** 2010. A *Candida albicans* cell wall-linked protein promotes invasive filamentation into semi-solid medium. *Mol Microbiol* **76:**733–748.
73. **Kumamoto CA.** 2008. Molecular mechanisms of mechanosensing and their roles in fungal contact sensing. *Nat Rev Microbiol* **6:**667–673.
74. **Puri S, Kumar R, Chadha S, Tati S, Conti HR, Hube B, Cullen PJ, Edgerton M.** 2012. Secreted aspartic protease cleavage of *Candida albicans* Msb2 activates Cek1 MAPK signaling affecting biofilm formation and oropharyngeal candidiasis. *PLoS One* **7:**e46020. doi:10.1371/journal.pone.0046020.
75. **Yi S, Sahni N, Daniels KJ, Pujol C, Srikantha T, Soll DR.** 2008. The same receptor, G protein, and mitogen-activated protein kinase pathway activate different downstream regulators in the alternative white and opaque pheromone responses of *Candida albicans*. *Mol Biol Cell* **19:**957–970.
76. **Al-Fattani MA, Douglas LJ.** 2006. Biofilm matrix of *Candida albicans* and *Candida tropicalis*: chemical composition and role in drug resistance. *J Med Microbiol* **55:**999–1008.
77. **Nett J, Lincoln L, Marchillo K, Massey R, Holoyda K, Hoff B, VanHandel M, Andes D.** 2007. Putative role of beta-1,3 glucans in *Candida albicans* biofilm resistance. *Antimicrob Agents Chemother* **51:**510–520.
78. **Hawser SP, Baillie GS, Douglas LJ.** 1998. Production of extracellular matrix by *Candida albicans* biofilms. *J Med Microbiol* **47:**253–256.
79. **Srikantha T, Daniels KJ, Pujol C, Kim E, Soll DR.** 2013. Identification of genes upregulated by the transcription factor Bcr1 that are involved in impermeability, impenetrability, and drug resistance of *Candida albicans* a/alpha biofilms. *Eukaryot Cell* **12:**875–888.
80. **Nett JE, Sanchez H, Cain MT, Andes DR.** 2010. Genetic basis of *Candida* biofilm resistance due to drug-sequestering matrix glucan. *J Infect Dis* **202:**171–175.
81. **Thomas DP, Bachmann SP, Lopez-Ribot JL.** 2006. Proteomics for the analysis of the *Candida albicans* biofilm lifestyle. *Proteomics* **6:**5795–5804.
82. **Martins M, Uppuluri P, Thomas DP, Cleary IA, Henriques M, Lopez-Ribot JL, Oliveira R.** 2010. Presence of extracellular DNA in the *Candida albicans* biofilm matrix and its contribution to biofilms. *Mycopathologia* **169:**323–331.
83. **Ganguly S, Bishop AC, Xu W, Ghosh S, Nickerson KW, Lanni F, Patton-Vogt J, Mitchell AP.** 2011. Zap1 control of cell-cell signaling in *Candida albicans* biofilms. *Eukaryot Cell* **10:**1448–1454.

84. **Perez A, Pedros B, Murgui A, Casanova M, Lopez-Ribot JL, Martinez JP.** 2006. Biofilm formation by *Candida albicans* mutants for genes coding fungal proteins exhibiting the eight-cysteine-containing CFEM domain. *FEMS Yeast Res* **6:**1074–1084.
85. **Weissman Z, Kornitzer D.** 2004. A family of *Candida* cell surface haem-binding proteins involved in haemin and haemoglobin-iron utilization. *Mol Microbiol* **53:**1209–1220.
86. **Nett JE, Sanchez H, Cain MT, Ross KM, Andes DR.** 2011. Interface of *Candida albicans* biofilm matrix-associated drug resistance and cell wall integrity regulation. *Eukaryot Cell* **10:**1660–1669.
87. **Robbins N, Uppuluri P, Nett J, Rajendran R, Ramage G, Lopez-Ribot JL, Andes D, Cowen LE.** 2011. Hsp90 governs dispersion and drug resistance of fungal biofilms. *PLoS Pathog* **7:** e1002257. doi:10.1371/journal.ppat.1002257.
88. **Rossignol T, Ding C, Guida A, d'Enfert C, Higgins DG, Butler G.** 2009. Correlation between biofilm formation and the hypoxic response in *Candida parapsilosis*. *Eukaryot Cell* **8:**550–559.
89. **Zhu Z, Wang H, Shang Q, Jiang Y, Cao Y, Chai Y.** 2013. Time course analysis of *Candida albicans* metabolites during biofilm development. *J Proteome Res* **12:**2375–2385.
90. **Askew C, Sellam A, Epp E, Hogues H, Mullick A, Nantel A, Whiteway M.** 2009. Transcriptional regulation of carbohydrate metabolism in the human pathogen *Candida albicans*. *PLoS Pathog* **5:**e1000612. doi:10.1371/journal.ppat.1000612.
91. **Stichternoth C, Ernst JF.** 2009. Hypoxic adaptation by Efg1 regulates biofilm formation by *Candida albicans*. *Appl Environ Microbiol* **75:**3663–3672.
92. **Mukherjee PK, Mohamed S, Chandra J, Kuhn D, Liu S, Antar OS, Munyon R, Mitchell AP, Andes D, Chance MR, Rouabhia M, Ghannoum MA.** 2006. Alcohol dehydrogenase restricts the ability of the pathogen *Candida albicans* to form a biofilm on catheter surfaces through an ethanol-based mechanism. *Infect Immun* **74:**3804–3816.
93. **Hornby JM, Jensen EC, Lisec AD, Tasto JJ, Jahnke B, Shoemaker R, Dussault P, Nickerson KW.** 2001. Quorum sensing in the dimorphic fungus *Candida albicans* is mediated by farnesol. *Appl Environ Microbiol* **67:**2982–2992.
94. **Ramage G, Saville SP, Wickes BL, Lopez-Ribot JL.** 2002. Inhibition of *Candida albicans* biofilm formation by farnesol, a quorum-sensing molecule. *Appl Environ Microbiol* **68:**5459–5463.
95. **Lindsay AK, Deveau A, Piispanen AE, Hogan DA.** 2012. Farnesol and cyclic AMP signaling effects on the hypha-to-yeast transition in *Candida albicans*. *Eukaryot Cell* **11:**1219–1225.
96. **Lu Y, Su C, Unoje O, Liu H.** 2014. Quorum sensing controls hyphal initiation in *Candida albicans* through Ubr1-mediated protein degradation. *Proc Natl Acad Sci USA* **111:**1975–1980.
97. **Hohmann S.** 2002. Osmotic stress signaling and osmo adaptation in yeasts. *Microbiol Mol Biol Rev* **66:**300–372.
98. **Nett JE, Crawford K, Marchillo K, Andes DR.** 2010. Role of Fks1p and matrix glucan in *Candida albicans* biofilm resistance to an echinocandin, pyrimidine, and polyene. *Antimicrob Agents Chemother* **54:**3505–3508.
99. **Martins M, Henriques M, Lopez-Ribot JL, Oliveira R.** 2012. Addition of DNase improves the *in vitro* activity of antifungal drugs against *Candida albicans* biofilms. *Mycoses* **55:**80–85.
100. **Mukherjee PK, Chandra J, Kuhn DM, Ghannoum MA.** 2003. Mechanism of fluconazole resistance in *Candida albicans* biofilms: phase-specific role of efflux pumps and membrane sterols. *Infect Immun* **71:**4333–4340.
101. **Ramage G, Bachmann S, Patterson TF, Wickes BL, Lopez-Ribot JL.** 2002. Investigation of multidrug efflux pumps in relation to fluconazole resistance in *Candida albicans* biofilms. *J Antimicrob Chemother* **49:**973–980.
102. **LaFleur MD, Kumamoto CA, Lewis K.** 2006. *Candida albicans* biofilms produce antifungal-tolerant persister cells. *Antimicrob Agents Chemother* **50:**3839–3846.
103. **Bink A, Vandenbosch D, Coenye T, Nelis H, Cammue BP, Thevissen K.** 2011. Superoxide dismutases are involved in *Candida albicans* biofilm persistence against miconazole. *Antimicrob Agents Chemother* **55:**4033–4037.
104. **Stewart PS, Franklin MJ.** 2008. Physiological heterogeneity in biofilms. *Nat Rev Microbiol* **6:**199–210.
105. **Nett JE, Andes DR.** 2015. Fungal Biofilms: *In Vivo* Models for Discovery of Anti-Biofilm Drugs. *In* Ghannoum M, Parsek M, Whiteley M, Mukherjee P (ed), *Microbial Biofilms,* 2nd ed. ASM Press, Washington, DC, in press.

Candida Biofilms: Development, Architecture, and Resistance

6

JYOTSNA CHANDRA[1] and PRANAB K. MUKHERJEE[1]

INTRODUCTION

The use of indwelling devices in current therapeutic practice is associated with hospital-acquired bloodstream and deep tissue infections (1). Transplantation medical procedures, immunosuppression, and prolonged intensive care unit stays have also increased the prevalence of nosocomial infections. Device-associated infections are commonly associated with the ability of bacteria and fungi to form biofilms, which are defined as communities of sessile organisms irreversibly associated with a surface, encased within a polysaccharide-rich extracellular matrix, and exhibiting enhanced resistance to antimicrobial drugs (2–5). Forming a biofilm provides the microbes protection from host immunity, environmental stresses due to contaminants, and nutritional depletion or imbalances, while being dangerous to human health due to biofilms' inherent robustness and elevated resistance.

Fungal infections are the fourth most common cause of nosocomial bloodstream infection (6), with *Candida* spp. being the most common fungi associated with these infections. Among *Candida* spp. *Candida albicans* is the most prevalent species causing both superficial and systemic disease (although

[1]Center for Medical Mycology and Mycology Reference Laboratory, Department of Dermatology, University Hospitals of Cleveland and Case Western Reserve University, Cleveland, OH 44106.

Microbial Biofilms 2nd Edition
Edited by Mahmoud Ghannoum, Matthew Parsek, Marvin Whiteley, and Pranab K. Mukherjee

doi:10.1128/microbiolspec.MB-0020-2015

infections due to non-*albicans* species are increasing). Even with current antifungal therapy, mortality associated with candidiasis can be as high as 50% in adults and up to 30% in children (7–10). In one of the earliest studies documenting the ability of *Candida* to form biofilms, Marrie and Costerton reported formation of *Candida parapsilosis* biofilms on vascular catheters (11). Initial studies also reported that *Candida* biofilms formed on different surfaces including Hickman catheters (12), soft contact lenses, ureteral stents (13), and corneas (14). Subsequent studies have demonstrated that *Candida* biofilms can form on a wide variety of indwelling medical devices including dentures, central venous catheters (CVCs), and urinary catheters.

Recent technological advances have facilitated the development of novel approaches to investigate the formation of biofilms and identify specific markers for biofilms. These studies have provided extensive knowledge of the effect of different variables, including growth time, nutrients, and physiological conditions, on biofilm formation, morphology, and architecture (15). In this chapter, we will focus on *Candida* biofilms (biofilms caused by *Aspergillus* are covered in reference 155) and provide an update on their development, architecture, and resistance mechanisms.

EXPERIMENTAL MODELS OF *CANDIDA* BIOFILMS

Microbial biofilms undergo multistep growth processes involving physical, chemical, and biological changes (16). Due to the versatility with which *Candida* biofilms can develop in human hosts, it is necessary to develop reproducible *in vitro* and *in vivo* models that could mimic these forms/situations. It is also necessary to develop models that can establish common and specific characteristics of *Candida* biofilm morphology. In this respect, various model systems have been studied to investigate the properties of microbial biofilms *in vitro* (17). These range from simple assays with catheter discs to more complex flow systems, such as the perfused biofilm fermenter or reactors and shear stress rotating disc systems (18, 19). Subsequent *in vitro* model systems have included forming biofilms on a variety of different plastics, microtiter plates, biofilm chips formed on glass slides, Calgary biofilm devices, microporous membrane cellulose filters, acrylic strips, voice prostheses, catheter discs, contact lenses, and tissue culture flasks (20–28). Although a variety of substrates support the formation of biofilms, those formed on clinically relevant substrates such as catheters, denture acrylic strips, voice prostheses, and contact lenses under physiological conditions are likely to be closer to the clinical setting than those formed on nonphysiologically relevant substrates.

Biofilm formation *in vitro* generally proceeds through three sequential steps: (i) pretreatment of the substrate, (ii) cell attachment, and (iii) colonization of cells and matrix formation. Various models have been evaluated to study detailed development, architecture, and morphology of biofilms (summarized in Tables 1 and 2). (These studies are described in greater detail in reference 156, and only a brief summary, relevant to *Candida* biofilms, is presented here.)

In vitro Models

In one of the first *in vitro* models of *Candida* biofilms, Hawser and Douglas (22) formed *C. albicans* biofilms on discs cut from a variety of catheters including latex urinary catheters, polyvinyl chloride CVCs, silicone elastomer-coated latex urinary Foley catheters, silicone urinary Foley catheters, and polyurethane CVCs. These investigators quantified biofilm growth using a colorimetric assay based on reduction of a tetrazolium salt (3-[4, 5-dimethylthiazol-2-yl]-2,5-diphenyltetrazolium bromide [MTT]) or incorporation of ^{3}H-leucine (22). This study reported an increase in MTT values and ^{3}H-leucine incorporation

TABLE 1 Summary of *in vitro Candida*-associated biofilm models

In vitro models	Substrate	Features
Plastic/microtiter plates	Polystyrene, flat-bottom 96-well plates, plastic slides	Biofilm for 96 strains/species can be tested at one time
Calgary Biofilm Device (CBD)	96-well polystyrene pegs/plates	CBD is a useful, simple, low-cost miniature device that has utility for parallel study of *Candida* biofilms and for elucidating factors modulating this phenomenon
Microporous membrane filters	Cellulose	Useful model to study antimycotic perfusion through biofilms and complex interactions between biofilm-antifungal interphase
Voice prostheses	Silicone rubber	Mimic clinical conditions
Catheters	Latex urinary catheters, polyvinyl chloride, CVCs, silicone elastomer-coated latex urinary Foley catheters, silicone urinary Foley catheters, and polyurethane CVCs; silicone elastomers	Clinically relevant substrates; mimic intravascular catheter-associated infections
Denture acrylic strips	Polymethylmethacrylate, acrylic resins	Mimic denture stomatitis, oral clinical conditions
Contact lenses	Lotrafilcon A, etafilcon A galyfilcon A, balafilcon A, alphafilcon A	Mimic keratitis and other eye-associated clinical conditions
Flow system biofilm models	Glass microfermentors	Provide a continuous flow of media or fluid, mimicking the physiological conditions present at the infection site (e.g., mimicking the flow of saliva, blood, or urine). Such flow of liquids can influence nutrient exchange and the structural integrity of biofilms
High-throughput biofilm models utilizing biofilm chip system (CaBChip)	Microarray platform with nano-biofilms encapsulated in a collagen matrix	Miniaturization and automation of chip cut reagent use and analysis time; minimize labor-intensive steps and reduces assay costs. Also accelerate the antifungal drug discovery process by enabling rapid, convenient, and inexpensive screening of hundreds to thousands of compounds simultaneously

levels with the maturation of biofilms and showed that both quantification methods resulted in strong correlation with biofilm dry weight (22). An *in vitro* voice prosthesis biofilm model was described by Everaert et al. (27), who evaluated biofilm formation on argon plasma-treated silicone rubber voice prostheses.

Our group investigated the development and characterization of *C. albicans* biofilms formed on two common bioprosthetic materials: (i) silicone elastomer, a commonly used catheter material (20), and (ii) polymethylmethacrylate, used to form denture acrylic (21). Briefly, cells were adhered on these substrates and then transferred to the specific media for biofilms to mature (20, 21, 29). We also used various soft contact lenses to analyze differences in biofilm architecture (23). Measurement of biofilm growth was performed using two quantitative methods: (i) colorimetric assays that involved the reduction of 2,3-bis (2-methoxy-4-nitro-5-sulfophenyl)-5-[(phenyl amino) carbonyl-2H-tetrazolium hydroxide] (XTT) by mitochondrial dehydrogenase in the living cells into a colored water-soluble product measured spectrophotometrically and (ii) dry weight determination, in which biofilms were scraped off the substrate surface and

TABLE 2 Summary of *in vivo Candida*-associated biofilm models

In vivo models	Animal species	Features
Catheter-associated *in vivo* models, models utilizing amphogel coating on catheters and subcutaneous catheter models	Rat, mouse, rabbit	Rat and mouse models have advantage over rabbit models because they have a relatively low cost in setting, are easy to handle, and mimic the clinical conditions of rabbit models
Denture-associated models	Rat	Low cost, mimic clinical conditions
Contact lens *in vivo* models	Mouse	Low cost and clinically relevant
Models using biotic surfaces such as oral cavity, oropharyngeal mucosa, tongue, vaginal mucosa	Mouse	Low setting cost, mimic clinical conditions

filtered through a preweighed membrane filter under vacuum (20–22). Our results showed that dry weight and XTT values increased with the formation of biofilms (20). The study showed that there was amorphous granular material covering yeast and hyphal forms and identified the developmental phases (20) associated with the biofilm growth: early (0 to 11 h), intermediate (~12 to 24 h), and mature phases (24 to 48 h) (20).

High-Throughput Models

To facilitate screening of compounds for their antibiofilm activity it is necessary to develop high-throughput biofilm models. Ramage et al. (30) used a microtiter plate model to assess the variability between *C. albicans* biofilms formed in independent wells of the same microtiter plate. All biofilms formed on the microtiter plates over a 24-h period displayed consistent metabolic activity (30). Our group developed a microtiter plate–based assay using catheter discs, in which biofilms are formed on catheter discs placed in the wells of a microtiter plate (31, 32). The advantage of this model is that biofilms are formed on actual catheter material instead of the plastic surface of a microtiter plate, which allows concomitant quantitative (XTT, dry weight) and microscopic (fluorescence, electron, confocal) evaluation of *Candida* biofilms at the same time (31, 32). Another microtiter plate model is the Calgary Biofilm Device model, developed by Ceri et al. (33) at the University of Calgary. This device has been used by several investigators in different studies, including evaluation of the ability of *Candida glabrata* to form biofilms (34), susceptibility of *Candida* biofilms to metal ions (35), interspecies variations (36), and identification of persister cells in *Candida* biofilms (37). More recently, Srinivasan et al. (38) developed a *C. albicans* biofilm chip microarray system (CaBChip), which comprises more than 700 independent and uniform nano-biofilms encapsulated in a collagen matrix and represents the first miniature biofilm model for *C. albicans*. Despite several-fold miniaturization, the biofilms formed on the chip had similar phenotypic characteristics as *in vitro* biofilms, including a mixture of yeast, pseudohyphae, and hyphal cells, and a high level of antifungal drug resistance (38). The models represent exciting advances in the field and are likely to facilitate rapid and in-depth analysis of *Candida* biofilms and allow the identification of potential antibiofilm drugs.

In vivo Models

Several investigators have developed *in vivo* models to characterize and delineate the role of biofilms in animals (reviewed in depth in reference 156). In this regard, catheter-associated *Candida in vivo* biofilm models have been developed in rodents which provided information on biofilm architecture and antifungal resistance (39). These catheter-based *in vivo* biofilm models showed similar biofilm structures as seen using *in vitro* models after 24 h, with layers of yeast, pseudohyphae, and elongated hyphal cells embedded in an extracellular matrix (40). Our group developed a rabbit model of catheter-associated *C. albicans* biofilm infection (41) and showed that 7 days postinfection, quantitative catheter cultures consistently yielded >2 log CFU/catheter segment, which is considered the threshold for catheter-related infections. We also used a subcutaneous mouse model to evaluate the effect of coating a catheter substrate with amphogels (amphotericin B–based gel) on *Candida* biofilms (42) and demonstrated that the subcutaneous model has utility in studies evaluating catheter surface modification on the ability of *Candida* to form biofilms.

While the majority of biofilm models have focused on *Candida,* our group developed a murine model of contact lens–associated *Fusarium* keratitis (43, 44). This model was prompted by the association of fungal keratitis and biofilm noted in an outbreak of this disease in humans (45–47). *Fusarium*-infected mice had severe corneal opacification within 24 h, which progressed with unimpaired fungal growth in the cornea and with hyphae penetrating into the anterior chamber (43, 44).

Host Tissue–Associated *Candida* Biofilm Models

Biofilms formed on host surfaces are not well characterized, since tissue samples are sparse and not easily available (48). This gap was partially addressed in a mucosal model of oropharyngeal candidiasis *in situ* in mice, which demonstrated for the first time that epithelial cells, neutrophils, and commensal oral bacteria coexist within fungal biofilms formed on mouse tongue (48, 49). *C. albicans* can also form biofilms on the vaginal mucosa, with typical biofilm composition of yeast and hyphal cells embedded in an extracellular matrix; this vaginal biofilm model was replicated in immunocompetent estradiol-treated mice (50).

Development of these *in vitro* and *in vivo* models has allowed detailed investigation, microscopic evaluation, and gene/protein profiling of *Candida* biofilms. The availability of *in vivo* models is especially encouraging since this allows the conduct of studies aiming to elucidate host-pathogen interactions occurring on biofilms as they exist on bodily tissues.

BIOFILM MORPHOLOGY AND ARCHITECTURE

Candida biofilms formed *in vitro* comprise fungal cells embedded in a polysaccharide-rich extracellular matrix. When formed *in vivo* or in samples obtained from patients (e.g., used intravascular catheters, urinary catheters), these biofilms also contain host-derived biomolecules such as fibrinogen, dead cells, etc. In this regard, Lazarus et al. (51) reported catheter- and drug-induced occlusion in CVCs inserted into patients with malignancies before administration of intensive cytotoxic therapy; these CVC-associated occlusions (biofilms) were noted in Gram-positive and -negative bacteria (78%) and fungi (22%). Marrie and Costerton (11) performed scanning electron microscopy of *C. parapsilosis* biofilms formed on vascular catheters and reported *Candida* biofilm to contain fungal cells in a fibrous matrix resembling fibrin. Other investigators (12) reported similar morphology for *Candida* biofilms by scanning and transmission electron microscopy. Hawser and Douglas (22) used scanning electron microscopy to demonstrate that mature *C. albicans* biofilms (grown for 48 h) consisted of a dense network of yeasts, germ tubes, pseudohyphae, and hyphae, with extracellular polymeric material on the surfaces of some of these morphological forms.

Our group characterized the surface topography and three-dimensional architecture of *Candida* biofilms formed on denture strips and catheter discs (21, 52). Initially, scanning electron microscopy analyses of *C. albicans* biofilms formed on denture strips revealed a dense layer of coaggregating blastospores, and few hyphal elements, embedded in an extracellular, granular, polymeric matrix (21). Subsequent analyses of denture- and catheter-associated *Candida* biofilms with fluorescence microscopy and confocal microscopy revealed important differences in the surface topography and three-dimensional architecture of biofilms formed on these two substrates (52). Fluorescence microscopy showed that *C. albicans* biofilm formation on denture strips proceeds in three distinct developmental phases: early (≈0 to 11 h), intermediate (≈12 to 30 h), and maturation (≈38 to 72 h) phases. In the early phase, *C. albicans* cells grew as blastospores (yeast forms) adhering to the denture surface, which continued to grow as distinct colonies. In the intermediate phase, the fungal cells coaggregated into thick "tracks" due to growth along areas of surface irregularities and produced a noncellular, polysaccharide-rich "hazy" film covering the aggregating colonies. As the biofilms matured with time, the amount of extracellular material increased, until *C. albicans* communities were completely encased within this extracellular matrix. Biofilm formation on catheter (silicone elastomer) substrate exhibited similar phases, with one key difference: these biofilms had abundant hyphal elements. The difference in biofilm morphology was associated with the presence of a salivary conditioning film on denture biofilms, while the catheter biofilms contained a conditioning film of serum, a known inducer of filamentation in *C. albicans*.

Confocal microscopy analyses revealed a highly heterogeneous architecture of mature *C. albicans* biofilms in terms of the distribution of fungal cells and extracellular material. These analyses also underscored the key differences in architecture between the denture and catheter biofilms. Denture biofilms were 20 to 30 μm thick, comprised mostly yeast cells in confluent layers, and had irregular topography. In contrast, catheter biofilms were much thicker (up to 450 μm thick), with a 10- to 12-μm-thick basal layer of yeast cells overlaid with a hypha-rich layer and uniform thickness. These differences could be attributed to differences in nutrient conditions and substrate properties for the two models.

FACTORS INFLUENCING BIOFILM FORMATION AND ARCHITECTURE

Biofilm formation is influenced by several host and *Candida*-derived variables, including fluid flow, nutrients, host receptor, and microbial products.

Fluid Flow

Physiological conditions including fluid flow at the infection site are important modulators of biofilm, since the flow of liquids can influence the nutrient exchange and structural integrity of biofilms (53–56). Efforts have been made to mimic these conditions *in vitro,* including mimicking the flow of saliva, blood, and urine, and the use of continuous flow cells to evaluate fungal biofilms. In this regard, Busscher et al. (57) investigated the ability of *C. albicans* and *Candida tropicalis* to form biofilms on silicone rubber voice prostheses with or without a salivary conditioning film in a parallel-plate flow chamber, and showed that biofilms formed under flow in the presence of salivary film tended to detach faster than those formed directly on the substrate. Other investigators used the parallel-plate flow chamber to evaluate formation of *Candida*-bacteria mixed biofilms on glass (58, 59) and acrylic (60). Zimmermann et al. (61) used the continuous flow culture to show that when tested under anaerobic conditions, fluconazole and voriconazole exhibit cidal activity, while under aerobic conditions, these agents were static against *Candida* biofilms. Our group investigated the effect of liquid flow shear on *Candida* biofilms using a rotating disc system (19), mimicking catheters placed intravenously that are exposed to shear stress caused by blood flow. Briefly, biofilms were formed on catheter discs and exposed to physiological levels of shear stress using a rotating disc system. Control biofilms were grown under conditions of no flow (19).

Tetrazolium assay and dry weight measurements were used to quantify metabolic activity and biofilm mass, respectively (19). Suci and Tyler (62) described an *in situ* method for assessment of the activity of chlorhexidine against *Candida* biofilms in a flow cell system by monitoring the kinetics of propidium iodide (PI) penetration into the cytoplasm of individual cells during dosing with chlorhexidine. This model allowed monitoring of the rate of PI penetration into the different subpopulations (yeast vs. hyphae) of the biofilm. Hawser et al. (63) showed that *Candida* biofilms formed under flow produced increased levels of extracellular matrix compared to those formed under static conditions. These results were confirmed in a subsequent study by the same group (64). Investigators have also used airflow models to evaluate voice prostheses, since obstruction of airflow is a major cause of early, premature replacement of these devices (65, 66).

Substrate

The role of substrate in modulating the ability of *Candida* to form biofilm has been demonstrated in several studies, which show that different substrates can greatly influence the architecture, morphology, and thickness of biofilms. Hawser and Douglas (22) evaluated various catheter materials and showed that biofilm formation by *C. albicans* was slightly increased on latex or silicone elastomer ($P < 0.05$) compared with polyvinyl chloride but substantially decreased on polyurethane or 100% silicone ($P < 0.001$). Scanning electron microscopy demonstrated that after 48 h, *C. albicans* biofilms consisted of a dense network of yeasts, germ tubes, pseudohyphae, and hyphae; extracellular polymeric material was visible on the surfaces of some of these morphological forms. Our group investigated whether surface modifications of polyetherurethane (Elasthane 80A [E80A]), polycarbonateurethane, and poly (ethyleneterephthalate) can influence fungal biofilm formation (67). We found that biofilm formation by *C. albicans* was significantly reduced on 6PEO-E80A (by 78%) compared to biofilms formed on the nonmodified E80A (optical densities of 0.054 to 0.020 and 0.24 to 0.10, respectively; $P = 0.037$) (67). The total biomass of *Candida* biofilm formed on 6PEO-E80A was 74% lower than that on the unmodified E80A surface (0.46 to 0.15 versus 1.76 to 0.32 mg, respectively; $P = 0.003$). More recently, Estivill et al. (68) evaluated biofilm formation by 84 strains of five *Candida* species on three clinical materials and reported that all tested *Candida* strains were able to form biofilms and that all species showed greater biofilm formation capacity on Teflon, with the exception of *C. glabrata* which displayed higher biofilm formation capacity on polyvinyl chloride.

Taken together, these studies showed that the ability of *Candida* to form biofilms is greatly influenced by the type of material on which it grows and on the species and strain of *Candida*.

Nutrients

Nutrients in the growth media, including sugars, lipids, and serum, are crucial determinants of the biofilm-forming ability of *Candida*. Richards and Russell (69) investigated the effect of sucrose on the colonization of acrylic by *C. albicans* in pure and mixed culture in an artificial mouth and showed that the number of *Candida* cells was significantly increased on acrylic exposed to sucrose, while the number of salivary bacteria was unaffected by sucrose. In a separate study, the growth of *C. albicans* biofilms in medium containing 500 mM galactose or 50 mM glucose reached a maximum after 48 h and then declined;

however, the cell yield was lower in low-glucose medium (22). Swindell et al. (70) determined the effect of parenteral lipid emulsion on *Candida* biofilms formed on medical catheter surfaces. Biofilms were formed on silicone-elastomer catheter discs and analyzed by scanning electron microscopy and confocal laser microscopy. Addition of lipid emulsion to a standard growth medium increased *C. albicans* biofilm production and resulted in changes in biofilm morphology and architecture. Furthermore, lipid emulsion induced germination and supported the growth of *C. albicans*. These findings may explain the increased risk of candidemia in patients receiving lipid emulsion via medical catheters. In a recent study, Samaranayake et al. (71) reported that human serum promotes *C. albicans* biofilm growth on silicone biomaterial and induces the expression of genes associated with adhesion (ALS3 and HWP1) and hydrolase-production (SAP, PLB1, and PLB2).

Species Variability

The ability to form biofilms may vary widely among and between strains of *Candida*. In this regard, in an early study, Branchini et al. (72) used electrophoretic karyotyping and pulsed-field gel electrophoresis to demonstrate genotypic variation and slime production among 31 isolates of *C. parapsilosis* obtained from patients with bloodstream or catheter infections. A total of 14 DNA subtypes were identified among the 31 isolates, of which 80% produced biofilms; biofilm-forming ability among the strains ranged from moderate to strong (67%) to weak (13%). Hawser and Douglas (22) compared biofilm formation by 15 different isolates of *C. albicans* and reported some correlation with pathogenicity: isolates of the less pathogenic *C. parapsilosis* (Glasgow), *Candida pseudotropicalis*, and *C. glabrata* formed significantly less biofilm ($P < 0.001$) than the more pathogenic *C. albicans*. Pfaller et al. (73) reported wide variability in the ability of clinical isolates of *C. parapsilosis* to form biofilms ("slime"). These investigators showed that 65% of the isolates tested produced biofilms (37% were moderately to strongly positive; 28% were weakly positive), and 35% did not form biofilms. A vast majority (83%) of the biofilm-forming isolates were blood and catheter isolates, suggesting that biofilm formation was closely associated with catheter-related bloodstream infections of *Candida*. Kuhn et al. (74) compared biofilms formed by *C. albicans* and *C. parapsilosis* on catheter surfaces using XTT and dry weight assays, followed by fluorescence microscopy and confocal scanning laser microscopy. These investigators reported significant differences in biofilm formation between invasive and noninvasive isolates of *C. albicans* ($P < 0.001$); *C. albicans* isolates produced more biofilm than *C. parapsilosis*, *C. glabrata*, and *C. tropicalis* isolates ($P < 0.001$ for all comparisons). Moreover, *C. albicans* biofilms consisted of a basal blastospore layer with a dense overlying matrix composed of exopolysaccharides and hyphae, while *C. parapsilosis* biofilms were comprised exclusively of clumped blastospores and had less volume than *C. albicans* biofilms. Unlike planktonically grown cells, *Candida* biofilms rapidly (within 6 h) developed fluconazole resistance (MIC > 128 μg/ml).

In a subsequent study, Silva et al. (75) characterized biofilms formed by three non-*albicans Candida* species (*C. parapsilosis*, *C. tropicalis*, and *C. glabrata*) recovered from different sources, using crystal violet staining. All non-*albicans Candida* species were able to form biofilms, although these were less extensive for *C. glabrata* than *C. parapsilosis* and *C. tropicalis*, and *C. parapsilosis* biofilm production was highly strain dependent. Scanning electron microscopy revealed that *C. parapsilosis* biofilm matrix had large amounts of carbohydrate with less protein. Conversely, matrix extracted from *C. tropicalis* biofilms had low amounts of carbohydrate and protein. Interestingly, *C. glabrata* biofilm matrix was high in both protein and carbohydrate content. Parahitiyawa et al. (36) used the Calgary Biofilm Device to evaluate biofilms formed by different *Candida* species and showed that *Candida krusei* developed the largest biofilm mass ($p < 0.05$) relative to *C. albicans*, *C. glabrata*, *Candida dubliniensis*, and *C. tropicalis*. These investigators also reported that *C. krusei* produced a thick multi-layered biofilm of pseudohyphal forms embedded within the polymer matrix, whereas *C. albicans*, *C. dubliniensis*, and *C. tropicalis* biofilms consisted of clusters or chains of cells with sparse extracellular matrix material (34). Lattif et al. (76) characterized biofilm formation by 10 clinical isolates each of *C. parapsilosis*, *Candida orthopsilosis*, and *Candida metapsilosis* and reported that these three species formed biofilms to the same extent, as measured by XTT and biomass assays. However, strain-dependent variations in the metabolic activity of formed biofilms was noted for all three species tested. Scanning electron and confocal microscopy revealed that while the three species formed biofilms with similar topography and architecture, *C. metapsilosis* biofilms showed a trend of lower biofilm thickness compared to *C. parapsilosis* and *C. orthopsilosis*. Estivill et al. (68) demonstrated similar trends of species-dependent biofilm formation by five different *Candida* species.

Taken together, these results demonstrated that biofilm-forming ability, structure, and matrix composition are highly species dependent. In general, *C. albicans* produces quantitatively larger and qualitatively more complex biofilms than other species.

Microbial Cohabitants

The ability of *Candida* to form biofilm is also affected by the presence of additional *Candida* species or of different bacterial cohabitants. In this regard, Holmes et al. (77) reported that *C. albicans* and *C. tropicalis*, two common oral fungi, bind to *Streptococcus gordonii*, while two other *Candida* species (*C. krusei* and *Candida kefyr*) do not. Moreover, there was a positive correlation between the ability of *Candida* to adhere to *S. gordonii* and adherence to experimental salivary pellicle. Whole saliva either stimulated or slightly inhibited adherence of *C. albicans* to *S. gordonii* depending on the streptococcal growth conditions. Reid et al. (78) showed that the ability of *Candida* to form biofilms on fibers and uroepithelial cells is affected by *Lactobacillus*. Fibers precoated with lactobacilli inhibited *Candida* adhesion by 0 to 67%, while lactobacilli exposure resulted in up to 91% displacement of preformed *C. albicans* biofilms. Experiments with uroepithelial cells also showed that the lactobacilli could significantly interfere with the adhesion of *Candida* to the cells, suggesting that members of the normal female urogenital flora might interfere with infections caused by *Candida*.

Webb et al. (79) showed that *S. gordonii* biofilms reduced the adhesion of *Candida* species to polystyrene. However, *Candida* species were able to coaggregate with *S. gordonii* in suspension, with one strain of *C. albicans* (GDH 2346, a denture stomatitis isolate) showing greater coaggregation than the other strains or species. Adam et al. (80) reported that extracellular polymer produced by *S. epidermidis* could inhibit fluconazole penetration in mixed *C. albicans*–bacterial biofilms. Conversely, the presence of *C. albicans* in a biofilm appeared to protect the slime-negative *Staphylococcus* against vancomycin. In a subsequent study, El-Azizi et al. (81) evaluated the interactions between *C. albicans* and 12 other species of *Candida* and bacteria in biofilms and reported reduced biofilm formation by *C. albicans* when the fungus was added to preformed biofilms of non-*albicans Candida* and bacteria. However, when *C. parapsilosis*, *Staphylococcus epidermidis* (a nonglycocalyx producer), or *Serratia marcescens* was added to preformed biofilms of *C. albicans*, the number of cells of the added microbes increased in the growing biofilms, demonstrating a dynamic interaction between *C. albicans* biofilms and other bacteria and fungi. In separate studies, Hogan et al. (82, 83) reported a pathogenic interaction between *Pseudomonas aeruginosa* and *C. albicans*. These investigators showed that *P. aeruginosa* formed a dense biofilm on *C. albicans* filaments and killed the fungus. In contrast, *P. aeruginosa* neither bound nor killed yeast-form *C. albicans*. Park et al. (84) recently reported that coculturing with bacteria decreased the biofilm-forming ability of *C. albicans*. van der Mei et al. (85) evaluated the ability of *C. albicans* and *C. tropicalis* to form biofilms on silicone voice prostheses in the absence and presence of various commensal bacterial strains and *Lactobacillus* strains, and reported that biofilms consisting of combinations of *C. albicans* and a bacterial strain comprised significantly fewer viable organisms than combinations comprising *C. tropicalis*. Moreover, high percentages of *Candida* were found in biofilms grown in combination with lactobacilli.

The mechanisms underlying these interactions within *Candida* biofilms have been proposed to involve host products (e.g., salivary adhesins) as well as microbial proteins (e.g., *Candida* proteins and those produced by bacteria). Holmes et al. (86) reported that binding of *C. albicans* to *S. gordonii* involves multiple adhesin-receptor interactions, including the *S. gordonii* *csh*A and *csh*B genes (encoding high-molecular-mass cell surface polypeptides) and *ssp*A and *ssp*B genes (encoding antigen I/II salivary adhesins). Vilchez et al. (87) reported that *S. mutans* produces trans-2-decenoic acid (SDSF), a fatty acid signaling molecule, which inhibits *HWP1* expression in *C. albicans*, thus affecting fungal biofilm architecture. Studies have also shown that several *P. aeruginosa* virulence factors, including homoserine lactones and phenazine (e.g., pyocyanin), are involved in the inhibition of *Candida* biofilms (82, 83, 88–90).

These studies demonstrate that fungal-fungal and fungal-bacterial interactions play critical roles in modulating the ability of *Candida* to form biofilms. How these interactions relate to differences in microbial communities (bacteriome and mycobiome) within a biofilm is an area that has not been investigated and holds promise for future research efforts.

Candida Products

Studies performed using targeted gene disruptions, microarray-based transcriptomics, proteomics, and genomics have shown that several genes, proteins, and metabolites play critical roles in the maintenance of biofilm phenotype by *Candida* (see reference 157). In the first proteomic analysis of *Candida* biofilms, we identified alcohol dehydrogenase as one of the proteins that can modulate biofilms, by controlling the ethanol-acetaldehyde conversion (91). In a subsequent study, we also performed proteomics analysis of the extracellular matrix of *Candida* biofilms (92). Initially, we compared five methods to isolate the matrix and showed that treatment with EDTA followed by ultrasonication was the optimal method to isolate this component of *Candida* biofilms. Proteomics analysis of biofilm matrix isolated using this optimized method revealed the presence of specific proteins (including glyceraldehyde 3-phosphate dehydrogenase and pyruvate kinase) in the biofilm matrix. Additional *Candida* genes implicated in biofilm formation include *ACE2* (93), *YWP1* (94), *HWP1* (95), *LL34* (*RIX7*) (96), *ALS3* (97, 98), *GAL10* (99), *VPS1* (100), *SUR7* (101), *GUP1* (102), *PEP12* (103), *TPK1/2* (104), *NRG1* (transcriptional repressor) and its target *BRG1* (GATA family transcription factor) (105), *UME6* (transcriptional regulator), *HGC1* (a cyclin-related protein), *SUN41* (a putative cell wall glycosidase), *EFG1* (106, 107), *STV1* and *VPH1* (Golgi/vacuolar subunits of vacuolar proton-translocating ATPase isoforms) (108), *CEK1* (map kinase) (109), *CDK8* (88), *BCR1* (110), *SPT20* (111), and *SAC1* (PIP phosphatase) (112). In addition, quorum sensing molecules (such as 3R-hydroxy-tetradecaenoic acid [3R-HTDE, a beta-oxidation metabolite of endogenously present linoleic acid] [113]), farnesol (114–117), and *cis*-2-dodecenoic acid (BDSF) (118) and metabolic processes (e.g., carbohydrate assimilation, amino acid metabolism, and intracellular transport) (119) and glycolytic flux and hypoxia adaptation (120) have been suggested to play critical roles in *Candida* biofilm formation. The mechanism by which these genes and proteins modulate *Candida* biofilm formation and resistance phenotypes is currently being investigated.

ANTIFUNGAL SUSCEPTIBILITY PROFILE OF *CANDIDA* BIOFILMS

Candida biofilms are well documented to be resistant to commonly used antifungals, including azoles and polyenes (18, 121–124). Chandra et al. (21) evaluated the antifungal susceptibility of *Candida* biofilms formed on denture acrylic *in vitro* and showed that *C. albicans* biofilms exhibited resistance to amphotericin B, nystatin, chlorhexidine, and fluconazole. In contrast, planktonically grown *C. albicans* were susceptible to these agents. Separate studies have reported similar resistance profiles for *C. albicans* and *C. dubliniensis* biofilms formed in microtiter plates (30, 125). Kuhn et al. (126) evaluated *C. albicans* and *C. parapsilosis* biofilms formed on catheter discs and found that lipid formulations of amphotericin B and echinocandins showed activity against *Candida* biofilms. Confocal analyses revealed that treatment with voriconazole, caspofungin, and a lipid formulation of amphotericin B resulted in drug-specific morphological alterations. Bernhardt et al. (127) also reported that voriconazole stopped growth and colonization of *C. albicans* on cover slips in microtiter plates, and treated fungal cells exhibited short, swollen, deformed mycelia. Bachmann et al. (128) evaluated the *in vitro* activity of caspofungin against *C. albicans* biofilms and showed that this echinocandin displayed potent *in vitro* activity against *C. albicans* biofilms.

Scanning electron microscopy and confocal scanning laser microscopy indicated that caspofungin affected the cellular morphology and the metabolic status of cells within the biofilms. Coating of biomaterials with caspofungin had an inhibitory effect on subsequent biofilm development by *C. albicans*. Aminocandin, a newer echinocandin, has also been shown to exhibit antibiofilm properties (129). Recently, Kaneko et al. (130) performed time-lapse microscopic observation of the effect of micafungin (an echinocandin) and fluconazole on *Candida* biofilms formed for up to 24 h on silicon disks in RPMI medium under flow (20 ml/h). These investigators showed that *Candida* biofilms grew at a uniform rate in the absence of drugs (17.2 ± 1.3 μm/h) and observed detachment of clusters of fungal cells from the hyphal tips in mature biofilms. Moreover, although neither drug eradicated biofilms, fluconazole exhibited an antibiofilm effect against early-phase (5-h grown) biofilms after 15 h of incubation. In contrast, micafungin suppressed biofilm growth within minutes after addition of the drug, with disruption of cells in the biofilms and release of undefined extracellular string-like structures from the burst hyphae.

Echinocandins, especially caspofungin, may also exhibit a paradoxical effect on *Candida* biofilms, defined as a resurgence of growth at drug concentrations above the MIC (131). These investigators reported that all *Candida* isolates (except *C. tropicalis*) displayed a paradoxical effect more frequently when grown as biofilms compared to planktonic cells. A paradoxical effect of echinocandin can also be discerned in the study by Kaneko et al. (130), who compared the antibiofilm activity of micafungin against early-phase biofilms after continued exposure for up to 24 h (described above). These investigators reported that after 15 h of incubation, micafungin-exposed biofilms exhibited some regrowth, compared to exposure for 5 min, when almost all the biofilm was inhibited. A paradoxical effect was associated with microscopic changes in cell morphology, manifested as the accumulation of enlarged, globose cells, suggesting drug-induced changes in cell wall composition as the mechanism underlying the paradoxical effect of echinocandins.

Several experimental agents are currently under investigation as potential antibiofilm drugs for *Candida* biofilms. These include chlorhexidine, sodium hypochlorite, zosteric acid, filastatin, EDTA/ethanol catheter lock solutions, gentian violet, and essential oils (62, 79, 132–142). In addition, physical interventions such as low-level laser (143), photodynamic therapy (144–147), and antimicrobial coating of catheters (148–151) have also been proposed as possible therapeutic alternatives. More detailed investigations are warranted to determine the efficacy of these agents against *Candida* biofilms.

MECHANISMS OF RESISTANCE OF *CANDIDA* BIOFILMS

The development of various models has allowed detailed evaluation and understanding of the mechanisms underlying *C. albicans* biofilm resistance. These methods include studying alterations in drug targets involving changes in membrane sterol, membrane localized drug efflux pump assays at the functional and transcriptional level, and reduced or limited drug penetration through biofilms.

The cellular target for azoles is a 14-α demethylase enzyme involved in the ergosterol biosynthetic pathway. Alterations in sterol composition are linked to antifungal resistance. Our group for the first time developed methods involving isolation of membrane sterols from biofilms (20). Briefly, total membrane sterols were isolated from biofilms and planktonic cells and were analyzed by gas liquid chromatography (20). These studies show that the ergosterol levels of biofilms grown to the intermediate and mature phases were reduced by 41 and 50%, respectively, compared to early-phase *C. albicans* biofilm. These results showed that the level of sterols is modulated during *C. albicans* biofilm formation and suggested that such modulation may contribute to drug resistance in a phase-specific manner (20). We also standardized an assay based on the efflux of Rhodamine 123 (Rh123), a fluorescent substrate for drug efflux proteins, to evaluate the functionality of efflux pump proteins (CDR/MDR proteins). Resistant cells overexpressing functional efflux pumps do not retain Rh123, while susceptible cells, which lack or have a low number of these pumps, retain the fluorescent dye, which is quantified by fluorescence measurements (152). Our results showed that in early-phase biofilms, efflux pumps contributed to antifungal resistance, while in mature-phase biofilms, resistance was associated with changes in the levels of ergosterol biosynthesis intermediates (152). The role of efflux pumps in biofilm-associated resistance was confirmed in a separate study by Mateus et al. (153), who evaluated efflux pump activity at the transcriptional level and showed that adherence of *C. albicans* to silicone induces immediate enhanced tolerance to fluconazole and that expression of *MDR1* and *CDR1* genes was significantly lower in daughter cells from 48-h biofilms than in firmly adherent cells (2 h after attachment), suggesting that efflux pump expression in adherent cultures is transient.

Next, our group investigated whether drug binding/penetration plays a role in the resistance of *C. albicans* biofilms against fluconazole. We performed preliminary studies by using equilibrium dialysis and diffusion bioassay methods (154). Briefly, the ability of fluconazole to bind/penetrate *Candida* biofilms formed on cellulose membrane was determined by using equilibrium dialysis equipment, which consists of two chambers (1 ml volume each) separated by the membrane. To form biofilm on the membrane, a fungal cell suspension (1 x 10^7 cells in yeast nitrogen base media) was added to chamber 1;

the other chamber was filled with 1 ml yeast nitrogen base media, and the apparatus was incubated at 37°C for 48 h (154). After biofilm formation on the membrane, fluconazole (4, 64, 256, or 1,024 μg/ml) was added to chamber 1 and allowed to equilibrate for 48 h, and the amount of free drug in each chamber was determined as inhibitory zones using a diffusion bioassay (154). At a concentration of 64 μg/ml, fluconazole was equally distributed in the two chambers of the equilibrium dialyzer, indicating that the drug freely penetrated. In contrast, when *C. albicans* biofilm was incubated with a higher concentration of fluconazole (256 μg/ml or 1,024 μg/ml), the free drug equilibrated between chambers 1 and 2 accounted for 200 μg/ml, while 56 μg/ml fluconazole was bound to the biofilm. Furthermore, incubation of *Candida* biofilm with 1,024 μg/ml fluconazole also resulted in binding of 56 μg/ml of the drug, indicating saturation of biofilm at high concentrations. These studies showed that at clinically relevant low concentrations, fluconazole did not bind to the biofilm, suggesting that drug binding/penetration does not play a major role in azole resistance of *C. albicans* biofilms (154).

CONCLUSION

Recent advances have resulted in the development of an array of new tools and techniques to analyze *Candida* biofilms at the morphological, physiological, biochemical, and molecular levels, providing in-depth insight into their biology and pathogenesis. This new knowledge will fuel future investigations that are likely to lead to better management of diseases associated with fungal biofilms. Finally, the findings that fungi-fungi and fungi-bacteria affect each other in a mixed biofilm environment point to the need to understand how biofilms are influenced and the role of these interactions as components of microbial communities such as the mycobiome and bacteriome.

ACKNOWLEDGMENTS

Funding support is acknowledged from the NIH/NIDCR (RO1DE17846 and the Oral HIV AIDS Research Alliance [BRS-ACURE-S-11-000049-110229 and AI-U01-68636]) to MAG; NIH/NIDCR (R01DE024228) to MAG and PKM; NIH/NEI and NIH/NIAID (R21EY021303 and R21AI074077), pilot funding from the Infectious Diseases Drug Development Center (IDDDC, Case), the National Eczema Association (Research Grant), and the National Psoriasis Foundation (Lozick Discovery Research Grant) to PKM; and the CWRU/UH Center for AIDS Research (CFAR, NIH grant number P30 AI036219).

Conflicts of interest: We disclose no conflicts.

CITATION

Chandra J, Mukherjee PK. 2015. *Candida* biofilms: development, architecture, and resistance. Microbiol Spectrum 3(4):MB-0020-2015.

REFERENCES

1. **Nicastri E, Petrosiillo N, Viale P, Ippolito G.** 2001. Catheter-related bloodstream infections in HIV-infected patients. *Ann NY Acad Sci* **946:**274–290.
2. **Costerton JW, Cheng KJ, Geesey GG, Ladd TI, Nickel JC, Dasgupta M, Marrie TJ.** 1987. Bacterial biofilms in nature and disease. *Annu Rev Microbiol* **41:**435–464.
3. **Costerton JW, Lewandowski Z, Caldwell DE, Korber DR, Lappin-Scott HM.** 1995. Microbial biofilms. *Annu Rev Microbiol* **49:** 711–745.
4. **Costerton JW, Stewart PS, Greenberg EP.** 1999. Bacterial biofilms: a common cause of persistent infections. *Science* **284:**1318–1322.
5. **Donlan RM.** 2002. Biofilms: microbial life on surfaces. *Emerg Infect Dis* **8:**881–890.
6. **Edmond MB, Wallace SE, McClish DK, Pfaller MA, Jones RN, Wenzel RP.** 1999. Nosocomial bloodstream infections in United States hospitals: a three-year analysis. *Clin Infect Dis* **29:**239–244.
7. **Andes DR, Safdar N, Baddley JW, Playford G, Reboli AC, Rex JH, Sobel JD, Pappas PG,**

Kullberg BJ. 2012. Impact of treatment strategy on outcomes in patients with candidemia and other forms of invasive candidiasis: a patient-level quantitative review of randomized trials. *Clin Infect Dis* **54:**1110–1122.

8. **Costa SF, Marinho I, Araujo EA, Manrique AE, Medeiros EA, Levin AS.** 2000. Nosocomial fungaemia: a 2-year prospective study. *J Hosp Infect* **45:**69–72.
9. **Moran C, Grussemeyer CA, Spalding JR, Benjamin DK Jr, Reed SD.** 2009. *Candida albicans* and non-*albicans* bloodstream infections in adult and pediatric patients: comparison of mortality and costs. *Pediatr Infect Dis J* **28:**433–435.
10. **Viudes A, Peman J, Canton E, Ubeda P, Lopez-Ribot JL, Gobernado M.** 2002. Candidemia at a tertiary-care hospital: epidemiology, treatment, clinical outcome and risk factors for death. *Eur J Clin Microbiol Infect Dis* **21:**767–774.
11. **Marrie TJ, Costerton JW.** 1984. Scanning and transmission electron microscopy of *in situ* bacterial colonization of intravenous and intraarterial catheters. *J Clin Microbiol* **19:**687–693.
12. **Tchekmedyian NS, Newman K, Moody MR, Costerton JW, Aisner J, Schimpff SC, Reed WP.** 1986. Special studies of the Hickman catheter of a patient with recurrent bacteremia and candidemia. *Am J Med Sci* **291:**419–424.
13. **Reid G, Denstedt JD, Kang YS, Lam D, Nause C.** 1992. Microbial adhesion and biofilm formation on ureteral stents *in vitro* and *in vivo*. *J Urol* **148:**1592–1594.
14. **Elder MJ, Matheson M, Stapleton F, Dart JK.** 1996. Biofilm formation in infectious crystalline keratopathy due to *Candida albicans*. *Cornea* **15:**301–304.
15. **Chandra J, Mukherjee PK, Ghannoum MA.** 2010. Fungal biofilms in the clinical lab setting. *Curr Rep Fungal Inf* **4:**137–144.
16. **Nett J, Andes DR.** 2006. *Candida albicans* biofilm development, modeling a host-pathogen interaction. *Curr Opin Microbiol* **9:**340–345.
17. **Douglas LJ.** 2003. *Candida* biofilms and their role in infection. *Trends Microbiol* **11:**30–36.
18. **Baillie GS, Douglas LJ.** 1999. *Candida* biofilms and their susceptibility to antifungal agents. *Methods Enzymol* **310:**644–656.
19. **Mukherjee PK, Chand DV, Chandra J, Anderson JM, Ghannoum MA.** 2009. Shear stress modulates the thickness and architecture of *Candida albicans* biofilms in a phase-dependent manner. *Mycoses* **52:**440–446.
20. **Chandra J, Kuhn DM, Mukherjee PK, Hoyer LL, McCormick T, Ghannoum MA.** 2001. Biofilm formation by the fungal pathogen *Candida albicans*: development, architecture and drug resistance. *J Bacteriol* **183:**5385–5394.
21. **Chandra J, Mukherjee PK, Leidich SD, Faddoul FF, Hoyer LL, Douglas LJ, Ghannoum MA.** 2001. Antifungal resistance of candidal biofilms formed on denture acrylic *in vitro*. *J Dent Res* **80:**903–908.
22. **Hawser SP, Douglas LJ.** 1994. Biofilm formation by *Candida* species on the surface of catheter materials *in vitro*. *Infect Immun* **62:**915–921.
23. **Imamura Y, Chandra J, Mukherjee PK, Abdul Lattif A, Szczotka-Flynn LB, Pearlman E, Lass JH, O'Donnell K, Ghannoum MA.** 2008. *Fusarium* and *Candida albicans* biofilms on soft contact lenses: model development, influence of lens type and susceptibility to lens care solutions. *Antimicrob Agents Chemother* **52:**171–182.
24. **Nikawa H, Yamamoto T, Hamada T.** 1995. Effect of components of resilient denture-lining materials on the growth, acid production and colonization of *Candida albicans*. *J Oral Rehabil* **22:**817–824.
25. **Ramage G, Vande WK, Wickes BL, Lopez-Ribot JL.** 2001. Biofilm formation by *Candida dubliniensis*. *J Clin Microbiol* **39:**3234–3240.
26. **Samaranayake YH, Ye J, Yau JYY, Cheung BPK, Samaranayake LP.** 2005. *In vitro* method to study antifungal perfusion in *Candida* biofilms. *J Clin Microbiol* **43:**818–825.
27. **Everaert EP, van de Belt-Gritter B, van der Mei HC, Busscher HJ, Verkerke GJ, Dijk F, Mahieu HF, Reitsma A.** 1998. *In vitro* and *in vivo* microbial adhesion and growth on argon plasma-treated silicone rubber voice prostheses. *J Mater Sci Mater Med* **9:**147–157.
28. **van der Mei HC, Free RH, Elving GJ, van Weissenbruch R, Albers FW, Busscher HJ.** 2000. Effect of probiotic bacteria on prevalence of yeasts in oropharyngeal biofilms on silicone rubber voice prostheses *in vitro*. *J Med Microbiol* **49:**713–718.
29. **Chandra J, Mukherjee PK, Ghannoum MA.** 2008. *In vitro* growth and analysis of *Candida* biofilms. *Nat Protoc* **3:**1909–1924.
30. **Ramage G, VandeWalle K, Wickes BL, Lopez-Ribot JL.** 2001. Standardized method for *in vitro* antifungal susceptibility testing of *Candida albicans* biofilms. *Antimicrob Agents Chemother* **45:**2475–2479.
31. **Nweze EI, Ghannoum A, Chandra J, Ghannoum MA, Mukherjee PK.** 2012.

Development of a 96-well catheter-based microdilution method to test antifungal susceptibility of *Candida* biofilms. *J Antimicrob Chemother* **67:**149–153.

32. **Nweze EI, Ghannoum A, Chandra J, Ghannoum MA, Mukherjee PK.** 2010. Microdilution method to test antifungal susceptibility of biofilms (BFs) formed by *Candida* (CA) on catheters. Presented at the 50th Interscience Conference on Antimicrobial Agents and Chemotherapy (ICAAC); Abstract Number M-1101; Sep 12-15. Boston, MA, American Society for Microbiology.
33. **Ceri H, Olson ME, Stremick C, Read RR, Morck D, Buret A.** 1999. The Calgary Biofilm Device: new technology for rapid determination of antibiotic susceptibilities of bacterial biofilms. *J Clin Microbiol* **37:**1771–1776.
34. **Almshawit H, Macreadie I, Grando D.** 2014. A simple and inexpensive device for biofilm analysis. *J Microbiol Methods* **98:**59–63.
35. **Harrison JJ, Ceri H, Yerly J, Rabiei M, Hu Y, Martinuzzi R, Turner RJ.** 2007. Metal ions may suppress or enhance cellular differentiation in *Candida albicans* and *Candida tropicalis* biofilms. *Appl Environ Microbiol.* [Epub ahead of print.] doi:AEM.02711-06.
36. **Parahitiyawa NB, Samaranayake YH, Samaranayake LP, Ye J, Tsang PW, Cheung BP, Yau JY, Yeung SK.** 2006. Interspecies variation in *Candida* biofilm formation studied using the Calgary biofilm device. *APMIS* **114:** 298–306.
37. **Harrison JJ, Turner RJ, Ceri H.** 2007. A subpopulation of *Candida albicans* and *Candida tropicalis* biofilm cells are highly tolerant to chelating agents. *FEMS Microbiol Lett* **272:** 172–181.
38. **Srinivasan A, Uppuluri P, Lopez-Ribot J, Ramasubramanian AK.** 2011. Development of a high-throughput *Candida albicans* biofilm chip. *PLoS One* **6:**e19036. doi:10.3791/3845.
39. **Coenye T.** 2010. Response of sessile cells to stress: from changes in gene expression to phenotypic adaptation. *FEMS Immunol Med Microbiol* **59:**239–252.
40. **Andes DR, Nett J, Oschel P, Albrecht R, Marchillo K, Pitula A.** 2004. Development and characterization of an *in vivo* central venous catheter *Candida albicans* biofilm model. *Infect Immun* **72:**6023–6031.
41. **Schinabeck MK, Long LA, Hossain MA, Chandra J, Mukherjee PK, Mohamed S, Ghannoum MA.** 2004. Rabbit model of *Candida albicans* biofilm infection: liposomal Amphotericin B antifungal lock therapy. *Antimicrob Agents Chemother* **48:**1727–1732.
42. **Zumbuehl A, Ferreira L, Kuhn D, Astashkina A, Long L, Yeo Y, Iaconis T, Ghannoum M, Fink GR, Langer R, Kohane DS.** 2007. Antifungal hydrogels. *Proc Natl Acad Sci USA* **104:**12994–12998.
43. **Mukherjee PK, Chandra J, Yu C, Sun Y, Pearlman E, Ghannoum MA.** 2012. Characterization of *Fusarium* keratitis outbreak isolates: contribution of biofilms to antimicrobial resistance and pathogenesis. *Invest Ophthalmol Vis Sci* **53:**4450–4457.
44. **Sun Y, Chandra J, Mukherjee PK, Szczotka-Flynn L, Ghannoum M, Pearlman E.** 2010. A murine model of contact lens associated *Fusarium* keratitis. *Invest Ophthalmol Vis Sci* **51:**1511–1516.
45. **Chang DC, Grant GB, O'Donnell K, Wannemuehler KA, Noble-Wang J, Rao CY, Jacobson LM, Crowell CS, Sneed RS, Lewis FM, Schaffzin JK, Kainer MA, Genese CA, Alfonso EC, Jones DB, Srinivasan A, Fridkin SK, Park BJ.** 2006. Multistate outbreak of *Fusarium* keratitis associated with use of a contact lens solution. *JAMA* **296:**953–963.
46. **Donnio A, Van Nuoi DNG, Catanese M, Desbois N, Ayeboua L, Merle H.** 2007. Outbreak of keratomycosis attributable to *Fusarium solani* in the French West Indies. *Am J Ophthalmol* **143:**356–358.
47. **Khor WB, Aung T, Saw SM, Wong TY, Tambyah PA, Tan AL, Beuerman R, Lim L, Chan WK, Heng WJ, Lim J, Loh RS, Lee SB, Tan DT.** 2006. An outbreak of *Fusarium* keratitis associated with contact lens wear in Singapore. *JAMA* **295:**2867–2873.
48. **Tournu H, Van Dijck P.** 2012. *Candida* biofilms and the host: models and new concepts for eradication. *Int J Microbiol.* doi:10.1155/2012/845352.
49. **Dongari-Bagtzoglou A, Kashleva H, Dwivedi P, Diaz P, Vasilakos J.** 2009. Characterization of mucosal *Candida albicans* biofilms. *PLoS ONE* **4:** e7967. doi:10.1371/journal.pone.0007967.
50. **Harriott MM, Lilly EA, Rodriguez TE, Fidel PL Jr, Noverr MC.** 2010. *Candida albicans* forms biofilms on the vaginal mucosa. *Microbiology* **156:**3635–3644.
51. **Lazarus HM, Lowder JN, Herzig RH.** 1983. Occlusion and infection in Broviac catheters during intensive cancer therapy. *Cancer* **52:** 2342–2348.
52. **Chandra J, Kuhn DM, Mukherjee PK, Hoyer LL, McCormick T, Ghannoum MA.** 2001. Biofilm formation by the fungal pathogen *Candida albicans*: development, architecture, and drug resistance. *J Bacteriol* **183:**5385–5394.

53. **Hall-Stoodley L, Costerton JW, Stoodley P.** 2004. Bacterial biofilms: from the natural environment to infectious diseases. *Nat Rev Microbiol* **2:**95–108.
54. **Hall-Stoodley L, Stoodley P.** 2002. Developmental regulation of microbial biofilms. *Curr Opin Biotechnol* **13:**228–233.
55. **Kolenbrander PE.** 2000. Oral microbial communities: biofilms, interactions, and genetic systems. *Annu Rev Microbiol* **54:**413–437.
56. **Kolenbrander PE, Palmer RJ, Periasamy S, Jakubovics NS.** 2010. Oral multispecies biofilm development and the key role of cell–cell distance. *Nat Rev Microbiol* **8:**471–480.
57. **Busscher HJ, Geertsema-Doornbusch GI, van der Mei HC.** 1997. Adhesion to silicone rubber of yeasts and bacteria isolated from voice prostheses: influence of salivary conditioning films. *J Biomed Mater Res* **34:**201–209.
58. **Millsap KW, Bos R, Busscher HJ, van der Mei HC.** 1999. Surface aggregation of *Candida albicans* on glass in the absence and presence of adhering *Streptococcus gordonii* in a parallel-plate flow chamber: a surface thermodynamical analysis based on acid-base interactions. *J Colloid Interface Sci* **212:**495–502.
59. **Roosjen A, Boks NP, van der Mei HC, Busscher HJ, Norde W.** 2005. Influence of shear on microbial adhesion to PEO-brushes and glass by convective-diffusion and sedimentation in a parallel plate flow chamber. *Colloids Surf B Biointerfaces* **46:**1–6.
60. **Millsap KW, Bos R, van der Mei HC, Busscher HJ.** 1999. Adhesion and surface-aggregation of *Candida albicans* from saliva on acrylic surfaces with adhering bacteria as studied in a parallel plate flow chamber. *Antonie Van Leeuwenhoek* **75:**351–359.
61. **Zimmermann K, Bernhardt J, Knoke M, Bernhardt H.** 2002. Influence of voriconazole and fluconazole on *Candida albicans* in long-time continuous flow culture. *Mycoses* **45:**41–46.
62. **Suci PA, Tyler BJ.** 2002. Action of chlorhexidine digluconate against yeast and filamentous forms in an early-stage *Candida albicans* biofilm. *Antimicrob Agents Chemother* **46:**3522–3531.
63. **Hawser SP, Baillie GS, Douglas LJ.** 1998. Production of extracellular matrix by *Candida albicans* biofilms. *J Med Microbiol* **47:**253–256.
64. **Al-Fattani MA, Douglas LJ.** 2006. Biofilm matrix of *Candida albicans* and *Candida tropicalis*: chemical composition and role in drug resistance. *J Med Microbiol* **55:**999–1008.
65. **Elving GJ, van der Mei HC, Busscher HJ, van Weissenbruch R, Albers FW.** 2001. Air-flow resistances of silicone rubber voice prostheses after formation of bacterial and fungal biofilms. *J Biomed Mater Res* **58:**421–426.
66. **Elving GJ, van der Mei HC, Busscher HJ, van Weissenbruch R, Albers FW.** 2002. Comparison of the microbial composition of voice prosthesis biofilms from patients requiring frequent versus infrequent replacement. *Ann Otol Rhinol Laryngol* **111:**200–203.
67. **Chandra J, Patel JD, Li J, Zhou G, Mukherjee PK, McCormick TS, Anderson JM, Ghannoum MA.** 2005. Modification of surface properties of biomaterials influences the ability of *C. albicans* to form biofilms. *Appl Environ Microbiol* **71:**8795–8801.
68. **Estivill D, Arias A, Torres-Lana A, Carrillo-Munoz AJ, Arevalo MP.** 2011. Biofilm formation by five species of *Candida* on three clinical materials. *J Microbiol Methods* **86:**238–242.
69. **Richards S, Russell C.** 1987. The effect of sucrose on the colonization of acrylic by *Candida albicans* in pure and mixed culture in an artificial mouth. *J Appl Bacteriol* **62:**421–427.
70. **Swindell K, Lattif AA, Chandra J, Mukherjee PK, Ghannoum MA.** 2009. Parenteral lipid emulsion induces germination of *Candida albicans* and increases biofilm formation on medical catheter surfaces. *J Infect Dis* **200:**473–480.
71. **Samaranayake YH, Cheung BP, Yau JY, Yeung SK, Samaranayake LP.** 2013. Human serum promotes *Candida albicans* biofilm growth and virulence gene expression on silicone biomaterial. *PLoS One* **8:**e62902. doi:10.1371/journal.pone.0062902.
72. **Branchini ML, Pfaller MA, Rhine-Chalberg J, Frempong T, Isenberg HD.** 1994. Genotypic variation and slime production among blood and catheter isolates of *Candida parapsilosis*. *J Clin Microbiol* **32:**452–456.
73. **Pfaller MA, Messer SA, Hollis RJ.** 1995. Variations in DNA subtype, antifungal susceptibility, and slime production among clinical isolates of *Candida parapsilosis*. *Diagn Microbiol Infect Dis* **21:**9–14.
74. **Kuhn DM, Chandra J, Mukherjee PK, Ghannoum MA.** 2002. Comparison of biofilms formed by *Candida albicans* and *Candida parapsilosis* on bioprosthetic surfaces. *Infect Immun* **70:**878–888.
75. **Silva S, Henriques M, Martins A, Oliveira R, Williams D, Azeredo J.** 2009. Biofilms of non-*Candida albicans Candida* species:

quantification, structure and matrix composition. *Med Mycol* **47:**681–689.

76. **Lattif AA, Mukherjee PK, Chandra J, Swindell K, Lockhart SR, Diekema DJ, Pfaller MA, Ghannoum MA.** 2010. Characterization of biofilms formed by *Candida parapsilosis, C. metapsilosis,* and *C. orthopsilosis. Int J Med Microbiol* **300:**265–270.
77. **Holmes AR, Cannon RD, Jenkinson HF.** 1995. Interactions of *Candida albicans* with bacteria and salivary molecules in oral biofilms. *J Ind Microbiol* **15:**208–213.
78. **Reid G, Tieszer C, Lam D.** 1995. Influence of lactobacilli on the adhesion of *Staphylococcus aureus* and *Candida albicans* to fibers and epithelial cells. *J Ind Microbiol* **15:**248–253.
79. **Webb BC, Willcox MD, Thomas CJ, Harty DW, Knox KW.** 1995. The effect of sodium hypochlorite on potential pathogenic traits of *Candida albicans* and other *Candida* species. *Oral Microbiol Immunol* **10:**334–341.
80. **Adam B, Baillie GS, Douglas LJ.** 2002. Mixed species biofilms of *Candida albicans* and *Staphylococcus epidermidis. J Med Microbiol* **51:**344–349.
81. **El-Azizi MA, Starks SE, Khardori N.** 2004. Interactions of *Candida albicans* with other *Candida* spp. and bacteria in the biofilms. *J Appl Microbiol* **96:**1067–1073.
82. **Hogan DA, Kolter R.** 2002. *Pseudomonas - Candida* interactions: an ecological role for virulence factors. *Science* **296:**2229–2232.
83. **Hogan DA, Vik A, Kolter R.** 2004. A *Pseudomonas aeruginosa* quorum-sensing molecule influences *Candida albicans* morphology. *Mol Microbiol* **54:**1212–1223.
84. **Park SJ, Han KH, Park JY, Choi SJ, Lee KH.** 2014. Influence of bacterial presence on biofilm formation of *Candida albicans. Yonsei Med J* **55:**449–458.
85. **van der Mei HC, Buijssen KJ, van der Laan BF, Ovchinnikova E, Geertsema-Doornbusch GI, Atema-Smit J, van de Belt-Gritter B, Busscher HJ.** 2014. Voice prosthetic biofilm formation and *Candida morphogenic* conversions in absence and presence of different bacterial strains and species on silicone-rubber. *PLoS One* **9:**e104508. doi:10.1371/journal.pone.0104508.
86. **Holmes AR, McNab R, Jenkinson HF.** 1996. *Candida albicans* binding to the oral bacterium *Streptococcus gordonii* involves multiple adhesin-receptor interactions. *Infect Immun* **64:**4680–4685.
87. **Vilchez R, Lemme A, Ballhausen B, Thiel V, Schulz S, Jansen R, Sztajer H, Wagner-Dobler I.** 2010. *Streptococcus mutans* inhibits *Candida albicans* hyphal formation by the fatty acid signaling molecule trans-2-decenoic acid (SDSF). *ChemBioChem* **11:**1552–1562.
88. **Lindsay AK, Morales DK, Liu Z, Grahl N, Zhang A, Willger SD, Myers LC, Hogan DA.** 2014. Analysis of *Candida albicans* mutants defective in the Cdk8 module of mediator reveal links between metabolism and biofilm formation. *PLoS Genet* **10:**e1004567. doi:10.1371/journal.pgen.1004567.
89. **Mear JB, Kipnis E, Faure E, Dessein R, Schurtz G, Faure K, Guery B.** 2013. *Candida albicans* and *Pseudomonas aeruginosa* interactions: more than an opportunistic criminal association? *Med Mal Infect* **43:**146–151.
90. **Morales DK, Grahl N, Okegbe C, Dietrich LE, Jacobs NJ, Hogan DA.** 2013. Control of *Candida albicans* metabolism and biofilm formation by *Pseudomonas aeruginosa* phenazines. *MBio* **4:**e00526-12. doi:10.1128/mBio.00526-12.
91. **Mukherjee PK, Mohamed S, Chandra J, Kuhn D, Liu S, Antar OS, Munyon R, Mitchell AP, Andes D, Chance MR, Rouabhia M, Ghannoum MA.** 2006. Alcohol dehydrogenase restricts the ability of the pathogen *Candida albicans* to form a biofilm on catheter surfaces through an ethanol-based mechanism. *Infect Immun* **74:**3804–3816.
92. **Lattif AA, Chandra J, Chang J, Liu S, Zhou G, Chance MR, Ghannoum MA, Mukherjee PK.** 2008. Proteomic and pathway analyses reveal phase-dependent over-expression of proteins associated with carbohydrate metabolic pathways in *Candida albicans* biofilms. *Open Proteom J* **1:**5–26.
93. **Kelly MT, MacCallum DM, Clancy SD, Odds FC, Brown AJ, Butler G.** 2004. The *Candida albicans* CaACE2 gene affects morphogenesis, adherence and virulence. *Mol Microbiol* **53:**969–983.
94. **Granger BL, Flenniken ML, Davis DA, Mitchell AP, Cutler JE.** 2005. Yeast wall protein 1 of *Candida albicans. Microbiology* **151:**1631–1644.
95. **Orsi CF, Borghi E, Colombari B, Neglia RG, Quaglino D, Ardizzoni A, Morace G, Blasi E.** 2014. Impact of *Candida albicans* hyphal wall protein 1 (HWP1) genotype on biofilm production and fungal susceptibility to microglial cells. *Microb Pathog* **69–70:**20–27.
96. **Melo AS, Padovan AC, Serafim RC, Puzer L, Carmona AK, Juliano Neto L, Brunstein A, Briones MR.** 2006. The *Candida albicans* AAA ATPase homologue of *Saccharomyces cerevisiae* Rix7p (YLL034c) is essential for proper morphology, biofilm formation and

activity of secreted aspartyl proteinases. *Genet Mol Res* **5:**664–687.

97. **Dranginis AM, Rauceo JM, Coronado JE, Lipke PN.** 2007. A biochemical guide to yeast adhesins: glycoproteins for social and antisocial occasions. *Microbiol Mol Biol Rev* **71:**282–294.
98. **Zhao X, Daniels KJ, Oh SH, Green CB, Yeater KM, Soll DR, Hoyer LL.** 2006. *Candida albicans* Als3p is required for wild-type biofilm formation on silicone elastomer surfaces. *Microbiology* **152:**2287–2299.
99. **Singh V, Satheesh SV, Raghavendra ML, Sadhale PP.** 2007. The key enzyme in galactose metabolism, UDP-galactose-4-epimerase, affects cell-wall integrity and morphology in *Candida albicans* even in the absence of galactose. *Fungal Genet Biol* **44:**563–574.
100. **Bernardo SM, Khalique Z, Kot J, Jones JK, Lee SA.** 2008. *Candida albicans* VPS1 contributes to protease secretion, filamentation, and biofilm formation. *Fungal Genet Biol* **45:**861–877.
101. **Bernardo SM, Lee SA.** 2010. *Candida albicans* SUR7 contributes to secretion, biofilm formation, and macrophage killing. *BMC Microbiol* **10:**133.
102. **Ferreira C, Silva S, Faria-Oliveira F, Pinho E, Henriques M, Lucas C.** 2010. *Candida albicans* virulence and drug-resistance requires the O-acyltransferase Gup1p. *BMC Microbiol* **10:**238.
103. **Palanisamy SK, Ramirez MA, Lorenz M, Lee SA.** 2010. *Candida albicans* PEP12 is required for biofilm integrity and *in vivo* virulence. *Eukaryot Cell* **9:**266–277.
104. **Giacometti R, Kronberg F, Biondi RM, Passeron S.** 2011. *Candida albicans* Tpk1p and Tpk2p isoforms differentially regulate pseudohyphal development, biofilm structure, cell aggregation and adhesins expression. *Yeast* **28:**293–308.
105. **Cleary IA, Lazzell AL, Monteagudo C, Thomas DP, Saville SP.** 2012. BRG1 and NRG1 form a novel feedback circuit regulating *Candida albicans* hypha formation and virulence. *Mol Microbiol* **85:**557–573.
106. **Banerjee M, Uppuluri P, Zhao XR, Carlisle PL, Vipulanandan G, Villar CC, Lopez-Ribot JL, Kadosh D.** 2013. Expression of UME6, a key regulator of *Candida albicans* hyphal development, enhances biofilm formation via Hgc1- and Sun41-dependent mechanisms. *Eukaryot Cell* **12:**224–232.
107. **Connolly LA, Riccombeni A, Grozer Z, Holland LM, Lynch DB, Andes DR, Gacser A, Butler G.** 2013. The APSES transcription factor Efg1 is a global regulator that controls morphogenesis and biofilm formation in *Candida parapsilosis. Mol Microbiol* **90:**36–53.
108. **Raines SM, Rane HS, Bernardo SM, Binder JL, Lee SA, Parra KJ.** 2013. Deletion of vacuolar proton-translocating ATPase V(o)a isoforms clarifies the role of vacuolar pH as a determinant of virulence-associated traits in *Candida albicans. J Biol Chem* **288:**6190–6201.
109. **Herrero-de-Dios C, Alonso-Monge R, Pla J.** 2014. The lack of upstream elements of the Cek1 and Hog1 mediated pathways leads to a synthetic lethal phenotype upon osmotic stress in *Candida albicans. Fungal Genet Biol* **69:**31–42.
110. **Pannanusorn S, Ramirez-Zavala B, Lunsdorf H, Agerberth B, Morschhauser J, Romling U.** 2014. Characterization of biofilm formation and the role of BCR1 in clinical isolates of *Candida parapsilosis. Eukaryot Cell* **13:**438–451.
111. **Tan X, Fuchs BB, Wang Y, Chen W, Yuen GJ, Chen RB, Jayamani E, Anastassopoulou C, Pukkila-Worley R, Coleman JJ, Mylonakis E.** 2014. The role of *Candida albicans* SPT20 in filamentation, biofilm formation and pathogenesis. *PLoS One* **9:**e94468. doi:10.1371/journal.pone.0094468.
112. **Zhang B, Yu Q, Jia C, Wang Y, Xiao C, Dong Y, Xu N, Wang L, Li M.** 2015. The actin-related protein Sac1 is required for morphogenesis and cell wall integrity in *Candida albicans. Fungal Genet Biol.* [Epub ahead of print.] doi:10.1016/j.fgb.2014.12.007.
113. **Nigam S, Ciccoli R, Ivanov I, Sczepanski M, Deva R.** 2011. On mechanism of quorum sensing in Candida albicans by 3(R)-hydroxy-tetradecaenoic acid. *Curr Microbiol* **62:**55–63.
114. **Deveau A, Hogan DA.** 2011. Linking quorum sensing regulation and biofilm formation by *Candida albicans. Methods Mol Biol* **692:**219–233.
115. **Pammi M, Liang R, Hicks JM, Barrish J, Versalovic J.** 2011. Farnesol decreases biofilms of *Staphylococcus epidermidis* and exhibits synergy with nafcillin and vancomycin. *Pediatr Res* **70:**578–583.
116. **Weber K, Schulz B, Ruhnke M.** 2010. The quorum-sensing molecule E,E-farnesol: its variable secretion and its impact on the growth and metabolism of *Candida species. Yeast* **27:**727–739.
117. **Yu LH, Wei X, Ma M, Chen XJ, Xu SB.** 2012. Possible inhibitory molecular mechanism of farnesol on the development of fluconazole resistance in *Candida albicans* biofilm. *Antimicrob Agents Chemother* **56:**770–775.

118. **Tian J, Weng LX, Zhang YQ, Wang LH.** 2013. BDSF inhibits *Candida albicans* adherence to urinary catheters. *Microb Pathog* **64:**33–38.
119. **Yeater KM, Chandra J, Cheng G, Mukherjee PK, Zhao X, Rodriguez-Zas SL, Kwast KE, Ghannoum MA, Hoyer LL.** 2007. Temporal analysis of *Candida albicans* gene expression during biofilm development. *Microbiology* **153:**2373–2385.
120. **Bonhomme J, Chauvel M, Goyard S, Roux P, Rossignol T, d'Enfert C.** 2011. Contribution of the glycolytic flux and hypoxia adaptation to efficient biofilm formation by *Candida albicans*. *Mol Microbiol* **80:**995–1013.
121. **Baillie GS, Douglas LJ.** 1998. Effect of growth rate on resistance of *Candida albicans* biofilms to antifungal agents. *Antimicrob Agents Chemother* **42:**1900–1905.
122. **Baillie GS, Douglas LJ.** 1998. Iron-limited biofilms of *Candida albicans* and their susceptibility to amphotericin B. *Antimicrob Agents Chemother* **42:**2146–2149.
123. **Hawser SP, Douglas LJ.** 1995. Resistance of *Candida albicans* biofilms to antifungal agents *in vitro*. *Antimicrob Agents Chemother* **39:**2128–2131.
124. **Kalya AV, Ahearn DG.** 1995. Increased resistance to antifungal antibiotics of *Candida* spp. adhered to silicone *J Ind Microbiol* **14:**451–455.
125. **Ramage G, VandeWalle K, Wickes BL, Lopez-Ribot JL.** 2001. Standardized method for *in vitro* antifungal susceptibility testing of *Candida albicans* biofilms. *Antimicrob Agents Chemother* **45:**2475–2479.
126. **Kuhn DM, George T, Chandra J, Mukherjee PK, Ghannoum MA.** 2002. Antifungal susceptibility of *Candida* biofilms: unique efficacy of amphotericin B lipid formulations and echinocandins. *Antimicrob Agents Chemother* **46:**1773–1780.
127. **Bernhardt H, Knoke M, Bernhardt J.** 2003. Changes in *Candida albicans* colonization and morphology under influence of voriconazole. *Mycoses* **46:**370–374.
128. **Bachmann SP, VandeWalle K, Ramage G, Patterson TF, Wickes BL, Graybill JR, Lopez-Ribot JL.** 2002. *In vitro* activity of caspofungin against *Candida albicans* biofilms. *Antimicrob Agents Chemother* **46:**3591–3596.
129. **Cateau E, Levasseur P, Borgonovi M, Imbert C.** 2007. The effect of aminocandin (HMR 3270) on the *in-vitro* adherence of *Candida albicans* to polystyrene surfaces coated with extracellular matrix proteins or fibronectin. *Clin Microbiol Infect* **13:**311–315.
130. **Kaneko Y, Miyagawa S, Takeda O, Hakariya M, Matsumoto S, Ohno H, Miyazaki Y.** 2013. Real-time microscopic observation of *Candida biofilm* development and effects due to micafungin and fluconazole. *Antimicrob Agents Chemother* **57:**2226–2230.
131. **Melo AS, Colombo AL, Arthington-Skaggs BA.** 2007. Paradoxical growth effect of caspofungin observed on biofilms and planktonic cells of five different *Candida* species. *Antimicrob Agents Chemother* **51:**3081–3088.
132. **Bersan SM, Galvao LC, Goes VF, Sartoratto A, Figueira GM, Rehder VL, Alencar SM, Duarte RM, Rosalen PL, Duarte MC.** 2014. Action of essential oils from Brazilian native and exotic medicinal species on oral biofilms. *BMC Complement Altern Med* **14:**451.
133. **da Silva PM, Acosta EJ, Pinto Lde R, Graeff M, Spolidorio DM, Almeida RS, Porto VC.** 2011. Microscopical analysis of *Candida albicans* biofilms on heat-polymerised acrylic resin after chlorhexidine gluconate and sodium hypochlorite treatments. *Mycoses* **54:**e712–e717.
134. **Fazly A, Jain C, Dehner AC, Issi L, Lilly EA, Ali A, Cao H, Fidel PL Jr, Rao RP, Kaufman PD.** 2013. Chemical screening identifies filastatin, a small molecule inhibitor of *Candida albicans* adhesion, morphogenesis, and pathogenesis. *Proc Natl Acad Sci USA* **110:**13594–13599.
135. **Furletti VF, Teixeira IP, Obando-Pereda G, Mardegan RC, Sartoratto A, Figueira GM, Duarte RM, Rehder VL, Duarte MC, Hofling JF.** 2011. Action of *Coriandrum sativum* L. essential oil upon oral *Candida albicans* biofilm formation. *Evid Based Complement Alternat Med* **2011:**985832.
136. **Lee HJ, Park HS, Kim KH, Kwon TY, Hong SH.** 2011. Effect of garlic on bacterial biofilm formation on orthodontic wire. *Angle Orthod* **81:**895–900.
137. **de Freitas Lima R, Alves EP, Rosalen PL, Ruiz AL, Teixeira Duarte MC, Goes VF, de Medeiros AC, Pereira JV, Godoy GP, Melo de Brito Costa EM.** 2014. Antimicrobial and antiproliferative potential of *Anadenanthera colubrina* (Vell.) Brenan. *Evid Based Complement Alternat Med* **2014:**802696.
138. **Palmeira-de-Oliveira A, Gaspar C, Palmeira-de-Oliveira R, Silva-Dias A, Salgueiro L, Cavaleiro C, Pina-Vaz C, Martinez-de-Oliveira J, Queiroz JA, Rodrigues AG.** 2012. The anti-*Candida* activity of *Thymbra capitata* essential oil: effect upon pre-formed biofilm. *J Ethnopharmacol* **140:**379–383.
139. **Raut JS, Shinde RB, Chauhan NM, Karuppayil SM.** 2013. Terpenoids of plant origin inhibit morphogenesis, adhesion, and biofilm formation by *Candida albicans*. *Biofouling* **29:**87–96.

140. **Sudjana AN, Carson CF, Carson KC, Riley TV, Hammer KA.** 2012. *Candida albicans* adhesion to human epithelial cells and polystyrene and formation of biofilm is reduced by sub-inhibitory *Melaleuca alternifolia* (tea tree) essential oil. *Med Mycol* **50:**863–870.
141. **Traboulsi RS, Mukherjee PK, Chandra J, Salata RA, Jurevic R, Ghannoum MA.** 2011. Gentian violet exhibits activity against biofilms formed by oral *Candida* isolates obtained from HIV-infected patients. *Antimicrob Agents Chemother* **55:**3043–3045.
142. **Villa F, Pitts B, Stewart PS, Giussani B, Roncoroni S, Albanese D, Giordano C, Tunesi M, Cappitelli F.** 2011. Efficacy of zosteric acid sodium salt on the yeast biofilm model *Candida albicans*. *Microb Ecol* **62:**584–598.
143. **Basso FG, Oliveira CF, Fontana A, Kurachi C, Bagnato VS, Spolidorio DM, Hebling J, de Souza Costa CA.** 2011. *In vitro* effect of low-level laser therapy on typical oral microbial biofilms. *Braz Dent J* **22:**502–510.
144. **Chabrier-Rosello Y, Foster TH, Perez-Nazario N, Mitra S, Haidaris CG.** 2005. Sensitivity of *Candida albicans* germ tubes and biofilms to photofrin-mediated phototoxicity. *Antimicrob Agents Chemother* **49:**4288–4295.
145. **Lopes M, Alves CT, Rama Raju B, Goncalves MS, Coutinho PJ, Henriques M, Belo I.** 2014. Application of benzo[a]phenoxazinium chlorides in antimicrobial photodynamic therapy of *Candida albicans* biofilms. *J Photochem Photobiol B* **141:**93–99.
146. **Machado-de-Sena RM, Correa L, Kato IT, Prates RA, Senna AM, Santos CC, Picanco DA, Ribeiro MS.** 2014. Photodynamic therapy has antifungal effect and reduces inflammatory signals in *Candida albicans*-induced murine vaginitis. *Photodiagnosis Photodyn Ther* **11:**275–282.
147. **Rossoni RD, Barbosa JO, de Oliveira FE, de Oliveira LD, Jorge AO, Junqueira JC.** 2014. Biofilms of *Candida albicans* serotypes A and B differ in their sensitivity to photodynamic therapy. *Lasers Med Sci* **29:**1679–1684.
148. **Farber BF, Wolff AG.** 1993. Salicylic acid prevents the adherence of bacteria and yeast to silastic catheters. *J Biomed Mater Res* **27:**599–602.
149. **Maki DG, Stolz SM, Wheeler S, Mermel LA.** 1997. Prevention of central venous catheter-related bloodstream infection by use of an antiseptic-impregnated catheter. A randomized, controlled trial. *Ann Intern Med* **127:**257–266.
150. **Raad I, Darouiche R, Hachem R, Sacilowski M, Bodey GP.** 1995. Antibiotics and prevention of microbial colonization of catheters. *Antimicrob Agents Chemother* **39:**2397–2400.
151. **Zhou L, Tong Z, Wu G, Feng Z, Bai S, Dong Y, Ni L, Zhao Y.** 2010. Parylene coating hinders *Candida albicans* adhesion to silicone elastomers and denture bases resin. *Arch Oral Biol* **55:**401–409.
152. **Mukherjee PK, Chandra J, Kuhn DM, Ghannoum MA.** 2003. Mechanism of fluconazole resistance in *Candida albicans* biofilms: phase-specific role of efflux pumps and membrane sterols. *Infect Immun* **71:**4333–4340.
153. **Mateus C, Crow SA Jr, Ahearn DG.** 2004. Adherence of *Candida albicans* to silicone induces immediate enhanced tolerance to fluconazole. *Antimicrob Agents Chemother* **48:**3358–3366.
154. **Chandra J, Mukherjee PK, Mohamed S, Schinabeck MK, Ghannoum MA.** 2003. Role of antifungal binding in fluconazole resistance of *Candida albicans* biofilms. Presented at the 43rd ICAAC; Abstract number PMArchive.
155. **Beauvais A, Latgé J-P.** 2015. *Aspergillus* biofilm *in vitro* and *in vivo*. *In* Ghannoum M, Parsek M, Whiteley M, Mukherjee P (ed), *Microbial Biofilms*. 2nd ed. ASM Press, Washington, DC, in press.
156. **Nett JE, Andes DR.** 2015. Fungal biofilms: *in vivo* models for discovery of anti-biofilm drugs. *In* Ghannoum M, Parsek M, Whiteley M, Mukherjee P (ed), *Microbial Biofilms*. 2nd ed. ASM Press, Washington, DC, in press.
157. **Desai JV, Mitchell AP.** 2015. *Candida albicans* biofilm development and its genetic control. *In* Ghannoum M, Parsek M, Whiteley M, Mukherjee P (ed), *Microbial Biofilms*. 2nd ed. ASM Press, Washington, DC, in press.

Biofilm Formation by *Cryptococcus neoformans*

7

LUIS R. MARTINEZ[1] and ARTURO CASADEVALL[2]

INTRODUCTION

Historically, microbiologists have studied microbes that cause infectious diseases by analyzing microbial cells grown in suspension (planktonic) in the laboratory. This tradition derives in great part from the early influences of Koch's postulates, which emphasized working with pure cultures. Unfortunately, this growth in pure cultures has little to do with the growth of microbes in "natural" or host environments. Advances in confocal microscopy and molecular genetics in the last two decades have provided evidence that biofilm formation represents the most common mode of growth of microorganisms in nature. This growth form presumably allows microbial cells to survive in hostile environments, enhances their resistance to physical and chemical pressures, and promotes metabolic cooperation (1). In fact, it is estimated that approximately 80% of all bacteria in the environment exist in biofilm communities, and more than 65% of human microbial infections involve biofilm formation (2). Microbial biofilms are dynamic communities of microorganisms strongly attached to biological and nonbiological substrata that are enclosed in a self-produced protective exopolymeric matrix (EPM) (3).

[1]New York Institute of Technology, College of Osteopathic Medicine, Department of Biomedical Sciences, Old Westbury, NY 11568; [2]Department of Molecular Microbiology and Immunology, Johns Hopkins Bloomberg School of Public Health, Baltimore, MD 21205.

Microbial Biofilms 2nd Edition
Edited by Mahmoud Ghannoum, Matthew Parsek, Marvin Whiteley, and Pranab K. Mukherjee

doi:10.1128/microbiolspec.MB-0006-2014

Researchers' interest in studying the role of microbial biofilms in human disease stems from the observation that microbes within biofilms display unique phenotypic characteristics that increase resistance to host immune mechanisms and antimicrobial therapy (4–7). For instance, the successful eradication of biofilms *in vivo* usually requires concentrations of antimicrobial drugs that are usually toxic to the host (8). Similarly, microbial biofilms are more resistant to antimicrobial molecules produced by the host immune system compared to their planktonic counterparts (5). Although bacterial biofilms have been extensively studied since the mid-1980s, little attention was paid to medically relevant fungal biofilms until the past decade.

Cryptococcus neoformans is an encapsulated, environmental fungus that frequently causes life-threatening meningoencephalitis in immunocompromised patients and also occasionally causes disease in apparently normal individuals. *C. neoformans* capsular polysaccharide is mainly composed of glucuronoxylomannan (GXM), and the capsule is a critical virulence phenotype since acapsular strains are not pathogenic (9). Copious amounts of GXM are released during cryptococcal infection, causing deleterious effects for the host immune response (9). In addition, *C. neoformans* forms biofilms on polystyrene plates and medical devices including ventriculoatrial shunt catheters (10, 11). Similarly, there are several reports of *C. neoformans* infection of polytetrafluoroethylene peritoneal dialysis fistula (12) and prosthetic cardiac valves (13) that highlight the ability of this organism to adhere to medical devices. The increasing use of ventriculoperitoneal shunts to manage intracranial hypertension associated with cryptococcal meningoencephalitis suggests the importance of investigating the biofilm-forming properties of this organism (11). This chapter describes the current knowledge of the biology of *C. neoformans* biofilms, the role of the polysaccharide capsule in biofilm formation, development of therapeutic strategies in preventing and treating cryptococcal biofilms, and recent advances in the field.

C. neoformans BIOFILM FORMATION

C. neoformans biofilm-related infections were reported clinically in the mid-1980s (10), but until we visited this problem approximately a decade ago there was no information on the dynamics of this process (14). The characteristics of *C. neoformans* biofilm development were described using a microtiter plate model, microscopic examinations, and a colorimetric XTT reduction assay to observe the metabolic activity of cryptococci within a biofilm. Biofilm formation by this fungus exhibited coordinated phases such as surface attachment, microcolony formation, EPM production, and maturation (8).

During the adhesion period (2 to 4 h) or early stages of biofilm formation, the cryptococcal cells were metabolically active and became firmly attached to the plastic surface of the microtiter plate in a monolayer arrangement. However, adhesion of fungi to a surface can also be facilitated by formation of an organic conditioning layer, which may include compounds released by the host inflammatory response in serum, saliva, or vaginal excretions (15). For instance, cerebrospinal fluid surrounding a ventriculoperitoneal shunt contains high concentrations of cations that may promote interactions of the microbe with the support surface. Furthermore, constant motion of cerebrospinal fluid across the solid surface influences the adhesion of microorganisms to biomaterials. These variables may affect the rate and the extent of fungal attachment (16). The cryptococcal cells adherent to the plastic support consisted of growing cells as indicated by the presence of many budding cells. At the intermediate stage (4 to 16 h), the fungal population had increased significantly and consisted of yeast cells spread uniformly throughout the plastic support, forming microcolonies. The proximity of cells within the microcolony could present a formidable environment for the establishment of nutrient gradients, genetic exchange, and quorum sensing. During the maturation stage (24 to 48 h), the metabolic activity of the cryptococcal cells on the biofilms remained high and steady. Scanning electron and confocal microscopy examinations have demonstrated that the microarchitecture of mature *C. neoformans* biofilms became more complex due to an increasing amount of extracellular material surrounding the cells and producing compact structures that tenaciously adhered to the plastic support (14, 16, 17) (Fig. 1). The structural organization of biofilms and the presence of flowing water channels (18) may allow nutrient and gas exchange while providing the fungal cells with a sheltered niche for protection against environmental predators, immune cells, shear forces, and antimicrobial drugs.

Because *C. neoformans* is an environmental fungus found ubiquitously in association with pigeon excreta in urban environments, and only an accidental pathogen, it is not surprising that biofilm formation constitutes an important survival strategy in hostile environmental conditions (e.g., ultraviolet light) and against predation (16, 19). Conditions that mimic the external environment have been suggested to be permissive for biofilm formation (16). Perhaps biofilm formation is a survival strategy used by *C. neoformans* that emerged and developed through environmental interactions due to the constant selection by predation (20). In this scenario, cryptococcal survival strategies are the result of environmental selection, suggesting that the fungus undergoes accidental adaptation to the host. Additional evidence for this theory comes from the finding that during nonlytic exocytosis from macrophages (21), fungal cells that were internalized by antibody (Ab)-opsonization emerge in biofilm-like microcolonies (22).

The addition of conditioned medium also stimulated the cryptococcal polysaccharide capsule and melanin production, mechanisms needed by this eukaryotic pathogen to thrive

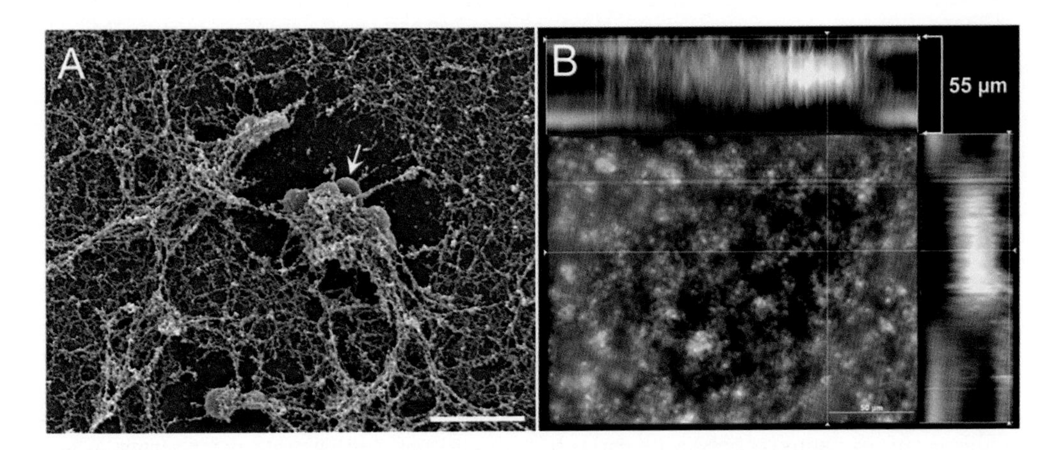

FIGURE 1 Images of a mature *C. neoformans* biofilm grown on polystyrene plates reveal a highly organized architecture. (A) Scanning electron microscopy image of a *C. neoformans* biofilm shows fungal cells (white arrow) surrounded by large amounts of EPM. Scale bar: 10 µm. This scanning electron microscopy image was originally published elsewhere (17). (B) Confocal microscopy image of a cryptococcal biofilm demonstrates a complex structure with internal regions of metabolically active cells interwoven with extracellular polysaccharide material. The thickness of a mature biofilm is approximately 55 µm. This confocal microscopy image was originally published elsewhere (14). doi:10.1128/microbiolspec.MB-0006-2014.f1

in the environment and cause disease in the host. These investigators provided evidence that *C. neoformans* cells can act in concert when expressing the necessary genes to survive in the host and cross-talk to other fungal species in the environment. Likewise, Cfl1, an adhesion protein, was identified to be responsible for the paracrine communication in *C. neoformans* (24). Consistent with its role in communication, Cfl1 is highly induced during mating colony differentiation, and some of the Cfl1 proteins undergo shedding and are released from the cell wall. However, its role in cryptococcal biofilm formation and pathogenesis remain to be elucidated.

POLYSACCHARIDE CAPSULE IN BIOFILM FORMATION

The composition of the microbial cell surface, which may exhibit fimbriae (25), flagella (26), or a capsule (14), greatly influences the rate and extent of attachment. The role of *C. neoformans* polysaccharide capsule in biofilm formation was elucidated (14). Studies with the acapsular *cap59* mutant C536 revealed no biofilm formation relative to the parental strain 3501 or the encapsulated complemented mutant 538 (14). This indicated that capsular polysaccharide was necessary for biofilm formation and implied a critical role for capsular polysaccharide in this process. Since the *C. neoformans* capsule is composed primarily of GXM (16), which is a constituent of the cryptococcal biofilm EPM, it is reasonable to conclude that the inability of the acapsular mutant strain C536 to form a biofilm reflects a failure to shed polysaccharide and form a matrix. Furthermore, the addition of exogenous polysaccharide to acapsular cell cultures was not sufficient to compensate for a lack of capsular production in biofilm formation (14).

The importance of *C. neoformans* capsular polysaccharide release in biofilm formation was also investigated (14). Using enzyme-linked immunosorbent assay (ELISA) spot assays, it was shown that *C. neoformans* biofilm is established via the local release of capsular polysaccharide by attached cryptococcal cells. In addition, the binding of shed polysaccharide to the solid surface created an EPM. However, it was also established that biofilm formation was dependant on environmental conditions such as surface support differences, conditioning films on the surface, characteristics of the medium, and properties of the microbial cell (16).

C. neoformans copiously releases capsular polysaccharide in the supernatant of liquid cultures (27) and in tissues (28). Previous studies have shown that monoclonal antibodies (mAbs) to *C. neoformans* GXM significantly reduced serum GXM levels *in vitro* (29) and *in vivo* (30). GXM-specific Ab inhibits polysaccharide release from encapsulated cells by cross-linking the carbohydrate molecules in the capsule (29). Since *C. neoformans* can form biofilms on medical devices that presumably contain polysaccharide components, this finding raised the intriguing possibility that Ab to GXM would also interfere with cryptococcal biofilm formation. Addition of GXM-binding mAbs to *C. neoformans* cultures in microtiter plates did not affect fungal cell adhesion but prevented biofilm formation by interference with release of GXM (14) (Fig. 2). This function had not been previously described for specific Abs to polysaccharides, and it suggests a new role for humoral immunity in defense against biofilm-forming microbes. Moreover, Ab-mediated inhibition of biofilm formation was associated with protective Abs, suggesting the tantalizing possibility that this effect is involved in Ab protective efficacy against *C. neoformans* (14). Subsequent studies established that Ab binding to the capsule triggered changes in gene expression (31), suggesting the possibility that the inhibition of biofilm formation was also a reflection of Ab-mediated alterations in fungal physiology.

Preparations of cryptococcal biofilm matrix material were analyzed for carbohydrate composition by combined gas chromatography/

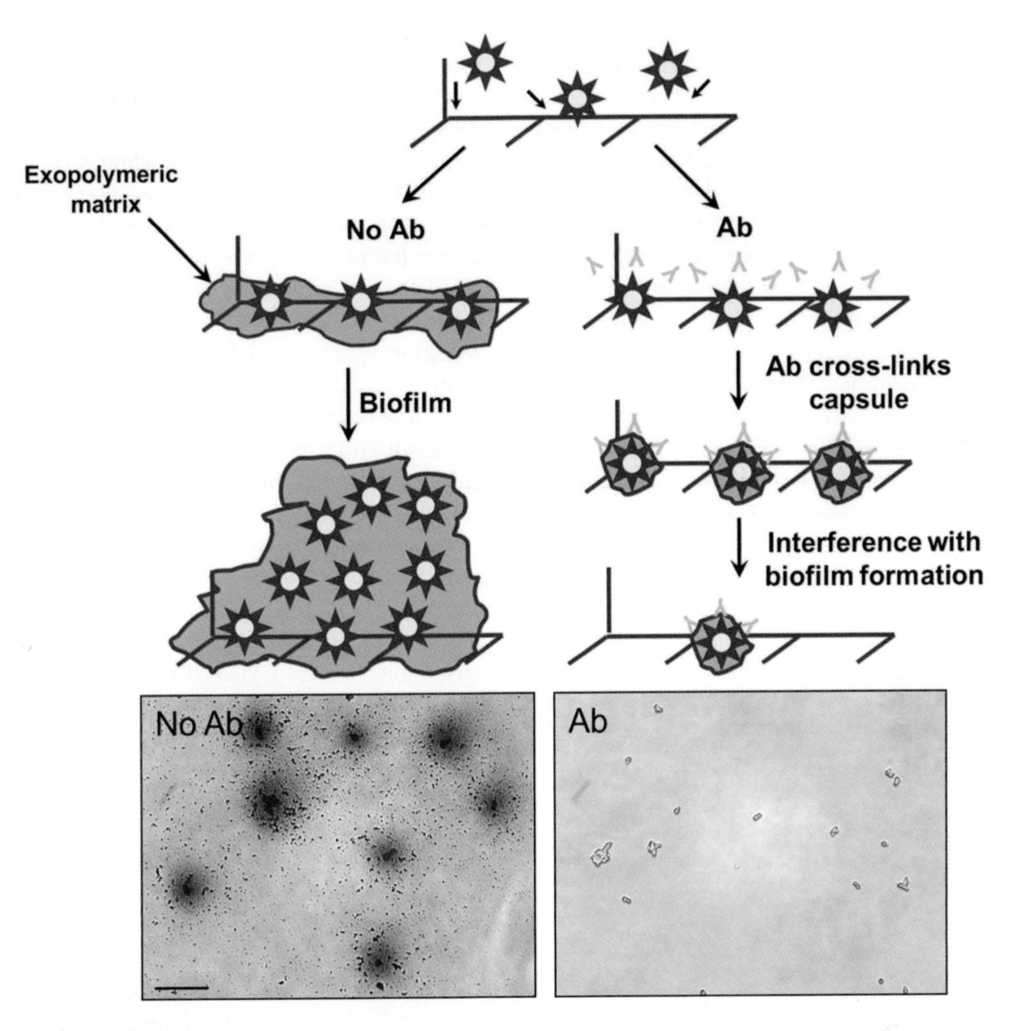

FIGURE 2 Model of antibody-mediated inhibition of *C. neoformans* biofilm formation. In the absence of mAb, *C. neoformans* cells release capsular polysaccharide which is involved in attachment to the plastic surface. In the presence of a mAb specific to *C. neoformans* polysaccharide capsule, the immunoglobulin prevents capsular polysaccharide release, which blocks the adhesion of the yeast cells to the surface. Light microscopic images of spots formed by *C. neoformans* during ELISA spot assay. Images were obtained after 2 h of incubation of fungal cells in the absence and presence of GXM-binding mAb in a polystyrene microtiter plates. Scale bar: 50 μm. The model and light microscopy images in this figure were originally published elsewhere (14). doi:10.1128/microbiolspec.MB-0006-2014.f2

mass spectrometry (16). The glycosyl composition of EPM isolated from biofilms was consistent with the presence of GXM. However, significant quantities of sugars not found in GXM, such as glucose, ribose, and fucose, were also detected, implying that the EPM of *C. neoformans* biofilms includes polysaccharides other than GXM. Based on the observation that GXM may be the main component of the extracellular polysaccharide surrounding fungal cells within a mature *C. neoformans* biofilm, a specific Ab to the capsular polysaccharide of this fungus was employed successfully as a reagent to stain the extracellular polysaccharide matrix of the fungal biofilms (16). Using light microscopy, investigators showed that *C. neoformans* GXM was copiously released to the medium and built up around attached cells, and it encased fungal cells within an EPM that could not be removed by shear forces (Fig. 3). Similarly, confocal microscopy revealed mature *C. neoformans* biofilm with a complex structure, which included internal regions of metabolically active cells interwoven with extracellular polysaccharide material and interspersed with water channels. These findings suggest that most of the extracellular polysaccharide comprising the matrix enclosing cryptococci within a mature biofilm is shed GXM and can be stained by specific mAbs. Therefore, the use of specific mAbs as a simple and effective method to study microbial biofilm by microscopy was introduced.

RESISTANCE TO HOST IMMUNE MECHANISMS AND ANTIFUNGAL THERAPY

Biofilm formation is associated with persistent infection because biofilms increase resistance to host immune mechanisms and antimicrobial therapy. Given that microbial cells within biofilms are highly resistant to standard concentrations of antimicrobial agents, the antifungal activity of commonly used and newly developed drugs against *C. neoformans* biofilms has been evaluated. Biofilms were more resistant than planktonic cells to amphotericin B and caspofungin and completely resistant to the two azole compounds, fluconazole and voriconazole (5). These results correlate with other reports suggesting that biofilm phenotype confers resistance against antifungal drug therapy (32, 33). Amphotericin B and caspofungin mediated a significant reduction of the metabolic activity of *C. neoformans* cells consistent with their fungicidal properties, but the concentrations used were high and above achievable levels *in vivo* after systemic administration and their susceptibility to these drugs was further reduced if cryptococcal cells contained melanin. Moreover, an antifungal surfactant-like lipopeptide, kannurin, produced by *Bacillus cereus* showed moderate effects against cryptococcal biofilms (34).

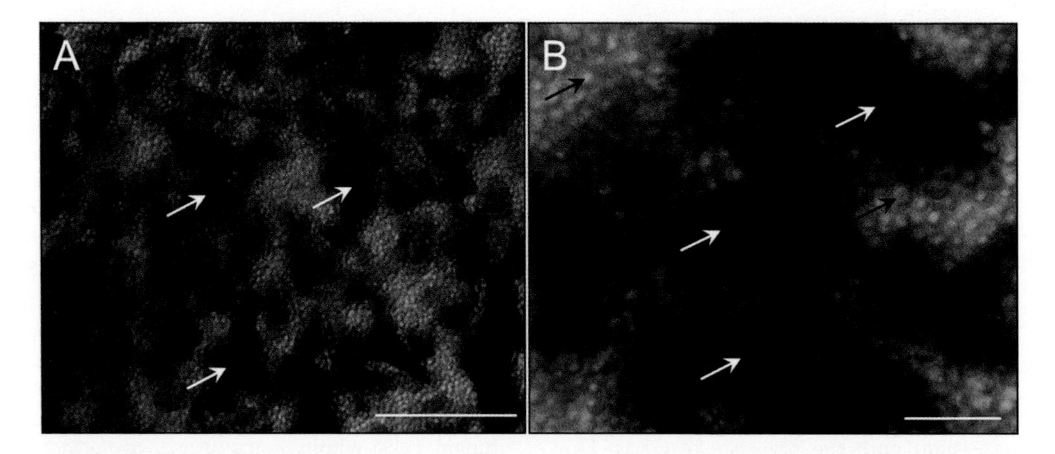

FIGURE 3 Light microscopy images of the EPM of a mature *C. neoformans* biofilm stained with GXM-specific mAb. Images of a mature biofilm show that capsular-binding mAb binds and darkly stains shed capsular polysaccharide. (A) Picture was taken using a 10× power field. Scale bar: 50 µm. (B) Picture was taken using a 40× power field. Scale bar: 10 µm. Black and white arrows denote yeast cells and EPM, respectively. These light microscopy images were originally published elsewhere (16). doi:10.1128/microbiolspec.MB-0006-2014.f3

Although biofilms are known to be less susceptible to antimicrobial drugs, little is known about their susceptibility to antimicrobial molecules produced by the innate immune system. Lactoferrin, a component of the innate immune system, was unable to prevent fungal biofilm formation (14), despite its reported efficacy against bacterial biofilms (35). *C. neoformans* cells within biofilms were more resistant than planktonic cells to oxidative stress but remained vulnerable to cationic antimicrobial peptides (4). Nevertheless, melanin production protected fungal biofilms against antimicrobial peptides (5).

Numerous studies have established that specific Abs can enhance the efficacy of antifungal therapy in animal models of fungal infection (36, 37) and *in vitro* (38). However, none of these models involved the formation of fungal biofilms. Since biofilm formation results in the formation of a physical barrier against host immune mechanisms and antimicrobial therapy, the efficacy of combining GXM-binding mAbs and antifungal drugs against *C. neoformans* biofilms was investigated (39). The presence of GXM-specific IgG1 protected cryptococcal biofilms from amphotericin B or caspofungin, presumably by creating a protein layer upon binding to extracellular polysaccharides. This effect was not observed with an irrelevant or nonspecific IgG1. Confocal microscopy revealed that GXM-specific IgG1 binds through all the EPM surrounding metabolically active yeast cells within *C. neoformans* biofilms. Given that amphotericin B and caspofungin are relatively large molecules of 924 and 1,213 Da, respectively, the antagonism observed may be a result of Ab-mediated interference with drug penetration. These findings suggest the possibility of antagonistic effects when combining Ab and drug therapy for those clinical situations where the presence of established biofilms can be expected, such as with infected prosthetic devices. Hence, Abs may be useful in preventing biofilm formation, but once the biofilms are formed, the binding of additional protein to the matrix could produce antagonistic effects with antimicrobial drugs. Furthermore, these results raised the possibility that products of the immune response contribute to drug resistance for biofilms formed *in vivo*. One can anticipate that microbial biofilms formed on prosthetic devices in tissues contain Abs and other molecules such as complement that may contribute to acquired drug resistance *in vivo*.

Ab-mediated agglutination and biofilm formation can also mediate resistance to and escape from phagocytosis inside macrophages (22). Both *Cryptococcus gattii* and *C. neoformans* cells exited macrophages in biofilm-like microcolonies where the yeast cells were aggregated in a polysaccharide matrix that contained bound Ab. In contrast, complement-opsonized *C. neoformans* was released from macrophages dispersed as individual cells. Hence, both Ab- and complement-mediated phagocytosis resulted in intracellular replication, but the mode of opsonization affected the outcome of exocytosis. The biofilm-like microcolony exit strategy of cryptococcal species following Ab opsonization reduced fungal cell dispersion. This finding suggests that Ab effects inside phagocytic cells might mediate physiological effects on fungal cells.

These are examples of biofilm phenotype increasing resistance against host immune mechanisms, a phenomenon that could contribute to the ability of biofilm-forming microbes to establish persistent infections. Various mechanisms of biofilm resistance to antimicrobial agents and molecules have been proposed, including the presence of physical barriers that prevent the penetration of the antimicrobial compounds into the biofilm (40), slow growth or regulation of the metabolic activity of the biofilm due to nutrient limitation (41), phenotypic switching (42), activation of the general stress response (43), changes in temperature (44), and the existence of a subpopulation of cells within the biofilm (known as persisters) that are preserved by antimicrobial pressure (45).

THERAPEUTIC APPROACHES

Currently, strategies to prevent microbial colonization of catheters have included impregnation of the catheter material with antimicrobial drugs, altering the chemical composition of the polymer, use of nanotechnology, and changing the physical surface properties. Unfortunately, these approaches have not been very effective in reducing the problem of biofilm formation by microbes. In fact, some antibiotics may contribute to the problem (46). For instance, in certain organisms, aminoglycoside antibiotics can induce bacterial biofilm formation.

Based on the urgent need to develop novel therapeutic strategies to combat microbial biofilms in patients, it was demonstrated using confocal microscopy that a capsular polysaccharide-binding IgG1 penetrated the EPM of a biofilm and bound to metabolically active *C. neoformans* cells, which were susceptible to alpha-radiation (47) (Fig. 4). Unlabeled IgG1, alpha-radiation-labeled nonspecific IgG1, and gamma and beta types of radiation did not have any effect on biofilms. The lack of efficacy of gamma and beta radiation probably reflects the radioprotective properties of EPM. In contrast, when GXM-specific IgM labeled with alpha-radiation was

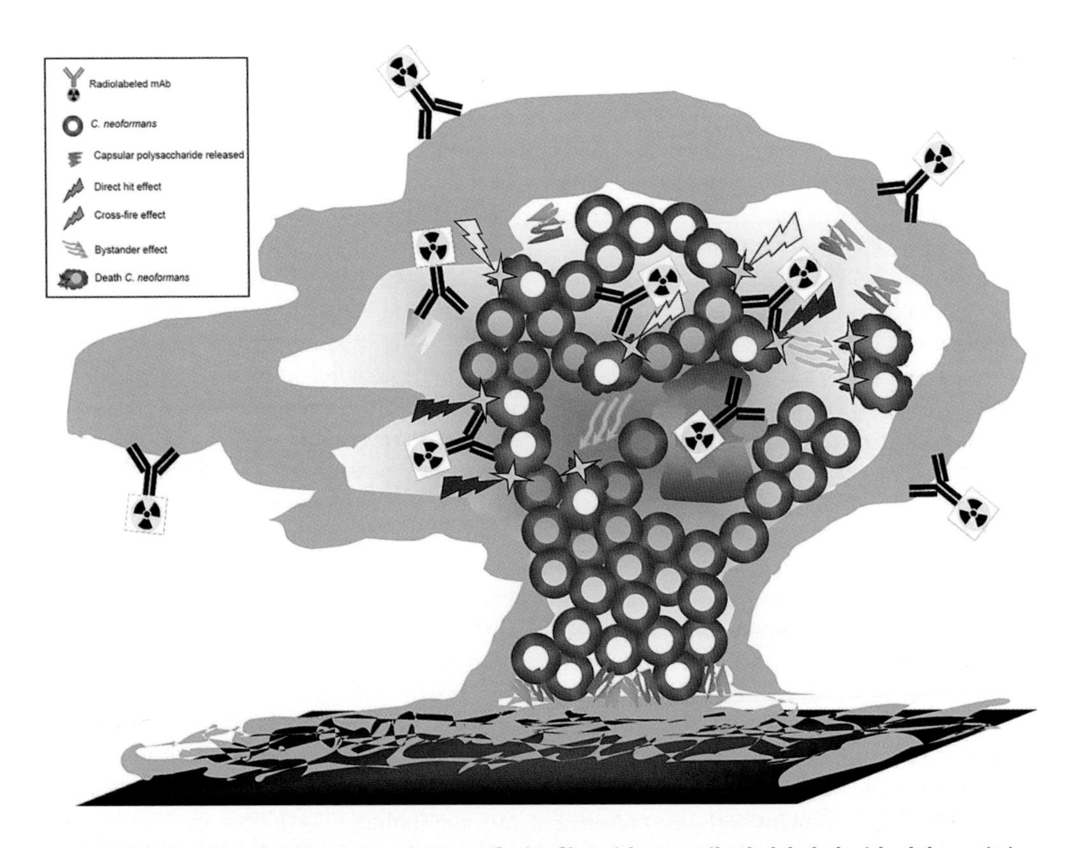

FIGURE 4 Schematic of radioimmunotherapy of a biofilm with an antibody labeled with alpha-emitting radionuclide. The "direct hit" effect is the killing of a cell by radiation emanating from a radiolabeled antibody molecule bound to this cell. "Cross-fire" is the killing of a cell by radiation emanating from a radiolabeled antibody bound to an adjacent or a distant cell. "Bystander" denotes the death of an unirradiated cell through the signaling from irradiated cells. doi:10.1128/microbiolspec.MB-0006-2014.f4

used, there was no penetration of the fungal biofilm and no damage, most likely due to the large size of the pentameric molecule. These results suggested a novel option for the prevention or treatment of microbial biofilms on indwelling medical devices. Since removing certain types of indwelling devices is difficult, one can imagine situations where it may be possible to treat infected devices *in situ* with radioimmunotherapy by local administration of radiolabeled mAbs in close proximity to the infected device. Alternatively, since mAbs may have a role in preventing biofilm formation, a prophylactic dose of unlabeled and radiolabeled Ab may be administered immediately after insertion of the device. In this regard, successful clinical experience has been accumulated in oncology in locoregional administration of radiolabeled mAbs. Novel therapeutic strategies against biofilm-related microbial infections may also be designed by combining radioimmunotherapy and conventional antimicrobial therapy.

Chitosan, a polymer isolated from crustacean exoskeletons, may offer a flexible, biocompatible platform for designing coatings to protect surfaces from infection. As recently demonstrated, chitosan in biofilms significantly reduced the metabolic activity and the cell viability of *C. neoformans* (17). Notably, melanization, an important virulence determinant of *C. neoformans*, did not protect cryptococcal biofilms against chitosan. This phenomenon was attributed to the ability of cationic chitosan to disrupt negatively charged cell membranes as microbes settle on the surface (48). Chitosan has a profound effect on the negative charge of the fungal cellular membrane and thus may interfere with surface colonization or adhesion and cell-cell interactions during biofilm formation (49). Charging the fungal surfaces may keep yeast cells in suspension, preventing biofilm formation (50), or may increase phagocytosis and killing of fungal cells by macrophages. Binding of chitosan to DNA and inhibition of messenger RNA synthesis occurs through enhanced chitosan penetration (51). Thus, it is likely that the interaction between positively charged chitosan molecules and negatively charged microbial cell membranes leads to the leakage of proteinaceous and other intracellular constituents, causing cell death (52). The chitosan concentrations used in these studies to evaluate the effect on biofilm formation were not toxic to human endothelial cells, suggesting an option for preventing or treating fungal biofilms on indwelling medical devices.

To improve the efficacy of antifungal drugs against fungal biofilms and biofilm-associated infections, treatment of early biofilm stages, novel formulations of antifungal drugs, and antifungal drug combinations have been recently developed and successfully applied. One of these examples is that the inhibitory effect of the antifolate combinations sulfamethoxazole–trimethoprim (SMX/TMP) and sulfadiazine–pyrimethamine (SDZ/PYR) against planktonic cells and biofilms of multiple environmental or clinical *C. neoformans* or *C. gattii* strains was recently evaluated (53). SMX/TMP and SDZ/PYR showed antifungal activity against free living cells and sessile cells of *Cryptococcus* spp. The drug combinations SMX/TMP and SDZ/PYR were able to prevent biofilm formation and showed an inhibitory effect against mature biofilms of both species. Additionally, the study showed that antifolate drugs reduced the ergosterol content in *C. neoformans* and *C. gattii* planktonic cells, highlighting the antifungal potential of these drug combinations.

Several groups are assessing the modulatory and synergistic effects of chemical compounds with traditional antimicrobials. For example, EDTA inhibits biofilm growth by *C. neoformans*, and the inhibition could be reversed by the addition of magnesium or calcium, implying that the inhibitory effect is by divalent cation starvation (54). EDTA also reduces GXM released into the EPM, thus providing a potential mechanism for the inhibitory effect of this cation-chelating compound. Unfortunately, the addition of EDTA does not make biofilms more susceptible to antifungal drugs such as fluconazole and voriconazole. Similarly, butyrate is a short-chain fatty acid that is produced by several human commensal bacteria, such as *Clostridium* and *Lactobacillus* species, and acts by inhibiting histone deacetylase (55). In contrast, sodium butyrate strongly inhibited yeast growth in a concentration-dependent manner, interfered with virulence traits such as melanization and capsule formation in *C. neoformans* and, importantly, significantly decreased yeast biofilm formation (56). This fatty acid also enhanced the antifungal activity of azole drugs. Additionally, sodium butyrate augmented the antifungal activity of macrophages by enhancing the production of reactive oxygen species.

Although there have been important advances for combating fungal biofilms, the majority of these prospective strategies are in preclinical development. There is hope that these promising approaches will progress into single or innovative combinatorial therapies to combat microbial biofilms in industrial and medical settings.

CONCLUSIONS AND FUTURE PERSPECTIVES

We have reviewed the available information and synthesized it to provide evidence that biofilm formation by *C. neoformans* may play an important role in fungal pathogenesis. However, the understanding of fungal biofilms is still in its infancy. New technological advances not existent before have made available better tools to study microbial cells within biofilms. Hence, the continuing study of *C. neoformans* biofilms would emphasize the elucidation of the genes expressed during biofilm development. For instance, recent proteomic data comparing cryptococcal biofilms and planktonic cells suggest general changes in metabolism, protein turnover, and global stress responses (43). Many changes in metabolic enzymes were identified in studies of bacterial biofilm, potentially revealing a conserved strategy in biofilm lifestyle. There seems to be a striking parallel between biofilm formation in bacteria, which results in chronic dormant infection with the potential for acute outbreaks, and the dormant state of primary infection followed by secondary outbreaks in *C. neoformans*. In this regard, Moranova et al. proposed a provocative hypothesis that the cryptococcal response to hypoxia might be the driving force for developing a biofilm-like state of dormant infection which is characterized by slowed proliferation and extensive changes in transcriptome and phenotype (57). This dormant state might facilitate *C. neoformans* survival in the host and possibly develop life-threatening acute outbreaks later.

An observation that supports this school of thought was recently documented in a murine model of methamphetamine abuse and *C. neoformans* infection (58). This drug of abuse stimulated colonization and biofilm formation in the lungs, followed by dissemination of the fungus to the central nervous system, where cryptococcomas harbored fungal cells surrounded by vast amounts of capsular material in a biofilm-like structure. It is particularly interesting that *C. neoformans* modifies its capsular polysaccharide after methamphetamine exposure, highlighting the fungus's ability to adapt to environmental stimuli, a possible explanation for its pathogenesis. In contrast, Lei et al. discovered that the NLRP3 inflammasome preferably responds to *C. neoformans* biofilms over planktonic cells, suggesting that this multiprotein oligomer is an important arm for host defense against *C. neoformans* infection (59). This group proposed that manipulating NLRP3 signaling may help to oppose *C. neoformans* infection in patients. Perhaps it is unproductive to begin a debate of whether cryptococcal biofilms are either a dormant infection or an infection that induces a strong immune response; the important point is that these thoughts opened novel avenues for investigation and expansion of this exciting field.

Additional studies would focus on understanding how microbial cells in a mature biofilm communicate by the isolation and characterization of the molecules and mechanisms involved in quorum sensing. Although a quorum sensing system (23) and putative molecules such as Cfl1 (24) have been identified, the use of genomic and proteomic techniques should help to elucidate the phenotypes associated with quorum sensing and the mechanisms by which these pathways work in cryptococcal biofilms. Therapeutics that antagonize quorum sensing molecules may potentially be useful in inhibiting fungal growth, capsular production, and biofilm formation by this opportunistic fungus.

Finally, major efforts should be concentrated on the development of strategies for either preventing or eradicating microbial colonization of medical prosthetic devices and development of new methods for assessing the efficacy of these treatments. In particular, studies of biofilms from *in vivo* and clinical sources, by using genome sequencing, comparative genomics, proteomics, and other multidimensional technologies, might open the door to a new frontier in dissecting the molecular mechanisms of cryptococcal biofilm formation and of designing and developing novel antibiofilm agents.

ACKNOWLEDGMENTS

Conflicts of interest: We disclose no conflicts.

CITATION

Martinez LR, Casadevall A. 2015. Biofilm formation by *Cryptococcus neoformans*. Microbiol Spectrum 3(3):MB-0006-2014.

REFERENCES

1. **Jabra-Rizk MA, Falkler WA, Meiller TF.** 2004. Fungal biofilms and drug resistance. *Emerg Infect Dis* **10:**14–19.
2. **Donlan RM.** 2002. Biofilms: microbial life on surfaces. *Emerg Infect Dis* **8:**881–890.
3. **Costerton JW, Lewandowski Z, Caldwell DE, Korber DR, Lappin-Scott HM.** 1995. Microbial biofilms. *Annu Rev Microbiol* **49:**711–745.
4. **Martinez LR, Casadevall A.** 2006. *Cryptococcus neoformans* cells in biofilms are less susceptible than planktonic cells to antimicrobial molecules produced by the innate immune system. *Infect Immun* **74:**6118–6123.
5. **Martinez LR, Casadevall A.** 2006. Susceptibility of *Cryptococcus neoformans* biofilms to antifungal agents *in vitro*. *Antimicrob Agents Chemother* **50:**1021–1033.
6. **Mowat E, Butcher J, Lang S, Williams C, Ramage G.** 2007. Development of a simple model for studying the effects of antifungal agents on multicellular communities of *Aspergillus fumigatus*. *J Med Microbiol* **56:**1205–1212.
7. **Kuhn DM, Ghannoum MA.** 2004. *Candida* biofilms: antifungal resistance and emerging therapeutic options. *Curr Opin Investig Drugs* **5:**186–197.
8. **Rasmussen TB, Givskov M.** 2006. Quorum-sensing inhibitors as anti-pathogenic drugs. *Int J Med Microbiol* **296:**149–161.
9. **Vecchiarelli A.** 2000. Immunoregulation by capsular components of *Cryptococcus neoformans*. *Med Mycol* **38:**407–417.
10. **Walsh TJ, Schlegel R, Moody MM, Costerton JW, Salcman M.** 1986. Ventriculoatrial shunt infection due to *Cryptococcus neoformans*: an ultrastructural and quantitative microbiological study. *Neurosurgery* **18:**373–375.
11. **Bach MC, Tally PW, Godofsky EW.** 1997. Use of cerebrospinal fluid shunts in patients having acquired immunodeficiency syndrome with cryptococcal meningitis and uncontrollable intracranial hypertension. *Neurosurgery* **41:**1280–1283.
12. **Braun DK, Janssen DA, Marcus JR, Kauffman CA.** 1994. Cryptococcal infection of a prosthetic dialysis fistula. *Am J Kidney Dis* **24:**864–867.
13. **Banerjee U, Gupta K, Venugopal P.** 1997. A case of prosthetic valve endocarditis caused by *Cryptococcus neoformans* var. neoformans. *J Med Vet Mycol* **35:**139–141.
14. **Martinez LR, Casadevall A.** 2005. Specific antibody can prevent fungal biofilm formation and this effect correlates with protective efficacy. *Infect Immun* **73:**6350–6362.
15. **Mittelman MW.** 1996. *Adhesion to Biomaterials*. Wiley-Liss, New York, NY.
16. **Martinez LR, Casadevall A.** 2007. *Cryptococcus neoformans* biofilm formation depends on

surface support and carbon source and reduces fungal cell susceptibility to heat, cold, and UV light. *Appl Environ Microbiol* **73:**4592–4601.
17. **Martinez LR, Mihu MR, Han G, Frases S, Cordero RJ, Casadevall A, Friedman AJ, Friedman JM, Nosanchuk JD.** 2010. The use of chitosan to damage *Cryptococcus neoformans* biofilms. *Biomaterials* **31:**669–679.
18. **de Beer D, Stoodley P, Lewandowski Z.** 1994. Liquid flow in heterogeneous biofilms. *Biotechnol Bioeng* **44:**636–641.
19. **Joubert LM, Wolfaardt GM, Botha A.** 2006. Microbial exopolymers link predator and prey in a model yeast biofilm system. *Microb Ecol* **52:**187–197.
20. **Steenbergen JN, Shuman HA, Casadevall A.** 2001. *Cryptococcus neoformans* interactions with amoebae suggest an explanation for its virulence and intracellular pathogenic strategy in macrophages. *Proc Natl Acad Sci USA* **98:** 15245–15250.
21. **Alvarez M, Casadevall A.** 2006. Phagosome extrusion and host-cell survival after *Cryptococcus neoformans* phagocytosis by macrophages. *Curr Biol* **16:**2161–2165.
22. **Alvarez M, Saylor C, Casadevall A.** 2008. Antibody action after phagocytosis promotes *Cryptococcus neoformans* and *Cryptococcus gattii* macrophage exocytosis with biofilm-like microcolony formation. *Cell Microbiol* **10:**1622–1633.
23. **Albuquerque P, Nicola AM, Nieves E, Paes HC, Williamson PR, Silva-Pereira I, Casadevall A.** 2013. Quorum sensing-mediated, cell density-dependent regulation of growth and virulence in *Cryptococcus neoformans. MBio* **5:**e00986-00913. doi:10.1128/mBio.00986-13.
24. **Wang L, Tian X, Gyawali R, Lin X.** 2013. Fungal adhesion protein guides community behaviors and autoinduction in a paracrine manner. *Proc Natl Acad Sci USA* **110:**11571–11576.
25. **Froeliger EH, Fives-Taylor P.** 2001. *Streptococcus parasanguis* fimbria-associated adhesin fap1 is required for biofilm formation. *Infect Immun* **69:**2512–2519.
26. **Gavin R, Rabaan AA, Merino S, Tomas JM, Gryllos I, Shaw JG.** 2002. Lateral flagella of *Aeromonas* species are essential for epithelial cell adherence and biofilm formation. *Mol Microbiol* **43:**383–397.
27. **Cherniak R, Sundstrom JB.** 1994. Polysaccharide antigens of the capsule of *Cryptococcus neoformans. Infect Immun* **62:**1507–1512.
28. **Goldman DL, Lee SC, Casadevall A.** 1995. Tissue localization of *Cryptococcus neoformans* glucuronoxylomannan in the presence and absence of specific antibody. *Infect Immun* **63:** 3448–3453.
29. **Martinez LR, Moussai D, Casadevall A.** 2004. Antibody to *Cryptococcus neoformans* glucuronoxylomannan inhibits the release of capsular antigen. *Infect Immun* **72:**3674–3679.
30. **MacGill TC, MacGill RS, Casadevall A, Kozel TR.** 2000. Biological correlates of capsular (quellung) reactions of *Cryptococcus neoformans. J Immunol* **164:**4835–4842.
31. **McClelland EE, Nicola AM, Prados-Rosales R, Casadevall A.** 2010. Ab binding alters gene expression in *Cryptococcus neoformans* and directly modulates fungal metabolism. *J Clin Invest* **120:**1355–1361.
32. **Theraud M, Bedouin Y, Guiguen C, Gangneux JP.** 2004. Efficacy of antiseptics and disinfectants on clinical and environmental yeast isolates in planktonic and biofilm conditions. *J Med Microbiol* **53:**1013–1018.
33. **Chandra J, Mukherjee PK, Leidich SD, Faddoul FF, Hoyer LL, Douglas LJ, Ghannoum MA.** 2001. Antifungal resistance of candidal biofilms formed on denture acrylic *in vitro. J Dent Res* **80:**903–908.
34. **Ajesh K, Sudarslal S, Arunan C, Sreejith K.** 2013. Kannurin, a novel lipopeptide from *Bacillus cereus* strain AK1: isolation, structural evaluation and antifungal activities. *J Appl Microbiol* **115:**1287–1296.
35. **Singh PK, Parsek MR, Greenberg EP, Welsh MJ.** 2002. A component of innate immunity prevents bacterial biofilm development. *Nature* **417:**552–555.
36. **Mukherjee J, Zuckier LS, Scharff MD, Casadevall A.** 1994. Therapeutic efficacy of monoclonal antibodies to *Cryptococcus neoformans* glucuronoxylomannan alone and in combination with amphotericin B. *Antimicrob Agents Chemother* **38:**580–587.
37. **Dromer F, Charreire J.** 1991. Improved amphotericin B activity by a monoclonal anti-*Cryptococcus neoformans* antibody: study during murine cryptococcosis and mechanisms of action. *J Infect Dis* **163:**1114–1120.
38. **Nooney L, Matthews RC, Burnie JP.** 2005. Evaluation of Mycograb, amphotericin B, caspofungin, and fluconazole in combination against *Cryptococcus neoformans* by checkerboard and time-kill methodologies. *Diagn Microbiol Infect Dis* **51:**19–29.
39. **Martinez LR, Christaki E, Casadevall A.** 2006. Specific antibody to *Cryptococcus neoformans* glucurunoxylomannan antagonizes antifungal drug action against cryptococcal biofilms *in vitro. J Infect Dis* **194:**261–266.

40. **Tseng BS, Zhang W, Harrison JJ, Quach TP, Song JL, Penterman J, Singh PK, Chopp DL, Packman AI, Parsek MR.** 2013. The extracellular matrix protects *Pseudomonas aeruginosa* biofilms by limiting the penetration of tobramycin. *Environ Microbiol* **15:**2865–2878.
41. **Resch A, Rosenstein R, Nerz C, Gotz F.** 2005. Differential gene expression profiling of *Staphylococcus aureus* cultivated under biofilm and planktonic conditions. *Appl Environ Microbiol* **71:**2663–2676.
42. **Martinez LR, Ibom DC, Casadevall A, Fries BC.** 2008. Characterization of phenotypic switching in *Cryptococcus neoformans* biofilms. *Mycopathologia* **166:**175–180.
43. **Santi L, Beys-da-Silva WO, Berger M, Calzolari D, Guimaraes JA, Moresco JJ, Yates JR 3rd.** 2014. Proteomic profile of *Cryptococcus neoformans* biofilm reveals changes in metabolic processes. *J Proteome Res* **13:**1545–1559.
44. **Pettit RK, Repp KK, Hazen KC.** 2010. Temperature affects the susceptibility of *Cryptococcus neoformans* biofilms to antifungal agents. *Med Mycol* **48:**421–426.
45. **LaFleur MD, Kumamoto CA, Lewis K.** 2006. *Candida albicans* biofilms produce antifungal-tolerant persister cells. *Antimicrob Agents Chemother* **50:**3839–3846.
46. **Hoffman LR, D'Argenio DA, MacCoss MJ, Zhang Z, Jones RA, Miller SI.** 2005. Aminoglycoside antibiotics induce bacterial biofilm formation. *Nature* **436:**1171–1175.
47. **Martinez LR, Bryan RA, Apostolidis C, Morgenstern A, Casadevall A, Dadachova E.** 2006. Antibody-guided alpha radiation effectively damages fungal biofilms. *Antimicrob Agents Chemother* **50:**2132–2136.
48. **Rabea EI, Badawy ME, Stevens CV, Smagghe G, Steurbaut W.** 2003. Chitosan as antimicrobial agent: applications and mode of action. *Biomacromolecules* **4:**1457–1465.
49. **Miyake Y, Tsunoda T, Minagi S, Akagawa Y, Tsuru H, Suginaka H.** 1990. Antifungal drugs affect adherence of *Candida albicans* to acrylic surfaces by changing the zeta-potential of fungal cells. *FEMS Microbiol Lett* **57:**211–214.
50. **Savard T, Beaulieu C, Boucher I, Champagne CP.** 2002. Antimicrobial action of hydrolyzed chitosan against spoilage yeasts and lactic acid bacteria of fermented vegetables. *J Food Prot* **65:**828–833.
51. **Sudarshan NR, Hoover DG, Knorr D.** 1992. Antibacterial action of chitosan. *Food Biotechnol* **6:**257–272.
52. **Jung B, Kim C, Choi K, Lee YM, Kim J.** 1999. Preparation of amphiphilic chitosan and their antimicrobial activities. *J App Polym Sci* **72:**1713–1719.
53. **de Aguiar Cordeiro R, Mourao CI, Rocha MF, de Farias Marques FJ, Teixeira CE, de Oliveira Miranda DF, Neto LV, Brilhante RS, de Jesus Pinheiro Gomes Bandeira T, Sidrim JJ.** 2013. Antifolates inhibit *Cryptococcus* biofilms and enhance susceptibility of planktonic cells to amphotericin B. *Eur J Clin Microbiol Infect Dis* **32:**557–564.
54. **Robertson EJ, Wolf JM, Casadevall A.** 2012. EDTA inhibits biofilm formation, extracellular vesicular secretion, and shedding of the capsular polysaccharide glucuronoxylomannan by *Cryptococcus neoformans*. *Appl Environ Microbiol* **78:**7977–7984.
55. **Sekhavat A, Sun JM, Davie JR.** 2007. Competitive inhibition of histone deacetylase activity by trichostatin A and butyrate. *Biochem Cell Biol* **85:**751–758.
56. **Nguyen LN, Lopes LC, Cordero RJ, Nosanchuk JD.** 2011. Sodium butyrate inhibits pathogenic yeast growth and enhances the functions of macrophages. *J Antimicrob Chemother* **66:**2573–2580.
57. **Moranova Z, Kawamoto S, Raclavsky V.** 2009. Hypoxia sensing in *Cryptococcus neoformans*: biofilm-like adaptation for dormancy? *Biomed Pap Med Fac Univ Palacky Olomouc Czech Repub* **153:**189–193.
58. **Patel D, Desai GM, Frases S, Cordero RJ, DeLeon-Rodriguez CM, Eugenin EA, Nosanchuk JD, Martinez LR.** 2013. Methamphetamine enhances *Cryptococcus neoformans* pulmonary infection and dissemination to the brain. *MBio* **4.** doi:10.1128/mBio.00400-13.
59. **Lei G, Chen M, Li H, Niu JL, Wu S, Mao L, Lu A, Wang H, Chen W, Xu B, Leng Q, Xu C, Yang G, An L, Zhu LP, Meng G.** 2013. Biofilm from a clinical strain of *Cryptococcus neoformans* activates the NLRP3 inflammasome. *Cell Res* **23:**965–968.

Aspergillus Biofilm *In Vitro* and *In Vivo*

8

ANNE BEAUVAIS[1] and JEAN-PAUL LATGÉ[1]

INTRODUCTION

Aspergillus fumigatus is the most important airborne fungal pathogen in the world. The conidia are inhaled by the entire population and cause a wide range of diseases from simple rhinitis to fatal invasive aspergillosis (IA) in immunocompromised patients (1). The number of *A. fumigatus* chronic infections is constantly increasing in immunocompetent patients suffering from respiratory problems such as chronic obstructive pulmonary disease (22%), asthma (1 to 5%), and cystic fibrosis (5 to 10%), along with 15% allergic bronchopulmonary aspergillosis. *Aspergillus* is also the cause of lung and sinus aspergilloma and serious fungal keratitis infections (2, 3).

Until a few years ago, most studies undertaken to understand *Aspergillus* physiology and virulence were performed with the fungus growing in shaken liquid flasks or fermentors. Such an experimental set-up was the most appropriate to obtain an important biomass to undertake biochemical studies and to purify secreted molecules or antigens from the culture filtrates or mycelial extracts. In contrast, in all *Aspergillus* infections, as well as in nature on a solid substratum, *A. fumigatus* grows as a colony characterized by multicellular and multilayered hyphae which are embedded in an extracellular

[1]Unité des Aspergillus, Institut Pasteur, 75015 Paris, France.

Microbial Biofilms 2nd Edition
Edited by Mahmoud Ghannoum, Matthew Parsek, Marvin Whiteley, and Pranab K. Mukherjee

doi:10.1128/microbiolspec.MB-0017-2015

matrix (ECM) (4, 5). This type of growth is consistent in general with the definition of a biofilm: a structural microbial community of cells enclosed in an ECM. However, *A. fumigatus* biofilms are very different from yeast biofilms (6). In this regard, biofilms formed by filamentous fungi contain septate hyphae that are structurally attached to form microbial colonies. Thus, a better understanding of the infectious process should be based on the study of the biofilm colonies rather than on cells grown in planktonic form in shaken flasks. This article will summarize our present knowledge of biofilms formed by *A. fumigatus* and the role of ECM components during biofilm growth *in vivo* and *in vitro*. In addition, we will discuss the molecular resistance of *A. fumigatus* biofilms to external threats.

BIOFILM CHARACTERISTICS *IN VIVO*

Ultrastructural studies of *Aspergillus* biofilms have been investigated in aspergilloma and IA only (4). In aspergilloma, *A. fumigatus* grows as a typical biofilm characterized by hyphae strongly attached within an ECM (Fig. 1A). Similar biofilms formed by filamentous fungi on contact lenses have been described in fungal keratitis (7, 8). In addition, mycelial "grains" within biofilm structures (referred to as mycetoma) have also been described in bronchoalvelolar lavages of chronic pulmonary aspergillosis and in neutropenic cancer patients with IA (4, 9). In IA, although hyphae grow separately in the lung, they tend to be covered by an ECM (Fig. 1B).

The composition of ECM of *Aspergillus* biofilms formed in human lung aspergilloma and mouse lung with IA was analyzed by immunocytochemistry (4). These studies showed that the ECM of both lungs contained the polysaccharides galactomannan and GAG. Interestingly, while the polysaccharide α1,3 glucan and melanin were detected in the ECM of aspergilloma biofilms, in IA, the α1,3 glucans were only found in the inner layer of the hyphal cell wall, and melanin could not be determined (Table 1, Fig. 2) (4). Surprisingly, the two major antigens, dipeptidylpeptase V (DppV, AFUA_2G09030) and catalase B (CatB, AFUA_3G02270), and the major allergen, AspF1 (AFUA_5G02330), were not detected in the ECMs of either aspergilloma or IA (4).

BIOFILM CHARACTERISTICS *IN VITRO*

Two *in vitro* biofilm models that mimic the *in vivo* situation were established to study

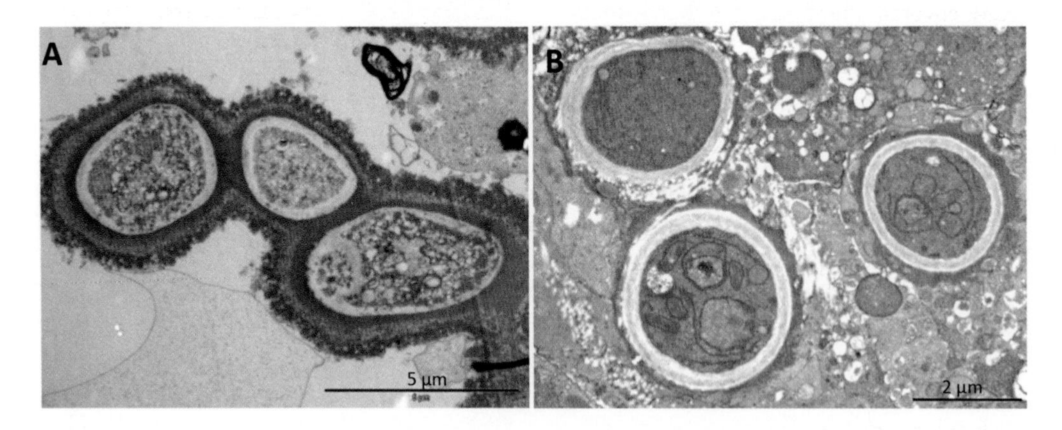

FIGURE 1 Ultrastructure of a human aspergilloma (A) and IA in mouse lung (B) showing the network of hyphae embedded in an ECM. doi:10.1128/microbiolspec.MB-0017-2015.f1

TABLE 1 Composition of the extracellular matrix *in vivo* and *in vitro* in model 2 biofilm conditions

	Aspergilloma	Invasive aspergillosis	Biofilm *in vitro*
Galactosaminogalactan	+	+	+
Galactomannan	+	+	+
α1,3 Glucan	+	–	+
Melanin	+	Unknown	+
Major antigens (DppV, CatB, AspF1)	–	–	+

biofilms formed by *Aspergillus* (Fig. 3). Model 1 used polystyrene microtiter plates containing liquid medium cultured with or without human bronchial epithelial cells obtained from healthy or cystic fibrosis patients and grown in static condition (10–12), while model 2 used agar medium (static conditions) or shaking conditions (13). Both models used an optimal inoculum of 10^6 conidia per milliliter of *A. fumigatus*, which allowed the formation of robust biofilms. Formation of biofilms involved the adhesion of swollen conidia, emergence of germ tubes, and establishment of a monolayer of hyphae. This was followed by an increase in the thickness of the biofilm with a concomitant increase in the complexity of the formed three-dimensional structure. With this model, it was noted that ECM production occurred as soon as the conidia started to swell and increased with the maturity of the biofilm. The biofilm reached the maturation phase when the complexity of the three-dimensional structure was maximal and ECM covered the entire biofilm. Although biofilms formed using the two models have some similarities, they also exhibit some differences. For example, in model 1, biofilms grow slowly and do not sporulate for 72 h at 37°C. The distribution of the ECM is heterogeneous, between packed threads of hyphae as well as surrounding them (Fig. 3) (11). In model 2, biofilms grow faster and begin sporulation after 16 h at 37°C. The ECM is uniformly distributed between hyphae, gluing them together and forming a continuous sheath on the surface of the biofilm (13). Molecular biology and functional genomics studies of *A. fumigatus* biofilms *in vitro* showed that early-phase biofilms are energy-dependent, while mature biofilms are characterized by a reduction of metabolic activity linked to downregulation of glycolysis/glucogenesis and tricarboxylic acid cycle (1, 14, 15). In addition, these studies showed that maturation of biofilms acquires energy not by fermentation but by oxidative phosphorylation.

The composition of ECM in *Aspergillus* biofilms was studied *in vitro* using model 2 biofilms and employing biochemical, molecular, and immunological techniques (Table 1) (13). Similar to the *in vivo* ECM, the ECM produced *in vitro* was shown to contain the polysaccharides galactomannan, GAG, and α1,3 glucans and melanin. However, in contrast to the *in vivo* situation, the three major antigens DppV, CatB, and AspF1 were detected in the ECM. Furthermore, extracellular DNA (eDNA) was not detected in biofilms produced in model 2. The biochemical composition of the ECM of *Aspergillus* biofilms formed in model 1 is still unknown, with the exception that eDNA was shown to be present in these biofilms (16).

Transcriptomic analysis of *Aspergillus* biofilms using microarrays and RNA sequencing (RNAseq) methods showed that biofilms formed in both models had upregulation of genes encoding for hydrophobins and proteins involved in the biosynthesis of secondary metabolites such as fumitremorgin and gliotoxin and in adhesion such as glycophosphatidylinositol (GPI)-anchored cell wall and cell surface proteins (1, 14, 15).

The biological function of the identified molecules in *A. fumigatus* biofilm is discussed below.

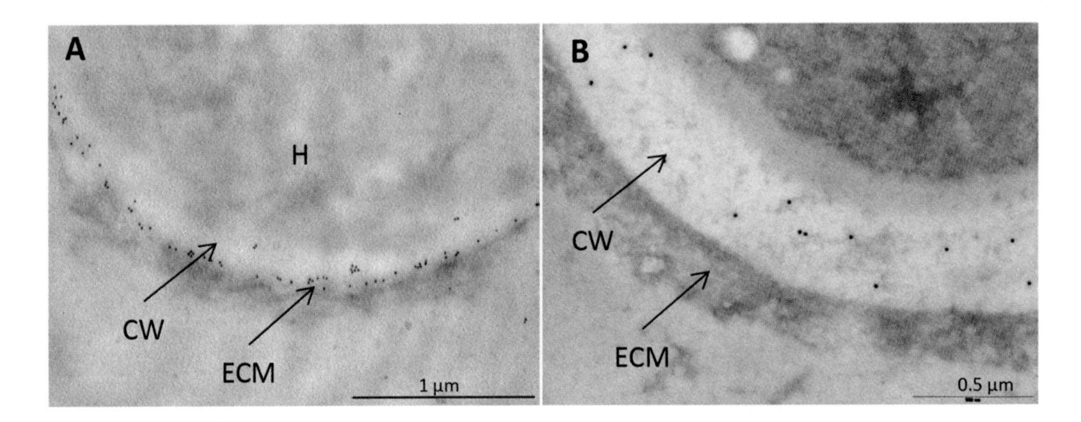

FIGURE 2 Immunolabeling of α1,3 glucan on an ultrathin section of an aspergilloma (A) and invasive aspergillosis in mouse lung (B) with a polyclonal rabbit antibody against anti-α1,3 glucan (1/50 diluted; kind gift of Dr. Ohno, Tokyo University of Pharmacy and Life Science, Japan) and an anti-IgG (whole molecule) conjugated with gold 10 nm. doi:10.1128/microbiolspec.MB-0017-2015.f2

ROLE OF BIOFILM MOLECULES

Polysaccharides

Galactomannan

Galactomannan, a β-(1,5)-galacto-α-(1,2)-α-(1,6)-mannan, is the most useful diagnostic marker in patients with IA (17). DC-SIGN, a C-type lectin receptor present on the surface of both macrophages and dendritic cells, contains a mannan-binding domain that has been shown to bind to the galactomannan of *A. fumigatus* (18). This interaction reduces the T-cell stimulatory activity of dendritic cells. Galactomannan activates Th2/Th17 cells in mice and humans, which results in an increase of interleukin 10 (IL-10) secretion. Elevated levels of IL-10, with no concomitant activation of Treg cells, were observed in response to galactomannan, a finding suggesting that silencing inflammation via IL-10 production may facilitate the initial establishment of *A. fumigatus* in the lung (19).

GAG

GAG is a heterogeneous linear polymer composed of galacto-pyranose (in contrast to galacto-furanose found in the galactomannan polymer) and (*N*-acetyl)-galactosamine linked with α-(1,4)-linkages. It is absent from the conidia and secreted by the mycelium (20). GAG is synthesized as soon as the conidia swell, and it is a dominant adhesin of *A. fumigatus* and mediates adherence to plastic, fibronectin, and epithelial cells (21). A mutant deleted in UDP-Glc epimerase, Δuge3 (*UGE3*, AFUA_3G07910), which cannot convert UDP-*N*-acetyl-glucosamine into UDP-(*N*-acetyl)-galactosamine, is unable to produce GAG and is impaired in the formation of adhesive hyphae and, accordingly, biofilm (21). GAG also plays a role in the virulence of the fungus *in vivo*. Murine experiments have shown that GAG favored fungal growth *in vivo* (20). In GAG-treated mice, no inflammatory pathology was seen in the lung where actual fungal growth was observed. This is associated with the fact that GAG inhibited protective Th1/Treg cells and promoted a Th2 response. Accordingly, Δuge3 that is not producing GAG is less virulent in experimental murine infections (21). The mechanisms responsible for the anti-inflammatory property of GAG were investigated (22). These studies showed (i) that GAG induces the production of IL-1 receptor antagonist (IL-1Ra), a potent anti-inflammatory cytokine that blocks IL-1

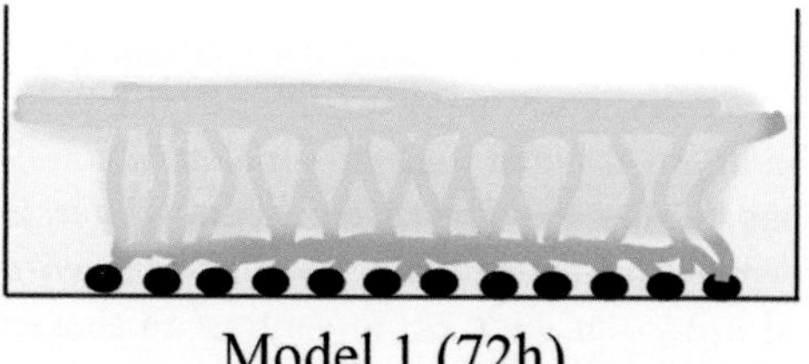

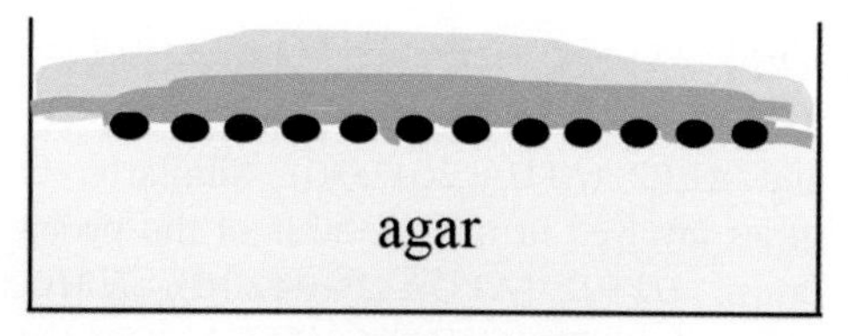

FIGURE 3 Schematic representation of model 1 and 2 biofilms. Model 1 is in liquid medium, and model 2 is in solid agar medium. Both models are static. Maturation is achieved in 72 h in model 1 and 16 h in model 2 at 37°C. doi:10.1128/microbiolspec.MB-0017-2015.f3

signaling and (ii) that GAG induces neutrophil apoptosis via an NK-cell-dependent mechanism (23). Taken together, these results demonstrate that GAG in the ECM of *A. fumigatus* biofilms represents a new virulence factor that allows the fungus to escape the innate immune system.

α1,3-glucan

A. fumigatus α1,3 glucan is composed of linear chains of α-1,3-glucan with intrachain α-(1,4)-linked glucose units for every hundred α-(1,3)-linked glucose units. As soon as the conidia germinate (*in vivo* as well as *in vitro*), melanin and rodlets disappear, exposing α1,3-glucans at the surface of the swollen conidia (24). Like GAG, α1,3-glucan functions as an adhesin and mediates agglutination between swollen conidia or hyphae through interactions between α1,3-glucan chains (24). This property explains the differential localization of this polysaccharide in aspergilloma or *in vitro* biofilm and in IA. In aspergilloma and *in vitro* biofilm, α1,3-glucan in the ECM mediates agglutination of hyphae, whereas in IA, α1,3-glucan is present in the inner layer of the cell wall, which leads to hyphae being separated.

The role of α-1,3-glucan during infection varies depending on whether this molecule is tested separately or in conjunction with other polysaccharides of the cell wall. Injected into mice in a vaccine model, α-(1,3)-glucan is protective because it induces an anti–*A. fumigatus* response associated with the production of a Th1 response that is directed against the fungus (19). However, a mutant devoid of α-(1,3)-glucan (Δags) is less virulent than the parental strain in an experimental model of murine aspergillosis, showing that α-(1,3)-glucan is essential for proper assembly of the cell wall to get the maximal protection against host defense reactions (25). In the modified α-(1,3)-glucan-less cell wall, pathogen-associated molecular patterns such as chitin and β-(1,3)-glucan as well as glycoproteins, are exposed on the fungal surface and promote the host immune response that is responsible for the increased killing of the fungus.

Melanin

Melanin has a protective role, especially against reactive oxidants and drugs, and contributes to the virulence of pathogens (26). Moreover, melanin increases both the negative charges and hydrophobicity of the cells, which would favor a role in biofilm formation. Two types of melanin are synthesized by *A. fumigatus*. DHN-melanin, synthesized from acetyl- and malonyl-CoA, is responsible for the characteristic gray-green color of conidia, and pyomelanin, synthesized from L-tyrosine is secreted by the mycelium (27, 28). The genes responsible for the synthesis of DHN-melanin belong to a cluster of six genes in the genome: *PKSP* (AFUA_2G17600), *AYG1* (AFUA_2G17550),

ARP1 (AFUA_2G17580), *ARP2* (AFUA_2G17560), *ABR1* (AFUA_2G17540), and *ABR2* (AFUA_2G17530). Similarly, the genes involved in the synthesis of the pyomelanin, *HPPD* (AFUA_2G04200), *HMGX* (AFUA_2G04210), *HMGA* (AFUA_2G04220), *FAHA* (AFUA_2G04230), and *MAIA* (AFUA_2G04240), are also organized in a cluster.

It is still unknown which type of melanin is found in biofilm *in vivo* and *in vitro*. The only molecular data available show that the transcription of the pyomelanin gene cluster is increased under cell wall stress and when confronting human neutrophils (27, 29). The global gene expression profiles of *A. fumigatus* grown *in vitro* under model 2 biofilm conditions, in nonbiofilm conditions (shaken flasks), and *in vivo* in the lungs of experimentally infected mice were recently analyzed by RNAseq. We observed that the DHN-melanin cluster is slightly more expressed in biofilm conditions *in vitro* than in nonbiofilm conditions (Table 2). The pyomelanin cluster is downregulated in biofilm conditions *in vitro* and *in vivo* in mice IA, in contrast to shaken flasks *in vitro* (planktonic hyphae), in which the cluster is highly expressed (Table 2). These results suggest that the DHN-melanin cluster is expressed in conditions which promote the formation of highly agglutinated hyphae such as in model 2 biofilm, whereas the pyomelanin cluster is expressed when hyphae grow separately without ECM such as *in vitro* in nonbiofilm conditions. This result was in agreement with the study of Langfelder et al. (30), which suggests that the *PKSP* gene contributed to invasive growth of the fungus *in vivo*.

TABLE 2 DHN melanin (*ABR2* to *PKSP* rows), pyomelanin (*HPPD* to *MAIA* rows) and hydrophobin (*RODA* to *RODG* rows) gene expression profiles obtained by RNAseq analysis in various conditions

			Study A (RPKM values)[a]		Study B (normalized count values)[b]		Study C (normalized count values)[b]	
Gene name	Strain A1163	Strain Af293	Biofilm	Planktonic	Resting conidia	Germinating conidia (8 h)	*In vitro* (planktonic)	*In vivo*
ABR2	AFUB_033220	AFUA_2G17530	NA	NA	636	9	NA	NA
ABR1	AFUB_033230	AFUA_2G17540	4	0	242	4	NA	NA
AYG1	AFUB_033240	AFUA_2G17550	20	3	86	8	21	7
ARP2	AFUB_033250	AFUA_2G17560	16	2	226	22	71	40
ARP1	AFUB_033270	AFUA_2G17580	15	1	61	17	26	38
PKSP	AFUB_033290	AFUA_2G17600	3	3	1,832	735	84	296
HPPD	AFUB_021270	AFUA_2G04200	2	7	2,159	54	16,518	484
HGMX	AFUB_021280	AFUA_2G04210	3	16	274	32	3,584	113
HMGA	AFUB_021290	AFUA_2G04220	10	16	2,200	273	7,016	381
FAHA	AFUB_021300	AFUA_2G04230	19	1	255	530	1,629	298
MAIA	AFUB_021310	AFUA_2G04240	6	6	271	40	2,679	124
RODA	AFUB_057130	AFUA_5G09580	76	1	670	42	NA	NA
RODB	AFUB_016640	AFUA_1G17250	3512	1	0	0	2298	321
RODC	AFUB_080740	AFUA_8G07060	NA	NA	3	3	NA	NA
RODD	AFUB_050030	AFUA_5G01490	26	2	7	34	217	70
RODE	AFUB_081650	AFUA_8G05890	NA	NA	138	5	NA	NA
RODF	AFUB_051810	AFUA_5G03280	18	5	607	220	9	77
RODG		AFUA_2G14661	+					

[a]NA, not included because the RPKM values in both conditions were <2 (17); +, positive by RT-PCR; strain ATCC46645 was used in this study.
[b]NA, not included because in both conditions normalized count values were <20 (J-P Latgé et al., unpublished); strain akuB ku80 was used in studies B and C.

Proteins

Secreted proteins

By immunolabeling, two major antigens, DppV and CatB, and one allergen, AspF1, are detected in the ECM *in vitro*. DppV, CatB, and AspF1 are normally secreted by *A. fumigatus* during its vegetative life *in vivo* and *in vitro* since they induce the production of high levels of specific antibodies which are used in the serodiagnosis of aspergillosis (31). DppV hydrolyzes the dipeptides Ala-Ala, Lys-Ala, His-Ser, and Ser-Tyr from the N terminus. CatB is a mycelial catalase that detoxifies H_2O_2. In a vaccination protocol in mice, maximal immunoprotection against IA induced by *A. fumigatus* conidia was correlated to high levels of *A. fumigatus* antibodies against these antigens (32). RNAseq data showed that genes coding for 38 (on 81 total allergens identified in *A. fumigatus*) are also upregulated in model 2 biofilm (15).

Many genes that encode proteins involved in the biosynthesis of secondary metabolites are also upregulated in biofilm (14, 15, 33). Among them, the most expressed clusters are the fumitremorgin B (*FTM*, AFUA_8G00170 to AFUA_8G00260) and gliotoxin (*GLI*, AFUA_6G09630 to AFUA_6G09740) gene clusters. The upregulation of the secondary metabolism regulator gene *LAEA* in biofilm is in agreement with these data since LaeA is the major regulator of secondary metabolite gene clusters (34).

Cell wall proteins

The major fungal cell wall proteins are the GPI-anchored proteins. The role of some of these proteins in biofilm formation is suggested by RNAseq data as well as mutant phenotypes. The genes encoding the cell wall GPI-galactomannoprotein Mp1 (AFUA_4G03240) and Mp2p (AFUA_2G05150), which are specific to *A. fumigatus*, are upregulated in model 2 biofilms (15). The repeat-rich GPI–cell surface protein CspAp (AFUA_3G08990) is involved in the first step of biofilm formation, the adhesion of the swollen conidia, since deletion of *CSPA* results in decreased adhesion to ECM derived from human alveolar epithelial cells (35). In *Candida albicans*, a family of GPI-anchored surface-bound proteins containing an eight-cysteine domain (CFEM domain) are essential for the formation of a normal biofilm (36). However, the homologous CFEM family in *A. fumigatus*, CfmAp to CfmCp (AFUA_6G14090, AFUA_6G10580, AFUA_6G06690), have no role in biofilm formation (37).

The expression of a gene encoding a non-GPI protein, the cell surface laminin-binding protein CalAp (AFUA_3G09690), is induced in biofilm conditions (38). Recombinant *A. fumigatus* CalAp shows significant binding with laminin and murine lung cells. The upregulation of this protein could increase the adhesion of biofilm to biotic or abiotic surfaces.

Hydrophobins

Hydrophobins are small hydrophobic proteins (<20 kDa) described only in filamentous fungi (39, 40). They are characterized by eight conserved cysteine residues organized in four doublets (C1-C6, C2-C5, C3-C4, C7-C8). They exist as gene families of two to eight members and are classified into two classes based on hydropathy plots, solubility, and the kind of layer they form. Class I hydrophobins assemble into highly insoluble polymeric layers (rodlets) which can be solubilized only with harsh acids. Class II hydrophobins assemble into polymers that lack the rodlet morphology and can be solubilized with organic solvents and detergents. *A. fumigatus* has seven hydrophobins (RodA [AFUA_5G09580], RodB [AFUA_1G17250], RodC [AFUA_8G07060], RodD [AFUA_5G01490] RodE [AFUA_8G05890], RodF [AFUA_5G03280], and RodG [AFUA_2G14661]); they belong to class I, except for RodD, RodF, and RodG, which are classified in an intermediate class (39, 41). Hydrophobins mediate adhesion in filamentous fungi (42). In *A. fumigatus*, only

RodAp has been studied, because this hydrophobin is responsible for the formation of the hydrophobic and immunologically inert rodlet layer on the surface of the conidia (43). In an RNAseq study, *RODA* and the other hydrophobins, *RODB*, *RODD*, and *RODF*, are more expressed in biofilm (hydrophobic hyphae) compared to shaken conditions (hydrophilic planktonic hyphae) (Table 2) (13, 14). *RODB* is the most expressed gene in all biofilm models *in vitro* (15). *RODC* and *RODE* were not differentially expressed in model 2 biofilm conditions in this study (15). However, RT-PCR studies have shown that *RODC* and *RODE* are indeed expressed in mature biofilms (13, 33). RodG is a new hydrophobin identified recently using three programs: (i) Perl to identify the hydrophobin common cysteine residues pattern, (ii) Pfam to verify their function as hydrophobins and (iii) SignalP 3.0 to identify the signal peptide as described by Jensen et al. (39). The expression of *RODG* was evaluated only by RT-PCR. We demonstrated that *RODG* was highly expressed in model 2 biofilm compared to shaken conditions. In contrast to biofilm results, RNAseq quantification of hydrophobin gene expression in *A. fumigatus in vivo* in mice with IA shows that all hydrophobins are poorly expressed, although *RODB*, *RODD*, and *RODF* are the most expressed *in vitro* in biofilm and are also the most expressed *in vivo* (Table 2).

To better understand the role of hydrophobins in biofilm formation, we individually deleted each *ROD* gene in *A. fumigatus* and quantified the biofilm biomass in model 2 conditions (unpublished results). The mutants ΔrodD and ΔrodF presented a lower biofilm biomass than the wild type (WT) and other Δrod strains (Fig. 4). Surprisingly, ΔrodB did not show any phenotype in biofilm conditions, in spite of being the most expressed gene. The composition of the ECM in WT and Δrod biofilms was analyzed as previously described (13). No composition difference was seen in the WT or any of the Δrod mutants. Confocal scanning laser microscopy did not show a

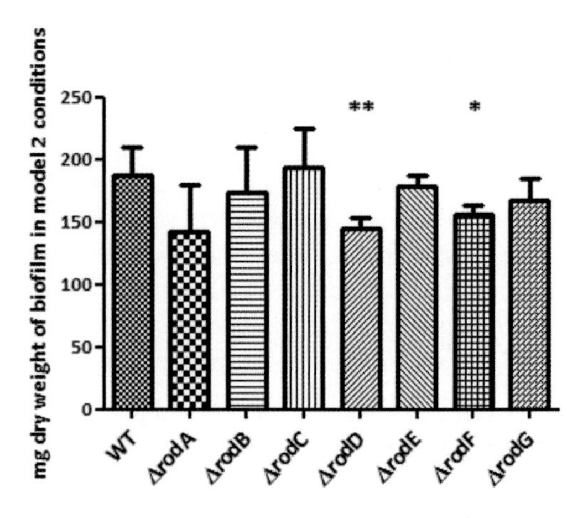

FIGURE 4 Biomass of hydrophobin mutants in model 2 biofilm conditions showing the lower biomass of ΔrodD and ΔrodF mutants. *, $P < 0.05$. doi:10.1128/microbiolspec.MB-0017-2015.f4

difference in the structure of WT and Δrod biofilms. These data showed that single *ROD* deletion did not alter the biofilm structure. In the future, multiple Δrod mutants will be constructed and analyzed for their capacity to make biofilm.

eDNA

It was demonstrated *in vitro* that the presence of eDNA increases the maturation of fungal or bacterial biofilms, and ECM production provides structural integrity and antifungal resistance to the biofilm (12, 44). ECM produced by model 1 biofilms contains genomic eDNA released during autolysis of hyphae, mediated by *A. fumigatus*–secreted chitinase, whereas ECM of model 2 biofilm does not contain eDNA (16). *In vivo*, host DNA released by neutrophils binds to fungi in neutrophil extracellular traps (45). The formation of neutrophil extracellular trap structures against *A. fumigatus* was observed *in vivo* in a mouse model of early IA (10-h lung infection), by two-photon microscopy in infected lungs (45). It has been repeatedly shown that fungal DNA is found in serum (46). However, the mechanisms responsible for the secretion of DNA are totally unknown.

All these data suggest that *in vivo*, as *in vitro*, biofilms integrate eDNA, coming from the fungus or the host, in the ECM to elaborate a more resistant structural biofilm.

BIOFILM DRUG RESISTANCE

Fungal biofilms increase resistance to antimicrobial agents and act as reservoirs for persistent sources of infection in a patient. In *A. fumigatus*, increased resistance of biofilms to polyenes, azoles and echinocandins has been reported (10, 11, 13, 47). The resistance of *A. fumigatus* biofilms to drugs increases with biofilm maturation. Young biofilms are significantly more susceptible to voriconazole and caspofungin than mature biofilms in model 1 conditions (47). Nothing is known regarding the role of *A. fumigatus* biofilm ECM in antifungal resistance.

In fungi and bacteria, multidrug resistance (MDR) pumps include ATP-binding-cassette (ABC) and the major facilitator superfamily. They are involved in the active exclusion of antimicrobial molecules. In *A. fumigatus*, MDR pumps have been associated with increased resistance to azoles. *In silico* analysis of the *A. fumigatus* genome suggests that *A. fumigatus* has 278 different major facilitator superfamilies and 49 ABC transporters. Among them, 140 efflux pump genes including the ABC transporters *MDR1* (AFUA_5G06070), *MDR2* (AFUA_4G10000), and *MDR4* (AFUA_1G12690) are upregulated in biofilm conditions *in vitro* (15, 48). Maximal expression of *MDR4* is correlated with the highest increase in resistance in *A. fumigatus* mature biofilms. *In vivo* in subcutaneously implanted *A. fumigatus* biofilm in mice, the expression of *MDR4* was upregulated in response to treatment with voriconazole (49). In addition, these data are confirmed by the significant increase in efflux pump activity in mature biofilm compared to young biofilm in model 1 conditions (49).

An additional mode of action for polyenes and azoles is the production of high levels of reactive oxidants (ROSs) inducing apoptosis of the fungal cells (50). Antioxidant defenses are specifically induced in biofilm cells (33, 50). This increase in ROS scavenging capacity can be responsible for an increased resistance of biofilm to drugs.

Hsp90 (AspF12, AFUA_5G04170) is also involved in the resistance of biofilms of *A. fumigatus* to drugs. Inhibition of Hsp90 by geldanamycin reduced resistance of *A. fumigatus* biofilms to echinocandins (51). However, in the absence of drugs, *HSP90* is downregulated in model 2 biofilm in RNAseq studies (14). The expression of *HSP90* in the presence of drugs has not been investigated in biofilm.

The mechanisms of resistance of *A. fumigatus* biofilms to antifungal drugs described above seem to be general mechanisms developed by fungal and bacterial biofilms to resist to drugs. Genes encoding for ABC transporters (*CDR1*, *CDR2*, and *MDR1*) are overexpressed in *C. albicans* biofilms independent of the exposure to antifungal agents. Similarly, impairment of Hsp90 function genetically or pharmacologically transformed fluconazole from ineffective to highly effective in eradicating biofilms in a rat venous catheter infection model (51, 52). In *Pseudomonas aeruginosa*, inactivation of efflux pumps renders biofilms significantly more susceptible to five different classes of antibiotics (53).

Another way for fungal and bacterial biofilms to resist drugs is the production of persister cells. Persister cells are not mutants but phenotypic variants of the WT which enter dormancy (54, 55). The presence of persister cells in biofilms treated with drugs has been described for bacteria and *Candida* spp. The percent of persister cells in yeast and bacterial biofilms varies from 0.9 to 9%. In *A. fumigatus*, persister cells have also been observed after treatment of model 2 biofilm by amphotericin B (56). However, though it is relatively easy to study the characteristics of bacteria or yeast persister cells, it has not been possible to isolate the persister cells in *A. fumigatus* biofilm because it is not possible to disaggregate the biofilm.

PERSPECTIVES

Biofilms can become a problematic clinical entity due to their resistance to antifungal agents. Most data which have shown why *Aspergillus* biofilms allow the fungus to resist external threats have been obtained *in vitro*. The next step will be to verify the *in vitro* results on *A. fumigatus* biofilms *in vivo*. Although very few *in vivo* models exist now, many are under current investigation. They involve *in vivo* animal models such as an *A. fumigatus* biofilm diffusion chamber implanted subcutaneously in a mouse and a murine model of contact lens–associated fungal keratitis (7, 49) and *ex vivo* models such as human primary bronchial epithelial cells at the air-liquid interface (J. Fernandes, personal communication) and the human airway epithelium, MucilAir™ (Epithelix Sarl) (7). These models will be most useful to understand the establishment of a biofilm in a patient and to analyze the host response during the development of *A. fumigatus* biofilm. These *ex* and *in vivo* models will also be used to validate new therapeutic strategies and, especially, combinatorial therapies. ROS-inducing antimicrobial agents (lactoferrin, defensins, the antimicrobial peptides arenicin and pleurocidin, derivatives of 2-aminotetralin, etc.) are not active directly against the pathogen, although they induce ROS accumulation in the fungal cell (50). However, they demonstrate synergistic interactions with the currently used antifungal azoles, polyenes, and echinocandins. Several well-known molecules such as anti-inflammatory drugs (diclofenac, aspirin) have synergistic interactions with amphotericin B or echinocandins against *Candida* spp. biofilm (50). In *A. fumigatus*, geldanamycin, an inhibitor of Hsp90, is very effective in reducing biofilm development in combination with echinocandins (51).

Another need is to study the *A. fumigatus* biofilm *in vivo* in its microbiota. The formation of a microbiotal lung biofilm formed by common commensal and pathogenic flora occupying the respiratory system has not been reported yet. *P. aeruginosa* and *A. fumigatus* are the two most common bacterial and fungal inhabitants of the lung microbiota. In cystic fibrosis patients, 80% are infected by *P. aeruginosa*, and 60% are also colonized by *A. fumigatus* (57, 58). It is hypothesized that *P. aeruginosa* colonization under biofilm structures favors *A. fumigatus* infections because the bacterial colonization takes place earlier than *A. fumigatus* infections (59). Our preliminary unpublished data show that *P. aeruginosa* initially produces toxins that inhibit fungal growth but subsequently modifies the physiology of the fungus, making it more stress resistant. However, *A. fumigatus* secondary metabolites stimulate biofilm formation by *P. aeruginosa*. An experimental murine model currently developed in our laboratory will permit us to study chronic bacterial and fungal co-infections. Microbiota will add another layer of complexity to our understanding of the role of *A. fumigatus* biofilm during lung invasion.

ACKNOWLEDGMENTS

Conflicts of interest: We disclose no conflicts.

CITATION

Beauvais A, Latgé J-P. 2015. *Aspergillus* biofilm *in vitro* and *in vivo*. Microbiol Spectrum 3(4):MB-0017-2015.

REFERENCES

1. **Latgé JP, Steinbach WJ.** 2009. A perspective view of *Aspergillus fumigatus* research for the next ten years, p 549–558. *In* Latge JP, Steinbach WJ (ed), *Aspergillus fumigatus and Aspergillosis*. ASM Press, Washington, DC.
2. **Taylor PR, Leal SM Jr, Sun Y, Pearlman E.** 2014. *Aspergillus* and *Fusarium* corneal infections are regulated by Th17 cells and IL-17-producing neutrophils. *J Immunol* **192:**3319–3327.
3. **Xu Y, Gao C, Li X, He Y, Zhou L, Pang G, Sun S.** 2013. *In vitro* antifungal activity of silver nanoparticles against ocular pathogenic

filamentous fungi. *J Ocul Pharmacol Ther* **29**:270–274.

4. **Loussert C, Schmitt C, Prevost MC, Balloy V, Fadel E, Philippe B, Kauffmann-Lacroix C, Latgé JP, Beauvais A.** 2010. *In vivo* biofilm composition of *Aspergillus fumigatus*. *Cell Microbiol* **12**:405–410.
5. **Muller FM, Seidler M, Beauvais A.** 2011. *Aspergillus fumigatus* biofilms in the clinical setting. *Med Mycol* **49**(Suppl 1):S96–S100.
6. **Fanning SMitchell AP.** 2012. Fungal biofilms. *PLoS Pathog* **8**:e1002585. doi:10.1371/journal.ppat.1002585.
7. **Chandra J, Pearlman E, Ghannoum MA.** 2014. Animal models to investigate fungal biofilm formation. *Methods Mol Biol* **1147**:141–157.
8. **Mukherjee PK, Chandra J, Yu C, Sun Y, Pearlman E, Ghannoum MA.** 2012. Characterization of fusarium keratitis outbreak isolates: contribution of biofilms to antimicrobial resistance and pathogenesis. *Invest Ophthalmol Vis Sci* **53**:4450–4457.
9. **Kaur S, Singh S.** 2013. Biofilm formation by *Aspergillus fumigatus*. *Med Mycol* **52**:2–9.
10. **Mowat E, Butcher J, Lang S, Williams C, Ramage G.** 2007. Development of a simple model for studying the effects of antifungal agents on multicellular communities of *Aspergillus fumigatus*. *J Med Microbiol* **56**:1205–1212.
11. **Seidler MJ, Salvenmoser S, Muller FM.** 2008. *Aspergillus fumigatus* forms biofilms with reduced antifungal drug susceptibility on bronchial epithelial cells. *Antimicrob Agents Chemother* **52**:4130–4136. doi:10.1128/AAC.00234-08.
12. **Shopova I, Bruns S, Thywissen A, Kniemeyer O, Brakhage AA, Hillmann F.** 2013. Extrinsic extracellular DNA leads to biofilm formation and colocalizes with matrix polysaccharides in the human pathogenic fungus *Aspergillus fumigatus*. *Front Microbiol* **4**:141. doi:10.3389/fmicb.2013.00141.
13. **Beauvais A, Schmidt C, Guadagnini S, Roux P, Perret E, Henry C, Paris S, Mallet A, Prevost MC, Latgé JP.** 2007. An extracellular matrix glues together the aerial-grown hyphae of *Aspergillus fumigatus*. *Cell Microbiol* **9**:1588–1600.
14. **Gibbons JG, Beauvais A, Beau R, McGary KL, Latgé JP, Rokas A.** 2012. Global transcriptome changes underlying colony growth in the opportunistic human pathogen *Aspergillus fumigatus*. *Eukaryot Cell* **11**:68–78.
15. **Muszkieta L, Beauvais A, Pahtz V, Gibbons JG, Anton Leberre V, Beau R, Shibuya K, Rokas A, Francois JM, Kniemeyer O, Brakhage AA, Latgé JP.** 2013. Investigation of *Aspergillus fumigatus* biofilm formation by various "omics" approaches. *Front Microbiol* **4**:13.
16. **Rajendran R, Williams C, Lappin DF, Millington O, Martins M, Ramage G.** 2013. Extracellular DNA release acts as an antifungal resistance mechanism in mature *Aspergillus fumigatus* biofilms. *Eukaryot Cell* **12**:420–429.
17. **Leeflang MM, Debets-Ossenkopp YJ, Visser CE, Scholten RJ, Hooft L, Bijlmer HA, Reitsma JB, Bossuyt PM, Vandenbroucke-Grauls CM.** 2008. Galactomannan detection for invasive aspergillosis in immunocompromized patients. *Cochrane Database Syst Rev* **Oct 8**(4):CD007394.
18. **Latgé JP, Beauvais A.** 2007. Interactions of *Aspergillus fumigatus* with its host during invasive pulmonary infections, p 331–358. *In* Brown GD, Netea MG (ed), *Immunology of Fungal Infections*. Springer, Dordrecht, The Netherlands.
19. **Bozza S, Clavaud C, Giovannini G, Fontaine T, Beauvais A, Sarfati J, D'Angelo C, Perruccio K, Bonifazi P, Zagarella S, Moretti S, Bistoni F, Latgé JP, Romani L.** 2009. Immune sensing of *Aspergillus fumigatus* proteins, glycolipids, and polysaccharides and the impact on Th immunity and vaccination. *J Immunol* **183**:2407–2414.
20. **Fontaine T, Delangle A, Simenel C, Coddeville B, van Vliet SJ, van Kooyk Y, Bozza S, Moretti S, Schwarz F, Trichot C, Aebi M, Delepierre M, Elbim C, Romani L, Latgé JP.** 2011. Galactosaminogalactan, a new immunosuppressive polysaccharide of *Aspergillus fumigatus*. *PLoS Pathog* **7**:e1002372. doi:10.1371/journal.ppat.1002372.
21. **Gravelat FN, Beauvais A, Liu H, Lee MJ, Snarr BD, Chen D, Xu W, Kravtsov I, Hoareau CM, Vanier G, Urb M, Campoli P, Al Abdallah Q, Lehoux M, Chabot JC, Ouimet MC, Baptista SD, Fritz JH, Nierman WC, Latgé JP, Mitchell AP, Filler SG, Fontaine T, Sheppard DC.** 2013. *Aspergillus* galactosaminogalactan mediates adherence to host constituents and conceals hyphal beta-glucan from the immune system. *PLoS Pathog* **9**:e1003575. doi:10.1371/journal.ppat.1003575.
22. **Gresnigt MS, Bozza S, Becker KL, Joosten LA, Abdollahi-Roodsaz S, van der Berg WB, Dinarello CA, Netea MG, Fontaine T, De Luca A, Moretti S, Romani L, Latgé JP, van de Veerdonk FL.** 2014. A polysaccharide virulence factor from *Aspergillus fumigatus* elicits anti-inflammatory effects through induction of interleukin-1 receptor antagonist. *PLoS Pathog* **10**:e1003936. doi:10.1371/journal.ppat.1003936.

23. **Robinet P, Baychelier F, Fontaine T, Picard C, Debre P, Vieillard V, Latgé JP, Elbim C.** 2014. A polysaccharide virulence factor of a human fungal pathogen induces neutrophil apoptosis via NK cells. *J Immunol* **192:**5332–5342.
24. **Fontaine T, Beauvais A, Loussert C, Thevenard B, Fulgsang CC, Ohno N, Clavaud C, Prevost MC, Latgé JP.** 2010. Cell wall alpha1-3 glucans induce the aggregation of germinating conidia of *Aspergillus fumigatus. Fungal Genet Biol* **47:**707–712.
25. **Beauvais A, Bozza S, Kniemeyer O, Formosa C, Balloy V, Henry C, Roberson RW, Dague E, Chignard M, Brakhage AA, Romani L, Latgé JP.** 2013. Deletion of the alpha-(1,3)-glucan synthase genes induces a restructuring of the conidial cell wall responsible for the avirulence of *Aspergillus fumigatus. PLoS Pathog* **9:** e1003716. doi:10.1371/journal.ppat.1003716.
26. **Eisenman HC, Casadevall A.** 2012. Synthesis and assembly of fungal melanin. *Appl Microbiol Biotechnol* **93:**931–940.
27. **Schmaler-Ripcke J, Sugareva V, Gebhardt P, Winkler R, Kniemeyer O, Heinekamp T, Brakhage AA.** 2009. Production of pyomelanin, a second type of melanin, via the tyrosine degradation pathway in *Aspergillus fumigatus. Appl Environ Microbiol* **75:**493–503.
28. **Youngchim S, Morris-Jones R, Hay RJ, Hamilton AJ.** 2004. Production of melanin by *Aspergillus fumigatus. J Med Microbiol* **53:** 175–181.
29. **Heinekamp T, Thywissen A, Macheleidt J, Keller S, Valiante V, Brakhage AA.** 2013. *Aspergillus fumigatus* melanins: interference with the host endocytosis pathway and impact on virulence. *Front Microbiol* **3:**440. doi:10.3389/fmicb.2012.00440.
30. **Langfelder K, Philippe B, Jahn B, Latgé JP, Brakhage AA.** 2001. Differential expression of the *Aspergillus fumigatus* pksP gene detected *in vitro* and *in vivo* with green fluorescent protein. *Infect Immun* **69:**6411–6418.
31. **Sarfati J, Boucias DG, Latgé JP.** 1995. Antigens of *Aspergillus fumigatus* produced *in vivo. J Med Vet Mycol* **33:**9–14.
32. **Beauvais A, Monod M, Debeaupuis JP, Diaquin M, Kobayashi H, Latgé JP.** 1997. Biochemical and antigenic characterization of a new dipeptidyl-peptidase isolated from *Aspergillus fumigatus. J Biol Chem* **272:**6238–6244.
33. **Bruns S, Seidler M, Albrecht D, Salvenmoser S, Remme N, Hertweck C, Brakhage AA, Kniemeyer O, Muller FM.** 2010. Functional genomic profiling of *Aspergillus fumigatus* biofilm reveals enhanced production of the mycotoxin gliotoxin. *Proteomics* **10:**3097–3107.
34. **Perrin RM, Fedorova ND, Bok JW, Cramer RA, Wortman JR, Kim HS, Nierman WC, Keller NP.** 2007. Transcriptional regulation of chemical diversity in *Aspergillus fumigatus* by LaeA. *PLoS Pathog* **3:**e50.
35. **Levdansky E, Kashi O, Sharon H, Shadkchan Y, Osherov N.** 2010. The *Aspergillus fumigatus* cspA gene encoding a repeat-rich cell wall protein is important for normal conidial cell wall architecture and interaction with host cells. *Eukaryot Cell* **9:**1403–1415.
36. **Perez A, Ramage G, Blanes R, Murgui A, Casanova M, Martinez JP.** 2011. Some biological features of *Candida albicans* mutants for genes coding fungal proteins containing the CFEM domain. *FEMS Yeast Res* **11:**273–284.
37. **Vaknin Y, Shadkchan Y, Levdansky E, Morozov M, Romano J, Osherov N.** 2014. The three *Aspergillus fumigatus* CFEM-domain GPI-anchored proteins (CfmA-C) affect cell-wall stability but do not play a role in fungal virulence. *Fungal Genet Biol* **63:**55–64.
38. **Upadhyay SK, Mahajan L, Ramjee S, Singh Y, Basir SF, Madan T.** 2009. Identification and characterization of a laminin-binding protein of *Aspergillus fumigatus*: extracellular thaumatin domain protein (AfCalAp). *J Med Microbiol* **58:**714–722.
39. **Jensen BG, Andersen MR, Pedersen MH, Frisvad JC, Sondergaard I.** 2010. Hydrophobins from *Aspergillus* species cannot be clearly divided into two classes. *BMC Res Notes* **3:**344.
40. **Sunde M, Kwan AH, Templeton MD, Beever RE, Mackay JP.** 2008. Structural analysis of hydrophobins. *Micron* **39:**773–784.
41. **Littlejohn KA, Hooley P, Cox PW.** 2012. Bioinformatics predicts diverse *Aspergillus* hydrophobins with novel properties. *Food Hydrocolloids* **27:**503–516.
42. **Kershaw MJ, Talbot NJ.** 1998. Hydrophobins and repellents: proteins with fundamental roles in fungal morphogenesis. *Fungal Genet Biol* **23:** 18–33.
43. **Aimanianda V, Bayry J, Bozza S, Kniemeyer O, Perruccio K, Elluru SR, Clavaud C, Paris S, Brakhage AA, Kaveri SV, Romani L, Latgé JP.** 2009. Surface hydrophobin prevents immune recognition of airborne fungal spores. *Nature* **460:**1117–1121.
44. **Gloag ES, Turnbull L, Huang A, Vallotton P, Wang H, Nolan LM, Mililli L, Hunt C, Lu J, Osvath SR, Monahan LG, Cavaliere R, Charles IG, Wand MP, Gee ML, Prabhakar R, Whitchurch CB.** 2013. Self-organization of bacterial biofilms is facilitated by extracellular

DNA. *Proc Natl Acad Sci USA* **110:**11541–11546. doi:10.1073/pnas.1218898110.

45. **Bruns S, Kniemeyer O, Hasenberg M, Aimanianda V, Nietzsche S, Thywissen A, Jeron A, Latgé JP, Brakhage AA, Gunzer M.** 2010. Production of extracellular traps against *Aspergillus fumigatus in vitro* and in infected lung tissue is dependent on invading neutrophils and influenced by hydrophobin RodA. *PLoS Pathog* **6:**e1000873. doi:10.1371/journal.ppat.1000873.
46. **White PL, Perry MD, Barnes RA.** 2009. An update on the molecular diagnosis of invasive fungal disease. *FEMS Microbiol Lett* **296:** 1–10.
47. **Mowat E, Lang S, Williams C, McCulloch E, Jones B, Ramage G.** 2008. Phase-dependent antifungal activity against *Aspergillus fumigatus* developing multicellular filamentous biofilms. *J Antimicrob Chemother* **62:**1281–1284.
48. **Ramage G, Rajendran R, Sherry L, Williams C.** 2012. Fungal biofilm resistance. *Int J Microbiol* **2012:**528521.
49. **Rajendran R, Mowat E, McCulloch E, Lappin DF, Jones B, Lang S, Majithiya JB, Warn P, Williams C, Ramage G.** 2011. Azole resistance of *Aspergillus fumigatus* biofilms is partly associated with efflux pump activity. *Antimicrob Agents Chemother* **55:**2092–2097.
50. **Delattin N, Cammue BP, Thevissen K.** 2014. Reactive oxygen species-inducing antifungal agents and their activity against fungal biofilms. *Future Med Chem* **6:**77–90.
51. **Robbins N, Uppuluri P, Nett J, Rajendran R, Ramage G, Lopez-Ribot JL, Andes D, Cowen LE.** 2011. Hsp90 governs dispersion and drug resistance of fungal biofilms. *PLoS Pathog* **7:**e1002257. doi:10.1371/journal.ppat.1002257.
52. **d'Enfert C.** 2006. Biofilms and their role in the resistance of pathogenic *Candida* to antifungal agents. *Curr Drug Targets* **7:**465–470.
53. **Liao J, Schurr MJ, Sauer K.** 2013. The MerR-like regulator BrlR confers biofilm tolerance by activating multidrug efflux pumps in *Pseudomonas aeruginosa* biofilms. *J Bacteriol* **195:** 3352–3363.
54. **Lafleur MD, Qi Q, Lewis K.** 2010. Patients with long-term oral carriage harbor high-persister mutants of *Candida albicans. Antimicrob Agents Chemother* **54:**39–44.
55. **Schumacher MA, Piro KM, Xu W, Hansen S, Lewis K, Brennan RG.** 2009. Molecular mechanisms of HipA-mediated multidrug tolerance and its neutralization by HipB. *Science* **323:**396–401.
56. **Beauvais A, Müller FM.** 2009. Biofilm formation in *Aspergillus fumigatus*, p 149–158. *In* Latgé JP, Steinbach WJ (ed), *Aspergillus fumigatus and Aspergillosis.* ASM Press, Washington DC.
57. **Baxter CG, Rautemaa R, Jones AM, Webb AK, Bull M, Mahenthiralingam E, Denning DW.** 2013. Intravenous antibiotics reduce the presence of *Aspergillus* in adult cystic fibrosis sputum. *Thorax* **68:**652–657.
58. **Paugam A, Baixench MT, Demazes-Dufeu N, Burgel PR, Sauter E, Kanaan R, Dusser D, Dupouy-Camet J, Hubert D.** 2010. Characteristics and consequences of airway colonization by filamentous fungi in 201 adult patients with cystic fibrosis in France. *Med Mycol* **48**(Suppl 1):S32–S36.
59. **Kraemer R, Delosea N, Ballinari P, Gallati S, Crameri R.** 2006. Effect of allergic bronchopulmonary aspergillosis on lung function in children with cystic fibrosis. *Am J Respir Crit Care Med* **174:**1211–1220.

Adhesins Involved in Attachment to Abiotic Surfaces by Gram-Negative Bacteria

9

CÉCILE BERNE,[1] ADRIEN DUCRET,[1] GAIL G. HARDY,[1] and YVES V. BRUN[1]

INTRODUCTION

The ability of bacterial cells to adhere to and interact with surfaces to eventually form a biofilm is a crucial trait for the survival of any microorganism in a complex environment. As a result, different strategies aimed at providing specific or nonspecific interactions between the bacterial cell and the surface have evolved. While adhesion to abiotic surfaces is usually mediated by nonspecific interactions, adhesion to biotic surfaces typically requires a specific receptor-ligand interaction (1). In both cases, these interactions usually originate from the same fundamental physicochemical forces: covalent bonds, Van der Waals forces, electrostatic forces, and acid-base interactions (2). Strong adhesion occurs if a bacterium and a surface are capable of forming either covalent, ionic, or metallic bonds, but weaker forces, such as polar, hydrogen bonding, or Van der Waals interactions, can also strengthen or achieve strong interactions when a high number of contacts are involved (2, 3). Due the net negative charge of their cell envelopes, bacteria are subjected to repulsive electrostatic forces when approaching surfaces. Bacterial cells also encounter repulsive hydrodynamic forces near the surface in a liquid environment. To overcome these two repulsive barriers, bacteria typically use

[1]Department of Biology, Jordan Hall JH142, Indiana University, Bloomington, IN 47405.

Microbial Biofilms 2nd Edition
Edited by Mahmoud Ghannoum, Matthew Parsek, Marvin Whiteley, and Pranab K. Mukherjee

doi:10.1128/microbiolspec.MB-0018-2015

organelles, such as flagella or pili, which act either as an active propeller or a grappling hook (4–6). Once on the surface, the cell can enhance attachment to the surface via specific and/or nonspecific adhesins to eventually trigger irreversible attachment. This irreversible attachment is strongly influenced by environmental factors (i.e., pH, salinity, etc.) and the physicochemical properties of the surface (i.e., rugosity, hydrophobicity, charge, etc.) but also by the presence of the conditioning film, a layer of organic and inorganic contaminants adsorbed on the surface which changes its physicochemical properties (7). To achieve permanent adhesion under such variable conditions, bacterial cells have developed a series of adhesins able to facilitate adhesion under various environmental conditions (8, 9). In this article, we will focus exclusively on nonspecific adhesins, which are primarily responsible for biofilm formation and bacterial adhesion to abiotic surfaces. We will review the current knowledge of fimbrial, nonfimbrial, and discrete polysaccharide adhesins involved in adhesion to abiotic surfaces and cell aggregation in Gram-negative bacteria.

FIMBRIAL ADHESINS/PILI

Fimbrial adhesins are a varied yet ubiquitous group of adhesins in both Gram-positive and Gram-negative bacteria. Also referred to as attachment pili, these polymeric fibers are involved in an array of functions, including attachment to both biotic and abiotic surfaces, motility, DNA transfer, and biofilm formation. Visible on the cell surface via electron microscopy, fimbrial adhesins are complex appendages that often require a large number of proteins for proper assembly.

The role of pili in the biofilm is multifaceted. First, they are important for the initial stage of bacterial cell attachment to both biotic and abiotic surfaces. Pili are often involved in the transition between motility and irreversible attachment, as seen with chaperone-usher pili (CUP) of *Escherichia coli* and tight adherence (Tad) pili of *Caulobacter crescentus* (10, 11). Second, they facilitate intercellular interactions through aggregation or microcolony formation, which is demonstrated with both Tad pili and curli. Finally, they also play a role in the secondary structure of the biofilm through their function in twitching motility. The type IVa pili of *Pseudomonas aeruginosa* help generate a mature biofilm structure through type IV pili–mediated migration of a key subpopulation of cells within the biofilm (12–14). Many bacterial species have more than one type of pili, and these can be divided into four subgroups, generally defined by their secretion and assembly processes: (i) CUP, (ii) type IV pili, (iii) alternative chaperone-usher pathway pili, and (iv) pili assembled by the extracellular nucleation-precipitation pathway, also known as curli. The structure and the role of Gram-negative fimbrial adhesins in biofilm formation will be discussed in this section.

The CUP

The CUP are the most ubiquitous type of pili and are generally made of either long and thin or heavier rod-like filaments. CUP are assembled by the concerted action of a periplasmic chaperone and a pore-forming protein, called the usher, which gives this type of pilus its name. Although best described in *Enterobacteriaceae*, CUP are also found in a variety of other Gram-negative bacteria, including *Acinetobacter* spp., *Haemophilus influenzae*, *Burkholderia cenocepacia*, *Rhizobium* spp., and *Xylella fastidiosa* (Table 1) (15–18). Based on the phylogeny of the usher protein sequences, CUP are subdivided into six main clades: α, β, γ, κ, π, and σ (19). Although there are as many as 38 different CUP in *E. coli* (20), the best studied are type I (γ1 clade) and P pili (π clade), which provide a good model for CUP assembly. Here we will use the *E. coli* type I pilus as the example for assembly (for more detail, see reference 21 or 22).

TABLE 1 Examples of fimbrial adhesins involved in biofilm formation

Pili type	Major pilus proteins	Minor proteins and assembly proteins	Bacteria	Reference
Chaperone/usher	EcpA or MatA (ECP pili)	EcpC, EcpD, EcpE	*E. coli*	196
	FimA (type I)	FimC, FimD, FimF, FimG, FimH	*E. coli, Klebsiella pneumoniae, X. fastidiosa, Enterobacter amylovora, Serratia marcescans*	21, 197–200
	CsuA/B (type I)	CsuC, CsuD, CsuE	*Acinetobacter baummanii*	201
	MrkA (type 3)	MrkB, MrkC, MrkD	*K. pneumoniae, E. coli (UPEC), Citrobacter koseri*	202, 203
Type IV pili				
Type IVa	PilA,	PilE, PilD, PilV, PilW, PilX	*P. aeruginosa*	14, 45, 46
	PilE	PilD, PilH, PilI, PilJ, PilK, PilX , PilV, ComP, PilD	*Neiserria* spp.	204, 205
	MshA (MSHA)	MshB, C, D, E, F, G, I, J, K, L, M, N, O, P, and MshQ	*Vibrio parahemolyticus* and *Vibrio cholerae*	49, 206
	PilA (ChiRP)	PilB,PilC,PilD	*V. parahemolyticus*	49, 207
Type IVb	BfpA (bundle forming)	BfpP, BfpI, BfpJ, BfpK,	*E. coli* (EPEC)	208
	TcpA (Tcp)	TcpB, C, D, E, F, TcpJ	*V. cholerae*	50
Tad	Flp	TadA, TadB/C, TadD, TadE, TadF, TadG, TadV, RcpA, RcpB, TadZ,	*Aggregatibacter actinomycetemcomitans*	58, 64
	PilA	CpaA, CpaB, CpaC, CpaD, CpaE, CpaF	*C. crescentus*	11, 66
	Flp	TadA, TadB, TadC, TadD, TadF, TadG, FppA, RcpA, RcpC, TadZ	*P. aeruginosa*	56, 209
	CtpA (common pili)	CtpA, CtpB, CtpC, CtpD, CtpE, CtpF, CtpG, CtpH, CtpI	*Agrobacterium tumefaciens*	252
Alternative CU	CooA (CS1)	CooB, CooC, CooD	*E. coli* (ETEC)	26, 28
	CblA (cable pilus)	CblB, CblC, CblD	*Burkholderia cepacia* complex	36
Nucleation/precipitation	CsgA(Curli)	CsgB, CsgG, CsgE, CsgF, CsgD	*E. coli, Enterobacter cloacae, Citrobacter* spp.	72, 77
	AgfA (Tafi)	AgfB, AgfC, AgfD, AgfE, AgfF	*Salmonella enteritidis*	210, 211
	FapC	FapA, FapB, FapD, FapE, FapF	*Pseudomonas* spp.	212

The type I pili consist of six subunits that are first translocated into the periplasm via the Sec translocation pathway (Fig. 1A). After crossing the inner membrane, the pilus subunits associate with a periplasmic chaperone, FimC, which facilitates association with FimD, the outer membrane usher protein. FimD, an 800–amino acid β-barrel protein, is referred to as the type 1 pilus assembly platform. The pilus subunit, chaperone, and usher proteins participate in a process called donor strand exchange and donor strand complementation to facilitate pilus filament assembly (21). Briefly, each pilus subunit has the structure of an incomplete immunoglobulin fold such that it is missing a C-terminal β-strand within the fold. Each subunit also contains an N-terminal extension. During folding and association with the chaperone subunit (FimC), the chaperone complements the missing β-strand within the pilus subunit with hydrophobic residues from the G1 strand of the chaperone. Then, the chaperone (FimC) and the pilus subunit associate with the usher (FimD) (Fig. 1A). The N-terminal extension of the pilus subunit already bound

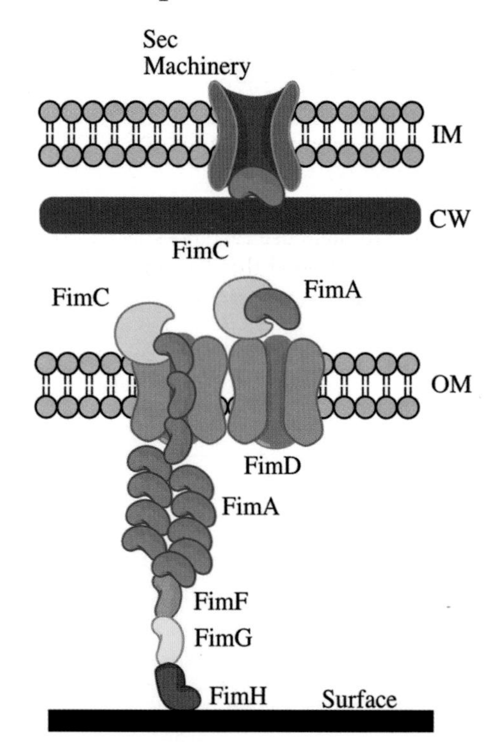
A. Chaperone-Usher
Sec
Machinery
IM
CW
FimC
FimC
FimA
OM
FimD
FimA
FimF
FimG
FimH
Surface

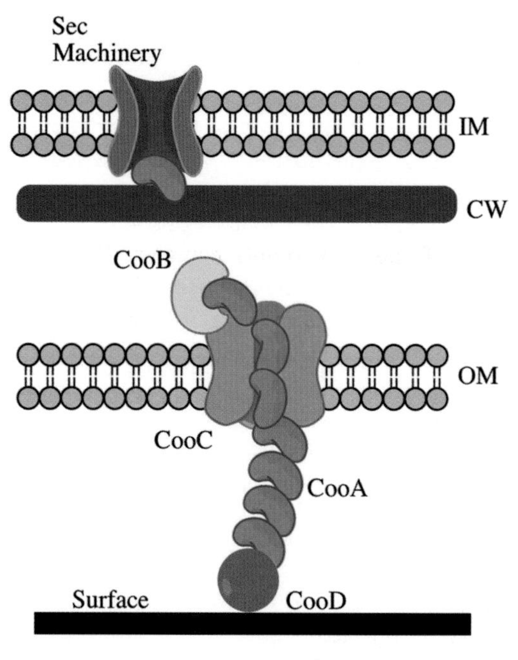
B. Alternate Chaperone-Usher
Sec
Machinery
IM
CW
CooB
OM
CooC
CooA
Surface
CooD

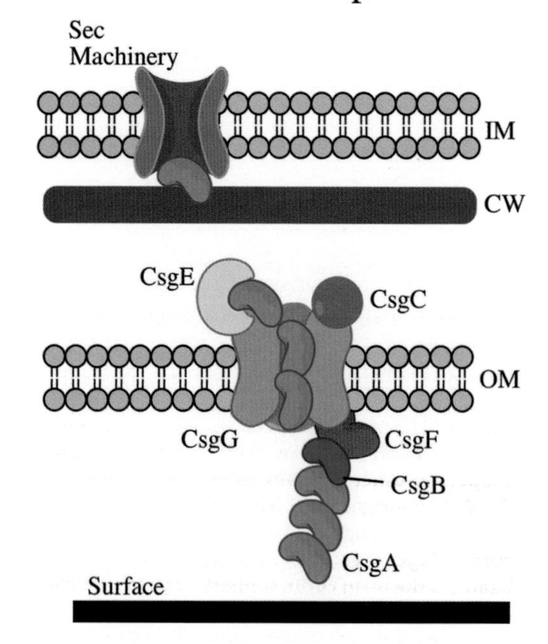
C. Nucleation-Precipitation
Sec
Machinery
IM
CW
CsgE
CsgC
OM
CsgG
CsgF
CsgB
CsgA
Surface

to FimD moves into the groove of the pilus subunit occupied by the chaperone (FimC) G1 strand. The donor strand exchange occurs subsequently by a zip-in zip-out mechanism to mediate transfer of the incoming pilus subunit onto the growing pilus rod. The chaperone (FimC) is then recycled to bind newly formed pilus subunits (Fig. 1A). The initiation of pilus biogenesis occurs when the FimC-FimH complex associates with the usher FimD. Elongation begins with the arrival of the FimC-FimG complex. FimH and FimG comprise the 3-nm tip fibrillum, a small fiber associated with the tip of the main pilus (Fig. 1A). Then, FimF is connected to the tip adhesin and subsequently initiates assembly of the body of the main rod that is composed of FimA (Fig. 1A). P pili assembly is then terminated with the addition of PapH at the proximal end of the pilus. PapH is unable to undergo donor-strand exchange and thus terminates pilus growth. By comparison, no homologues of PapH have been identified in the type I pilus assembly.

The *E. coli* type I pilus is composed of 500 to 3,000 copies of FimA, resulting in a fiber that is 1 to 2 μm long and 6.9 nm wide. The absence of the tip adhesin FimH reduces adherence and biofilm formation, suggesting that FimH is the business end of the type I pilus (10). FimH is specific for mannose-containing receptors of host cells but can also bind to abiotic surfaces in a mannose-dependent manner (10, 13). Mannose was found to specifically inhibit biofilm formation in *E. coli* on a variety of abiotic surfaces, including various plastic polymers and glass (10). This specificity resulted in the development of mannose drug analogs to inhibit bacterial attachment (23). In addition to being involved in attachment to abiotic surfaces, FimH primarily mediates strong adherence to mannose receptors using catch bonds, a type of bond that becomes stronger over time or becomes activated when the receptor and ligand are pulled apart (24, 25). Catch bonds are particularly effective for organisms that experience shear stress, such as those that colonize host mucosal surfaces or air-water interfaces. A variety of other CUP have been reported to play a role in biofilm formation (Table 1).

The Alternative CUP

The alternative CUP (or class 5 pilus assembly) requires "usher" and "chaperone" proteins and has been classified in the α-clade of the CU family of fimbriae (19). However, the proteins in the alternative CUP pathway have little to no similarity to those in the CUP pathway, and only four proteins are known to be required for proper assembly of the alternative CUP (19). This clade includes

FIGURE 1 Assembly and secretion of fimbrial adhesins. All the assembly pathways are oriented such that the inside of the cell is at the top and the surface to which the adhesin is binding, represented by the thick black line, is at the bottom. The subunits for the three described systems are believed to be transported across the inner membrane by the Sec machinery. (A) A schematic of the CUP pathway represented by the assembly of the *E. coli* type I pilus. FimC (green moon) is a chaperone. FimD (blue-gray) is the outer membrane usher shown as a dimeric channel. FimA (blue bean) is the main pilus subunit. FimF (orange bean) links the tip fibrillum to the main fiber. FimG (yellow bean) is the tip fibrillum. FimH (red bean) is the mannose-specific tip fibrillum adhesin. (B) A schematic of the alternative chaperone-usher pathway using the *E. coli* CS1 pilus as a model. CooB (green moon) is the chaperone. CooC (blue-gray) is the outer membrane usher. CooA (blue bean) is the main pilus subunit. CooD (red circle) is the pilus tip adhesin. (C) Model of *E. coli* curlin assembly as a nucleation-precipitation pathway model. CsgE (green moon) is the chaperone. CsgG (blue-gray) is the outer membrane usher. CsgA (blue beans) is the main curlin subunit. CsgB (dark blue bean) is the minor curlin subunit. CsgF (red bean) is the outer membrane protein needed for curlin polymerization and CsgB localization. CsgC (red ball) may be important for CsgG localization. Abbreviations: IM, inner membrane; CW, cell wall; OM, outer membrane.
doi:10.1128/microbiolspec.MB-0018-2015.f1

several coli surface (CS) antigens (CS) of enterotoxigenic *E. coli* (ETEC), which are involved in the intestinal epithelium colonization or in biofilm formation on indwelling devices such as bladder catheters (26), and the cable pili of *Burkholderia cepacia* that facilitate bacterial aggregation, microcolony formation, and association with mucin, a prerequisite in the development of cystic fibrosis (Table 1) (27). In this section, the CS antigen 1 (CS1) pilus of ETEC will serve as our model (Fig. 1B) (Table 1) (for more detail on the alternative CU pathway see references 28–30).

It is believed that the CS1 pilus proteins rely on the Sec pathway for translocation across the inner membrane, as also observed for the proteins of the CUP system. The 15.2-kDa CooA is the major pilin subunit. CooA can spontaneously form multimers in the periplasm, but the multimerization is greatly improved in the presence of CooB, which acts as a chaperone for CooA pilin subunits (Fig. 1B) (28, 31, 32). CooC is a 94-kDa integral outer membrane protein (33) predicted to be the usher protein required for transport and assembly of the pili on the cell surface (Fig. 1B). CooD is a 38-kDa minor pilin associated with the tip of the CS1 pili (Fig. 1B) (33). The ratio of CooD to CooA is 1:1,800, indicating that one CooD could be present at the tip of the pilus. CS1 pili are not assembled in the absence of CooD, suggesting that CooD could initiate pilus assembly as observed for the FimH tip adhesin (34). Similar to the CUP pilin subunits, the pilin of the alternative CUP pathway also undergoes donor strand exchange to facilitate transfer of the pilin subunits from the chaperone to the growing pilus fiber.

ETEC is a leading cause of diarrhea in children in developing countries. These bacteria use a variety of surface adhesins to bind to the intestinal epithelium (35). ETEC can also form biofilms on indwelling devices such as bladder catheters. Both an ETEC CS2 *cotD* mutant and a colonization factor antigen II (CFA/II) *cfaE* mutant had reduced biofilm formation relative to the wild-type parental strains (26), suggesting that alternative CUP are important for adhesion.

B. cenocepacia is an important pathogen associated with cystic fibrosis. The cable pili of *B. cenocepacia* facilitate bacterial cell-cell interactions or microcolony formation and association with mucin (Table 1) (27). However, these cell-cell interactions actually prevent nonspecific aggregation that can result in clearance by the host. A pilin subunit (*cblA*) mutant had increased aggregation relative to wild-type (36).

Type IV Pili

The type IV pili are multifunctional organelles involved in diverse functions such as bacterial twitching motility, auto-aggregation, attachment, and DNA uptake (37). They are generally long thin filaments (6 to 8 nm wide and 1 to 4 μm long) composed of pilin subunits that in some cases assemble into bundles.

The type IV pili are often homopolymers of a 15- to 20-kDa pilin protein, and they can also have an additional adhesive tip subunit. Both the CUP and type IV pilus assembly requires the general secretory pathway to translocate subunits across the inner membrane as well as a variety of outer membrane components to assemble the pilus on the cell surface. However, by comparison with CUP, type IV pili also require additional inner membrane components to facilitate fiber assembly. The dynamic assembly/disassembly of type IV pilin into a fiber is powered by ATP and requires at least 10 proteins: (i) the major pilin subunit and sometimes minor pilin subunits, (ii) an assembly-specific ATPase, (iii) a prepilin peptidase that cleaves the N-terminal signal peptide, (iv) an assembly platform, (v) an outer membrane protein that recruits the ATPase, and (vi) an outer membrane secretin that is necessary for the presentation of the pili on the cell surface (Fig. 2) (13). Based on the length of the prepilin signal peptides, type IV pili can be

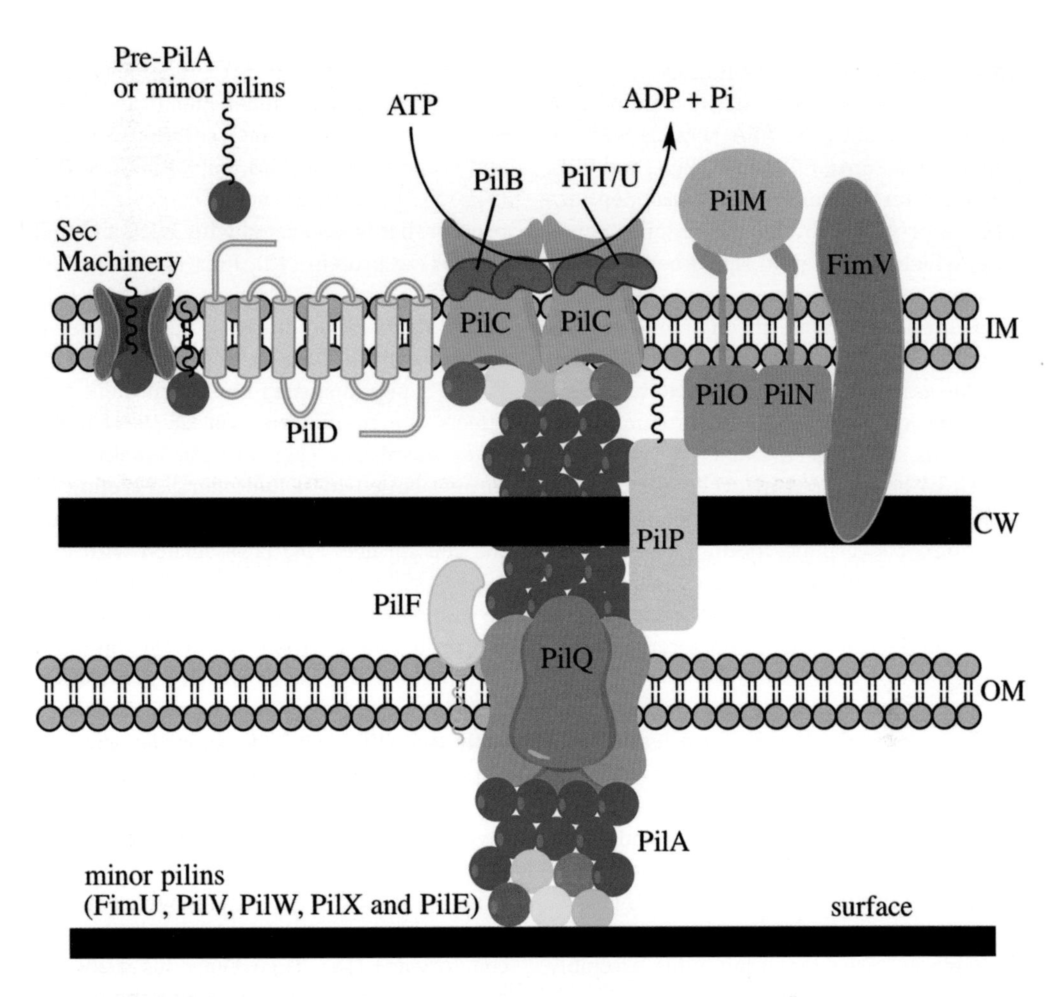

FIGURE 2 Type IV assembly and secretion pathway. Given that the type IV pili have similar elements, we are using the *P. aeruginosa* type IVa pilus as the model for biogenesis. Many type IVa proteins utilize the Sec machinery to translocate the inner membrane (aqua pore). PilA (blue sphere) is the main pilus subunit. FimU, PilE, PilX, PilW, and PilV are minor pilins (red, yellow, light blue, green, and purple spheres, respectively). The prepilins are processed by PilD (orange integral IM protein), the prepilin protease. PilB (red bean) is the ATPase that supplies energy for pilus assembly, and PilU/PilT (purple bean) is the ATPase for pilus retraction. PilC (green porin) is an inner membrane protein of the motor complex for assembly of the pilus. PilM, PilN, PilO, PilP, and FimV are the alignment complex. PilQ is the multimeric secretin in the outer membane that translocates the pilus outside the cell. PilF is a pilotin needed for localization of the PilQ in the OM. FimV is a peptidoglycan binding protein needed for multimerization of PilQ. Abbreviations: IM, inner membrane; CW, cell wall; OM, outer membrane. doi:10.1128/microbiolspec.MB-0018-2015.f2

subdivided into two categories: type IVa, characterized by prepilins with a short signal peptide (5 to 10 amino acids) and type IVb, characterized by prepilins with a long signal peptide (15 to 30 amino acids) (13). In this section we will focus on individual examples of type IVa, type IVb, and a subset of type IVb called the Tad pili.

Type IVa pili

We will use *P. aeruginosa* as a model for type IVa pilus assembly and secretion (Fig. 2,

Tables 1 and 2) (for more detail see reference 14). The pilus itself is primarily composed of the 15-kDa main pilin PilA (Fig. 2) and the minor pilins FimU, PilV, PilW, PilX, and PilE. All the pilin subunits have signal peptides and are processed by PilD, a prepilin peptidase, which cleaves pilin signal peptides and is bound to the inner membrane. The minor pilins are incorporated throughout the pilus structure and are important in the initiation of pilus assembly (38). There have been several other roles suggested for the minor type IVa pilins. PilX, PilY1, and PilW have been shown to be involved in cyclic diguanosine monophosphate (c-di-GMP)–mediated suppression of swarming motility and biofilm formation which is facilitated through a decrease in cellular c-di-GMP levels in cells grown on an agar surface compared to liquid medium (39). In addition, PilY1 is needed for both twitching and swarming motility and facilitates attachment to host cells in a calcium-dependent manner (40).

The inner membrane components of the type IVa pilus include the motor and alignment subcomplexes. The motor complex is composed of PilB, PilC, PilT, and PilU (Fig. 2, Table 2). PilB, PilT, and PilU are putative ATPases required for either pilus assembly (PilB) or disassembly (PilT and PilU). PilC is an integral inner membrane protein thought to play a role in regulation of pilus assembly and depolymerization (41). The alignment complex bridges the inner membrane motor and the outer membrane secretion complex and is composed of PilM, PilN, PilO, and PilP (Fig. 2). PilM is an actin-like cytoplasmic protein that is associated with PilN, an inner membrane protein (42). PilN also associates with PilO and the lipoprotein PilP to subsequently interact with PilQ, the outer membrane secretin. PilN also associates with FimV, a peptidoglycan binding protein that affects the multimerization of PilQ in the outer membrane (Fig. 2) (43). Finally, PilQ forms a large 1-MDa multimeric secretin that facilitates the presentation of the pilus on the cell surface. PilQ is associated with PilF via several tetratricopeptide repeats. PilF is an outer membrane lipoprotein required for the proper localization of PilQ in the outer membrane and is referred to as a pilotin (44).

While type IVa pili are usually associated with twitching motility, they also play an important role in attachment and biofilm structure. Twitching motility or crawling involves the ability of bacteria to move over semisolid surfaces using type IV pili. Cells extend individual pili that attach to a surface and then retract the pili, pulling the cell forward (14). Type IVa pilus–mediated twitching motility of *P. aeruginosa* is important for not only the initial attachment of the bacteria, but also for the formation of

TABLE 2 Type IV pilus components

	Type IVa	Type IVb	Tad or Flp
Bacteria	*P. aeruginaosa*	*V. cholerae*	*A. actinomycetemcomitans*
Components			
Pilin subunit	PilA	TcpA	Flp-1
Minor pilins	FimU, PilV, PilW, PilX, PilE	TcpB	Flp-2, TadE, TadF
Prepilin peptidase	PilD	TcpJ	TadV
Assembly ATPase	PilB	TcpT	TadA
Retraction ATPase	PilT/PilU	NF[a]	NF[a]
Secretin	PilQ	TcpC	RcpA
Platform proteins	PilC	TcpE	TadB, TadC, TadG
Pilotin	PilF	TpcQ	TadD
Alignment complex	PilM, PilN, PilO, PilP, FimV	TcpD, TcpR	RcpB, RcpC
Secreted proteins		TcpF	
Localization proteins			TadZ

[a]NF, not found

microcolonies and three-dimensional structures in the biofilm architecture (45, 46). Hyperpiliated strains of *P. aeruginosa*, where twitching motility is inhibited, form dense flat monolayers (47). Type IVa pili, and more specifically the MSHA (mannose-sensitive hemagglutinating) pilus of *Vibrio cholerae* and *Vibrio parahaemolyticus* facilitate proper biofilm formation on biotic and abiotic surfaces, including both chitin and nonchitinaceaous surfaces, but remain dispensable for the process of initial attachment (48, 49).

Type IVb pili

While most type IVa pili are associated with twitching motility, the type IVb and Tad pili are more often associated with attachment, since the PilT ATPase homologue required for pilus retraction is absent. There are seven type IVb pili that have been characterized in Gram-negative bacteria, including the Tad pili. To examine the structure and assembly of type IVb pili, we will look at the *V. cholerae* toxin coregulated pilus (TCP) (Tables 1 and 2). TCP plays a role in microcolony formation and mediates bacterial interactions on chitinaceous surfaces (50) that are abundant as part of shellfish in aquatic environments. The TCP results in a more stable and fit biofilm on chitin-rich surfaces (50). *V. cholerae* can degrade chitin and utilize it as a source of food, which can facilitate survival and spread of *V. cholerae* in the environment (51).

To be properly assembled, TCP require about 10 to 14 genes that are usually organized into one long operon called the TCP locus (52). TcpA and TcpB are the major and minor pilin of TCP pili, respectively (Table 2). TcpC is an outer membrane secretin interacting with a small protein, TcpQ, which provides proper TcpC localization and stability (53). TcpJ is the prepilin peptidase, belonging to the aspartic acid protease family, which cleaves the prepilin signal sequence between a charged and hydrophobic region within the N-terminus. The type IVb prepilin signal sequences are approximately 25 residues (Fig. 2) (54). TcpT, the putative ATPase required for pili biogenesis, is localized to the inner membrane in a TcpR-dependent manner (Table 2) (55).

Tad pili

The Tad loci of bacteria are a subset or distinct clade of the type IVb pili and are found in both Gram-negative and Gram-positive bacteria. Tad pili are very important for initial adhesion, cellular aggregation, and biofilm formation (56). Due to their small size, pili of the Tad system are also referred to as Flp (fimbrial-low-molecular weight protein) (Table 2) (56). We will review the Flp pili of *Aggregatibacter actinomycetemcomitans* as an example of Tad pili. Many proteins of the *tad* loci in *A. actinomycetemcomitans* share similarities with proteins involved in the type II, type III, and type IV secretion pathways (Flp1, TadV, RcpA, TadA, TadB, TadC, TadE, and TadF), but some of the Tad proteins (RcpB, RcpC, TadZ, TadD, and TadG) remain exclusively associated with the Tad system (Table 2). The pilin Flp1 is the major structural component of the Flp pili (57). TadE and TadF are important for Flp pili biogenesis (58) but are classified as pseudopilins, because they have conserved prepilin processing sites and have never been found associated with the pilus (59). Thus far, it is believed that Tad pili are not retracted, because no homologue of PilT, the ATPase required for pilus retraction, has been identified (Table 2). As observed for the *P. aeruginosa* type IVa ATPase PilB, the putative ATPase TadA forms hexomeric complexes to power pilus assembly (60). The localization of TadA is facilitated by TadZ, a cytoplasmic multidomain protein localized to the cell pole (61). TadB and TadC share similarity with PilC-like proteins and may associate to facilitate the passage of other Tad components across the inner membrane. RcpA is strongly suspected to form a channel required for the translocation of the pilus to the cell surface, but the ability of

RcpA to function as a secretin is not well characterized (62, 63) (Table 2). RcpB is an outer membrane protein that may function as a gating protein (63). TadD is a predicted lipoprotein and may be important for the assembly and targeting of the other outer membrane proteins (63).

A. actinomycetemcomitans is a strong biofilm former on a variety of surfaces and is a causative agent of infective endocarditis. The Tad pili of *A. actinomycetemcomitans* form bundles of fibers about 5 nm in diameter, composed of the 6.5-kDa main pilin Flp1, which are involved in adhesion to abiotic surfaces (58). The lack of processing of pre-Flp1, pre-TadE, or pre-TadF or loss of pili results in decreased cell aggregation and biofilm formation (59, 64, 65).

Unlike *A. actinomycetemcomitans*, the Tad pili in *C. crescentus* are not bundled but are expressed as individual pili at the flagellar pole (66). The Tad pili in *C. crescentus* are important for the initiation of adherence and biofilm formation (Table 1). They are crucial for the initial reversible attachment to surfaces in conjunction with flagella since the loss of pili decreases initial adherence by about 50% (67). In addition, pili are involved in surface contact stimulation of holdfast secretion. Holdfast is the powerful polysaccharide adhesin responsible for irreversible adhesion and biofilm formation, as described below (see "Holdfast" below) (68). Pili also play an important role in biofilm formation and maturation, because they are responsible for the integrity of the biofilm architecture (11).

P. aeruginosa makes a Tad or Flp pilus in addition to type IVa and type IVb pili (Table 1). The Flp pilus of *P. aeruginosa* has also been shown to be involved in biofilm formation and facilitates attachment to epithelial cells (69).

The plant pathogen *Agrobacterium tumefaciens* has a Tad pilus encoded by the *ctp* locus. Mutations in *ctpA*, *ctpB*, or *ctpG* resulted in decreased adherence on abiotic surfaces in both short-term assays and biofilm formation relative to wild type. However, no difference was readily observed between a *ctpA* mutant and wild type during root biofilm formation (252).

The Nucleation-Precipitation Pili

The final group of fimbrial adhesins is assembled via the nucleation-precipitation pathway. The curli of *E. coli* and *Salmonella* are the best characterized (Table 1). This group of adhesins is covered in depth in reference 251, so it will only be touched on briefly here. Curli are members of a growing group of proteins called functional amyloids. Aggregation is the basis of amyloid formation. A protein is in an "amyloid state" when it forms bundles of fibers made of β-sheets. Uncontrolled aggregation of proteins can result in damage or disease, such as may be the case in Alzheimer's disease. Functional amyloids in bacteria arise from aggregation producing an ordered β-sheet structure in a controlled process used to generate fibrils, spore coats, or surface coats (70). Curli systems are present in a wide variety of bacteria (71). Curli are thin fimbriae (6 to 12 nm wide) with variable lengths (72). The main curlin monomer is CsgA, which when secreted via CsgG, self-assembles into fibers on the cell surface in association with the minor curlin nucleator CsgB (Fig. 1C). CsgE and CsgF function as chaperones, and CsgD is a positive regulator of the curlin system. For an in-depth understanding of curli biogenesis and regulation, see reference 251 or the review by Evans and Chapman (73).

Curli are associated with both initial adherence and biofilm formation in Shigatoxin-producing *E. coli* (74). Gastrointestinal commensal *E. coli* express curli that were shown to contribute to biofilm formation (75). The curli and cellulose of enterohemorragic *E. coli* and ETEC work in concert in adherence and biofilm formation (76). As a consequence of curli protein self-assembly on the cell surface, cross-seeding of curli proteins has been documented between multiple species, including *E. coli*, *Salmonella*, and *Citrobacter*. Enhanced adherence and pellicle biofilm formation upon cross-seeding indicates that curli can facilitate multispecies biofilm formation (77).

NONFIMBRIAL ADHESINS

Unlike the long, polymeric adhesins discussed above, the nonfimbrial adhesins are short monomeric or trimeric structures. This type of adhesin is widely spread among bacteria and is involved in cell attachment to abiotic surfaces and/or host cells. In this section, only nonfimbrial adhesins involved in biofilm formation will be presented. The different roles in nonspecific adhesion played by these adhesins will be discussed: nonfimbrial adhesins are directly anchored to the outer cell membrane via covalent or noncovalent interactions and, due to their relatively short size, are usually involved in close contact between the bacterial cell and the substrate. Nonfimbrial adhesins are also involved in cell-cell interactions and aggregation. Furthermore, this type of adhesin is also shown to interact with various components of the biofilm extracellular matrix, linking the bacteria to the matrix and maintaining the biofilm architecture.

Although highly diverse in terms of structure and/or adhesive properties, nonfimbrial adhesins in Gram-negative bacteria are usually grouped into two main categories: the nonfimbrial adhesins secreted through a type 1 secretion system (T1SS) and the nonfimbrial adhesins secreted through one of the type 5 secretion systems (T5SSs). Detailed examples of nonfimbrial adhesins from both categories will be discussed in this section (for more detail on nonfimbrial adhesins see references 78 and 79).

Adhesin Secreted by the T1SS

In Gram-negative bacteria, a very large group of nonfimbrial adhesins is secreted by the T1SS. This secretion system is one of the simplest described to date in Gram-negative bacteria. It is a heteromeric complex composed of three components: (i) an inner membrane ABC (ATP binding cassette) transporter, (ii) a membrane fusion protein in the periplasmic space, and (iii) an outer membrane pore (Fig. 3). These three proteins interact and form a channel that specifically transports T1SS substrates through the bacterial envelope in a single step, directly from the cytoplasm to the extracellular space. Proteins targeted by the T1SS share a C-terminal domain of lightly conserved secondary structure that is not cleaved off during the secretion process. Proteins are secreted in an unfolded state and first interact with the ABC transporter, triggering a conformational change of the channel and subsequent ATP hydrolysis, followed by the transport of the protein through the entire secretion system and subsequent folding of the secreted protein (reviewed in reference 80).

The most studied example of nonfimbrial adhesins secreted via T1SS is the Bap (biofilm-associated protein) family of proteins. The adhesins belonging to this group are high molecular weight multidomain proteins and contain a core domain of long repeated units of variable length and sequence. All Baps share three distinct features (Fig. 4): (i) an N-terminal secretion signal for transport from the cytoplasm to the periplasm, (ii) a core domain composed of highly repeated motifs, and (iii) a glycine-rich C-terminal domain. The first Bap was identified in *Staphylococcus aureus* (81), and since then, numerous other Bap family members have been shown to be involved in cell adhesion to abiotic surfaces and biofilm formation in both Gram-positive and Gram-negative bacteria (for a review, see reference 82). A computational study revealed that large adhesins sharing the Bap family features are widespread among bacteria (83). However, only a few of these proteins have been characterized experimentally. Table 3 shows the Bap proteins that have

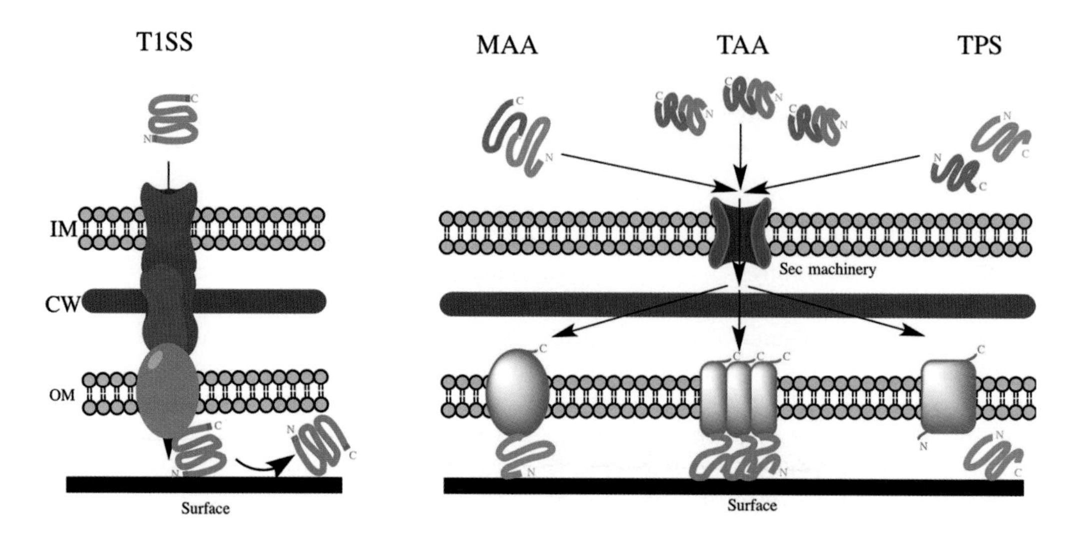

FIGURE 3 Schematic overview of the various secretion systems of nonfimbrial adhesins. The type 1 secretion system (T1SS) and three classes of type 5 secretion system (T5SS) (monomeric autotransporter adhesins [MAA], trimeric autotransporter adhesins [TAA], and two-partner secretion [TPS] systems) are represented. In T1SS, the adhesin is exported directly from the cytoplasm to the extracellular milieu via a pore comprised of three proteins. In T5SS, the adhesin is translocated from the cytoplasm to the periplasm by the Sec machinery and auto-assembled in the outer membrane. See text for more details. Abbreviations: IM, inner membrane; CW, cell wall; OM, outer membrane. doi:10.1128/microbiolspec.MB-0018-2015.f3

been reported to be involved in biofilm formation by Gram-negative bacteria.

In *P. fluorescens* and *P. putida*, the Bap protein LapA (large adhesion protein A) is crucial for biofilm formation (84, 85). This surface-associated protein, the largest one expressed by both organisms, is responsible for the transition between reversible adhesion via a single pole to irreversible adhesion along the entire cell length (84, 85). Indeed, *lapA* or the ABC transporter for LapA (required for the export of LapA to the cell surface) mutants is able to first attach to surfaces via their pole but fails to undergo later irreversible adhesion and to develop the typical biofilm architecture (85). LapA is conserved between many *P. fluorescens* and *P. putida* strains, but the length of the protein in different strains is highly variable due to the flexible number of amino acid repeats (86). LapA mediates cell adhesion to a wide array of abiotic surfaces, from various plastics to glass or quartz, suggesting that the interactions between this protein and the surface are nonspecific (84, 85). Single cell force spectroscopy experiments recently showed that different domains of LapA are involved in different adhesion processes: while the repeated units in the core domain are mainly responsible for adhesion on hydrophobic surfaces, the C-terminal domain is involved in adhesion to hydrophilic surfaces, allowing LapA adhesion to a wider range of substrates and increasing the versatility of *P. fluorescens* colonization in diverse environments (87). In addition, single cell force spectroscopy analysis of the footprint left behind by *P. fluorescens* cells detached from a surface reveals a local accumulation of LapA occurring at the cell-surface interface and the presence of multiple adhesion peaks with extended rupture lengths, highlighting the critical role of LapA in mediating the irreversible cell adhesion to surfaces (88).

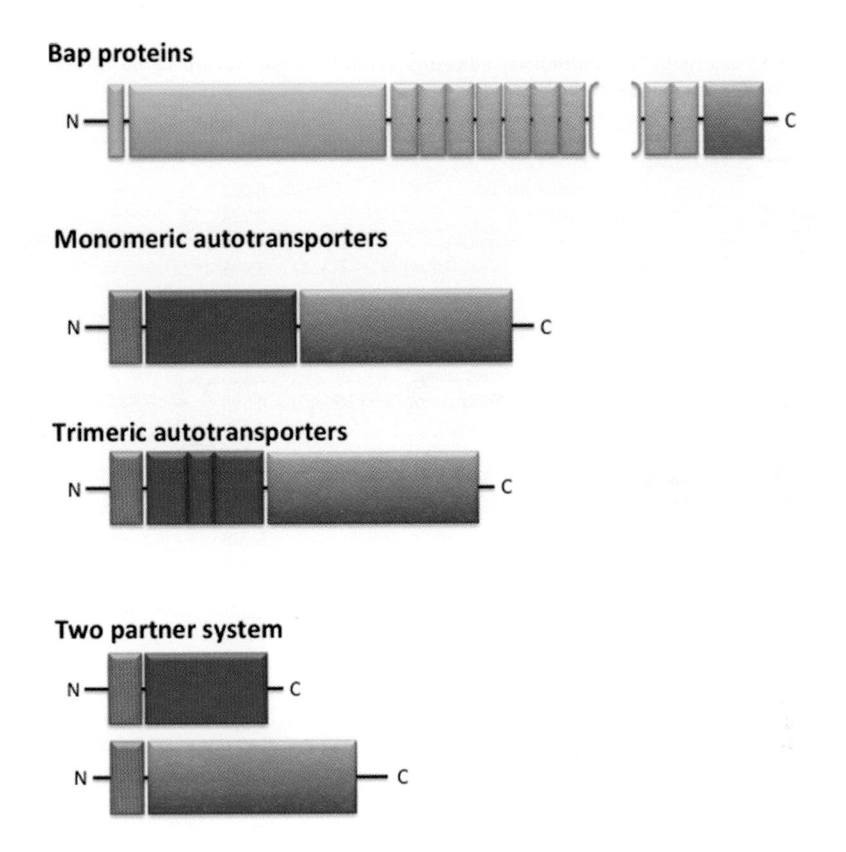

FIGURE 4 Nonfimbrial adhesin organization. See text for details. Green, signal sequence for proper localization and processing; turquoise, core domain; orange, glycine-rich repeated domain (brackets depict the variable number of repeats); magenta, serine-rich C-terminal region; navy blue, passenger domain; red, translocator domain. doi:10.1128/microbiolspec.MB-0018-2015.f4

Another large adhesion protein found in *P. putida* that participates in biofilm formation is LapF (89). LapF is a surface-associated protein and shares the key features of Baps (89). While LapA is responsible for irreversible adhesion of single cells to the surface, LapF is believed to meditate the cell-cell interaction in the biofilm and to play a major role during later development phases in biofilm maturation and stabilization (86, 89). It is worth noting that LapA is not present in pathogenic pseudomonads, such as *P. aeruginosa* and *Pseudomonas syringae* (85, 90), suggesting that while LapA is crucial for biofilm formation in environmental pseudomonads, pathogenic strains have developed other mechanisms for adhesion. *P. aeruginosa* strains have a *lapF* gene, whereas no LapF ortholog has been found in *P. syringae* (90). A similar observation is reported for Bap in *S. aureus*, where *bap* genes are absent from the genomes of human pathogenic strains (91).

Once secreted via the T1SS, all Baps remain loosely anchored to the outer membrane and are frequently released into the extracellular space. The nature of the interaction between the cell envelope and Baps remains unknown, but various hypotheses have been proposed: (i) Baps could transiently interact with the T1SS apparatus upon secretion; (ii) Baps could interact with an unknown surface-exposed outer membrane protein; or (iii) released Baps could

TABLE 3 Selected examples of nonfimbrial adhesins experimentally shown to be involved in biofilm formation by Gram-negative bacteria

Protein	Organism	Size (aa)	Reference
Biofilm associated proteins (Bap) – T1SS			
LapA	*Pseudomonas putida*	8,682	84
BapA / AdhA	*B. cenocepacia*	2,924	213
LapA	*Pseudomonas fluorescens*	4,920	85
BapA	*S. enterica*	3,825	214
YeeJ	*E. coli*	2,358	102
Bap	*Acinetobacter baumannii*	8,621	215
LapF	*P. putida*	6,310	89
BfpA	*Shewanella oneidensis*	2,768	216
MRP	*Pectobacterium atrosepticum*	4,558	217
BfpA	*Shewanella putrefaciens*	4,220	218
Cat-1	*Psychrobacter articus*	6,715	219
Monomeric autotransporter adhesins – T5SS			
Ag43	*E. coli*	1,039	104
Cah	*E. coli*	2,850	220
AIDA	*E. coli*	1,237	107
TibA	*E. coli*	989	221
YfaL/EhaC	*E. coli*	1,250	102
YpjA/EhaD	*E. coli*	1,526	102
YcgV	*E. coli*	955	102
Hap	*H. influenzae*	1,392	127
EhaA	*E. coli*	1,328	222
EhaB	*E. coli*	980	223
UpaH	*E. coli*	2,845	224
UpaC	*E. coli*	996	225
UpaI	*E. coli*	1,254	226
MisL	*S. enterica*	955	227
Trimeric autotransporter adhesins – T5SS			
YadA	*Yersinia pseudotuberculosis*	434	119
UspA1	*Moraxella catarrhalis*	955	228
Hap/MID	*M. catarrhalis*	2,090	228
UpaG	*E. coli*	1,779	120
SadA	*S. enterica*	1,461	121
AtaA	*Acinetobacter* sp. Tol5	3,630	229
EhaG	*E. coli*	1,589	230
BbfA	*Burkolderia pseudomallei*	1,527	122
Hemagglutin-like adhesins – T5SS			
HxfB	*X. fastidiosa*	3,376	231
HxfA	*X. fastidiosa*	3,458	231
HMW1	*H. influenzae*	1,536	127
HMW2	*H. influenzae*	1,477	127
XadA	*X. fastidiosa*	763	125
YapH	*Xanthomonas fuscans*	3,397	124
FhaB	*X. fuscans*	4,490	124
XacFhaB	*Xanthomonas axonopodis*	4,753	232
CdrA	*P. aeruginosa*	2,154	129
FHA	*B. pertussis*	3,590	128
BcpA	*Burkholderia thailandensis*	3,147	233

eventually engage in homotypic interaction with the residual surface-associated Baps (78, 92). Recent studies of the large adhesion SiiE from *Salmonella enterica* revealed that the retention of SiiE on the cell surface is modulated by pH, ionic strength, and osmolarity and is probably due to the interaction of the SiiE coil-coiled domains with the proteins forming the T1SS pore (93). In addition, extensive studies of the Baps LapA and LapF from *P. putida* show that these proteins are important components of the biofilm extracellular matrix and could be a link between the cells and matrix exopolysaccharides (94–96). During biofilm dispersal, LapA is cleaved from the cell surface by the specific protease LapG and released into the extracellular medium, freeing the bacteria from the biofilm matrix (94). Thus, the balance between retention on the cell surface and the release of Baps into the extracellular space may play a role in the regulation of cell adhesion to surfaces and eventually biofilm formation and dispersion.

T5SS Autotransporter Adhesins

The other major class of nonfimbrial adhesins is secreted by the T5SS. One of the largest groups of secreted proteins in Gram-negative bacteria, T5SS proteins display a two-step process: proteins are first transported from the cytoplasm to the periplasm via the Sec machinery and are then secreted across the outer membrane through a channel formed by a β-barrel pore (Fig. 3) (97). Originally thought to be self-sufficient for proper assembly on the cell surface, this type of protein was called "autotransporter." However, more recent studies suggest that the Bam complex, the machinery responsible for proper folding and proper insertion of outer membrane proteins into the outer membrane, is required for proper assembly of T5SS (98, 99).

All T5SS proteins share common structural and functional features (Fig. 4): (i) an N-terminal Sec-dependent signal peptide, (ii) a passenger domain providing the protein function, and (iii) a C-terminal β-barrel domain that allows secretion of the passenger domain. Once the protein is processed, the β-barrel domain integrates into the outer membrane to form a pore through which the passenger domain is secreted to the cell surface (for recent reviews, see references 79, 100).

The T5SS autotransporter family is represented by three distinct groups: (i) the monomeric autotransporters, (ii) the trimeric autotransporters, and (iii) the two-partner secretion systems.

Monomeric autotransporter adhesins (MAAs)

This group is the largest group of T5SS proteins. In monomeric autotransporters, passenger and secretion domains are integrated into a single multidomain protein (Fig. 3). The secretion β-barrel domain (250 to 300 amino acids) has high sequence similarity among monomeric autotransporters and is predicted to form a pore comprised of 14 antiparallel β-strands (97). Conversely, the passenger regions consist of repetitive amino acid motifs, highly variable in sequence and size, responsible for the difference in specificity of each protein (101). The properties of autotransporters are very diverse, because they can be virulence factors, toxins, or adhesins, among other functions. The majority of the studies of these autotransporter adhesins focus on specific interactions and host cell-bacterial adhesion, and only a few monomeric autotransporter adhesins have been shown to play a role in biofilm formation and nonspecific adhesion to abiotic surfaces (Table 3).

One of the most studied monomeric autotransporters is Ag43 from *E. coli*. Based on sequence similarity with Ag43, several other monomeric autotransporter adhesins have been identified in *E. coli*, and a few have been shown experimentally to be involved in cell aggregation and biofilm formation (Table 3) (102). Ag43 is present in the majority of commensal and pathogenic *E. coli* strains (103) and promotes autoaggregation and biofilm formation (104, 105). The exact role of monomeric autotransporters in biofilm formation has not been entirely elucidated yet, but recent experiments suggest that Ag43, like other related autotransporters including AIDA (adhesin involved in diffuse adherence), may mediate cell-cell aggregation by homotypic interactions that could eventually play a role in biofilm formation (106, 107). In addition, it has been shown that different monomeric autotransporter adhesins, like Ag43 and AIDA, can cross-interact, leading to the formation of mixed cell aggregates (79, 101, 107). It has been shown that many autotransporters are O-glycosylated proteins, like Ag43, AIDA, and TibA in *E. coli* (108–110). However, this glycosylation does not play a role in cell-cell aggregation or biofilm formation (109, 112). Ag43 expression is controlled by phase variation and switches from Ag43^{on} to Ag43^{off} at a high rate (112). A recent study highlighted the importance of Ag43 phase variation during biofilm formation mediated by both Ag43 and other adhesins in *E. coli* (113). Indeed, Ag43^{on} bacteria are physically selected for in the biofilm, probably due to their inherent ability to auto-aggregate, which may subsequently modulate interaction with surfaces via the production of other adhesins (113). The recent resolution of the crystal structure of Ag43 reveals a "Velcro-like" organization similar to the self-oligomerization of the Hap autotransporter from *H. influenzae* (114), mediating Ag43-Ag43 interactions and probably Ag43 interactions with other autotransporters and promoting cell-cell aggregation (115).

Trimeric autotransporter adhesions (TAAs)

All the members of the TAA family described so far are nonfimbrial adhesin proteins (92). Numerous TAAs have been reported to specifically bind to biotic surfaces on host cells, or host extracellular matrix components including fibronectin or collagen, and only a few studies have focused on the role of these adhesins in biofilm formation and attachment to abiotic surfaces (Table 3).

While the overall organization of these autotransporters is similar to the monomeric autotransporters, TAAs usually have a shorter C-terminal secretion domain (50 to 100 amino acids) with four β-strands. The formation of a full-size 12-β-strand pore is achieved by trimerization (116). The passenger domain contains conserved elements designated as head, connector, and stalk domains (117). The head harbors the adhesive properties, while the connector and stalk primarily function to extend the head domain away from the bacterial cell body (117). Unlike the monomeric autotransporters described above, TAA are not cleaved or released into the extracellular space (106).

The most studied TAA in the multifunctional nonfimbrial adhesins is YadA from *Yersinia* spp. (118). This adhesin has been linked to virulence, adhesion to host cells, and autoaggregation (118). However, the putative role of YadA in adhesion to abiotic surfaces and biofilm formation is poorly understood. A specific 31–amino acid motif, present only in the N-terminal sequence of YadA from *Yersinia pseudotuberculosis*, is required for cell aggregation and biofilm formation, but the mechanism of adhesion is unknown (119). Only a few other TAAs have been reported to play a role in biofilm formation (Table 3). UpaG is involved in cell-cell interactions, promoting biofilm formation to abiotic surfaces in *E. coli* (120). SadA, an ortholog of UpaG in *S. enterica* and *S. typhimirium*, has been shown to mediate cell aggregation and biofilm formation in cells unable to express long O-antigen on the cell surface (121), suggesting that bacteria need to be in close proximity for SadA to mediate cell-cell interactions. Recently, BbfA (*Burkolderia*biofilm factor A) has been shown to play a major role in biofilm formation in *Burkolderia pseudomallei*, probably by mediating cell aggregation (122).

Two-partners secretion system (TPS)

TPS proteins form another group of autotransporters utilizing T5SS. In this family, the secretion mechanism is very similar to the ones described above, but the passenger and membrane anchor domains are encoded by two distinct genes, usually organized in a single operon. One gene encodes the secretion signal and the adhesin domain, and the other encodes the outer membrane β-barrel pore (79) (Fig. 3). The pores are highly conserved proteins of typically 500 to 800 amino acids (123). The effectors are large proteins and can be highly variable, but they all contain two common features: (i) a β-helix or β-solenoid secondary structure and (ii) a conserved N-terminal domain, the TPS domain (123). The TPS domain provides specific recognition of the appropriate pore for translocation through the outer membrane (116).

Filamentous hemagglutinin proteins are typical adhesins exported by the TPS (Table 3) and are usually involved in tight interaction between the bacterium and a specific receptor on the host cell. Their involvement in biofilm formation and nonspecific adhesion to abiotic surfaces has been studied only sparsely. Filamentous hemagglutinins have been shown to play a role in biofilm formation by different plant pathogens, including *X. fastidiosa* or various *Xanthomonas* spp. (124–126). Unlike the two other types of autrotransporter adhesins described above, it seems that filamentous hemagglutinins from these species do not mediate cell-cell interactions and autoaggregation, but rather attach cells to abiotic surfaces, thereby facilitating their interaction with each other (125). Filamentous hemagglutinins also play a role in biofilm formation in the human pathogens *H. influenzae* (127) and *Bordetella pertussis* (128). Filamentous hemagglutinin from *B. pertussis* is a critical virulence factor and is involved in interactions between bacterial cells, as well as interactions between the bacteria and abiotic surfaces (128). In addition, it has been shown that the hemagglutinins HMW (high molecular weight protein) 1 and HMW 2 are found around *H. influenzae* cells grown in a biofilm, as well as interacting with the biofilm extracellular matrix (127), strongly suggesting that these proteins play an important role in adhesion to abiotic surfaces by linking the bacteria to the biofilm matrix and stabilizing the overall architecture.

Another example of a nonfimbrial adhesin belonging to the TPS group is CdrA (c-di-GMP regulated TPS A) in *P. aeruginosa*. This adhesin is transported by CdrB (c-di-GMP regulated TPS B) and is expressed when the level of c-di-GMP in the cell is high or in biofilm cultures (129). Once secreted outside the cell, CdrA is proposed to bind to the exopolysaccharide Psl to promote cell aggregation and biofilm formation. This adhesin is part of the biofilm matrix and plays a role in maintaining biofilm structural integrity, probably by bundling individual Psl strands together and/or by anchoring Psl fibers to the bacteria in the biofilm (129).

POLYSACCHARIDE ADHESINS

In addition to the various protein adhesins described above, many bacterial species produce extracellular polysaccharides that promote adhesion. These polysaccharides can be firmly associated with the cell surface, forming a capsule (capsular polysaccharide), or be loosely associated or even released from the cell surface (extracellular polysaccharide [EPS]). The difference between capsular polysaccharide and EPS is usually experimentally defined and may have limited physiological relevance. From an adhesive point of view, it is more appropriate to distinguish protective from aggregative polysaccharides. While protective polysaccharides provide a protective barrier around the cell, the aggregative polysaccharides have adhesive and/or cohesive properties that are exploited by bacteria to mediate adherence to surfaces and/or reinforce structural integrity of the biofilm (130, 131). Although protective polysaccharides could also contribute to the adhesive properties of bacterial cells, only the aggregative polysaccharides that serve as adhesins will be discussed in this section. For simplicity, aggregative polysaccharides will be referred to as EPS.

Structure and Function

EPSs often have long chains (molecular weight > 10^6 Da) composed of a repetition of the same sugar residue (homopolysaccharide) or a mixture of neutral and/or charged sugar residues (heteropolysaccharide). These polymers are mostly negatively charged, but neutral and even positively charged EPSs also exist (6, 132–134). The adhesiveness of EPSs strongly depends on chain conformation and is greatly influenced by substituents that modify interchain and intrachain interactions (135). The polar and hydrogen bonding functional groups of polysaccharides, such as ethers and hydroxyls, provide good adhesion to polar surfaces. The adhesiveness of EPSs can be strongly influenced by the presence carboxylic acid groups that promote the association of divalent cations such as Ca^{2+} and Mg^{2+} (6, 132–134, 136).

The degree of acetylation of a polysaccharide can drastically modify its cohesiveness and adhesiveness (137–141). Partial deacetylation of the well-characterized poly-β-1,6-*N*-acetyl-D-glucosamine EPS (PGA) is required to initiate biofilm formation by *E. coli*, *S. aureus*, and *Staphylococcus epidermidis* (143–145). Similarly, the selective O-acetylation of mannuronate residues on alginate, a polysaccharide secreted by *P. aeruginosa*, seems to improve biofilm adhesion and increase adhesion to host surfaces (137, 139, 146). Furthermore, partial deacetylation of *E. coli* PGA is required to facilitate the secretion of the polymer through the outer membrane (143). On the other hand, partial deacetylation of holdfast in *C. crescentus* and *Asticcacaulis biprosthecum* appears to be essential to maintain the structure, the adhesiveness, and the retention of the polymer on the cell surface in both species (141). In this case, the deacetylation of polysaccharides might facilitate the conformational transition of the polymer chain from random coils to ordered helices, promoting better interchain interactions and more stiffness of the polymer (138).

Biosynthesis, Secretion, and Anchoring

Although chemically diverse, the molecular mechanisms by which EPSs are assembled and exported to the cell surface can be categorized into three distinct mechanisms: (i) the Wzx/Wzy transporter-dependent pathway, (ii) the ABC transporter–dependent pathway, and (iii) the synthase-dependent pathway (Fig. 5). These three mechanisms have been extensively reviewed, and only a brief overview will be presented in this section (for more detail see reference 131).

In the Wzx/Wzy-dependent pathway, each polysaccharide repeat unit is first assembled on an undecaprenyl phosphate acceptor moiety in the cytoplasm, which is then transported across the inner membrane before being polymerized into a high molecular weight polysaccharide in the periplasm. In contrast, the entire polysaccharide chain is assembled in the cytoplasm on a lipid acceptor in the ABC transporter–dependent pathway. Although mechanistically different, these two pathways use a similar mechanism to facilitate EPS export across the periplasm and through the outer membrane. Both pathways use a large protein complex spanning the periplasmic space, formed by an outer membrane polysaccharide export pore and a polysaccharide copolymerase. In the synthase-dependent pathway, both the polymerization and the transport of the polymer across the inner membrane are carried out by the same membrane-embedded glycosyl transferase (131, 147). The polymer is protected by a molecular chaperone, a tetratricopeptide repeat–containing protein, and then exported across the outer membrane through a β-barrel porin (147).

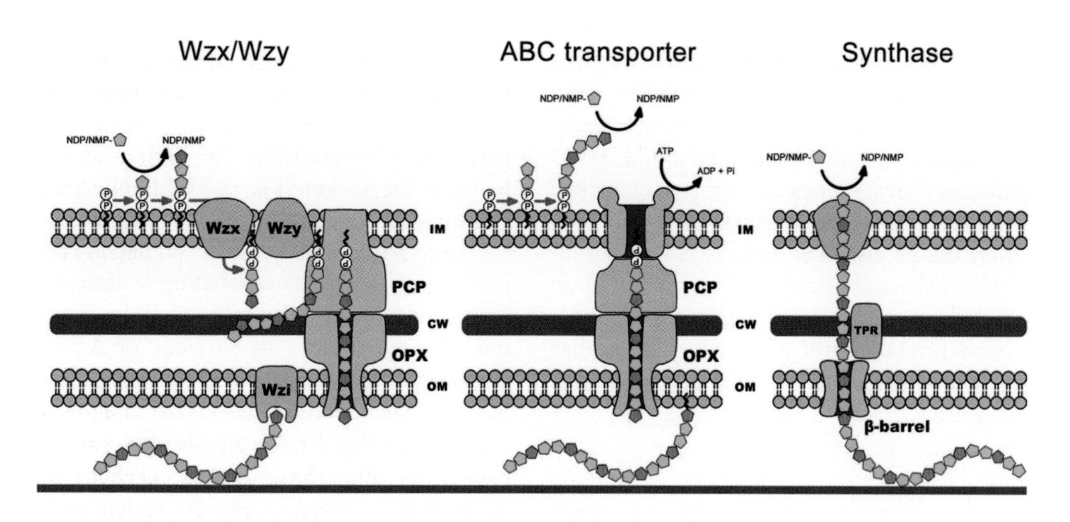

FIGURE 5 Polysaccharide biosynthesis pathways. Overview of the Wzx/Wzy-, ABC-transporter-, and synthase-dependent exopolysaccharide biosynthesis pathways. Only the key components for each pathway are indicated on the diagram. In the Wzx/Wzy-dependent pathway, the polysaccharide repeat unit assembly is initiated on an undecaprenyl phosphate acceptor moiety located in the inner leaflet of the inner membrane, which is then transported across the inner membrane by the flippase, Wzx. The polymerization into high–molecular weight polysaccharide occurs in the periplasm by the action of the polymerase Wzy. The export and secretion of the polysaccharide through the outer membrane are facilitated by the outer membrane polysaccharide export (OPX) and the polysaccharide copolymerase (PCP) protein families. Depending on the polysaccharide being synthesized, the nascent polymer could be anchored to the outer membrane via a specific protein, such as Wzi. In the ABC transporter–dependent pathway, the entire polysaccharide chain is assembled into the cytoplasm on a lipid acceptor that is then transported across the inner membrane by the ABC transporter. As observed for the Wzx/Wzy-dependant pathway, the export and secretion of the polysaccharide through the outer membrane also involve the OPX and PCP protein families. In the synthase-dependent pathway, both the polymerization and the transport of the polymer across the inner membrane are carried out by the same membrane-embedded glycosyl transferase. The export and secretion of the polysaccharide through the outer membrane are facilitated by a molecular chaperone and a β-barrel porin. Abbreviations: IM, inner membrane; CW, cell wall; OM, outer membrane. doi:10.1128/microbiolspec.MB-0018-2015.f5

When associated with the cell surface, the exact connection between EPS and the outer membrane is not always known and usually involves a linker, which can be a sugar, lipid, or protein. For EPS exported through the ABC transporter–dependent pathway, the conserved phospholipid terminus is attached at the reducing end of the polysaccharide via a poly-3-deoxy-D-manno-oct-2-ulosonic acid (also known as KDO) linker that is responsible for proper translocation and contributes to the attachment of the polymer to the cell surface (148), although ionic interactions between the core region of the lipopolysaccharide (LPS) and the polymer may also be involved (159). As observed for LPS O-antigen, carbohydrate chains forming the capsular or K-antigen (derived from the German word *Kapsel* [150]) may also be linked to the lipid A core of an LPS molecule, but the distinction from a traditional O-antigen remains subtle and purely operational (151, 152). Finally, polysaccharides can be anchored to the cell surface via dedicated proteins (153–155). In *E. coli* and *K. pneumoniae*, the outer membrane protein Wzi is a critical factor in the anchoring of the K30 type I capsule to the cell surface (153, 155). The 2.6Å resolution structure of Wzi suggests that Wzi may act as a lectin binding to the nascent polymer once translocated through the Wza translocase, but other cell surface components may also play a role (153). In *C. crescentus*, the HfaA, HfaB, and HfaD proteins provide an extremely strong anchor for the holdfast polysaccharide to the cell envelope, as discussed below (154). Alternatively, polysaccharides can be anchored to the cell surface by glycosylation of proteins, a system primarily utilized to decorate flagellar proteins and protein adhesins (for review see references 156, 157).

Polysaccharide Diversity

The composition and the structure of EPS can vary both between and within species (148), and it is now well accepted that bacteria are able to produce different types of EPS to provide adhesive adaptability under varying conditions (Table 4). Most of these carbohydrate polymers generate a highly hydrated and extensive layer surrounding the cell, but a few bacterial species adhere to surfaces by using a clearly defined, discrete patch of polysaccharide (158–160). We will focus on the discrete aggregative polysaccharides in this section.

There are diverse instances of discrete adhesive polysaccharide mediating surface attachment. Among these is the well-described unipolar attachment to surfaces of *Alphaproteobacteria*, such as *C. crescentus* and *A. tumefaciens*, and the less common discrete slime deposition by the motile *Deltaproteobacterium*, *Myxococcus xanthus* (Fig. 6). In both cases, the discrete polysaccharide provides a specific function for different phases of the life cycle, supporting a transient or an irreversible attachment with the surface. In this section, the unipolar polysaccharide adhesin produced by *C. crescentus* and by some *Rhizobiales* and the adhesive slime produced by *M. xanthus* will be discussed.

TABLE 4 Selected examples of aggregative polysaccharides experimentally shown to be involved in biofilm formation by Gram-negative bacteria

Polysaccharide	Organism	Composition/structure	Reference
Alginate	*P. aeruginosa*	β-1,4-linked mannuronic acids and guluronic acids	234
Cellulose	*Gluconacetobacter xylinus*, *A. tumefaciens*, *Rhizobium leguminosarum* bv. Trifolii, *Sarcina ventriculli*, *Salmonella* spp., *E. coli*, *K. pneumoniae*	β-1,4-linked D-glucose	235–239
Holdfast	*Caulobacter* spp., *Asticcacaulis biprosthecum*, *Hyphomonas adherens*, *Hyphomonas rosenbergii*, *Hyphomicrobium zavarzinii*, *Maricaulis maris*, *Oceanicaulis alexandrii*	Suspected to contain β-1,4-linked N-acetyl-D-glucosamine, but the exact composition and structure remain unknown	160, 163, 166, 240–243
PGA	*E. coli*, *Yersinia pestis*, *Bordetella* spp., *Actinobacillus* spp., *P. fluorescens*	β-1,6-linked N-acetyl-D-glucosamine	244, 245
Psl	*P. aeruginosa*	Repeating pentasaccharide of 3 mannose, 1 rhamnose, and 1 glucose	246–248
Pel	*P. aeruginosa*, *P. fluorescens*	Unknown, but reported to be a glucose-rich polysaccharide polymer	246, 249, 250
Slime	*M. xanthus*	Suspected to contain α-D-mannose or α-D-glucose residues, but the exact composition and structure remain unknown	192
UPP	*A. tumefaciens*	Suspected to contain N-acetyl-D-glucosamine residues, but the exact composition and structure remain unknown	68, 179

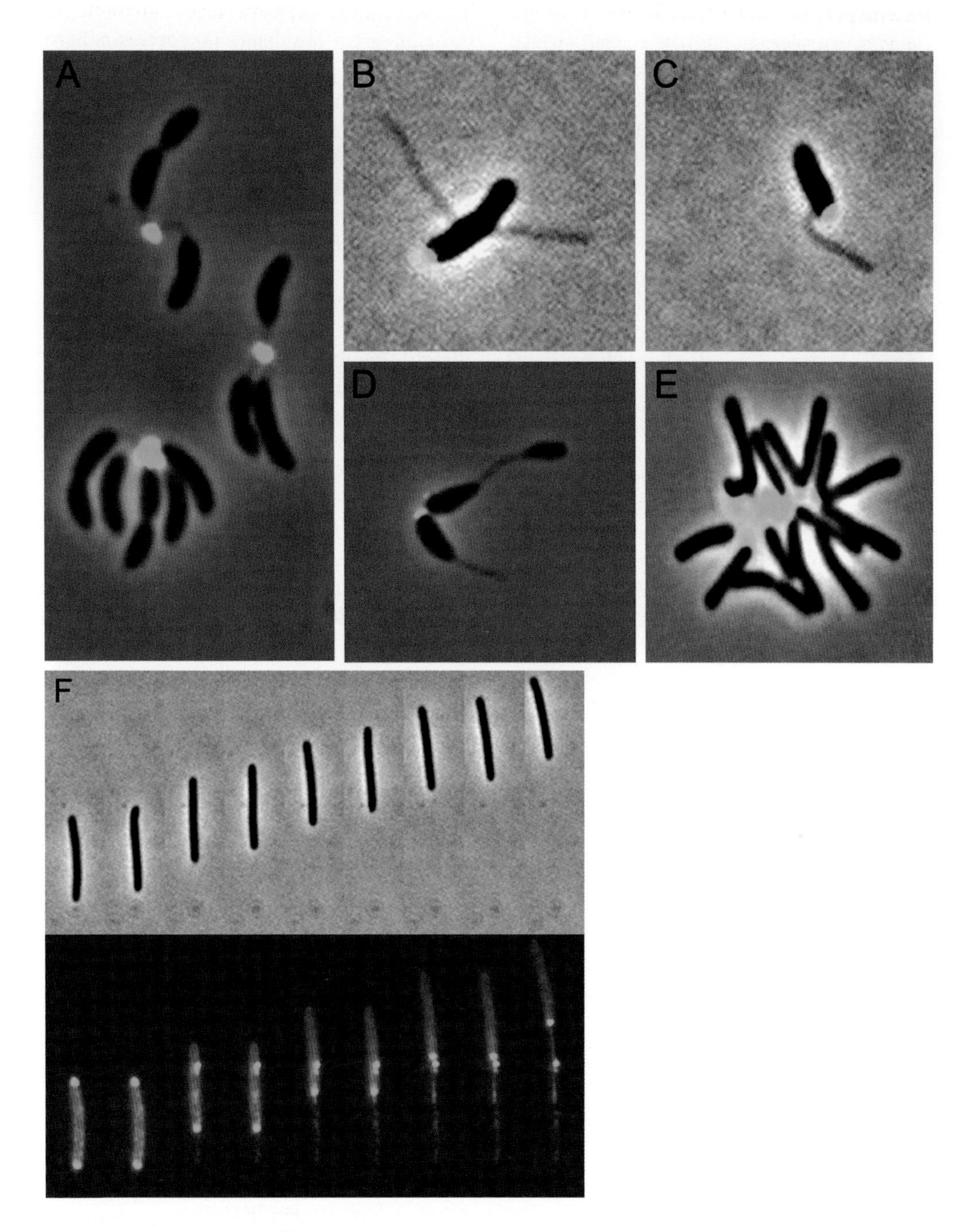

FIGURE 6 Selected examples of discrete polysaccharides. AF488-conjugated wheat germ agglutinin lectin labelling of the holdfast in (A) *C. crescentus*, (B) *A. biprosthecum* (courtesy of Chao Jiang), (C) *Asticcacaulis excentricus* (courtesy of Chao Jiang), and (D) *Hyphomicrobium vulgare* (courtesy of Ellen Quardokus). (E) AF488-conjugated wheat germ agglutinin lectin labelling of the UPP in *A. tumefaciens*. (F) FITC-conjugated ConA lectin labelling of the slime in *M. xanthus*. doi:10.1128/microbiolspec.MB-0018-2015.f6

Holdfast

Caulobacter spp. and other stalked *Alphaproteobacteria* (161) synthesize a polar polysaccharide called holdfast (162) (Fig. 6). Only a small amount of holdfast is produced at a discrete site on the cell surface and can be observed either at one of the cell poles or at the tip of a thin cylindrical extension of the cell envelope referred to as the stalk. The holdfast is responsible for permanent adhesion to a surface and can maintain attachment for many generations, presumably to provide better access to limited nutrients under flow (163). The holdfast is also crucial for biofilm formation (162). *C. crescentus* holdfast secretion is regulated both developmentally and by contact dependence (68). *C. crescentus* exhibits a dimorphic life cycle that begins as a motile swarmer cell. Unless they come in contact with a surface, swarmer cells are able to swim for a period equivalent to one-third of the cell cycle, after which they differentiate into stalked cells. During this developmentally programmed transition, swarmer cells release their flagellum, retract their pili, and synthesize a holdfast and a stalk at the same pole. The small protein regulator HfiA is crucial for holdfast production at the correct time during the cell cycle, as well as under the correct nutritional conditions (164). In addition to the developmental timing of holdfast secretion, if a swarmer cell encounters the surface, the coordinated action of the flagellum and pili stimulates early holdfast secretion, thereby driving the rapid, just-in-time transition from reversible to irreversible attachment (68). The addition of crowding agents that impede the rotation of the flagellum also stimulates the production of holdfast, suggesting that increased load on the flagellum triggers holdfast synthesis (68).

Two main gene clusters in *C. crescentus* have been identified that are required for holdfast synthesis (*hfs*) and holdfast anchoring to the tip of the stalk (*hfa*). The *hfs* cluster encodes proteins involved in the biosynthesis and secretion of the polysaccharide in a process similar to the Wzx/Wzy transporter-dependent pathway (160, 165). The *hfa* cluster encodes proteins crucial for attaching the holdfast to the cell envelope. Mutations in the *hfa* locus result in inefficient attachment of the holdfast to the tip of the stalk, causing holdfast shedding into the culture supernatant (154). The *hfa* locus is a single operon comprised of *hfaA*, *hfaB*, and *hfaD*. The HfaA, HfaB, and HfaD proteins are associated with the outer membrane and are polarly localized. HfaA and HfaD form high molecular weight complexes that share properties with amyloid proteins, and HfaB plays a role in the secretion of both proteins (154). Although the exact anchoring mechanism between the polysaccharide and the high molecular weight complexes formed by HfaA and HfaD remains unknown, the HfaA, HfaB, and HfaD proteins provide an extremely strong anchor for the holdfast polysaccharide to the cell envelope of *C. crescentus* (154).

Holdfast is bound specifically by wheat germ agglutinin, a lectin that specifically recognizes *N*-acetyl-D-glucosamine (GlcNAc)–containing polymers. In addition, holdfast is sensitive to lysozyme, a glycolytic enzyme specific for cleavage of β-1,4 linkages in oligomers of GlcNAc, suggesting that the holdfast is mostly comprised of β-1,4-linked GlcNAc (163, 166). The exact composition and structure of the holdfast is still unknown, due in large part to the small quantity synthesized by cells and its insoluble nature. The holdfast of *C. crescentus* has a gel-like nature and exhibits an impressive adhesive force in the ×Newton range, the equivalent of 10,000 psi, making holdfast one of the strongest known biological adhesives characterized (166, 167). GlcNAc polymers have been proposed to provide elastic properties to the holdfast (166). Recent force spectroscopy experiments suggest that additional adhesive components within the holdfast might be responsible for the bulk of adhesion (168). The measure of time-dependence of holdfast rupture force on a variety of substrates suggests the existence of discrete and cooperative events of single adhesin molecules that may act in concert with the GlcNAc polymers (168). As mentioned earlier, the degree of deacetylation strongly influences the physical properties of the holdfast polysaccharide (141). The deletion of *hfsH*, a polysaccharide deacetylase, strongly affects the adhesiveness and the cohesiveness of holdfast (141).

Although *C. crescentus* holdfast remains the best studied, other stalked bacteria including *Hyphomonas*, *Hyphomicrobium*, *Asticcacaulis*, other freshwater *Caulobacter*, and a variety of marine bacteria also produce a discrete holdfast polysaccharide (Fig. 6) (for a review see reference 169). These polysaccharides localize to the pole of the cell and provide strong adhesion to surfaces in marine and freshwater environments. The holdfast biosynthesis gene clusters of the other species often contain additional genes for gycosyltransferases or polysaccharide modification enzymes as compared to *C. crescentus* (169), suggesting that holdfast composition may vary depending on the environment in which the different species attach.

Polar surface polysaccharide from *Rhizobiales*

In addition to the *Caulobacterales* holdfast, other members of the *Alphaproteobacteria* class exhibit similar polar polysaccharides that are involved in adhesion. The *Rhizobiales* *A. tumefaciens*, *Rhizobium leguminosarum*, and *Bradyrhizobium japonicum*, can form a pathogenic or endosymbiotic association with certain plants (170). This association starts with the attachment of the bacteria to the host plant tissues, a multistep process mediated by bacteria-secreted polysaccharides, bacteria-secreted adhesins, or in some cases, by plant-produced lectins (170–173). Once in contact with the plant surface, bacteria may produce cellulose fibrils to form a stable biofilm around the tips of root hairs or the plant tissues (174). Several different surface polysaccharides, including LPS and capsular acidic polysaccharides have been shown to play a role in plant infection and colonization. In addition, *A. tumefaciens*, *R. leguminosarum*, and *B. japonicum* produce a unipolar polysaccharide that appears to be essential for the initial step of adhesion (170, 174–176).

A. tumefaciens forms complex biofilms on abiotic surfaces and plant roots that eventually cause "crown gall" disease. Unlike *R. leguminosarum* and *B. japonicum*, *A. tumefaciens* is pathogenic and does not benefit the plant (177). The infection involves first the attachment of the bacteria to the plant surface and then the transfer of a DNA segment carried on the tumor-inducing (Ti) plasmid into the host plant genome. It has been shown that *A. tumefaciens* produces several different EPSs: succinoglycan, cellulose, β1-2 glucans, and β1-3 glucans. However, more recent studies indicate that the initial attachment to the plant root surface and abiotic surfaces is mediated by a coordinated action of flagella and the unipolar polysaccharide (UPP) (68, 178, 179). UPP is crucial for permanent attachment, but by comparison with the holdfast of *C. crescentus*, UPP is only observed upon surface contact (68). More recently, it has been proposed that secretion of UPP is stimulated by an increase of the intracellular level of c-di-GMP, suggesting that surface contact may stimulate the increase of c-di-GMP through the activity of diguanylate cyclases or phosphodiesterases by an as yet unknown mechanism (179).

The secretion of UPP is abolished by deletion of the *uppABCDEF* locus (178, 179). The genes within the *uppABCDEF* locus share similarity with several genes within the *hfs* locus in *C. crescentus*, most notably UppC with HfsD, the predicted Wza outer membrane export porin (175). No homologues of the *hfa* genes, which are involved in holdfast anchoring in *C. crescentus*, have been identified in agrobacterial genome sequences (175), suggesting that UPP and holdfast are anchored to the cell surface via different mechanisms. As observed for holdfast, UPP reacts specifically with wheat germ agglutinin, suggesting that the UPP also contains GlcNAc residues, but its exact composition and structure remain unknown (68, 179).

R. leguminosarum forms an endosymbiotic association with certain leguminous plants such as *Trifolium* (clover), *Pisum* (pea), *Vicia* (vetch), and *Phaseolus* (bean) to potentially nodulate and differentiate into nitrogen fixing bacteroids (180). This bacterium also synthesizes a discrete polar polysaccharide that is responsible for specific adhesion to plant receptors. The primary attachment of *R. leguminosarum* to host plant root hairs involves bacteria-secreted polysaccharides and plant-produced lectins (171, 172). The plant-produced lectins, such as the pea (*P. sativum*) lectin and the vetch (*V. sativa*) lectin are hypothesized to recognize and bind to a polarly localized glucomannan exopolysaccharide present at the surface of *R. leguminosarum*, promoting bacterial attachment at one pole (171–173). The glucomannan exopolysaccharide is mainly composed of mannose and glucose with minor amounts of galactose and rhamnose (171). Terminal mannose and glucose were identified, in addition to 1,2-mannose, 1,4-glucose, and 1,3,4-linked glucose, but the exact structure remains unknown (171). At least one gene (*gmsA*), whose deletion abolishes glucomannan secretion and biofilm formation on root hairs, has been identified as being involved in glucomannan biosynthesis (174). Interestingly, *gmsA* belongs to a conserved gene locus among other rhizobia and, more specifically, *A. tumefaciens*. This locus shares sequence similarity with the *uppABCDEF* locus in *A. tumefaciens*, suggesting that glucomannan could be biosynthesized through a similar pathway as UPP (178). Glucomannan is required for initial attachment only under moderate acidic conditions. Under neutral or alkaline conditions, attachment to root hairs is mediated by Rhicadhesin, a calcium-binding protein, or rhizobium-adhering proteins, named "Rap" (171, 173, 181).

B. japonicum forms an endosymbiotic association with soybean plants. It was proposed that initial attachment was mediated by the plant lectin SBA (soybean agglutinin), but another carbohydrate-binding protein, called BJ38, has been subsequently identified (176). Both SBA and BJ38 bind to a polarly localized surface polysaccharide produced by *B. japonicum*, but surprisingly, SBA and BJ38 recognize polysaccharides at opposite ends of the cell (176). It was suggested that the pole bound by BJ38 is probably involved in cell-cell and cell-host attachment since BJ38 labeled the point of bacterial attachment (176, 182). However, the exact nature of this interaction and the composition of this polar polysaccharide remain to be determined (176, 182). More recently, SBA has been shown to promote biofilm formation even in the absence of plant roots, suggesting that SBA also contributes to adhesion on abiotic surfaces (183).

In conclusion, the use of a polar adhesin appears to be a common feature shared by some *Alphaproteobacteria* to mediate adhesion of the bacterial cell to biotic or abiotic surfaces. The conservation of this feature suggests that the resulting asymmetric adhesion is important for the fitness of these species. However, future studies will be needed to identify the adhesive mechanism behind these strongly adhesive unipolar polysaccharides, their regulation, and their biosynthesis.

Slime

By comparison with the discrete unipolar polysaccharide of *C. crescentus*, the deltaproteobacterium *M. xanthus* exhibits an unusual patchy localization of an adhesive polysaccharide referred to as slime (Fig. 6). *M. xanthus* moves across solid surfaces by both twitching and gliding motility. While twitching motility results from polar retractile type IV pili, gliding motility results from the action of internal molecular motors that exert traction against the substrate through a large envelope-spanning complex (184–187).

Under low-nutrient conditions, *M. xanthus* cells coordinate their motility to merge into multicellular structures called fruiting bodies, where cells follow a developmental program that eventually triggers their differentiation into stress-resistant spores. Under high-nutrient conditions, spores germinate to form predatory swarms that secrete degradative enzymes to lyse prey bacteria (188). While moving over surfaces, slime is observed in the wake of motile bacteria. Slime has been observed for more than three decades (189) and has been proposed to be directly responsible for gliding motility (190), but recent evidence suggests the slime most likely promotes adhesion of the gliding motors to the substrate by a "paste-release" mechanism (191).

When observed at high contrast by a noninvasive technique called Wet-SEEC (wet-surface enhanced ellipsometry contrast microscopy) COMP:, slime trails appear to be secreted inconsistently behind the cell as extended continuous segments separated by patchy and localized deposits (191, 192). On cells, slime is unevenly distributed, forming conspicuous patches distributed at regular intervals on the surface of the cell body (192). Careful examination suggests that slime is locally concentrated and deposited underneath the cell body at fixed and discrete positions that colocalize with the gliding machinery. It has been suggested that the gliding machinery itself may produce its own adhesion substrate, enhancing *Myxococcus* substrate adhesion and allowing cell gliding on a larger range of substrates, but the exact anchoring mechanism between the polymer and the tip of the machinery remains to be elucidated (192).

Although slime reacts with Concanavalin-A, a lectin specific to internal and nonreducing terminal α-D-mannose or α-D-glucose, its exact composition remains unknown (192). Moreover, slime secretion remains unaffected by the deletion of genes involved in the secretion of exopolysaccharides through the Wzx/Wzy transporter-dependent pathway or in the biosynthesis of the O-antigen, suggesting that slime is secreted by an alternative secretion pathway. More recently, slime has been shown to contain embedded outer membrane vesicles, which might be involved in cell-cell communication or cell-cell transfer of outer membrane proteins between *Myxococcus* cells (191). Slime may play multiple roles by not only providing specific adhesion for the gliding machinery, but also by potentially supporting a form of cell-cell communication between distant *Myxoccocus* cells sharing the same track.

CONCLUSION

During the last decade a large repertoire of bacterial adhesins has been identified. The recent advances in our understanding of the secretion, assembly, and regulation of these bacterial adhesins demonstrate that, even if they share some key structural and regulatory factors that determine their expression at the surface of cells, their level of specificity varies significantly depending on the targeted surface. This characteristic clearly illustrates the ability of bacteria to successfully adapt and adhere to virtually all natural and man-made materials. This article reviewed major protein and localized discrete polysaccharide adhesins, but the large variety of adhesins suggests that virtually any biological molecule made by the bacterium can be used to support or modulate its adhesion to surfaces. Even previously unsuspected biological molecules, such as extracellular DNA and lipids, have been shown to play a role in the process of adhesion or biofilm maturation (193, 194). Recent discoveries in the regulation and the modulation of adhesion suggest that this process is much more complex and dynamic than originally anticipated. We have just begun to understand the factors that govern the temporal regulation of the expression of adhesin molecules, but many questions concerning the nature of signals triggering the secretion of the proper adhesin or the interaction between the different types of adhesins remain unanswered. The recent advent of single-cell approaches will yield a better understanding of the detailed roles of each adhesive factor and the environment during this transition from reversible to irreversible attachment (195). Surprisingly, even with decades of research, the exact molecular mechanisms that support or modulate the interaction between adhesins and surfaces have seldom been characterized. Even if we can grasp the huge versatility of bacterial adhesives, we are lacking the most important knowledge about how this adhesion occurs.

A better understanding of the adhesive properties of these adhesins holds promise for future methods to control bacterial surface attachment, promoting it when beneficial and eradicating it when detrimental. In addition, an improved knowledge of this field will allow us to use certain bacterial adhesins as models of biological adhesives, which could offer an alternative strategy to their synthetic counterparts in industrial and medical applications. Indeed, biological adhesives demonstrate impressive performance in their natural context: they enable attachment to a broad variety of surfaces in aqueous environments and share desirable properties, such as sustainability, biodegradability, and biocompatibility, resulting in a much-reduced impact on the environment.

ACKNOWLEDGMENTS

The authors want to thank the members of the Brun lab and Clay Fuqua for their fruitful discussions and critical reading of the manuscript. Work on bacterial adhesion in the Brun lab is supported by National Institutes of Health Grant GM102841 and a grant from the Indiana METACyt Initiative of Indiana University (funded in part through a major grant from the Lilly Endowment, Inc.).

Cécile Berne, Adrien Ducret, and Gail G. Hardy contributed equally to this work.

Conflicts of interest: We disclose no conflicts.

CITATION

Berne C, Ducret A, Hardy GG, Brun YV. 2015. Adhesins involved in attachment to abiotic surfaces by Gram-negative bacteria. Microbiol Spectrum 3(4):MB-0018-2015.

REFERENCES

1. **Dunne WM.** 2002. Bacterial adhesion: seen any good biofilms lately? *Clin Microbiol Rev* **15:**155–166.
2. **van Oss CJ.** 2003. Long-range and short-range mechanisms of hydrophobic attraction and hydrophilic repulsion in specific and aspecific interactions. *J Mol Recognit* **16:**177–190.
3. **Stewart RJ.** 2011. Protein-based underwater adhesives and the prospects for their biotechnological production. *Appl Microbiol Biotechnol* **89:**27–33.
4. **O'Toole G, Kaplan HB, Kolter R.** 2000. Biofilm formation as microbial development. *Annu Rev Microbiol* **54:**49–79.
5. **Monds RD, O'Toole GA.** 2009. The developmental model of microbial biofilms: ten years of a paradigm up for review. *Trends Microbiol* **17:**73–87.
6. **Donlan RM.** 2002. Biofilms: microbial life on surfaces. *Emerg Infect Dis* **8:**881–890.
7. **Geng J, Henry N.** 2011. Short time-scale bacterial adhesion dynamics, p 315–331. *In* Link D, Goldman A (ed), *Bacterial Adhesion*. Springer, Dordrecht, The Netherlands.
8. **Beloin C, Roux A, Ghigo J-M.** 2008. *Escherichia coli* biofilms, p 249–289. *In* Romeo T (ed), *Bacterial Biofilms*. Springer, Dordrecht, The Netherlands.
9. **Karatan E, Watnick P.** 2009. Signals, regulatory networks, and materials that build and break bacterial biofilms. *Microbiol Mol Biol Rev* **73:**310–347.
10. **Pratt LA, Kolter R.** 1998. Genetic analysis of *Escherichia coli* biofilm formation: roles of flagella, motility, chemotaxis and type I pili. *Mol Microbiol* **30:**285–293.
11. **Entcheva-Dimitrov P, Spormann AM.** 2004. Dynamics and control of biofilms of the oligotrophic bacterium *Caulobacter crescentus*. *J Bacteriol* **186:**8254–8266.
12. **Van Houdt R, Michiels CW.** 2005. Role of bacterial cell surface structures in *Escherichia coli* biofilm formation. *Res Microbiol* **156:**626–633.
13. **Proft T, Baker EN.** 2009. Pili in Gram-negative and Gram-positive bacteria: structure, assembly and their role in disease. *Cell Mol Life Sci* **66:**613–635.

14. **Burrows LL.** 2012. *Pseudomonas aeruginosa* twitching motility: type IV pili in action. *Annu Rev Microbiol* **66:**493–520.
15. **St Geme JW, Pinkner JS 3rd, Krasan GP, Heuser J, Bullitt E, Smith AL, Hultgren SJ.** 1996. *Haemophilus influenzae* pili are composite structures assembled via the HifB chaperone. *Proc Natl Acad Sci USA* **93:**11913–11918.
16. **Gohl O, Friedrich A, Hoppert M, Averhoff B.** 2006. The thin pili of *Acinetobacter* sp. strain BD413 mediate adhesion to biotic and abiotic surfaces. *Appl Environ Microbiol* **72:**1394–1401.
17. **Inhulsen S, Aguilar C, Schmid N, Suppiger A, Riedel K, Eberl L.** 2012. Identification of functions linking quorum sensing with biofilm formation in *Burkholderia cenocepacia* H111. *Microbiologyopen* **1:**225–242.
18. **Ormeno-Orrillo E, Menna P, Almeida LG, Ollero FJ, Nicolas MF, Pains Rodrigues E, Shigueyoshi Nakatani A, Silva Batista JS, Oliveira Chueire LM, Souza RC, Ribeiro Vasconcelos AT, Megias M, Hungria M, Martinez-Romero E.** 2012. Genomic basis of broad host range and environmental adaptability of *Rhizobium tropici* CIAT 899 and *Rhizobium* sp. PRF 81 which are used in inoculants for common bean (*Phaseolus vulgaris* L.). *BMC Genomics* **13:**735.
19. **Nuccio SP, Baumler AJ.** 2007. Evolution of the chaperone/usher assembly pathway: fimbrial classification goes Greek. *Microbiol Mol Biol Rev* **71:**551–575.
20. **Wurpel DJ, Beatson SA, Totsika M, Petty NK, Schembri MA.** 2013. Chaperone-usher fimbriae of *Escherichia coli*. *PLoS One* **8:**e52835.
21. **Busch A, Waksman G.** 2012. Chaperone-usher pathways: diversity and pilus assembly mechanism. *Philos Trans R Soc Lond B Biol Sci* **367:**1112–1122.
22. **Waksman G, Hultgren SJ.** 2009. Structural biology of the chaperone-usher pathway of pilus biogenesis. *Nat Rev Microbiol* **7:**765–774.
23. **Han Z, Pinkner JS, Ford B, Obermann R, Nolan W, Wildman SA, Hobbs D, Ellenberger T, Cusumano CK, Hultgren SJ, Janetka JW.** 2010. Structure-based drug design and optimization of mannoside bacterial FimH antagonists. *J Med Chem* **53:**4779–4792.
24. **Hertig S, Vogel V.** 2012. Catch bonds. *Curr Biol* **22:**R823–R825.
25. **Rakshit S, Sivasankar S.** 2014. Biomechanics of cell adhesion: how force regulates the lifetime of adhesive bonds at the single molecule level. *Phys Chem Chem Phys* **16:**2211–2223.
26. **Liaqat I, Sakellaris H.** 2012. Biofilm formation and binding specificities of CFA/I, CFA/II and CS2 adhesions of enterotoxigenic *Escherichia coli* and Cfae-R181A mutant. *Braz J Microbiol* **43:**969–980.
27. **Ammendolia MG, Bertuccini L, Iosi F, Minelli F, Berlutti F, Valenti P, Superti F.** 2010. Bovine lactoferrin interacts with cable pili of *Burkholderia cenocepacia*. *Biometals* **23:**531–542.
28. **Sakellaris H, Scott JR.** 1998. New tools in an old trade: CS1 pilus morphogenesis. *Mol Microbiol* **30:**681–687.
29. **Starks AM, Froehlich BJ, Jones TN, Scott JR.** 2006. Assembly of CS1 pili: the role of specific residues of the major pilin, CooA. *J Bacteriol* **188:**231–239.
30. **Galkin VE, Kolappan S, Ng D, Zong Z, Li J, Yu X, Egelman EH, Craig L.** 2013. The structure of the CS1 pilus of enterotoxigenic *Escherichia coli* reveals structural polymorphism. *J Bacteriol* **195:**1360–1370.
31. **Perez-Casal J, Swartley JS, Scott JR.** 1990. Gene encoding the major subunit of CS1 pili of human enterotoxigenic *Escherichia coli*. *Infect Immun* **58:**3594–3600.
32. **Voegele K, Sakellaris H, Scott JR.** 1997. CooB plays a chaperone-like role for the proteins involved in formation of CS1 pili of enterotoxigenic *Escherichia coli*. *Proc Natl Acad Sci USA* **94:**13257–13261.
33. **Sakellaris H, Balding DP, Scott JR.** 1996. Assembly proteins of CS1 pili of enterotoxigenic *Escherichia coli*. *Mol Microbiol* **21:**529–541.
34. **Froehlich BJ, Karakashian A, Melsen LR, Wakefield JC, Scott JR.** 1994. CooC and CooD are required for assembly of CS1 pili. *Mol Microbiol* **12:**387–401.
35. **Macfarlane S, Dillon JF.** 2007. Microbial biofilms in the human gastrointestinal tract. *J Appl Microbiol* **102:**1187–1196.
36. **Tomich M, Mohr CD.** 2003. Adherence and autoaggregation phenotypes of a *Burkholderia cenocepacia* cable pilus mutant. *FEMS Microbiol Lett* **228:**287–297.
37. **Giltner CL, Nguyen Y, Burrows LL.** 2012. Type IV pilin proteins: versatile molecular modules. *Microbiol Mol Biol Rev* **76:**740–772.
38. **Giltner CL, Habash M, Burrows LL.** 2010. *Pseudomonas aeruginosa* minor pilins are incorporated into type IV pili. *J Mol Biol* **398:**444–461.
39. **Kuchma SL, Griffin EF, O'Toole GA.** 2012. Minor pilins of the type IV pilus system participate in the negative regulation of swarming motility. *J Bacteriol* **194:**5388–5403.
40. **Johnson MD, Garrett CK, Bond JE, Coggan KA, Wolfgang MC, Redinbo MR.** 2011. *Pseudomonas aeruginosa* PilY1 binds integrin in an RGD- and calcium-dependent manner. *PLoS One* **6:**e29629. doi:10.1371/journal.pone.0029629.

41. **Takhar HK, Kemp K, Kim M, Howell PL, Burrows LL.** 2013. The platform protein is essential for type IV pilus biogenesis. *J Biol Chem* **288:**9721–9728.
42. **Tammam S, Sampaleanu LM, Koo J, Manoharan K, Daubaras M, Burrows LL, Howell PL.** 2013. PilMNOPQ from the *Pseudomonas aeruginosa* type IV pilus system form a transenvelope protein interaction network that interacts with PilA. *J Bacteriol* **195:**2126–2135.
43. **Wehbi H, Portillo E, Harvey H, Shimkoff AE, Scheurwater EM, Howell PL, Burrows LL.** 2011. The peptidoglycan-binding protein FimV promotes assembly of the *Pseudomonas aeruginosa* type IV pilus secretin. *J Bacteriol* **193:**540–550.
44. **Koo J, Tang T, Harvey H, Tammam S, Sampaleanu L, Burrows LL, Howell PL.** 2013. Functional mapping of PilF and PilQ in the *Pseudomonas aeruginosa* type IV pilus system. *Biochemistry* **52:**2914–2923.
45. **O'Toole GA, Kolter R.** 1998. Flagellar and twitching motility are necessary for *Pseudomonas aeruginosa* biofilm development. *Mol Microbiol* **30:**295–304.
46. **Klausen M, Aaes-Jorgensen A, Molin S, Tolker-Nielsen T.** 2003. Involvement of bacterial migration in the development of complex multicellular structures in *Pseudomonas aeruginosa* biofilms. *Mol Microbiol* **50:**61–68.
47. **Chiang P, Burrows LL.** 2003. Biofilm formation by hyperpiliated mutants of *Pseudomonas aeruginosa. J Bacteriol* **185:**2374–2378.
48. **Watnick PI, Fullner KJ, Kolter R.** 1999. A role for the mannose-sensitive hemagglutinin in biofilm formation by *Vibrio cholerae* El Tor. *J Bacteriol* **181:**3606–3609.
49. **Shime-Hattori A, Iida T, Arita M, Park KS, Kodama T, Honda T.** 2006. Two type IV pili of *Vibrio parahaemolyticus* play different roles in biofilm formation. *FEMS Microbiol Lett* **264:**89–97.
50. **Reguera G, Kolter R.** 2005. Virulence and the environment: a novel role for *Vibrio cholerae* toxin-coregulated pili in biofilm formation on chitin. *J Bacteriol* **187:**3551–3555.
51. **Lutz C, Erken M, Noorian P, Sun S, McDougald D.** 2013. Environmental reservoirs and mechanisms of persistence of *Vibrio cholerae. Front Microbiol* **4:**375.
52. **Manning PA.** 1997. The tcp gene cluster of *Vibrio cholerae. Gene* **192:**63–70.
53. **Bose N, Taylor RK.** 2005. Identification of a TcpC-TcpQ outer membrane complex involved in the biogenesis of the toxin-coregulated pilus of *Vibrio cholerae. J Bacteriol* **187:**2225–2232.
54. **LaPointe CF, Taylor RK.** 2000. The type 4 prepilin peptidases comprise a novel family of aspartic acid proteases. *J Biol Chem* **275:**1502–1510.
55. **Tripathi SA, Taylor RK.** 2007. Membrane association and multimerization of TcpT, the cognate ATPase ortholog of the *Vibrio cholerae* toxin-coregulated-pilus biogenesis apparatus. *J Bacteriol* **189:**4401–4409.
56. **Tomich M, Planet PJ, Figurski DH.** 2007. The tad locus: postcards from the widespread colonization island. *Nat Rev Microbiol* **5:**363–375.
57. **Inoue T, Tanimoto I, Ohta H, Kato K, Murayama Y, Fukui K.** 1998. Molecular characterization of low-molecular-weight component protein, Flp, in *Actinobacillus actinomycetemcomitans* fimbriae. *Microbiol Immunol* **42:**253–258.
58. **Kachlany SC, Planet PJ, DeSalle R, Fine DH, Figurski DH.** 2001. Genes for tight adherence of *Actinobacillus actinomycetemcomitans*: from plaque to plague to pond scum. *Trends Microbiol* **9:**429–437.
59. **Tomich M, Fine DH, Figurski DH.** 2006. The TadV protein of *Actinobacillus actinomycetemcomitans* is a novel aspartic acid prepilin peptidase required for maturation of the Flp1 pilin and TadE and TadF pseudopilins. *J Bacteriol* **188:**6899–6914.
60. **Bhattacharjee MK, Kachlany SC, Fine DH, Figurski DH.** 2001. Nonspecific adherence and fibril biogenesis by *Actinobacillus actinomycetemcomitans*: TadA protein is an ATPase. *J Bacteriol* **183:**5927–5936.
61. **Perez-Cheeks BA, Planet PJ, Sarkar IN, Clock SA, Xu Q, Figurski DH.** 2012. The product of tadZ, a new member of the parA/minD superfamily, localizes to a pole in *Aggregatibacter actinomycetemcomitans. Mol Microbiol* **83:**694–711.
62. **Haase EM, Zmuda JL, Scannapieco FA.** 1999. Identification and molecular analysis of rough-colony-specific outer membrane proteins of *Actinobacillus actinomycetemcomitans. Infect Immun* **67:**2901–2908.
63. **Clock SA, Planet PJ, Perez BA, Figurski DH.** 2008. Outer membrane components of the Tad (tight adherence) secreton of *Aggregatibacter actinomycetemcomitans. J Bacteriol* **190:**980–990.
64. **Inoue T, Shingaki R, Sogawa N, Sogawa CA, Asaumi J, Kokeguchi S, Fukui K.** 2003. Biofilm formation by a fimbriae-deficient mutant of *Actinobacillus actinomycetemcomitans. Microbiol Immunol* **47:**877–881.
65. **Saito T, Ishihara K, Ryu M, Okuda K, Sakurai K.** 2010. Fimbriae-associated genes

are biofilm-forming factors in *Aggregatibacter actinomycetemcomitans* strains. *Bull Tokyo Dent Coll* **51:**145–150.

66. **Skerker JM, Shapiro L.** 2000. Identification and cell cycle control of a novel pilus system in *Caulobacter crescentus*. *EMBO J* **19:**3223–3234.
67. **Bodenmiller D, Toh E, Brun YV.** 2004. Development of surface adhesion in *Caulobacter crescentus*. *J Bacteriol* **186:**1438–1447.
68. **Li G, Brown PJ, Tang JX, Xu J, Quardokus EM, Fuqua C, Brun YV.** 2012. Surface contact stimulates the just-in-time deployment of bacterial adhesins. *Mol Microbiol* **83:**41–51.
69. **Ruer S, Stender S, Filloux A, de Bentzmann S.** 2007. Assembly of fimbrial structures in *Pseudomonas aeruginosa*: functionality and specificity of chaperone-usher machineries. *J Bacteriol* **189:**3547–3555.
70. **Bednarska NG, Schymkowitz J, Rousseau F, Van Eldere J.** 2013. Protein aggregation in bacteria: the thin boundary between functionality and toxicity. *Microbiology* **159:**1795–1806.
71. **Dueholm MS, Albertsen M, Otzen D, Nielsen PH.** 2012. Curli functional amyloid systems are phylogenetically widespread and display large diversity in operon and protein structure. *PLoS One* **7:**e51274. doi:10.1371/journal.pone.0051274.
72. **Chapman MR, Robinson LS, Pinkner JS, Roth R, Heuser J, Hammar M, Normark S, Hultgren SJ.** 2002. Role of *Escherichia coli* curli operons in directing amyloid fiber formation. *Science* **295:**851–855.
73. **Evans ML, Chapman MR.** 2013. Curli biogenesis: order out of disorder. *Biochim Biophys Acta* [Epub ahead of print.] doi:10.1016/j.bbamcr.2013.09.010.
74. **Cookson AL, Cooley WA, Woodward MJ.** 2002. The role of type 1 and curli fimbriae of Shiga toxin-producing *Escherichia coli* in adherence to abiotic surfaces. *Int J Med Microbiol* **292:**195–205.
75. **Bokranz W, Wang X, Tschape H, Romling U.** 2005. Expression of cellulose and curli fimbriae by *Escherichia coli* isolated from the gastrointestinal tract. *J Med Microbiol* **54:**1171–1182.
76. **Saldana Z, Xicohtencatl-Cortes J, Avelino F, Phillips AD, Kaper JB, Puente JL, Giron JA.** 2009. Synergistic role of curli and cellulose in cell adherence and biofilm formation of attaching and effacing *Escherichia coli* and identification of Fis as a negative regulator of curli. *Environ Microbiol* **11:**992–1006.
77. **Zhou Y, Smith D, Leong BJ, Brannstrom K, Almqvist F, Chapman MR.** 2012. Promiscuous cross-seeding between bacterial amyloids promotes interspecies biofilms. *J Biol Chem* **287:**35092–35103.
78. **Gerlach RG, Hensel M.** 2007. Protein secretion systems and adhesins: the molecular armory of Gram-negative pathogens. *Int J Med Microbiol* **297:**401–415.
79. **Chagnot C, Zorgani MA, Astruc T, Desvaux M.** 2013. Proteinaceous determinants of surface colonization in bacteria: bacterial adhesion and biofilm formation from a protein secretion perspective. *Front Microbiol* **4:**303.
80. **Delepelaire P.** 2004. Type I secretion in Gram-negative bacteria. *Biochim Biophys Acta* **1694:**149–161.
81. **Cucarella C, Solano C, Valle J, Amorena B, Lasa I, Penades JR.** 2001. Bap, a *Staphylococcus aureus* surface protein involved in biofilm formation. *J Bacteriol* **183:**2888–2896.
82. **Lasa I, Penades JR.** 2006. Bap: a family of surface proteins involved in biofilm formation. *Res Microbiol* **157:**99–107.
83. **Yousef F, Espinosa-Urgel M.** 2007. *In silico* analysis of large microbial surface proteins. *Res Microbiol* **158:**545–550.
84. **Espinosa-Urgel M, Salido A, Ramos JL.** 2000. Genetic analysis of functions involved in adhesion of *Pseudomonas putida* to seeds. *J Bacteriol* **182:**2363–2369.
85. **Hinsa SM, Espinosa-Urgel M, Ramos JL, O'Toole GA.** 2003. Transition from reversible to irreversible attachment during biofilm formation by *Pseudomonas fluorescens* WCS365 requires an ABC transporter and a large secreted protein. *Mol Microbiol* **49:**905–918.
86. **Fuqua C.** 2010. Passing the baton between laps: adhesion and cohesion in *Pseudomonas putida* biofilms. *Mol Microbiol* **77:**533–536.
87. **El-Kirat-Chatel S, Beaussart A, Boyd CD, O'Toole GA, Dufrêne YF.** 2013. Single-cell and single-molecule analysis deciphers the localization, adhesion, and mechanics of the biofilm adhesin LapA. *ACS Chem Biol* **9:**485–494.
88. **El-Kirat-Chatel S, Boyd CD, O'Toole GA, Dufrêne YF.** 2014. Single-molecule analysis of *Pseudomonas fluorescens* footprints. *ACS Nano* **8:**1690–1698.
89. **Martinez-Gil M, Yousef-Coronado F, Espinosa-Urgel M.** 2010. LapF, the second largest *Pseudomonas putida* protein, contributes to plant root colonization and determines biofilm architecture. *Mol Microbiol* **77:**549–561.
90. **Duque E, de la Torre J, Bernal P, Molina-Henares MA, Alaminos M, Espinosa-Urgel M, Roca A, Fernandez M, de Bentzmann S, Ramos JL.** 2013. Identification of reciprocal adhesion genes in pathogenic and non-pathogenic *Pseudomonas*. *Environ Microbiol* **15:**36–48.

91. **Valle J, Latasa C, Gil C, Toledo-Arana A, Solano C, Penades JR, Lasa I.** 2012. Bap, a biofilm matrix protein of *Staphylococcus aureus* prevents cellular internalization through binding to GP96 host receptor. *PLoS Pathog* **8:** e1002843. doi:10.1371/journal.ppat.1002843.
92. **Wagner C, Hensel M.** 2011. Adhesive mechanisms of *Salmonella enterica*. *Adv Exp Med Biol* **715:**17–34.
93. **Wagner C, Polke M, Gerlach RG, Linke D, Stierhof YD, Schwarz H, Hensel M.** 2011. Functional dissection of SiiE, a giant non-fimbrial adhesin of *Salmonella enterica*. *Cell Microbiol* **13:**1286–1301.
94. **Gjermansen M, Nilsson M, Yang L, Tolker-Nielsen T.** 2010. Characterization of starvation-induced dispersion in *Pseudomonas putida* biofilms: genetic elements and molecular mechanisms. *Mol Microbiol* **75:**815–826.
95. **Martinez-Gil M, Quesada JM, Ramos-Gonzalez MI, Soriano MI, de Cristobal RE, Espinosa-Urgel M.** 2013. Interplay between extracellular matrix components of *Pseudomonas putida* biofilms. *Res Microbiol* **164:**382–389.
96. **Martínez-Gil M, Ramos-González MI, Espinosa-Urgel M.** 2014. Role of c-di-GMP and the Gac system in the transcriptional control of the genes coding for the *Pseudomonas putida* adhesins LapA and LapF. *J Bacteriol* **196:**1484–1495.
97. **Desvaux M, Parham NJ, Henderson IR.** 2004. Type V protein secretion: simplicity gone awry? *Curr Issues Mol Biol* **6:**111–124.
98. **Bernstein HD.** 2007. Are bacterial 'autotransporters' really transporters? *Trends Microbiol* **15:**441–447.
99. **Leyton DL, Rossiter AE, Henderson IR.** 2012. From self-sufficiency to dependence: mechanisms and factors important for autotransporter biogenesis. *Nat Rev Microbiol* **10:**213–225.
100. **Leo JC, Grin I, Linke D.** 2012. Type V secretion: mechanism(s) of autotransport through the bacterial outer membrane. *Philos Trans R Soc B Biol Sci* **367:**1088–1101.
101. **Klemm P, Vejborg RM, Sherlock O.** 2006. Self-associating autotransporters, SAATs: functional and structural similarities. *Int J Med Microbiol* **296:**187–195.
102. **Roux A, Beloin C, Ghigo JM.** 2005. Combined inactivation and expression strategy to study gene function under physiological conditions: application to identification of new *Escherichia coli* adhesins. *J Bacteriol* **187:**1001–1013.
103. **Owen P, Meehan M, de Loughry-Doherty H, Henderson I.** 1996. Phase-variable outer membrane proteins in *Escherichia coli*. *FEMS Immunol Med Microbiol* **16:**63–76.
104. **Hasman H, Chakraborty T, Klemm P.** 1999. Antigen-43-mediated autoaggregation of *Escherichia coli* is blocked by fimbriation. *J Bacteriol* **181:**4834–4841.
105. **Danese PN, Pratt LA, Dove SL, Kolter R.** 2000. The outer membrane protein, antigen 43, mediates cell-to-cell interactions within *Escherichia coli* biofilms. *Mol Microbiol* **37:**424–432.
106. **Grijpstra J, Arenas J, Rutten L, Tommassen J.** 2013. Autotransporter secretion: varying on a theme. *Res Microbiol* **164:**562–582.
107. **Sherlock O, Schembri MA, Reisner A, Klemm P.** 2004. Novel roles for the AIDA adhesin from diarrheagenic *Escherichia coli*: cell aggregation and biofilm formation. *J Bacteriol* **186:**8058–8065.
108. **Benz I, Schmidt MA.** 2001. Glycosylation with heptose residues mediated by the aah gene product is essential for adherence of the AIDA-I adhesin. *Mol Microbiol* **40:**1403–1413.
109. **Sherlock O, Dobrindt U, Jensen JB, Vejborg RM, Klemm P.** 2006. Glycosylation of the self-recognizing *Escherichia coli* Ag43 autotransporter protein. *J Bacteriol* **188:**1798–1807.
110. **Lindenthal C, Elsinghorst EA.** 1999. Identification of a glycoprotein produced by enterotoxigenic *Escherichia coli*. *Infect Immun* **67:**4084–4091.
111. **Côté J-P, Charbonneau M-È, Mourez M.** 2013. Glycosylation of the *Escherichia coli* TibA self-associating autotransporter influences the conformation and the functionality of the protein. *PloS One* **8:**e80739. doi:10.1371/journal.pone.0080739.
112. **Wallecha A, Munster V, Correnti J, Chan T, van der Woude M.** 2002. Dam- and OxyR-dependent phase variation of agn43: essential elements and evidence for a new role of DNA methylation. *J Bacteriol* **184:**3338–3347.
113. **Chauhan A, Sakamoto C, Ghigo JM, Beloin C.** 2013. Did I pick the right colony? Pitfalls in the study of regulation of the phase variable antigen 43 adhesin. *PLoS One* **8:**e73568. doi:10.1371/journal.pone.0073568.
114. **Meng G, Spahich N, Kenjale R, Waksman G, St Geme JW.** 2011. Crystal structure of the *Haemophilus influenzae* Hap adhesin reveals an intercellular oligomerization mechanism for bacterial aggregation. *EMBO J* **30:**3864–3874.
115. **Heras B, Totsika M, Peters KM, Paxman JJ, Gee CL, Jarrott RJ, Perugini MA, Whitten AE, Schembri MA.** 2014. The antigen 43 structure reveals a molecular Velcro-like mechanism of autotransporter-mediated bacterial clumping. *Proc Natl Acad Sci USA* **111:**457–462.

116. Henderson IR, Navarro-Garcia F, Desvaux M, Fernandez RC, Ala'Aldeen D. 2004. Type V protein secretion pathway: the autotransporter story. *Microbiol Mol Biol Rev* **68:**692–744.

117. Lyskowski A, Leo JC, Goldman A. 2011. Structure and biology of trimeric autotransporter adhesins. *Adv Exp Med Biol* **715:**143–158.

118. El Tahir Y, Skurnik M. 2001. YadA, the multifaceted *Yersinia* adhesin. *Int J Med Microbiol* **291:**209–218.

119. Heise T, Dersch P. 2006. Identification of a domain in *Yersinia* virulence factor YadA that is crucial for extracellular matrix-specific cell adhesion and uptake. *Proc Natl Acad Sci USA* **103:**3375–3380.

120. Valle J, Mabbett AN, Ulett GC, Toledo-Arana A, Wecker K, Totsika M, Schembri MA, Ghigo JM, Beloin C. 2008. UpaG, a new member of the trimeric autotransporter family of adhesins in uropathogenic *Escherichia coli. J Bacteriol* **190:**4147–4161.

121. Raghunathan D, Wells TJ, Morris FC, Shaw RK, Bobat S, Peters SE, Paterson GK, Jensen KT, Leyton DL, Blair JM, Browning DF, Pravin J, Flores-Langarica A, Hitchcock JR, Moraes CT, Piazza RM, Maskell DJ, Webber MA, May RC, MacLennan CA, Piddock LJ, Cunningham AF, Henderson IR. 2011. SadA, a trimeric autotransporter from *Salmonella enterica* serovar Typhimurium, can promote biofilm formation and provides limited protection against infection. *Infect Immun* **79:** 4342–4352.

122. Lazar Adler NR, Dean RE, Saint RJ, Stevens MP, Prior JL, Atkins TP, Galyov EE. 2013. Identification of a predicted trimeric autotransporter adhesin required for biofilm formation of *Burkholderia pseudomallei. PLoS One* **8:**e79461. doi:10.1371/journal.pone.0079461.

123. Mazar J, Cotter PA. 2007. New insight into the molecular mechanisms of two-partner secretion. *Trends Microbiol* **15:**508–515.

124. Darsonval A, Darrasse A, Durand K, Bureau C, Cesbron S, Jacques MA. 2009. Adhesion and fitness in the bean phyllosphere and transmission to seed of *Xanthomonas fuscans* subsp. fuscans. *Mol Plant Microbe Interact* **22:**747–757.

125. Feil H, Feil WS, Lindow SE. 2007. Contribution of fimbrial and afimbrial adhesins of *Xylella fastidiosa* to attachment to surfaces and virulence to grape. *Phytopathology* **97:**318–324.

126. Ryan RP, Vorholter FJ, Potnis N, Jones JB, Van Sluys MA, Bogdanove AJ, Dow JM. 2011. Pathogenomics of *Xanthomonas*: understanding bacterium-plant interactions. *Nat Rev Microbiol* **9:**344–355.

127. Webster P, Wu S, Gomez G, Apicella M, Plaut AG, St Geme JW 3rd. 2006. Distribution of bacterial proteins in biofilms formed by non-typeable *Haemophilus influenzae. J Histochem Cytochem* **54:**829–842.

128. Serra DO, Conover MS, Arnal L, Sloan GP, Rodriguez ME, Yantorno OM, Deora R. 2011. FHA-mediated cell-substrate and cell-cell adhesions are critical for *Bordetella pertussis* biofilm formation on abiotic surfaces and in the mouse nose and the trachea. *PLoS One* **6:** e28811. doi:10.1371/journal.pone.0028811.

129. Borlee BR, Goldman AD, Murakami K, Samudrala R, Wozniak DJ, Parsek MR. 2010. *Pseudomonas aeruginosa* uses a cyclic-di-GMP-regulated adhesin to reinforce the biofilm extracellular matrix. *Mol Microbiol* **75:**827–842.

130. Guo H, Yi W, Song JK, Wang PG. 2008. Current understanding on biosynthesis of microbial polysaccharides. *Curr Top Med Chem* **8:**141–151.

131. Whitney JC, Howell PL. 2013. Synthase-dependent exopolysaccharide secretion in Gram-negative bacteria. *Trends Microbiol* **21:**63–72.

132. Ahimou F, Semmens MJ, Haugstad G, Novak PJ. 2007. Effect of protein, polysaccharide, and oxygen concentration profiles on biofilm cohesiveness. *Appl Environ Microbiol* **73:**2905–2910.

133. Davey ME, O'Toole GA. 2000. Microbial biofilms: from ecology to molecular genetics. *Microbiol Mol Biol Rev* **64:**847–867.

134. Sutherland I. 2001. Biofilm exopolysaccharides: a strong and sticky framework. *Microbiology* **147:**3–9.

135. Haag AP. 2006. Mechanical properties of bacterial exopolymeric adhesives and their commercial development, p 1–19. *In* Smith AM, Callow JA (ed), *Biological Adhesives.* Springer-Verlag, Berlin.

136. Korstgens V, Flemming HC, Wingender J, Borchard W. 2001. Influence of calcium ions on the mechanical properties of a model biofilm of mucoid *Pseudomonas aeruginosa. Water Sci Technol* **43:**49–57.

137. Franklin MJ, Ohman DE. 1993. Identification of algF in the alginate biosynthetic gene cluster of *Pseudomonas aeruginosa* which is required for alginate acetylation. *J Bacteriol* **175:**5057–5065.

138. Rinaudo M. 2004. Role of substituents on the properties of some polysaccharides. *Biomacromolecules* **5:**1155–1165.

139. Tielen P, Strathmann M, Jaeger KE, Flemming HC, Wingender J. 2005. Alginate acetylation influences initial surface colonization by mucoid *Pseudomonas aeruginosa*. *Microbiol Res* **160:**165–176.

140. **Villain-Simonnet A, Milas M, Rinaudo M.** 2000. A new bacterial polysaccharide (YAS34). I. Characterization of the conformations and conformational transition. *Int J Biol Macromol* **27:**65–75.
141. **Wan Z, Brown PJ, Elliott EN, Brun YV.** 2013. The adhesive and cohesive properties of a bacterial polysaccharide adhesin are modulated by a deacetylase. *Mol Microbiol* **88:**486–500.
142. **Cerca N, Jefferson KK, Maira-Litran T, Pier DB, Kelly-Quintos C, Goldmann DA, Azeredo J, Pier GB.** 2007. Molecular basis for preferential protective efficacy of antibodies directed to the poorly acetylated form of staphylococcal poly-N-acetyl-beta-(1-6)-glucosamine. *Infect Immun* **75:**3406–3413.
143. **Itoh Y, Rice JD, Goller C, Pannuri A, Taylor J, Meisner J, Beveridge TJ, Preston JF 3rd, Romeo T.** 2008. Roles of pgaABCD genes in synthesis, modification, and export of the *Escherichia coli* biofilm adhesin poly-beta-1,6-N-acetyl-D-glucosamine. *J Bacteriol* **190:** 3670–3680.
144. **Vuong C, Kocianova S, Voyich JM, Yao Y, Fischer ER, DeLeo FR, Otto M.** 2004. A crucial role for exopolysaccharide modification in bacterial biofilm formation, immune evasion, and virulence. *J Biol Chem* **279:**54881–54886.
145. **Pokrovskaya V, Poloczek J, Little DJ, Griffiths H, Howell PL, Nitz M.** 2013. Functional characterization of *Staphylococcus epidermidis* IcaB, a de-N-acetylase important for biofilm formation. *Biochemistry* **52:**5463–5471.
146. **Riley LM, Weadge JT, Baker P, Robinson H, Codee JD, Tipton PA, Ohman DE, Howell PL.** 2013. Structural and functional characterization of *Pseudomonas aeruginosa* AlgX: role of AlgX in alginate acetylation. *J Biol Chem* **288:**22299–22314.
147. **Whitney JC, Hay ID, Li C, Eckford PD, Robinson H, Amaya MF, Wood LF, Ohman DE, Bear CE, Rehm BH, Howell PL.** 2011. Structural basis for alginate secretion across the bacterial outer membrane. *Proc Natl Acad Sci USA* **108:**13083–13088.
148. **Willis LM, Stupak J, Richards MR, Lowary TL, Li J, Whitfield C.** 2013. Conserved glycolipid termini in capsular polysaccharides synthesized by ATP-binding cassette transporter-dependent pathways in Gram-negative pathogens. *Proc Natl Acad Sci USA* **110:**7868–7873.
149. **Jimenez N, Senchenkova SN, Knirel YA, Pieretti G, Corsaro MM, Aquilini E, Regue M, Merino S, Tomas JM.** 2012. Effects of lipopolysaccharide biosynthesis mutations on K1 polysaccharide association with the *Escherichia coli* cell surface. *J Bacteriol* **194:**3356–3367.
150. **Gaastra W, De Graaf FK.** 1982. Host-specific fimbrial adhesins of noninvasive enterotoxigenic *Escherichia coli* strains. *Microbiol Rev* **46:**129.
151. **Franco AV, Liu D, Reeves PR.** 1996. A Wzz (Cld) protein determines the chain length of K lipopolysaccharide in *Escherichia coli* O8 and O9 strains. *J Bacteriol* **178:**1903–1907.
152. **Jann K, Dengler T, Jann B.** 1992. Core-lipid A on the K40 polysaccharide of *Escherichia coli* O8:K40:H9, a representative of group I capsular polysaccharides. *Zentralbl Bakteriol* **276:** 196–204.
153. **Bushell SR, Mainprize IL, Wear MA, Lou H, Whitfield C, Naismith JH.** 2013. Wzi is an outer membrane lectin that underpins group 1 capsule assembly in *Escherichia coli*. *Structure* **21:**844–853.
154. **Hardy GG, Allen RC, Toh E, Long M, Brown PJ, Cole-Tobian JL, Brun YV.** 2010. A localized multimeric anchor attaches the *Caulobacter* holdfast to the cell pole. *Mol Microbiol* **76:**409–427.
155. **Rahn A, Beis K, Naismith JH, Whitfield C.** 2003. A novel outer membrane protein, Wzi, is involved in surface assembly of the *Escherichia coli* K30 group 1 capsule. *J Bacteriol* **185:**5882–5890.
156. **Iwashkiw JA, Vozza NF, Kinsella RL, Feldman MF.** 2013. Pour some sugar on it: the expanding world of bacterial protein O-linked glycosylation. *Mol Microbiol* **89:**14–28.
157. **Song MC, Kim E, Ban YH, Yoo YJ, Kim EJ, Park SR, Pandey RP, Sohng JK, Yoon YJ.** 2013. Achievements and impacts of glycosylation reactions involved in natural product biosynthesis in prokaryotes. *Appl Microbiol Biotechnol* **97:**5691–5704.
158. **Quintero EJ, Busch K, Weiner RM.** 1998. Spatial and temporal deposition of adhesive extracellular polysaccharide capsule and fimbriae by hyphomonas strain MHS-3. *Appl Environ Microbiol* **64:**1246–1255.
159. **Ong CJ, Wong ML, Smit J.** 1990. Attachment of the adhesive holdfast organelle to the cellular stalk of *Caulobacter crescentus*. *J Bacteriol* **172:**1448–1456.
160. **Smith CS, Hinz A, Bodenmiller D, Larson DE, Brun YV.** 2003. Identification of genes required for synthesis of the adhesive holdfast

in *Caulobacter crescentus*. *J Bacteriol* **185:** 1432–1442.
161. **Poindexter JS.** 2006. Dimorphic prosthecate bacteria: the genera *Caulobacter, Asticcacaulis, Hyphomicrobium, Pedomicrobium, Hyphomonas* and *Thiodendron*, p 72–90. *In* Rosenberg E, DeLong EF (ed), *The Prokaryotes*, **vol. 5**. Springer, New York.
162. **Poindexter JS.** 1964. Biological properties and classification of the *Caulobacter* group. *Bacteriol Rev* **28:**231–295.
163. **Merker RI, Smit J.** 1988. Characterization of the adhesive holdfast of marine and freshwater caulobacters. *Appl Environ Microbiol* **54:**2078–2085.
164. **Fiebig A, Herrou J, Fumeaux C, Radhakrishnan SK, Viollier PH, Crosson S.** 2014. A cell cycle and nutritional checkpoint controlling bacterial surface adhesion. *PLoS Genet* **10:**e1004101. doi:10.1371/journal.pgen.1004101.
165. **Toh E, Kurtz HD Jr, Brun YV.** 2008. Characterization of the *Caulobacter crescentus* holdfast polysaccharide biosynthesis pathway reveals significant redundancy in the initiating glycosyltransferase and polymerase steps. *J Bacteriol* **190:**7219–7231.
166. **Li G, Smith CS, Brun YV, Tang JX.** 2005. The elastic properties of the *Caulobacter crescentus* adhesive holdfast are dependent on oligomers of N-acetylglucosamine. *J Bacteriol* **187:**257–265.
167. **Tsang PH, Li G, Brun YV, Freund LB, Tang JX.** 2006. Adhesion of single bacterial cells in the micronewton range. *Proc Natl Acad Sci USA* **103:**5764–5768.
168. **Berne C, Ma X, Licata NA, Neves BR, Setayeshgar S, Brun YV, Dragnea B.** 2013. Physiochemical properties of *Caulobacter crescentus* holdfast: a localized bacterial adhesive. *J Phys Chem B* **117:**10492–10503.
169. **Brown PJ, Hardy GG, Trimble MJ, Brun YV.** 2008. Complex regulatory pathways coordinate cell-cycle progression and development in *Caulobacter crescentus*. *Adv Microbial Physiol* **54:**1–101.
170. **Rodriguez-Navarro DN, Dardanelli MS, Ruiz-Sainz JE.** 2007. Attachment of bacteria to the roots of higher plants. *FEMS Microbiol Lett* **272:**127–136.
171. **Laus MC, Logman TJ, Lamers GE, Van Brussel AA, Carlson RW, Kijne JW.** 2006. A novel polar surface polysaccharide from *Rhizobium leguminosarum* binds host plant lectin. *Mol Microbiol* **59:**1704–1713.
172. **Xie F, Williams A, Edwards A, Downie JA.** 2012. A plant arabinogalactan-like glycoprotein promotes a novel type of polar surface attachment by *Rhizobium leguminosarum*. *Mol Plant Microbe Interact* **25:**250–258.
173. **Ausmees N, Jacobsson K, Lindberg M.** 2001. A unipolarly located, cell-surface-associated agglutinin, RapA, belongs to a family of *Rhizobium*-adhering proteins (Rap) in *Rhizobium leguminosarum* bv. trifolii. *Microbiology* **147:**549–559.
174. **Williams A, Wilkinson A, Krehenbrink M, Russo DM, Zorreguieta A, Downie JA.** 2008. Glucomannan-mediated attachment of *Rhizobium leguminosarum* to pea root hairs is required for competitive nodule infection. *J Bacteriol* **190:**4706–4715.
175. **Tomlinson AD, Fuqua C.** 2009. Mechanisms and regulation of polar surface attachment in *Agrobacterium tumefaciens*. *Curr Opin Microbiol* **12:**708–714.
176. **Loh JT, Ho SC, de Feijter AW, Wang JL, Schindler M.** 1993. Carbohydrate binding activities of *Bradyrhizobium japonicum*: unipolar localization of the lectin BJ38 on the bacterial cell surface. *Proc Natl Acad Sci USA* **90:**3033–3037.
177. **Merritt PM, Danhorn T, Fuqua C.** 2007. Motility and chemotaxis in *Agrobacterium tumefaciens* surface attachment and biofilm formation. *J Bacteriol* **189:**8005–8014.
178. **Xu J, Kim J, Danhorn T, Merritt PM, Fuqua C.** 2012. Phosphorus limitation increases attachment in *Agrobacterium tumefaciens* and reveals a conditional functional redundancy in adhesin biosynthesis. *Res Microbiol* **163:**674–684.
179. **Xu J, Kim J, Koestler BJ, Choi JH, Waters CM, Fuqua C.** 2013. Genetic analysis of *Agrobacterium tumefaciens* unipolar polysaccharide production reveals complex integrated control of the motile-to-sessile switch. *Mol Microbiol* **89:**929–948.
180. **Fujishige NA, Kapadia NN, De Hoff PL, Hirsch AM.** 2006. Investigations of *Rhizobium* biofilm formation. *FEMS Microbiol Ecol* **56:**195–206.
181. **Abdian PL, Caramelo JJ, Ausmees N, Zorreguieta A.** 2013. RapA2 is a calcium-binding lectin composed of two highly conserved cadherin-like domains that specifically recognize *Rhizobium leguminosarum* acidic exopolysaccharides. *J Biol Chem* **288:**2893–2904.
182. **Ho SC, Wang JL, Schindler M, Loh JT.** 1994. Carbohydrate binding activities of *Bradyrhizobium japonicum*. III. Lectin expression, bacterial binding, and nodulation efficiency. *Plant J* **5:**873–884.
183. **Pérez-Giménez J, Mongiardini EJ, Althabegoiti MJ, Covelli J, Quelas JI, López-García SL, Lodeiro AR.** 2009. Soybean lectin enhances

biofilm formation by *Bradyrhizobium japonicum* in the absence of plants. *Int J Microbiol* **2009:**719367.
184. **Luciano J, Agrebi R, Le Gall AV, Wartel M, Fiegna F, Ducret A, Brochier-Armanet C, Mignot T.** 2011. Emergence and modular evolution of a novel motility machinery in bacteria. *PLoS Genet* **7:**e1002268. doi:10.1371/journal.pgen.1002268.
185. **Nan B, Chen J, Neu JC, Berry RM, Oster G, Zusman DR.** 2011. Myxobacteria gliding motility requires cytoskeleton rotation powered by proton motive force. *Proc Natl Acad Sci USA* **108:**2498–2503.
186. **Nan B, Mauriello EM, Sun IH, Wong A, Zusman DR.** 2010. A multi-protein complex from *Myxococcus xanthus* required for bacterial gliding motility. *Mol Microbiol* **76:**1539–1554.
187. **Sun M, Wartel M, Cascales E, Shaevitz JW, Mignot T.** 2011. Motor-driven intracellular transport powers bacterial gliding motility. *Proc Natl Acad Sci USA* **108:**7559–7564.
188. **Zhang Y, Ducret A, Shaevitz J, Mignot T.** 2012. From individual cell motility to collective behaviors: insights from a prokaryote, *Myxococcus xanthus. FEMS Microbiol Rev* **36:**149–164.
189. **Burchard RP.** 1982. Trail following by gliding bacteria. *J Bacteriol* **152:**495–501.
190. **Wolgemuth C, Hoiczyk E, Kaiser D, Oster G.** 2002. How myxobacteria glide. *Curr Biol* **12:** 369–377.
191. **Ducret A, Fleuchot B, Bergam P, Mignot T.** 2013. Direct live imaging of cell-cell protein transfer by transient outer membrane fusion in *Myxococcus xanthus*. *Elife* **2:**e00868. doi:10.7554/eLife.00868.
192. **Ducret A, Valignat MP, Mouhamar F, Mignot T, Theodoly O.** 2012. Wet-surface-enhanced ellipsometric contrast microscopy identifies slime as a major adhesion factor during bacterial surface motility. *Proc Natl Acad Sci USA* **109:**10036–10041.
193. **Flemming H-C, Wingender J.** 2010. The biofilm matrix. *Nat Rev Microbiol* **8:**623–633.
194. **Stoodley P, Sauer K, Davies D, Costerton JW.** 2002. Biofilms as complex differentiated communities. *Annu Rev Microbiol* **56:**187–209.
195. **Wessel AK, Hmelo L, Parsek MR, Whiteley M.** 2013. Going local: technologies for exploring bacterial microenvironments. *Nat Rev Microbiol* **11:**337–348.
196. **Garnett JA, Martinez-Santos VI, Saldana Z, Pape T, Hawthorne W, Chan J, Simpson PJ, Cota E, Puente JL, Giron JA, Matthews S.** 2012. Structural insights into the biogenesis and biofilm formation by the *Escherichia coli* common pilus. *Proc Natl Acad Sci USA* **109:** 3950–3955.
197. **Kalivoda EJ, Stella NA, O'Dee DM, Nau GJ, Shanks RM.** 2008. The cyclic AMP-dependent catabolite repression system of *Serratia marcescens* mediates biofilm formation through regulation of type 1 fimbriae. *Appl Environ Microbiol* **74:**3461–3470.
198. **Koczan JM, Lenneman BR, McGrath MJ, Sundin GW.** 2011. Cell surface attachment structures contribute to biofilm formation and xylem colonization by *Erwinia amylovora. Appl Environ Microbiol* **77:**7031–7039.
199. **Mhedbi-Hajri N, Jacques MA, Koebnik R.** 2011. Adhesion mechanisms of plant-pathogenic *Xanthomonadaceae. Adv Exp Med Biol* **715:**71–89.
200. **Stahlhut SG, Struve C, Krogfelt KA, Reisner A.** 2012. Biofilm formation of *Klebsiella pneumoniae* on urethral catheters requires either type 1 or type 3 fimbriae. *FEMS Immunol Med Microbiol* **65:**350–359.
201. **Tomaras AP, Dorsey CW, Edelmann RE, Actis LA.** 2003. Attachment to and biofilm formation on abiotic surfaces by *Acinetobacter baumannii*: involvement of a novel chaperone-usher pili assembly system. *Microbiology* **149:**3473–3484.
202. **Ong CL, Beatson SA, Totsika M, Forestier C, McEwan AG, Schembri MA.** 2010. Molecular analysis of type 3 fimbrial genes from *Escherichia coli, Klebsiella* and *Citrobacter* species. *BMC Microbiol* **10:**183.
203. **Di Martino P, Cafferini N, Joly B, Darfeuille-Michaud A.** 2003. *Klebsiella pneumoniae* type 3 pili facilitate adherence and biofilm formation on abiotic surfaces. *Res Microbiol* **154:**9–16.
204. **Lappann M, Haagensen JA, Claus H, Vogel U, Molin S.** 2006. Meningococcal biofilm formation: structure, development and phenotypes in a standardized continuous flow system. *Mol Microbiol* **62:**1292–1309.
205. **Carbonnelle E, Helaine S, Nassif X, Pelicic V.** 2006. A systematic genetic analysis in *Neisseria meningitidis* defines the Pil proteins required for assembly, functionality, stabilization and export of type IV pili. *Mol Microbiol* **61:**1510–1522.
206. **Marsh JW, Taylor RK.** 1999. Genetic and transcriptional analyses of the *Vibrio cholerae* mannose-sensitive hemagglutinin type 4 pilus gene locus. *J Bacteriol* **181:**1110–1117.
207. **Meibom KL, Li XB, Nielsen AT, Wu CY, Roseman S, Schoolnik GK.** 2004. The *Vibrio cholerae* chitin utilization program. *Proc Natl Acad Sci USA* **101:**2524–2529.

208. **Moreira CG, Palmer K, Whiteley M, Sircili MP, Trabulsi LR, Castro AF, Sperandio V.** 2006. Bundle-forming pili and EspA are involved in biofilm formation by enteropathogenic *Escherichia coli*. *J Bacteriol* **188:**3952–3961.

209. **Bernard CS, Bordi C, Termine E, Filloux A, de Bentzmann S.** 2009. Organization and PprB-dependent control of the *Pseudomonas aeruginosa* tad Locus, involved in Flp pilus biology. *J Bacteriol* **191:**1961–1973.

210. **Collinson SK, Clouthier SC, Doran JL, Banser PA, Kay WW.** 1996. *Salmonella enteritidis* agfBAC operon encoding thin, aggregative fimbriae. *J Bacteriol* **178:**662–667.

211. **Austin JW, Sanders G, Kay WW, Collinson SK.** 1998. Thin aggregative fimbriae enhance *Salmonella enteritidis* biofilm formation. *FEMS Microbiol Lett* **162:**295–301.

212. **Dueholm MS, Sondergaard MT, Nilsson M, Christiansen G, Stensballe A, Overgaard MT, Givskov M, Tolker-Nielsen T, Otzen DE, Nielsen PH.** 2013. Expression of Fap amyloids in *Pseudomonas aeruginosa, P. fluorescens,* and *P. putida* results in aggregation and increased biofilm formation. *Microbiologyopen* **2:**365–382.

213. **Huber B, Riedel K, Kothe M, Givskov M, Molin S, Eberl L.** 2002. Genetic analysis of functions involved in the late stages of biofilm development in *Burkholderia cepacia* H111. *Mol Microbiol* **46:**411–426.

214. **Latasa C, Roux A, Toledo-Arana A, Ghigo JM, Gamazo C, Penades JR, Lasa I.** 2005. BapA, a large secreted protein required for biofilm formation and host colonization of *Salmonella enterica* serovar Enteritidis. *Mol Microbiol* **58:**1322–1339.

215. **Loehfelm TW, Luke NR, Campagnari AA.** 2008. Identification and characterization of an *Acinetobacter baumannii* biofilm-associated protein. *J Bacteriol* **190:**1036–1044.

216. **Theunissen S, De Smet L, Dansercoer A, Motte B, Coenye T, Van Beeumen JJ, Devreese B, Savvides SN, Vergauwen B.** 2010. The 285 kDa Bap/RTX hybrid cell surface protein (SO4317) of *Shewanella oneidensis* MR-1 is a key mediator of biofilm formation. *Res Microbiol* **161:**144–152.

217. **Pérez-Mendoza D, Coulthurst SJ, Humphris S, Campbell E, Welch M, Toth IK, Salmond GP.** 2011. A multi-repeat adhesin of the phytopathogen, *Pectobacterium atrosepticum,* is secreted by a type I pathway and is subject to complex regulation involving a non-canonical diguanylate cyclase. *Mol Microbiol* **82:**719–733.

218. **Wu C, Cheng YY, Yin H, Song XN, Li WW, Zhou XX, Zhao LP, Tian LJ, Han JC, Yu HQ.** 2013. Oxygen promotes biofilm formation of *Shewanella putrefaciens* CN32 through a diguanylate cyclase and an adhesin. *Sci Rep* **3:**1945.

219. **Hinsa-Leasure SM, Koid C, Tiedje JM, Schultzhaus JN.** 2013. Biofilm formation by *Psychrobacter arcticus* and the role of a large adhesin in attachment to surfaces. *Appl Environ Microbiol* **79:**3967–3973.

220. **Torres AG, Perna NT, Burland V, Ruknudin A, Blattner FR, Kaper JB.** 2002. Characterization of Cah, a calcium-binding and heat-extractable autotransporter protein of enterohaemorrhagic *Escherichia coli*. *Mol Microbiol* **45:**951–966.

221. **Sherlock O, Vejborg RM, Klemm P.** 2005. The TibA adhesin/invasin from enterotoxigenic *Escherichia coli* is self recognizing and induces bacterial aggregation and biofilm formation. *Infect Immun* **73:**1954–1963.

222. **Wells TJ, Sherlock O, Rivas L, Mahajan A, Beatson SA, Torpdahl M, Webb RI, Allsopp LP, Gobius KS, Gally DL, Schembri MA.** 2008. EhaA is a novel autotransporter protein of enterohemorrhagic *Escherichia coli* O157:H7 that contributes to adhesion and biofilm formation. *Environ Microbiol* **10:**589–604.

223. **Wells TJ, McNeilly TN, Totsika M, Mahajan A, Gally DL, Schembri MA.** 2009. The *Escherichia coli* O157:H7 EhaB autotransporter protein binds to laminin and collagen I and induces a serum IgA response in O157:H7 challenged cattle. *Environ Microbiol* **11:**1803–1814.

224. **Allsopp LP, Totsika M, Tree JJ, Ulett GC, Mabbett AN, Wells TJ, Kobe B, Beatson SA, Schembri MA.** 2010. UpaH is a newly identified autotransporter protein that contributes to biofilm formation and bladder colonization by uropathogenic *Escherichia coli* CFT073. *Infect Immun* **78:**1659–1669.

225. **Allsopp LP, Beloin C, Ulett GC, Valle J, Totsika M, Sherlock O, Ghigo JM, Schembri MA.** 2012. Molecular characterization of UpaB and UpaC, two new autotransporter proteins of uropathogenic *Escherichia coli* CFT073. *Infect Immun* **80:**321–332.

226. **Zude I, Leimbach A, Dobrindt U.** 2013. Prevalence of autotransporters in *Escherichia coli*: what is the impact of phylogeny and pathotype? *Int J Med Microbiol.* [Epub ahead of print.] doi:10.1016/j.ijmm.2013.10.006.

227. **Kroupitski Y, Brandl MT, Pinto R, Belausov E, Tamir-Ariel D, Burdman S, Sela Saldinger S.** 2013. Identification of *Salmonella enterica* genes with a role in persistence on lettuce

leaves during cold storage by recombinase-based *in vivo* expression technology. *Phytopathology* **103:**362–372.

228. **Pearson MM, Laurence CA, Guinn SE, Hansen EJ.** 2006. Biofilm formation by *Moraxella catarrhalis in vitro*: roles of the UspA1 adhesin and the Hag hemagglutinin. *Infect Immun* **74:**1588–1596.
229. **Ishikawa M, Nakatani H, Hori K.** 2012. AtaA, a new member of the trimeric autotransporter adhesins from *Acinetobacter* sp. Tol 5 mediating high adhesiveness to various abiotic surfaces. *PLoS One* **7:**e48830. doi:10.1371/journal.pone.0048830.
230. **Totsika M, Wells TJ, Beloin C, Valle J, Allsopp LP, King NP, Ghigo JM, Schembri MA.** 2012. Molecular characterization of the EhaG and UpaG trimeric autotransporter proteins from pathogenic *Escherichia coli*. *Appl Environ Microbiol* **78:**2179–2189.
231. **Guilhabert MR, Kirkpatrick BC.** 2005. Identification of *Xylella fastidiosa* antivirulence genes: hemagglutinin adhesins contribute a biofilm maturation to *X. fastidios* and colonization and attenuate virulence. *Mol Plant Microbe Interact* **18:**856–868.
232. **Gottig N, Garavaglia BS, Garofalo CG, Orellano EG, Ottado J.** 2009. A filamentous hemagglutinin-like protein of *Xanthomonas axonopodis* pv. citri, the phytopathogen responsible for citrus canker, is involved in bacterial virulence. *PLoS One* **4:**e4358. doi:10.1371/journal.pone.0004358.
233. **Garcia EC, Anderson MS, Hagar JA, Cotter PA.** 2013. *Burkholderia* BcpA mediates biofilm formation independently of interbacterial contact-dependent growth inhibition. *Mol Microbiol* **89:**1213–1225.
234. **Evans LR, Linker A.** 1973. Production and characterization of the slime polysaccharide of *Pseudomonas aeruginosa*. *J Bacteriol* **116:**915–924.
235. **Ausmees N, Jonsson H, Hoglund S, Ljunggren H, Lindberg M.** 1999. Structural and putative regulatory genes involved in cellulose synthesis in *Rhizobium leguminosarum* bv. trifolii. *Microbiology* **145:**1253–1262.
236. **Matthysse AG, Thomas DL, White AR.** 1995. Mechanism of cellulose synthesis in *Agrobacterium tumefaciens*. *J Bacteriol* **177:**1076–1081.
237. **Matthysse AG, White S, Lightfoot R.** 1995. Genes required for cellulose synthesis in *Agrobacterium tumefaciens*. *J Bacteriol* **177:** 1069–1075.
238. **Ross P, Mayer R, Benziman M.** 1991. Cellulose biosynthesis and function in bacteria. *Microbiol Rev* **55:**35–58.
239. **Zogaj X, Nimtz M, Rohde M, Bokranz W, Romling U.** 2001. The multicellular morphotypes of *Salmonella typhimurium* and *Escherichia coli* produce cellulose as the second component of the extracellular matrix. *Mol Microbiol* **39:**1452–1463.
240. **MacRae JD, Smit J.** 1991. Characterization of caulobacters isolated from wastewater treatment systems. *Appl Environ Microbiol* **57:**751–758.
241. **Moore RL, Marshall KC.** 1981. Attachment and rosette formation by hyphomicrobia. *Appl Environ Microbiol* **42:**751–757.
242. **Quintero EJ, Weiner RM.** 1995. Evidence for the adhesive function of the exopolysaccharide of hyphomonas strain MHS-3 in its attachment to surfaces. *Appl Environ Microbiol* **61:**1897–1903.
243. **Yun C, Ely B, Smit J.** 1994. Identification of genes affecting production of the adhesive holdfast of a marine caulobacter. *J Bacteriol* **176:**796–803.
244. **Wang X, Preston JF 3rd, Romeo T.** 2004. The pgaABCD locus of *Escherichia coli* promotes the synthesis of a polysaccharide adhesin required for biofilm formation. *J Bacteriol* **186:**2724–2734.
245. **Itoh Y, Wang X, Hinnebusch BJ, Preston JF 3rd, Romeo T.** 2005. Depolymerization of beta-1,6-N-acetyl-D-glucosamine disrupts the integrity of diverse bacterial biofilms. *J Bacteriol* **187:**382–387.
246. **Ghafoor A, Hay ID, Rehm BH.** 2011. Role of exopolysaccharides in *Pseudomonas aeruginosa* biofilm formation and architecture. *Appl Environ Microbiol* **77:**5238–5246.
247. **Jackson KD, Starkey M, Kremer S, Parsek MR, Wozniak DJ.** 2004. Identification of psl, a locus encoding a potential exopolysaccharide that is essential for *Pseudomonas aeruginosa* PAO1 biofilm formation. *J Bacteriol* **186:**4466–4475.
248. **Byrd MS, Sadovskaya I, Vinogradov E, Lu H, Sprinkle AB, Richardson SH, Ma L, Ralston B, Parsek MR, Anderson EM, Lam JS, Wozniak DJ.** 2009. Genetic and biochemical analyses of the *Pseudomonas aeruginosa* Psl exopolysaccharide reveal overlapping roles for polysaccharide synthesis enzymes in Psl and LPS production. *Mol Microbiol* **73:**622–638.
249. **Friedman L, Kolter R.** 2004. Genes involved in matrix formation in *Pseudomonas aeruginosa* PA14 biofilms. *Mol Microbiol* **51:**675–690.
250. **Friedman L, Kolter R.** 2004. Two genetic loci produce distinct carbohydrate-rich structural components of the *Pseudomonas aeruginosa* biofilm matrix. *J Bacteriol* **186:**4457–4465.

251. **Hufnagel DA, DePas WH, Chapman MR.** 2015. The biology of the *Escherichia coli* extracellular matrix. *In* Ghannoum M, Parsek M, Whiteley M, Mukherjee P (ed), *Microbial Biofilms*. 2nd ed. ASM Press, Washington, DC, in press.

252. **Wang Y, Haitjema CH, Fuqua C.** 2015. The Ctp type IVb pilus locus of *Agrobacterium tumefaciens* directs formation of the common pili and contributes to reversible surface attachment. *J Bacteriol* **196:**2979–2988.

Biofilm Matrix Proteins

10

JIUNN N. C. FONG[1] and FITNAT H. YILDIZ[1]

INTRODUCTION

Microorganisms in the natural environment typically live on or in close association with surfaces and predominantly exist as biofilms, surface attached microbial communities composed of cells and extracellular matrix (1–4). The exact compositions of biofilm matrices differ greatly between different microorganisms and growth conditions under which biofilms are formed but generally consist of exopolysaccharides, proteins, and nucleic acids. Proteinaceous components include cell surface adhesins, protein subunits of flagella and pili, secreted extracellular proteins, and proteins of outer membrane vesicles.

Cell surface proteins, pili, and flagella participate in the initial attachment to surfaces and, in some microorganisms, are also involved in migration along the surfaces, thereby facilitating surface colonization. Matrix proteins contribute to biofilm structure and stability. Such proteins were identified mostly by mutational studies, which showed that the absence of matrix proteins results in reduced biofilm formation and stability, and altered biofilm architectures (5–14). Structural analysis and biofilm localization studies have

[1]Department of Microbiology and Environmental Toxicology, University of California, Santa Cruz, Santa Cruz, CA 95064.

Microbial Biofilms 2nd Edition
Edited by Mahmoud Ghannoum, Matthew Parsek, Marvin Whiteley, and Pranab K. Mukherjee

doi:10.1128/microbiolspec.MB-0004-2014

provided further insights into the functions and mechanisms of action of matrix proteins. Some matrix proteins exhibit enzymatic properties toward matrix components, such as the glycosyl hydrolase dispersin B that hydrolyzes polysaccharides (15), proteases that target matrix proteins (16), and DNases that degrade extracellular nucleic acids (17, 18), thus facilitating either biofilm matrix reorganization or biofilm matrix degradation and dispersal.

Several studies have been carried out to identify the matrix proteome of several microorganisms, including *Vibrio cholerae* (19), *Pseudomonas aeruginosa* (20), *Myxococcus xanthus* (21), and natural biofilm communities of acid mine drainage (22). These studies revealed that, in addition to secreted proteins, the biofilm matrix also contains large numbers of periplasmic, cytoplasmic, inner, and outer membrane proteins (OMPs). These results implicate the involvement of cell lysis and/or outer membrane vesicles (OMVs) in modulating biofilm proteome composition.

In this article, we focus on the matrix proteins that play structural roles in biofilm formation. We will discuss the functions and mechanisms of action of matrix proteins and lectins produced by *V. cholerae* and *P. aeruginosa*, the biofilm-associated proteins from *Staphylococcus aureus*, and the hydrophobin from *Bacillus subtilis* in biofilm formation. Finally, we will review matrix proteome studies of *V. cholerae* and *P. aeruginosa* and the roles of OMVs and nucleoid-binding proteins in biofilm formation.

V. CHOLERAE MATRIX PROTEINS

V. cholerae is a facultative human pathogen that colonizes the human intestine and survives for extended periods in natural aquatic environments. Both pathogenesis and environmental survival are closely linked to the microbe's ability to form biofilms. Mature biofilm formation in *V. cholerae* depends on the production of *Vibrio* exopolysaccharides (VPS) (23, 24). *V. cholerae* produces two types of VPS. The repeating unit of the major variant consists of [-4)-α-GulNAcAGly3OAc-(1-4)-β-D-Glc-(1-4)-α-Glc-(1-4)-α-D-Gal-(1-]. In the minor variant, α-D-Glc is replaced with α-D-GlcNAc (25). Three major biofilm matrix proteins (RbmA, Bap1, and RbmC) (5, 6) are important for biofilm formation on abiotic surfaces, and the extracellular chitin-binding protein GbpA mediates attachment to chitinous surfaces of zooplankton (26). The structure, function, and mechanistic roles of these matrix proteins in *V. cholerae* surface adhesion and biofilm formation are reviewed below.

Rugosity and Biofilm Structure Modulator A (RbmA)

RbmA is a 26-kDa matrix protein involved in facilitating intercellular adhesion during biofilm formation (5, 27). Studies carried out using a rugose variant of *V. cholerae,* which exhibits enhanced biofilm formation and colony corrugation due to increased production of VPS and matrix proteins, revealed that an *rbmA* mutant exhibits a decrease in colony corrugation (Fig. 1), forms a biofilm with altered biofilm architectures, and disperses easily by shear force (5). Similarly, pellicles, which are biofilms formed at the air-liquid interface, formed by the *rbmA* mutant are less wrinkled and more fragile and disintegrate upon force (5). Addition of exogenous purified RbmA rescues the altered pellicle phenotype of an *rbmA* mutant strain (19), indicating that extracellular provision of RbmA enhances intercellular interactions. Taken together, these studies point out the importance of RbmA in the development of mature biofilm architecture and in stabilization of biofilms.

The crystal structure of RbmA revealed that it consists of two tandem fibronectin type III (FnIII) domains and functions as a 49-kDa dimer (28). The approximately 100-aa FnIII domain is found widely in many proteins, including eukaryotic cell surface

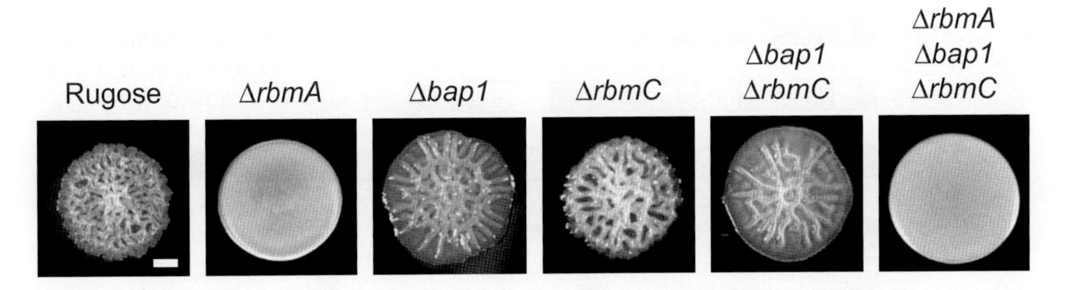

FIGURE 1 Colony morphology of *V. cholerae* rugose variant and mutant strains unable to produce RbmA, RbmC, and Bap1 matrix proteins. Bar = 0.5 mm. doi:10.1128/microbiolspec.MB-0004-2014.f1

receptors and prokaryotic carbohydrate-binding proteins (29), suggesting a possible role of RbmA in binding carbohydrates and in cell adhesion. The two tandem FnIII domains (Fig. 2) are not identical in peptide sequence but share 24% identity and 44% similarity (30). The FnIII domains of RbmA fold as a seven-strand β-sandwich, with the N-terminal of the FnIII domain of one monomer interacting tightly with a second monomer of the asymmetric unit (28). The crystal structure of RbmA also revealed a positively charged groove formed by the two adjacent FnIII domains (28). Three arginine residues (R116, R219, and R234) located within this groove, which are predicted to be involved in ligand binding, were found to be critical for RbmA function. Strains that produce mutated versions of RbmA, containing point mutations in these positively charged residues, exhibit a decrease in colony corrugation and/or pellicle formation when compared to the parental strain (28). RbmA also contains a negatively charged groove, formed between the two FnIII domains of the same monomer (28). However, site-directed mutagenesis resulting in either removing (E84A) or reversing (E84R) the negative charges did not affect RbmA function, suggesting that this negatively charged groove does not play a major role in RbmA-mediated biofilm formation under the conditions studied (28). The biological role of the negatively charged groove remains to be determined.

Possible ligands of RbmA were identified by glycan array studies (30). RbmA exhibits saccharide-binding specificity and multivalency toward sialic acid and fucose and has a lower binding preference to galactose, rhamnose, and *N*-acetylgalactosamine (GalNAc) (30). The ability of RbmA to bind to galactose, a component of VPS (23), suggests that RbmA-mediated biofilm formation may be, in part, due to RbmA-VPS interactions. The significance of RbmA binding to other sugars remains to be determined. It has also been speculated that RbmA may bind the lipopolysaccharide (LPS) found on the cell surface via interaction with sialic acid derivatives (30). Therefore, the positively charged groove may interact with negatively charged ligands such as carbohydrates on bacterial cell surfaces (30), VPS (28), and/or LPS (30). The unique flexibility of the FnIII domains (31, 32), together with the predicted interactions of RbmA with VPS, LPS, and/or cell surfaces, support a model where RbmA can act as an elastic scaffold in the biofilm matrix. RbmA may mediate flexible contacts with other matrix components and bacterial cell surfaces, thus increasing shear resistance and integrity of the biofilm matrix.

RbmA is secreted and contains a secretion signal sequence in its N-terminal region (Fig. 2) (5), indicating involvement of the general secretion (Sec) pathway in its secretion. However, the mechanism of RbmA secretion outside of the cell is currently unknown. Localization studies revealed that

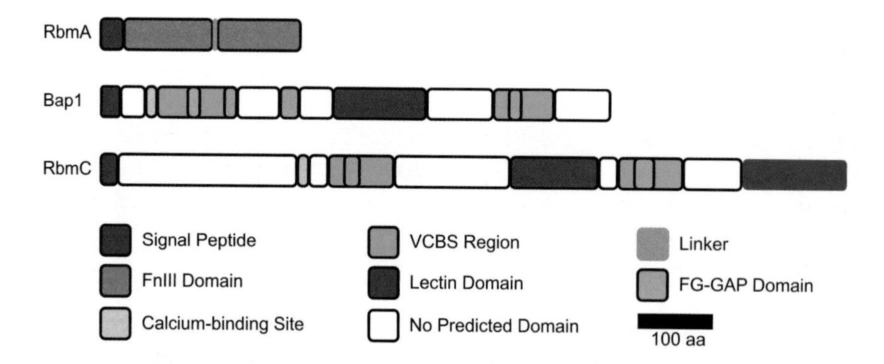

FIGURE 2 Domain organization of *V. cholerae* RbmA, Bap1, and RbmC. FnIII, fibronectin type III; VCBS, *Vibrio-Colwellia-Bradyrhizobium-Shewanella* repeat; FG-GAP, phenyl-alanyl-glycyl (FG) and glycyl-alanyl-prolyl (GAP). doi:10.1128/microbiolspec.MB-0004-2014.f2

RbmA is distributed throughout the biofilm (19, 27), specifically surrounding the bacterial cells (27), suggesting that RbmA facilitates cell-cell adhesion. In fact, RbmA was detected on cell surfaces 30 minutes after surface attachment and aids in the retention of newly divided daughter cells (27). Furthermore, retention of RbmA on cell surfaces is dependent on the presence of VPS. This indicates that interaction with VPS is essential for RbmA spatial distribution within the biofilm matrix and further reinforces the notion that RbmA binds VPS. The spatial and temporal localization of RbmA, as well as the dependency on VPS for retention on cell surfaces, strongly supports the role of RbmA as a scaffold that mediates cell-cell and cell-matrix interactions.

Biofilm-Associated Protein 1 (Bap1) and Rugosity and Biofilm Structure Modulator C (RbmC)

The 75-kDa Bap1 and 104-kDa RbmC matrix proteins share 47% peptide sequence similarity and are involved in biofilm formation in *V. cholerae* (6, 33). Despite their sequence similarity, Bap1 and RbmC affect biofilm formation differently. While the lack of either protein affects colony corrugation (Fig. 1), pellicle formation, and biofilm formation, the defects differ in magnitude (6). A mutant unable to produce both RbmC and Bap1 exhibits marked decreases in biofilm formation and the ability to stably attach to surfaces. Complementation analysis of *rbmC bap1* double mutant showed that Bap1 and RbmC can partially complement each other, but they are not functionally redundant (6, 27).

Sequence similarity database and Pfam motif searches, as well as sequence alignment analysis, showed that Bap1 and RbmC contain four overlapping *Vibrio-Colwellia-Bradyrhizobium-Shewanella* repeat (VCBS) domains that form two VCBS regions (Fig. 2). In addition, Bap1 contains four FG-GAP (phenyl-alanyl-glycyl [FG] and glycyl-alanyl-prolyl [GAP]) domains, while RbmC contains two FG-GAP domains. Although the VCBS domain (PF13517) is commonly found in multiple copies in proteins from several species of *Vibrio, Colwellia, Bradyrhizobium, and Shewanella*, very little is known about its function, except that it may be involved in cell adhesion. FG-GAP repeats (PF01839) are found in the N-terminal region of the eukaryotic integrin α-chain, which is important for ligand recognition and binding to the extracellular matrix or cell surface proteins (34–36). NCBI conserved domain searches also revealed that Bap1 and RbmC contain one and two jacalin-like lectin domains, respectively. The jacalin-like lectin domains (PF01419), or β-prism domains, have binding specificities to galactose, mannose, and/or glucose and their derivatives (37, 38). Since VPS contains these sugars (23), it is possible that Bap1 and RbmC may bind to the galactose, mannose, and/or glucose residues of VPS. The presence of VCBS, FG-GAP, and lectin domains suggests that Bap1 and RbmC likely mediate biofilm formation by facilitating adhesion and carbohydrate binding. Both Bap1 and RbmC contain a predicted EF-hand calcium-binding motif (Fig. 2), indicative of calcium binding. However, it remains to be determined whether these predicted sites bind calcium and if binding to calcium regulates protein function.

Both Bap1 and RbmC contain predicted secretion signal sequences at the N-terminal regions (Fig. 2), indicating that they are secreted from the cytoplasm via the Sec-dependent pathway. *V. cholerae* type II secretion system proteome analysis showed that RbmC is secreted by the type II secretion system (39), while the mechanism of secretion of Bap1 is currently unknown. Biofilm localization studies revealed that RbmC and Bap1 form envelopes around microcolonies/cell clusters and that Bap1, but not RbmC, localizes to the biofilm-surface interface (27). The localization of Bap1 and RbmC within the biofilm corroborates their functions and the biofilm phenotypes observed. While the presence of both Bap1 and RbmC in the cell cluster envelope likely allows partial functional redundancy, the additional role of Bap1 in anchoring developing biofilms onto surfaces cannot be fulfilled by RbmC. It is noteworthy that the specific spatial localization of Bap1 within the biofilm is not due to localized transcription of *bap1* within the cell population (19). The mechanisms by which Bap1 and RbmC are targeted to their specific location during biofilm formation are currently unknown. Retention of Bap1 and RbmC on cell surfaces is dependent on the presence of VPS, while the localization of Bap1 at the biofilm-substratum interface is VPS-independent (27). These observations indicate that Bap1 and RbmC could bind VPS, and these interactions may influence the spatial localization of these matrix proteins. The second lectin domain at the C-terminal region of RbmC has been reported to be nonessential for protein function, because a *bap1 rbmC* double mutant strain, harboring an *rbmC* allele with a truncation of the C-terminal lectin domain, is able to form biofilms (19). However, it is yet to be determined if the complemented strain can form biofilms with wild-type architectures.

It has been reported that Bap1 could associate with OMVs by binding to OmpT (40). Bap1-OmpT interaction requires the presence of the integrin-binding domain (leucine-aspartic acid–valine peptide) of OmpT and the FG-GAP domains of Bap1 (Fig. 2). It was also reported that RbmC does not interact with OmpT in the OMVs, although it is currently unclear why Bap1, but not RbmC, binds to OmpT. It is possible that Bap1 may exhibit higher binding affinity to OmpT because Bap1 contains four FG-GAP domains, while RbmC contains only two FG-GAP domains. Although it is yet to be determined if OMVs are part of the *V. cholerae* biofilm matrix, this observation highlights possible bifunctional roles of Bap1 in biofilm formation: facilitating cell attachment to surfaces and interactions among different biofilm components, i.e., VPS and OMVs. While the localization of Bap1 and RbmC in the biofilm matrix and their involvement in biofilm formation have been demonstrated, the importance of the various predicted adhesion and carbohydrate-binding domains is unknown. Further studies in determining the roles of these domains and critical residues for protein function will help provide better insights into the mechanistic functions of these matrix proteins.

V. cholerae GlcNAc-Binding Protein A (GbpA)

In the natural aquatic environment, *V. cholerae* adheres to chitinous surfaces of phytoplankton and zooplankton, including exoskeletons of copepods, and colonization of these surfaces enhances *V. cholerae* survival in the rapidly changing environment (41, 42). Chitin is a polymer of *N*-acetyl-D-glucosamine (GlcNAc) and is one of the most abundant polysaccharides in nature that can be used as a carbon and energy source by *V. cholerae* (43). The *V. cholerae* GbpA is a chitin-binding protein (26, 44) predicted to be critical for *V. cholerae* environmental survival by facilitating adhesion to chitinous surfaces. Indeed, *gbpA* mutant strains exhibited decreased attachment to zooplankton chitinous exoskeletons and egg sacs (26, 45). In addition, GbpA-deficient mutants exhibited reduced adherence to chitin and GlcNAc-coated beads (26, 44). Similarly, GbpA-chitin interaction was also observed in GbpA-containing cell lysates, and the interaction was abolished upon the addition of GlcNAc in a concentration-dependent manner, indicating that GbpA specifically binds to the chitin monomer GlcNAc (26). Although GbpA mediates adhesion to chitinous surfaces, it was reported that GbpA does not play a role in attachment and biofilm formation on abiotic surfaces. A *gbpA* mutant strain exhibited wild-type levels of attachment efficiency on inorganic quartz, quartzite, and calcium carbonate (marine sediment) surfaces (45) and was able to form biofilms on polyvinyl chloride surfaces (46), suggesting that the main role of GbpA is adhesion to GlcNAc/chitin-containing biotic surfaces. Furthermore, GbpA has also been reported to bind to intestinal mucin, which also contains GlcNAc (47). As such, GbpA functions as a virulence factor (26, 47, 48) and an adhesin in mediating intestinal attachment and adhesion on chitinous surfaces.

Crystal structures of GbpA revealed that it is an unusual, elongated protein. It consists of four domains (D1 to D4) that do not interact and are completely surface-exposed (48). The two terminal domains D1 and D4 exhibit structural similarity to chitin-binding protein 21 (CBP21) and sequence similarity to the C-terminal chitin-binding domain of chitinase B (ChiB) from *Serratia marcescens*, respectively (48). Domains D1 and D4 were shown to bind chitin in glycan array binding assays. However, in contrast to previous reports that showed binding of GbpA to GlcNAc using *V. cholerae* cells and cell lysates (26, 44), the purified full-length and various truncated versions of GbpA did not exhibit GlcNAc-binding ability (48), suggesting that additional *V. cholerae* factors may be required for GbpA-GlcNAc interaction. Domains D2 and D3 are essential for cell-surface interactions (48). Domain D2 shows distant structural similarity to the β-domain of protein p5 of *Sphingomonas* sp. A1 (48), which is predicted to be involved in cell-surface interactions (49, 50). Domain D3 exhibits an immunoglobulin fold that is associated with cell adhesion (48, 51, 52). Furthermore, D1 is essential for mucin binding, and D1 to D3 are required for intestinal colonization in infant mice (48). GbpA contains an N-terminal secretion signal sequence and has been reported to be secreted by the type II secretion system (26, 39). It appears that the unique elongated structure of GbpA results in modular binding, in which the two terminal domains, D1 and D4, bind to chitin, while the two central domains, D2 and D3, interact with cell surfaces. Although the binding targets of each domain have been elucidated, the essential residues for their specific binding are still unknown; such studies would provide mechanistic insights into GbpA-mediated adhesion.

Matrix Proteome of *V. cholerae*

The proteome of the *V. cholerae* biofilm matrix from biofilms grown in nutrient broth under static biofilm conditions has been reported (19). A total of 74 proteins with predicted extracytoplasmic localization were identified. These include known biofilm matrix proteins (RbmA and RbmC), several OMPs (OmpA, OmpU, OmpT, OmH, OmpK, OmpW, OmpS, and TolC), periplasmic proteins, flagellar proteins, mannose-sensitive hemagglutinin pili proteins, enzymes (hemagglutinin/protease HapA and chitinase ChiA-2), and the hemolysin HlyA. Protein localization studies, using immunofluorescence microscopy, further confirmed biofilm matrix localization of RbmA, Bap1, MshA, and HlyA (19). The cell-associated mannose-sensitive hemagglutinin type IV pili, which facilitates attachment to abiotic and chitinous biotic surfaces (53, 54), and the matrix proteins RbmA and Bap1, which contain adhesion and carbohydrate-binding lectin domains, are expected to be retained within the biofilm matrix. HlyA contains a jacalin-like lectin domain, suggesting that HlyA may be binding to the biofilm matrix via the lectin domain. The role of OMPs in biofilm formation is yet to be determined. Several hypothetical proteins were also identified in the study, and determining their functions may lead to the identification of new biofilm matrix proteins of *V. cholerae*.

P. AERUGINOSA MATRIX PROTEINS

P. aeruginosa is an opportunistic human pathogen capable of causing diverse infections in humans. The capacity of *P. aeruginosa* to form biofilms on various surfaces in the human body is critical for its pathogenicity. The *P. aeruginosa* biofilm matrix consists of exopolysaccharides (55–57), extracellular DNA (eDNA) (58, 59) and proteins (7, 8, 14, 60). In *P. aeruginosa*, three exopolysaccharides (alginate, Pel, and Psl) have been shown to be involved in biofilm formation in a strain-specific manner. Alginate is a high molecular weight acetylated exopolysaccharide that consists of uronic acids (mannuronic and guluronic acids) (55, 61) and is not essential for biofilm formation, as shown in the clinical alginate-overproducing mucoid strain (FRD1) and the laboratory nonmucoid (PAO1 and PA14) strains (62–64). However, overproduction of alginate affects the development of mature biofilm architectures (62, 64, 65). Pel is a glucose-rich polysaccharide and is required for pellicle and biofilm formation in *P. aeruginosa* PA14, while Psl is a mannose- and galactose-rich polysaccharide that is critical for pellicle and biofilm formation in PAO1 (56, 57, 66–69). eDNA has also been reported to be important in the initial stages of biofilm formation. The presence of DNase I prevents biofilm formation, and addition of DNase I on preformed biofilms at early stages of biofilm formation leads to dissolution of the biofilm matrix (58). While a wealth of knowledge is available on the exopolysaccharides produced by different strains of *P. aeruginosa*, only a few *P. aeruginosa* biofilm matrix proteins have been described. These include the lectins LecA and LecB, and the Psl-binding matrix protein CdrA.

LecA and LecB Lectins

P. aeruginosa produces two lectins, LecA and LecB (formerly known as PA-IL and PA-IIL, respectively) (70–72), that are involved in biofilm formation (7, 8, 60) and play a role during infections (73–75). Lectins are carbohydrate-binding proteins that exhibit sugar-binding specificity (76, 77). LecA is required for *P. aeruginosa* biofilm formation on polystyrene and stainless steel surfaces. A *lecA* mutant of PAO1 exhibits reduced substratum coverage, while a LecA-overproducing strain exhibits increased biofilm formation (7). LecA binds to galactose, *N*-acetyl-D-galactosamine, and glucose (60, 78). It is yet to be determined if LecA binds to the galactose-rich Psl and glucose-rich Pel; such an interaction would contribute to biofilm formation. LecB is involved in biofilm formation on glass surfaces. A *lecB* mutant of PAO1 forms a biofilm with reduced thickness and surface coverage compared to that formed by wild type (8). LecB exhibits high affinity for fucose and binds to several other monosaccharaides with the following preference: L-fucose > L-galactose > D-arabinose > D-fructose > D-mannose (79). LecB may mediate biofilm formation of *P. aeruginosa* via interactions with the galactose and mannose residues in Psl. Orthologs of LecB have also been identified in other bacteria (80–83), but their role in biofilm formation is unknown.

Crystal structures of LecA and LecB have been reported both in the native form and in complex with their binding saccharides (79, 84–86). LecA (51 kDa) is a tetrameric protein consisting of four 12.8-kDa subunits (84, 85, 87). In the LecA structure, each monomer adopts a small jelly-roll β-sandwich fold, consisting of two curved sheets, with a calcium-dependent ligand-binding site at the apex that binds one galactose ligand and one calcium ion (85). LecA also contains a secondary glucose-binding site in close proximity to the primary galactose-binding site, but the bound glucose residue does not interact with the amino acid residues of the galactose-binding site (78). LecA-mediated biofilm formation involves the galactose-binding site. Addition of galactosides that have high affinities to LecA, such as IPTG (isopropyl-β-D-thiogalactoside) and NPG (*p*-nitrophenyl-α-D-galactoside), reduce wild-type *P. aeruginosa* PAO1 biofilm formation and induce dispersion in mature biofilms (7).

LecB (47 kDa) is a tetrameric protein consisting of four 11.7-kDa subunits (88). In the LecB structure, each monomer is arranged as a nine-stranded antiparallel β-sandwich (84, 86). The amino acid residues N95 to D104 form a single loop, and together with S22, S23, and G114 make up the ligand-binding site in LecB that interacts with two calcium ions. The two calcium ions not only directly interact with the ligands but are also required to stabilize the N95-D104 loop, which forms the core ligand-binding site.

LecB is localized to the outer membrane (8). A mutation in the calcium-binding site (D104A) abolishes binding of LecB to the outer membrane. Furthermore, treatment of outer membrane fractions with NPF (*p*-nitrophenyl-α-L-fucose), which has a high affinity for LecB, resulted in dissociation of LecB from the outer membrane. In addition, pre-incubation of a fluorescently labeled LecB with L-fucose inhibited interaction of the lectin with cell surfaces (8). Collectively, these findings suggest that LecB likely interacts with a ligand on the cell surface in a calcium-dependent manner.

In a study designed to identify the cell surface ligand for LecB, direct interaction between LecB and the OMP OprF was demonstrated (89). Western blot analysis revealed that OprF was necessary for proper LecB localization, because LecB was released into the culture supernatant of the *oprF* mutant, instead of being retained on the outer membrane as observed in the wild-type strain. Specific LecB-OprF interaction was further demonstrated with copurification of OprF and LecB. In the same study, a far-Western blot analysis using purified LecB detected several positive protein bands from the membrane fraction of *P. aeruginosa* (89), suggesting that LecB may interact with other ligands on the membrane in addition to OprF. The mechanism of LecB-OprF interaction remains to be determined. It is also currently unclear if LecA interacts with any OMPs in *P. aeruginosa*. Although both LecA and LecB are localized in the cytosolic and outer membrane fractions in PAO1 (7, 8), they do not contain a predicted signal sequence as determined by SignalP 4.1. However, using SecretomeP 2.0, both LecA and LecB are predicted to be secreted via a nonclassical pathway (7). The exact mechanism of LecA and LecB secretion is currently unknown.

Cyclic Diguanylate-Regulated Two-Partner Secretion Partner A (CdrA)

P. aeruginosa produces a matrix protein, CdrA, which binds to the exopolysaccharide Psl and mediates biofilm formation on abiotic surfaces (14). *cdrA* is in an operon with *cdrB*, predicted to encode a putative outer membrane transporter. CdrA and CdrB are predicted to be members of the two-partner secretion systems. Western blot analysis demonstrated that CdrA exists as a full-length cell-associated protein that can be processed into a smaller fragment, which is released into culture supernatant. The exact mechanism of this proteolytic processing is not clear. A *cdrA* mutant forms biofilms that are thinner and less structured than biofilms formed by wild type (14). Overproduction of CdrA leads to an increase in cell auto-aggregation in liquid cultures, and CdrA-mediated auto-aggregation is dependent on Psl, but not Pel: a Psl-deficient strain is unable to auto-aggregate, while a Pel-deficient strain still exhibits auto-aggregation phenotype when CdrA is overproduced. Addition of mannose reduces the auto-aggregation phenotype, suggesting that CdrA binds to the mannose residues in Psl. CdrA also exhibits multivalency because the addition of fucose, fructose, or GlcNAc also reduces the auto-aggregation phenotype. Direct binding of CdrA to Psl exopolysaccharide was demonstrated with co-immunoprecipitation of CdrA with Psl. Therefore, CdrA mediates biofilm formation and cell auto-aggregation in *P. aeruginosa* PAO1 by binding to the Psl exopolysaccharide, leading to either cross-linking of the exopolysaccharide polymers and/or tethering of Psl to cells.

The secondary structure of CdrA is predicted to be dominated mainly by β-strands, and tertiary structure prediction revealed that CdrA forms a long, rod-shape structure with a β-helix structural motif (14). CdrA contains several putative adhesion domains (14), including a carbohydrate-dependent hemagglutination activity domain, a glycine-rich sugar-binding domain, and an RGD (Arg-Gly-Asp) sequence motif that may function as an integrin recognition site (51). Although the functions of these domains and motifs remain to be tested, it is likely that CdrA mediates biofilm formation via these putative adhesion domains.

Matrix Proteome of *P. aeruginosa*

A study designed to identify proteins associated with *P. aeruginosa* PAO1 biofilm matrix (matrix proteome) was recently carried out (20). Forty-five proteins that were not present in the whole-cell samples were identified exclusively in the matrix proteome, including the Psl-binding CdrA and the cognate transporter CdrB. Overall, the matrix proteome of *P. aeruginosa* contains secreted proteins (13.3%), cytoplasmic proteins (28.9%), periplasmic proteins (11.1%), cytoplasmic membrane proteins (2.2%), and most abundantly, OMPs (35.6%). The cellular locations of the rest of the proteins (8.9%) could not be identified with confidence. The presence of OMPs and cytoplasmic proteins in the matrix proteome could be due to cell lysis as well as the presence of OMVs in the biofilm matrix. When compared to proteins identified from biofilm OMVs, 53% of predicted OMPs were found in both matrix and OMV samples, indicating that a large portion of the matrix proteins are associated with OMVs. A large 362-kDa protein, predicted to be localized in the outer membrane, was also identified that may function as a surface protein for adhesion and biofilm formation. Several enzymes were also found, although not exclusively, in the biofilm matrix of *P. aeruginosa,* including alkaline protease, chitinase, protease IV, and a putative magnesium-dependent DNase (20), which are likely retained in the matrix by interaction with matrix components such as exopolysaccharides, eDNA, or proteins. The function of these proteins in biofilm formation is yet to be determined.

BIOFILM-ASSOCIATED PROTEIN (Bap)

The Bap protein family represents one of the most studied groups of matrix proteins involved in biofilm formation (Table 1). Members of the Bap family are usually very large secreted proteins of up to several hundred kilodaltons in molecular mass. The

TABLE 1 Members of the Bap family[a]

Microorganism[b]	Strain	Protein (residues)[c]	GenBank/ UniProtKB	Size (kDa)	A repeats	C repeats	D repeats	Amyloid peptide[d]	References
S. aureus	V329	Bap (2,276 aa)	Q79LN3	239	2	13 (2 partial)	3 (1 partial)	1 (STVTVTD)	11, 90, 91
Staphylococcus epidermidis	C533	Bap (2,742 aa)	AAY28519.1	284	2	16	6	1 (STVTVTD) 17 (STVTVTF)	90, 91
Staphylococcus chromogenes	C483	Bap (1,530 aa)	Q4ZHU4	162	1	5	4	1 (STVTVTD) 5 (STVTVTF)	90, 91
Staphylococcus hyicus	12	Bap (3,278 aa)	Q4ZHU0	338	2	25	3	1 (STVTVTD) 26 (STVTVTF)	90, 91
Staphylococcus xylosus	C482	Bap (3,271 aa)	AAY28517.1	337	0	24	9	23 (STVTVTF)	91
Staphylococcus simulans	ATCC1362	Bap (1,674 aa)	Q4ZHU2	177	2	6	4	1 (STVTVTD) 5 (STVTVTF)	90, 91
Salmonella enterica Typhimurium	3938	BapA (3,824 aa)	A9LS56	386	ND	29	ND	1 (STVTVTL)	9, 90

[a]aa, amino acids; ND, no data or no domains identified.
[b]Selected members of Bap family proteins.
[c]Protein lengths are in parentheses.
[d]Number of repetitions of the amyloid peptide within the protein were a combination of data reported by Lembre et al. (90) and from manual sequence gazing. The amyloid peptide sequences are in parentheses.

most unique feature of these proteins is the presence of multiple repeats of identical or near-identical amino acid residues in the core region. Members of the Bap family exhibit the following unique domain features (11). They contain an N-terminal secretion signal sequence, followed by a nonrepetitive N-terminal region B, which may be absent in some Bap orthologs (9). The central region, which makes up most of the protein, consists of multiple identical or near-identical repeats that may contain amyloid-like peptide sequences (90). The number of repeats in the central region can vary in different species and isolates, resulting in different extended structures and protein variants that likely aid in the evasion of host immune responses (91, 92). In Gram-positive bacteria, the protein ends with the C-terminal region carrying an LPxTG (leucine-proline-X-threonine-glycine, where X denotes any amino acid) cell-wall anchoring motif (11). Most of these Bap proteins can function both as virulence factors involved in pathogenesis and as matrix proteins mediating abiotic surface adhesion and subsequent biofilm formation. In the following section, we will discuss representative Bap proteins from *S. aureus, Staphylococcus epidermidis,* and *Salmonella enterica.*

S. aureus Bap

Bap, a 239-kDa cell surface protein (Table 1), was first identified in a study designed to identify genes involved in *Staphylococcal* biofilm formation. Mutants unable to produce Bap exhibit decreased colony corrugation, reduced intercellular adhesion (cell aggregation), and impaired biofilm formation on abiotic surfaces (11, 92–94).

Bap exhibits a unique domain organization (Fig. 3), containing repeats that can be grouped into four regions: A to D (11). The N-terminal region of Bap contains a putative secretion signal sequence and thus likely involves the general Sec-dependent pathway for its secretion. Following the N-terminal signal peptide is region A, which consists of two 32-amino acid repeats (A repeats)

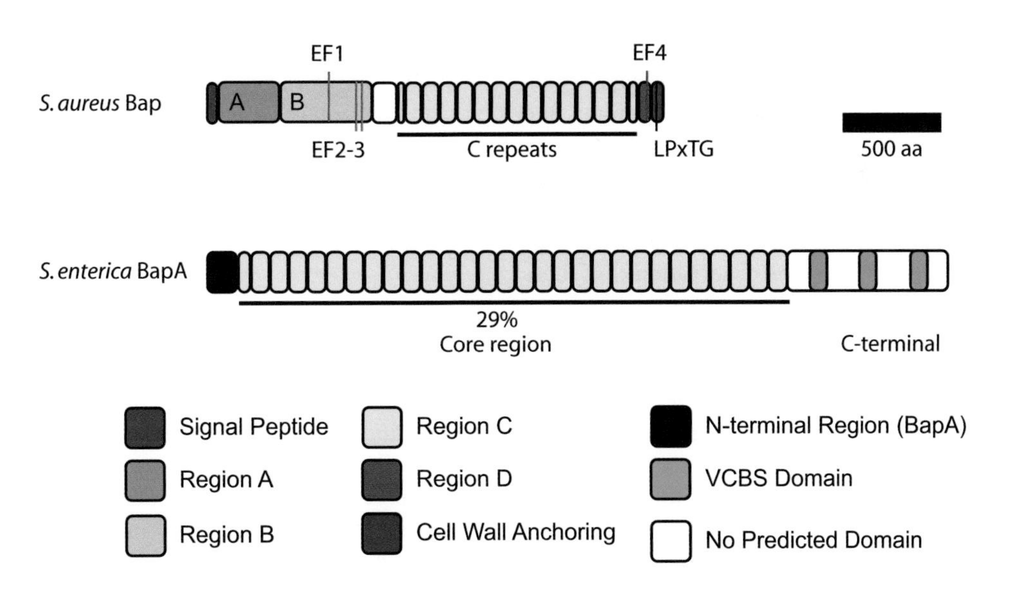

FIGURE 3 Domain organization of *S. aureus* Bap and *S. enterica* BapA. EF-hand calcium-binding motifs EF1 to 4 in Bap are indicated. LPxTG is the cell-wall anchoring motif. The repeats in the core regions of *S. enterica* BapA share 29% identity with the C repeats of *S. aureus* Bap. doi:10.1128/microbiolspec.MB-0004-2014.f3

separated by 26 amino acids (11). Region B does not contain any repeats. A dimerization domain is predicted in region A and region B, suggesting that Bap may form homodimers or heterodimers with other Bap orthologs (94). Formation of heterodimers of *S. aureus* Bap and Bap orthologs from other bacteria could facilitate mixed-species biofilm formation. The most distinctive region of Bap is the core region (region C), which is comprised of 13 near-identical repeats (C repeats) of 86 amino acids and two partial repeats at each end of region C. Region D consists of three short repeats of 18 amino acids (D repeats) in a stretch of sequence rich in serine and aspartic acid residues and the cell-wall anchoring LPxTG motif at the C-terminal region (11). The repeats in region C are predicted to fold as a seven-strand β-sandwich and exhibit similarity to members of the HYR (hyalin repeat) family that contain extracellular adhesion modules (51, 94). Thus, it is possible that the C repeats could mediate adhesion and be involved in biofilm formation.

However, differences in the number of repeats in region C do not appear to be critical for biofilm formation on abiotic surfaces or colony corrugation (92, 93): a mutant producing a shorter variant of Bap, which contains one C repeat, forms biofilms and exhibits colony morphology indistinguishable from those formed by the parental strain, which produces the wild-type Bap with 13 C repeats (93). This notion was further supported by the observation that there is no association between biofilm formation and natural *bap*-positive isolates containing a varying number of repeats in region C (92). Although the number of C repeats does not appear to affect biofilm formation, the presence of a single C repeat may be sufficient to mediate adherence and biofilm formation in *S. aureus*. Alternatively, region A and/or region B may be involved in biofilm formation. In fact, the N-terminal region of Esp, a Bap ortholog in *Enterococcus faecalis*, which exhibits 33% sequence identity to region B of *S. aureus* Bap, was found to be essential for biofilm formation (10, 95). Therefore, a more in-depth domain analysis would be critical to determine which region of Bap is required for biofilm formation on abiotic surfaces.

Posttranslational regulation of Bap function by calcium has been reported (96, 97). Calcium inhibits Bap-mediated biofilm formation, likely through induction of a conformational change (96, 97). Bap contains four predicted Ca_2^+-binding EF-hand motifs (PS00018) that exhibit $\geq$80% similarity to the loop consensus of EF-hand motifs (96). Three are found within region B, while the fourth one is located in region D (Fig. 3). Addition of calcium in the millimolar range reduced wild-type *S. aureus* biofilm formation on polystyrene plates and cell aggregation in liquid cultures, while no inhibitory effect on biofilm formation was observed with natural Bap-deficient *S. aureus* isolates (96). In addition, calcium also decreases *S. aureus* biofilm thickness (97). Mutations in two of the predicted EF-hand motifs abolish the biofilm-inhibitory effect of calcium (96). Western blot analysis of surface protein samples revealed that wild-type Bap harvested from cells grown in the presence of calcium is more resistant to protease degradation than that from cells grown in the absence of calcium. This suggests that binding to calcium, while inhibiting Bap function, renders the protein more resistant to proteolytic degradation (96), possibly due to conformational changes.

Bap Orthologs in Other Gram-Positive and Gram-Negative Bacteria

Bap orthologs have also been identified in other *Staphylococcus* species, including *S. epidermidis*. An *S. epidermidis* mutant lacking Bap is incapable of forming wild-type biofilms, while complementation with *bap in trans* increases the biofilm-forming capacity (91). In region C of Bap proteins, amyloid-like peptide sequences were identified (90). Amyloid proteins, such as *Escherichia coli* curli (98) and *B. subtilis* TasA (99), form extracellular fibers that are involved in biofilm formation. The common denominator of the amyloid consensus peptide sequences TVTVT are found in *S. aureus* and *S. epidermidis* Bap (Table 1) (90) and Bap orthologs from other Gram-negative bacteria, including BapA from *S. enterica* Typhimurium (Table 1). While the STVTVTF peptides form stiff fibers several microns long *in vitro*, as observed with atomic force microscopy (90), formation of amyloid-like fibers by Bap proteins has not been reported. It is predicted that the amyloid sequence repeats contribute to the adhesive properties of Bap in promoting cell-cell adhesion rather than participating in amyloid-like fiber formation. Site-directed mutagenesis or TVTVT motif deletion studies are likely to provide more insights into the role of these repeats in surface adhesion and biofilm formation by Bap and Bap orthologs.

In *S. enterica* Enteritidis, BapA was identified through sequence homology searches using *S. aureus* Bap (9). Phenotypic analyses using *bapA* deletion and overexpression strains revealed that BapA is involved in pellicle and biofilm formation (9). The core region of BapA contains 29 imperfect tandem repeats of 86 to 106 amino acid residues and shares 29% identity with C repeats of *S. aureus* Bap (Fig. 3, Table 1). Despite the absence of calcium-binding motifs, BapA-dependent biofilm formation was found to be mediated by calcium in the millimolar range, similar to that observed for *S. aureus* Bap. The mechanism of calcium regulation of BapA-mediated biofilm formation is currently unknown. Nonetheless, modulation of protein conformation by calcium may be a conserved regulatory mechanism in Bap-mediated biofilm formation in both Gram-positive and Gram-negative bacteria. BapA also contains three VCBS repeats at the C-terminal region (9), which have been implicated in cell adhesion, further corroborating the involvement of BapA in biofilm formation. BapA is secreted and loosely associated with the bacterial cell surface (9). Secretion of BapA likely occurs through a type I secretion system, because BapA contains three noninteracting α-helices at the C-terminal region that resemble the C-terminal secretion signal recognized by the type I secretion ABC transporter (9). In addition, deletion of downstream coding regions, including a putative ABC-type exporter, abolished BapA secretion and resulted in a strain unable to form pellicles (9). The presence of Bap orthologs with similar domain organization in both Gram-positive and Gram-negative bacteria highlights the importance of Bap-family proteins in biofilm formation in a diverse group of bacteria.

B. SUBTILIS BIOFILM-SURFACE LAYER PROTEIN (BslA)

The amphiphilic protein BslA, formerly known as YuaB, has been reported to be a major factor contributing to surface repellency and colony corrugation of biofilms formed by *B. subtilis*, a Gram-positive soil-dwelling bacterium. BslA plays a synergistic role with other matrix components, specifically TasA and exopolysaccharides, in the late stages of biofilm formation and confers enhanced repellency in the resultant biofilm with the formation of a unique hydrophobic layer on the biofilm surface (12, 13, 100, 101). Such increased repellency in biofilm surfaces may enhance *B. subtilis* survival in a natural soil habitat by repelling environmental pollutants and toxic compounds such as heavy metals and antimicrobial agents.

Strains lacking BslA form colonies and pellicles with decreased corrugation, altered surface microstructures, and loss of surface repellency (12, 13, 100, 101). Increased expression of *bslA* from an IPTG-inducible promoter in a BslA-deficient strain results in increased colony and pellicle corrugation (12), and exogenous addition of purified BslA complements the altered pellicle and colony phenotypes of a *bslA* mutant strain (13). Overproduction of exopolysaccharide or the amyloid protein TasA, the other two major biofilm matrix components, cannot complement the loss of colony and pellicle corrugation phenotypes due to *bslA* mutation (12), indicating that BslA functions synergistically with TasA and the exopolysaccharide in mediating biofilm formation in *B. subtilis*. Coculturing a BslA-deficient strain and a strain that is unable to produce TasA and exopolysaccharide complements the colony-biofilm phenotype (12). These observations reinforce the notion that the biofilm matrix is composed of secreted matrix components and that these extracellular components contribute collectively to biofilm formation by serving as communal resources. While strains unable to produce BslA, TasA, or exopolysaccharide exhibit altered pellicle structures, only TasA- and exopolysaccharide-deficient strains are incapable of forming cell clusters and aggregates (13), suggesting that BslA likely plays a role in biofilm formation after TasA- and exopolysaccharide-dependent cell clusters are formed.

As a bacterial hydrophobin, BslA also functions synergistically with TasA and the exopolysaccharide in mediating surface repellency, via its ability to form rough surface microstructures (13). BslA exhibits self-polymerization (13), similar to other hydrophobins, and forms a hydrophobic layer at the bottom liquid-cell interface of pellicles (101) and on both the top and bottom surfaces of colony biofilms (13, 101). The BslA hydrophobic layer localized at the bottom of the floating pellicle biofilm may form a protein raft carrying the biofilm mass. Using a *bslA* promoter–green fluorescent protein construct, it was shown that spatial localization of BslA at the surfaces of biofilms is not due to localized transcription of *bslA* (101). BslA therefore migrates, via an unknown mechanism, to the biofilm surface. BslA exhibits unique dual functions in biofilm formation: it functions as a hydrophobin in increasing biofilm repellency and a matrix protein critical for biofilm architecture formation.

BslA contains a secretion signal sequence (13) and has been reported to be part of the secretome of *B. subtilis* grown in liquid cultures (102); however, the mechanism of secretion is unclear. Using Western blot analyses (12, 13) and transmission electron microscopy (12), BslA was shown to be secreted and retained within the biofilm matrix. Although secretion is required for BslA-mediated biofilm formation, it is currently unclear if BslA is associated with the exopolysaccharides or cell wall, since BslA was found in different cellular fractions in different studies and was detected in the matrix only in standing, but not in shaking, cultures (12, 13).

Crystal structures reveal that BslA exhibits an Ig-like fold and contains a hydrophobic cap (101). The presence of the Ig-like fold corroborated the hypothesis that BslA plays an adhesive role in biofilm formation. Site-directed mutagenesis of the residues located in the hydrophobic cap (L76K, L77K, L79K, L121K, L123K, L153K, and I155K) resulted in mutant strains exhibiting altered colony biofilm phenotypes, but only two of these (L77K and L79K) show a loss of surface repellency, indicating that the loss of hydrophobicity is not merely due to a decrease in colony corrugation and complexity (101). However, the loss of surface repellency due to site-directed mutagenesis is always accompanied by loss of complex colony architectures. As such, BslA-mediated surface repellency in *B. subtilis* biofilms is dependent on the hydrophobic cap, which in turn affects biofilm formation, but loss of BslA function in biofilm formation does not necessarily affect biofilm repellency. Currently, the mechanism by which BslA affects biofilm formation and surface repellency is unknown. In addition, it is not clear how individual BslA proteins interact during the self-assembling process, how BslA localizes to the surface of the biofilm, or what type of interaction occurs between the BslA hydrophilic layer and the exopolysaccharide.

OMVs

OMVs are small spherical structures produced by Gram-negative bacteria through pinching off or blebbing from the outer membrane. They range from 10 to 300 nm in diameter and generally contain cytoplasmic and periplasmic contents such as proteases, alkaline phosphatase, lipases, and toxins, as well as OMPs and LPS. OMVs are involved in many biological processes, including biofilm formation, pathogenesis, quorum signaling, nutrient acquisition, and horizontal gene transfer (103–105). OMVs are produced by several Gram-negative bacteria, including *P. aeruginosa* (20, 106), *Helicobacter pylori* (107), *V. cholerae* (40, 108), and *Vibrio fischeri* (109), as observed with transmission electron microscopy of thin-section biofilms or extracted OMVs. A growing number of studies have focused on investigating the possible role of OMVs in biofilm formation.

P. aeruginosa OMVs are a component of the biofilm matrix (106). Using transmission electron microscopy, OMVs were observed to be present within the biofilm matrix formed by *P. aeruginosa* PAO1 (106). OMVs extracted from biofilm cultures of *P. aeruginosa* PAO1 differ in quantity, quality, and protein identity compared to those isolated from planktonic cultures (20, 106), suggesting that OMVs may play different roles in these different physiological growth states (i.e., biofilm versus planktonic). It is notable that there are more OMVs isolated from biofilms than from planktonic cultures (106). However, a direct role of OMVs in *P. aeruginosa* biofilm formation has not been demonstrated. Nonetheless, interactions of OMVs with secreted matrix proteins, through binding to OMPs present in the OMVs, have been reported, thus implicating OMVs in biofilm formation. The OMP OprF was identified in both the biofilm matrix and the OMVs (20). Since OprF interacts with the lectin LecB (89), it is possible that OMVs may be localized in the biofilm matrix via interaction of OMV-associated OprF and the potentially Psl exopolysaccharide–binding LecB, thus forming an OMV-OprF-LecB-Psl interaction. It has been documented that eDNA, a biofilm matrix component of *P. aeruginosa* PAO1, associates with OMVs in *P. aeruginosa* PAO1 (58, 110, 111), which suggests that OMVs may be involved in the process of biofilm formation by interacting with different biofilm matrix components, including eDNA.

A direct role of OMVs in biofilm formation has been demonstrated in *H. pylori*. Addition of purified OMV-fraction induces biofilm formation in a dose-dependent manner (107). OMV production in *H. pylori*, similar to that observed in other bacteria, depends on culture conditions and the physiological state of the cells, as increased OMV production and biofilm formation were observed with growth medium containing fetal calf serum in a dose-dependent manner (107). Although the protein compositions of OMVs produced by *H. pylori* have not been determined, protein profiles, as determined by SDS-PAGE, indicate that OMVs isolated at different stages of biofilm formation exhibit differences (112), suggesting that these OMVs may have distinct cargo.

Production of OMVs has also been reported in *V. cholerae* (40, 108); however, direct involvement of OMVs in *V. cholerae* biofilm formation has not been demonstrated. Nonetheless, OMVs isolated from planktonic *V. cholerae* cultures grown in the presence of the antimicrobial peptide polymyxin B, which induces cell envelope stress, have been shown to bind Bap1 (40), a matrix protein that is involved in biofilm formation (6, 27). As mentioned earlier, Bap1 associates with OMVs by binding to OmpT, and since Bap1 also functions as an adhesive protein, it is likely that Bap1 binding to OMVs results in adherence and localization of OMVs onto surfaces and/or exopolysaccharides, implicating OMVs in biofilm formation.

There is a clear connection between OMV production and biofilm formation in *V. fischeri* (109). Transmission electron microscopy analysis of colony biofilms showed that OMVs are present within the extracellular matrix. A *V. fischeri* strain that exhibits increased biofilm formation, due to overproduction of the RscS sensor kinase, produces approximately 2.5-fold more OMVs than the parental strain. Furthermore, the increase in OMV production is dependent on the production of the symbiosis polysaccharide.

OMV involvement in biofilm formation may be multifaceted, because OMVs could interact with various biofilm matrix components, including proteins, exopolysaccharides, and eDNA. These interactions likely involve proteins and/or LPS found on the surfaces of OMVs, and therefore could reinforce structural integrity of the biofilm matrix. OMVs may also be deposited onto surfaces, thus conditioning the substratum for subsequent bacterial attachment. Further investigations will help to gain greater insights into the role of OMVs in biofilm formation.

BACTERIAL NUCLEOID-BINDING PROTEINS

Recent studies revealed that nucleoid-binding proteins, besides being involved in the maintenance of DNA supercoiling and compaction, also play a "moonlighting" role in biofilm formation (113–115). DNABII family proteins, members of the nucleoid-associated protein superfamily, can be grouped into two subtypes: HU (histone-like protein from *E. coli* strain U39), which is ubiquitous in eubacteria, and IHF (integration host factor), which is found only in bacteria within the α- and γ-proteobacteria genera (113, 116). Extracellular-localized nucleoid-binding proteins have been found in association with eDNA in the biofilm matrix from sputum samples of cystic fibrosis patients (114, 115). eDNA contributes to biofilm stability in many bacterial species, including *P. aeruginosa* PAO1 (58). It has also been reported that addition of anti-IHF serum, which exhibits avidity to both IHF and HU, to preformed biofilms results in dissolution of *E. coli*, *Haemophilus influenza*, and *Burkholderia cenocepacia* biofilms (113, 114). Similar dissolutions of preformed biofilm with the addition of anti-IHF serum were also observed with other bacterial species, including *S. aureus*, *S. epidermidis*, uropathogenic *E. coli*, *Neisseria gonorrhoeae*, and *P. aeruginosa* (113). A study carried out to identify proteins found in the biofilm matrix of *P. aeruginosa* PAO1 identified the nucleoid-binding protein HU in the biofilm matrix (20), further suggesting that nucleoid-binding protein may be involved in biofilm formation. While several other nucleoid-binding proteins, including H-NS (histone-like nucleoid-structuring protein) and DPS (DNA-binding protein from starved cells), are produced in different bacteria, antisera directed against these proteins did not result in alteration of biofilm structures when added to the biofilms (113), suggesting that H-NS and DPS are unlikely to be involved in biofilm formation.

Currently, it is unclear how nucleoid-binding proteins are localized outside of the cells. Since they do not contain a secretion signal sequence, it is likely that they are released into the biofilm matrix via cell lysis or through a similar unknown mechanism involved in eDNA release. Since OMVs may be involved in eDNA release into the biofilm matrix, it is possible that the extracellular localization of these nucleoid-binding proteins may be OMV-mediated. Little is known about the identity of nucleoid-binding proteins in different bacterial biofilms and the role they play in biofilm formation.

CONCLUDING REMARKS

Matrix proteins play diverse and important roles in biofilm formation by mediating initial surface attachment, cell cluster and aggregate formation, and establishment and stabilization of elaborate three-dimensional biofilm architectures. Matrix proteins exhibit unique characteristics and function synergistically with each other and with other matrix components, such as exopolysaccharides and eDNA, in biofilm formation. A common feature of these matrix proteins is that they are organized into modules or domains. These include FnIII, FG-GAP, VCBS, lectin, glycine-rich sugar-binding domains, and RGD sequence motifs that participate in cell-cell adhesion and/or binding to extracellular matrix, cell surface proteins, or carbohydrates. Identification of residues within the domains that are critical for ligand interactions as well as identification of the target ligands will be crucial in deciphering the mechanisms of these matrix proteins in biofilm formation. The regulation of matrix proteins is complex, involving multiple positive and negative transcriptional regulators, alternative sigma factors, and small regulatory RNAs (5, 100, 117–125). Matrix protein production is commonly coordinated with the production of other matrix components, such as exopolysaccharides, leading to optimal biofilm structure and function. Better characterization of the biofilm matrix proteome, structure/function relationships of matrix proteins, and regulatory circuits controlling biofilm matrix production will provide further mechanistic insights into biofilm formation and facilitate development of antibiofilm therapeutics.

ACKNOWLEDGMENTS

We thank Karen Visick and Holger Sondermann as well as members of the Yildiz laboratory for their valuable comments on the manuscript. This work is supported by NIH R01 AI055987.

Conflicts of interest: We disclose no conflicts.

CITATION

Fong JNC, Yildiz FH. 2015. Biofilm matrix proteins. Microbiol Spectrum 3(2):MB-0004-2014.

REFERENCES

1. **Costerton JW, Lewandowski Z, Caldwell DE, Korber DR, Lappin-Scott HM.** 1995. Microbial biofilms. *Annu Rev Microbiol* **49:** 711–745.
2. **O'Toole G, Kaplan HB, Kolter R.** 2000. Biofilm formation as microbial development. *Annu Rev Microbiol* **54:**49–79.
3. **Branda SS, Vik S, Friedman L, Kolter R.** 2005. Biofilms: the matrix revisited. *Trends Microbiol* **13:**20–26.
4. **Flemming HC, Wingender J.** 2010. The biofilm matrix. *Nat Rev Microbiol* **8:**623–633.
5. **Fong JC, Karplus K, Schoolnik GK, Yildiz FH.** 2006. Identification and characterization of RbmA, a novel protein required for the development of rugose colony morphology and biofilm structure in *Vibrio cholerae*. *J Bacteriol* **188:**1049–1059.
6. **Fong JC, Yildiz FH.** 2007. The *rbmBCDEF* gene cluster modulates development of rugose colony morphology and biofilm formation in *Vibrio cholerae*. *J Bacteriol* **189:**2319–2330.
7. **Diggle SP, Stacey RE, Dodd C, Camara M, Williams P, Winzer K.** 2006. The galactophilic lectin, LecA, contributes to biofilm development in *Pseudomonas aeruginosa*. *Environ Microbiol* **8:**1095–1104.
8. **Tielker D, Hacker S, Loris R, Strathmann M, Wingender J, Wilhelm S, Rosenau F, Jaeger KE.** 2005. *Pseudomonas aeruginosa* lectin LecB is located in the outer membrane and is involved in biofilm formation. *Microbiology* **151:**1313–1323.
9. **Latasa C, Roux A, Toledo-Arana A, Ghigo JM, Gamazo C, Penades JR, Lasa I.** 2005. BapA, a large secreted protein required for biofilm formation and host colonization of *Salmonella enterica* serovar Enteritidis. *Mol Microbiol* **58:**1322–1339.
10. **Toledo-Arana A, Valle J, Solano C, Arrizubieta MJ, Cucarella C, Lamata M, Amorena B, Leiva J, Penades JR, Lasa I.** 2001. The enterococcal surface protein, Esp, is involved in *Enterococcus faecalis* biofilm formation. *Appl Environ Microbiol* **67:**4538–4545.
11. **Cucarella C, Solano C, Valle J, Amorena B, Lasa I, Penades JR.** 2001. Bap, a *Staphylococcus aureus* surface protein involved in biofilm formation. *J Bacteriol* **183:**2888–2896.
12. **Ostrowski A, Mehert A, Prescott A, Kiley TB, Stanley-Wall NR.** 2011. YuaB functions synergistically with the exopolysaccharide and TasA amyloid fibers to allow biofilm formation by *Bacillus subtilis*. *J Bacteriol* **193:**4821–4831.

13. **Kobayashi K, Iwano M.** 2012. BslA(YuaB) forms a hydrophobic layer on the surface of *Bacillus subtilis* biofilms. *Mol Microbiol* **85:**51–66.
14. **Borlee BR, Goldman AD, Murakami K, Samudrala R, Wozniak DJ, Parsek MR.** 2010. *Pseudomonas aeruginosa* uses a cyclic-di-GMP-regulated adhesin to reinforce the biofilm extracellular matrix. *Mol Microbiol* **75:**827–842.
15. **Kaplan JB, Ragunath C, Ramasubbu N, Fine DH.** 2003. Detachment of *Actinobacillus actinomycetemcomitans* biofilm cells by an endogenous beta-hexosaminidase activity. *J Bacteriol* **185:**4693–4698.
16. **Marti M, Trotonda MP, Tormo-Mas MA, Vergara-Irigaray M, Cheung AL, Lasa I, Penades JR.** 2010. Extracellular proteases inhibit protein-dependent biofilm formation in *Staphylococcus aureus*. *Microbes Infect* **12:**55–64.
17. **Mann EE, Rice KC, Boles BR, Endres JL, Ranjit D, Chandramohan L, Tsang LH, Smeltzer MS, Horswill AR, Bayles KW.** 2009. Modulation of eDNA release and degradation affects *Staphylococcus aureus* biofilm maturation. *PLoS One* **4:** e5822. doi:10.1371/journal.pone.0005822.
18. **Nijland R, Hall MJ, Burgess JG.** 2010. Dispersal of biofilms by secreted, matrix degrading, bacterial DNase. *PLoS One* **5:**e15668. doi:10.1371/journal.pone.0015668.
19. **Absalon C, Van Dellen K, Watnick PI.** 2011. A communal bacterial adhesin anchors biofilm and bystander cells to surfaces. *PLoS Pathog* **7:** e1002210. doi:10.1371/journal.ppat.1002210.
20. **Toyofuku M, Roschitzki B, Riedel K, Eberl L.** 2012. Identification of proteins associated with the *Pseudomonas aeruginosa* biofilm extracellular matrix. *J Proteome Res* **11:**4906–4915.
21. **Curtis PD, Atwood J 3rd, Orlando R, Shimkets LJ.** 2007. Proteins associated with the *Myxococcus xanthus* extracellular matrix. *J Bacteriol* **189:**7634–7642.
22. **Jiao Y, D'Haeseleer P, Dill BD, Shah M, Verberkmoes NC, Hettich RL, Banfield JF, Thelen MP.** 2011. Identification of biofilm matrix-associated proteins from an acid mine drainage microbial community. *Appl Environ Microbiol* **77:**5230–5237.
23. **Yildiz FH, Schoolnik GK.** 1999. *Vibrio cholerae* O1 El Tor: identification of a gene cluster required for the rugose colony type, exopolysaccharide production, chlorine resistance, and biofilm formation. *Proc Natl Acad Sci USA* **96:**4028–4033.
24. **Fong JC, Syed KA, Klose KE, Yildiz FH.** 2010. Role of *Vibrio* polysaccharide (*vps*) genes in VPS production, biofilm formation and *Vibrio cholerae* pathogenesis. *Microbiology* **156:**2757–2769.
25. **Yildiz F, Fong J, Sadovskaya I, Grard T, Vinogradov E.** 2014. Structural Characterization of the extracellular polysaccharide from *Vibrio cholerae* O1 El-Tor. *PLoS One* **9:**e86751. doi:10.1371/journal.pone.0086751.
26. **Kirn TJ, Jude BA, Taylor RK.** 2005. A colonization factor links *Vibrio cholerae* environmental survival and human infection. *Nature* **438:**863–866.
27. **Berk V, Fong JC, Dempsey GT, Develioglu ON, Zhuang X, Liphardt J, Yildiz FH, Chu S.** 2012. Molecular architecture and assembly principles of *Vibrio cholerae* biofilms. *Science* **337:**236–239.
28. **Giglio KM, Fong JC, Yildiz FH, Sondermann H.** 2013. Structural basis for biofilm formation via the *Vibrio cholerae* matrix protein RbmA. *J Bacteriol* **195:**3277–3286.
29. **Bork P, Doolittle RF.** 1992. Proposed acquisition of an animal protein domain by bacteria. *Proc Natl Acad Sci USA* **89:**8990–8994.
30. **Maestre-Reyna M, Wu WJ, Wang AH.** 2013. Structural insights into RbmA, a biofilm scaffolding protein of *Vibrio cholerae*. *PLoS One* **8:** e82458. doi:10.1371/journal.pone.0082458.
31. **Oberhauser AF, Marszalek PE, Erickson HP, Fernandez JM.** 1998. The molecular elasticity of the extracellular matrix protein tenascin. *Nature* **393:**181–185.
32. **Craig D, Gao M, Schulten K, Vogel V.** 2004. Tuning the mechanical stability of fibronectin type III modules through sequence variations. *Structure* **12:**21–30.
33. **Moorthy S, Watnick PI.** 2005. Identification of novel stage-specific genetic requirements through whole genome transcription profiling of *Vibrio cholerae* biofilm development. *Mol Microbiol* **57:**1623–1635.
34. **Baneres JL, Roquet F, Martin A, Parello J.** 2000. A minimized human integrin alpha(5) beta(1) that retains ligand recognition. *J Biol Chem* **275:**5888–5903.
35. **Hynes RO.** 2002. Integrins: bidirectional, allosteric signaling machines. *Cell* **110:**673–687.
36. **Hynes RO.** 1992. Integrins: versatility, modulation, and signaling in cell adhesion. *Cell* **69:**11–25.
37. **Raval S, Gowda SB, Singh DD, Chandra NR.** 2004. A database analysis of jacalin-like lectins: sequence-structure-function relationships. *Glycobiology* **14:**1247–1263.
38. **Sankaranarayanan R, Sekar K, Banerjee R, Sharma V, Surolia A, Vijayan M.** 1996. A novel mode of carbohydrate recognition in

jacalin, a *Moraceae* plant lectin with a beta-prism fold. *Nat Struct Biol* **3**:596–603.

39. **Sikora AE, Zielke RA, Lawrence DA, Andrews PC, Sandkvist M.** 2011. Proteomic analysis of the *Vibrio cholerae* type II secretome reveals new proteins, including three related serine proteases. *J Biol Chem* **286**:16555–16566.
40. **Duperthuy M, Sjostrom AE, Sabharwal D, Damghani F, Uhlin BE, Wai SN.** 2013. Role of the *Vibrio cholerae* matrix protein Bap1 in cross-resistance to antimicrobial peptides. *PLoS Pathog* **9**:e1003620. doi:10.1371/journal.ppat.1003620.
41. **Huq A, Small EB, West PA, Huq MI, Rahman R, Colwell RR.** 1983. Ecological relationships between *Vibrio cholerae* and planktonic crustacean copepods. *Appl Environ Microbiol* **45**:275–283.
42. **Chowdhury MA, Huq A, Xu B, Madeira FJ, Colwell RR.** 1997. Effect of alum on free-living and copepod-associated *Vibrio cholerae* O1 and O139. *Appl Environ Microbiol* **63**:3323–3326.
43. **Hunt DE, Gevers D, Vahora NM, Polz MF.** 2008. Conservation of the chitin utilization pathway in the *Vibrionaceae*. *Appl Environ Microbiol* **74**:44–51.
44. **Meibom KL, Li XB, Nielsen AT, Wu CY, Roseman S, Schoolnik GK.** 2004. The *Vibrio cholerae* chitin utilization program. *Proc Natl Acad Sci USA* **101**:2524–2529.
45. **Stauder M, Huq A, Pezzati E, Grim CJ, Ramoino P, Pane L, Colwell RR, Pruzzo C, Vezzulli L.** 2012. Role of GbpA protein, an important virulence-related colonization factor, for *Vibrio cholerae*'s survival in the aquatic environment. *Environ Microbiol Rep* **4**:439–445.
46. **Stauder M, Vezzulli L, Pezzati E, Repetto B, Pruzzo C.** 2010. Temperature affects *Vibrio cholerae* O1 El Tor persistence in the aquatic environment via an enhanced expression of GbpA and MSHA adhesins. *Environ Microbiol Rep* **2**:140–144.
47. **Bhowmick R, Ghosal A, Das B, Koley H, Saha DR, Ganguly S, Nandy RK, Bhadra RK, Chatterjee NS.** 2008. Intestinal adherence of *Vibrio cholerae* involves a coordinated interaction between colonization factor GbpA and mucin. *Infect Immun* **76**:4968–4977.
48. **Wong E, Vaaje-Kolstad G, Ghosh A, Hurtado-Guerrero R, Konarev PV, Ibrahim AF, Svergun DI, Eijsink VG, Chatterjee NS, van Aalten DM.** 2012. The *Vibrio cholerae* colonization factor GbpA possesses a modular structure that governs binding to different host surfaces. *PLoS Pathog* **8**:e1002373. doi:10.1371/journal.ppat.1002373.
49. **Maruyama Y, Momma M, Mikami B, Hashimoto W, Murata K.** 2008. Crystal structure of a novel bacterial cell-surface flagellin binding to a polysaccharide. *Biochemistry* **47**:1393–1402.
50. **Hashimoto W, He J, Wada Y, Nankai H, Mikami B, Murata K.** 2005. Proteomics-based identification of outer-membrane proteins responsible for import of macromolecules in *Sphingomonas* sp. A1: alginate-binding flagellin on the cell surface. *Biochemistry* **44**:13783–13794.
51. **Callebaut I, Gilges D, Vigon I, Mornon JP.** 2000. HYR, an extracellular module involved in cellular adhesion and related to the immunoglobulin-like fold. *Protein Sci* **9**:1382–1390.
52. **Kelly G, Prasannan S, Daniell S, Fleming K, Frankel G, Dougan G, Connerton I, Matthews S.** 1999. Structure of the cell-adhesion fragment of intimin from enteropathogenic *Escherichia coli*. *Nat Struct Biol* **6**:313–318.
53. **Chiavelli DA, Marsh JW, Taylor RK.** 2001. The mannose-sensitive hemagglutinin of *Vibrio cholerae* promotes adherence to zooplankton. *Appl Environ Microbiol* **67**:3220–3225.
54. **Watnick PI, Fullner KJ, Kolter R.** 1999. A role for the mannose-sensitive hemagglutinin in biofilm formation by *Vibrio cholerae* El Tor. *J Bacteriol* **181**:3606–3609.
55. **Evans LR, Linker A.** 1973. Production and characterization of the slime polysaccharide of *Pseudomonas aeruginosa*. *J Bacteriol* **116**:915–924.
56. **Friedman L, Kolter R.** 2004. Two genetic loci produce distinct carbohydrate-rich structural components of the *Pseudomonas aeruginosa* biofilm matrix. *J Bacteriol* **186**:4457–4465.
57. **Friedman L, Kolter R.** 2004. Genes involved in matrix formation in *Pseudomonas aeruginosa* PA14 biofilms. *Mol Microbiol* **51**:675–690.
58. **Whitchurch CB, Tolker-Nielsen T, Ragas PC, Mattick JS.** 2002. Extracellular DNA required for bacterial biofilm formation. *Science* **295**:1487.
59. **Allesen-Holm M, Barken KB, Yang L, Klausen M, Webb JS, Kjelleberg S, Molin S, Givskov M, Tolker-Nielsen T.** 2006. A characterization of DNA release in *Pseudomonas aeruginosa* cultures and biofilms. *Mol Microbiol* **59**:1114–1128.
60. **Garber N, Guempel U, Belz A, Gilboa-Garber N, Doyle RJ.** 1992. On the specificity of the D-galactose-binding lectin (PA-I) of *Pseudomonas aeruginosa* and its strong binding to hydrophobic derivatives of D-galactose and thiogalactose. *Biochim Biophys Acta* **1116**:331–333.

61. **Schurks N, Wingender J, Flemming HC, Mayer C.** 2002. Monomer composition and sequence of alginates from *Pseudomonas aeruginosa*. *Int J Biol Macromol* **30:**105–111.
62. **Nivens DE, Ohman DE, Williams J, Franklin MJ.** 2001. Role of alginate and its O acetylation in formation of *Pseudomonas aeruginosa* microcolonies and biofilms. *J Bacteriol* **183:** 1047–1057.
63. **Stapper AP, Narasimhan G, Ohman DE, Barakat J, Hentzer M, Molin S, Kharazmi A, Hoiby N, Mathee K.** 2004. Alginate production affects *Pseudomonas aeruginosa* biofilm development and architecture, but is not essential for biofilm formation. *J Med Microbiol* **53:**679–690.
64. **Wozniak DJ, Wyckoff TJ, Starkey M, Keyser R, Azadi P, O'Toole GA, Parsek MR.** 2003. Alginate is not a significant component of the extracellular polysaccharide matrix of PA14 and PAO1 *Pseudomonas aeruginosa* biofilms. *Proc Natl Acad Sci USA* **100:**7907–7912.
65. **Hentzer M, Teitzel GM, Balzer GJ, Heydorn A, Molin S, Givskov M, Parsek MR.** 2001. Alginate overproduction affects *Pseudomonas aeruginosa* biofilm structure and function. *J Bacteriol* **183:**5395–5401.
66. **Ma L, Lu H, Sprinkle A, Parsek MR, Wozniak DJ.** 2007. *Pseudomonas aeruginosa* Psl is a galactose- and mannose-rich exopolysaccharide. *J Bacteriol* **189:**8353–8356.
67. **Colvin KM, Gordon VD, Murakami K, Borlee BR, Wozniak DJ, Wong GC, Parsek MR.** 2011. The Pel polysaccharide can serve a structural and protective role in the biofilm matrix of *Pseudomonas aeruginosa*. *PLoS Pathog* **7:** e1001264. doi:10.1371/journal.ppat.1001264.
68. **Jackson KD, Starkey M, Kremer S, Parsek MR, Wozniak DJ.** 2004. Identification of *psl*, a locus encoding a potential exopolysaccharide that is essential for *Pseudomonas aeruginosa* PAO1 biofilm formation. *J Bacteriol* **186:**4466–4475.
69. **Ma L, Jackson KD, Landry RM, Parsek MR, Wozniak DJ.** 2006. Analysis of *Pseudomonas aeruginosa* conditional Psl variants reveals roles for the Psl polysaccharide in adhesion and maintaining biofilm structure post-attachment. *J Bacteriol* **188:**8213–8221.
70. **Gilboa-Garber N.** 1972. Purification and properties of hemagglutinin from *Pseudomonas aeruginosa* and its reaction with human blood cells. *Biochim Biophys Acta* **273:**165–173.
71. **Gilboa-Garber N, Mizrahi L, Garber N.** 1977. Mannose-binding hemagglutinins in extracts of *Pseudomonas aeruginosa*. *Can J Biochem* **55:**975–981.
72. **Gilboa-Garber N.** 1982. *Pseudomonas aeruginosa* lectins. *Methods Enzymol* **83:**378–385.
73. **Adam EC, Mitchell BS, Schumacher DU, Grant G, Schumacher U.** 1997. *Pseudomonas aeruginosa* PA-II lectin stops human ciliary beating: therapeutic implications of fucose. *Am J Respir Crit Care Med* **155:**2102–2104.
74. **Avichezer D, Gilboa-Garber N.** 1991. Antitumoral effects of *Pseudomonas aeruginosa* lectins on Lewis lung carcinoma cells cultured *in vitro* without and with murine splenocytes. *Toxicon* **29:**1305–1313.
75. **Bajolet-Laudinat O, Girod-de Bentzmann S, Tournier JM, Madoulet C, Plotkowski MC, Chippaux C, Puchelle E.** 1994. Cytotoxicity of *Pseudomonas aeruginosa* internal lectin PA-I to respiratory epithelial cells in primary culture. *Infect Immun* **62:**4481–4487.
76. **Gabius HJ, Andre S, Kaltner H, Siebert HC.** 2002. The sugar code: functional lectinomics. *Biochim Biophys Acta* **1572:**165–177.
77. **Barondes SH, Gitt MA, Leffler H, Cooper DN.** 1988. Multiple soluble vertebrate galactoside-binding lectins. *Biochimie* **70:**1627–1632.
78. **Blanchard B, Imberty A, Varrot A.** 2013. Secondary sugar binding site identified for LecA lectin from *Pseudomonas aeruginosa*. *Proteins* **82:**1060–1065.
79. **Imberty A, Wimmerova M, Mitchell EP, Gilboa-Garber N.** 2004. Structures of the lectins from *Pseudomonas aeruginosa*: insight into the molecular basis for host glycan recognition. *Microbes Infect* **6:**221–228.
80. **Sudakevitz D, Kostlanova N, Blatman-Jan G, Mitchell EP, Lerrer B, Wimmerova M, Katcoff DJ, Imberty A, Gilboa-Garber N.** 2004. A new *Ralstonia solanacearum* high-affinity mannose-binding lectin RS-IIL structurally resembling the *Pseudomonas aeruginosa* fucose-specific lectin PA-IIL. *Mol Microbiol* **52:**691–700.
81. **Zinger-Yosovich K, Sudakevitz D, Imberty A, Garber NC, Gilboa-Garber N.** 2006. Production and properties of the native *Chromobacterium violaceum* fucose-binding lectin (CV-IIL) compared to homologous lectins of *Pseudomonas aeruginosa* (PA-IIL) and *Ralstonia solanacearum* (RS-IIL). *Microbiology* **152:**457–463.
82. **Lameignere E, Malinovska L, Slavikova M, Duchaud E, Mitchell EP, Varrot A, Sedo O, Imberty A, Wimmerova M.** 2008. Structural basis for mannose recognition by a lectin from opportunistic bacteria *Burkholderia cenocepacia*. *Biochem J* **411:**307–318.
83. **Pokorna M, Cioci G, Perret S, Rebuffet E, Kostlanova N, Adam J, Gilboa-Garber N,**

Mitchell EP, Imberty A, Wimmerova M. 2006. Unusual entropy-driven affinity of *Chromobacterium violaceum* lectin CV-IIL toward fucose and mannose. *Biochemistry* **45:**7501–7510.

84. **Mitchell E, Houles C, Sudakevitz D, Wimmerova M, Gautier C, Perez S, Wu AM, Gilboa-Garber N, Imberty A.** 2002. Structural basis for oligosaccharide-mediated adhesion of *Pseudomonas aeruginosa* in the lungs of cystic fibrosis patients. *Nat Struct Biol* **9:**918–921.
85. **Cioci G, Mitchell EP, Gautier C, Wimmerova M, Sudakevitz D, Perez S, Gilboa-Garber N, Imberty A.** 2003. Structural basis of calcium and galactose recognition by the lectin PA-IL of *Pseudomonas aeruginosa*. *FEBS Lett* **555:** 297–301.
86. **Loris R, Tielker D, Jaeger KE, Wyns L.** 2003. Structural basis of carbohydrate recognition by the lectin LecB from *Pseudomonas aeruginosa*. *J Mol Biol* **331:**861–870.
87. **Avichezer D, Katcoff DJ, Garber NC, Gilboa-Garber N.** 1992. Analysis of the amino acid sequence of the *Pseudomonas aeruginosa* galactophilic PA-I lectin. *J Biol Chem* **267:** 23023–23027.
88. **Gilboa-Garber N, Katcoff DJ, Garber NC.** 2000. Identification and characterization of *Pseudomonas aeruginosa* PA-IIL lectin gene and protein compared to PA-IL. *FEMS Immunol Med Microbiol* **29:**53–57.
89. **Funken H, Bartels KM, Wilhelm S, Brocker M, Bott M, Bains M, Hancock RE, Rosenau F, Jaeger KE.** 2012. Specific association of lectin LecB with the surface of *Pseudomonas aeruginosa*: role of outer membrane protein OprF. *PLoS One* **7:**e46857. doi:10.1371/journal.pone.0046857.
90. **Lembre P, Vendrely C, Martino PD.** 2014. Identification of an amyloidogenic peptide from the Bap protein of *Staphylococcus epidermidis*. *Protein Pept Lett* **21:**75–79.
91. **Tormo MA, Knecht E, Gotz F, Lasa I, Penades JR.** 2005. Bap-dependent biofilm formation by pathogenic species of *Staphylococcus*: evidence of horizontal gene transfer? *Microbiology* **151:**2465–2475.
92. **Cucarella C, Tormo MA, Ubeda C, Trotonda MP, Monzon M, Peris C, Amorena B, Lasa I, Penades JR.** 2004. Role of biofilm-associated protein Bap in the pathogenesis of bovine *Staphylococcus aureus*. *Infect Immun* **72:**2177–2185.
93. **Valle J, Latasa C, Gil C, Toledo-Arana A, Solano C, Penades JR, Lasa I.** 2012. Bap, a biofilm matrix protein of *Staphylococcus aureus* prevents cellular internalization through binding to GP96 host receptor. *PLoS Pathog* **8:**e1002843. doi:10.1371/journal.ppat.1002843.
94. **Cucarella C, Tormo MA, Knecht E, Amorena B, Lasa I, Foster TJ, Penades JR.** 2002. Expression of the biofilm-associated protein interferes with host protein receptors of *Staphylococcus aureus* and alters the infective process. *Infect Immun* **70:**3180–3186.
95. **Tendolkar PM, Baghdayan AS, Shankar N.** 2005. The N-terminal domain of enterococcal surface protein, Esp, is sufficient for Esp-mediated biofilm enhancement in *Enterococcus faecalis*. *J Bacteriol* **187:**6213–6222.
96. **Arrizubieta MJ, Toledo-Arana A, Amorena B, Penades JR, Lasa I.** 2004. Calcium inhibits bap-dependent multicellular behavior in *Staphylococcus aureus*. *J Bacteriol* **186:**7490–7498.
97. **Shukla SK, Rao TS.** 2013. Effect of calcium on *Staphylococcus aureus* biofilm architecture: a confocal laser scanning microscopic study. *Colloids Surf B Biointerfaces* **103:**448–454.
98. **Chapman MR, Robinson LS, Pinkner JS, Roth R, Heuser J, Hammar M, Normark S, Hultgren SJ.** 2002. Role of *Escherichia coli* curli operons in directing amyloid fiber formation. *Science* **295:**851–855.
99. **Romero D, Aguilar C, Losick R, Kolter R.** 2010. Amyloid fibers provide structural integrity to *Bacillus subtilis* biofilms. *Proc Natl Acad Sci USA* **107:**2230–2234.
100. **Verhamme DT, Murray EJ, Stanley-Wall NR.** 2009. DegU and Spo0A jointly control transcription of two loci required for complex colony development by *Bacillus subtilis*. *J Bacteriol* **191:**100–108.
101. **Hobley L, Ostrowski A, Rao FV, Bromley KM, Porter M, Prescott AR, MacPhee CE, van Aalten DM, Stanley-Wall NR.** 2013. BslA is a self-assembling bacterial hydrophobin that coats the *Bacillus subtilis* biofilm. *Proc Natl Acad Sci USA* **110:**13600–13605.
102. **Voigt B, Antelmann H, Albrecht D, Ehrenreich A, Maurer KH, Evers S, Gottschalk G, van Dijl JM, Schweder T, Hecker M.** 2009. Cell physiology and protein secretion of *Bacillus licheniformis* compared to *Bacillus subtilis*. *J Mol Microbiol Biotechnol* **16:**53–68.
103. **Beveridge TJ.** 1999. Structures of Gram-negative cell walls and their derived membrane vesicles. *J Bacteriol* **181:**4725–4733.
104. **Kulp A, Kuehn MJ.** 2010. Biological functions and biogenesis of secreted bacterial outer membrane vesicles. *Annu Rev Microbiol* **64:** 163–184.
105. **Bielig H, Dongre M, Zurek B, Wai SN, Kufer TA.** 2011. A role for quorum sensing in

regulating innate immune responses mediated by *Vibrio cholerae* outer membrane vesicles (OMVs). *Gut Microbes* **2:**274–279.

106. **Schooling SR, Beveridge TJ.** 2006. Membrane vesicles: an overlooked component of the matrices of biofilms. *J Bacteriol* **188:**5945–5957.
107. **Yonezawa H, Osaki T, Kurata S, Fukuda M, Kawakami H, Ochiai K, Hanawa T, Kamiya S.** 2009. Outer membrane vesicles of *Helicobacter pylori* TK1402 are involved in biofilm formation. *BMC Microbiol* **9:**197.
108. **Kondo K, Takade A, Amako K.** 1993. Release of the outer membrane vesicles from *Vibrio cholerae* and *Vibrio parahaemolyticus*. *Microbiol Immunol* **37:**149–152.
109. **Shibata S, Visick KL.** 2012. Sensor kinase RscS induces the production of antigenically distinct outer membrane vesicles that depend on the symbiosis polysaccharide locus in *Vibrio fischeri*. *J Bacteriol* **194:**185–194.
110. **Schooling SR, Hubley A, Beveridge TJ.** 2009. Interactions of DNA with biofilm-derived membrane vesicles. *J Bacteriol* **191:**4097–4102.
111. **Renelli M, Matias V, Lo RY, Beveridge TJ.** 2004. DNA-containing membrane vesicles of *Pseudomonas aeruginosa* PAO1 and their genetic transformation potential. *Microbiology* **150:**2161–2169.
112. **Yonezawa H, Osaki T, Woo T, Kurata S, Zaman C, Hojo F, Hanawa T, Kato S, Kamiya S.** 2011. Analysis of outer membrane vesicle protein involved in biofilm formation of *Helicobacter pylori*. *Anaerobe* **17:**388–390.
113. **Goodman SD, Obergfell KP, Jurcisek JA, Novotny LA, Downey JS, Ayala EA, Tjokro N, Li B, Justice SS, Bakaletz LO.** 2011. Biofilms can be dispersed by focusing the immune system on a common family of bacterial nucleoid-associated proteins. *Mucosal Immunol* **4:**625–637.
114. **Novotny LA, Amer AO, Brockson ME, Goodman SD, Bakaletz LO.** 2013. Structural stability of *Burkholderia cenocepacia* biofilms is reliant on eDNA structure and presence of a bacterial nucleic acid binding protein. *PLoS One* **8:**e67629. doi:10.1371/journal.pone.0067629
115. **Gustave JE, Jurcisek JA, McCoy KS, Goodman SD, Bakaletz LO.** 2013. Targeting bacterial integration host factor to disrupt biofilms associated with cystic fibrosis. *J Cyst Fibros* **12:**384–389.
116. **Swinger KK, Rice PA.** 2004. IHF and HU: flexible architects of bent DNA. *Curr Opin Struct Biol* **14:**28–35.
117. **Yildiz FH, Liu XS, Heydorn A, Schoolnik GK.** 2004. Molecular analysis of rugosity in a *Vibrio cholerae* O1 El Tor phase variant. *Mol Microbiol* **53:**497–515.
118. **Beyhan S, Bilecen K, Salama SR, Casper-Lindley C, Yildiz FH.** 2007. Regulation of rugosity and biofilm formation in *Vibrio cholerae*: comparison of VpsT and VpsR regulons and epistasis analysis of *vpsT*, *vpsR*, and *hapR*. *J Bacteriol* **189:**388–402.
119. **Fong JC, Yildiz FH.** 2008. Interplay between cyclic AMP-cyclic AMP receptor protein and cyclic di-GMP signaling in *Vibrio cholerae* biofilm formation. *J Bacteriol* **190:**6646–6659.
120. **Kovacs AT, Kuipers OP.** 2011. Rok regulates *yuaB* expression during architecturally complex colony development of *Bacillus subtilis* 168. *J Bacteriol* **193:**998–1002.
121. **Goodman AL, Kulasekara B, Rietsch A, Boyd D, Smith RS, Lory S.** 2004. A signaling network reciprocally regulates genes associated with acute infection and chronic persistence in *Pseudomonas aeruginosa*. *Dev Cell* **7:**745–754.
122. **Jones CJ, Newsom D, Kelly B, Irie Y, Jennings LK, Xu B, Limoli DH, Harrison JJ, Parsek MR, White P, Wozniak DJ.** 2014. ChIP-Seq and RNA-Seq reveal an AmrZ-mediated mechanism for cyclic di-GMP synthesis and biofilm development by *Pseudomonas aeruginosa*. *PLoS Pathog* **10:** e1003984. doi:10.1371/journal.ppat.1003984.
123. **Irie Y, Starkey M, Edwards AN, Wozniak DJ, Romeo T, Parsek MR.** 2010. *Pseudomonas aeruginosa* biofilm matrix polysaccharide Psl is regulated transcriptionally by RpoS and post-transcriptionally by RsmA. *Mol Microbiol* **78:**158–172.
124. **Burrowes E, Baysse C, Adams C, O'Gara F.** 2006. Influence of the regulatory protein RsmA on cellular functions in *Pseudomonas aeruginosa* PAO1, as revealed by transcriptome analysis. *Microbiology* **152:**405–418.
125. **Moscoso JA, Mikkelsen H, Heeb S, Williams P, Filloux A.** 2011. The *Pseudomonas aeruginosa* sensor RetS switches type III and type VI secretion via c-di-GMP signalling. *Environ Microbiol* **13:**3128–3138.

Bacterial Extracellular Polysaccharides in Biofilm Formation and Function

11

DOMINIQUE H. LIMOLI,[1] CHRISTOPHER J. JONES,[2] and DANIEL J. WOZNIAK[1]

INTRODUCTION

The ability to construct and maintain a structured multicellular bacterial community depends critically on the production of extracellular matrix components (1, 2). While the biofilm matrix may be composed of various molecules, the focus of this chapter is on the extracellular polysaccharides (PSs) important for biofilm formation.

The PSs synthesized by microbial cells vary greatly in their composition and hence in their chemical and physical properties. Many are polyanionic, but others are neutral or polycationic (see Table 1 and Fig. 1) (2). In most natural and experimental environments, PSs are found in ordered compositions, with long, thin molecular chains, ranging in mass from 0.5 to 2.0 x 10^6 Da. PSs can be elaborated in a multitude of ways, influenced by the environment and association with other molecules such as lectins, proteins, lipids, and bacterial and host extracellular DNA (eDNA). Moreover, many biofilms are composed of multiple bacterial or even fungal species, whereby a range of PSs may interact to generate further permutations of unique architectures (3).

[1]Department of Microbial Infection and Immunity, Ohio State University, Columbus, OH 43210; [2]Department of Microbiology and Environmental Toxicology, University of California, Santa Cruz, Santa Cruz, CA 95064.

Microbial Biofilms 2nd Edition
Edited by Mahmoud Ghannoum, Matthew Parsek, Marvin Whiteley, and Pranab K. Mukherjee

doi:10.1128/microbiolspec.MB-0011-2014

TABLE 1 Summary of the cellular location, chemical composition, and functions of bacterial polysaccharides important for biofilm formation

	Localization	Charge	Functions		
			Aggregative	Protective	Architectural
Pel	Secreted	NA	X	X	X
Psl	Secreted/cell associated	Neutral	X	X	X
PIA	Secreted	Polycationic	X		X
Cellulose	Secreted	Neutral	X	X	
Alginate	Cell associated	Polyanionic		X	X
CPS	Covalently attached	Polyanionic		X	
Levan	Cell associated	Neutral	X	X	
Colanic acid	Cell associated	Polyanionic			X
VPS	Secreted	NA	X	X	X
***Bacillus* EPS**	Secreted	Neutral			X

The diversity in PS structure also provides a range of functional roles for PSs in microbial biofilms. For many bacteria, structural and physical consequences of PS expression confer unique colony morphology phenotypes (Fig. 2). The PS in the biofilm matrix dictates a framework for the biofilm landscape. Inhabitants of the biofilm need to be protected from the environment (host cells, antimicrobials, desiccation, temperature, competing microbes, etc.) while maintaining access to nutrients and the ability to respond to changes in the environment. Bacteria generate multiple PSs to cope with these needs in a variety of different ways. PSs can help bacteria adhere to a multitude of different surfaces and host and bacterial cells, provide protection from the onslaught of antimicrobials in the environment, provide reservoirs for nutrient acquisition, and aid in the creation of distinct architectures, which further potentiate an environment suitable for microbes to persist. In this chapter, we will discuss PSs that are known to be important for microbial biofilm formation. For strictly organizational purposes, PSs are divided here into three functional categories to highlight their importance and diversity in biofilm biology. While these PSs are subjectively categorized into aggregative, protective, and architectural, these divisions are by no means exclusive. Several PSs have roles in each of these categories (see Table 1), which will also be discussed below.

AGGREGATIVE POLYSACCHARIDES

The formation of biofilms occurs in multiple stages: initial attachment, microcolony and macrocolony formation, and detachment or disassembly (4–6). Aggregative PSs play essential roles in each of these steps: aiding in adhesion to surfaces, formation of complex structures by promoting microbial interactions, and relief of these interactions promoting dissolution of the biofilm. Bacteria can elaborate multiple PSs, which are important in different strains and varying environmental conditions, including surface substrate, nutrition, and flow rate (7). The redundancy of aggregative PSs produced by many bacteria highlights the essentiality of bacteria remaining associated with the biofilm community. Moreover, the ability to modify PS production provides compensatory mechanisms to adapt to changing environments. The PSs described in this section highlight the importance of aggregation in the biofilm community lifestyle and demonstrate the range of functions of these PSs.

Polysaccharide Intercellular Adhesion

Significance, structure, and regulation

The PS intercellular adhesion (PIA) is the primary PS involved in biofilm formation of *Staphylococcus aureus* and *Staphylococcus epidermidis*, which contribute significantly to endocarditis, osteomyelitis, and infections

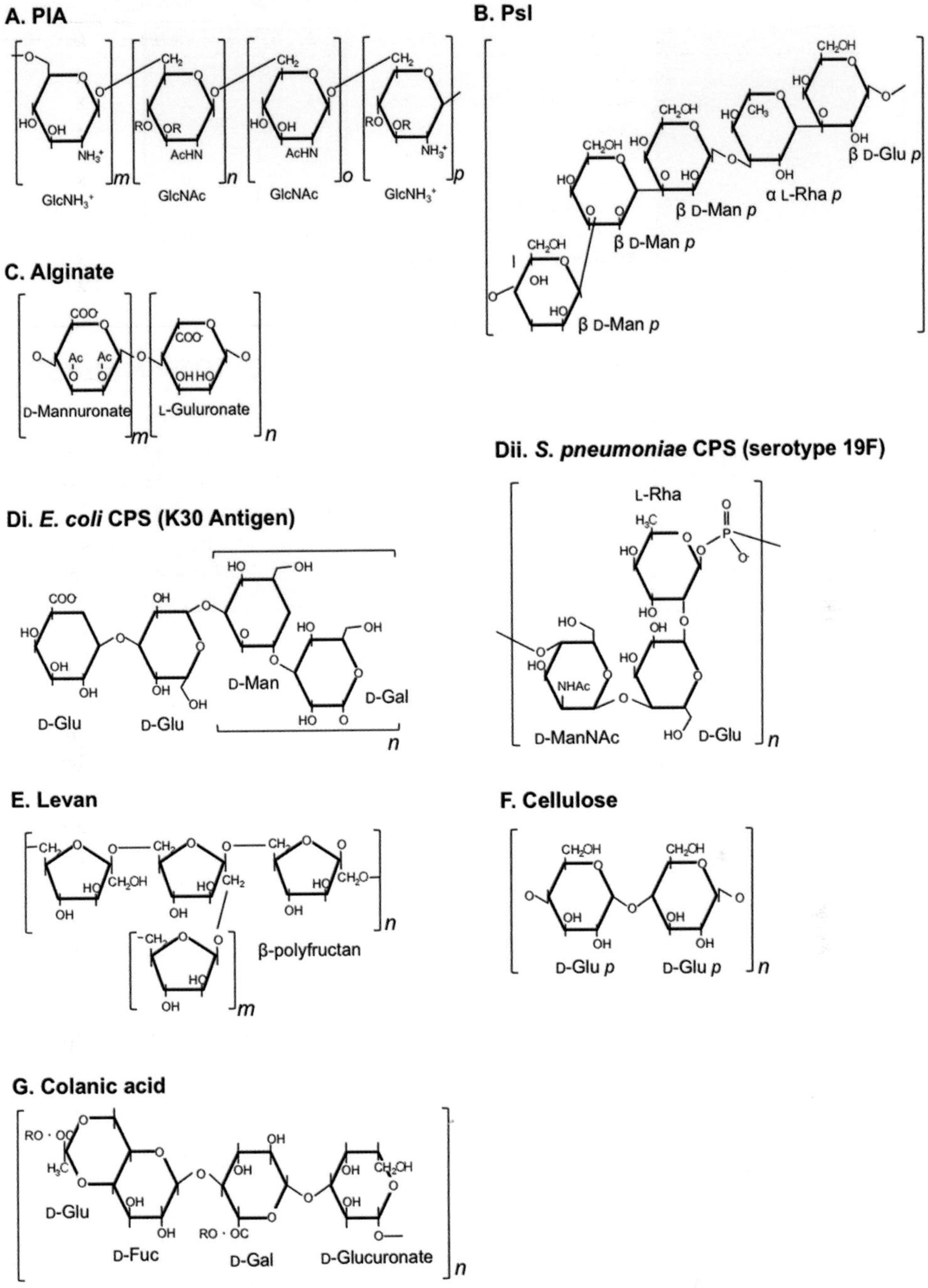

FIGURE 1 Adapted representative chemical structures of polysaccharides which participate in biofilm formation including (A) polysaccharide intercellular adhesin (PIA), (B) Psl, (C) alginate, capsular polysaccharide (CPS) from (Di) *E. coli* and (Dii) *S. pneumoniae*, (E) levan, (F) cellulose, and (G) colanic acid. Brackets depict repeating units. doi:10.1128/microbiolspec.MB-0011-2014.f1

A. *S. aureus* PIA

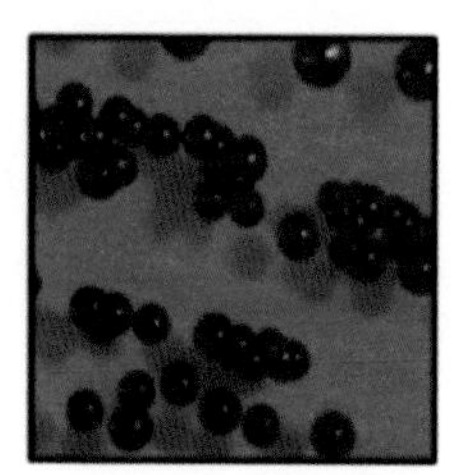

B. *P. aeruginosa* Pel

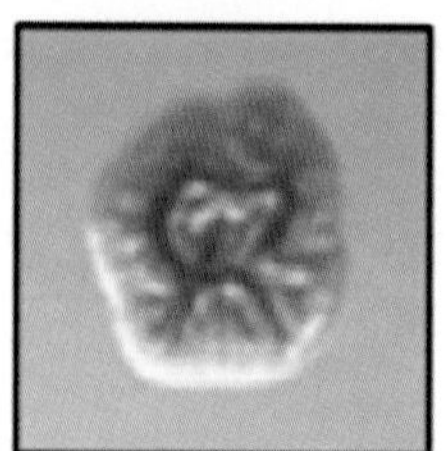

C. *P. aeruginosa* Psl

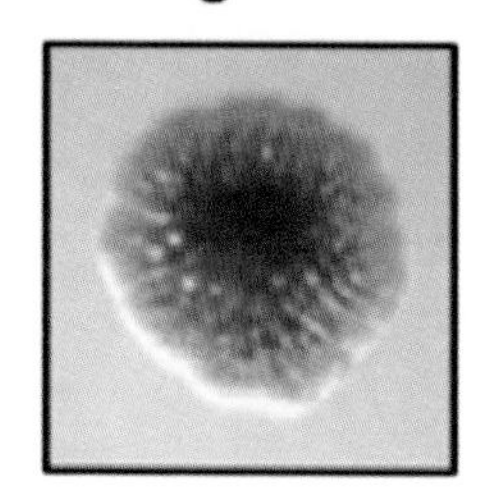

D. *P. aeruginosa* Alginate

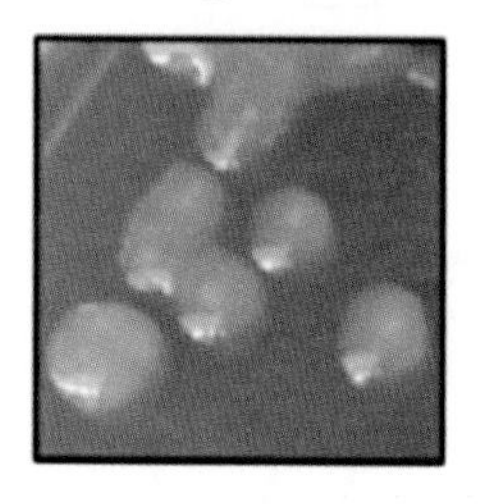

E. *E. coli* Colanic acid

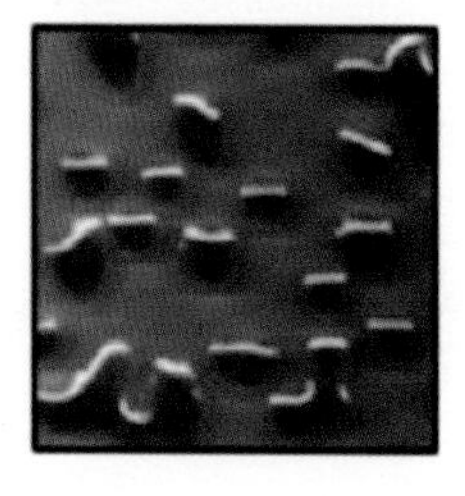

F. Vibrio VPS

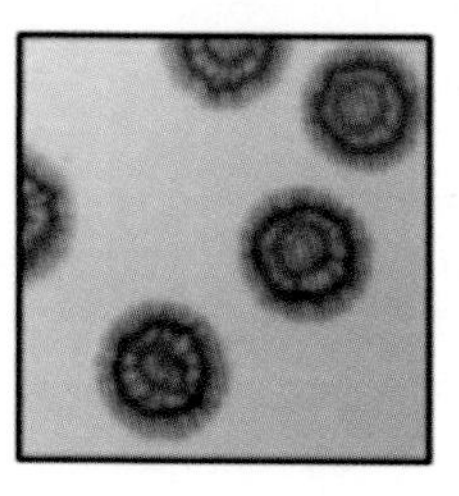

G. *B. subtilis* EPS

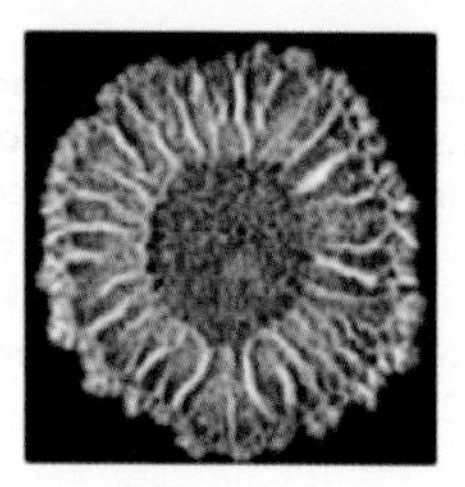

H. *E. coli* Cellulose

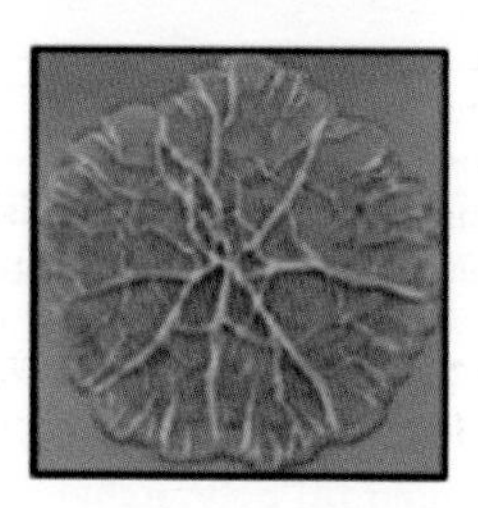

associated with indwelling medical devices (4, 8). PIA was originally identified in *S. epidermidis* and named as such due its role in mediating cell-cell interactions during biofilm formation (9–11). Similar PSs were subsequently identified in *S. aureus* and referred to as polysaccharide/adhesion (PS/A or PSA) (12), poly-N-acetylglucosamine (13) or *S. aureus* exopolysaccharide (14). It has since been verified that despite possible variations in the degree of *N*-acetylation, *O*-succinylation, and molecular weight, each PS represents the same homopolymer composed of β-1-6 linked 2-amino-2-deoxy-D-glucopyranosyl residues (Fig. 1) (11, 13, 15). PIA is synthesized by enzymes encoded by the *icaADBC* locus, which is negatively regulated by *icaR* located upstream of *icaA* (16–18). Importantly, *icaADBC* orthologous genes have also been identified in other human pathogens such as *Escherichia coli* (19), *Yersinia pestis* (20), *Aggregatibacter actinomycetemcomitans, Actinobacillus pleuropneumoniae* (21), *Bordetella* spp. (22), *Klebsiella pneumoniae, Enterobacter cloacae, Stenotrophomonas maltophilia,* the *Burkholderia cepacia* complex (BCC) (23), and *Acinetobacter baumannii* (24). The *ica* operon is tightly regulated *in vitro* and is induced by a number of environmental conditions. Although there is strain-to-strain variation, PIA can be induced in the presence of high NaCl, glucose, temperature and ethanol, anaerobiosis, and subinhibitory concentrations of some antibiotics (25–27). In *S. epidermidis* the quorum sensing LuxS system negatively regulates PIA expression, where a *luxS* mutant demonstrates increased biofilm formation and enhanced virulence in a rat model of biofilm-associated infection (28). σ^B and RsbU also positively regulate PIA. σ^B activates the *ica* operon by repressing transcription of *icaR* though an unknown intermediate (29). In *S. aureus* the Spx protein, which directly interacts with RNA polymerase to regulate transcription, negatively regulates the *ica* operon and biofilm formation (30, 31), while the global regulator SarA positively regulates PIA production and biofilms (32, 33). The importance of PIA during staphylococci infections has been extensively studied, and several roles have been attributed to PIA, such as promoting cellular aggregation and biofilm formation, increased virulence, and protection from host innate immune responses (34–36; reviewed in references 25, 37–40).

Role in biofilm biology

In *S. aureus* and *S.epidermidis*, initial attachment is mediated primarily by cell surface proteins that bind to mammalian extracellular matrix/plasma proteins (41). The micro- and macro- colony formation stage requires intercellular bacterial aggregation, and in staphylococci, this is mediated primarily by PIA. This has been illustrated by mutants in the *ica* operon retaining the ability to adhere

FIGURE 2 Colony phenotypes conferred upon expression or overexpression of PS by representative bacteria. (A) PS intercellular adhesion producing *Staphylococcus aureus*. Reprinted from *World Journal of Microbiology and Biotechnology* (214) with permission from the publisher. (B) Pel producing *Pseudomonas aeruginosa* (Δ*wspF*Δ*psl*). Reprinted from *Molecular Microbiology* (215) with permission from the publisher. (C) Psl producing *P. aeruginosa* (Δ*wspF*Δ*pel*). Reprinted from *Molecular Microbiology* (215) with permission from the publisher. (D) Alginate overproducing *P. aeruginosa* (*mucA22*). Not previously published. Credit: Daniel Wozniak. (E) Colanic acid producing *Escherichia coli*. Reprinted from *PLoS One* (216) with permission from the publisher. (F) VPS producing rugose variant of *Vibrio cholerae*. Reprinted from *The Proceedings of the National Academy of Sciences of the USA* (66) with permission from the publisher. (G) EPS producing *Bacillus subtilis*. Reprinted from *The Proceedings of the National Academy of Sciences of the USA* (217) with permission from the publisher. (H) Cellulose producing *E. coli* (*csgD::cm*). Reprinted from *The Journal of Medical Microbiology* (218) with permission from the publisher. doi:10.1128/microbiolspec.MB-0011-2014.f2

to surfaces while being incapable of forming multilayered biofilms (16). In the detachment stage of the biofilm lifestyle, matrix components are often degraded; however, staphylococci do not seem to possess PIA hydrolytic enzymes (37). Instead, PIA is dispersed by elevating expression of detergent-like peptides, which disrupt noncovalent interactions between PIA and the bacterial cell surface (38). The importance of PIA for staphylococci biofilm formation has been demonstrated *in vivo* in several animal models where the *ica* operon is upregulated and required for biofilm formation during infection (42–46). However, staphylococcal strains demonstrate a wide range of biofilm phenotypes, and many strains maintain the ability to form biofilms in the absence of PIA (32, 47, 48), where several PIA-independent protein factors and/or eDNA seem to be more important (8, 49, 50). At least some strains are also able to switch between PIA-dependent and -independent phenotypes, which may aid in adaptation to changing environments (51). Collectively, these data demonstrate that biofilm formation and PIA production are integral to the staphylococcal lifestyle and its ability to cause a range of diseases.

Pseudomonas aeruginosa Pel

Significance, structure, and regulation

Pel is an aggregative PS produced by *P. aeruginosa*. It derives its name from the thick pellicle observed in strains overexpressing the *pel* operon. This operon encodes seven enzymes with homology to other PS synthesis proteins; however, several genes predicted to be essential for Pel production are missing from this operon (52–54). This indicates that there may be reliance on several other PS synthesis enzymes encoded elsewhere on the chromosome. The structure and composition of Pel remain unknown, but studies are underway to identify the sugars and linkages present in this PS. The common strain PA14 relies exclusively on Pel production for aggregation, because it is missing the *pslABCD* genes (see below) (52).

Pel production is increased by elevated amounts of the intracellular second messenger cyclic diguanylate (c-di-GMP) (55). Recent studies have described a role for the flagellum regulator FleQ in the regulation of the *pel* operon, where the cellular pool of c-di-GMP is sensed by FleQ, which causes FleQ to change from a repressor of the *pel* promoter to an activator (56, 57). In addition, c-di-GMP functions post-translationally in Pel synthesis by modulating activity of PelD (55).

Role in biofilm biology

The contribution of Pel to biofilm structure is most evident when observing biofilms that form at the air-liquid interface, called pellicles. The production of Pel is strongly correlated with increased pellicle formation, adherence to culture tubes, and formation of aggregates in broth culture (52, 58, 59). The biofilm properties of Pel and Psl (below) are closely linked, because there is evidence of functional overlap in these two PSs (59). It appears that the structural requirements of these two PSs for biofilm formation are strain specific, because there are examples of strains which are incapable of producing one or the other yet still can efficiently form biofilms (59).

Pseudomonas aeruginosa Psl

Significance, structure, and regulation

The polysaccharide synthesis locus produces another important *P. aeruginosa* PS, Psl (60). The *psl* operon consists of 15 genes (*pslA-O*), which are cotranscribed (61). Mutagenesis studies revealed that 11 of these genes (*pslACDEFGHIJKL*) are essential for Psl production and surface attachment (62). There are two forms of Psl, a high molecular weight cell-associated form and a low molecular weight form that appears to be released from cells (62). Released Psl consists of D-mannose, D-glucose, and L-rhamnose in a 3:1:1 ratio, respectively (Fig. 1) (62, 63). The structure of cell-associated Psl is unknown but is believed to be a polymer of mannose, glucose, and rhamnose and possibly galactose (63). Psl is produced primarily in nonmucoid strains, because expression of the *psl* operon is repressed in mucoid strains (64).

The effect of Psl is most easily studied in rugose small colony variant strains such as the Δ*wspF* mutant, where Psl is overproduced (65). The overproduction of Psl leads to a rough, wrinkled colony morphology and hyperbiofilm phenotype, which is consistent with overexpression of adhesive PSs in other bacteria (66, 67). As with Pel, high expression of Psl in this strain is due to elevated levels of the second messenger cyclic di-GMP (65, 68, 69). There are several pathways that lead to regulation of Psl production by altering the c-di-GMP level of the cell. The first of these regulatory pathways to be described was the regulation of the *psl* operon by the master flagellum regulator, FleQ (56, 57). In this instance, c-di-GMP relieves the repressive effect of FleQ at the *psl* operon promoter, leading to expression of the *psl* operon (56, 65).

Another important observation is that patients who survive systemic infections with *P. aeruginosa* often produce neutralizing antibodies against Psl (70). Passive immunization with these antibodies is protective against *P. aeruginosa* infection in several murine models (70). Psl also increases adherence to epithelial cell layers and evasion of phagocytosis by neutrophils, emphasizing that Psl plays important roles during the initiation and persistence of chronic infections (71, 72).

Role in biofilm biology

PAO1 Δ*psl* mutants form thin, diffuse biofilms lacking structure compared to the wild type PAO1, while the *psl* overexpressing strains form thicker biofilms with significantly more biomass and microcolony height (73). While this suggests that Psl is essential for biofilm formation in *P. aeruginosa* strain PAO1, it is important to note that this requirement of Psl for biofilm formation varies from strain to strain. For example, the common strain PA14 does not synthesize Psl yet retains the ability to form biofilms (74). This discovery led to an investigation of the roles of the adhesive PSs, Psl, and Pel in various *P. aeruginosa* isolates. It was concluded that there are strain-specific requirements for PS production in biofilms. In some strains, Psl is exclusively required for biofilm formation, while in other strains, Psl and Pel are functionally redundant (59). Importantly, one of these two adhesive PSs is required in all strains tested to form a mature biofilm *in vitro* (59). Recent work indicates that Psl can also function as a signaling molecule that stimulates biofilm formation (75). Psl is also protective against many classes of antibiotics, serving yet another role in the biofilm matrix (76). Psl-producing cells were more resistant to a wide range of antibiotics than cells unable to produce Psl. This protective effect was transferable to cells unable to produce Psl when grown in coculture, including other species of bacteria such as *E. coli* and *S. aureus* (76). Collectively, these observations indicate that Psl is important for initial attachment, resistance, and biofilm structure.

PROTECTIVE POLYSACCHARIDES

Microbial protection from the onslaught of host and environmental factors is one of the most quintessential and frequently described attributes of the biofilm mode of growth. In fact, a current criteria used to define infectious biofilms is increased recalcitrance to antimicrobials and the host immune response (77, 78). While several factors including metabolic heterogeneity, altered chemical microenvironments, and persister cell populations contribute to biofilm antimicrobial resistance, an important role for PSs is clearly evident (79–81). It had been initially proposed that PSs in the matrix pose a diffusion barrier, which protects biofilm-grown cells from antibiotic penetration (82). However, studies have demonstrated that several antibiotics can readily penetrate the biofilm (83, 84). While diffusion may not be prevented, it may be delayed enough to induce expression of genes that mediate tolerance (85) or be neutralized by enzymes within the matrix before reaching the bacterial cells (86, 87).

Another prevalent and important role for PSs is protection from opsonic and nonopsonic phagocytosis. Several studies demonstrate that antibodies or inflammatory cells cannot efficiently penetrate biofilms embedded in a PS matrix (8, 49, 88–91). Protection from the immune response results in poor detection, promoting maintenance of stable chronic infections and decreased virulence. Importantly, PS-producing bacteria may confer resistance to nonproducing bacteria within the biofilm (76). PSs are also important for providing a hydrated biofilm environment, which can provide protection from desiccation and promote biofilm fluidity (2, 92). The PSs described in this section provide evidence for a diversity of protective functions toward biofilm bacteria.

Alginate

Significance, structure, and regulation

The first *P. aeruginosa* PS described was alginate due to its prevalence in cystic fibrosis (CF) pulmonary infections, whereby up to 90% of *P. aeruginosa* clinical isolates overproduce alginate (referred to as mucoid) (60, 93, 94). CF patients are initially colonized by nonmucoid *P. aeruginosa* (Psl and/or Pel producing) but convert to the mucoid phenotype during the course of infection. Mucoid conversion confers a selective advantage to *P. aeruginosa* in the CF lung (discussed further below) (95). Mucoid conversion is directly correlated with an increase in morbidity and mortality of CF patients and remains a significant clinical challenge for eradicating *P. aeruginosa* infections (96). While the majority of alginate research has focused on *P. aeruginosa*, alginate production has also been reported in other *Pseudomonads* and *Azotobacter vinelandii* (54, 97). Alginate is a random linear polymer of variably acetylated 1,4-linked β-D-mannuronic acid and its C5 epimer α-L-guluronic acid, originally identified in brown seaweeds (Fig. 1) (98, 99). This acetylated polymer structure yields a highly hygroscopic PS promoting biofilm fluidity and resistance to desiccation (2, 92).

Alginate synthesis is tightly regulated due to the high metabolic cost of production (alginate regulation has been extensively reviewed in references 95, 100–102). In brief, expression of the alginate biosynthetic operon (*algD-A*) is dependent on the alternative sigma factor AlgT (also called AlgU/σ^E/σ^{22}/RpoE). The activity of AlgT is antagonized by the anti–sigma factor MucA, which sequesters AlgT to the cell membrane, preventing interaction with AlgT-dependent promoters (100). Alginate is constitutively overproduced (as seen in mucoid CF isolates) upon acquisition of mutation(s) in the *mucA* gene, which results in a truncated, inactive protein. The mutant MucA is unable to inactivate AlgT, allowing expression of the AlgT regulon, including the alginate operon (95, 100, 103). Additional regulators of alginate production include AlgR, IHF, AlgB, and AmrZ (*algD* transcriptional regulators) and MucB to E and P, AlgW, KinB, and ClpX proteases (regulation of MucA stability) (100, 101, 104–106).

Role in biofilm biology

Early microscopy evidence demonstrated that mucoid *P. aeruginosa* forms microcolonies embedded in an extracellular matrix in the alveoli of CF patients (88, 107). Alginate was initially thought to be essential for *P. aeruginosa* adherence and biofilm formation, (108–111). However, this paradigm has been challenged by multiple groups who, by comparing isogenic mucoid and nonmucoid strains, demonstrated that alginate is not required for biofilm formation (112–15). These studies, corroborated by more recent work, demonstrate that the Psl and Pel PSs provide attachment and structural support for nonmucoid and mucoid biofilms (116, 117), while alginate contributes to generation of a unique architecture (112, 114, 115). Alginate acetylation further influences mucoid biofilm architecture by promoting cell-cell adhesion (112, 114, 115, 118).

The biofilm architecture constructed by mucoid *P. aeruginosa*, combined with the intrinsic properties of alginate, provide *P. aeruginosa* significant protection from antimicrobials in the microenvironment. Mucoid *P. aeruginosa* biofilms are more resistant to antibiotic treatment *in vitro* (115, 119, 120), and even the most aggressive antibiotic treatments are unable to eradicate mucoid *P. aeruginosa* infections in CF patients (96). Alginate also provides protection from the innate immune response by inhibiting both opsonic and nonopsonic phagocytosis, further influenced by the levels of alginate acetylation (121–125). Protection is additionally afforded by the ability of alginate to scavenge reactive oxygen intermediates (126, 127) and inhibit killing by cationic antimicrobial peptides (128–130). Although antibodies to alginate are found in the sera of chronically infected CF patients, these antibodies fail to mediate opsonic killing of *P. aeruginosa in vitro* (131). Additionally, the anionic nature of alginate promotes cation chelation (Ca^{2+} preferentially) (132). Calcium chelation by alginate biofilms induces type III secretion, whose effectors may provide further protection from host immune responses (133). Contemporary studies further emphasize the protective nature of *P. aeruginosa* alginate, demonstrating that mucoid biofilms contain more viable cells than nonmucoid biofilms and are more resistant to DNase treatment (116, 117). Moreover, compared to nonmucoid isolates, mucoid *P. aeruginosa* maintains *in vitro* biofilm formation capacity and gene expression profiles during chronic lung infection of CF patients (134). This suggests that protection from the CF microenvironment afforded by alginate results in a more stable population over time and may help to explain how clonal mucoid strains can persist for decades in CF patients.

Capsular Polysaccharide

Significance, structure, and regulation

Bacterial capsular polysaccharides (CPSs) are found on the cell surface of a broad range of species. The CPS is tightly associated with the bacterial cell surface via covalent attachments to either phospholipid or lipid-A molecules. CPSs are highly hydrated molecules composed of repeating monosaccharides joined by glycosidic linkages (135). CPSs are extremely diverse, not only in the monosaccharide constituents, but also in glycosidic linkages, branching, and substitutions with noncarbohydrate residues, leading to a nearly unlimited range of structures (two examples in Fig. 1) (136). For example, over 80 CPSs (K antigens) have been identified in *E. coli*, and 93 CPSs (serotypes) in *Streptococcus pneumoniae* (137, 138). Despite the variety of CPS structures, many of the details of polymer synthesis and regulation are shared among Gram-positive and Gram-negative bacteria. Capsule synthesis requires nucleotide diphosphosugar precursors available in the cytoplasm and concludes with assembly of the polymer at the periplasmic face of the plasma membrane (136). The three primary biosynthetic pathways are termed Wyz dependent, synthase dependent, and ATP-binding cassette transporter dependent (139). Many CPS-producing bacteria respond to environmental conditions by varying capsule production between a low CPS-producing phase and a high CPS-producing phase, whereby stable phenotypic variants may also be selected for under specific environmental conditions (138, 140, 141). Moreover, CPS plays pivotal roles in bacterial survival in the environment and infection in the host, providing protection from desiccation, opsonophagocytosis, and complement-mediated and cationic antimicrobial peptide-mediated killing (135).

Role in biofilm biology

The complexity of biofilm formation in varying environments and stages of infection, the ubiquity and diversity of CPSs among bacteria, and the propensity of bacteria to synthesize multiple PSs have created a complex and inconclusive view of the role of CPSs in biofilm biology. In general, CPS production is thought to inhibit bacterial adherence and biofilm formation. CPS-deficient strains in *S. pneumoniae, Neisseria meningitidis, S. aureus,* and *Vibrio vulnificans* demonstrate increased adherence to epithelial cells and surfaces and more robust biofilm formation (142–146). CPS is also often downregulated during biofilm formation and upon contact with epithelial cells (143, 147, 148). It is hypothesized that decreased CPS further enhances the quiescent nature often associated with biofilm formation: potentiating immune evasion, decreased virulence, and persistence (142, 143, 149). This is supported by observations that low-CPS-producing strains are more frequently isolated from chronic infections, and high-CPS strains, from acute infections (143, 146).

However, this simplistic, binary view may not illustrate the complete role of CPSs in biofilm formation. Studies have demonstrated that CPSs may be important in mature biofilms, aiding in the maintenance of biofilm size and dispersal. In *V. vulnificans,* CPS was found to be expressed in later stages of biofilm formation and was hypothesized to be synthesized and secreted after biofilms mature and reach a threshold of cell density (150). In support of this, CPS expression in mature biofilms and maintenance of biofilm size is regulated by quorum sensing molecules (151). Similarly, in *S. pneumoniae,* higher CPS expression has been demonstrated in biofilm towers, compared to cells at the biofilm surface (148). Further evidence supporting a role of CPS in biofilm maturation is seen during formation of intracellular biofilm-like communities within bladder epithelial cells during urinary tract infections by uropathogenic *E. coli* (149). Collectively, these data point toward an overall view for multiple pathogens that CPS plays an important, yet incompletely defined role in biofilm biology. Several lines of evidence support an inhibitory role during initial attachment and biofilm formation; however, the ability to then enhance CPS synthesis in later stages may prove to be critical during infection. This clearly requires further study.

Levan

Significance, structure, and regulation

Levans are high molecular mass, neutral homopolymers composed of ß-D-fructans with extensive and irregular branching (Fig. 1) (152). Levan production confers a mucoid phenotype to bacteria expressing this and has been described in several plant pathogens (*Pseudomonas syringae, Erwinia amylovora,* and *Bacillus subtilis*) and *Streptococcus mutans* oral biofilms (153–155). Levan is produced exclusively from extracellular sucrose catalyzed by excreted levansucrase (153, 156, 157). Levansucrase is activated in the presence of sucrose and is frequently regulated by two-component sensor kinases (LadS in *P. syringae*; SacX/SacY and DegS/DegU in *B. subtilis*; RscC/RscB, GacS/GacR (GrrS/GrrA), and EnvZ/OmpR in *E. amylovora*; and CovS/CovR [VicK/VicR] in *S. mutans* [152, 158, 159]). In *B. subtilis sacB* and *sacC* are involved in levansucrase production and modification. Levansucrase is secreted from the cell by the SecA pathway (160, 161). In *P. syringae* and *E. amylovora* the levansucrase biosynthetic operon is composed of three genes: *iscA-C* (153, 162). Proposed functions for levan include protection from desiccation and bacteriophages, nutrient storage, and virulence (157, 163, 164).

Role in biofilm biology

The role of levan production in biofilms has predominately been described in *S. mutans* dental biofilms. Upon exposure to sucrose, levan accumulates on dental plaques and is catabolized to acid when the environmental sugar is depleted. Prolonged exposure promotes dental carries (165). While levan is not required for *S. mutans* biofilm formation, supplementation of sucrose to biofilms *in vitro* yields thicker structures, which contain more levan and are more resistant to shear stress (92, 166, 167). Interestingly, levan is a water-soluble PS, and levan produced by *B. subtilis* possesses a particularly low viscosity. At typical concentrations where other PSs display gel behavior, levan remains fluid, suggesting that it may not form an adhesive framework (168). In *P. syringae*, levan is not required for biofilm formation and is not responsible for maintenance of biofilm structure. Instead, levan may function as a storage molecule, utilized during periods of starvation (164). Levan was identified in the voids and blebs of *P. syringae* biofilms, whose formation required supplementation of sucrose in the media (164). This theory of nutrient retention by levan was previously suggested for *S. mutans* in the oral cavity (163). Collectively, these data suggest that levan contributes to the physio-chemical properties of bacterial biofilms but is not required for its formation.

ARCHITECTURAL POLYSACCHARIDES

Some of the first biofilm-related PSs identified were studied due to the role that they play in biofilm structure. These PSs have provided extensive insight into the regulation of biofilm formation and structure, primarily due to the readily observable phenotypes associated with mutants or overproducers of these products. For example, through studying the rugose phenotype associated with PS produced by *Vibrio cholerae* (*Vibrio* polysaccharide [VPS]), important studies in gene regulation, second messenger signaling, and biofilm matrix assembly have been reported (169–171).

Colanic Acid

Significance, structure, and regulation

Colanic acid (CA), or M antigen, is a branched PS composed of glucose, galactose, and glucuronic acid (Fig. 1) (172). The 19-gene *wca* cluster is responsible for producing CA in several species of *Enterobacteriaceae* (172–174). This cluster was formerly called the *cps* cluster, but it was renamed when it was discovered that these genes encode enzymes that produce CA (172). As with most PSs, several of the precursors of CA, such as UDP-D-glucose and UDP-D-galactose are produced by enzymes that are located in other loci, while many of the enzymes dedicated to CA synthesis are included in the *wca* cluster (172). For many years, the study of CA was focused on its effect on colony morphology, but advances in molecular techniques allowed a genetic approach to the regulation, structure, and role of CA in the biofilm matrix. The *wca* gene cluster is regulated by the well-characterized Rcs phosphorelay system. In this variation of a two-component system, RcsC transfers a phosphate to RcsD, which subsequently transfers it to RcsB. RcsB is a LuxR family helix-turn-helix response regulator, which binds the promoter of the *wca* operon and activates transcription (175–178). The highly unstable accessory protein RcsA can enhance this interaction. Collectively, this signaling cascade leads to upregulation of the *wca* cluster and production of CA, which allows the bacterium to endure environmental stresses such as desiccation (179). CA is primarily produced at lower temperatures, indicating that it is most beneficial to the bacterium in the environmental phase of its life cycle (175).

Role in biofilm biology

CA was one of the first PSs studied with respect to its role in biofilm formation. Early *in vitro* work demonstrated that an *E. coli* mutant incapable of producing CA had two interesting phenotypes. First, the CA mutant demonstrated attachment capabilities similar to the wild type strain. This was interesting to the biofilm field, because previous work on the role of PSs in biofilms indicated that the PS was involved in the initial attachment to surfaces (12, 180). Second, though the CA mutant was able to attach at wild type levels, it formed biofilms with less three-dimensional structure than the biofilms formed by the wild type strain (181). These biofilms were described as "collapsed," with cells densely packed against the substrate. It was concluded that CA was critical for the development of mature, three-dimensional biofilms by *E. coli* (181). This work was integral to the biofilm field, because it was one of the first observations that implicated PSs as structural members of the biofilm matrix.

In *E. coli*, CA is produced at low temperatures and therefore is more important for environmental survival than during infection (182). In contrast, the role of CA produced by the pathogen *Salmonella enterica* serovar *Typhimurium* is dispensable for binding abiotic surfaces (183). Further characterization of the role of CA in *S. enterica* biofilms indicated that the CA mutants were unable to build a three-dimensional biofilm on either HEp-2 cells or chicken intestinal epithelium, suggesting that CA is involved in pathogenesis (183). These divergent roles for CA in two relatively closely related bacteria demonstrate how PSs can enhance the diversity of bacterial colonization niches and biofilm phenotypes.

Vibrio Polysaccharide

Significance, structure, and regulation

V. cholerae is a Gram-negative aquatic bacterium that is the causative agent of the diarrheal disease cholera. This bacterium must endure many environmental stresses as it transits from an aquatic environment to the human gastrointestinal system. One of the mechanisms that are important in maintaining this infectious cycle is the production of the VPS. High expression of this PS is the cause of the rugose variant of an El Tor *V. cholerae* strain (66, 184). VPS was demonstrated to be essential for biofilm formation and resistance to chlorine. The NtrC family response regulator VpsR activates transcription of the *vps* operon, allowing for production of the PS (185).

The VPS is encoded by two operons, *vpsA* through *vpsK*, and *vpsL* through *vpsQ*. These operons are positively regulated by the response regulators VpsR and VpsT (185, 186). Expression analysis indicates that VpsR is epistatic to the positive regulator VpsT (187). Additionally, both VpsT and VpsR induce expression of each other, presenting a rapid positive feedback loop resulting in production of VPS (186). The negative regulator HapR represses VpsR, VpsT, and the *vps* operons. Collectively, these three regulators affect the expression of VPS and therefore biofilm phenotypes and colony rugosity. Deletion of *hapR* in a smooth strain relieves the negative regulation of *vps* expression, resulting in the transition to a rugose colony. Conversely, deletion of *vpsR* or *vpsT* in a rugose strain produces a smooth strain (187).

Several studies have indicated that *vps* genes are expressed during intestinal colonization and infection. VPS could aid in the formation of hyper-infectious aggregates of cells, as well as attachment and colonization of the intestine. In support of this, *vps* operon mutants demonstrated a significant defect in colonization of infant mouse intestines, supporting VPS being involved in host colonization (188).

Recent work indicates that the expression of the *vps* gene clusters is dependent on elevated intracellular levels of the second messenger c-di-GMP. VpsT binds c-di-GMP, resulting in oligomerization and upregulation of the *vps* operons (169, 170). Several diguanylate cyclases and phosphodiesterases affect VPS production by altering the cellular pool of c-di-GMP. Specifically, the phosphodiesterase VieA reduces *vps* expression by degrading c-di-GMP (189), while the diguanylate cyclases CdgA, H, K, L, and M all contribute to the activation of VpsT by producing elevated levels of c-di-GMP (170).

VPS consists mainly of glucose (52.6%) and galactose (37.0%), with small amounts of

N-acetylglucosamine, mannose, and xylose (5.1%, 3.8%, and 1.5%, respectively). The primary linkages in this PS are 4-linked glucose and 4-linked galactose, suggesting that these sugars form the linear backbone. Branching of the PS is suggested by the detection of small amounts of 3,4- and 4,6-linked galactose and glucose, as well as 2,4-linked galactose (66).

Role in biofilm biology

Much of what is known regarding the role of VPS in biofilm biology stems from the observation of *V. cholerae* rugose variants. These colonies have a highly structured, wrinkled morphology due to the overproduction of VPS. When the VPS genes are deleted from a rugose strain, the colony reverts to a smooth morphology. In addition to the rough colony morphology, rugose strains demonstrate enhanced attachment and biofilm formation phenotypes (66, 185). Both *vps* operons are essential for VPS production and, therefore, biofilm development (188). Super-resolution confocal microscopy revealed that VPS promotes the retention of daughter cells in the biofilm, as well as the accumulation of the biofilm matrix proteins RbmA, Bap1, and RbmC (171). In the absence of VPS, biofilms were restricted to the monolayer stage. During biofilm growth, VPS was extruded from the cell and formed spherical foci on the cell surface. It was proposed that the VPS, along with the proteins Bap1 and RbmC, forms the biofilm matrix envelope around cell clusters. Collectively, VPS is regarded as an essential part of the *V. cholerae* biofilm matrix, which is important in both the environmental and host phases of the infectious cycle.

B. subtilis Polysaccharide

Significance, structure, and regulation

B. subtilis produces several PSs that affect biofilm structure and function, but the prevalent architectural PS is EPS (157, 190, 191). During a study of the formation of fruiting bodies, it was observed that Δ*yveQ* (*epsG*) and Δ*yveR* (*epsH*) mutants were unable to form structured pellicles or produce multicellular fruiting bodies (190). Sixteen genes encoding EPS are in an operon, and most are homologous to those involved in PS production and modification in other bacteria (190). A subsequent report updated the operon annotation to 15 genes and renamed it *epsA-O* (67). Further characterization of the genes involved in biofilm formation indicated that an unlinked gene, *yhxB*, is essential for biofilm formation (191). The Δ*yhxB* mutant forms pellicles that are similar to the EPS-deficient Δ*yveR* mutant. *yhxB* has sequence similarity to both phosphoglucomutase and phosphomannomutase sugar-modifying enzymes that are involved in PS synthesis in other bacteria. It was concluded that *yhxB* along with the *epsA-O* operon are essential for the production of EPS and biofilms (191). Though this PS has been extensively studied, the structure is unknown (157).

The *epsA-O* operon is repressed by SinR (67), which binds to an operator upstream of the operon (67, 192). PS repression is relieved by SinI, which forms a complex with SinR and antagonizes its activity (67). As a result, SinR is deemed the master biofilm regulator of *B. subtilis* (67).

Role in biofilm biology

EPS is essential for the formation of pellicles and surface structures by *B. subtilis* (191). A Δ*epsH* mutant forms weak pellicles and smooth colonies, in contrast to the robust pellicle and rugose colonies formed by the wild type strain (67). Increased EPS expression in a Δ*sinR* mutant exhibits enhanced cell aggregation, pellicle formation, colony rugosity, and impaired swarming motility (67). The importance of PS in biofilm formation and cell aggregation is highlighted with the observation that addition of norspermidine can induce biofilm dissolution by targeting PS and interfering with its ability to aggregate and provide biofilm structure (193), though recent work has contradicted some of the claims about the effect of norspermidine (194), indicating that further research is required on this topic. Collectively, these data demonstrate the importance of the structural role of EPS in biofilm formation and stability.

Cellulose

Significance, structure, and regulation

Cellulose is one of the most abundant PSs in nature. The structure of cellulose fibrils is conserved throughout evolution, with bacterial cellulose indistinguishable from cellulose produced by higher-order algae, fungi, and plants. This polymer consists of repeating chains of β-1,4 linked D-glucose that form fibrils, which resemble cables (Fig. 1). Fibrils range from 1 to 25 nm in width and 1 to 9 μm in length and are assembled in the extracellular space immediately adjacent to the cell. These cables can align to form sheets, introducing the structural rigidity that plant cellulose provides due to its insoluble and inelastic nature. Cellulose fibrils have a tensile strength similar to steel (195). This structure is most evident in wood and paper products, which are largely composed of cellulose.

The role of cellulose in bacteria was first described in the Gram-negative bacterium *Gluconacetobacter xylinus* (formerly *Acetobacter xylinum*) (196–198). This bacterium produces large quantities of highly pure cellulose that shares many properties with cellulose produced by higher-order organisms, which aided in research on the role of plant and algal cellulose. Static cultures of *G. xylinus* produce a robust pellicle that is rich in cellulose, facilitating the isolation of the polymer from the matrix. Subsequently, many bacteria have been found to produce cellulose, including commensal organisms such as *E. coli*, pathogens such as *S. enterica*, and environmental organisms such as many *Pseudomonas* spp. (199–201). The diversity in the niches of the organisms that produce cellulose highlights the variety of functions that it can play in the biofilm matrix.

The cellulose production genes have several names, depending on the species. These include *acs* (*Acetobacter* cellulose synthesis), *bcs* (bacterial cellulose synthesis), and *cel* (cellulose) (200). In general, the operon consists of two conserved genes and several accessory genes. The first conserved gene is *bcsA* (*acsA* or *celA*), which encodes the cellulose synthase enzyme. This is followed by *bcsB* (*acsB* or *celB*), which encodes a c-di-GMP binding protein (200). Additional genes in the operon can vary but generally include genes including a cellulase (*bcsZ* or *celC*) (200). The most thorough characterization of the cellulose biosynthetic machinery has been conducted in *G. xylinus*. In this bacterium, the cellulose synthase enzyme complexes are linked to the cytoplasmic membrane by 8 to 10 transmembrane domains. Approximately 50 synthase complexes form in a row along the long axis of the bacterium, excreting cellulose fibers that rapidly aggregate with nearby fibers to form a ribbon (200, 202).

Regulation of cellulose production is linked to c-di-GMP levels. In fact, the initial characterization of c-di-GMP and the diguanylate cyclase and phosphodiesterase enzymes responsible for this molecule were made in *G. xylinus* as a result of studies investigating factors that promote cellulose production (203, 204). c-di-GMP binds to BcsB and promotes the production of cellulose. In addition, cellulose production in *E. coli* and *S. enterica* is dependent on AdrA, which contains a GGDEF domain (199, 205). The cellulose synthase is expressed constitutively but is not active without the c-di-GMP inducer. It has been proposed that AdrA is the diguanylate cyclase that provides the c-di-GMP required for cellulose production to the cellulose synthase. This is supported by the evidence that expression of *adrA* from a plasmid is sufficient to induce cellulose production (199). This system provides rapid production of cellulose, since the biosynthetic machinery is present, requiring only production of AdrA to provide the allosteric inducer c-di-GMP.

Role in biofilm biology

Cellulose provides both structure and protection to bacterial cells in biofilms. The first reports of the contribution of cellulose to *G. xylinus* pellicles indicated that cellulose was responsible for cell aggregation in the pellicle, allowing the biofilm to float to the surface of the culture where oxygen was readily available to the bacteria (198). In addition to this role, cellulose provides protection from the mutagenic effects of ultraviolet light (195, 206).

The role of cellulose in biofilm formation has expanded across many species. One report describes the production of cellulose in many *Enterobacteriaceae*, including *E. coli*, *Salmonella enteritidis*, *Salmonella typhimurium*, and *K. pneumoniae* (199). In these species, cellulose often interacts with the curli fimbriae to develop the biofilm matrix that is responsible for the *rdar* (rough, dry, and red) colony phenotype (199, 200). Other groups have observed that cellulose production results in pellicle formation in many *Pseudomonas* spp. (201). Cellulose production enhances binding of epithelial cells and reduces immune response to the bacteria in the probiotic strain of *E. coli* Nissle 1917 (207). These functions are important for establishing a commensal relationship with the host epithelium. Cellulose is produced by many bacteria to utilize the properties of this simple polymer for a wide variety of functions, providing a good example of the functional diversity of PSs.

FUTURE PERSPECTIVES

A wealth of knowledge has been gained regarding the composition and functions of PS components of the biofilm matrix by utilizing methods such as traditional genetic techniques combined with biochemistry and immunochemistry, lectin/carbohydrate stains, and microscopy (208). While necessary for initial observations, this reductionist point of view leaves significant gaps in our understanding of biofilm biology. PS isolation is difficult to achieve, especially from environmental or host sources, which contain a diverse range of components. Moreover, the paucity of animal models, particularly for chronic infections, presents a significant challenge for correlating *in vitro* findings to infection. Many bacteria produce multiple PSs, and *in vivo* infections are rarely comprised of a single microbe. Our current understanding of how PSs interact in polymicrobial communities in the environment or the host is extremely limited. Bridging this gap will require the development of more extensive *in situ* techniques to visualize, measure, and fully define PSs in real-world situations. Further development and utilization of microscopy techniques such as super-resolution and Raman scattering microscopy, combined with fully hydrated living models, will be essential to provide a complete interpretation of the complexity of microbial biofilms and the contribution of PS production (209).

While PSs are a predominant and well-studied biofilm component, the matrix is clearly composed of a range of macromolecules, including eDNA, proteins, and lipids. It stands to reason and is frequently speculated that these molecules interact in the matrix to form specific and dynamic interactions to fine-tune the biofilm community to varying environments. However, information regarding the biological significance, regulation, and direct visualization of these interactions are currently lacking. For example, since the seminal discovery by Whitchurch and colleagues in 2002 that eDNA is an important component of *P. aeruginosa* biofilms (210), only a few studies have defined how eDNA interacts with PS in the matrix. It has been suggested that in *Myxococcus xanthus* biofilms, eDNA directly interacts with PS, enhancing the physical strength and resistance to biological stress (211). Accordingly, a mathematical model of biofilm stress relaxation reveals evidence that filamentous eDNA may interact with PS to produce well-defined agglomerate structures with unique physio-elastic properties (212). For protein-EPS interactions, an additional observation in *M. xanthus* illustrated the first direct interaction of EPS and type IV pili (213). However, the biological significance and the nature of these interactions remain unclear, representing a significant gap in our understanding of the biofilm matrix.

In addition to interacting with other components of the biofilm matrix, it is likely that various PSs, produced by either the same bacteria or by other species in the environment, interact with each other to produce unique and/or compensatory functions. Some PSs may be exclusively expressed at specific stages of biofilm formation or under certain environmental conditions. Interrogating the spatial and temporal production of PSs and how they interact with other molecules in the environment will significantly enhance our understanding of how biofilms develop and how they may be modulated.

ACKNOWLEDGMENTS

Dominique H. Limoli and Christopher J. Jones contributed equally to this work.

Conflicts of interest: We disclose no conflicts.

CITATION

Limoli DH, Jones CJ, Wozniak DJ. 2015. Bacterial extracellular polysaccharides in biofilm formation and function. Microbiol Spectrum 3(3):MB-0011-2014.

REFERENCES

1. **Branda SS, Vik Å, Friedman L, Kolter R.** 2005. Biofilms: the matrix revisited. *Trends Microbiol* **13:**20–26.
2. **Sutherland IW.** 2001. The biofilm matrix: an immobilized but dynamic microbial environment. *Trends Microbiol* **9:**222–227.
3. **Jenkinson HF, Lamont R.** 1997. *Streptococcal* adhesion and colonization. *Crit Rev Oral Biol Med* **8:**175–200.
4. **Boles BR, Horswill AR.** 2011. *Staphylococcal* biofilm disassembly. *Trends Microbiol* **19:**449–455.
5. **Otto M.** 2008. *Staphylococcal* biofilms. *Curr Top Microbiol Immunol* **322:**207–228.
6. **Fey PD, Olson ME.** 2010. Current concepts in biofilm formation of *Staphylococcus epidermidis*. *Future Microbiol* **5:**917–933.
7. **Kostakioti M, Hadjifrangiskou M, Hultgren SJ.** 2013. Bacterial biofilms: development, dispersal, and therapeutic strategies in the dawn of the postantibiotic era. *Cold Spring Harbor Perspect Med* **3:**a010306. doi:10.1101/cshperspect.a010306.
8. **Rohde H, Burandt EC, Siemssen N, Frommelt L, Burdelski C, Wurster S, Scherpe S, Davies AP, Harris LG, Horstkotte MA, Knobloch JKM, Ragunath C, Kaplan JB, Mack D.** 2007. Polysaccharide intercellular adhesin or protein factors in biofilm accumulation of *Staphylococcus epidermidis* and *Staphylococcus aureus* isolated from prosthetic hip and knee joint infections. *Biomaterials* **28:**1711–1720.
9. **Mack D, Siemssen N, Laufs R.** 1992. Parallel induction by glucose of adherence and a polysaccharide antigen specific for plastic-adherent *Staphylococcus epidermidis*: evidence for functional relation to intercellular adhesion. *Infect Immun* **60:**2048–2057.
10. **Mack D, Nedelmann M, Krokotsch A, Schwarzkopf A, Heesemann J, Laufs R.** 1994. Characterization of transposon mutants of biofilm-producing *Staphylococcus epidermidis* impaired in the accumulative phase of biofilm production: genetic identification of a hexosamine-containing polysaccharide intercellular adhesin. *Infect Immun* **62:**3244–3253.
11. **Mack D, Fischer W, Krokotsch A, Leopold K, Hartmann R, Egge H, Laufs R.** 1996. The intercellular adhesin involved in biofilm accumulation of *Staphylococcus epidermidis* is a linear beta-1,6-linked glucosaminoglycan: purification and structural analysis. *J Bacteriol* **178:**175–183.
12. **McKenney D, Hübner J, Muller E, Wang Y, Goldmann DA, Pier GB.** 1998. The ica locus of *Staphylococcus epidermidis* encodes production of the capsular polysaccharide/adhesin. *Infect Immun* **66:**4711–4720.
13. **Maira-Litran T, Kropec A, Abeygunawardana C, Joyce J, Mark G III, Goldmann DA, Pier**

GB. 2002. Immunochemical properties of the *Staphylococcal* poly-N-acetylglucosamine surface polysaccharide. *Infect Immun* **70:**4433–4440.

14. **Joyce JG, Abeygunawardana C, Xu Q, Cook JC, Hepler R, Przysiecki CT, Grimm KM, Roper K, Ip CCY, Cope L, Montgomery D, Chang M, Campie S, Brown M, McNeely TB, Zorman J, Maira-Litrán T, Pier GB, Keller PM, Jansen KU, Mark GE III.** 2003. Isolation, structural characterization, and immunological evaluation of a high-molecular-weight exopolysaccharide from *Staphylococcus aureus*. *Carbohydr Res* **338:**903–922.
15. **Sadovskaya I, Vinogradov E, Flahaut S, Kogan G, Jabbouri S.** 2005. Extracellular carbohydrate-containing polymers of a model biofilm-producing strain, *Staphylococcus epidermidis* RP62A. *Infect Immun* **73:**3007–3017.
16. **Cramton SE, Gerke C, Schnell NF, Nichols WW, Götz F.** 1999. The intercellular adhesion (ica) locus is present in *Staphylococcus aureus* and is required for biofilm formation. *Infect Immun* **67:**5427–5433.
17. **Heilmann C, Schweitzer O, Gerke C, Vanittanakom N, Mack D, Götz F.** 1996. Molecular basis of intercellular adhesion in the biofilm-forming *Staphylococcus epidermidis*. *Mol Microbiol* **20:**1083–1091.
18. **Conlon KM, Humphreys H, O'Gara JP.** 2002. *icaR* encodes a transcriptional repressor involved in environmental regulation of *ica* operon expression and biofilm formation in *Staphylococcus epidermidis*. *J Bacteriol* **184:**4400–4408.
19. **Wang X, Preston JF, Romeo T.** 2004. The *pgaABCD* locus of *Escherichia coli* promotes the synthesis of a polysaccharide adhesin required for biofilm formation. *J Bacteriol* **186:**2724–2734.
20. **Bobrov AG, Kirillina O, Forman S, Mack D, Perry RD.** 2008. Insights into *Yersinia pestis* biofilm development: topology and co-interaction of Hms inner membrane proteins involved in exopolysaccharide production. *Environ Microbiol* **10:**1419–1432.
21. **Kaplan JB, Velliyagounder K, Ragunath C, Rohde H, Mack D, Knobloch JKM, Ramasubbu N.** 2004. Genes involved in the synthesis and degradation of matrix polysaccharide in *Actinobacillus actinomycetemcomitans* and *Actinobacillus pleuropneumoniae* biofilms. *J Bacteriol* **186:**8213–8220.
22. **Parise G, Mishra M, Itoh Y, Romeo T, Deora R.** 2007. Role of a putative polysaccharide locus in *Bordetella* biofilm development. *J Bacteriol* **189:**750–760.
23. **Skurnik D, Davis MR, Benedetti D, Moravec KL, Cywes-Bentley C, Roux D, Traficante DC, Walsh RL, Maira-Litran T, Cassidy SK, Hermos CR, Martin TR, Thakkallapalli EL, Vargas SO, McAdam AJ, Lieberman TD, Kishony R, LiPuma JJ, Pier GB, Goldberg JB, Priebe GP.** 2012. Targeting pan-resistant bacteria with antibodies to a broadly conserved surface polysaccharide expressed during infection. *J Infect Dis* **205:**1709–1718.
24. **Choi AHK, Slamti L, Avci FY, Pier GB, Maira-Litran T.** 2009. The *pgaABCD* locus of *Acinetobacter baumannii* encodes the production of poly-1-6-N-acetylglucosamine, which is critical for biofilm formation. *J Bacteriol* **191:** 5953–5963.
25. **Cue D, Lei MG, Lee CY.** 2012. Genetic regulation of the intercellular adhesion locus in *Staphylococci*. *Front Cell Infect Microbiol* **2:**38.
26. **Rachid S, Ohlsen K, Witte W, Hacker J, Ziebuhr W.** 2000. Effect of subinhibitory antibiotic concentrations on polysaccharide intercellular adhesin expression in biofilm-forming *Staphylococcus epidermidis*. *Antimicrob Agents Chemother* **44:**3357–3363.
27. **Nuryastuti T, Krom BP, Aman AT, Busscher HJ, van der Mei HC.** 2010. Ica-expression and gentamicin susceptibility of *Staphylococcus epidermidis* biofilm on orthopedic implant biomaterials. *J Biomed Mater Res* **96A:**365–371.
28. **Xu L, Li H, Vuong C, Vadyvaloo V, Wang J, Yao Y, Otto M, Gao Q.** 2005. Role of the *luxS* quorum-sensing system in biofilm formation and virulence of *Staphylococcus epidermidis*. *Infect Immun* **74:**488–496.
29. **Knobloch JKM, Bartscht K, Sabottke A, Rohde H, Feucht HH, Mack D.** 2001. Biofilm formation by *Staphylococcus epidermidis* depends on functional RsbU, an activator of the *sigB* operon: differential activation mechanisms due to ethanol and salt stress. *J Bacteriol* **183:**2624–2633.
30. **Frees D, Chastanet A, Qazi S, Sørensen K, Hill P, Msadek T, Ingmer H.** 2004. Clp ATPases are required for stress tolerance, intracellular replication and biofilm formation in *Staphylococcus aureus*. *Mol Microbiol* **54:** 1445–1462.
31. **Pamp SJ, Frees D, Engelmann S, Hecker M, Ingmer H.** 2006. Spx is a global effector impacting stress tolerance and biofilm formation in *Staphylococcus aureus*. *J Bacteriol* **188:**4861–4870.
32. **Beenken KE, Blevins JS, Smeltzer MS.** 2003. Mutation of *sarA* in *Staphylococcus aureus* limits biofilm formation. *Infect Immun* **71:**4206–4211.

33. **Valle J, Toledo-Arana A, Berasain C, Ghigo J-M, Amorena B, Penadés JR, Lasa I.** 2003. SarA and not sigmaB is essential for biofilm development by *Staphylococcus aureus*. *Mol Microbiol* **48:**1075–1087.
34. **Vuong C, Kocianova S, Voyich JM, Yao Y, Fischer ER, DeLeo FR, Otto M.** 2004. A crucial role for exopolysaccharide modification in bacterial biofilm formation, immune evasion, and virulence. *J Biol Chem* **279:**54881–54886.
35. **Kristian SA, Birkenstock TA, Sauder U, Mack D, Götz F, Landmann R.** 2008. Biofilm formation induces C3a release and protects *Staphylococcus epidermidis* from IgG and complement deposition and from neutrophil-dependent killing. *J Infect Dis* **197:**1028–1035.
36. **França A, Vilanova M, Cerca N, Pier GB.** Monoclonal antibody raised against PNAG has variable effects on static *S. epidermidis* biofilm accumulation *in vitro*. *Int J Biol Sci* **9:**518–520.
37. **Otto M.** 2009. *Staphylococcus epidermidis:* the "accidental" pathogen. *Nat Rev Microbiol* **7:**555–567.
38. **Kong K-F, Vuong C, Otto M.** 2006. *Staphylococcus* quorum sensing in biofilm formation and infection. *Int J Med Microbiol* **296:**133–139.
39. **Mack D, Becker P, Chatterjee I, Dobinsky S, Knobloch JKM, Peters G, Rohde H, Herrmann M.** 2004. Mechanisms of biofilm formation in *Staphylococcus epidermidis* and *Staphylococcus aureus*: functional molecules, regulatory circuits, and adaptive responses. *Int J Med Microbiol* **294:**203–212.
40. **Rohde H, Frankenberger S, Zähringer U, Mack D.** 2010. Structure, function and contribution of polysaccharide intercellular adhesin (PIA) to *Staphylococcus epidermidis* biofilm formation and pathogenesis of biomaterial-associated infections. *Eur J Cell Biol* **89:**103–111.
41. **Patti JM, Allen BL, McGavin MJ, Höök M.** 1994. MSCRAMM-mediated adherence of microorganisms to host tissues. *Annu Rev Microbiol* **48:**585–617.
42. **Rupp ME, Fey PD, Heilmann C, Götz F.** 2001. Characterization of the importance of *Staphylococcus epidermidis* autolysin and polysaccharide intercellular adhesin in the pathogenesis of intravascular catheter-associated infection in a rat model. *J Infect Dis* **183:**1038–1042.
43. **Rupp ME, Ulphani JS, Fey PD, Bartscht K, Mack D.** 1999. Characterization of the importance of polysaccharide intercellular adhesin/hemagglutinin of *Staphylococcus epidermidis* in the pathogenesis of biomaterial-based infection in a mouse foreign body infection model. *Infect Immun* **67:**2627–2632.
44. **Rupp ME, Ulphani JS, Fey PD, Mack D.** 1999. Characterization of *Staphylococcus epidermidis* polysaccharide intercellular adhesin/hemagglutinin in the pathogenesis of intravascular catheter-associated infection in a rat model. *Infect Immun* **67:**2656–2659.
45. **McKenney D, Pouliot KL, Wang Y, Murthy V, Ulrich M, Doring G, Lee JC, Goldmann DA, Pier GB.** 1999. Broadly protective vaccine for *Staphylococcus aureus* based on an *in vivo*-expressed antigen. *Science* **284:**1523–1527.
46. **Fluckiger U, Ulrich M, Steinhuber A, Doring G, Mack D, Landmann R, Goerke C, Wolz C.** 2005. Biofilm formation, *icaADBC* transcription, and polysaccharide intercellular adhesin synthesis by *Staphylococci* in a device-related infection model. *Infect Immun* **73:**1811–1819.
47. **Boles BR, Horswill AR.** 2008. *agr*-mediated dispersal of *Staphylococcus aureus* biofilms. *PLoS Pathog* **4:**e1000052. doi:10.1371/journal.ppat.1000052.
48. **Lauderdale KJ, Boles BR, Cheung AL, Horswill AR.** 2009. Interconnections between sigma B, *agr*, and proteolytic activity in *Staphylococcus aureus* biofilm maturation. *Infect Immun* **77:**1623–1635.
49. **Boles BR, Thoendel M, Roth AJ, Horswill AR.** 2010. Identification of genes involved in polysaccharide-independent *Staphylococcus aureus* biofilm formation. *PLoS One* **5:**e10146. doi:10.1371/journal.pone.0010146.
50. **Izano EA, Amarante MA, Kher WB, Kaplan JB.** 2008. Differential roles of poly-N-acetylglucosamine surface polysaccharide and extracellular DNA in *Staphylococcus aureus* and *Staphylococcus epidermidis* biofilms. *Appl Environ Microbiol* **74:**470–476.
51. **Hennig S, Nyunt Wai S, Ziebuhr W.** 2007. Spontaneous switch to PIA-independent biofilm formation in an *ica*-positive *Staphylococcus epidermidis* isolate. *Int J Med Microbiol* **297:**117–122.
52. **Friedman L, Kolter R.** 2004. Two genetic loci produce distinct carbohydrate-rich structural components of the *Pseudomonas aeruginosa* biofilm matrix. *J Bacteriol* **186:**4457–4465.
53. **Franklin MJ, Nivens DE, Weadge JT, Howell PL.** 2011. Biosynthesis of the *Pseudomonas aeruginosa* extracellular polysaccharides, alginate, Pel, and Psl. *Front Microbiol* **2:**167.
54. **Mann EE, Wozniak DJ.** 2012. *Pseudomonas* biofilm matrix composition and niche biology. *FEMS Microbiol Rev* **36:**893–916.
55. **Lee VT, Matewish JM, Kessler JL, Hyodo M, Hayakawa Y, Lory S.** 2007. A cyclic-di-GMP receptor required for bacterial exopolysaccharide production. *Mol Microbiol* **65:**1474–1484.

56. **Hickman JW, Harwood CS.** 2008. Identification of FleQ from *Pseudomonas aeruginosa* as a c-di-GMP-responsive transcription factor. *Mol Microbiol* **69:**376–389.
57. **Baraquet C, Murakami K, Parsek MR, Harwood CS.** 2012. The FleQ protein from *Pseudomonas aeruginosa* functions as both a repressor and an activator to control gene expression from the pel operon promoter in response to c-di-GMP. *Nucleic Acids Res* **40:** 7207–7218.
58. **Colvin KM, Gordon VD, Murakami K, Borlee BR, Wozniak DJ, Wong GCL, Parsek MR.** 2011. The pel polysaccharide can serve a structural and protective role in the biofilm matrix of *Pseudomonas aeruginosa. PLoS Pathog* **7:**e1001264. doi:10.1371/journal.ppat.1001264.
59. **Colvin KM, Irie Y, Tart CS, Urbano R, Whitney JC, Ryder C, Howell PL, Wozniak DJ, Parsek MR.** 2012. The Pel and Psl polysaccharides provide *Pseudomonas aeruginosa* structural redundancy within the biofilm matrix. *Environ Microbiol* **14:**1913–1928.
60. **Jackson KD, Starkey M, Kremer S, Parsek MR, Wozniak DJ.** 2004. Identification of *psl*, a locus encoding a potential exopolysaccharide that is essential for *Pseudomonas aeruginosa* PAO1 biofilm formation. *J Bacteriol* **186:**4466–4475.
61. **Matsukawa M, Greenberg EP.** 2004. Putative exopolysaccharide synthesis genes influence *Pseudomonas aeruginosa* biofilm development. *J Bacteriol* **186:**4449.
62. **Byrd MS, Sadovskaya I, Vinogradov E, Lu H, Sprinkle AB, Richardson SH, Ma L, Ralston B, Parsek MR, Anderson EM, Lam JS, Wozniak DJ.** 2009. Genetic and biochemical analyses of the *Pseudomonas aeruginosa* Psl exopolysaccharide reveal overlapping roles for polysaccharide synthesis enzymes in Psl and LPS production. *Mol Microbiol* **73:**622–638.
63. **Ma L, Lu H, Sprinkle A, Parsek MR, Wozniak DJ.** 2007. *Pseudomonas aeruginosa* Psl is a galactose- and mannose-rich exopolysaccharide. *J Bacteriol* **189:**8353–8356.
64. **Jones CJ, Ryder CR, Mann EE, Wozniak DJ.** 2013. AmrZ modulates *Pseudomonas aeruginosa* biofilm architecture by directly repressing transcription of the *psl* operon. *J Bacteriol* **195:**1637–1644.
65. **Hickman JW, Tifrea DF, Harwood CS.** 2005. A chemosensory system that regulates biofilm formation through modulation of cyclic diguanylate levels. *Proc Natl Acad Sci USA* **102:**14422–14427.
66. **Yildiz FH, Schoolnik GK.** 1999. *Vibrio cholerae* O1 El Tor: identification of a gene cluster required for the rugose colony type, exopolysaccharide production, chlorine resistance, and biofilm formation. *Proc Natl Acad Sci USA* **96:**4028–4033.
67. **Kearns DB, Chu F, Branda SS, Kolter R, Losick R.** 2005. A master regulator for biofilm formation by *Bacillus subtilis. Mol Microbiol* **55:**739–749.
68. **Kirisits MJ, Prost L, Starkey M, Parsek MR.** 2005. Characterization of colony morphology variants isolated from *Pseudomonas aeruginosa* biofilms. *Appl Environ Microbiol* **71:**4809–4821.
69. **Starkey M, Hickman JH, Ma L, Zhang N, De Long S, Hinz A, Palacios S, Manoil C, Kirisits MJ, Starner TD, Wozniak DJ, Harwood CS, Parsek MR.** 2009. *Pseudomonas aeruginosa* rugose small-colony variants have adaptations that likely promote persistence in the cystic fibrosis lung. *J Bacteriol* **191:**3492–3503.
70. **DiGiandomenico A, Warrener P, Hamilton M, Guillard S, Ravn P, Minter R, Camara MM, Venkatraman V, MacGill RS, Lin J, Wang Q, Keller AE, Bonnell JC, Tomich M, Jermutus L, McCarthy MP, Melnick DA, Suzich JA, Stover CK.** 2012. Identification of broadly protective human antibodies to *Pseudomonas aeruginosa* exopolysaccharide Psl by phenotypic screening. *J Exp Med* **209:**1273–1287.
71. **Byrd MS, Pang B, Mishra M, Swords WE, Wozniak DJ.** 2010. The *Pseudomonas aeruginosa* exopolysaccharide Psl facilitates surface adherence and NF-kappaB activation in A549 cells. *MBio* **1:**e00140-10. doi:10.1128/mBio.00140-10.
72. **Mishra M, Byrd MS, Sergeant S, Azad AK, Parsek MR, McPhail L, Schlesinger LS, Wozniak DJ.** 2012. *Pseudomonas aeruginosa* Psl polysaccharide reduces neutrophil phagocytosis and the oxidative response by limiting complement-mediated opsonization. *Cell Microbiol* **14:**95–106.
73. **Ma L, Conover M, Lu H, Parsek MR, Bayles K, Wozniak DJ.** 2009. Assembly and development of the *Pseudomonas aeruginosa* biofilm matrix. *PLoS Pathog* **5:**e1000354. doi:10.1371/journal.ppat.1000354.
74. **Friedman L, Kolter R.** 2004. Genes involved in matrix formation in *Pseudomonas aeruginosa* PA14 biofilms. *Mol Microbiol* **51:**675–690.
75. **Irie Y, Borlee BR, O'Connor JR, Hill PJ, Harwood CS, Wozniak DJ, Parsek MR.** 2012. Self-produced exopolysaccharide is a signal that stimulates biofilm formation in *Pseudomonas aeruginosa. Proc Natl Acad Sci USA* **109:**20632–20636.
76. **Billings N, Millan M, Caldara M, Rusconi R, Tarasova Y, Stocker R, Ribbeck K.** 2013. The

extracellular matrix component Psl provides fast-acting antibiotic defense in *Pseudomonas aeruginosa* biofilms. *PLoS Pathog* **9:**e1003526. doi:10.1371/journal.ppat.1003526.
77. **Parsek MR, Singh PK.** 2003. Bacterial biofilms: an emerging link to disease pathogenesis. *Annu Rev Microbiol* **57:**677–701.
78. **Hall-Stoodley L, Stoodley P.** 2009. Evolving concepts in biofilm infections. *Cell Microbiol* **11:**1034–1043.
79. **Stewart PS, Costerton JW.** 2001. Antibiotic resistance of bacteria in biofilms. *Lancet* **358:** 135–138.
80. **Brown MR, Allison DG, Gilbert P.** 1988. Resistance of bacterial biofilms to antibiotics: a growth-rate related effect? *J Antimicrob Chemother* **22:**777–780.
81. **Mah TF, O'Toole GA.** 2001. Mechanisms of biofilm resistance to antimicrobial agents. *Trends Microbiol* **9:**34–39.
82. **Costerton JW, Cheng KJ, Geesey GG, Ladd TI, Nickel JC, Dasgupta M, Marrie TJ.** 1987. Bacterial biofilms in nature and disease. *Annu Rev Microbiol* **41:**435–464.
83. **Walters MC, Roe F, Bugnicourt A, Franklin MJ, Stewart PS.** 2003. Contributions of antibiotic penetration, oxygen limitation, and low metabolic activity to tolerance of *Pseudomonas aeruginosa* biofilms to ciprofloxacin and tobramycin. *Antimicrob Agents Chemother* **47:**317–323.
84. **de Beer D, Stoodley P, Lewandowski Z.** 1997. Measurement of local diffusion coefficients in biofilms by microinjection and confocal microscopy. *Biotechnol Bioeng* **53:**151–158.
85. **Jefferson KK, Goldmann DA, Pier GB.** 2005. Use of confocal microscopy to analyze the rate of vancomycin penetration through *Staphylococcus aureus* biofilms. *Antimicrob Agents Chemother* **49:**2467–2473.
86. **Dibdin GH, Assinder SJ, Nichols WW, Lambert PA.** 1996. Mathematical model of beta-lactam penetration into a biofilm of *Pseudomonas aeruginosa* while undergoing simultaneous inactivation by released beta-lactamases. *J Antimicrob Chemother* **38:**757–769.
87. **Høiby N, Bjarnsholt T, Givskov M, Molin S, Ciofu O.** 2010. Antibiotic resistance of bacterial biofilms. *Int J Antimicrob Agents* **35:**322–332.
88. **Lam J, Chan R, Lam K, Costerton JW.** 1980. Production of mucoid microcolonies by *Pseudomonas aeruginosa* within infected lungs in cystic fibrosis. *Infect Immun* **28:**546–556.
89. **Leid JG, Shirtliff ME, Costerton JW, Stoodley AP.** 2002. Human leukocytes adhere to, penetrate, and respond to *Staphylococcus aureus* biofilms. *Infect Immun* **70:**6339–6345.
90. **Kocianova S, Vuong C, Yao Y, Voyich JM, Fischer ER, DeLeo FR, Otto M.** 2005. Key role of poly-gamma-DL-glutamic acid in immune evasion and virulence of *Staphylococcus epidermidis*. *J Clin Invest* **115:**688–694.
91. **Jesaitis AJ, Franklin MJ, Berglund D, Sasaki M, Lord CI, Bleazard JB, Duffy JE, Beyenal H, Lewandowski Z.** 2003. Compromised host defense on *Pseudomonas aeruginosa* biofilms: characterization of neutrophil and biofilm interactions. *J Immunol* **171:**4329–4339.
92. **Sutherland I.** 2001. Biofilm exopolysaccharides: a strong and sticky framework. *Microbiology* **147:**3–9.
93. **Doggett RG, Harrison GM, Stillwell RN, Wallis ES.** 1966. An atypical *Pseudomonas aeruginosa* associated with cystic fibrosis of the pancreas. *J Pediatr* **68:**215–221.
94. **Elston HR, Hoffman KC.** 1967. Increasing incidence of encapsulated *Pseudomonas aeruginosa* strains. *Am J Clin Pathol* **48:**519–523.
95. **Govan JR, Deretic V.** 1996. Microbial pathogenesis in cystic fibrosis: mucoid *Pseudomonas aeruginosa* and *Burkholderia cepacia*. *Microbiol Rev* **60:**539–574.
96. **Gaspar MC, Couet W, Olivier JC, Pais AACC, Sousa JJS.** 2013. *Pseudomonas aeruginosa* infection in cystic fibrosis lung disease and new perspectives of treatment: a review. *Eur J Clin Microbiol Infect Dis* **32:**1231–1252.
97. **Clementi F.** 1997. Alginate production by *Azotobacter vinelandii*. *Crit Rev Biotechnol* **17:** 327–361.
98. **Evans LR, Linker A.** 1973. Production and characterization of the slime polysaccharide of *Pseudomonas aeruginosa*. *J Bacteriol* **116:** 915–924.
99. **Franklin MJ, Ohman DE.** 1993. Identification of *algF* in the alginate biosynthetic gene cluster of *Pseudomonas aeruginosa* which is required for alginate acetylation. *J Bacteriol* **175:**5057–5065.
100. **Ramsey DM, Wozniak DJ.** 2005. Understanding the control of *Pseudomonas aeruginosa* alginate synthesis and the prospects for management of chronic infections in cystic fibrosis. *Mol Microbiol* **56:**309–322.
101. **Damron FH, Goldberg JB.** 2012. Proteolytic regulation of alginate overproduction in *Pseudomonas aeruginosa*. *Mol Microbiol* **84:**595–607.
102. **Hay ID, Wang Y, Moradali M, Rehman ZU, Rehm BHA.** 2014. Genetics and regulation of bacterial alginate production. *Environ Microbiol.* [Epub ahead of print.] doi:10.1111/1462-2920.12389.
103. **Martin DW, Schurr MJ, Mudd MH, Govan JR, Holloway BW, Deretic V.** 1993. Mechanism of conversion to mucoidy in *Pseudomonas*

aeruginosa infecting cystic fibrosis patients. *Proc Natl Acad Sci USA* **90:**8377–8381.

104. **Wood LF, Ohman DE.** 2012. Identification of genes in the σ22 regulon of *Pseudomonas aeruginosa* required for cell envelope homeostasis in either the planktonic or the sessile mode of growth. *MBio* **3:**e00094-12. doi:10.1128/mBio.00094-12.
105. **Damron FH, Owings JP, Okkotsu Y, Varga JJ, Schurr JR, Goldberg JB, Schurr MJ, Yu HD.** 2011. Analysis of the *Pseudomonas aeruginosa* regulon controlled by the sensor kinase KinB and sigma factor RpoN. *J Bacteriol* **194:**1317–1330.
106. **Damron FH, Yu HD.** 2011. *Pseudomonas aeruginosa* MucD regulates the alginate pathway through activation of MucA degradation via MucP proteolytic activity. *J Bacteriol* **193:**286–291.
107. **Singh PK, Schaefer AL, Parsek MR, Moninger TO, Welsh MJ, Greenberg EP.** 2000. Quorum-sensing signals indicate that cystic fibrosis lungs are infected with bacterial biofilms. *Nature* **407:**762–764.
108. **Ramphal R, Pier GB.** 1985. Role of *Pseudomonas aeruginosa* mucoid exopolysaccharide in adherence to tracheal cells. *Infect Immun* **47:**1–4.
109. **Davies DG, Chakrabarty AM, Geesey GG.** 1993. Exopolysaccharide production in biofilms: substratum activation of alginate gene expression by *Pseudomonas aeruginosa. Appl Environ Microbiol* **59:**1181–1186.
110. **Davies DG, Geesey GG.** 1995. Regulation of the alginate biosynthesis gene *algC* in *Pseudomonas aeruginosa* during biofilm development in continuous culture. *Appl Environ Microbiol* **61:**860–867.
111. **Hoyle BD, Williams LJ, Costerton JW.** 1993. Production of mucoid exopolysaccharide during development of *Pseudomonas aeruginosa* biofilms. *Infect Immun* **61:**777–780.
112. **Nivens DE, Ohman DE, Williams J, Franklin MJ.** 2001. Role of alginate and its O acetylation in formation of *Pseudomonas aeruginosa* microcolonies and biofilms. *J Bacteriol* **183:**1047–1057.
113. **Wozniak DJ, Wyckoff TJO, Starkey M, Keyser R, Azadi P, O'Toole GA, Parsek MR.** 2003. Alginate is not a significant component of the extracellular polysaccharide matrix of PA14 and PAO1 *Pseudomonas aeruginosa* biofilms. *Proc Natl Acad Sci USA* **100:**7907–7912.
114. **Stapper AP, Narasimhan G, Ohman DE, Barakat J, Hentzer M, Molin S, Kharazmi A, Høiby N, Mathee K.** 2004. Alginate production affects *Pseudomonas aeruginosa* biofilm development and architecture, but is not essential for biofilm formation. *J Med Microbiol* **53:**679–690.
115. **Hentzer M, Teitzel GM, Balzer GJ, Heydorn A, Molin S, Givskov M, Parsek MR.** 2001. Alginate overproduction affects *Pseudomonas aeruginosa* biofilm structure and function. *J Bacteriol* **183:**5395–5401.
116. **Ghafoor A, Hay ID, Rehm BHA.** 2011. Role of exopolysaccharides in *Pseudomonas aeruginosa* biofilm formation and architecture. *Appl Environ Microbiol* **77:**5238–5246.
117. **Yang L, Hengzhuang W, Wu H, Damkiær S, Jochumsen N, Song Z, Givskov M, Høiby N, Molin S.** 2012. Polysaccharides serve as scaffold of biofilms formed by mucoid *Pseudomonas aeruginosa. FEMS Immunol Med Microbiol* **65:**366–376.
118. **Tielen P, Strathmann M, Jaeger K-E, Flemming H-C, Wingender J.** 2005. Alginate acetylation influences initial surface colonization by mucoid *Pseudomonas aeruginosa. Microbiol Res* **160:**165–176.
119. **Hengzhuang W, Wu H, Ciofu O, Song Z, Hoiby N.** 2011. Pharmacokinetics/pharmacodynamics of colistin and imipenem on mucoid and nonmucoid *Peudomonas aeruginosa* biofilms. *Antimicrob Agents Chemother* **55:**4469–4474.
120. **Alkawash MA, Soothill JS, Schiller NL.** 2006. Alginate lyase enhances antibiotic killing of mucoid *Pseudomonas aeruginosa* in biofilms. *APMIS* **114:**131–138.
121. **Pier GB, Coleman F, Grout M, Franklin M, Ohman DE.** 2001. Role of alginate O acetylation in resistance of mucoid *Pseudomonas aeruginosa* to opsonic phagocytosis. *Infect Immun* **69:**1895–1901.
122. **Leid JG, Willson CJ, Shirtliff ME, Hassett DJ, Parsek MR, Jeffers AK.** 2005. The exopolysaccharide alginate protects *Pseudomonas aeruginosa* biofilm bacteria from IFN-gamma-mediated macrophage killing. *J Immunol* **175:**7512–7518.
123. **Schwarzmann S, Boring JR.** 1971. Antiphagocytic effect of slime from a mucoid strain of *Pseudomonas aeruginosa. Infect Immun* **3:**762–767.
124. **Simpson JA, Smith SE, Dean RT.** 1988. Alginate inhibition of the uptake of *Pseudomonas aeruginosa* by macrophages. *J Gen Microbiol* **134:**29–36.
125. **Mai GT, Seow WK, Pier GB, McCormack JG, Thong YH.** 1993. Suppression of lymphocyte and neutrophil functions by *Pseudomonas aeruginosa* mucoid exopolysaccharide (alginate): reversal by physicochemical, alginase, and specific monoclonal antibody treatments. *Infect Immun* **61:**559–564.
126. **Simpson JA, Smith SE, Dean RT.** 1989. Scavenging by alginate of free radicals released by macrophages. *Free Radic Biol Med* **6:**347–353.
127. **Learn DB, Brestel EP, Seetharama S.** 1987. Hypochlorite scavenging by *Pseudomonas aeruginosa* alginate. *Infect Immun* **55:**1813–1818.

128. **Chan C, Burrows LL, Deber CM.** 2004. Helix induction in antimicrobial peptides by alginate in biofilms. *J Biol Chem* **279:**38749–38754.
129. **Chan C, Burrows LL, Deber CM.** 2005. Alginate as an auxiliary bacterial membrane: binding of membrane-active peptides by polysaccharides. *J Pept Res* **65:**343–351.
130. **Limoli DH, Rockel AB, Host KM, Jha A, Kopp BT, Hollis T, Wozniak DJ.** 2014. Cationic antimicrobial peptides promote microbial mutagenesis and pathoadaptation in chronic infections. *PLoS Pathog* **10:**e1004083. doi:10.1371/journal.ppat.1004083.
131. **Pier GB, Boyer D, Preston M, Coleman FT, Llosa N, Mueschenborn-Koglin S, Theilacker C, Goldenberg H, Uchin J, Priebe GP, Grout M, Posner M, Cavacini L.** 2004. Human monoclonal antibodies to *Pseudomonas aeruginosa* alginate that protect against infection by both mucoid and nonmucoid strains. *J Immunol* **173:**5671–5678.
132. **Lattner D, Flemming H-C, Mayer C.** 2003. 13C-NMR study of the interaction of bacterial alginate with bivalent cations. *Int J Biol Macromol* **33:** 81–88.
133. **Horsman SR, Moore RA, Lewenza S.** 2012. Calcium chelation by alginate activates the type III secretion system in mucoid *Pseudomonas aeruginosa* biofilms. *PLoS One* **7:**e46826. doi:10.1371/journal.pone.0046826.
134. **Lee B, Schjerling CK, Kirkby N, Hoffmann N, Borup R, Molin S, Høiby N, Ciofu O.** 2011. Mucoid *Pseudomonas aeruginosa* isolates maintain the biofilm formation capacity and the gene expression profiles during the chronic lung infection of CF patients. *APMIS* **119:**263–274.
135. **Roberts IS.** 1996. The biochemistry and genetics of capsular polysaccharide production in bacteria. *Annu Rev Microbiol* **50:**285–315.
136. **Whitfield C, Roberts IS.** 1999. Structure, assembly and regulation of expression of capsules in *Escherichia coli. Mol Microbiol* **31:**1307–1319.
137. **Bratcher PE, Kim KH, Kang JH, Hong JY, Nahm MH.** 2010. Identification of natural *Pneumococcal* isolates expressing serotype 6D by genetic, biochemical and serological characterization. *Microbiology* **156:**555–560.
138. **van der Woude MW.** 2011. Phase variation: how to create and coordinate population diversity. *Curr Opin Microbiol* **14:**205–211.
139. **Yother J.** 2011. Capsules of *Streptococcus pneumoniae* and other bacteria: paradigms for polysaccharide biosynthesis and regulation. *Annu Rev Microbiol* **65:**563–581.
140. **Lukáčová M, Barák I, Kazár J.** 2008. Role of structural variations of polysaccharide antigens in the pathogenicity of Gram-negative bacteria. *Clin Microbiol Infect* **14:**200–206.
141. **Kim JO, Weiser JN.** 1998. Association of intrastrain phase variation in quantity of capsular polysaccharide and teichoic acid with the virulence of *Streptococcus pneumoniae. J Infect Dis* **177:**368–377.
142. **Qin L, Kida Y, Imamura Y, Kuwano K, Watanabe H.** 2013. Impaired capsular polysaccharide is relevant to enhanced biofilm formation and lower virulence in *Streptococcus pneumoniae. J Infect Chemother* **19:**261–271.
143. **Sanchez CJ, Kumar N, Lizcano A, Shivshankar P, Dunning Hotopp JC, Jorgensen JH, Tettelin H, Orihuela CJ.** 2011. *Streptococcus pneumoniae* in biofilms are unable to cause invasive disease due to altered virulence determinant production. *PLoS One* **6:**e28738. doi:10.1371/journal.pone.0028738.
144. **Joseph LA, Wright AC.** 2004. Expression of *Vibrio vulnificus* capsular polysaccharide inhibits biofilm formation. *J Bacteriol* **186:**889–893.
145. **Yi K, Rasmussen AW, Gudlavalleti SK, Stephens DS, Stojiljkovic I.** 2004. Biofilm formation by *Neisseria meningitidis. Infect Immun* **72:**6132–6138.
146. **Tuchscherr LPN, Buzzola FR, Alvarez LP, Caccuri RL, Lee JC, Sordelli DO.** 2005. Capsule-negative *Staphylococcus aureus* induces chronic experimental mastitis in mice. *Infect Immun* **73:**7932–7937.
147. **Deghmane A-E, Giorgini D, Larribe M, Alonso J-M, Taha M-K.** 2002. Down-regulation of pili and capsule of *Neisseria meningitidis* upon contact with epithelial cells is mediated by CrgA regulatory protein. *Mol Microbiol* **43:** 1555–1564.
148. **Hall-Stoodley L, Nistico L, Sambanthamoorthy K, Dice B, Nguyen D, Mershon WJ, Johnson C, Hu FZ, Stoodley P, Ehrlich GD, Post JC.** 2008. Characterization of biofilm matrix, degradation by DNase treatment and evidence of capsule downregulation in *Streptococcus pneumoniae* clinical isolates. *BMC Microbiol* **8:**173.
149. **Goller CC, Seed PC.** 2010. Revisiting the *Escherichia coli* polysaccharide capsule as a virulence factor during urinary tract infection: contribution to intracellular biofilm development. *Virulence* **1:**333–337.
150. **Kim H-S, Park S-J, Lee K-H.** 2009. Role of NtrC-regulated exopolysaccharides in the biofilm formation and pathogenic interaction of *Vibrio vulnificus. Mol Microbiol* **74:**436–453.
151. **Lee K-J, Kim J-A, Hwang W, Park S-J, Lee K-H.** 2013. Role of capsular polysaccharide

(CPS) in biofilm formation and regulation of CPS production by quorum-sensing in *Vibrio vulnificus*. *Mol Microbiol* **90:**841–857.

152. **Velázquez-Hernández ML, Baizabal-Aguirre VM, Bravo-Patiño A, Cajero-Juárez M, Chávez-Moctezuma MP, Valdez-Alarcón JJ.** 2009. Microbial fructosyltransferases and the role of fructans. *J Appl Microbiol* **106:**1763–1778.
153. **Osman SF, Fett WF, Fishman ML.** 1986. Exopolysaccharides of the phytopathogen *Pseudomonas syringae pv. glycinea*. *J Bacteriol* **166:**66–71.
154. **Bereswill S, Geider K.** 1997. Characterization of the *rcsB* gene from *Erwinia amylovora* and its influence on exoploysaccharide synthesis and virulence of the fire blight pathogen. *J Bacteriol* **179:**1354–1361.
155. **Dedonder R, Peaud-Lenoel C.** 1957. Studies on the levansucrase of *Bacillus subtilis*. I. Production of levans and levansucrase (levan-sucharotransfructosidase) by cultures of *Bacillus subtilis*. *Bull Soc Chim Biol (Paris)* **39:**483–497.
156. **Li H, Ullrich MS.** 2001. Characterization and mutational analysis of three allelic *isc* genes encoding levansucrase in *Pseudomonas syringae*. *J Bacteriol* **183:**3282–3292.
157. **Marvasi M, Visscher PT, Casillas Martinez L.** 2010. Exopolymeric substances (EPS) from *Bacillus subtilis*: polymers and genes encoding their synthesis. *FEMS Microbiol Lett* **313:**1–9.
158. **Records AR, Gross DC.** 2010. Sensor kinases RetS and LadS regulate *Pseudomonas syringae* type VI secretion and virulence factors. *J Bacteriol* **192:**3584–3596.
159. **Li J, Kim IH.** 2013. Effects of levan-type fructan supplementation on growth performance, digestibility, blood profile, fecal microbiota, and immune responses after lipopolysaccharide challenge in growing pigs. *J Anim Sci* **91:**5336–5343.
160. **Shida T, Mukaijo K, Ishikawa S, Yamamoto H, Sekiguchi J.** 2002. Production of long-chain levan by a *sacC* insertional mutant from *Bacillus subtilis* 327UH. *Biosci Biotechnol Biochem* **66:**1555–1558.
161. **Leloup L, Driessen AJ, Freudl R, Chambert R, Petit-Glatron MF.** 1999. Differential dependence of levansucrase and alpha-amylase secretion on SecA (Div) during the exponential phase of growth of *Bacillus subtilis*. *J Bacteriol* **181:**1820–1826.
162. **Hettwer U, Jaeckel FR, Boch J, Meyer M, Rudolph K, Ullrich MS.** 1998. Cloning, nucleotide sequence, and expression in *Escherichia coli* of levansucrase genes from the plant pathogens *Pseudomonas syringae pv. glycinea* and *P. syringae pv. phaseolicola*. *Appl Environ Microbiol* **64:**3180–3187.
163. **Burne RA, Chen YYM, Wexler DL, Kuramitsu H, Bowen WH.** 1996. Cariogenicity of *Streptococcus mutans* strains with defects in fructan metabolism assessed in a program-fed specific-pathogen-free rat model. *J Dent Res* **75:**1572–1577.
164. **Laue H, Schenk A, Li H, Lambertsen L, Neu TR, Molin S, Ullrich MS.** 2006. Contribution of alginate and levan production to biofilm formation by *Pseudomonas syringae*. *Microbiology* **152:** 2909–2918.
165. **Kiska DL, Macrina FL.** 1994. Genetic analysis of fructan-hyperproducing strains of *Streptococcus mutans*. *Infect Immun* **62:**2679–2686.
166. **Dogsa I, Brloznik M, Stopar D, Mandic-Mulec I.** 2013. Exopolymer diversity and the role of levan in *Bacillus subtilis* biofilms. *PLoS One* **8:** e62044. doi:10.1371/journal.pone.0062044.
167. **Kolenbrander PE, London J.** 1993. Adhere today, here tomorrow: oral bacterial adherence. *J Bacteriol* **175:**3247–3252.
168. **Arvidson SA, Rinehart BT, Gadala-Maria F.** 2006. Concentration regimes of solutions of levan polysaccharide from *Bacillus* sp. *Carbohydr Polym* **65:**144–149.
169. **Krasteva PV, Fong JCN, Shikuma NJ, Beyhan S, Navarro MVAS, Yildiz FH, Sondermann H.** 2010. *Vibrio cholerae* VpsT regulates matrix production and motility by directly sensing cyclic di-GMP. *Science* **327:**866–868.
170. **Shikuma NJ, Fong JCN, Yildiz FH.** 2012. Cellular levels and binding of c-di-GMP control subcellular localization and activity of the *Vibrio cholerae* transcriptional regulator VpsT. *PLoS Pathog* **8:**e1002719. doi:10.1371/journal.ppat.1002719.
171. **Berk V, Fong JCN, Dempsey GT, Develioglu ON, Zhuang X, Liphardt J, Yildiz FH, Chu S.** 2012. Molecular architecture and assembly principles of *Vibrio cholerae* biofilms. *Science* **337:**236–239.
172. **Stevenson G, Andrianopoulos K, Hobbs M, Reeves PR.** 1996. Organization of the *Escherichia coli* K-12 gene cluster responsible for production of the extracellular polysaccharide colanic acid. *J Bacteriol* **178:**4885–4893.
173. **Anderson ES, Rogers AH.** 1963. Slime polysaccharides of the *Enterobacteriaceae*. *Nature* **198:**714–715.
174. **Grant WD, Sutherland IW, Wilkinson JF.** 1969. Exopolysaccharide colanic acid and its occurrence in the *Enterobacteriaceae*. *J Bacteriol* **100:**1187–1193.
175. **Gottesman S, Trisler P, Torres-Cabassa A.** 1985. Regulation of capsular polysaccharide synthesis in *Escherichia coli* K-12: characterization of three regulatory genes. *J Bacteriol* **162:**1111–1119.

176. **Brill JA, Quinlan-Walshe C, Gottesman S.** 1988. Fine-structure mapping and identification of two regulators of capsule synthesis in *Escherichia coli* K-12. *J Bacteriol* **170:**2599–2611.
177. **Stout V, Gottesman S.** 1990. RcsB and RcsC: a two-component regulator of capsule synthesis in *Escherichia coli*. *J Bacteriol* **172:**659–669.
178. **Majdalani N, Gottesman S.** 2005. The Rcs phosphorelay: a complex signal transduction system. *Annu Rev Microbiol* **59:**379–405.
179. **Ophir T, Gutnick DL.** 1994. A role for exopolysaccharides in the protection of microorganisms from desiccation. *Appl Environ Microbiol* **60:** 740–745.
180. **Watnick PI, Kolter R.** 1999. Steps in the development of a *Vibrio cholerae* El Tor biofilm. *Mol Microbiol* **34:**586–595.
181. **Danese PN, Pratt LA, Kolter R.** 2000. Exopolysaccharide production is required for development of *Escherichia coli* K-12 biofilm architecture. *J Bacteriol* **182:**3593–3596.
182. **Navasa N, Rodríguez-Aparicio L, Martínez-Blanco H, Arcos M, Ferrero MÁ.** 2009. Temperature has reciprocal effects on colanic acid and polysialic acid biosynthesis in *E. coli* K92. *Appl Microbiol Biotechnol* **82:**721–729.
183. **Ledeboer NA, Jones BD.** 2005. Exopolysaccharide sugars contribute to biofilm formation by *Salmonella enterica* serovar typhimurium on HEp-2 cells and chicken intestinal epithelium. *J Bacteriol* **187:**3214–3226.
184. **Ali A, Mahmud ZH, Morris JG, Sozhamannan S, Johnson JA.** 2000. Sequence analysis of TnphoA insertion sites in *Vibrio cholerae* mutants defective in rugose polysaccharide production. *Infect Immun* **68:**6857–6864.
185. **Yildiz FH, Dolganov NA, Schoolnik GK.** 2001. VpsR, a member of the response regulators of the two-component regulatory systems, is required for expression of *vps* biosynthesis genes and EPS(ETr)-associated phenotypes in *Vibrio cholerae* O1 El Tor. *J Bacteriol* **183:**1716–1726.
186. **Casper-Lindley C, Yildiz FH.** 2004. VpsT is a transcriptional regulator required for expression of *vps* biosynthesis genes and the development of rugose colonial morphology in *Vibrio cholerae* O1 El Tor. *J Bacteriol* **186:**1574–1578.
187. **Beyhan S, Bilecen K, Salama SR, Casper-Lindley C, Yildiz FH.** 2007. Regulation of rugosity and biofilm formation in *Vibrio cholerae*: comparison of VpsT and VpsR regulons and epistasis analysis of *vpsT, vpsR,* and *hapR*. *J Bacteriol* **189:**388–402.
188. **Fong JCN, Syed KA, Klose KE, Yildiz FH.** 2010. Role of *Vibrio* polysaccharide (*vps*) genes in VPS production, biofilm formation and *Vibrio cholerae* pathogenesis. *Microbiology* **156:**2757–2769.
189. **Tischler AD, Camilli A.** 2004. Cyclic diguanylate (c-di-GMP) regulates *Vibrio cholerae* biofilm formation. *Mol Microbiol* **53:**857–869.
190. **Branda SS, González-Pastor JE, Ben-Yehuda S, Losick R, Kolter R.** 2001. Fruiting body formation by *Bacillus subtilis*. *Proc Natl Acad Sci USA* **98:**11621–11626.
191. **Branda SS, González-Pastor JE, Dervyn E, Ehrlich SD, Losick R, Kolter R.** 2004. Genes involved in formation of structured multicellular communities by *Bacillus subtilis*. *J Bacteriol* **186:**3970–3979.
192. **Chu F, Kearns DB, Branda SS, Kolter R, Losick R.** 2006. Targets of the master regulator of biofilm formation in *Bacillus subtilis*. *Mol Microbiol* **59:**1216–1228.
193. **Kolodkin-Gal I, Cao S, Chai L, Böttcher T, Kolter R, Clardy J, Losick R.** 2012. A self-produced trigger for biofilm disassembly that targets exopolysaccharide. *Cell* **149:**684–692.
194. **Hobley L, Kim SH, Maezato Y, Wyllie S, Fairlamb AH, Stanley-Wall NR, Michael AJ.** 2014. Norspermidine is not a self-produced trigger for biofilm disassembly. *Cell* **156:**844–854.
195. **Ross P, Mayer R, Benziman M.** 1991. Cellulose biosynthesis and function in bacteria. *Microbiol Rev* **55:**35–58.
196. **Brown AJ.** 1886. XLIII.—On an acetic ferment which forms cellulose. *J Chem Soc* **49:**432–439.
197. **Hestrin S, Schramm M.** 1954. Synthesis of cellulose by *Acetobacter xylinum*. *Biochem J* **58:** 345–352.
198. **Cook KE, Colvin JR.** 1980. Evidence for a beneficial influence of cellulose production on growth of *Acetobacter xylinum* in liquid medium. *Curr Microbiol* **3:**203–205.
199. **Zogaj X, Nimtz M, Rohde M, Bokranz W, Römling U.** 2001. The multicellular morphotypes of *Salmonella typhimurium* and *Escherichia coli* produce cellulose as the second component of the extracellular matrix. *Mol Microbiol* **39:**1452–1463.
200. **Römling U.** 2002. Molecular biology of cellulose production in bacteria. *Res Microbiol* **153:**205–212.
201. **Ude S, Arnold DL, Moon CD, Timms-Wilson T, Spiers AJ.** 2006. Biofilm formation and cellulose expression among diverse environmental *Pseudomonas* isolates. *Environ Microbiol* **8:**1997–2011.
202. **Kimura S, Chen HP, Saxena IM, Brown RM, Itoh T.** 2001. Localization of c-di-GMP-binding protein with the linear terminal complexes of *Acetobacter xylinum*. *J Bacteriol* **183:**5668–5674.

203. **Ross P, Weinhouse H, Aloni Y, Michaeli D, Weinberger-Ohana P, Mayer R, Braun S, de Vroom E, van der Marel GA, van Boom JH, Benziman M.** 1987. Regulation of cellulose synthesis in *Acetobacter xylinum* by cyclic diguanylic acid. *Nature* **325:**279–281.
204. **Tal R, Wong HC, Calhoon R, Gelfand D, Fear AL, Volman G, Mayer R, Ross P, Amikam D, Weinhouse H, Cohen A, Sapir S, Ohana P, Benziman M.** 1998. Three *cdg* operons control cellular turnover of cyclic di-GMP in *Acetobacter xylinum*: genetic organization and occurrence of conserved domains in isoenzymes. *J Bacteriol* **180:**4416–4425.
205. **Römling U.** 2001. Genetic and phenotypic analysis of multicellular behavior in *Salmonella typhimurium*. *Methods Enzymol* **336:**48–59.
206. **Williams WS, Cannon RE.** 1989. Alternative environmental roles for cellulose produced by *Acetobacter xylinum*. *Appl Environ Microbiol* **55:**2448–2452.
207. **Monteiro C, Saxena I, Wang X, Kader A, Bokranz W, Simm R, Nobles D, Chromek M, Brauner A, Brown RM, Römling U.** 2009. Characterization of cellulose production in *Escherichia coli* Nissle 1917 and its biological consequences. *Environ Microbiol* **11:**1105–1116.
208. **Flemming H-C, Wingender J.** 2010. The biofilm matrix. *Nat Rev Microbiol* **8:**623–633.
209. **Roche Y, Cao-Hoang L, Perrier-Cornet J-M, Waché Y.** 2012. Advanced fluorescence technologies help to resolve long-standing questions about microbial vitality. *Biotechnol J* **7:**608–619.
210. **Whitchurch CB, Tolker-Nielsen T, Ragas PC, Mattick JS.** 2002. Extracellular DNA required for bacterial biofilm formation. *Science* **295:**1487.
211. **Hu W, Li L, Sharma S, Wang J, McHardy I, Lux R, Yang Z, He X, Gimzewski JK, Li Y, Shi W.** 2012. DNA builds and strengthens the extracellular matrix in *Myxococcus xanthus* biofilms by interacting with exopolysaccharides. *PLoS One* **7:**e51905. doi:10.1371/journal.pone.0051905.
212. **Peterson BW, van der Mei HC, Sjollema J, Busscher HJ, Sharma PK.** 2013. A distinguishable role of eDNA in the viscoelastic relaxation of biofilms. *MBio* **4:**e00497-13. doi:10.1128/mBio.00497-13.
213. **Hu W, Yang Z, Lux R, Zhao M, Wang J, He X, Shi W.** 2011. Direct visualization of the interaction between pilin and exopolysaccharides of *Myxococcus xanthus* with eGFP-fused PilA protein. *FEMS Microbiol Lett* **326:**23–30.
214. **Zmantar T, Bettaieb F, Chaieb K, Ezzili B, Mora-Ponsonnet L, Othmane A, Jaffrézic N, Bakhrouf A.** 2011. Atomic force microscopy and hydrodynamic characterization of the adhesion of *Staphylococcus aureus* to hydrophilic and hydrophobic substrata at different pH values. *J Microbiol Biotechnol* **27:**887–896.
215. **Irie Y, Starkey M, Edwards AN, Wozniak DJ, Romeo T, Parsek MR.** 2010. *Pseudomonas aeruginosa* biofilm matrix polysaccharide Psl is regulated transcriptionally by RpoS and post-transcriptionally by RsmA. *Mol Microbiol* **78:**158–172.
216. **Lacour S, Bechet E, Cozzone AJ, Mijakovic I, Grangeasse C.** 2008. Tyrosine phosphorylation of the UDP-glucose dehydrogenase of *Escherichia coli* is at the crossroads of colanic acid synthesis and polymyxin resistance. *PLoS One* **3:**e3053. doi:10.1371/journal.pone.0003053.
217. **Romero D, Aguilar C, Losick R, Kolter R.** 2010. Amyloid fibers provide structural integrity to *Bacillus subtilis* biofilms. *Proc Nat Acad Sci USA* **107:**2230–2234.
218. **Bokranz W, Wang X, Tschäpe H, Römling U.** 2005. Expression of cellulose and curli fimbriae by *Escherichia coli* isolated from the gastrointestinal tract. *J Med Micrbiol* **54:**1171–1182.

The Biology of the *Escherichia coli* Extracellular Matrix

12

DAVID A. HUFNAGEL,[1] WILLIAM H. DEPAS,[2] and MATTHEW R. CHAPMAN[1]

BASIC HISTORY, PHYLOGENY, AND HABITATS OF *ESCHERICHIA COLI*

The *Enterobacteriaceae* bacterial family includes a variety of intestinal symbionts as well as notable pathogens such as *Salmonella enterica*, *Serratia marcescens*, *Klebsiella*, and *Yersinia pestis* (1). Also included among the *Enterobacteriaceae* family is the most well-documented bacterial species on Earth, *E. coli*. *E. coli* is a fascinatingly diverse bug, featuring a cadre of different strains that have adapted to diverse environmental conditions and lifestyles. While the typical *E. coli* genome contains roughly 4,800 genes, only approximately 1,700 are shared by every *E. coli* strain (2). In total, there are over 15,000 genes that make up the *E. coli* pangenome (2, 3). The genomic plasticity of various *E. coli* isolates provides *E. coli* the ability to proliferate and survive in an array of environments (4, 5).

A major niche of *E. coli* is the lower intestinal tract of mammals, birds, and reptiles (6). Indeed, *E. coli* was first isolated by Theodor Escherich from a human stool sample in 1886 (7). Among the first bacteria to colonize the intestinal tract of human infants, *E. coli* establishes a stable population of roughly 10^8 CFU/g of feces by adulthood (5, 8, 9). The intestinal tract is an

[1]Department of Molecular, Cellular, and Developmental Biology; [2]Microbiology and Immunology, University of Michigan, Ann Arbor, MI 48109.

Microbial Biofilms 2nd Edition
Edited by Mahmoud Ghannoum, Matthew Parsek, Marvin Whiteley, and Pranab K. Mukherjee

doi:10.1128/microbiolspec.MB-0014-2014

oxygen limiting environment, and *E. coli* is a facultative anaerobe that can reduce several alternate terminal electron acceptors such as nitrate, fumarate, dimethyl sulfoxide (DMSO), and trimethylamine N-oxide (TMAO) (10). Mouse colonization studies have revealed that in the intestine, fumarate reductase, nitrate reductase, and *bd* oxidase (high-affinity oxygen cytochrome) are particularly important for *E. coli* fitness (11, 12). The physiological flexibility of *E. coli* renders it well suited for the diverse environments encountered in the intestinal tract.

While the lower digestive tract is the primary habitat of *E. coli*, fecal dissemination leads to the passage of *E. coli* to its secondary environment outside of the host. The host and nonhost environments can be dynamic and differ in nutrient availability, temperature, and the number and nature of competitors, among other things. Once *E. coli* is excreted in stool, there is generally a net-negative fitness cost compared to growth in most host environments (9, 13). *E. coli* doubles around every two days in a human host, but outside of the host, instead of dividing, *E. coli* perishes after an average of four days depending on the environment (9). However, as *E. coli* is constantly being excreted, it has been estimated that half of the *E. coli* cells on Earth exist outside the host (9). Interestingly, studies have found that *E. coli* can grow in soil not only in tropical conditions, but also in colder temperatures (14).

Many studies have investigated how enterohemorrhagic *E. coli* (EHEC) strains, particularly EHEC O157:H7, make their way from environmental reservoirs to humans (15). *E. coli* is transmitted from host to host by the fecal-oral route. The ability to survive outside of the host therefore facilitates the host-to-host transmission of all *E. coli* variants, including pathogenic strains. EHEC strains cause diarrhea, abdominal cramps, and in certain cases the life-threatening hemolytic-uremic syndrome (16). The intestinal tract of domesticated cattle serves as the primary reservoir for EHEC in the United States, and EHEC can contaminate meat during the slaughter process (17, 18). Additionally, EHEC is shed in feces and can survive in manure for months (17, 19–21), making contact with animal feces a risk factor for EHEC infections. Even using untreated manure as fertilizer can result in contaminated produce (15, 22–24). While contaminated manure or water can foul the surface of plants, there is also evidence that EHEC can invade plant tissue (23–25). *E. coli*'s ability to survive outside of the intestinal tract is integral to EHEC's ability to cause outbreaks.

Extraintestinal pathogenic *E. coli* (ExPEC) are strains that cause disease outside of the intestinal tract of animals, including uropathogenic, or UPEC, strains. Urinary tract infections (UTIs) (11) are one of the most common bacterial infections, costing over $3 billion in health care expenses in the United States alone (26). UPEC can live in the intestine without causing disease, and recent work has shown that UPEC has no fitness defects in the gut environment when compared with commensal *E. coli* strains (27). How ExPEC strains like UPEC transmit from the intestine to the site of disease is not completely understood, but at least some strains can spread via the fecal-oral route (28). In fact, the primary source of UPEC that colonizes the urethra is a patient's own intestinal tract (29). ExPEC, including UPEC strains, can also be found on food products, and clonal UPEC outbreaks have been reported (30–32). Taken together, these data suggest that UPEC strains are at least partly dependent on survival outside of the host before they can recolonize a second host and cause disease.

UPEC has specifically adapted to cause disease in the urinary tract (26). It ascends the urinary tract to the bladder and causes infection (26). UPEC invades bladder epithelia cells in a type 1 pili-dependent mechanism, where it forms tight-knit and aggregated intracellular bacterial communities (IBCs) (33, 34). IBCs are drug resistant and can evade host immune responses, allowing for cells within the IBC to proliferate and to further infect additional bladder epithelial cells (35, 36).

INTRODUCTION TO BIOFILMS AND THE EXTRACELLULAR MATRIX

Biofilm formation can increase bacterial fitness in both host and nonhost environments (37, 38). In this review we will use the general definition of a biofilm as a group of surface-associated bacteria enveloped in a self-produced extracellular matrix (39). The *E. coli* extracellular matrix contains a major protein polymer called curli and the carbohydrate polymer, cellulose (40–42). Although curli and cellulose are typically the most abundant biofilm constituents, the extracellular matrix of *E. coli* can also include type 1 pili, flagella, antigen 43, DNA, β-1,6-N-acetylglucosamine (β-1,6-GlcNAc), capsule sugars, and colanic acid (43). Most pathogenic strains of *E. coli* form robust biofilms, but some laboratory strains of *E. coli* are attenuated in their ability to produce biofilms. The K12 strain of *E. coli* was first isolated from Stanford in 1922, and was subsequently passaged for more than 50 years (44). This passaging led to evolutionary adaptation to the laboratory growth conditions and to the loss of certain traits that influence biofilms (45). K12 *E. coli* therefore requires extended periods of time to adhere to surfaces and form biofilms (40). On the other hand, a host of pathogenic, environmental, and commensal *E. coli* isolates readily form biofilms in the laboratory and therefore make great model organisms for biofilm formation studies (46–50).

Biofilm formation correlates with resistance to a variety of environmental stresses, including antibiotics, the immune system, and predation (38). Resistance is conferred through at least two distinct mechanisms. First, the extracellular matrix forms a physical barrier that can resist shear stress and recognition and phagocytosis by immune cells (38). Second, bacteria within biofilms often assemble into subpopulations that have distinct physiological characteristics (46, 51, 52). Subpopulation development can be triggered by mutations, stochastic gene expression, or chemical gradients that develop during biofilm formation (37, 52–54). For instance, bacteria at the biofilm surface are exposed to more oxygen, stimulating a higher rate of aerobic respiration (37, 55, 56). Metabolic changes often coincide with resistance to different stresses (54, 57). A biofilm community with multiple subpopulations, each resistant to different stresses, therefore demonstrates resistance to a broader range of environmental pressures to the biofilm community as a whole (37, 54, 57).

LABORATORY BIOFILM MODELS

Laboratory biofilm models have been used to determine most of the spatiotemporal and molecular characteristics of biofilms (Fig. 1). *E. coli* forms at least two distinct types of biofilm in static liquid cultures; both form at the air liquid interface and both require production of extracellular polymers (41, 43, 47, 58, 59) (Fig. 1A,B,C). The accumulation of biomass at the air-exposed edges of polyvinyl chloride wells or glass culture tubes with cultures grown in lysogeny broth (LB) media is reliant on type 1 pili, poly-β-1,6-GlcNAc, and flagella (47, 58, 60) (Fig. 1A, top image). The crystal violet–stained biofilm rings can be visualized in the top image in Figure 1A, whereas the flagella mutant (*fliC*::kan) grown in the tube shown in the bottom image did not form rings on the glass culture tube.

E. coli can also form a pellicle biofilm that floats at the air liquid interface. Pellicles are films of curli/cellulose-encased cells that span the entire air-liquid interface of a single well (47, 59, 61). Pellicle formation can be quantified by crystal violet staining of the biofilm biomass (62, 63) (Fig. 1B). The culture in the top well in Figure 1B formed a robust

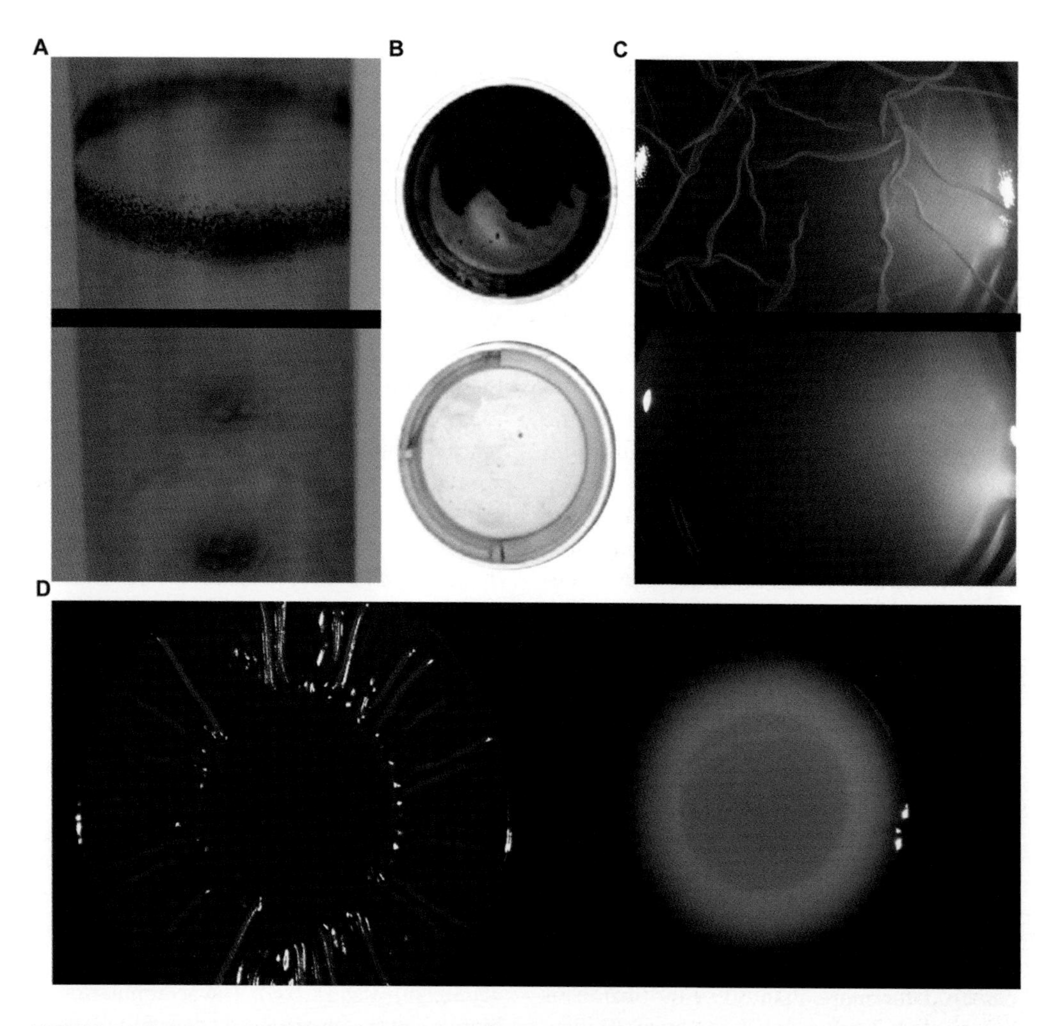

FIGURE 1 Laboratory *E. coli* biofilm models. (A) Ring biofilm stained by crystal violet (CV). Cultures were grown in LB media in glass tubes at 26°C for 48 hours. Liquid culture was removed and the tube was stained with 0.1% (w/v) CV for 5 minutes. Tubes were subsequently washed with water. The top image is a WT strain, and the lower image is a flagella mutant (*fliC*::kan). (B) Pellicle biofilms grown in a 24-well plate for 48 hours at 26°C. Liquid media was removed followed by 5 minutes of staining with 0.1% CV. Stained pellicles were washed three times with water prior to imaging. The top image is a CV-stained WT UTI89 pellicle, whereas the lower picture is a culture of a *ΔcsgD* mutant that did not produce a pellicle. (C) Pellicle biofilms grown in 1:7500 (Congo red:YESCA) media in a 24-well dish for 48 hours at 26°C. The top image shows a WT UTI89 culture that produced a pellicle, whereas the lower image is a culture of a *ΔcsgD* mutant that did not form a pellicle. (D) 4-μL spots of 1-OD_{600}*E. coli* were grown at 26°C for 48 hours on YESCA CR plates. The colony on the left is UTI89 WT; on the right is a *csgD* mutant colony.
doi:10.1128/microbiolspec.MB-0014-2014.f1

pellicle at the air-liquid interface of the culture, whereas the culture in the bottom well did not.

Pellicle biofilm formation relies on the biofilm master regulator, CsgD. CsgD regulates expression of the matrix components curli and cellulose (41, 64). CsgD-mediated biofilms can be inhibited by the presence of glucose, temperatures greater than 30°C, and high osmolarity, because all of these conditions repress CsgD activity (47, 65–67). When *E. coli* is grown in conditions that are conducive to *csgD* expression, curli and cellulose-dependent biofilms can manifest in a variety of ways. The culture grown in the bottom image of Figure 1B is a *csgD* mutant that is unable to produce curli or cellulose. The pellicle architecture at the air-biofilm interface can be highlighted by growing the static culture in media amended with the diazo dye Congo red (61, 68) (Fig. 1C). Congo red stains both curli and cellulose, allowing for visualization of the pellicle biofilms and the ornate wrinkled morphology at the surface of the culture (Fig. 1C, top image).

Confocal laser scanning microscopy and electron microscopy analysis revealed that two separate populations exist at the pellicle biofilm air-liquid interface (59, 61, 69). At the air-biofilm interface cells are encased in a thick fibrous extracellular matrix, whereas cells at the liquid interface are more evenly spaced and often not surrounded by a fibrous extracellular matrix (59). While curli and cellulose are the chief components of pellicle biofilms, pili are involved in pellicle development and maturation, because deletion of type-1 pili leads to a less robust pellicle (59). Flagella are also required for pellicle development, since motile cells are necessary for colonization at the top of the static culture (47, 59).

The CsgD-induced matrix components curli and cellulose are also required for colony biofilm formation in *E. coli* and *Salmonella* spp. A variety of bacterial species, including *E. coli*, produce wrinkled colony biofilms on agar plates (Fig. 1D). The nomenclature for the wrinkled colony phenotype varies between species, but some common names are rugose biofilms, wrinkled colony biofilms, and red dry and rough (rdar) biofilms. Here, we will use the term rugose biofilms to describe all wrinkled colony biofilms on agar plates. The mechanics of rugose biofilm formation have been studied extensively in multiple species including *E. coli*, *Vibrio cholera*, *Pseudomonas aeruginosa*, *Bacillus subtilis*, and *S. enterica* serovar *Typhimurium* (41, 70–78).

When grown under CsgD-inducing conditions on agar plates, *E. coli* will form rugose colony biofilms (41, 79) (Fig. 1D, left image). Curli and cellulose are required for rugose development and are the major extracellular structures in the rugose biofilm matrix (41, 46, 80). Curli and cellulose mutant strains form nonspreading and unwrinkled colonies that do not bind Congo red (Fig. 1D, right image). While flagella are not required for the rugose colony morphotype, including colony spreading, the interior cells in a rugose colony are heavily flagellated (51). Since *E. coli* K12 strains have acquired a mutation that prevents cellulose production, they do not form rugose biofilms unless the cellulose synthesis defect is repaired (81). Indeed, repeated culturing tends to select for genetic suppressors of biofilm formation in a variety of bacteria (45). However, more recently isolated commensal *E. coli*, ExPEC, and intestinal pathogenic *E. coli* are able to form rugose biofilms (62, 82, 83) (Fig. 1D). Curli, cellulose, and rugose biofilms are produced not only in *E. coli* and *Salmonella* spp., but also in other *Enterobacteriaceae*, such as *Citrobacter* spp. (46, 84). Because curli and cellulose are required for rugose biofilm development, the environmental signals leading to rugose development coincide with those affecting CsgD expression (78, 85).

CsgD-MEDIATED CONTROL OF THE EXTRACELLULAR MATRIX

CsgD is a FixJ/LuxR/UhpA–type response regulator that contains a typical C-terminal helix-turn-helix DNA-binding domain (64, 86). The N-terminus of CsgD is thought to process environmental signals that affect the protein's ability to bind DNA and regulate transcription (86–88). The N-termini of FixJ-type response regulators are typically modified by phosphorylation on highly conserved aspartic acid residues (89, 90). However,

CsgD is missing the conserved phosphorylation sites, which has confused the role that phosphorylation plays in regulating CsgD (88, 91). However, acetyl phosphate can phosphorylate CsgD *in vitro* (89), which reduces the ability of CsgD to bind particular promoters, suggesting that post-translational phosphorylation regulation occurs (89).

The expression of *csgD* is controlled by a large number of transcriptional regulators and small RNAs (92, 93). In general, low salt, low temperature, and low glucose conditions trigger *csgD* expression (64, 66, 92, 94, 95). Additionally, *csgD* expression requires the stationary phase sigma factor RpoS (67, 77), and CsgD activity is regulated by the small molecule, cyclic-di-GMP (96). Diguanylate cyclases contain a GGDEF domain that promotes cyclic-di-GMP production, and phosphodiesterases contain an EAL domain that promotes the breakdown of c-di-GMP (96). *E. coli* and *S. enterica* encode a number of diguanylate cyclases and phosphodiesterases to regulate cytoplasmic levels of cyclic-di-GMP (97–99). The GGDEF- and EAL-containing proteins, STM2123 (YegE) and STM3388, are both required for wild type (WT) CsgD protein levels in *S. enterica* serovar Typhimurium (99). Alternatively, the diguanylate cyclases, YdaM and YfiN, are required for WT CsgD protein levels in *E. coli* (68, 98). The mechanism by which these diguanylate cyclases control CsgD protein levels is unknown. C-di-GMP can be bound by RNA (riboswitches) and by multiple binding motifs in proteins (100). In addition, the diguanylate cyclases themselves can have a secondary function outside of c-di-GMP production, leading to a wide variety of potential causes of CsgD protein level suppression (100).

CsgD controls a modest regulon of roughly 13 genes/operons (64, 101–103). Included in the regulon of genes that CsgD upregulates is *iraP*, which leads to a relay system where the IraP protein stabilizes RpoS, which results in more CsgD expression (102). CsgD directly represses the flagella biosynthesis genes *fliE* and *fliF* (103) and induces expression of curli by directly binding to the *csgBAC* promoter (64). CsgD induces cellulose production indirectly by positively regulating the diguanylate cyclase AdrA (41) (Fig. 2). Cyclic-di-GMP produced by AdrA activates the cellulose synthase and promotes cellulose production (41) (Fig. 2). The complex regulation of CsgD and the promotion of curli and cellulose go hand in hand (Fig. 2), suggesting that only

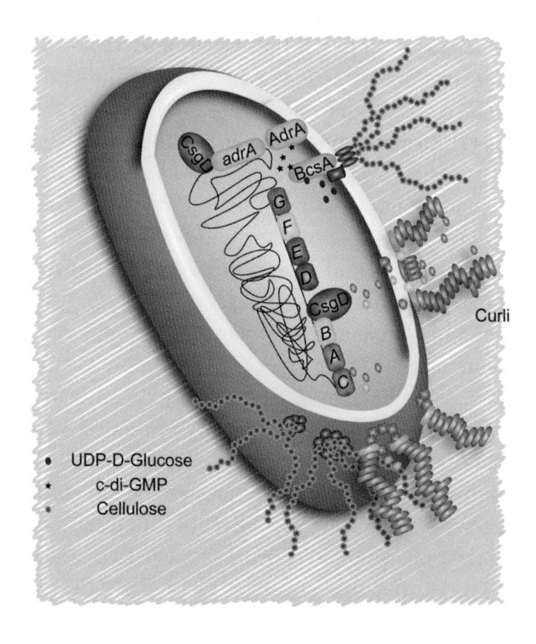

FIGURE 2 ECM production model. CsgD is the master regulator of the biofilm extracellular matrix. CsgD transcriptionally upregulates the *csgB* and *csgA* genes, which encode the minor and major curli fiber subunits, respectively. CsgA and CsgB are secreted through an outer membrane pore formed by CsgG. CsgE is thought to facilitate translocation of curli subunits across the outer membrane by capping the periplasmic side of the secretion vestibule so that movement in the channel is unidirectional. CsgB associates with the cell surface and templates amyloid polymerization of CsgA. CsgD also transcriptionally upregulates *adrA*. AdrA is an inner membrane diguanylate cyclase, which produces the secondary messenger, c-di-GMP. c-di-GMP binds and activates BcsA, which then produces cellulose fibers via the building block UDP-glucose. C-di-GMP that activates BcsA can also be produced via YedQ and YfiN. doi:10.1128/microbiolspec.MB-0014-2014.f2

under particular conditions is it advantageous for the cell to decrease motility and produce an extracellular matrix.

HISTORY OF CURLI

Curli were first described in 1989 when Staffan Normark and colleagues were investigating bovine mastitis isolates for the ability to bind to host cell matrix components (94). The authors noticed that half of the isolates bound fibronectin when grown under conditions that we now know to favor curli production (94). Electron microscopy revealed that the fibronectin-binding isolates produced fibrous coiled surface structures that they called curli (94). The presence of curli fibers in *Salmonella* was discovered a few years later by Collinson et al. (104). These fibers were originally termed thin aggregative fimbriae, or tafi, and the authors did not initially think that they were related to curli (104) because *Salmonella* produced tafi fibers when grown at both 30°C and 37°C (104). However, Arnqvist et al. found that both *Salmonella enteritidis* tafi and *E. coli* curli are the same fiber and that they are both primarily composed of the protein monomer, CsgA (105).

THE CURLI BIOGENESIS PATHWAY

Curli fibers were the first described extracellular fibers to polymerize by the nucleation-precipitation mechanism, also known as the type VII secretion system (106–110). Curli fiber subunits and their assembly machinery are encoded on two divergently transcribed operons, *csgDEFG* and *csgBAC* (Fig. 2) (64). The *csgBAC* operon encodes the major and minor curli subunits, CsgA and CsgB, respectively. Both CsgA and CsgB are secreted across the outer membrane. On the cell surface, CsgB associates with the outer membrane and provides a template that promotes CsgA to adopt an amyloid fold and form cell-associated curli fibers (111) (Fig. 2). Electron microscopy analysis has led to the estimate that 30% of *E. coli* cells produce curli under curli-inducing conditions (112). A *csgB* mutant secretes unfolded CsgA that can polymerize on the surface of a *csgA* mutant cell (which presents a surface-associated CsgB), in a process called "interbacterial complementation" (106, 108). Curli subunits from *E. coli, S. enterica* serovar Typhimurium, and *Citrobacter* can cross-seed and polymerize into amyloid fibers *in vitro* (113). Furthermore, interbacterial complementation between *E. coli* and *S. enterica* serovar Typhimurium has been observed *in vivo*. Interspecies curli production restores agar adherence to the mixed species bacterial communities (113), supporting the idea that curli subunits may be a community resource that can be utilized indiscriminately by all cells in a localized community.

The *csgDEFG* operon encodes the CsgD regulator and three curli assembly accessory proteins that are each required for WT levels of curli production (Fig. 2) (64). CsgG forms a nonameric pore in the outer membrane that facilitates curli secretion (Fig. 2) (112, 114–116) and has an approximately 2-nm central pore that spans the outer membrane (112, 116). During curli assembly CsgG forms discrete puncta that associate with two other curli accessory proteins, CsgE and CsgF (Fig. 2) (114). Electron microscopy analysis revealed that CsgG, CsgE, and CsgF cluster together around curli fibers (112). CsgE is a periplasmic protein that provides substrate specificity to CsgG-mediated secretion and also has chaperone-like activity because it can prevent CsgA from assembling amyloid fibers *in vitro* (108, 117). CsgE caps the secretion vestibule on the periplasmic side of CsgG, which facilitates unidirectional migration of substrates to the cell surface (116). CsgE antiamyloid activity can arrest CsgA fiber formation at different stages of CsgA polymerization, although it cannot disassemble preformed fibers (118). When CsgE is added exogenously to the media, pellicle biofilm formation is interrupted (118). CsgC is a periplasmic protein that inhibits CsgA amyloid polymerization at very low sub-stoichiometric ratios (163). When CsgA isn't properly secreted to the cell surface, CsgC is required to hold CsgA in a non-aggregated state that allows for proteolytic degradation of CsgA (163). Interestingly, CsgC can also function to inhibit polymerization of other disease-associated amyloids like α-synuclein (163). Finally, CsgF is necessary for both cell surface association of curli and CsgA polymerization into fibers *in vivo* (108, 111) (Fig. 2).

CHEMICAL AND PROTEIN MODULATION OF CsgA AMYLOID FORMATION

Curli are part of a growing class of β-rich, ordered protein fibers called amyloids. Amyloids have a villainous history, because the amyloid fold is associated with type 2 diabetes, Alzheimer's disease, and Parkinson's disease, among many others (119). The signature and well-studied role of amyloid formation in human diseases has led to the search for drugs that can prevent amyloid formation. However, progress in the amyloid formation field has been somewhat hampered by a lack of robust, reproducible and tractable cellular model systems. The curli biogenesis system in *E. coli* is ideally situated as a platform for screening drugs that might interrogate the amyloid formation process. The curli system in *E. coli* provides sophisticated genetics and a suite of different assays to measure *in vivo* amyloid formation, including curli-dependent biofilm formation and Congo red binding. We have used the curli system to begin screening and characterizing a library of designer peptidomimetic compounds called 2-pyridones (61, 118, 120). Pyridone compounds were initially recognized as antiamyloid compounds in a screen of small organic molecules testing for binding strength to the amyloid associated with Alzheimer's disease, $A\beta^{1\text{-}42}$ (121). The 2-pyridone compounds have a rigid bicyclic structure, which maintains the compounds in a straight conformation that mimics a β-strand (118). Altered 2-pyridone compounds were constructed that had increased solubility and were used to inhibit the type 1 pilus assembly in UTI89 via interruption of the chaperone-usher interaction (122, 123).

The CF_3-phenyl substituted 2-pyridone, FN075, is a potent amyloid inhibitor (61, 120, 123). FN075 inhibits CsgA amyloid polymerization *in vitro*, and it also blocks biofilm formation by preventing curli formation and inhibiting type 1 pilus assembly (61). FN075 inhibits CsgA amyloid fiber formation by directing CsgA into soluble oligomers that are not on a pathway to the amyloid fiber (120). Interestingly, the CF_3-phenyl substitution does not abrogate 2-pyridone activity against the type 1 pilus. Thus, FN075 functions as a dual-function pilus inhibitor (pilicide) and curli inhibitor (curlicide). FN075 co-inoculation with UPEC in a mouse UTI model caused decreases in initial titers of bacteria and IBC formation (61). Recently, FN075 was used as a scaffold for the production of new, chemically distinct compounds (118). Compounds from this library are being screened for their ability to inhibit CsgA amyloid formation, and several interesting leads have been identified (118). Two of these compounds were found to accelerate CsgA amyloid formation (118). Interestingly, acceleration of amyloid formation may be one route to prevent amyloid-related cellular toxicity, because conformationally dynamic amyloid intermediates are hypothesized to be the root of amyloid toxicity (124). Therefore, small molecule "accelerators" should facilitate the conversion of dynamic oligomers into the more stable and less cytotoxic amyloid state, and thus have promising therapeutic value.

CELLULOSE PRODUCTION

The other major polymer present in the *E. coli* extracellular matrix is cellulose, a linear chain of β-(1,4)-linked glucose monomers (41, 80). Bacterial cellulose production was first described in 1887, and in recent years *Gluconacetobacter xylinus* has been the model organism for studies of bacterial cellulose synthesis (125–127). Cellulose production by the *Enterobacteriaceae* family was first described in 2001 (41), and this initiated the discovery of cellulose production in many additional *E. coli* strains (41, 46, 62, 84, 128). Most non-K12 *E. coli* strains can produce cellulose as a component of the biofilm matrix, while K12 strains do not (41, 128). In the case of K12 strains W3110 and MG1655, the cellulose defect is due to a single nucleotide polymorphism in *bcsQ* that results in a premature stop codon (81). Repairing this single nucleotide polymorphism in *bcsQ* restores cellulose synthesis in *E. coli* W3110 and induces rugose colony formation (81).

Expression of the cellulose synthesis genes is usually dependent on the master biofilm regulator, CsgD (41, 46). CsgD modulates cellulose synthesis by activating transcription of *adrA* (41, 46, 85, 128) (Fig. 2). AdrA contains a GGDEF domain typical of diguanylate cyclases (85, 97), and AdrA-produced cyclic-di-GMP binds to the PilZ domain of the cellulose synthase, BcsA, which stimulates cellulose synthesis (129, 130) (Fig. 2). BcsA covalently links UDP-D-glucose monomers into a growing glucan chain (127, 131) (Fig. 2). AdrA is not the only diguanylate cyclase that can produce cyclic-di-GMP to drive BcsA activation. *E. coli* 1094 can produce cellulose independently of CsgD, and does so via the diguanylate cyclase YedQ that works in place of AdrA (128). Furthermore, mutants in the disulfide bonding system produce cellulose independently of CsgD in the uropathogenic strain UTI89 (68). The YfiN negative regulator protein, YfiR, is unstable and degraded in disulfide bonding system mutants and can no longer repress activation of the diguanylate cyclase, YfiN (68, 132). Constitutive activation of YfiN leads to accumulation of cyclic-di-GMP and activation of the cellulose synthase in the presence of high salt, high temperature, and glucose, conditions in which CsgD is normally repressed (68). Additionally, YfiN and cellulose production are activated by reducing conditions, suggesting that under these conditions cellulose can be produced independently of curli (68).

The biophysics of BcsA activation and cellulose production have been recently elucidated (131, 133, 134). The BcsA glycosyl transferase domain is on the cytoplasmic C-terminus located next to the PilZ domain (133). In *Rhodobacter sphaeroides*, a salt bridge between an arginine of the PilZ domain and a glutamine in the gating loop is broken once ci-di-GMP is bound, which allows the glycosyl transferase domain to engage UDP-glucose (133, 134). Mutation of the glutamine in the gating loop results in constitutive cellulose expression, because the inhibitory salt bridge is not formed (134). The UDP-binding site allows space for a single addition to the nascent glucan chain (133). A glucan chain of approximately 10 glucose monomers aligns on the periplasmic side of BcsA, perpendicular to the axis of the BcsA β-barrel (133). Additionally, a by-product of cellulose formation, UDP, acts as a competitive feedback inhibitor of the glycosyl transferase (131, 134). *E. coli* and *R. sphaeroides* BcsA and BcsB are sufficient for production of cellulose *in vitro*, but the reaction requires c-di-GMP, UDP-glucose, and both BcsA and BcsB to be colocalized in the same liposome to function (131).

ENTEROBACTERIACEAE EXTRACELLULAR MATRIX IN THE HOST AND DISEASE

Despite the fact that many *E. coli* strains express curli and cellulose maximally at 26°C, as might be the case outside of the host, both components are expressed in the host environment and play a role in commensal and disease interactions. Both *E. coli* and *Salmonella* clinical isolates express CsgD, curli, and cellulose at 37°C (62, 77). *Salmonella* produces cellulose inside macrophages, which causes a decrease in virulence (164). Cellulose null strains and cellulose inhibition through the MgtC protein were found to increase virulence during infection, and it is hypothesized that cellulose can act as a mediator between pathogenicity and long-term survival to increase *Salmonella* transmission (164). UPEC curli production is part of early-stage bladder colonization in a mouse model, because curli mutants have decreased bladder titers (61). Curli promote interactions with epithelial cells, and they can inhibit the polymerization of the antimicrobial factor LL-37, which plays a major role in host defense of the urinary tract (135). Untreated UPEC infections can lead to dissemination and infections of the bloodstream, also known as bacteremia (136). Interestingly, UPEC strains isolated from bacteremic patients produce more curli at 37°C than UPEC strains isolated from nonbacteremic patients (137). Additionally, over 50% of *E. coli* isolates from patients with sepsis produce curli at 37°C (138). Serum samples from patients that had sepsis had antibodies against the major curli subunit, CsgA, whereas control patients did not (138). Because curli and cellulose are expressed by clinical isolates under host-colonization conditions, it is imperative to understand how these matrix components shape the host-pathogen interaction and the molecular pathways that induce their production in these nonlaboratory conditions.

In addition to curli and cellulose, the extracellular matrix components type 1 pili, K1 capsule, and antigen 43 also have roles in promoting UPEC bladder colonization and biofilm formation. Antigen 43 is an outer-membrane protein required for biofilm formation in glucose minimal media, but not LB media (139). Antigen 43 drives aggregation and cell-surface and cell-cell interactions of *E. coli* biofilms via antigen 43–antigen 43 interaction (139, 140). Antigen 43 production increases interspecies biofilm formation between *E. coli* and *Pseudomonas fluorescens* (140). IBCs that form in the superficial facet cells that line the bladder epithelium are composed of densely compact and coccoid-shaped cells. Type 1 pili mutant strains have cells that are very dispersed and retain their rod shape (34). The polysaccharide K1 capsule is produced in IBCs and helps UPEC evasion of neutrophils in the host (141). K1 capsule mutants form dispersed and disordered intracellular bacterial populations that have neutrophils within the confines of their communities (141).

Beyond causing disease in the urinary tract, most *E. coli* in the human body persists and associates within the lower intestinal tract by interacting with gut and colon epithelia (9, 13). Curli and cellulose are required for proper attaching and effacing of *E. coli* to host colon cancer cells and to bovine cow colon explants (142). Curli also increase the intestinal colon cell internalization of *E. coli* commensal isolates (143).

The immune system and host cells interact with the extracellular matrix components produced by *E. coli* and *Salmonella*. Colon cells have increased Il-8 production in the presence of both flagellated and "curliated" *E. coli*, while macrophages have increased nitrous oxide and IL-6 production in the presence of purified *Salmonella* curli fibers (143, 144). Host cells recognize amyloids and also *Salmonella* curli fibers through TLR1/2 heterodimer (145). CD14 on the host cell surface complexes with TLR1/2 and increases the colon cell response to curli fibers (144). Polarized epithelial cells recognize curli and decrease the permeability of junctions between cells, thus decreasing dissemination of *Salmonella* past gut cells (146). The permeability of host gut epithelial cells was visualized via the migration of labeled beads through a polarized cell epithelia monolayer in the presence and absence of curli (146). In agreement with these findings, infection with curli-positive *Salmonella in vivo* also led to increased barrier formation in gut epithelial cells, and less curli-positive *Salmonella* was recovered from extraintestinal tissues in a mouse model (146).

ENTEROBACTERIACEAE EXTRACELLULAR MATRIX OUTSIDE OF THE HOST

E. coli and *Salmonella* also utilize biofilms in their life cycle outside of the host. The strategy of gut microbes outside of the host shifts from growth and nutrient acquisition to a lifestyle of attachment, aerobiosis, and prolonged survival (13). Conditions associated with life outside of the host such as low temperature, low glucose and nutrients, and oxidative stress induce extracellular matrix production (46, 147–149). Curli and cellulose are generally coexpressed due to their mutual dependence on the transcriptional regulator CsgD (41, 64, 95). The extracellular matrix components cellulose and curli contribute to the survival of *S. enterica* serovar *Typhimurium* to both desiccation and bleach stress (150). Expression of both curli fibers and cellulose in the extracellular matrix fraction of rugose biofilms correlates with H_2O_2 resistance (46, 150). Cellulose protects *E. coli* from feeding by the nematode Caenorhabditis elegans, and curli protects *E. coli* from the predatory bacteria *Myxococcus xanthus* (165). Unsurprisingly, matrix encased cells in the biofilm were more protected against predation. Curli and cellulose were also found to be produced by environmental isolates of *E. coli*, *Salmonella*, and *Citrobacter koseri* in extra-host mimicking environments like pig dung, produce, and meat (165). Conditions outside of the host induce curli and cellulose, which protect *Enterobacteriaceae* from these harsh conditions.

The extracellular matrix helps bacteria remain in close proximity with one another and also facilitates attachment to the surfaces colonized in the extra-host environment (40, 151). EHEC curli expression increases attachment to produce and abiotic surfaces (152, 153). Curli are important for *E. coli* attachment to alfalfa sprouts and seed coats after 3 days of coincubation (154). *E. coli* O157:H7–expressing curli strains are more hydrophobic and have increased binding to produce surfaces (155). Interaction with the lettuce rhizosphere upregulates a curli regulator, Crl, and attachment to the rhizosphere is decreased in a curli mutant (156).

Colanic acid is an extracellular polysaccharide that has varying uses in different *E. coli* strains and biofilms. Colanic acid is upregulated in conditions that occur outside of the host (157). In *E. coli* K12, colanic acid does not contribute to surface adherence but is necessary for development of three-dimensional architecture in biofilms on glass slides and for robust LB biofilm development in static culture (158, 159). In contrast to other surfaces, colanic acid does increase adherence of *E. coli* to alfalfa sprouts and to certain plastic surfaces (160).

CONCLUSIONS AND THOUGHTS FOR FUTURE STUDIES

The fantastic understanding we have of *E. coli* physiology, combined with the wealth of genetic tools afforded by *E. coli*, provides an exciting platform for understanding biofilm biology. We must better define the role that curli and cellulose play in the life cycle of enteric bacteria like *E. coli*. To date, only a few studies have attempted to assess the prevalence of curli and cellulose in environmental soil and water samples. Initial studies have estimated that some 5 to 40% of environmental biofilms contain microbial curli-like amyloids, highlighting the prevalence of amyloids in natural systems. Furthermore, *csg* homologues have been found in a wide array of bacterial species inside and outside of *Enterobacteriaceae* (161, 162). Fully understanding the range of curli in natural environments will allow a greater appreciation of the importance of the ECM to Earth's microbiota.

We must also better understand the role that the ECM plays in mediating interactions with plants and animals. Mounting evidence suggests that curli are required for interactions with plants, and curliated *E. coli* adheres robustly to the rhizosphere and the leaves of various plants (152, 155, 156). Advancing the biochemical knowledge of curli and plant interaction could foster various treatments for the prevention and removal of contaminating bacteria on fresh produce.

How the ECM shapes the infectious cycle of important human pathogens has yet to be defined. Antibodies are produced against curli in patients suffering from sepsis, so the host is clearly exposed to curli during an infection (138, 146). Curli have been shown to augment epithelial cell barriers and to decrease titers in extraintestinal tissue of mice, suggesting that maybe curliated bacteria in the intestinal tract lead to a commensal relationship between resident flora and mammalian gut cells (146). An important part of this equation will be to determine how certain isolates of *E. coli* and *Salmonella* produce cellulose and curli at human body temperature, while other strains only produce curli at temperatures less than 30°C. The study of various ECM activation pathways of *E. coli* at human body temperatures could potentially lead to powerful therapeutic molecular targets for ECM-related *E. coli* infections.

ACKNOWLEDGMENTS

Thank you to the Chapman and Blaise Boles Labs for useful discussions on the manuscript. Thanks to the Ann Miller Lab for the use of their dissecting scope. This work was supported by National Institutes of Health Grant RO1 A1073847-6 and University of Michigan funding through the M-cubed program.

Conflicts of interest: We disclose no conflicts.

CITATION

Hufnagel DA, DePas WH, Chapman MR. 2015. The biology of the *Escherichia coli* extracellular matrix. Microbiol Spectrum 3(3):MB-0014-2014.

REFERENCES

1. **Farmer JJ, Davis BR, Hickmanbrenner FW, McWhorter A, Huntleycarter GP, Asbury MA, Riddle C, Wathen-Grady HG, Elias C, Fanning GR.** 1985. Biochemical-identification of new species and biogroups of *Enterobacteriaceae* isolated from clinical specimens. *J Clin Microbiol* **21:**46–76.
2. **Kaas RS, Friis C, Ussery DW, Aarestrup FM.** 2012. Estimating variation within the genes and inferring the phylogeny of 186 sequenced diverse *Escherichia coli* genomes. *BMC Genomics* **13:**577.
3. **Touchon M, Hoede C, Tenaillon O, Barbe V, Baeriswyl S, Bidet P, Bingen E, Bonacorsi S, Bouchier C, Bouvet O, Calteau A, Chiapello H, Clermont O, Cruveiller S, Danchin A, Diard M, Dossat C, Karoui ME, Frapy E, Garry L, Ghigo JM, Gilles AM, Johnson J, Le Bouguénec C, Lescat M, Mangenot S, Martinez-Jéhanne V, Matic I, Nassif X, Oztas S, Petit MA, Pichon C, Rouy Z, Ruf CS, Schneider D, Tourret J, Vacherie B, Vallenet D, Médigue C, Rocha EP, Denamur E.** 2009. Organised genome dynamics in the *Escherichia coli* species results in highly diverse adaptive paths. *PLoS Genet* **5:** e1000344. doi:10.1371/journal.pgen.1000344.
4. **Kaper JB, Nataro JP, Mobley HL.** 2004. Pathogenic *Escherichia coli. Nat Rev Microbiol* **2:**123–140.
5. **Tenaillon O, Skurnik D, Picard B, Denamur E.** 2010. The population genetics of commensal *Escherichia coli. Nat Rev Microbiol* **8:**207–217.
6. **Smith HW.** 2965. Observations on the flora of the alimentary tract of animals and factors affecting its composition. *J Pathol Bacteriol* **89:** 95–122.
7. **Shulman ST, Friedmann HC, Sims RH.** 2007. Theodor Escherich: the first pediatric infectious diseases physician? *Clin Infect Dis* **45:**1025–1029.
8. **Mitsuoka T, Hayakawa K, Kimura N.** 1975. The fecal flora of man. III. Communication: the composition of *Lactobacillus* flora of different age groups (author's transl). *Zentralbl Bakteriol Orig A* **232:**499–511.
9. **Savageau MA.** 1983. *Escherichia-coli* habitats, cell-types, and molecular mechanisms of gene-control. *Am Nat* **122:**732–744.
10. **Unden G, Bongaerts J.** 1997. Alternative respiratory pathways of *Escherichia coli*: energetics and transcriptional regulation in response to electron acceptors. *Biochim Biophys Acta* **1320:**217–234.
11. **Jones SA, Chowdhury FZ, Fabich AJ, Anderson A, Schreiner DM, House AL, Autieri SM, Leatham MP, Lins JJ, Jorgensen M,**

Cohen PS, Conway T. 2007. Respiration of *Escherichia coli* in the mouse intestine. *Infect Immun* **75:**4891–4899.

12. **Jones SA, Gibson T, Maltby RC, Chowdhury FZ, Stewart V, Cohen PS, Conway T.** 2100. Anaerobic respiration of *Escherichia coli* in the mouse intestine. *Infect Immun* **79:** 4218–4226.
13. **Winfield MD, Groisman EA.** 2003. Role of nonhost environments in the lifestyles of *Salmonella* and *Escherichia coli*. *Appl Environ Microbiol* **69:**3687–3694.
14. **Ishii S, Ksoll WB, Hicks RE, Sadowsky MJ.** 2006. Presence and growth of naturalized *Escherichia coli* in temperate soils from Lake Superior watersheds. *Appl Environ Microbiol* **72:**612–621.
15. **Ferens WA, Hovde CJ.** *Escherichia coli* O157: H7: animal reservoir and sources of human infection. *Foodborne Pathog Dis* **8:**465–487.
16. **Orth D, Grif K, Zimmerhackl LB, Wurzner R.** 2008. Prevention and treatment of enterohemorrhagic *Escherichia coli* infections in humans. *Expert Rev Anti Infect Theor* **6:**101–108.
17. **Gyles CL.** 2007. Shiga toxin-producing *Escherichia coli*: an overview. *J Anim Sci* **85**(Suppl): E45–E62.
18. **Laven RA, Ashmore A, Stewart CS.** 2003. *Escherichia coli* in the rumen and colon of slaughter cattle, with particular reference to *E. coli* O157. *Vet J* **165:**78–83.
19. **Widiasih DA, Ido N, Omoe K, Sugii S, Shinagawa K.** 2004. Duration and magnitude of faecal shedding of Shiga toxin-producing *Escherichia coli* from naturally infected cattle. *Epidemiol Infect* **132:**67–75.
20. **Fegan N, Vanderlinde P, Higgs G, Desmarchelier P.** 2004. The prevalence and concentration of *Escherichia coli* O157 in faeces of cattle from different production systems at slaughter. *J Appl Microbiol* **97:**362–370.
21. **Kudva IT, Blanch K, Hovde CJ.** 1998. Analysis of *Escherichia coli* O157:H7 survival in ovine or bovine manure and manure slurry. *Appl Environ Microbiol* **64:**3166–3174.
22. **Locking ME, O'Brien SJ, Reilly WJ, Wright EM, Campbell DM, Coia JE, Browning LM, Ramsay CN.** 2001. Risk factors for sporadic cases of *Escherichia coli* O157 infection: the importance of contact with animal excreta. *Epidemiol Infect* **127:**215–220.
23. **Solomon EB, Yaron S, Matthews KR.** 2002. Transmission of *Escherichia coli* O157:H7 from contaminated manure and irrigation water to lettuce plant tissue and its subsequent internalization. *Appl Environ Microbiol* **68:**397–400.
24. **Islam M, Doyle MP, Phatak SC, Millner P, Jiang X.** 2004. Persistence of enterohemorrhagic *Escherichia coli* O157:H7 in soil and on leaf lettuce and parsley grown in fields treated with contaminated manure composts or irrigation water. *J Food Prot* **67:**1365–1370.
25. **Jablasone J, Warriner K, Griffiths M.** 2005. Interactions of *Escherichia coli* O157:H7, *Salmonella* typhimurium and *Listeria monocytogenes* plants cultivated in a gnotobiotic system. *Int J Food Microbiol* **99:**7–18.
26. **Sivick KE, Mobley HL.** 2010. Waging war against uropathogenic *Escherichia coli*: winning back the urinary tract. *Infect Immun* **78:** 568–585.
27. **Chen SL, Wu M, Henderson JP, Hooton TM, Hibbing ME, Hultgren SJ, Gordon JI.** 2013. Genomic diversity and fitness of *E. coli* strains recovered from the intestinal and urinary tracts of women with recurrent urinary tract infection. *Sci Transl Med* **5:**184ra60.
28. **Manges AR, Johnson JR.** 2012. Food-borne origins of *Escherichia coli* causing extraintestinal infections. *Clin Infect Dis* **55:**712–719.
29. **Yamamoto S, Tsukamoto T, Terai A, Kurazono H, Takeda Y, Yoshida O.** 1997. Genetic evidence supporting the fecal-perineal-urethral hypothesis in cystitis caused by *Escherichia coli*. *J Urol* **157:**1127–1129.
30. **Johnson JR, Kuskowski MA, Smith K, O'Bryan TT, Tatini S.** 2005. Antimicrobial-resistant and extraintestinal pathogenic *Escherichia coli* in retail foods. *J Infect Dis* **191:** 1040–1049.
31. **Phillips I, Eykyn S, King A, Gransden WR, Rowe B, Frost JA, Gross RJ.** 1988. Epidemic multiresistant *Escherichia coli* infection in West Lambeth Health District. *Lancet* **1:** 1038–1041.
32. **Manges AR, Johnson JR, Foxman B, O'Bryan TT, Fullerton KE, Riley LW.** 2001. Widespread distribution of urinary tract infections caused by a multidrug-resistant *Escherichia coli* clonal group. *N Engl J Med* **345:**1007–1013.
33. **Mulvey MA, Schilling JD, Hultgren SJ.** 2001. Establishment of a persistent *Escherichia coli* reservoir during the acute phase of a bladder infection. *Infect Immun* **69:**4572–4579.
34. **Wright KJ, Seed PC, Hultgren SJ.** 2007. Development of intracellular bacterial communities of uropathogenic *Escherichia coli* depends on type 1 pili. *Cell Microbiol* **9:**2230–2241.
35. **Blango MG, Mulvey MA.** 2010. Persistence of uropathogenic *Escherichia coli* in the face of multiple antibiotics. *Antimicrob Agents Chemother* **54:**1855–1863.

36. **Jorgensen I, Seed PC.** 2012. How to make it in the urinary tract: a tutorial by *Escherichia coli*. *PLoS Pathog* **8:**e1002907. doi:10.1371/journal.ppat.1002907.
37. **Stewart PS, Franklin MJ.** 2008. Physiological heterogeneity in biofilms. *Nat Rev Microbiol* **6:**199–210.
38. **Hall-Stoodley L, Costerton JW, Stoodley P.** 2004. Bacterial biofilms: from the natural environment to infectious diseases. *Nat Rev Microbiol* **2:**95–108.
39. **Wilson M.** 2001. Bacterial biofilms and human disease. *Sci Prog* **84:**235–254.
40. **Vidal O, Longin R, Prigent-Combaret C, Dorel C, Hooreman M, Lejeune P.** 1998. Isolation of an *Escherichia coli* K-12 mutant strain able to form biofilms on inert surfaces: involvement of a new *ompR* allele that increases curli expression. *J Bacteriol* **180:**2442–2449.
41. **Zogaj X, Nimtz M, Rohde M, Bokranz W, Romling U.** 2001. The multicellular morphotypes of *Salmonella* typhimurium and *Escherichia coli* produce cellulose as the second component of the extracellular matrix. *Mol Microbiol* **39:**1452–1463.
42. **Whitchurch CB, Tolker-Nielsen T, Ragas PC, Mattick JS.** 2002. Extracellular DNA required for bacterial biofilm formation. *Science* **295:** 1487.
43. **Beloin C, Roux A, Ghigo JM.** 2008. *Escherichia coli* biofilms. *Curr Top Microbiol Immunol* **322:** 249–289.
44. **Bachmann BJ.** 1972. Pedigrees of some mutant strains of *Escherichia-coli* K-12. *Bacteriol Rev* **36:**525–557.
45. **Fux CA, Shirtliff M, Stoodley P, Costerton JW.** 2005. Can laboratory reference strains mirror 'real-world' pathogenesis? *Trends Microbiol* **13:**58–63.
46. **DePas WH, Hufnagel DA, Lee JS, Blanco LP, Bernstein HC, Fisher ST, James GA, Stewart PS, Chapman MR.** 2013. Iron induces bimodal population development by *Escherichia coli*. *Proc Natl Acad Sci USA* **110:**2629–2634.
47. **Hadjifrangiskou M, Gu AP, Pinkner JS, Kostakioti M, Zhang EW, Greene SE, Hultgren SJ.** 2012. Transposon mutagenesis identifies uro pathogenic *Escherichia coli* biofilm factors. *J Bacteriol* **194:**6195–6205.
48. **Uhlich GA, Cooke PH, Solomon EB.** 2006. Analyses of the red-dry-rough phenotype of an *Escherichia coli* O157:H7 strain and its role in biofilm formation and resistance to antibacterial agents. *Appl Environ Microbiol* **72:**2564–2572.
49. **Liu NT, Nou X, Lefcourt AM, Shelton DR, Lo YM.** 2014. Dual-species biofilm formation by *Escherichia coli* O157:H7 and environmental bacteria isolated from fresh-cut processing facilities. *Int J Food Microbiol* **171:**15–20.
50. **Reisner A, Krogfelt KA, Klein BM, Zechner EL, Molin S.** 2006. *In vitro* biofilm formation of commensal and pathogenic *Escherichia coli* strains: impact of environmental and genetic factors. *J Bacteriol* **188:**3572–3581.
51. **Serra DO, Richter AM, Klauck G, Mika F, Hengge R.** 2013. Microanatomy at cellular resolution and spatial order of physiological differentiation in a bacterial biofilm. *mBio* **4:** e00103-13. doi:10.1128/mBio.00103-13.
52. **Boles BR, Singh PK.** 2008. Endogenous oxidative stress produces diversity and adaptability in biofilm communities. *Proc Natl Acad Sci USA* **105:**12503–12508.
53. **Chai Y, Chu F, Kolter R, Losick R.** 2008. Bistability and biofilm formation in *Bacillus subtilis*. *Mol Microbiol* **67:**254–263.
54. **Boles BR, Thoendel M, Singh PK.** 2004. Self-generated diversity produces "insurance effects" in biofilm communities. *Proc Natl Acad Sci USA* **101:**16630–16635.
55. **Xu KD, Stewart PS, Xia F, Huang CT, McFeters GA.** 1998. Spatial physiological heterogeneity in *Pseudomonas aeruginosa* biofilm is determined by oxygen availability. *Appl Environ Microbiol* **64:**4035–4039.
56. **Williamson KS, Richards LA, Perez-Osorio AC, Pitts B, McInnerney K, Stewart PS, Franklin MJ.** 2012. Heterogeneity in *Pseudomonas aeruginosa* biofilms includes expression of ribosome hibernation factors in the antibiotic-tolerant subpopulation and hypoxia-induced stress response in the metabolically active population. *J Bacteriol* **194:**2062–2073.
57. **Lewis K.** 2008. Multidrug tolerance of biofilms and persister cells. *Curr Top Microbiol Immunol* **322:**107–131.
58. **Pratt LA, Kolter R.** 1998. Genetic analysis of *Escherichia coli* biofilm formation: roles of flagella, motility, chemotaxis and type I pili. *Mol Microbiol* **30:**285–293.
59. **Hung C, Zhou YZ, Pinkner JS, Dodson KW, Crowley JR, Heuser J, Chapman MR, Hadjifrangiskou M, Henderson JP, Hultgren SJ.** 2013. *Escherichia coli* biofilms have an organized and complex extracellular matrix structure. *mBio* **4**(5)**:**e00645-13. doi:10.1128/mBio.00645-13.
60. **Wang X, Preston JF 3rd, Romeo T.** 2004. The *pgaABCD* locus of *Escherichia coli* promotes the synthesis of a polysaccharide adhesin required for biofilm formation. *J Bacteriol* **186:**2724–2734.
61. **Cegelski L, Pinkner JS, Hammer ND, Cusumano CK, Hung CS, Chorell E, Åberg V,**

Walker JN, Seed PC, Almqvist F, Chapman MR, Hultgren SJ. 2009. Small-molecule inhibitors target *Escherichia coli* amyloid biogenesis and biofilm formation. *Nat Chem Biol* **5:**913–919.

62. **Bokranz W, Wang X, Tschape H, Romling U.** 2005. Expression of cellulose and curli fimbriae by *Escherichia coli* isolated from the gastrointestinal tract. *J Med Microbiol* **54:**1171–1182.
63. **Zhou Y, Smith DR, Hufnagel DA, Chapman MR.** 2013. Experimental manipulation of the microbial functional amyloid called curli. *Methods Mol Biol* **966:**53–75.
64. **Hammar M, Arnqvist A, Bian Z, Olsen A, Normark S.** 1995. Expression of two *csg* operons is required for production of fibronectin- and Congo red-binding curli polymers in *Escherichia coli* K-12. *Mol Microbiol* **18:**661–670.
65. **Jubelin G, Vianney A, Beloin C, Ghigo JM, Lazzaroni JC, Lejeune P, Dorel C.** 2005. CpxR/OmpR interplay regulates curli gene expression in response to osmolarity in *Escherichia coli*. *J Bacteriol* **187:**2038–2049.
66. **Zheng D, Constantinidou C, Hobman JL, Minchin SD.** 2004. Identification of the CRP regulon using *in vitro* and *in vivo* transcriptional profiling. *Nucleic Acids Res* **32:**5874–5893.
67. **Olsen A, Arnqvist A, Hammar M, Sukupolvi S, Normark S.** 1993. The RpoS sigma factor relieves H-NS-mediated transcriptional repression of csgA, the subunit gene of fibronectin-binding curli in *Escherichia coli*. *Mol Microbiol* **7:**523–536.
68. **Hufnagel DA, DePas WH, Chapman MR.** 2014. The disulfide bonding system suppresses CsgD-independent cellulose production in *E. coli*. *J Bacteriol* [Epub ahead of print.] doi:10.1128/JB.02019-14.
69. **Wu C, Lim JY, Fuller GG, Cegelski L.** 2012. Quantitative analysis of amyloid-integrated biofilms formed by uropathogenic *Escherichia coli* at the air-liquid interface. *Biophys J* **103:**464–471.
70. **Wai SN, Mizunoe Y, Takade A, Kawabata SI, Yoshida SI.** *Vibrio cholerae* O1 strain TSI-4 produces the exopolysaccharide materials that determine colony morphology, stress resistance, and biofilm formation. *Appl Environ Microbiol* **64:**3648–3655.
71. **Yildiz FH, Dolganov NA, Schoolnik GK.** 2001. VpsR, a member of the response regulators of the two-component regulatory systems, is required for expression of *vps* biosynthesis genes and EPS(ETr)-associated phenotypes in *Vibrio cholerae* O1 El Tor. *J Bacteriol* **183:**1716–1726.
72. **Yildiz FH, Liu XS, Heydorn A, Schoolnik GK.** 2004. Molecular analysis of rugosity in a *Vibrio cholerae* O1 El Tor phase variant. *Mol Microbiol* **53:**497–515.
73. **Asally M, Kittisopikul M, Rue P, Du Y, Hu Z, Cagatay T, Robinson AB, Lu H, Garcia-Ojalvo J, Süel GM.** 2012. Localized cell death focuses mechanical forces during 3D patterning in a biofilm. *Proc Natl Acad Sci USA* **109:**18891–18896.
74. **Dietrich LE, Okegbe C, Price-Whelan A, Sakhtah H, Hunter RC, Newman DK.** 2013. Bacterial community morphogenesis is intimately linked to the intracellular redox state. *J Bacteriol* **195:**1371–1380.
75. **Epstein AK, Pokroy B, Seminara A, Aizenberg J.** 2011. Bacterial biofilm shows persistent resistance to liquid wetting and gas penetration. *Proc Natl Acad Sci USA* **108:**995–1000.
76. **Kolodkin-Gal I, Elsholz AK, Muth C, Girguis PR, Kolter R, Losick R.** 2013. Respiration control of multicellularity in *Bacillus subtilis* by a complex of the cytochrome chain with a membrane-embedded histidine kinase. *Gene Dev* **27:**887–899.
77. **Romling U, Sierralta WD, Eriksson K, Normark S.** 1998. Multicellular and aggregative behaviour of *Salmonella* typhimurium strains is controlled by mutations in the *agfD* promoter. *Mol Microbiol* **28:**249–264.
78. **Romling U.** 2005. Characterization of the rdar morphotype, a multicellular behaviour in *Enterobacteriaceae*. *Cell Mol Life Sci* **62:**1234–1246.
79. **Lim JY, May JM, Cegelski L.** 2012. Dimethyl sulfoxide and ethanol elicit increased amyloid biogenesis and amyloid-integrated biofilm formation in *Escherichia coli*. *Appl Environ Microbiol* **78:**3369–3378.
80. **McCrate OA, Zhou X, Reichhardt C, Cegelski L.** 2013. Sum of the parts: composition and architecture of the bacterial extracellular matrix. *J Mol Biol* **425:**4286–4294.
81. **Serra DO, Richter AM, Hengge R.** 2013. Cellulose as an architectural element in spatially structured *Escherichia coli* biofilms. *J Bacteriol* **195:**5540–5554.
82. **Weiss-Muszkat M, Shakh D, Zhou Y, Pinto R, Belausov E, Chapman MR, Sela S.** 2010. Biofilm formation by and multicellular behavior of *Escherichia coli* O55:H7, an atypical enteropathogenic strain. *Appl Environ Microbiol* **76:**1545–1554.
83. **Lim JY, Pinkner JS, Cegelski L.** 2014. Community behavior and amyloid-associated phenotypes among a panel of uropathogenic *E. coli*. *Biochem Biophys Res Commun* **443:**345–350.
84. **Zogaj X, Bokranz W, Nimtz M, Romling U.** 2003. Production of cellulose and curli fimbriae

by members of the family *Enterobacteriaceae* isolated from the human gastrointestinal tract. *Infect Immun* **71:**4151–4158.
85. **Romling U, Rohde M, Olsen A, Normark S, Reinkoster J.** 2000. AgfD, the checkpoint of multicellular and aggregative behaviour in *Salmonella* typhimurium regulates at least two independent pathways. *Mol Microbiol* **36:**10–23.
86. **Gao R, Mack TR, Stock AM.** 2007. Bacterial response regulators: versatile regulatory strategies from common domains. *Trends Biochem Sci* **32:**225–234.
87. **Skerker JM, Perchuk BS, Siryaporn A, Lubin EA, Ashenberg O, Goulian M, Laub MT.** 2008. Rewiring the specificity of two-component signal transduction systems. *Cell* **133:**1043–1054.
88. **Stock AM, Robinson VL, Goudreau PN.** 2000. Two-component signal transduction. *Annu Rev Biochem* **69:**183–215.
89. **Zakikhany K, Harrington CR, Nimtz M, Hinton JCD, Romling U.** 2010. Unphosphorylated CsgD controls biofilm formation in *Salmonella enterica* serovar Typhimurium. *Mol Microbiol* **77:**771–786.
90. **Wang L, Tian X, Wang J, Yang H, Fan K, Xu G, Yang K, Tan H.** 2009. Autoregulation of antibiotic biosynthesis by binding of the end product to an atypical response regulator. *Proc Natl Acad Sci USA* **106:**8617–8622.
91. **Sitnikov DM, Schineller JB, Baldwin TO.** 1995. Transcriptional regulation of bioluminesence genes from *Vibrio fischeri. Mol Microbiol* **17:** 801–812.
92. **Evans ML, Chapman MR.** 2013. Curli biogenesis: order out of disorder. *Biochim Biophys Acta* **1843:**1551–1558.
93. **Mika F, Hengge R.** Small regulatory RNAs in the control of motility and biofilm formation in *E. coli* and *Salmonella. Int J Mol Sci* **14:**4560–4579.
94. **Olsen A, Jonsson A, Normark S.** 1989. Fibronectin binding mediated by a novel class of surface organelles on *Escherichia coli. Nature* **338:**652–655.
95. **Romling U, Bian Z, Hammar M, Sierralta WD, Normark S.** 1998. Curli fibers are highly conserved between *Salmonella typhimurium* and *Escherichia coli* with respect to operon structure and regulation. *J Bacteriol* **180:**722–731.
96. **Romling U, Gomelsky M, Galperin MY.** 2005. C-di-GMP: the dawning of a novel bacterial signalling system. *Mol Microbiol* **57:**629–639.
97. **Simm R, Morr M, Kader A, Nimtz M, Romling U.** 2004. GGDEF and EAL domains inversely regulate cyclic di-GMP levels and transition from sessility to motility. *Mol Microbiol* **53:** 1123–1134.
98. **Sommerfeldt N, Possling A, Becker G, Pesavento C, Tschowri N, Hengge R.** 2009. Gene expression patterns and differential input into curli fimbriae regulation of all GGDEF/EAL domain proteins in *Escherichia coli. Microbiology* **155:**1318–1331.
99. **Weber H, Pesavento C, Possling A, Tischendorf G, Hengge R.** 2006. Cyclic-di-GMP-mediated signalling within the sigma network of *Escherichia coli. Mol Microbiol* **62:**1014–1034.
100. **Hengge R.** 2009. Principles of c-di-GMP signalling in bacteria. *Nat Rev Microbiol* **7:**263–273.
101. **Brombacher E, Dorel C, Zehnder AJ, Landini P.** 2003. The curli biosynthesis regulator CsgD co-ordinates the expression of both positive and negative determinants for biofilm formation in *Escherichia coli. Microbiology* **149:**2847–2857.
102. **Gualdi L, Tagliabue L, Landini P.** 2007. Biofilm formation-gene expression relay system in *Escherichia coli*: modulation of sigmaS-dependent gene expression by the CsgD regulatory protein via sigmaS protein stabilization. *J Bacteriol* **189:**8034–8043.
103. **Ogasawara H, Yamamoto K, Ishihama A.** 2011. Role of the biofilm master regulator CsgD in cross-regulation between biofilm formation and flagellar synthesis. *J Bacteriol* **193:**2587–2597.
104. **Collinson SK, Emody L, Muller KH, Trust TJ, Kay WW.** 1991. Purification and characterization of thin, aggregative fimbriae from *Salmonella enteritidis. J Bacteriol* **173:**4773–4781.
105. **Arnqvist A, Olsen A, Pfeifer J, Russell DG, Normark S.** 1992. The Crl protein activates cryptic genes for curli formation and fibronectin binding in *Escherichia coli* HB101. *Mol Microbiol* **6:**2443–2452.
106. **Hammar M, Bian Z, Normark S.** Nucleator-dependent intercellular assembly of adhesive curli organelles in *Escherichia coli. Proc Natl Acad Sci USA* **93:**6562–6566.
107. **Bian Z, Normark S.** 1997. Nucleator function of CsgB for the assembly of adhesive surface organelles in *Escherichia coli. EMBO J* **16:** 5827–5836.
108. **Chapman MR, Robinson LS, Pinkner JS, Roth R, Heuser J, Hammar M, Normark S, Hultgren SJ.** 2002. Role of *Escherichia coli* curli operons in directing amyloid fiber formation. *Science* **295:**851–855.
109. **Abdallah AM, Van Pittius NCG, Champion PAD, Cox J, Luirink J, Vandenbroucke-Grauls CMJE, Appelmelk BJ, Bitter W.** 2007. Type VII secretion: mycobacteria show the way. *Nat Rev Microbiol* **5:**883–891.
110. **Desvaux M, Hebraud M, Talon R, Henderson IR.** 2009. Secretion and subcellular localizations

of bacterial proteins: a semantic awareness issue. *Trends Microbiol* **17:**139–145.

111. **Hammer ND, Schmidt JC, Chapman MR.** 2007. The curli nucleator protein, CsgB, contains an amyloidogenic domain that directs CsgA polymerization. *Proc Natl Acad Sci USA.* **104:**12494–12499.
112. **Epstein EA, Reizian MA, Chapman MR.** 2009. Spatial clustering of the curlin secretion lipoprotein requires curli fiber assembly. *J Bacteriol* **191:**608–615.
113. **Zhou Y, Smith D, Leong BJ, Brannstrom K, Almqvist F, Chapman MR.** 2012. Promiscuous cross-seeding between bacterial amyloids promotes interspecies biofilms. *J Biol Chem* **287:**35092–35103.
114. **Robinson LS, Ashman EM, Hultgren SJ, Chapman MR.** 2006. Secretion of curli fibre subunits is mediated by the outer membrane-localized CsgG protein. *Mol Microbiol* **59:**870–881.
115. **Loferer H, Hammar M, Normark S.** 1997. Availability of the fibre subunit CsgA and the nucleator protein CsgB during assembly of fibronectin-binding curli is limited by the intracellular concentration of the novel lipoprotein CsgG. *Mol Microbiol* **26:**11–23.
116. **Goyal G, Van Gerven N, Gubellini F, Van den Broeck I, Troupoiotis-Tsailaki A, Jonkheere W, Pejau-Arnaudet G, Pinkner JS, Chapman MR, Hultgren SJ, Howorka S, Fronzes R, Remaut H.** 2014. Structural and mechanistic insights into the bacterial amyloid secretion channel CsgG. *Nature* **516:**250–253.
117. **Nenninger AA, Robinson LS, Hammer ND, Epstein EA, Badtke MP, Hultgren SJ, Chaman MR.** 2011. CsgE is a curli secretion specificity factor that prevents amyloid fibre aggregation. *Mol Microbiol* **81:**486–499.
118. **Andersson EK, Bengtsson C, Evans ML, Chorell E, Sellstedt M, Lindgren AE, Hufnagel DA, Bhattacharya M, Tessier PM, Wittung-Stafshede P, Almqvist F, Chapman MR.** 2013. Modulation of curli assembly and pellicle biofilm formation by chemical and protein chaperones. *Chem Biol* **20:**1245–1254.
119. **Fowler DM, Koulov AV, Balch WE, Kelly JW.** 2007. Functional amyloid: from bacteria to humans. *Trends Biochem Sci* **32:**217–224.
120. **Horvath I, Weise CF, Andersson EK, Chorell E, Sellstedt M, Bengtsson C, Olofsson A, Hultgren SJ, Chapman M, Wolf-Watz M, Almqvist F, Wittung-Stafshede P.** 2012. Mechanisms of protein oligomerization: inhibitor of functional amyloids templates alpha-synuclein fibrillation. *J Am Chem Soc* **134:** 3439–3444.
121. **Kuner P, Bohrmann B, Tjernberg LO, Naslund J, Huber G, Celenk S, Grüninger-Leitch F, Richards JG, Jakob-Roetne R, Kemp JA, Nordstedt C.** 2000. Controlling polymerization of beta-amyloid and prion-derived peptides with synthetic small molecule ligands. *J Biol Chem* **275:**1673–1678.
122. **Pinkner JS, Remaut H, Buelens F, Miller E, Aberg V, Pemberton N, Hedenström M, Larsson A, Seed P, Waksman G, Hultgren SJ, Almqvist F.** 2006. Rationally designed small compounds inhibit pilus biogenesis in uropathogenic bacteria. *Proc Natl Acad Sci USA* **103:** 17897–17902.
123. **Aberg V, Norman F, Chorell E, Westermark A, Olofsson A, Sauer-Eriksson AE, Almqvist F.** 2005. Microwave-assisted decarboxylation of bicyclic 2-pyridone scaffolds and identification of Abeta-peptide aggregation inhibitors. *Org Biomol Chem* **3:**2817–2823.
124. **Haass C, Selkoe DJ.** 2007. Soluble protein oligomers in neurodegeneration: lessons from the Alzheimer's amyloid beta-peptide. *Nat Rev Mol Cell Biol* **8:**101–112.
125. **Brown AJ.** 1887. Note on the cellulose formed by *Bacterium xylinum. J Chem Soc* **51:**643.
126. **Romling U.** 2002. Molecular biology of cellulose production in bacteria. *Res Microbiol* **153:**205–212.
127. **Ross P, Mayer R, Benziman M.** 1991. Cellulose biosynthesis and function in bacteria. *Microbiol Rev* **55:**35–58.
128. **Da Re S, Ghigo JM.** 2006. A CsgD-independent pathway for cellulose production and biofilm formation in *Escherichia coli. J Bacteriol* **188:** 3073–3087.
129. **Ryjenkov DA, Simm R, Romling U, Gomelsky M.** 2006. The PilZ domain is a receptor for the second messenger c-di-GMP: the PilZ domain protein YcgR controls motility in enterobacteria. *J Biol Chem* **281:**30310–30314.
130. **Amikam D, Galperin MY.** 2006. PilZ domain is part of the bacterial c-di-GMP binding protein. *Bioinformatics* **22:**3–6.
131. **Omadjela O, Narahari A, Strumillo J, Melida H, Mazur O, Bulone V, Zimmer J.** 2013. BcsA and BcsB form the catalytically active core of bacterial cellulose synthase sufficient for *in vitro* cellulose synthesis. *Proc Natl Acad Sci USA* **110:**17856–17861.
132. **Malone JG, Jaeger T, Spangler C, Ritz D, Spang A, Arrieumerlou C, Kaever V, Landmann R, Jenal U.** 2010. YfiBNR mediates cyclic di-GMP dependent small colony variant formation and persistence in *Pseudomonas aeruginosa. PLoS Pathog* **6:**e1000804. doi:10.1371/journal.ppat.1000804.

133. **Morgan JLW, Strumillo J, Zimmer J.** 2013. Crystallographic snapshot of cellulose synthesis and membrane translocation. *Nature* **493:**181–186.
134. **Morgan JL, McNamara JT, Zimmer J.** 2014. Mechanism of activation of bacterial cellulose synthase by cyclic di-GMP. *Nat Struct Mol Biol* **21:**489–496.
135. **Kai-Larsen Y, Luthje P, Chromek M, Peters V, Wang X, Holm A, Kádas L, Hedlund KO, Johansson J, Chapman MR, Jacobson SH, Römling U, Agerberth B, Brauner A.** 2010. Uropathogenic *Escherichia coli* modulates immune responses and its curli fimbriae interact with the antimicrobial peptide LL-37. *PLoS Pathog* **6:**e1001010. doi:10.1371/journal.ppat.1001010.
136. **Al-Hasan MN, Eckel-Passow JE, Baddour LM.** 2010. Bacteremia complicating Gram-negative urinary tract infections: a population-based study. *J Infect* **60:**278–285.
137. **Hung C, Marschall J, Burnham CAD, Byun AS, Henderson JP.** 2014. The bacterial amyloid curli is associated with urinary source bloodstream infection. *PloS One* **9:**e86009. doi:10.1371/journal.pone.0086009.
138. **Bian Z, Brauner A, Li Y, Normark S.** 2000. Expression of and cytokine activation by *Escherichia coli* curli fibers in human sepsis. *J Infect Dis* **181:**602–612.
139. **Danese PN, Pratt LA, Dove SL, Kolter R.** 2000. The outer membrane protein, antigen 43, mediates cell-to-cell interactions within *Escherichia coli* biofilms. *Mol Microbiol* **37:**424–432.
140. **Kjaergaard K, Schembri MA, Ramos C, Molin S, Klemm P.** 2000. Antigen 43 facilitates formation of multispecies biofilms. *Environ Microbiol* **2:**695–702.
141. **Anderson GG, Goller CC, Justice S, Hultgren SJ, Seed PC.** 2010. Polysaccharide capsule and sialic acid-mediated regulation promote biofilm-like intracellular bacterial communities during cystitis. *Infect Immun* **78:**963–975.
142. **Saldana Z, Xicohtencatl-Cortes J, Avelino F, Phillips AD, Kaper JB, Puente JL, Girón JA.** 2009. Synergistic role of curli and cellulose in cell adherence and biofilm formation of attaching and effacing *Escherichia coli* and identification of Fis as a negative regulator of curli. *Environ Microbiol* **11:**992–1006.
143. **Wang X, Rochon M, Lamprokostopoulou A, Lunsdorf H, Nimtz M, Romling U.** 2006. Impact of biofilm matrix components on interaction of commensal *Escherichia coli* with the gastrointestinal cell line HT-29. *Cell Mol Life Sci* **63:**2352–2363.
144. **Rapsinski GJ, Newman TN, Oppong GO, van Putten JP, Tukel C.** 2013. CD14 protein acts as an adaptor molecule for the immune recognition of *Salmonella* curli fibers. *J Biol Chem* **288:**14178–14188.
145. **Tukel C, Nishimori JH, Wilson RP, Winter MG, Keestra AM, van Putten JP, Bäumler AJ.** 2010. Toll-like receptors 1 and 2 cooperatively mediate immune responses to curli, a common amyloid from enterobacterial biofilms. *Cell Microbiol* **12:**1495–1505.
146. **Oppong GO, Rapsinski GJ, Newman TN, Nishimori JH, Biesecker SG, Tukel C.** 2013. Epithelial cells augment barrier function via activation of the Toll-like receptor 2/phosphatidylinositol 3-kinase pathway upon recognition of *Salmonella enterica* serovar Typhimurium curli fibrils in the gut. *Infect Immun* **81:**478–486.
147. **Gerstel U, Romling U.** 2001. Oxygen tension and nutrient starvation are major signals that regulate *agfD* promoter activity and expression of the multicellular morphotype in *Salmonella* typhimurium. *Environ Microbiol* **3:**638–648.
148. **Olsen A, Arnqvist A, Hammar M, Normark S.** 1993. Environmental regulation of curli production in *Escherichia coli*. *Infect Agents Dis* **2:**272–274.
149. **Reshamwala SM, Noronha SB.** 2011. Biofilm formation in *Escherichia coli cra* mutants is impaired due to down-regulation of curli biosynthesis. *Arch Microbiol* **193:**711–722.
150. **White AP, Gibson DL, Kim W, Kay WW, Surette MG.** 2006. Thin aggregative fimbriae and cellulose enhance long-term survival and persistence of *Salmonella*. *J Bacteriol* **188:**3219–3227.
151. **Cookson AL, Cooley WA, Woodward MJ.** 2002. The role of type 1 and curli fimbriae of Shiga toxin-producing *Escherichia coli* in adherence to abiotic surfaces. *Int J Med Microbiol* **292:**195–205.
152. **Macarisin D, Patel J, Bauchan G, Giron JA, Sharma VK.** 2012. Role of curli and cellulose expression in adherence of *Escherichia coli* O157:H7 to spinach leaves. *Foodborne Pathog Dis* **9:**160–167.
153. **Pawar DM, Rossman ML, Chen J.** 2005. Role of curli fimbriae in mediating the cells of enterohaemorrhagic *Escherichia coli* to attach to abiotic surfaces. *J Appl Microbiol* **99:**418–425.
154. **Jeter C, Matthysse AG.** 2005. Characterization of the binding of diarrheagenic strains of *E. coli* to plant surfaces and the role of curli in the interaction of the bacteria with alfalfa sprouts. *Mol Plant Microbe Interact* **18:**1235–1242.

155. **Patel J, Sharma M, Ravishakar S.** 2011. Effect of curli expression and hydrophobicity of *Escherichia coli* O157:H7 on attachment to fresh produce surfaces. *J Appl Microbiol* **110:**737–745.
156. **Hou Z, Fink RC, Black EP, Sugawara M, Zhang Z, Diez-Gonzalez F, Sadowdky MJ.** 2012. Gene expression profiling of *Escherichia coli* in response to interactions with the lettuce rhizosphere. *J Appl Microbiol* **113:**1076–1086.
157. **Gottesman S, Stout V.** 1991. Regulation of capsular polysaccharide synthesis in *Escherichia coli* K12. *Mol Microbiol* **5:**1599–1606.
158. **Hanna A, Berg M, Stout V, Razatos A.** 2003. Role of capsular colanic acid in adhesion of uropathogenic *Escherichia coli. Appl Environ Microbiol* **69:**4474–4481.
159. **Danese PN, Pratt LA, Kolter R.** 2000. Exopolysaccharide production is required for development of *Escherichia coli* K-12 biofilm architecture. *J Bacteriol* **182:**3593–3596.
160. **Matthysse AG, Deora R, Mishra M, Torres AG.** 2008. Polysaccharides cellulose, poly-beta-1,6-N-acetyl-D-glucosamine, and colanic acid are required for optimal binding of *Escherichia coli* O157:H7 strains to alfalfa sprouts and K-12 strains to plastic but not for binding to epithelial cells. *Appl Environ Microbiol* **74:**2384–2390.
161. **Larsen P, Nielsen JL, Dueholm MS, Wetzel R, Otzen D, Nielsen PH.** 2007. Amyloid adhesins are abundant in natural biofilms. *Environ Microbiol* **9:**3077–3090.
162. **Dueholm MS, Albertsen M, Otzen D, Nielsen PH.** 2012. Curli functional amyloid systems are phylogenetically widespread and display large diversity in operon and protein structure. *PloS One* 7(12):e51274. doi:10.1371/journal.pone.0051274.
163. **Evans ML, Chorell E, Taylor JD, Aden J, Gotheson A, Li F.** 2015. The bacterial curli system possesses a potent and selective inhibitor of amyloid formation. *Molec Cell* **57:**445–455.
164. **Pontes MH, Lee EJ, Choi J, Groisman EA.** 2015. *Salmonella* promotes virulence by repressing cellulose production. *Proc Natl Acad Sci USA* **112:**5183–5188.
165. **DePas WH, Syed AK, Sifuentes M, Lee JS, Warshaw D, Saggar V.** 2014. Biofilm formation protects *Escherichia coli* against killing by Caenorhabditis elegans and *Myxococcus xanthus. Appl Environ Microbiol* **80:**7079–7087.

Antimicrobial Tolerance in Biofilms 13

PHILIP S. STEWART[1]

EXAMPLES OF REDUCED BIOFILM SUSCEPTIBILITY

Tolerance to antimicrobial agents is a common feature of microbial biofilm formation (1–7). Table 1 presents a few examples of biofilm tolerance to biocides and antiseptics, and Table 2 summarizes some examples of antibiotic tolerance in biofilms. Neither of these listings is comprehensive, but these two data sets can be analyzed to gain insight into the factors that influence biofilm tolerance. The examples have been selected to illustrate the wide variety of microbial species, growth environments, and antimicrobial chemistries for which biofilm reduced susceptibility has been reported. The short list in Table 1 encompasses studies designed to mimic biofilms in dental plaque, hot tubs, paper mills, drinking water, household drains, urinary catheters, food processing plants, cooling water systems, and hospitals. These examples employ a range of individual and mixed species biofilms and diverse biocidal chemistries including halogens, phenolics, quaternary ammonium compounds, aldehydes, a plant essential oil, and peroxides. The studies captured in Table 2 cover 19 antibiotics and 9 organisms that include aerobic bacteria, strict anaerobes, and a fungus.

[1]Center for Biofilm Engineering, Montana State University - Bozeman, Bozeman, MT 59717.

Microbial Biofilms 2nd Edition
Edited by Mahmoud Ghannoum, Matthew Parsek, Marvin Whiteley, and Pranab K. Mukherjee

doi:10.1128/microbiolspec.MB-0010-2014

TABLE 1 Selected examples of tolerance of bacteria in biofilms to biocides and antiseptics

Organisms	Agent	Molecular weight (g mole^{-1})	Substratum[a]	(log$_{10}$ cfu cm^{-2})X_o	*TF*	References
Pseudomonas aeruginosa,	Hypochlorite, pH 11	52.5	SS	9.9	767	37
Klebsiella pneumoniae	Chlorosulfamate	131			272	
P. aeruginosa	Peracetic acid	76.1	PS	8	6.7	67
Aggregatibacter	Chlorhexidine	506	CN	7.48	2.7	68
actinomycetocomitans	Cetylpyridinium chloride	340			3.4	
Legionella in	Glutaraldehyde	100	RR	6.1	2.0	69
mixed species	Bromo-nitropropane-diol	200			1.0	
P. aeruginosa	Hydrogen peroxide	34	PE	6.65	2.8	70
Mixed drinking water	Chlorine dioxide	67.5	G	5.3	1.0	71
Staphylococcus aureus	Benzalkonium chloride	360	G	7.9	52	38
P. aeruginosa	Bromine	96.9	PC	5.3	1.4	72
P. aeruginosa	Benzylchlorophenol,	195	G	8.5	7.4	39
S. aureus	phenylphenol				4.2	
Salmonella typhimurium	Triclosan	290	Pellicle	7.2	20	73
Citrobacter diversus	Povidone-iodine	365	S	8.8	11	74
Listeria monocytogenes	Iodine	254	SS	5.2	1.7	8
Mixed paper mill white water	Thymol	150	SS	7.7	1.1	75
Mixed oral	Chlorhexidine	506	HAP	9	13.5	76

[a]Abbreviations: SS, stainless steel; PS, polystyrene; CN, cellulose nitrate; RR, red rubber; PE, polyester; G, glass; PC, polycarbonate; S, silicone; BR, Buna rubber; HAP, hydroxyapatite.

Biofilm reduced susceptibility is quantified in Tables 1 and 2 by a tolerance factor, *TF*, defined as:

$$TF = (LR_P * t_B * C_B / LR_B * t_P * C_P)$$

where C_P and C_B denote planktonic and biofilm dose concentration, respectively, t_P and t_B denote planktonic and biofilm dose duration, respectively, and LR_P and LR_B denote the measured log reduction in planktonic and biofilm populations, respectively.

TF compares the rate of killing in the planktonic and biofilm states. For example, a value of $TF = 10$ means that biofilm killing is 10 times slower than in the planktonic condition. A quick inspection of Tables 1 and 2 reveals that the tolerance factor ranges widely, from a value of 1.0 (no difference at all between suspended and sessile susceptibility) to a value of more than 1,000.

FACTORS INFLUENCING BIOFILM SUSCEPTIBILITY

One of the challenges of understanding biofilm tolerance is the large number of factors that likely influence the susceptibility in a particular biofilm. Some of the factors that could be important are antimicrobial chemistry, substratum material, areal cell density or thickness, biofilm age, microbial speciation, and medium composition. Here I attempt to shed some light on some of these factors by meta-analyses of the literature.

Antimicrobial Chemistry

When the tolerance factors for biocides reported in Table 1 are regressed against the molecular weight of the antimicrobial agent, no correlation whatsoever is apparent (Fig. 1A; $R^2 = 0.0007$). Indeed, this value of R^2 suggests that none of the considerable variation in *TF* can be attributed to the

TABLE 2 Selected examples of tolerance of bacteria or fungi in biofilms to antibiotics

Organisms	Agent	Molecular weight (g mole^{-1})	Substratum[a]	($\log_{10}$ cfu cm^{-2})X_o	*TF*	References
Propionibacterium acnes	Rifampin	823	G		4	77
	Daptomycin	1620			16	
	Vancomycin	1468			16	
	Penicillin G	334			2	
Corynebacterium urealyticum	Ciprofloxacin	330	PS		2048	78
	Moxifloxacin	401			512	
	Vancomycin	1468			512	
Pseudomonas aeruginosa	Gentamicin	478			4	79
	Tobramycin	468			4	
	Ciprofloxacin	330			8	
	Ofloxacin	361			4	
P. aeruginosa	Tobramycin	468	SS	9.6	4.4	80
	Ciprofloxacin	330			3.5	
K. pneumoniae	Ciprofloxacin	330	PC	10.3	90	64
	Ampicillin	371		10.2	14	
Staphylococcus epidermidis	Ciprofloxacin	330	SS	8.9	14	81
	Rifampin	823			7	
P. aeruginosa	Tobramycin	468	PC	10.4	265	82
	Ciprofloxacin	330			104	
P. aeruginosa	Tobramycin	468	S	8.3	208	45
				7.4	1.5	
S. aureus	Nisin	3354	PS	7.5	5.3	83
	Vancomycin	1468		7.8	55	
S. epidermidis	Levofloxacin	350	G	10.3	12	48
	Vancomycin	1468			157	
Porphyromonas gingivalis	Amoxicillin	365	CA		3.3	84
	Doxycycline	444			21	
	Metronidazole	171			4.2	
Staphylococcus lugdunensis	Cefazolin	456	PS		256	85
	Rifampin	823			4	
	Daptomycin	1620			64	
	Moxifloxacin	401			4	
	Naficillin	414			16	
Candida albicans	Amphotericin B	923	PVC		3.4	86
C. albicans	Fluconazole	306	PVC		4.4	87

[a]Abbreviations: as for Table 1; CA, cellulose acetate; PVC, polyvinyl choride.

size of the antimicrobial molecule itself. A similar analysis of the tolerance factors for antibiotics (Table 2) also reveals no correlation (Fig. 1B; R^2 = 0.012). There are no reports in the literature demonstrating that antimicrobial size is a predictor of efficacy against biofilms.

TF ranges widely even for a single antimicrobial agent. For example, values of *TF* for tobramycin measured against just one bacterium, *Pseudomonas aeruginosa*, run from 1.5 to 265. *TF* values for ciprofloxacin, measured against four different bacteria, range from 3.5 to 2,048. It will be seen shortly that the rate of biofilm killing by chlorine ranges over three orders of magnitude even when scaled for the dose concentration. These observations suggest that the numerical value of *TF* is not specific to a particular antimicrobial agent. Put another way, at least at this point in time, the chemistry of a particular antimicrobial does not allow us to predict its relative efficacy against a biofilm.

At the risk of redundancy, it is important not to extrapolate a *TF* value pulled from the tables compiled here to some other

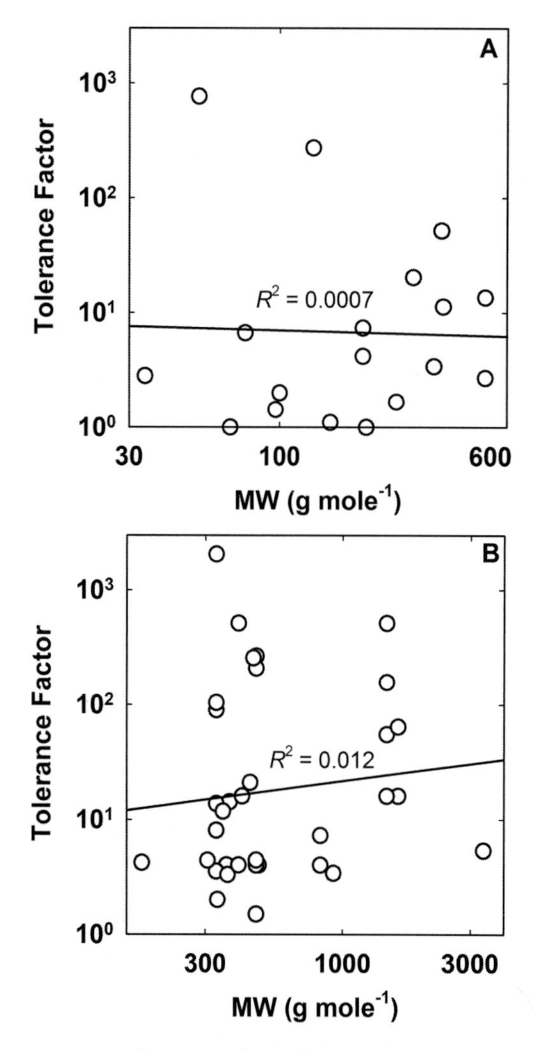

FIGURE 1 Tolerance factors versus antimicrobial agent molecular weight for the data on (A) biocides and antiseptics from Table 1 and (B) antibiotics from Table 2. doi:10.1128/microbiolspec.MB-0010-2014.f1

system. It is to be expected that if more measurements were available, we would find that *TF* values for every antimicrobial range widely.

Substratum Material

The data compiled in Tables 1 and 2 reflect measurements made using biofilms formed on a wide variety of materials: polystyrene, glass, stainless steel, cellulose acetate or nitrate, polycarbonate, silicone, polyvinyl chloride, rubber, polyester, and hydroxyapatite. Though analysis of variance of these data (plotted in Fig. 2) indicates a borderline statistically significant difference between the five groups (p = 0.053), I suspect that the root of this difference is in methodology rather than material. The polystyrene group, which has somewhat higher *TF* values, is all data from multiwell plates. Most of the data collected in plate assays derive from a series of antimicrobial concentrations as opposed to kill data in time. This method can produce very high *TF* values with antibiotics when delivered at extremely high (and not physiologically relevant) concentrations. Inspection of the data shows that *TF* ranges by two orders of magnitude for a given material (Fig. 2). For example, tolerance factors reported for biofilms grown on stainless steel (n = 8) range from 1.1 to 767.

There may be occasional situations in which the substratum material does influence

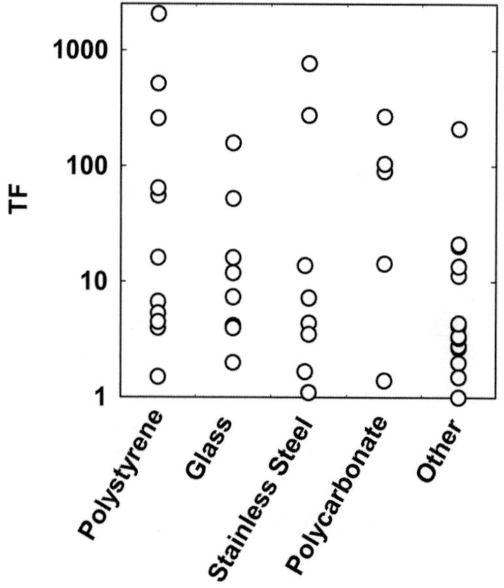

FIGURE 2 Tolerance factors grouped and compared by substratum material. doi:10.1128/microbiolspec.MB-0010-2014.f2

biofilm accumulation and antimicrobial tolerance. For example, whereas iodine was relatively effective at killing *Listeria* on stainless steel (*TF* = 1.7), it was ineffective against the same strain when biofilms were formed on Buna rubber (*TF* = 70) (8). Buna rubber was shown to have an independent bacteriostatic effect. Biofilms formed on mild steel, in which some corrosion of the metal was evident, were less susceptible to killing by monochloramine than biofilms on stainless steel (9). These examples suggest that the substratum material is most likely to influence biofilm susceptibility when it leaches or corrodes.

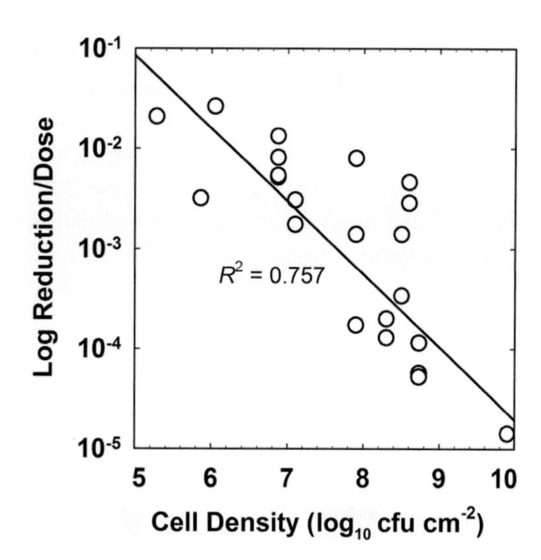

FIGURE 4 Efficacy of chlorine treatment against biofilms as a function of the untreated control biofilm areal cell density. The y-axis is the reported log reduction divided by the product of dose concentration and duration ($C_B t_B$). The line is the least squares regressed fit. Sources: references 8, 37–44. doi:10.1128/microbiolspec.MB-0010-2014.f4

Cell Density

When the tolerance factors for biocides and antiseptics tabulated in Table 1 are regressed against the untreated control biofilm areal cell density (measured in units of $\log_{10}$ CFU cm^{-2}), a clear correlation emerges (Fig. 3; R^2 = 0.629). To put these values in terms of the approximate thickness of the biofilm, a biofilm of 6.0 $\log_{10}$ CFU cm^{-2} corresponds roughly to a sparse monolayer, whereas the most massive biofilm in this data set (9.9 $\log_{10}$ CFU cm^{-2}) was nearly 1 mm thick. This result shows that tolerance to biocides depends on the extent of biofilm accumulation.

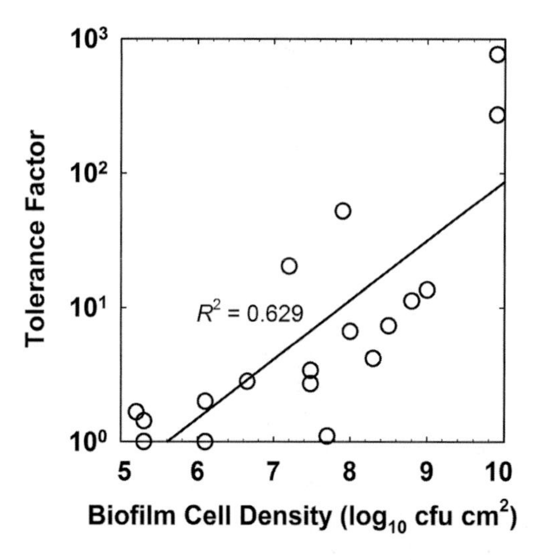

FIGURE 3 Tolerance factor versus biofilm cell density for the data in Table 1. The line is the least squares regressed fit. doi:10.1128/microbiolspec.MB-0010-2014.f3

There are few biocides for which there is sufficient data available to perform an agent-specific analysis of the role of biofilm cell density in susceptibility. Chlorine is one such agent, and Figure 4 presents a correlation (R^2 = 0.757) that reinforces an important role for the extent of biofilm accumulation prior to treatment in determining the efficacy of a chlorine dose. This analysis includes data from nine independent investigations using *Staphylococcus*, *Pseudomonas*, *Listeria*, *Salmonella*, and mixed-species biofilms.

There is also an important dependence of biofilm antibiotic tolerance on the cell density of the biofilm. This is most clearly demonstrated in investigations which have challenged biofilms at different stages of development with the same antibiotic dose. Older, thicker biofilms are invariably less susceptible than younger, less dense biofilms (Fig. 5). The overall correlation of log reduction with cell density is poor in this case (R^2 = 0.125), but the effect within an investigation is obvious and consistent.

Age

Several investigations have compared the efficacy of identical antimicrobial challenges against biofilms of different ages. Within a given experimental system, biofilms tend to become less susceptible as they age (Fig. 6), though here again the overall correlation is not strong (R^2 = 0.217). Assuming a first order process, the characteristic time (expressed as a half-life) for biofilm tolerance to develop as determined from these data sets was 2.7 ± 2.0 days (n = 12). This suggests that, at least *in vitro*, biofilm tolerance manifests over a timescale of a few days. This result also provides an important clue that the biological state of the organisms in a biofilm is a key factor in determining their susceptibility.

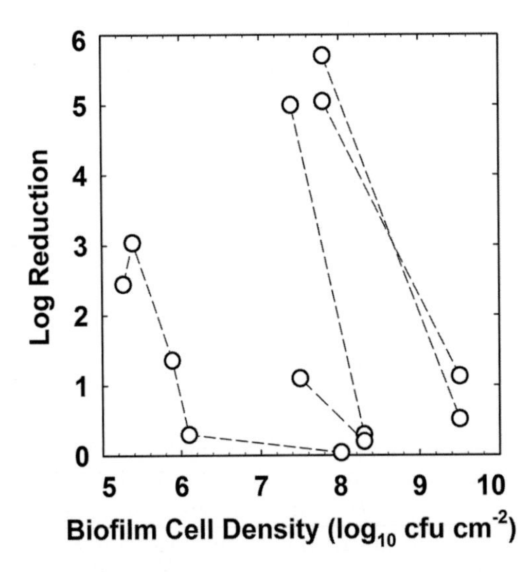

FIGURE 5 Antibiotic efficacy against *Pseudomonas aeruginosa* biofilms as a function of the untreated control biofilm areal cell density. Dashed lines connect data points from the same investigation. The antibiotics used include tobramycin, ciprofloxacin, and gentamicin. Sources: references 11, 45–47. doi:10.1128/microbiolspec.MB-0010-2014.f5

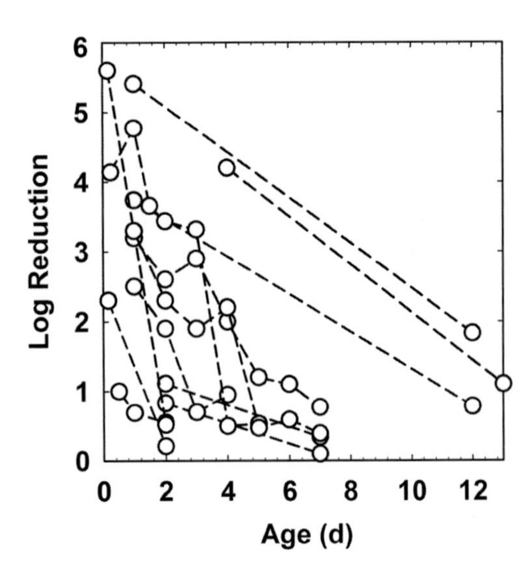

FIGURE 6 Antimicrobial efficacy as a function of biofilm age. Dashed lines connect data points from the same investigation. Sources: references 11, 25, 45, 47, 48–51. doi:10.1128/microbiolspec.MB-0010-2014.f6

Biofilm age and biofilm cell density are usually strongly correlated. The effects of these two parameters are therefore easily confounded. Is the difference in susceptibility between 2-day-old and 7-day-old biofilms (10) a function of age or a function of the substantial difference in biofilm accumulation at these two time points? Here I analyze one data set where, fortuitously, it is possible to separate these two parameters. Wolcott et al. (11) reported on the challenge of *Staphylococcus aureus* biofilm with gentamicin. During more than 100 h of the maturation of the biofilm there was little change in the biofilm cell density (Fig. 7A). Biofilms removed at different ages were treated with gentamicin, and the log reduction in viable counts was determined. This log reduction did not correlate with the untreated control biofilm cell density (R^2 = 0.087). There was correlation between the biofilm susceptibility to gentamicin and biofilm age (Fig. 7B; R^2 = 0.470).

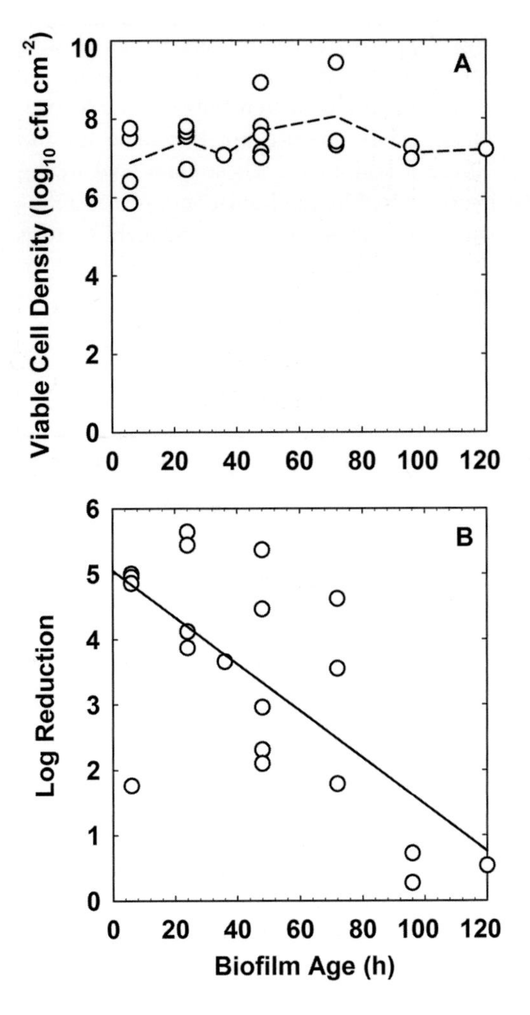

FIGURE 7 (A) Maturation of *S. aureus* biofilm and (B) change in gentamicin susceptibility with age. The dashed line in panel A connects the mean values at each time point. The solid line in panel B is the least squares regressed fit. Source: reference 11. doi:10.1128/microbiolspec.MB-0010-2014.f7

Though the preceding example indicates a more important role for biofilm age than for cell density, in general it is very difficult to uncouple the individual contributions of age and density to biofilm tolerance.

Species Composition

In this section I explore the role of the microbial composition of a biofilm on its antimicrobial tolerance. Tolerance factors, for both biocides and antibiotics, are grouped by phylum in Figure 8. There is no statistically significant difference between the mean values of *TF* for any of the phyla ($p = 0.26$ by analysis of variance). For the three phyla for which there are four or more data points (*Firmicutes, Proteobacteria, Actinobacteria*), *TF* ranges over at least two orders of magnitude. One thing these data suggest is that tolerance is not specific to any particular subgroup of microorganisms. Indeed, reduced biofilm susceptibility appears to be a broadly distributed capability across the microbial world.

Medium Composition

Antimicrobial susceptibility can be very sensitive to the composition of the medium used in the assay. I was not able to devise an informative way to test for effects of medium composition on *TF* values. To under-

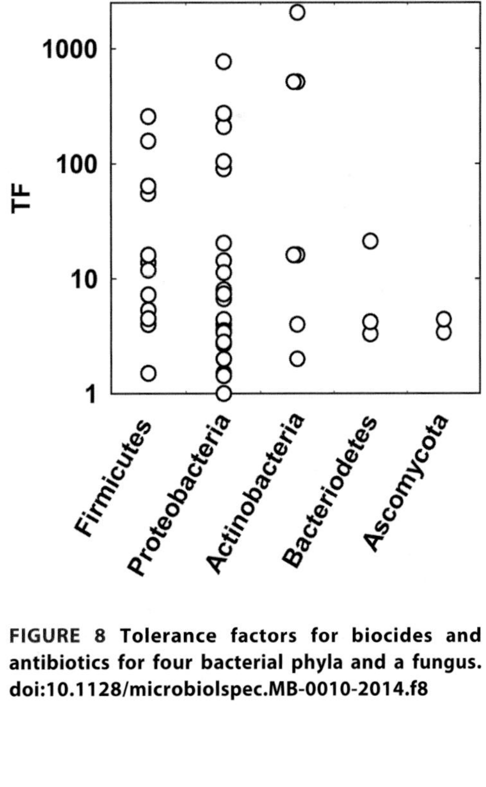

FIGURE 8 Tolerance factors for biocides and antibiotics for four bacterial phyla and a fungus. doi:10.1128/microbiolspec.MB-0010-2014.f8

score the dramatic influence medium composition can play, Figure 9 presents some measurements made with young *Escherichia coli* or *P. aeruginosa* biofilms. At this early stage of development, antibiotics can be very effective against the bacteria under certain culture conditions. However, changes in the medium can drastically alter bacterial susceptibility. For example, 6-h-old *E. coli* biofilms are decimated by kanamycin when challenged on LB medium (8.4 log reduction) but scarcely affected when the medium is supplemented with glucose (1.2 log reduction). A similarly dramatic effect is seen for ampicillin treatment, except that it is exactly the reverse: on LB medium ampicillin is ineffective (1.4 log reduction), whereas the addition of glucose greatly enhances killing (7.6 log reduction). Analogous alterations in antibiotic susceptibility can be seen in 4-h-old *P. aeruginosa* biofilms exposed to tetracycline or tobramycin on different media (Fig. 9). For each agent there are conditions under which they are very effective and conditions under which they are ineffective. These conditions are not the same for the different antibiotics. These data lead to the hypothesis that medium composition influences microbial physiology, which in turn alters antimicrobial susceptibility.

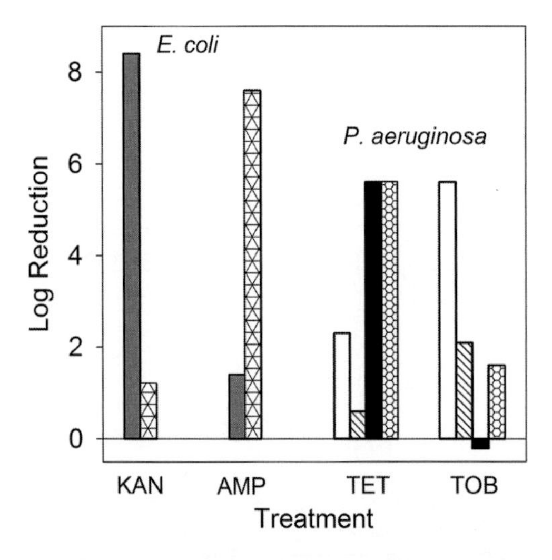

FIGURE 9 Medium effects on biofilm susceptibility to antibiotics. The different bar fills denote various media: LB (gray); LB + glucose (triangles); TSA, aerobic (white); TSA, anaerobic (hatched); noble agar, aerobic (black); noble agar, anaerobic (honeycomb). Sources: reference 52 for *E. coli* and unpublished data of Borriello and Stewart for *P. aeruginosa*. doi:10.1128/microbiolspec.MB-0010-2014.f9

Summary

What has been shown so far is that there is no discernable generalized role of antimicrobial size, antimicrobial chemistry, substratum material, or microbial species composition on the quantitative level of tolerance established during biofilm formation. Only areal cell density and biofilm age partially correlate with antimicrobial tolerance. This suggests that there is something that happens during biofilm maturation, either physical or physiological, that is essential for full biofilm tolerance. Case study results also point to an important role for medium composition, and hence physiology, in biofilm tolerance. Another way to say this is that the details of how the biofilm is grown for a particular test are likely to be more important than the choice of antimicrobial agent or microorganism.

MECHANISMS OF BIOFILM ANTIMICROBIAL TOLERANCE

Antimicrobial Depletion

One simple and possibly underappreciated mechanism of biofilm protection is depletion of the antimicrobial agent in the fluid bathing the biofilm. The antimicrobial could be depleted either by reaction in the fluid phase, by reaction with the biofilm or attachment substratum, or by sorption to constituents of the biofilm or substratum material. This mechanism is especially plausible in systems with a relatively high surface area to volume ratio, such as a 96-well microtiter plate. In this type of system, the demand exhibited by the biofilm could quickly reduce the dissolved concentration of antimicrobial.

The obvious way to control for antimicrobial depletion is to assay the bulk fluid during the course of treatment, or at least before and after the exposure period, to test whether the antimicrobial concentration is sustained. This is not typically done.

Since the surface area to volume ratio is a critical physical characteristic of a system, determining the potential for antimicrobial depletion, and since most of the data sets in Tables 1 and 2 include details permitting calculation of this ratio, a quantitative analysis can be conducted. When the tolerance factors in Table 1 are regressed against the surface area to volume ratio, no correlation is apparent ($R^2 = 0.022$). Neither do the biofilm *TF*s for antibiotics in Table 2 correlate with the surface area to volume ratio of the biofilm test system ($R^2 = 0.010$). What these analyses indicate is that antimicrobial depletion is probably not a general cause of biofilm tolerance in *in vitro* models.

Penetration

The extent of antimicrobial penetration into a biofilm is expected to depend on biofilm thickness, effective diffusivity of the agent in the biofilm, reactivity of the agent in the biofilm, the sorptive capacity of the biofilm for the agent, the dose concentration and dose duration, and external mass transfer properties (12). In other words, this is a complex interaction and problem. A good starting place is to examine actual measurements of antimicrobial penetration in biofilms.

A survey of experimentally measured penetration times of antimicrobial agents in biofilms is presented in Figure 10. This data set excludes measurements made using diffusion chambers in which the biofilm is sandwiched between two compartments. These approaches can be useful for determining whether penetration occurs but are not appropriate for determining absolute penetration times, because the time constants are dependent on the device geometry. The measurements reported in Figure 10 were made using microelectrodes, time lapse microscopy of fluorescent-tagged drugs, total internal reflection spectroscopy, and time lapse microscopy of fluorescence loss from cells preloaded with a fluorophore.

The penetration times in Figure 10 range from a fraction of a minute to almost a full day. It is tempting to judge some of these as fast and others as slow, but keep in mind that the important comparison to be made is between the dose duration and the penetration time. A penetration time of 30 min could be fast if the dose duration is 8 h and slow if the dose duration is 3 min.

Penetration times do not increase with the molecular weight of the antimicrobial as intuition might suggest. Indeed, one thing

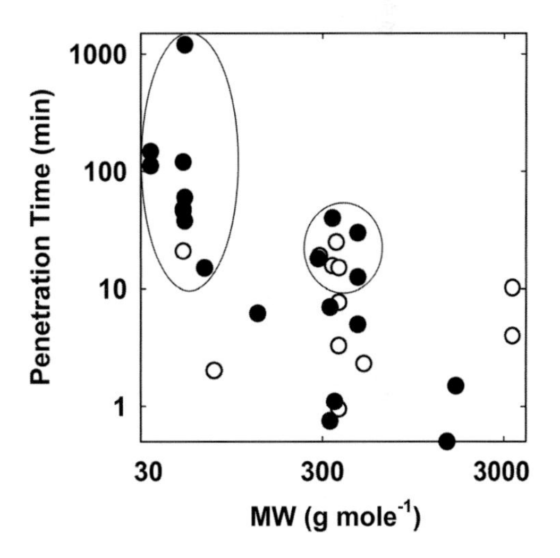

FIGURE 10 Experimentally measured antimicrobial penetration times in biofilms versus molecular weight of the antimicrobial. The penetration time was determined as the time to attain, at the base or center of the biofilm, 50% of the equilibrium concentration of the antimicrobial agent either through a direct measurement of the antimicrobial agent (solid circles) or by loss of membrane integrity detected with a fluorescent probe (open circles). Penetration times greater than 12 min are circled. Sources: references 15, 37, 53–63. doi:10.1128/microbiolspec.MB-0010-2014.f10

that can be inferred from Figure 10 is that even large antibiotics or antimicrobial peptides can penetrate a biofilm within a few minutes. Some examples of large agents that access the interior of a biofilm relatively quickly include vancomycin (0.5 min), daptomycin (1.5 min), and nisin (4 to 10 min).

There are two groups of agents, circled in Figure 10, with measured penetration times longer than 12 min. The antimicrobials in the first group (lower molecular weight) are all reactive oxidants: chlorine, chlorine dioxide, monochloramine, and hydrogen peroxide. The agents in the second group (higher molecular weight) are mostly cationic molecules including quaternary ammonium compounds, such as cetylpyridinium chloride and benzalkonium chloride, and an aminoglycoside antibiotic. The retarded penetration of these agents into the biofilm derives from the reaction or sorption of the agent in the biofilm as it diffuses. Halogens react with uncharacterized components of biomass and are neutralized. Hydrogen peroxide is destroyed by the action of catalase. Agents with a positive charge likely bind to negatively charged polymers or to cell surfaces, delaying penetration. Retarded penetration due to reaction and sorption has been analyzed by mathematical models (12, 13).

When considering agents that are subject to reaction or sorption in the biofilm, it is anticipated that the rate of penetration will depend on the applied concentration. This prediction is borne out by the subset of data plotted in Figure 11. This analysis shows that agents such as chlorine, peracetic acid, and tobramycin (all members of the circled groups in Fig. 10) penetrate a given biofilm faster as the applied concentration is increased. The slope of the regressed line in Figure 11 is close to −1. This tells us that penetration time for these agents is inversely proportional to dose concentration. For example, a dose concentration that is 10 times higher will result in a penetration time that is one 10th as long. This is not expected to be true of antimicrobials that do not react or sorb in the biofilm. The 50% penetration time for a noninteracting solute is predicted to be independent of the applied concentration.

The preceding analysis and discussion is helpful for gaining insight into the fundamental phenomenon of antimicrobial penetration in a biofilm, but it does not tell us if retarded penetration actually limits antimicrobial efficacy in practice. The most likely situation for incomplete penetration to occur is when reactive oxidants are delivered at relatively low concentrations to thick biofilms for brief dose durations. Antibiotics likely penetrate biofilms *in vivo* because dose durations are relatively long. Another argument for penetration of antibiotics, including the sticky aminoglycosides, is that they result in log reductions *in vivo* that indicate access to most of the bacteria. For example, a classic clinical study of inhaled tobramycin in cystic fibrosis patients reported log reductions of *P. aeruginosa* in sputum of slightly greater than 2 after 2 weeks of therapy (14). This tells us that

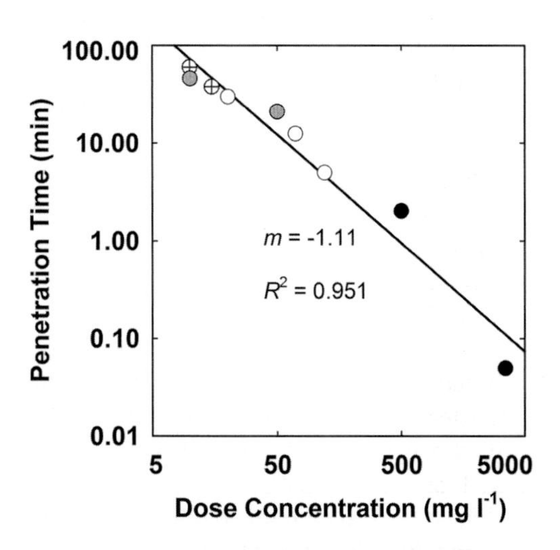

FIGURE 11 Experimentally measured antimicrobial penetration times in biofilms versus dose concentration. The line is the least squares regressed fit. Symbols indicate data for chlorine (cross, 55), chlorine (gray, 54), tobramycin (white, 62), peracetic acid (black, 53). doi:10.1128/microbiolspec.MB-0010-2014.f11

the drug reached 99% of the bacteria. Even in applications in which the dose duration is brief, for example, a mouth rinse treating dental plaque, penetration of the antimicrobial may not be limiting. Corbin (15) found no correlation between the clinical efficacy of mouth rinse active ingredients and their *in vitro* penetration time.

Physiology

Microorganisms in biofilms may be tolerant to antimicrobial agents because they enter less susceptible physiological states. For example, it is widely appreciated that microorganisms in the stationary phase of a batch planktonic culture, which may be slow-growing or nongrowing and may be less metabolically active than growing cells, can be less sensitive to killing by antimicrobials. A few research studies have compared killing in exponential phase, stationary phase, and biofilms (Fig. 12). Though this analysis lacks sufficient data to make a strong conclusion, it suggests that whereas exponential phase planktonic cells are clearly less susceptible than biofilms cells, stationary phase planktonic cells are not consistently different from biofilm cells. The terminology of batch planktonic cultures is probably inadequate as a basis for characterizing the physiological heterogeneity within a biofilm.

A wide variety of terms have been used to characterize the physiological state of a microbial cell: exponential phase, stationary phase, lag phase, nongrowing, stressed, adapted, inactive, viable but nonculturable, persister, dormant. Figure 13 presents a simplified categorization of physiological states based on discrimination of three features: (i) growth, (ii) metabolic and anabolic activity, and (iii) deployment of specific stress-adaptive responses. Though the schematic in Figure 13 presents these as discrete states, it may be more realistic to think of them as stations along continua. The susceptibility of a cell will depend on both the physiological state and the particular antimicrobial agent. Here are a few examples to illustrate the diversity of protected states.

In general, when biofilm microorganisms are compared to planktonic cells for antimicrobial susceptibility, the comparison is to a growing batch culture (Fig. 13A). These are cells that may be relatively sensitive to antimicrobial attack because their current environment is growth permissive and their current investment is in cell growth and replication rather than survival. Cells that transition to a nongrowing but still active state (Fig. 13B) may quickly acquire tolerance to some agents. In the conceptualization of Figure 13, this state is conceived of as cells with an active membrane potential and capacity for generation of some ATP along with submaximal capacity for transcription and translation. These cells do not exhibit DNA replication, cell wall synthesis, or

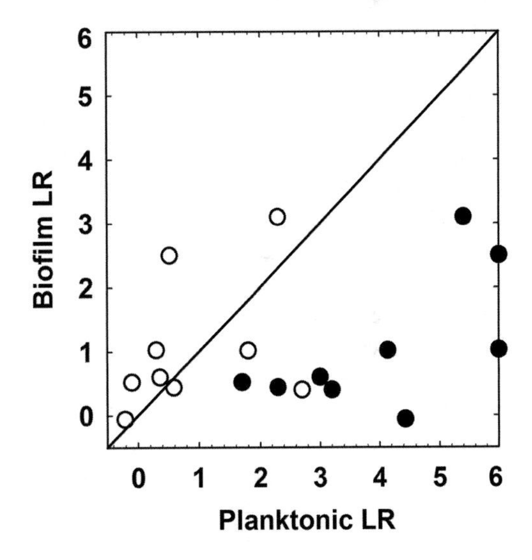

FIGURE 12 Comparison of antimicrobial susceptibility of exponential phase planktonic (solid symbols) or stationary phase planktonic (open symbols) to biofilm cells. The solid line is the line of equality. Points below the line indicate that biofilm cells were less susceptible than planktonic cells. Points above the line indicate that planktonic cells were less susceptible than biofilm cells. Sources: references 38, 64–66. doi:10.1128/microbiolspec.MB-0010-2014.f12

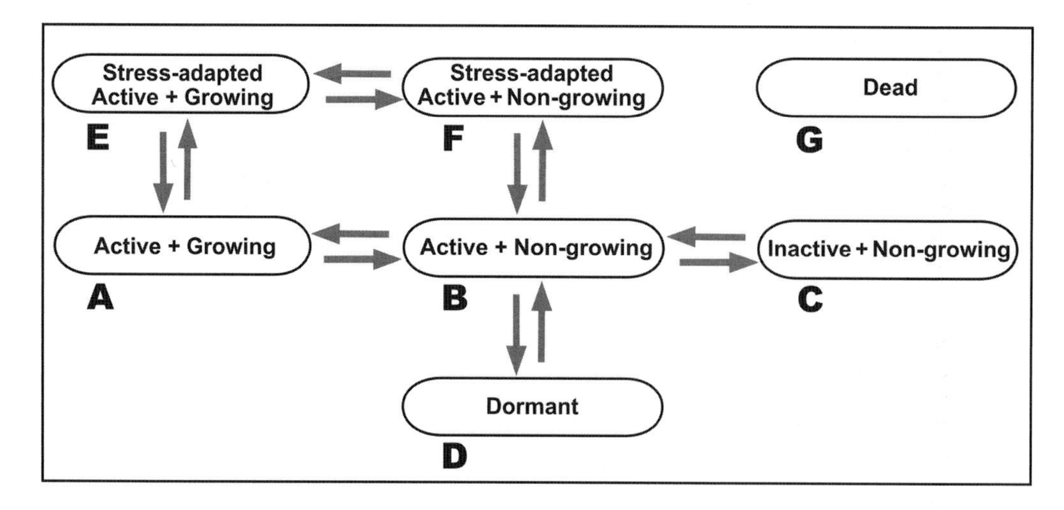

FIGURE 13 Conceptual diagram of distinct cell states important for antimicrobial sensitivity. The dead cell state can presumably be accessed from any of the other states. doi:10.1128/microbiolspec.MB-0010-2014.f13

balanced translation of all of the proteins required to make a new cell. In such a state, bacteria become insensitive to β-lactam antibiotics, which lyse cells by inhibiting cell wall synthesis as the cell continues to expand (16). Cells that transition to the inactive, non-growing state (Fig. 13C) lack any catabolism or anabolism. Such a cell cannot maintain a membrane potential and thus may become insensitive to aminoglycoside antibiotics, which depend on active transport to reach their intracellular targets (17). The dormant state (Fig. 13D) is conceived of as distinct from the inactive, nongrowing state (Fig. 13C).

The dormant state is also metabolically inactive and nongrowing. To enter the dormant state, however, the cell has implemented protective modifications. Such modifications could include, hypothetically, alteration of membrane lipid and porin composition to reduce permeability, hibernation of ribosomes, inhibition of transcription and replication machinery, and deployment of enzymes that protect against oxidative stress without consuming ATP (e.g., catalase). In contrast, the nongrowing, inactive state (Fig. 13C) is an energetically disabled cell that has no other protective modifications. By way of an analogy, the state in Figure 13C could be compared to a car that has run out of gas by the side of the road and been abandoned, whereas a vehicle analogous to the cell state in Figure 13D, while also out of gas, has had the windows rolled up, the radiator drained, the battery disconnected, and a cover tied over it. Such a dormant cell state could confer tolerance to a wide variety of antimicrobial challenges. The much-discussed persister cell may represent such a dormant state (18, 19). Metabolically active bacteria are able to sense their environment and actively respond to the presence of an antimicrobial stress. In the schema of Figure 13, either growing cells (Fig. 13A) or nongrowing yet active cells (Fig. 13B) have the capacity to deploy active stress responses (resulting in the states shown in Figs. 13E and 13F, respectively). Examples of stress responses that have been demonstrated in bacterial biofilms include catalase induction upon treatment with hydrogen peroxide (20), β-lactamase induction upon treatment with imipenem (21), and induction of the lipopolysaccharide-modifying *pmr* operon upon treatment with colistin (22). In each of these examples, the induced gene or genes enhance the capacity of the biofilm to tolerate the antimicrobial either by augmenting destruction of the antimicrobial agent or by modifying the cell to make it less susceptible.

Because biofilms are known to contain niches of varying environmental chemistry and biological activity, it is important to recognize that a biofilm could harbor cells in more than one, possibly all, of the states shown in Figure 13 (23). This physiological heterogeneity or diversification is likely an important factor in the tolerance of the biofilm state. Note that none of these states is necessarily exclusively associated with either a planktonic or biofilm cell.

One difficulty with analyzing the physiological variety diagrammed in Figure 13 is a lack of standard quantitative measures of most of the physiological characteristics. There is an excellent quantitative parameter to characterize microbial growth: specific growth rate. Techniques to measure local growth rates within biofilms could offer insight into the spectrum of physiological states that influence antimicrobial susceptibility. In addition, it would be helpful to have quantitative measures of the overall cellular capacity for transcription or translation, the relative expression of adaptive stress responses, and some quantitative definition of dormancy.

The cell states diagrammed in Figure 13 are surely associated with the differential expression of specific sets of genes in a particular organism. One issue to keep in mind in interpreting the analysis of antimicrobial susceptibility of genetic mutants grown as biofilms is that a mutation that affects the areal cell density of the biofilm could indirectly alter its susceptibility. Indeed, this effect is to be expected, as discussed above and presented in Figures 3 to 5. Some of the systems that have been reported to contribute to biofilm antimicrobial tolerance include the stringent response (24), the SOS response (25), efflux pumps (26, 27), quorum sensing (28), toxin-antitoxin modules (29, 30), the elaboration of periplasmic or extracellular polysaccharides (31, 32, 33), and others (34–36). At this time it is still too early to be able to identify a consensus genetic basis for biofilm antimicrobial tolerance, but these details are certain to follow.

ACKNOWLEDGMENTS

Conflicts of interest: I declare no conflicts.

CITATION

Stewart PS. 2015. Antimicrobial tolerance in biofilms. Microbiol Spectrum 3(3):MB-0010-2014.

REFERENCES

1. **Stewart PS, McFeters GA, Huang CT.** 2000. Biofilm control by antimicrobial agents, p 373–405. *In* Bryers JD (ed), *Biofilms*, 2nd ed. John Wiley & Sons, New York.
2. **Mah TF, O'Toole GA.** 2001. Mechanisms of biofilm resistance to antimicrobial agents. *Trends Microbiol* **9:**34–39.
3. **Stewart PS, Costerton JW.** 2001. Antibiotic resistance of bacteria in biofilms. *Lancet* **358:** 135–138.
4. **Lewis K.** 2001. Riddle of biofilm resistance. *Antimicrob Agents Chemother* **45:**999–1007.
5. **Davies D.** 2003. Understanding biofilm resistance to antibacterial agents. *Nat Rev Drug Dis* **2:**114–122.
6. **Stewart PS, Mukherjee PK, Ghannoum MA.** 2004. Biofilm antimicrobial resistance, p 250–268. *In* Ghannoum M, O'Toole GA (ed), *Microbial Biofilms*, 1st ed. ASM Press, Washington, DC.
7. **Bridier A, Briandet R, Thomas V, Dubois-Brissonnet F.** 2011. Resistance of bacterial biofilms to disinfectants: a review. *Biofouling* **27:**1017–1032.
8. **Ronner AB, Wong AC.** 1993. Biofilm development and sanitizer inactivation of *Listeria monocytogenes* and *Salmonella typhimurium* on stainless steel and Buna-n rubber. *J Food Prot* **56:**750–758.
9. **Chen CI, Griebe T, Srinivasan R, Stewart PS.** 1993. Effects of various metal substrata on accumulation of *Pseudomonas aeruginosa* biofilms and efficacy of monochloramine as a biocide. *Biofouling* **7:**421–251.
10. **Anwar H, Strap JL, Costerton JW.** 1992. Establishment of aging biofilms: possible

mechanism of bacterial resistance to antimicrobial therapy. *Antimicrob Agents Chemother* **36:**1347–1351.

11. **Wolcott RD, Rumbaugh KP, James G, Schultz G, Phillips P, Yang Q, Watters C, Stewart PS, Dowd SE.** 2010. Biofilm maturity studies indicate sharp debridement opens a time-dependent therapeutic window. *J Wound Care* **19:**320–328.
12. **Stewart PS.** 1996. Theoretical aspects of antibiotic diffusion into microbial biofilms. *Antimicrob Agents Chemother* **40:**2517–2522.
13. **Stewart PS, Raquepas JB.** 1995. Implications of reaction-diffusion theory for the disinfection of microbial biofilms by reactive antimicrobial agents. *Chem Eng Sci* **50:**3099–3104.
14. **Ramsey BW, Pepe MS, Quan JM, Otto KL, Montgomery AB, Williams-Warren J, Vasiljev-K M, Borowitz D, Bowman CM, Marshall BC, Marshall S, Smith AL.** 1999. Intermittent administration of inhaled tobramycin in patients with cystic fibrosis. *New Engl J Med* **340:**23–30.
15. **Corbin A, Pitts B, Parker A, Stewart PS.** 2011. Antimicrobial penetration and efficacy in an *in vitro* oral biofilm model. *Antimicrob Agents Chemother* **55:**3338–3344.
16. **Tuomanen E, Cozens R, Tosch W, Zak O, Tomasz A.** 1986. The rate of killing of *Escherichia coli* by [beta]-lactam antibiotics is strictly proportional to the rate of bacterial growth. *J Gen Microbiol* **132:**1297–1304.
17. **Taber HW, Mueller JP, Miller PF, Arrow AS.** 1987. Bacterial uptake of aminoglycoside antibiotics. *Microbiol Rev* **51:**439–457.
18. **Lewis K.** 2007. Persister cells, dormancy and infectious disease. *Nat Rev Microbiol* **5:**48–56.
19. **Lewis K.** 2010. Persister cells. *Annu Rev Microbiol* **64:**357–372.
20. **Elkins JG, Hassett DJ, Stewart PS, Schweizer HP, McDermott TR.** 1999. *Pseudomonas aeruginosa* biofilm resistance to hydrogen peroxide: protective role of catalase. *Appl Environ Microbiol* **65:**4594–4600.
21. **Bagge N, Schuster M, Hentzer M, Ciofu O, Givskov M, Greenberg EP, Høiby N.** 2004. *Pseudomonas aeruginosa* biofilms exposed to imipenem exhibit changes in global gene expression and beta-lactamase and alginate production. *Antimicrob Agents Chemother* **48:** 1175–1187.
22. **Pamp SJ, Gjermansen M, Johansen HK, Tolker-Nielsen T.** 2008. Tolerance to the antimicrobial peptide colistin in *Pseudomonas aeruginosa* biofilms is linked to metabolically active cells, and depends on the *pmr* and *mexAB-oprM* genes. *Mol Microbiol* **68:**223–240.
23. **Stewart PS, Franklin MJ.** 2008. Physiological heterogeneity in biofilms. *Nat Rev Microbiol* **6:**199–210.
24. **Nguyen D, Joshi-Datar A, Lepine F, Bauerle E, Olakanmi O, Beer K, McKay G, Siehnel R, Schafhauser J, Wang Y, Britigan BE, Singh PK.** 2011. Active starvation responses mediate antibiotic tolerance in biofilms and nutrient-limited bacteria. *Science* **334:**982–986.
25. **Bernier SP, Lebeaux D, DeFrancesco AS, Valomon A, Soubigou G, Coppée JY, Ghigo JM, Beloin C.** 2013. Starvation, together with the SOS response, mediates high biofilm-specific tolerance to the fluoroquinolone ofloxacin. *PLoS Genet* **9:**e1003144. doi:10.1371/journal.pgen.1003144.
26. **Liao J, Schurr MJ, Sauer K.** 2013. The MerR-like regulator BrlR confers biofilm tolerance by activating multidrug-efflux pumps in *Pseudomonas aeruginosa* biofilms. *J Bacteriol* **195:**3352–3363.
27. **Zhang L, Mah TF.** 2008. Involvement of a novel efflux system in biofilm-specific resistance to antibiotics. *J Bacteriol* **190:**4447–4452.
28. **Bjarnsholt T, Jensen PØ, Burmølle M, Hentzer M, Haagensen JA, Hougen HP, Calum H, Madsen KG, Moser C, Molin S, Høiby N, Givskov M.** 2005. *Pseudomonas aeruginosa* tolerance to tobramycin, hydrogen peroxide and polymorphonuclear leukocytes is quorum-sensing dependent. *Microbiology* **151:**373–383.
29. **Harrison JJ, Wade WD, Akierman S, Vacchi-Suzzi C, Stremick CA, Turner RJ, Ceri H.** 2009. The chromosomal toxin gene *yafQ* is a determinant of multidrug tolerance for *Escherichia coli* growing as a biofilm. *Antimicrob Agents Chemother* **53:**2253–2258.
30. **Van Acker H, Sass A, Dhondt I, Nelis HJ, Coenye T.** 2014. Involvement of toxin-antitoxin modules in *Burkholderia cenocepacia* biofilm persistence. *Pathog Dis.* [Epub ahead of print.] doi:10.1111/2049-632X.12177.
31. **Mah TF, Pitts B, Pellock B, Walker GC, Stewart PS, O'Toole GA.** 2003. A genetic basis for *Pseudomonas aeruginosa* biofilm antibiotic resistance. *Nature* **426:**306–310.
32. **Colvin KM, Gordon VD, Murakami K, Borlee BR, Wozniak DJ, Wong GC, Parsek MR.** 2011. The Pel polysaccharide can serve a structural and protective role in the biofilm matrix of *Pseudomonas aeruginosa*. *PLoS Pathog* **7:** e10012164. doi:10.1371/journal.ppat.1001264.
33. **Billings N, Millan MR, Caldara M, Rusconi R, Tarasova Y, Stocker R, Ribbeck K.** 2013. The extracellular matrix component Psl provides

fast-acting antibiotic defense in *Pseudomonas aeruginosa* biofilms. *PLoS Pathog* **9:**e1003526. doi:10.1371/journal.ppat.1003526.

34. **Lynch SV, Dixon L, Benoit MR, Brodie EL, Keyhan M, Hu P, Ackerley DF, Andersen GL, Matin A.** 2007. Role of the *rapA* gene in controlling antibiotic resistance of *Escherichia coli* biofilms. *Antimicrob Agents Chemother* **51:**3650–3658.
35. **Zhang L, Hinz AJ, Nadeau JP, Mah TF.** 2011. *Pseudomonas aeruginosa tssC1* links type VI secretion and biofilm-specific antibiotic resistance. *J Bacteriol* **193:**5510–5513.
36. **Zhang L, Chiang WC, Gao Q, Givskov M, Tolker-Nielsen T, Yang L, Zhang G.** 2012. The catabolite repression control protein Crc plays a role in the development of antimicrobial-tolerant subpopulations in *Pseudomonas aeruginosa* biofilms. *Microbiology* **158:**3014–3019.
37. **Stewart PS, Rayner J, Roe F, Rees WM.** 2001. Biofilm penetration and disinfection efficacy of alkaline hypochlorite and chlorosulfmates. *J Appl Microbiol* **91:**525–532.
38. **Luppens SB, Reij MW, van der Heijden RW, Rombouts FM, Abee T.** 2002. Development of a standard test to assess the resistance of *Staphylococcus aureus* biofilm cells to disinfectants. *Appl Environ Microbiol* **68:**4194–4200.
39. **Buckingham-Meyer K, Goeres DM, Hamilton MA.** 2007. Comparative evaluation of biofilm disinfectant efficacy tests. *J Microbiol Methods* **70:**236–244.
40. **Byun MW, Kim JH, Kim DH, Kim HJ, Jo C.** 2007. Effects of irradiation and sodium hypochlorite on the micro-organisms attached to a commercial food container. *Food Microbiol* **24:**544–548.
41. **Griebe T, Chen CI, Srinivasan R, Stewart PS.** 1993. Analysis of biofilm disinfection by monochloramine and free chlorine, p 151–161. *In* Geesey GG, Lewandowski Z, Flemming HC (ed), *Biofouling and Biocorrosion in Industrial Water Systems.* Lewis Publishers, Boca Raton, FL.
42. **Norwood DE, Gilmour A.** 2000. The growth and resistance to sodium hypochlorite of *Listeria monocytogenes* in a steady-state multispecies biofilm. *J Appl Microbiol* **88:**512–520.
43. **Oie S, Huang Y, Kamiya A, Konishi H, Nakazawa T.** 1996. Efficacy of disinfectants against biofilm cells of methicillin-resistant *Staphylococcus aureus. Microbios* **85:**223–230.
44. **Kim J, Pitts B, Stewart PS, Camper A, Yoon J.** 2008. Comparison of the antimicrobial effects of chlorine, silver ion, and tobramycin on biofilm. *Antimicrob Agents Chemother* **52:** 1446–1453.
45. **Anwar H, van Biesen T, Dasgupta M, Costerton JW.** 1989. Interaction of biofilm bacteria with antibiotics in a novel *in vitro* chemostat system. *Antimicrob Agents Chemother* **33:**1824–1826.
46. **Jass J, Costerton JW, Lappin-Scott HM.** 1995. The effect of electrical currents and tobramycin on *Pseudomonas aeruginosa* biofilms. *J Indust Microbiol* **15:**234–242.
47. **Borriello G, Werner EM, Roe F, Kim AM, Ehrlich GD, Stewart PS.** 2004. Oxygen limitation contributes to antibiotic tolerance of *Pseudomonas aeruginosa* in biofilms. *Antimicrob Agents Chemother* **48:**2659–2664.
48. **Shapiro JA, Nguyen VL, Chamberlain NR.** 2011. Evidence for persisters in *Staphylococcus epidermidis* RP62a planktonic cultures and biofilms. *J Med Microbiol* **60:**950–960.
49. **Tré-Hardy M, Macé C, Manssouri NE, Vanderbist F, Traore H, Devleeschouwer MJ.** 2009. Effect of antibiotic co-administration on young and mature biofilms of cystic fibrosis clinical isolates: the importance of the biofilm model. *Int J Antimicrob Agents* **33:**40–45.
50. **Singla S, Harjai K, Chhibber S.** 2013. Susceptibility of different phases of biofilm of *Klebsiella pneumoniae* to three different antibiotics. *J Antibiot* **66:**61–66.
51. **Corcoran M, Morris D, De Lappe N, O'Connor J, Lalor P, Dockery P, Cormican M.** 2014. Commonly used disinfectants fail to eradicate *Salmonella enterica* biofilms from food contact surface materials. *Appl Environ Microbiol* **80:**1507–1514.
52. **Zuroff TR, Bernstein H, Lloyd-Randolfi J, Jimenez-Taracido L, Stewart PS, Carlson RP.** 2010. Robustness analysis of culturing perturbations on *Escherichia coli* colony biofilm beta-lactam and aminoglycoside antibiotic tolerance. *BMC Microbiol* **10:**185.
53. **Bridier A, Dubois-Brissonnet F, Greub G, Thomas V, Briandet R.** 2011. Dynamics of the action of biocides in *Pseudomonas aeruginosa* biofilms. *Antimicrob Agents Chemother* **55:** 2648–2654.
54. **Davison WM, Pitts B, Stewart PS.** 2010. Spatial and temporal patterns of biocide action against *Staphylococcus epidermidis* biofilms. *Antimicrob Agents Chemother* **54:**2920–2927.
55. **de Beer D, Srinivasan R, Stewart PS.** 1994. Direct measurement of chlorine penetration into biofilms during disinfection. *Appl Environ Microbiol* **60:**4339–4344.
56. **Jang A, Szabo J, Hosni AA, Coughlin M, Bishop PL.** 2006. Measurement of chlorine dioxide penetration in dairy process pipe

biofilms during disinfection. *Appl Microbiol Biotechnol* **72:**368–376.

57. **Lee WH, Wahman DG, Bishop PL, Pressman JG.** 2011. Free chlorine and monochloramine application to nitrifying biofilm: comparison of biofilm penetration, activity, and viability. *Environ Sci Technol* **45:**1412–1419.
58. **Liu X, Roe F, Jesaitis A, Lewandowski Z.** 1998. Resistance of biofilms to the catalase inhibitor 3-amino-1,2,4-triazole. *Biotechnol Bioeng* **59:**156–162.
59. **Dabbi-Oubekka S, Briandet R, Fontaine-Aupart MP, Steenkeste K.** 2012. Correlative time-resolved fluorescence microscopy to assess antibiotic diffusion-reaction in biofilms. *Antimicrob Agents Chemother* **56:**3349–3358.
60. **Sandt C, Barbeau J, Gagnon MA, LaFleur M.** 2007. Role of the ammonium group in the diffusion of quaternary ammonium compounds in *Streptococcus mutans* biofilms. *J Antimicrob Chemother* **60:**1281–1287.
61. **Stewart PS, Roe F, Rayner J, Elkins JG, Lewandowski Z, Ochsner UA, Hassett DJ.** 2000. Effect of catalase on hydrogen peroxide penetration into *Pseudomonas aeruginosa* biofilms. *Appl Environ Microbiol* **66:**836–838.
62. **Tseng BS, Zhang W, Harrison JJ, Quach TP, Song JL, Penterman J, Singh PK, Chopp DL, Packman AI, Parsek MR.** 2013. The extracellular matrix protects *Pseudomonas aeruginosa* biofilms by limiting the penetration of tobramycin. *Environ Microbiol* **15:**2865–2878.
63. **Vrany JD, Stewart PS, Suci PA.** 1997. Comparison of recalcitrance to ciprofloxacin and levofloxacin exhibited by *Pseudomonas aeruginosa* biofilms displaying rapid-transport characteristics. *Antimicrob Agents Chemother* **41:**1352–1358.
64. **Anderl JN, Franklin MJ, Stewart PS.** 2000. Role of antibiotic penetration limitation in *Klebsiella pneumoniae* biofilm resistance to ampicillin and ciprofloxacin. *Antimicrob Agents Chemother* **44:**1818–1824.
65. **Liao J, Sauer K.** 2012. The MerR-like transcriptional regulator BrlR contributes to *Pseudomonas aeruginosa* biofilm tolerance. *J Bacteriol* **194:**4823–4836.
66. **Podos SD, Thanassi JA, Leggio M, Pucci MJ.** 2012. Bactericidal activity of ACH-702 against nondividing and biofilm staphylococci. *Antimicrob Agents Chemother* **56:**3812–3818.
67. **Spoering AL, Lewis K.** 2001. Biofilms and planktonic cells of *Pseudomonas aeruginosa* have similar resistance to killing by antimicrobials. *J Bacteriol* **183:**6746–6751.
68. **Thrower Y, Pinney RJ, Wilson M.** 1997. Susceptibilities of *Actinobacillus actinomycetemcomitans* biofilms to oral antiseptics. *J Med Microbiol* **46:**425–429.
69. **Green PN, Pirrie RS.** 1993. A laboratory apparatus for the generation and biocide efficacy testing of *Legionella* biofilms. *J Appl Bacteriol* **74:**388–393.
70. **Wood P, Jones M, Bhakoo M, Gilbert P.** 1996. A novel strategy for control of microbial biofilms through generation of biocide at the biofilm-surface interface. *Appl Environ Microbiol* **62:**2598–2602.
71. **Walker JT, Morales M.** 1997. Evaluation of chlorine dioxide (ClO_2) for the control of biofilms. *Wat Sci Technol* **35:**319–323.
72. **Goeres DM, Loetterle LR, Hamilton MA.** 2007. A laboratory hot tub model of disinfectant efficacy evaluation. *J Microbiol Meth* **68:**184–192.
73. **Tabak M, Scher K, Hartog E, Römling U, Matthews KR, Chikindas ML, Yaron S.** 2007. Effect of triclosan on *Salmonella typhimurium* at different growth stages and in biofilms. *FEMS Microbiol Lett* **267:**200–206.
74. **Stickler D, Hewett P.** 1991. Activity of antiseptics against biofilms of mixed bacterial species growing on silicone surfaces. *Eur J Clin Microbiol Infect Dis* **10:**416–421.
75. **Neyret C, Herry JM, Meylheuc T, Dubois-Brissonnet F.** 2014. Plant-derived compounds as natural antimicrobials to control paper mill biofilms. *J Ind Microbiol Biotechnol* **41:**87–96.
76. **Blanc V, Isabal S, Sánchez MC, Llama-Palacios A, Herrera D, Sanz M, León R.** 2013. Characterization and application of a flow system for *in vitro* multispecies oral biofilm formation. *J Periodont Res* [Epub ahead of print.] doi:10.1111/jre.12110.
77. **Tafin UF, Corvec S, Betrisey B, Zimmerli W, Trampuz A.** 2012. Role of rifampin against *Propionibacterium acnes* biofilm *in vitro* and in an experimental foreign-body infection model. *Antimicrob Agents Chemother* **56:**1885–1891.
78. **Soriano F, Huelves L, Naves P, Rodriguez-Cerrato V, del Prado G, Ruiz V, Ponte C.** 2009. *In vitro* activity of ciprofloxacin, moxifloxacin, vancomycin, and erythromycin against planktonic and biofilm forms of *Corynebacterium urealyticum*. *J Antimicrob Chemother* **63:**353–356.
79. **Khan W, Bernier SP, Kuchma SL, Hammond JH, Hasan F, O'Toole GA.** 2010. Aminoglycoside resistance of *Pseudomonas aeruginosa* biofilms modulated by extracellular polysaccharide. *Int Microbiol* **13:**207–212.
80. **Folsom JP, Richards L, Pitts B, Roe F, Ehrlich GD, Parker A, Mazurie A, Stewart PS.** 2010. Physiology of *Pseudomonas aeruginosa* in

biofilms as revealed by transcriptome analysis. *BMC Microbiol* **10:**294.

81. **Zheng Z, Stewart PS.** 2004. Growth limitation of *Staphylococcus epidermidis* in biofilms contributes to rifampin tolerance. *Biofilms* **1:**31–35.
82. **Walters MC, Roe F, Bugnicourt A, Franklin MJ, Stewart PS.** 2003. Contributions of antibiotic penetration, oxygen limitation, and low metabolic activity to tolerance of *Pseudomonas aeruginosa* biofilms to ciprofloxacin and tobramycin. *Antimicrob Agents Chemother* **47:**317–323.
83. **Okuda K, Zendo T, Sugimoto S, Iwase T, Tajima A, Yamada S, Sonomoto K, Mizunoe Y.** 2013. Effects of bacteriocins on methicillin-resistant *Staphylococcus aureus* biofilm. *Antimicrob Agents Chemother* **57:**5572–5579.
84. **Larsen T.** 2014. Susceptibility of *Porphyromonas gingivalis* in biofilms to amoxicillin, doxycycline and metronidazole. *Oral Microbiol Immunol* **17:**267–271.
85. **Frank KL, Reichert EJ, Piper KE, Patel R.** 2007. *In vitro* effects of antimicrobial agents on planktonic and biofilm forms of *Staphylococcus lugdunensis* clinical isolates. *Antimicrob Agents Chemother* **51:**888–895.
86. **Al-Dhaheri RS, Douglas LJ.** 2008. Absence of amphotericin B-tolerant persister cells in biofilms of some *Candida* species. *Antimicrob Agents Chemother* **52:**1884–1887.
87. **Hawser SP, Douglas LJ.** 1995. Resistance of *Candida albicans* biofilms to antifungal agents *in vitro*. *Antimicrob Agents Chemother* **39:** 2128–2131.

How Biofilms Evade Host Defenses 14

EMMANUEL ROILIDES,[1] MARIA SIMITSOPOULOU,[1]
ASPASIA KATRAGKOU,[1,2] and THOMAS J. WALSH[2]

INTRODUCTION

Historically, microbial organisms have been grown in pure liquid cultures as free-floating "planktonic" cells, promoting the general theory of the unicellular lifestyle. However, in the late 1970s, Costerton et al. (1) demonstrated that groups of bacteria were embedded in a highly hydrated polysaccharide matrix that mediated adhesion to solid aquatic surfaces. Several years later, the same research team called these cellular communities "biofilms," defined as a functionally heterogeneous aggregate of microcolonies or single cells encased in a matrix of self-produced extracellular polymeric molecules that could adhere either to organic, abiotic surfaces or to each other (2). Microbial biofilms can develop into highly organized structures containing channels in which water, nutrients, and metabolic waste can be transported. Adhesion to substrates or surfaces induces expression of a large number of genes, while cell aggregates in different regions in a biofilm exhibit different gene expression profiles that regulate biofilm development and maturation processes (3, 4). A large amount of research since the 1980s has brought to light the theory that

[1]Infectious Diseases Unit, 3rd Department of Pediatrics, Faculty of Medicine, Aristotle University School of Health Sciences, Hippokration Hospital, 54642 Thessaloniki, Greece; [2]Transplantation-Oncology Infectious Diseases Program, Weill Cornell Medical Center of Cornell University, New York, NY 14850.

Microbial Biofilms 2nd Edition
Edited by Mahmoud Ghannoum, Matthew Parsek, Marvin Whiteley, and Pranab K. Mukherjee

doi:10.1128/microbiolspec.MB-0012-2014

most, if not all, bacteria and fungi can form biofilms as a survival mechanism in hostile environments, providing protection from biotic and abiotic stresses (5–8). Prime candidates for cell attachment and biofilm growth are surfaces exposed to or containing moisture and some nutrients. Natural or man-made substrates for cell attachment and biofilm growth include river stones, oil and gas installations, ship hulls, water pipes, food-processing surfaces, contaminated surgical instruments, indwelling medical devices, human teeth, and infected wounds (9–11). In this chapter, we will present an overview of the life cycle of biofilms and their diversity, detection methods for biofilm development, and host immune responses to pathogens. We will then focus on current concepts in bacterial and fungal biofilm immune evasion mechanisms.

BIOFILM FORMATION

Although different model systems have been described to define biofilm development in bacteria and fungi, the steps involved in the biofilm growth cycle are fairly universal with many common characteristics (Table 1). In the human body, attachment to accessible human host proteins is the first step of biofilm formation and is dependent either on the hydrophobic nature of microbial surfaces or on specific cell surface molecules that enable adherence to host proteins. For example, *Staphylococcus epidermidis* and *Staphylococcus aureus* express MSCRAMMs (microbial surface components recognizing adhesive matrix molecules) that have domains for noncovalent attachment to peptidoglycan moieties and also harbor binding sites for fibronectin, fibrinogen, laminin, collagen IV, and other human matrix proteins (12, 13). For bacteria with flagellar motility, swimming cells attach to the surface first transiently and then permanently, expressing conditionally synthesized adhesive proteins to form a monolayer biofilm (14, 15). Like the exopolysaccharide (EPS) matrix of bacterial biofilms, the polysaccharide capsule of *Cryptococcus neoformans,* a human pathogenic fungus, promotes the attachment process to prosthetic medical devices, whereas cell-surface glycoproteins facilitate adhesion of *Candida albicans* and *Candida glabrata* (16–19). Attachment is promoted by several environmental signals, such as changes in nutrient concentrations, pH, flow velocity of surrounding body fluids (urine, blood, saliva), temperature, oxygen concentration, osmolality, and iron.

TABLE 1 Steps in the biofilm growth cycle

Steps
Attachment of microorganisms to surfaces
Adherence
Microcolony arrangement
Detachment and dissemination of single or clustered cells to other organ systems

The next phase of biofilm formation is characterized by a microcolony arrangement on the attached surface produced by the multiplication of free-floating cells; motility is reduced and exopolysaccharide production is activated to trap nutrients and planktonic cells, whereas signal molecules are secreted in a cell-density-dependent manner to communicate and coordinate the cellular responses through a process called quorum sensing. The presence of farnesol, a quorum sensing molecule first described in *C. albicans,* has been found to inhibit the yeast-to-mycelium conversion, enhancing active budding yeast production without compromising cellular growth rates (20). In *Pseudomonas aeruginosa,* rhamnolipids maintain biofilm architecture by affecting the attachment of bacterial cells to substrates and cell-to-cell-interactions. When cell density increases during the later stages of the maturation phase, rhamnolipids maintain open channels between cellular aggregates to distribute certain types and amounts of nutrients to support cell growth (21).

Another way to trap both nutrients and planktonic cells once a critical biofilm mass has been achieved is the production and secretion of extracellular matrix predominantly composed of polysaccharides, proteins, and extracellular DNA (eDNA). This matrix promotes initial cell adhesion, triggers polysaccharide formation, and serves as a support that links molecules together in the biofilm matrix, thus influencing the structure and organization of mature biofilms. Extracellular DNA is an important component of *Aspergillus fumigatus* biofilms that originates from either fungal autolysis (22) or is externally supplied from human neutrophils attracted to the infected site (23, 24).

The last phase of biofilm formation is cell detachment, which is required for the dissemination of single or clustered cells to other organ systems. Factors that could contribute to cell dispersal include sheer mechanical forces (blood flow), enzymes that digest the extracellular matrix, and nutrient limitation in a mature biofilm. The dispersed cell population of *C. albicans* displays yeast morphology with different phenotypic characteristics from their planktonic counterparts, including increased adherence capacity, filamentation, biofilm formation, and increased pathogenicity to establish new foci of infection (25). In staphylococci, quorum sensing–controlled phenol-soluble modulins (PSMs) participate in establishing biofilm architecture as well as detachment processes using a mechanism similar to that used by *P. aeruginosa* but with different effector molecules. PSMs are part of a novel toxin family with multiple roles in staphylococcal pathogenesis, because they promote formation of biofilm channels and control biofilm expansion by detachment and regrowth, contributing to bacterial dissemination. During the dispersal phase, PSMs act as surfactant-like peptides that inhibit cell-to-cell interactions at the surface of the biofilm, leading to the detachment of bacterial cells at the fluid-biofilm interphase and the subsequent systemic spread of biofilm-associated infections (26, 27).

Natural biofilms in most environments, including human disease, tend to coexist forming polymicrobial communities (28, 29). The interactions that take place between fungal and/or bacterial species can either be synergistic or antagonistic in nature (30). An example of a mutually beneficial interaction is a phenomenon called "coaggregation symbiosis" that is observed between *C. albicans* and *S. aureus*, whereby *Candida* hyphal penetration through epithelial layers provides a route of entry for staphylococci. Moreover, the observed hyphal-mediated increased pathogenicity of *S. aureus* may be attributed not only to the physical interactions but also to the differential regulation of virulence factors produced during polymicrobial growth (31). Different microbial species in a single biofilm community could offer passive resistance, metabolic cooperation, quorum sensing systems, and genotypic variability that give an advantage to counteract adverse environmental conditions (32). In contrast, the interactions between *A. fumigatus* and *P. aeruginosa* (both present in the cystic fibrosis [CF] lung microbiome) is described as being antagonistic. *A. fumigatus* biofilm formation is inhibited by direct contact with a *P. aeruginosa*–secreted heat-stable soluble factor, suggesting that small diffusible molecules can interfere with filamentous fungal growth in polymicrobial environments (33). Although in recent years investigations have shifted to polymicrobial biofilms, there is considerable knowledge yet to be gained in our understanding of microbial cohabitation and microbes' interaction with the host to control the impact of polymicrobial biofilm-associated diseases.

DETECTION AND QUANTIFICATION METHODS OF BIOFILM FORMATION

Various methods have been developed for biofilm detection and quantitation (Table 2). They include staining and microscopic visualization of biofilm structure as well as quantitation of the numbers of biofilm-associated cells *in situ* or after detachment of microbial cells from the substrate.

TABLE 2 Methods for biofilm detection and quantitation

Direct methods	Indirect methods
Tissue culture plating	Dry cell weight assays
Tube method	Colony-forming-unit counting
Congo red agar method	DNA quantification
Crystal violet staining	XTT reduction assay
Visual assessment by scanning electron microscopy	
Visual assessment by confocal scanning laser microscopy	

Biofilm-producing microorganisms are also detected by methods that include growth in tissue culture wells or silicone tube biofilm reactors, followed by staining with Congo red or crystal violet stains. These stained biofilms can be visually assessed by scanning electron microscopy or confocal scanning laser microscopy, which is most effective for studying an intact biofilm's three-dimensional architecture (34–36). Quantitative measurement of biofilm growth is determined by using methods that include determining dry cell weight assays, assessing colony forming unit by plating on solid media, DNA quantification, or an XTT 2,3-bis(2-methoxy-4-nitro-5-sulfophenyl)-5-[(phenylamino) carbonyl]-2H-tetrazolium hydroxide reduction assay (34, 37, 38). The latter method is colorimetric and uses XTT, a colorless or slightly yellow compound that when reduced to a formazan product becomes bright orange. The oxidation-reduction process is initiated by mitochondrial dehydrogenases of metabolically active cells. The colorimetric change is proportional to the number of living cells and can be quantified spectrophotometrically. The biodiversity and abundance of biofilm-associated microorganisms in polymicrobial communities is detected using a combination of molecular diagnostics based on PCR, sequencing technologies, and advanced mathematical algorithms to identify the biofilm-producing microorganisms present and to analyze the copy number of each organism relative to the total number of copies for all organisms (39–41).

HOST DEFENSES AGAINST BIOFILMS

Upon infection, innate immune defense strategies are able to establish an immediate response through effector mechanisms mediated by immune cells, receptors, and several humoral factors (Table 3). The main role of humoral factors, such as mannose binding lectins, collectins, complement, and antibodies, is to bind invading microbes and promote receptor-mediated recognition and phagocytosis by cells of the innate immune system. Pathogen-associated molecular structures found on the surface of microbes are recognized by specific receptors on natural killer cells or professional phagocytes consisting of polymorphonuclear neutrophils (PMNs), mononuclear leukocytes (macrophages), and dendritic cells (DCs). Toll-like receptors (TLRs), the best-defined immune sensors of invading pathogens on the surface of phagocytes, initiate a cascade of signaling pathways that induce phagocytosis, secretion of antimicrobial products generated in phagocytes by oxidative and nonoxidative mechanisms, release of pro- and anti-inflammatory cytokines, or other factors that contribute to the activation, maturation, and immunoregulation of adaptive immune responses. DCs are the main cell populations that bridge innate and adaptive immunity, because recruitment by inflammatory signals leads to DC accumulation at injured or infected sites to capture, internalize, and present microbial cell fragments to naïve T cells in the lymph nodes. TLRs and C-type lectin receptors contribute to the recognition and activation of specific DC programs to decode the structural information of the antigens captured and to convert it into different T-cell immune responses (42–45). Thus, early phases of immune response help to control excessive propagation of pathogens, while activation of proper T cell subsets leads to long-lived protective immunity.

In most of the studies conducted to date, host-pathogen interactions concern bacteria and fungi in their free-living planktonic state.

TABLE 3 Relation of innate host defenses and specific microbial biofilms

Organism	Host defenses
Pseudomonas	Activation of oxidative burst by PMN
	Altered migratory behavior of phagocytes
	Maintenance of phagocytic functions
	Biofilm formation increases in the presence of PMNs
	Function of lactoferrin is severely compromised by both bacteria and PMNs
	Secreted rhamnolipids protect bacteria against PMN antibacterial activity
	Dislodged biofilm cells are susceptible to immune cells
Enterococcus	Effective phagocytosis of biofilm-embedded cells
	Infected monocytes with biofilm cells display reduced cytokine, chemokine expression
Candida	Monocytes exert a biofilm-enhancing effect associated with increased pro-inflammatory cytokine expression
	Biofilms have an immunosuppressive effect
	Mature biofilms fail to trigger a reactive oxygen species response
	β-glucan in the extracellular matrix evades host defenses
	Cytokine-primed PMNs do not enhance biofilm damage
	Biofilm glycosyl phosphatidylinositol–anchored proteins confer resistance to PMN killing
	Echinocandins have positive immunomodulatory effects on host cells against biofilms
	Dislodged biofilm cells are susceptible to immune cells
Aspergillus	Biofilms are resistant to the antifungal activity of phagocytes
	Gliotoxin downregulates vitamin D and cytokine expression

However, despite the paucity of information, recent studies have begun to investigate immune responses to biofilms. Antimicrobial factors, such as lactoferrin and the human cationic host defense peptide LL-37 found at mucosal surfaces or in secondary granules of PMN, when used *in vitro* at very low concentrations, strongly inhibit formation of *P. aeruginosa* biofilms. Lactoferrin, as an iron-binding glycoprotein, reduces the iron supply necessary for biofilm growth and promotes twitching motility (46), whereas LL-37 reduces the initial attachment phase, increases surface motility, and interferes with the quorum sensing system of *P. aeruginosa* (47). However, bacteria are known to counteract such host defense processes by secreting proteases able to degrade both lactoferrin and LL-37 (48, 49). Although specific immune receptors for the biofilm mode of growth have not been identified, the active involvement of PMNs against biofilms is demonstrated by the increased oxygen depletion caused by enhanced oxidative burst, by the increased glucose uptake by PMN in CF lungs, and by the high concentration of L-lactate in the sputum of CF patients with chronic lung infection (50). It has been shown that upon contact of PMNs with *S. aureus* biofilms, a decrease in biofilm mass is caused by phagocytosis, rapid degranulation of lactoferrin and elastase, and finally, DNA release (51). Time-lapse video microscopy has documented migration of PMNs into *S. aureus* biofilms, and clearance of bacterial cells within the biofilm by phagocytosis is shown to be dependent on biofilm maturation, because young biofilms are more susceptible to PMN antimicrobial functions compared to mature biofilms (52). In contrast to planktonic cells of *S. epidermidis*, adherence of PMNs to biofilms and phagocytosis seems to be opsonization-independent, suggesting that biofilms could contain PMN signaling molecules (53).

Confocal scanning laser microscopy shows that phagocytes enhance the ability of *C. albicans* to form biofilms and that these phagocytes appear as unstimulated rounded cells, inducing significantly less damage to biofilms than their planktonic counterparts (54, 55). *C. neoformans*, an encapsulated fungus that can cause meningoencephalitis in immunocompromised patients, shows resistance to oxidative stress, but it becomes considerably more susceptible to cationic antimicrobial peptides when switching to the biofilm mode of growth. Defensins with higher net positive charges (β-defensin-1, β-defensin-3) interact strongly with negatively charged biofilm surfaces, while their hydrophobic characteristics enable them to enter the cells' membranes and create temporary pores through which cell contents leak, leading to cell death. In contrast, defensins with lower net positive charges (α-defensin-3 and magainin-1) are less efficient against biofilms, implying that the greater affinity to biofilms results in increased biofilm susceptibility (56).

C. neoformans capsule is primarily composed of glucuronoxylomannan, a polysaccharide present in large amounts in the extracellular biofilm matrix and for this reason indispensible for biofilm formation. Moreover, it has been shown that anti–*C. neoformans* antibodies have a multitude of protective functions, including enhancement of animal survival against cryptococcosis, promotion of phagocytosis, antigen presentation, complement activation, and modulation of immune protein expression. Testing the hypothesis that antibodies could also interfere with biofilm production, Martinez and Casadevall demonstrated that monoclonal antibodies raised against glucuronoxylomannan inhibit biofilm formation, suggesting that humoral immune responses could be important to the prevention of biofilm formation. In the clinical setting, administration of protective antibodies could present a valuable alternative in the management of biofilm-associated diseases (17).

HOW BIOFILMS OF BACTERIA AND FUNGI EVADE HOST DEFENSES

Staphylococcus spp.

Staphylococcus spp. are among the leading causes of health care facility– and community-associated infections. *S. epidermidis* and *S. aureus* are frequent etiologic factors of biofilm-associated infections on indwelling medical devices as well as chronic and recalcitrant infections such as endocarditis, periodontitis, rhinosinusitis, and osteomyelitis (57). The emergence of drug-resistant strains, especially methicillin-resistant *S. aureus*, further accentuates their therapeutic difficulty. While the bulk of the available information on biofilms concerns biofilm development, the host immune responses against biofilm infections remain largely unidentified.

Initial studies of medically relevant *S. aureus* biofilms showed that human leukocytes are able to effectively penetrate the biofilm, possibly by using the nutrient channels that exist in mature biofilms; however, leukocytes exhibit impaired phagocytosis and show a decreased ability to kill the bacteria (58). Subsequent studies using more sophisticated methods demonstrated that PMNs are able to clear *Staphylococcus* spp. biofilms by phagocytosis (51, 52, 59). Furthermore, immature biofilms were more sensitive to phagocytosis compared to mature ones, although mature biofilms are not entirely immune to PMN attack (52). By comparison, macrophages did not show appreciable phagocytosis of *S. aureus* biofilms. Of note, macrophages were capable of engulfing disrupted biofilm material, suggesting that the size and/or physical complexity of biofilm ultrastructure are responsible for their recalcitrance to phagocytosis (60).

The opsonization of *S. aureus* biofilms with immunoglobulin G did not affect the adherence of PMNs to the biofilms; however, it increased the clearance of biofilm, possibly through upregulated oxygen radical production in PMNs (61). It seems that biofilm matrix can protect bacteria from antibody-mediated phagocytosis (62).

Another mechanism utilized by *S. aureus* biofilms to evade host immunity, which may explain their *in vivo* persistence, is the macrophage dysfunction and cell death caused upon contact with the biofilm itself. It was noted that macrophages showed a differential sensitivity to cell death based on their physical distance from the biofilm surface, with macrophages most intimately associated with biofilm being dead, while macrophages that remained above the biofilm surface remained viable (60). Among the potential mechanisms proposed to account for this phenomenon is the phenomenon of metabolic "layering" within the biofilm (63). Regions of anaerobic and aerobic microenvironments within the biofilm, bacterial-influenced fluctuations in pH, and macrophage survival may be affected due to the release of bacterially produced toxic byproducts of metabolism. Furthermore, the biofilm may contain lytic toxins, which may lead to cell death.

Using a mouse model of catheter-associated biofilm infection, it was demonstrated that several inflammatory signals responsible for macrophage and neutrophil recruitment (CCL2 and CXXL2, respectively) and activation (tumor necrosis factor alpha and interleukin [IL]-1β) were significantly reduced in biofilm-infected tissues compared to the wound healing response elicited by sterile catheters (60). The role of IL-1β is potentially mediated through the MyD88-dependent pathway and can reduce biofilm development but is not sufficient to eradicate staphylococcal biofilms (63). Additionally, inducible nitric oxide synthase (iNOS) expression was decreased and arginase-1, a key enzyme involved in collagen biosynthesis, was increased in macrophages surrounding the biofilm. These findings would be expected to skew the cellular response away from bacterial killing to favor fibrosis and could represent another mechanism to account for the ability of biofilms to evade clearance (60, 63). Further *in vivo* data showed that *S. aureus* biofilms could evade TLR2 and TLR9 recognition (60). While the mechanism responsible for TLR2/TLR9 evasion is not known, it could be attributed to ligand inaccessibility.

Current data collectively indicate that the persistence of staphylococcal biofilm could be attributed to its ability to skew the immune response to favor anti-inflammatory and profibrotic pathways. As such, it has been shown that targeting macrophage pro-inflammatory activity can represent a novel therapeutic strategy to overcome the local immune inhibitory environment created during biofilm infections (64).

From the currently available data, one can deduce that innate host defense cells may be able to recognize *Staphylococcus* spp. biofilms, migrate toward biofilms, and degrade biofilms *in vitro*. However, the latter depends greatly on the experimental conditions, namely the *Staphylococcus* strain, biofilm maturation phase, opsonization with immunoglobulin, and the immune cell population used. It is likely that the unique properties of *Staphylococcus* spp. biofilms circumvent traditional antimicrobial pathways. Additional studies are warranted to investigate the mechanisms that lead to host defense impairment upon contact with *Staphylococcus* spp. biofilms.

Pseudomonas spp.

The opportunistic pathogen *P. aeruginosa* has drawn special attention in the biofilm field due to its role in chronic and recalcitrant infections, particularly lung infection of CF patients, ventilator-associated pneumonia, chronic wounds, otitis media, and medical device–related infections. In humans, infection by *P. aeruginosa*, and subsequent biofilm formation, invariably occurs in association with an exuberant inflammatory response. *Pseudomonas* biofilm–related infections seem to follow the current dogma regarding host defense mechanisms where immune responses fail in eradicating biofilms (65).

Initial studies using a chemiluminescence assay showed that while biofilms of *P. aeruginosa* were capable of activating the oxidative burst response of PMNs, this response was reduced by about 25% compared to planktonic cells (66). *In vitro* microscopic evidence suggests that the normal migratory behavior of phagocytes is disrupted by *P. aeruginosa* biofilms. The innate cells settle into biofilms but do not appear to be capable of migrating from the point of contact even though they appear to mount a respiratory burst, degranulate, and retain their phagocytic and secretory activity. The mechanism by which the biofilms immobilize neutrophils (yet they are still capable of phagocytosing bacteria) is not known (67). It is possible that exotoxins or other components produced by *P. aeruginosa* and immune cells, respectively, may play a role (46, 67). Others suggest that *P. aeruginosa* biofilm tolerance to PMNs is quorum sensing dependent (68).

It has also been reported that *P. aeruginosa* biofilm formation is increased by 2- to 3.5-fold in the presence of PMNs (69, 70). The mechanism of biofilm enhancement by PMNs was attributed to PMN-generated polymers comprised of actin and DNA (69). This is consistent with the clinical finding of higher numbers of necrotic PMNs detected in CF lungs infected with *P. aeruginosa* compared with other bacterial infections (71). Lactoferrin, a common secretory component of human neutrophil granules, has been shown to inhibit *P. aeruginosa* biofilm production (46). However, lactoferrin inhibitory effect was not evident when the total content of neutrophil granules was combined with *P. aeruginosa* (69). In this regard, it is likely that the role of lactoferrin in microbial biofilms is modulated by scavenging and protease- or oxygen radical–mediated degradation by *P. aeruginosa* and neutrophils (46, 67).

The pathogen-beneficial effects of the biofilm matrix have been demonstrated in other studies (65, 72). The exopolysaccharide alginate protects *P. aeruginosa* biofilm bacteria from leukocyte phagocytosis. *In vitro* studies demonstrate that in the presence of the potent leukocyte activator interferon-γ, human phagocytes killed *P. aeruginosa* biofilm bacteria lacking the ability to produce the alginate exopolysaccharide (72). Of note, the inability of innate cells to eliminate biofilms is abolished as soon as biofilm cells are mechanically disrupted into individual cells (65, 73).

There is increasing evidence supporting the role of rhamnolipids (a class of glycolipids produced by *P. aeruginosa* that are considered a key virulence determinant acting under the control of the quorum sensing system) in the host defense against PMNs, especially in CF lung infection or other chronic infections (65, 74). The activity of rhamnolipids was shown *in vitro* and *in vivo* in pulmonary models, where *P. aeruginosa* biofilms produce a shield of excreted rhamnolipids that protects from the bactericidal activity of PMNs (65, 75, 76). Accordingly, disabling the biofilm shield of rhamnolipids either by mutation or by treatment with quorum sensing inhibitors leads to increased PMN-mediated clearance of the biofilm (65, 76). Rhamnolipids appear to be a crucial component of a vicious self-enhancing cycle in which rhamnolipids induce necrotic lysis of PMNs, and the lysed PMNs subsequently cause more inflammation and in turn attract more PMNs (65). The hypothesis of a "biofilm shield" of rhamnolipids which offers protection from the antibacterial activity of PMNs seems to contribute to biofilm infections of *P. aeruginosa* (65).

Candida spp.

While the interactions between the host immune system and planktonic *Candida* cells have been extensively examined, the corresponding biofilm interactions are largely unstudied (77). Initial studies showed that monocytes influence the ability of *C. albicans* to form biofilms. Monocytes exert a biofilm-enhancing effect, which has been associated with a modulated immune response of increased levels of pro-inflammatory cytokines such as IL-1β and decreased levels of IL-6, IL-10, MCP-1, I-309, and tumor necrosis factor alpha (54). Overall, *Candida* biofilms appear to have an immunosuppressive effect, because monocytes are unable to phagocytose fungal cells in the maturing biofilms and appear to be entrapped within the biofilm ultrastructure (54). Lack of monocyte-mediated phagocytosis within biofilms was also observed for different stages of *C. albicans* biofilm development. Confocal scanning microscopy showed that human phagocytes resemble unstimulated cells, presenting a rounded shape when in the presence of biofilms (55). This was also confirmed by reduced cytokine production in the biofilm-phagocyte co-culture compared to a planktonic cell-phagocyte mixture (55). However, the same phenomenon was not observed with dislodged biofilm cells (55). Consistently, mature *Candida* biofilms were more resistant to killing by leukocytes than early biofilms. Further, mature biofilms failed to trigger a reactive oxygen species response (78). The architecture of mature biofilms and the presence of β-glucans in the extracellular matrix represent an important innate immune evasion mechanism of *C. albicans* biofilms (55, 78).

Previous studies have unequivocally shown that interferon-γ and granulocyte colony-stimulating factor enhance the antifungal activity of human PMNs against planktonic *Candida* spp. (77). On the contrary, exposure of *C. albicans* biofilms to cytokine-activated PMNs does not enhance biofilm damage (77, 79). The increased biofilm damage induced by cytokine-activated PMNs can be explained by the upregulatory activity of the two cytokines to the number and/or to the affinity of the mannose and FcγR receptors on the surface of the phagocytes leading to a more efficient interaction between PMNs and fungal targets. Knowing that killing of *C. albicans* requires ligation of the various host immune receptors by *C. albicans* surface structures (mannoproteins, mannans, or β-glucans) and knowing the role of extracellular matrix to biofilm immune evasion, the lack of cytokine effectiveness was attributed to the inability of the effector cells to recognize the cells within the biofilm (79). Nevertheless, the notion of biofilm recognition by the immune system appears to be important. Large amounts of (1→3)-β-D-glucans are found in the extracellular matrix of *Candida* biofilms *in vitro* and *in vivo* (80–82). In addition, *Hyr1*, which encodes cell surface glycosyl phosphatidylinositol–anchored proteins, confers resistance to neutrophil killing *in vitro* and in an oral mucosal tissue biofilm model (81, 83).

In this regard, exposure of *C. albicans* biofilms to anidulafungin, an antifungal agent that causes heightened β-glucan exposure, was associated with a significant increase in phagocyte-mediated damage and a larger pro-inflammatory response from phagocytes (55). Furthermore, *Candida* spp. biofilms were more susceptible to the combined effect of anidulafungin with phagocytes compared to voriconazole with phagocytes (55, 79). These data underline the notion that echinocandins (i.e., caspofungin, micafungin, and anidulafungin) have immunomodulatory effects on host cells against biofilms.

Other Organisms

Enterococci constitute an important cause of hospital-acquired infections and become particularly pathogenic in intensive care settings, in debilitated patients with impaired immune systems, and in the elderly. Despite the clinical importance of enterococci, little is known about their interactions with immune cells (84, 85). The prevailing notion that biofilm cells encased in an exopolysaccharide matrix are resistant to phagocytosis by immune cells seems to contradict recent findings (86, 87). *In vitro* studies showed that immune cells are able to phagocytose enterococcal cells recovered from biofilms as effectively as their planktonic counterparts (86, 87). In addition, enterococcal biofilms seem to elicit a differential immune response relative to planktonic cells, in that infected monocytes with enterococcal biofilm cells invoke less pro-inflammatory cytokine and chemokine expression (86, 87). However, *in vivo* it is likely that the complex involvement of multiple host factors may allow biofilm cells to circumvent host defenses, thus explaining the persistence of biofilm-related infections.

Aspergillus spp. are frequently isolated from the respiratory tract of patients with CF (88), where it causes typical biofilms and may be associated with deterioration of lung function, invasive infection, or allergic bronchopulmonary aspergillosis. Host defenses against *Aspergillus* in the respiratory tract of CF patients have been reviewed (89, 90). Like all biofilms, *Aspergillus* biofilms are quite resistant to the antifungal activity of PMNs and macrophages, the two major airway phagocytes. Of note, *Aspergillus* downregulates vitamin D receptor gene and protein expression *in vitro* and *in vivo*. This is mediated by the fungal virulence factor gliotoxin, an immune-evader and promoter of cellular apoptosis. Reduced gliotoxin and IL-5 as well as IL-13 (Th2 cytokines) concentrations were detectable post–itraconazole treatment in patients with CF and *Aspergillus* with concomitant increases in vitamin D airway receptor expression (91).

HIGHLIGHTS OF RECENT FINDINGS AND FUTURE QUESTIONS

A large range of biofilm model systems has been developed; however, the technical challenges associated with the growth and assessment of biofilm development further complicate the generalization of the results of various studies. In assessing the effects of host defenses on microbial biofilms, the experimental conditions, the thickness, and the complexity of biofilms should be considered. A major challenge in biofilm research is the development of suitable models that effectively reproduce the host defense conditions at different infection sites and that permit evaluation of the efficacy of novel antifungal treatment strategies under *in vivo* conditions (92). Animal models of biofilm formation represent an invaluable alternative for improving our understanding of biofilm development inside the host, their resilience to immune cells, and their interaction with the host immune defense system. This appears to be of particular importance because the use of host immune responses may represent a novel therapeutic approach against biofilms (64).

ACKNOWLEDGMENTS

Conflicts of interest: We disclose no conflicts.

CITATION

Roilides E, Simitsopoulou M, Katragkou A, Walsh TJ. 2015. How biofilms evade host defenses. Microbiol Spectrum 3(3):MB-0012-2014.

REFERENCES

1. **Costerton JW, Geesey GG, Cheng KJ.** 1978. How bacteria stick. *Sci Am* **238:**86–95.
2. **Costerton JW, Lewandowski Z, Caldwell DE, Korber DR, Lappin-Scott HM.** 1995. Microbial biofilms. *Annu Rev Microbiol* **49:**711–745.
3. **Costerton JW, Stewart PS, Greenberg EP.** 1999. Bacterial biofilms: a common cause of persistent infections. *Science* **284:**1318–1322.
4. **Donlan RM, Costerton JW.** 2002. Biofilms: survival mechanisms of clinically relevant microorganisms. *Clin Microbiol Rev* **15:**167–193.
5. **Mack D, Becker P, Chatterjee I, Dobinsky S, Knobloch JK, Peters G, Rohde H, Herrmann M.** 2004. Mechanisms of biofilm formation in *Staphylococcus epidermidis* and *Staphylococcus aureus*: functional molecules, regulatory circuits, and adaptive responses. *Int J Med Microbiol* **294:**203–212.
6. **Ramage G, Mowat E, Jones B, Williams C, Lopez-Ribot J.** 2009. Our current understanding of fungal biofilms. *Crit Rev Microbiol* **35:** 340–355.
7. **Wei Q, Ma LZ.** 2013. Biofilm matrix and its regulation in *Pseudomonas aeruginosa*. *Int J Mol Sci* **14:**20983–21005.
8. **Palkova Z, Vachova L.** 2006. Life within a community: benefit to yeast long-term survival. *FEMS Microbiol Rev* **30:**806–824.
9. **Donlan RM.** 2001. Biofilms and device-associated infections. *Emerg Infect Dis* **7:**277–281.
10. **Salta M, Wharton JA, Blache Y, Stokes KR, Briand JF.** 2013. Marine biofilms on artificial surfaces: structure and dynamics. *Environ Microbiol.* [Epub ahead of print.] doi:10.1111/1462-2920.12186.

11. **Abdallah M, Benoliel C, Drider D, Dhulster P, Chihib NE.** 2014. Biofilm formation and persistence on abiotic surfaces in the context of food and medical environments. *Arch Microbiol.* [Epub ahead of print.] doi:10.1007/s00203-014-0983-1.
12. **Otto M.** 2008. Staphylococcal biofilms. *Curr Top Microbiol Immunol* **322:**207–228.
13. **Fey PD, Olson ME.** 2010. Current concepts in biofilm formation of *Staphylococcus epidermidis. Future Microbiol* **5:**917–933.
14. **Moorthy S, Watnick PI.** 2004. Genetic evidence that the *Vibrio cholerae* monolayer is a distinct stage in biofilm development. *Mol Microbiol* **52:**573–587.
15. **Caiazza NC, Merritt JH, Brothers KM, O'Toole GA.** 2007. Inverse regulation of biofilm formation and swarming motility by *Pseudomonas aeruginosa* PA14. *J Bacteriol* **189:** 3603–3612.
16. **Walsh TJ, Schlegel R, Moody M, Costerton JW, Salcman M.** 1986. Ventriculo-atrial shunt infection due to *Cryptococcus neoformans*: an ultrastructural and quantitative microbiological study. *Neurosurgery* **18:**373–375.
17. **Martinez LR, Casadevall A.** 2005. Specific antibody can prevent fungal biofilm formation and this effect correlates with protective efficacy. *Infect Immun* **73:**6350–6362.
18. **Zhao X, Oh SH, Yeater KM, Hoyer LL.** 2005. Analysis of the *Candida albicans* Als2p and Als4p adhesins suggests the potential for compensatory function within the Als family. *Microbiology* **151:**1619–1630.
19. **Castano I, Pan SJ, Zupancic M, Hennequin C, Dujon B, Cormack BP.** 2005. Telomere length control and transcriptional regulation of subtelomeric adhesins in *Candida glabrata. Mol Microbiol* **55:**1246–1258.
20. **Hornby JM, Jensen EC, Lisec AD, Tasto JJ, Jahnke B, Shoemaker R, Dussault P, Nickerson KW.** 2001. Quorum sensing in the dimorphic fungus *Candida albicans* is mediated by farnesol. *Appl Environ Microbiol* **67:**2982–2992.
21. **Davey ME, Caiazza NC, O'Toole GA.** 2003. Rhamnolipid surfactant production affects biofilm architecture in *Pseudomonas aeruginosa* PAO1. *J Bacteriol* **185:**1027–1036.
22. **Rajendran R, Williams C, Lappin DF, Millington O, Martins M, Ramage G.** 2013. Extracellular DNA release acts as an antifungal resistance mechanism in mature *Aspergillus fumigatus* biofilms. *Eukaryot Cell* **12:**420–429.
23. **Shopova I, Bruns S, Thywissen A, Kniemeyer O, Brakhage AA, Hillmann F.** 2013. Extrinsic extracellular DNA leads to biofilm formation and colocalizes with matrix polysaccharides in the human pathogenic fungus *Aspergillus fumigatus. Front Microbiol* **4:**141.
24. **Krappmann S, Ramage G.** 2013. A sticky situation: extracellular DNA shapes *Aspergillus fumigatus* biofilms. *Front Microbiol* **4:**159.
25. **Uppuluri P, Chaturvedi AK, Srinivasan A, Banerjee M, Ramasubramaniam AK, Kohler JR, Kadosh D, Lopez-Ribot JL.** 2010. Dispersion as an important step in the *Candida albicans* biofilm developmental cycle. *PLoS Pathog* **6:** e1000828. doi:10.1371/journal.ppat.1000828.
26. **Wang R, Khan BA, Cheung GY, Bach TH, Jameson-Lee M, Kong KF, Queck SY, Otto M.** 2011. *Staphylococcus epidermidis* surfactant peptides promote biofilm maturation and dissemination of biofilm-associated infection in mice. *J Clin Invest* **121:**238–248.
27. **Periasamy S, Joo HS, Duong AC, Bach TH, Tan VY, Chatterjee SS, Cheung GY, Otto M.** 2012. How *Staphylococcus aureus* biofilms develop their characteristic structure. *Proc Natl Acad Sci USA* **109:**1281–1286.
28. **Harriott MM, Noverr MC.** 2009. *Candida albicans* and *Staphylococcus aureus* form polymicrobial biofilms: effects on antimicrobial resistance. *Antimicrob Agents Chemother* **53:** 3914–3922.
29. **Demuyser L, Jabra-Rizk MA, Van Dijck P.** 2014. Microbial cell surface proteins and secreted metabolites involved in multispecies biofilms. *Pathog Dis* **70:**219–230.
30. **Shirtliff ME, Peters BM, Jabra-Rizk MA.** 2009. Cross-kingdom interactions: *Candida albicans* and bacteria. *FEMS Microbiol Lett* **299:**1–8.
31. **Peters BM, Jabra-Rizk MA, Scheper MA, Leid JG, Costerton JW, Shirtliff ME.** 2010. Microbial interactions and differential protein expression in *Staphylococcus aureus-Candida albicans* dual-species biofilms. *FEMS Immunol Med Microbiol* **59:**493–503.
32. **Wolcott R, Costerton JW, Raoult D, Cutler SJ.** 2013. The polymicrobial nature of biofilm infection. *Clin Microbiol Infect* **19:**107–112.
33. **Mowat E, Rajendran R, Williams C, McCulloch E, Jones B, Lang S, Ramage G.** 2010. *Pseudomonas aeruginosa* and their small diffusible extracellular molecules inhibit *Aspergillus fumigatus* biofilm formation. *FEMS Microbiol Lett* **313:**96–102.
34. **Christensen GD, Simpson WA, Younger JJ, Baddour LM, Barrett FF, Melton DM, Beachey EH.** 1985. Adherence of coagulase-negative staphylococci to plastic tissue culture plates: a quantitative model for the adherence of staphylococci to medical devices. *J Clin Microbiol* **22:** 996–1006.

35. **Chandra J, Kuhn DM, Mukherjee PK, Hoyer LL, McCormick T, Ghannoum MA.** 2001. Biofilm formation by the fungal pathogen *Candida albicans*: development, architecture, and drug resistance. *J Bacteriol* **183:**5385–5394.
36. **Hassan A, Usman J, Kaleem F, Omair M, Khalid A, Iqbal M.** 2011. Evaluation of different detection methods of biofilm formation in the clinical isolates. *Braz J Infect Dis* **15:**305–311.
37. **Taff HT, Nett JE, Andes DR.** 2012. Comparative analysis of *Candida* biofilm quantitation assays. *Med Mycol* **50:**214–218.
38. **Mowat E, Butcher J, Lang S, Williams C, Ramage G.** 2007. Development of a simple model for studying the effects of antifungal agents on multicellular communities of *Aspergillus fumigatus*. *J Med Microbiol* **56:**1205–1212.
39. **Sontakke S, Cadenas MB, Maggi RG, Diniz PP, Breitschwerdt EB.** 2009. Use of broad range 16S rDNA PCR in clinical microbiology. *J Microbiol Methods* **76:**217–225.
40. **Pontes DS, Lima-Bittencourt CI, Chartone-Souza E, Amaral Nascimento AM.** 2007. Molecular approaches: advantages and artifacts in assessing bacterial diversity. *J Ind Microbiol Biotechnol* **34:**463–473.
41. **Rhoads DD, Wolcott RD, Sun Y, Dowd SE.** 2012. Comparison of culture and molecular identification of bacteria in chronic wounds. *Int J Mol Sci* **13:**2535–2550.
42. **Banchereau J, Briere F, Caux C, Davoust J, Lebecque S, Liu YJ, Pulendran B, Palucka K.** 2000. Immunobiology of dendritic cells. *Annu Rev Immunol* **18:**767–811.
43. **Gantner BN, Simmons RM, Canavera SJ, Akira S, Underhill DM.** 2003. Collaborative induction of inflammatory responses by dectin-1 and Toll-like receptor 2. *J Exp Med* **197:**1107–1117.
44. **Romani L.** 2011. Immunity to fungal infections. *Nat Rev Immunol* **11:**275–288.
45. **Norrby-Teglund A, Johansson L.** 2013. Beyond the traditional immune response: bacterial interaction with phagocytic cells. *Int J Antimicrob Agents* **42**(Suppl)**:**S13–S16.
46. **Singh PK, Parsek MR, Greenberg EP, Welsh MJ.** 2002. A component of innate immunity prevents bacterial biofilm development. *Nature* **417:**552–555.
47. **Overhage J, Campisano A, Bains M, Torfs EC, Rehm BH, Hancock RE.** 2008. Human host defense peptide LL-37 prevents bacterial biofilm formation. *Infect Immun* **76:**4176–4182.
48. **Britigan BE, Hayek MB, Doebbeling BN, Fick RB Jr.** 1993. Transferrin and lactoferrin undergo proteolytic cleavage in the *Pseudomonas aeruginosa*-infected lungs of patients with cystic fibrosis. *Infect Immun* **61:**5049–5055.
49. **Schmidtchen A, Frick IM, Andersson E, Tapper H, Bjorck L.** 2002. Proteinases of common pathogenic bacteria degrade and inactivate the antibacterial peptide LL-37. *Mol Microbiol* **46:**157–168.
50. **Jensen PO, Givskov M, Bjarnsholt T, Moser C.** 2010. The immune system vs. *Pseudomonas aeruginosa* biofilms. *FEMS Immunol Med Microbiol* **59:**292–305.
51. **Meyle E, Stroh P, Gunther F, Hoppy-Tichy T, Wagner C, Hansch GM.** 2010. Destruction of bacterial biofilms by polymorphonuclear neutrophils: relative contribution of phagocytosis, DNA release, and degranulation. *Int J Artif Organs* **33:**608–620.
52. **Gunther F, Wabnitz GH, Stroh P, Prior B, Obst U, Samstag Y, Wagner C, Hansch GM.** 2009. Host defense against *Staphylococcus aureus* biofilms infection: phagocytosis of biofilms by polymorphonuclear neutrophils (PMN). *Mol Immunol* **46:**1805–1813.
53. **Meyle E, Brenner-Weiss G, Obst U, Prior B, Hansch GM.** 2012. Immune defense against *S. epidermidis* biofilms: components of the extracellular polymeric substance activate distinct bactericidal mechanisms of phagocytic cells. *Int J Artif Organs* **35:**700–712.
54. **Chandra J, McCormick TS, Imamura Y, Mukherjee PK, Ghannoum MA.** 2007. Interaction of *Candida albicans* with adherent human peripheral blood mononuclear cells increases *C. albicans* biofilm formation and results in differential expression of pro- and anti-inflammatory cytokines. *Infect Immun* **75:**2612–2620.
55. **Katragkou A, Kruhlak MJ, Simitsopoulou M, Chatzimoschou A, Taparkou A, Cotten CJ, Paliogianni F, Diza-Mataftsi E, Tsantali C, Walsh TJ, Roilides E.** 2010. Interactions between human phagocytes and *Candida albicans* biofilms alone and in combination with antifungal agents. *J Infect Dis* **201:**1941–1949.
56. **Martinez LR, Casadevall A.** 2006. *Cryptococcus neoformans* cells in biofilms are less susceptible than planktonic cells to antimicrobial molecules produced by the innate immune system. *Infect Immun* **74:**6118–6123.
57. **Archer NK, Mazaitis MJ, Costerton JW, Leid JG, Powers ME, Shirtliff ME.** 2011. *Staphylococcus aureus* biofilms: properties, regulation, and roles in human disease. *Virulence* **2:**445–459.
58. **Leid JG, Shirtliff ME, Costerton JW, Stoodley P.** 2002. Human leukocytes adhere to, penetrate, and respond to *Staphylococcus aureus* biofilms. *Infect Immun* **70:**6339–6345.
59. **Guenther F, Stroh P, Wagner C, Obst U, Hansch GM.** 2009. Phagocytosis of staphylococci biofilms by polymorphonuclear neutrophils: *S. aureus* and *S. epidermidis* differ with

regard to their susceptibility towards the host defense. *Int J Artif Organs* **32:**565–573.

60. **Thurlow LR, Hanke ML, Fritz T, Angle A, Aldrich A, Williams SH, Engebretsen IL, Bayles KW, Horswill AR, Kielian T.** 2011. *Staphylococcus aureus* biofilms prevent macrophage phagocytosis and attenuate inflammation *in vivo*. *J Immunol* **186:**6585–6596.
61. **Stroh P, Gunther F, Meyle E, Prior B, Wagner C, Hansch GM.** 2011. Host defense against *Staphylococcus aureus* biofilms by polymorphonuclear neutrophils: oxygen radical production but not phagocytosis depends on opsonisation with immunoglobulin G. *Immunobiology* **216:**351–357.
62. **Cerca N, Jefferson KK, Oliveira R, Pier GB, Azeredo J.** 2006. Comparative antibody-mediated phagocytosis of *Staphylococcus epidermidis* cells grown in a biofilm or in the planktonic state. *Infect Immun* **74:**4849–4855.
63. **Hanke ML, Kielian T.** 2012. Deciphering mechanisms of staphylococcal biofilm evasion of host immunity. *Front Cell Infect Microbiol* **2:**62.
64. **Hanke ML, Heim CE, Angle A, Sanderson SD, Kielian T.** 2013. Targeting macrophage activation for the prevention and treatment of *Staphylococcus aureus* biofilm infections. *J Immunol* **190:**2159–2168.
65. **Alhede M, Bjarnsholt T, Givskov M, Alhede M.** 2014. *Pseudomonas aeruginosa* biofilms: mechanisms of immune evasion. *Adv Appl Microbiol* **86:**1–40.
66. **Jensen ET, Kharazmi A, Lam K, Costerton JW, Hoiby N.** 1990. Human polymorphonuclear leukocyte response to *Pseudomonas aeruginosa* grown in biofilms. *Infect Immun* **58:**2383–2385.
67. **Jesaitis AJ, Franklin MJ, Berglund D, Sasaki M, Lord CI, Bleazard JB, Duffy JE, Beyenal H, Lewandowski Z.** 2003. Compromised host defense on *Pseudomonas aeruginosa* biofilms: characterization of neutrophil and biofilm interactions. *J Immunol* **171:**4329–4339.
68. **Bjarnsholt T, Jensen PO, Burmolle M, Hentzer M, Haagensen JA, Hougen HP, Calum H, Madsen KG, Moser C, Molin S, Hoiby N, Givskov M.** 2005. *Pseudomonas aeruginosa* tolerance to tobramycin, hydrogen peroxide and polymorphonuclear leukocytes is quorum-sensing dependent. *Microbiology* **151:**373–383.
69. **Walker TS, Tomlin KL, Worthen GS, Poch KR, Lieber JG, Saavedra MT, Fessler MB, Malcolm KC, Vasil ML, Nick JA.** 2005. Enhanced *Pseudomonas aeruginosa* biofilm development mediated by human neutrophils. *Infect Immun* **73:**3693–3701.
70. **Parks QM, Young RL, Poch KR, Malcolm KC, Vasil ML, Nick JA.** 2009. Neutrophil enhancement of *Pseudomonas aeruginosa* biofilm development: human F-actin and DNA as targets for therapy. *J Med Microbiol* **58:**492–502.
71. **Watt AP, Courtney J, Moore J, Ennis M, Elborn JS.** 2005. Neutrophil cell death, activation and bacterial infection in cystic fibrosis. *Thorax* **60:**659–664.
72. **Leid JG, Willson CJ, Shirtliff ME, Hassett DJ, Parsek MR, Jeffers AK.** 2005. The exopolysaccharide alginate protects *Pseudomonas aeruginosa* biofilm bacteria from IFN-gamma-mediated macrophage killing. *J Immunol* **175:**7512–7518.
73. **Jensen ET, Kharazmi A, Hoiby N, Costerton JW.** 1992. Some bacterial parameters influencing the neutrophil oxidative burst response to *Pseudomonas aeruginosa* biofilms. *APMIS* **100:**727–733.
74. **Jensen PO, Bjarnsholt T, Phipps R, Rasmussen TB, Calum H, Christoffersen L, Moser C, Williams P, Pressler T, Givskov M, Hoiby N.** 2007. Rapid necrotic killing of polymorphonuclear leukocytes is caused by quorum-sensing-controlled production of rhamnolipid by *Pseudomonas aeruginosa*. *Microbiology* **153:**1329–1338.
75. **Alhede M, Bjarnsholt T, Jensen PO, Phipps RK, Moser C, Christophersen L, Christensen LD, van Gennip M, Parsek M, Hoiby N, Rasmussen TB, Givskov M.** 2009. *Pseudomonas aeruginosa* recognizes and responds aggressively to the presence of polymorphonuclear leukocytes. *Microbiology* **155:**3500–3508.
76. **Van Gennip M, Christensen LD, Alhede M, Phipps R, Jensen PO, Christophersen L, Pamp SJ, Moser C, Mikkelsen PJ, Koh AY, Tolker-Nielsen T, Pier GB, Hoiby N, Givskov M, Bjarnsholt T.** 2009. Inactivation of the rhlA gene in *Pseudomonas aeruginosa* prevents rhamnolipid production, disabling the protection against polymorphonuclear leukocytes. *APMIS* **117:**537–546.
77. **Roilides E, Walsh T.** 2004. Recombinant cytokines in augmentation and immunomodulation of host defenses against *Candida* spp. *Med Mycol* **42:**1–13.
78. **Xie Z, Thompson A, Sobue T, Kashleva H, Xu H, Vasilakos J, Dongari-Bagtzoglou A.** 2012. *Candida albicans* biofilms do not trigger reactive oxygen species and evade neutrophil killing. *J Infect Dis* **206:**1936–1945.
79. **Katragkou A, Simitsopoulou M, Chatzimoschou A, Georgiadou E, Walsh TJ, Roilides E.** 2011. Effects of interferon-gamma and granulocyte colony-stimulating factor on antifungal activity of human polymorphonuclear neutrophils against *Candida albicans* grown as biofilms or planktonic cells. *Cytokine* **55:**330–334.

80. **Nett J, Lincoln L, Marchillo K, Andes D.** 2007. Beta -1,3 glucan as a test for central venous catheter biofilm infection. *J Infect Dis* **195:**1705–1712.
81. **Dongari-Bagtzoglou A, Kashleva H, Dwivedi P, Diaz P, Vasilakos J.** 2009. Characterization of mucosal *Candida albicans* biofilms. *PLoS One* **4:**e7967. doi:10.1371/journal.pone.0007967.
82. **Al-Fattani MA, Douglas LJ.** 2006. Biofilm matrix of *Candida albicans* and *Candida tropicalis*: chemical composition and role in drug resistance. *J Med Microbiol* **55:**999–1008.
83. **Luo G, Ibrahim AS, Spellberg B, Nobile CJ, Mitchell AP, Fu Y.** 2010. *Candida albicans* Hyr1p confers resistance to neutrophil killing and is a potential vaccine target. *J Infect Dis* **201:** 1718–1728.
84. **Giard JC, Riboulet E, Verneuil N, Sanguinetti M, Auffray Y, Hartke A.** 2006. Characterization of Ers, a PrfA-like regulator of *Enterococcus faecalis. FEMS Immunol Med Microbiol* **46:**410–418.
85. **Gentry-Weeks CR, Karkhoff-Schweizer R, Pikis A, Estay M, Keith JM.** 1999. Survival of *Enterococcus faecalis* in mouse peritoneal macrophages. *Infect Immun* **67:**2160–2165.
86. **Daw K, Baghdayan AS, Awasthi S, Shankar N.** 2012. Biofilm and planktonic *Enterococcus faecalis* elicit different responses from host phagocytes *in vitro. FEMS Immunol Med Microbiol* **65:**270–282.
87. **Mathew S, Yaw-Chyn L, Kishen A.** 2010. Immunogenic potential of *Enterococcus faecalis* biofilm under simulated growth conditions. *J Endod* **36:**832–836.
88. **Chotirmall SH, McElvaney NG.** 2014. Fungi in the cystic fibrosis lung: bystanders or pathogens? *Int J Biochem Cell Biol* **52:**161–173.
89. **Roilides E, Simitsopoulou M.** 2010. Local innate host response and filamentous fungi in patients with cystic fibrosis. *Med Mycol* **48**(Suppl 1)**:**S22–S31.
90. **Tasina E, Simitsopoulou M, Roilides E.** 2012. The innate immune response to filamentous fungi in patients with cystic fibrosis. *CML Cystic Fibrosis* **2:**29–39.
91. **Kreindler JL, Steele C, Nguyen N, Chan YR, Pilewski JM, Alcorn JF, Vyas YM, Aujla SJ, Finelli P, Blanchard M, Zeigler SF, Logar A, Hartigan E, Kurs-Lasky M, Rockette H, Ray A, Kolls JK.** 2010. Vitamin D3 attenuates Th2 responses to *Aspergillus fumigatus* mounted by CD4+ T cells from cystic fibrosis patients with allergic bronchopulmonary aspergillosis. *J Clin Invest* **120:**3242–3254.
92. **Nett J, Andes D.** 2006. *Candida albicans* biofilm development, modeling a host-pathogen interaction. *Curr Opin Microbiol* **9:**340–345.

c-di-GMP and its Effects on Biofilm Formation and Dispersion: a *Pseudomonas Aeruginosa* Review

15

DAE-GON HA[1] and GEORGE A. O'TOOLE[1]

BIOFILM FORMATION AND OTHER RELEVANT PHENOTYPES IN *PSEUDOMONAS AERUGINOSA*

P. aeruginosa is a Gram-negative bacterium that has become an indispensable model organism in our quest to understand the A-to-Z of bacterial biofilms (1). It is genetically tractable, has a sequenced and annotated genome (http://www.pseudomonas.com), and boasts a number of useful tools (2, 3) that facilitate both *in vivo* and *in vitro* studies. Although it is commonly found as an environmental isolate, it is also an opportunistic pathogen capable of colonizing plants and mammalian hosts and is particularly significant for its efficient colonization of the lungs of cystic fibrosis patients (4–6). The versatility of *P. aeruginosa* is in large part attributed to a battery of traits that provide it with selective advantage(s) across diverse environments. Like many other bacteria, *P. aeruginosa* is capable of transitioning between motile and sessile/biofilm lifestyles, which is believed to contribute to this bacterium's versatility.

Bacteria were once perceived to be simple, single-celled organisms; however, it is quite clear that microbes can participate in a broad range of

[1]Departments of Microbiology and Immunonology, Geisel School of Medicine at Dartmouth, Hanover, NH 03755.

Microbial Biofilms 2nd Edition
Edited by Mahmoud Ghannoum, Matthew Parsek, Marvin Whiteley, and Pranab K. Mukherjee

doi:10.1128/microbiolspec.MB-0003-2014

complex multicellular behaviors, including quorum sensing, the formation of complex spore-forming aggregates by *Myxococcus* and *Bacillus*, swarming motility, and the formation of bacterial biofilms (7–13). Biofilms are defined as a structured community of bacterial cells enclosed in a self-produced polymeric matrix and adherent to inert or living surfaces (14). The ability to form a biofilm is a common trait of a diverse array of microbes, including lower order eukaryotes. We now understand that in bacteria the intracellular second messenger molecule (3′,5′-cyclic diguanylic acid, or c-di-GMP [Fig. 1]) appears to control many facets of group behavior, including biofilm formation by *P. aeruginosa*.

c-di-GMP is synthesized from two GTP molecules by diguanylate cyclases (DGC) and is degraded by phosphodiesterases (PDE). The genome of *P. aeruginosa* PAO1 encodes 41 c-di-GMP proteins predicted to participate in c-di-GMP metabolism, while *P. aeruginosa* PA14 has 40 such proteins (15, 16). Interestingly, most of these proteins are linked to various sensory input domains on their N-terminus, including PAS (first discovered in **P**er, **A**rnt, and **S**im proteins), GAF (found in c**G**MP-specific phosphodiesterases, **a**denylyl cyclases, and **F**hlA), and REC (**REC**eiver) domains (17, 18), presumably transducing environmental stimuli to cellular response(s). In fact, c-di-GMP has been implicated in numerous cellular functions including regulation of cell cycle, differentiation, biofilm formation and dispersion, motility, and virulence (19–28), adding credence to this prediction. With regard to biofilm formation in particular, the current general model associates high intracellular levels of c-di-GMP with biofilm formation or a sessile lifestyle, and low c-di-GMP levels are associated with a motile or planktonic existence (29). While this general pattern holds true in many instances, the relationship between c-di-GMP levels and phenotypic outputs is likely to be much more complex (15, 29).

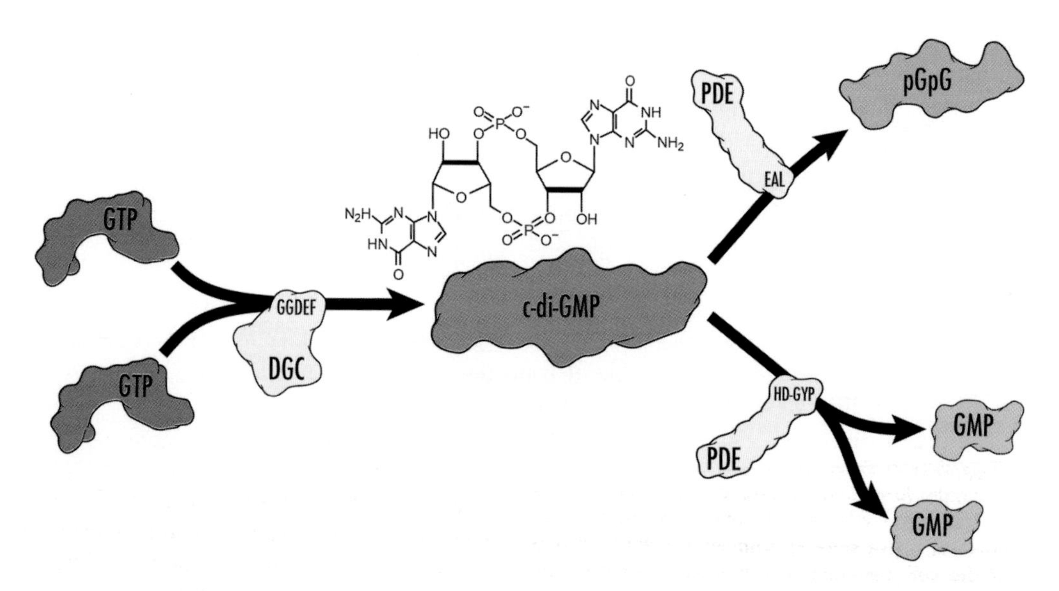

FIGURE 1 c-di-GMP: A central regulator of biofilms. The structure of c-di-GMP is shown (center). This molecule is synthesized from two molecules of GTP by enzymes known as diguanylate cyclases (DGCs), which carry a conserved GGDEF domain. c-di-GMP can be degraded by two families of phosphodiesterases (PDEs); those with an EAL domain linearize the molecule to produce pGpG, and proteins with an HD-GYP domain generate two molecules of GMP from the signal. Illustration ©2014 William Scavone, Kestrel Studio, reprinted with permission. doi:10.1128/microbiolspec.MB-0003-2014.f1

In the laboratory, biofilm formation is a cyclical process wherein a free-swimming planktonic cell encounters a surface—biotic or abiotic—and initiates cell-to-surface attachment (Fig. 2). The cell initially attaches reversibly by its polar flagellum (30) and then subsequently attaches irreversibly along the cell's longitudinal axis, which we believe is the first committed step in biofilm formation. Irreversibly attached cells form the basis of the monolayer upon which mature biofilms are established, which vary in morphology from mushroom-shaped macrocolonies to uniform, thick layers of bacteria. Lastly, biofilm-associated cells are dispersed from the mature biofilm to resume a planktonic lifestyle, which completes the cycle. At this point, it is worth remembering that biofilm formation is dependent on a bevy of cellular factors and coordinated pathways. For example, previous reports have identified flagellar motility (31, 32), twitching motility (mediated by type IV pili) (31), and EPS (exopolysaccharides) (33, 34) as prerequisites to *P. aeruginosa* biofilms. Therefore, understanding the full scope of biofilm formation requires a comprehensive view into how each component, and its regulation, contributes to the overall process.

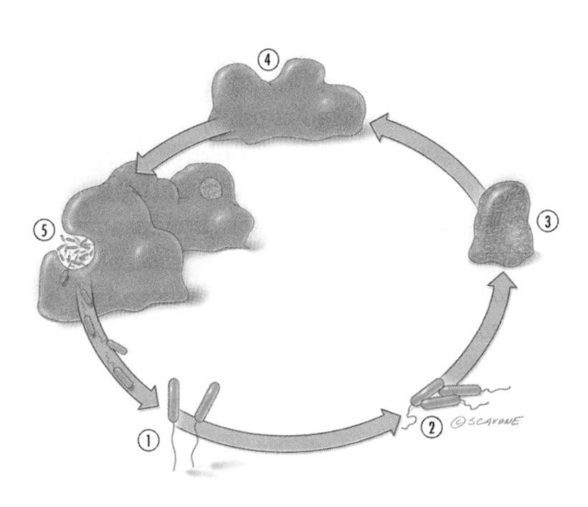

FIGURE 2 A model for biofilm formation and dispersion in *Pseudomonas aeruginosa*. The steps of biofilm formation, as described in this article are (1) reversible attachment, likely via the flagellar pole, (2) irreversible attachment via the long axis of the cell, resulting in a monolayer of cells, (3) microcolony formation, (4) macrocolony formation, and (5) dispersion. See the text for more detailed explanations of each step, which are based largely on laboratory studies. Illustration ©2014 William Scavone, Kestrel Studio, reprinted with permission. doi:10.1128/microbiolspec.MB-0003-2014.f2

Here, we aim to highlight key findings within the context of *P. aeruginosa* biofilm formation, focusing on how second messenger molecule c-di-GMP regulates the numerous factors and pathways that contribute to the formation of a mature biofilm by this organism. This article is divided into the various stages of biofilm formation: initial attachment (reversible, then irreversible), maturation, and finally, dispersal. It is important to keep in mind that biofilm formation studies using *P. aeruginosa* as a model have used a variety of different strains, including PA14, PAO1, PAK, and others. Thus, while we have tried to integrate the literature regarding *P. aeruginosa* biofilm formation into a single narrative, it is important to note that there may be examples of strain-specific findings, which we will highlight as appropriate.

INITIAL ATTACHMENT: REVERSIBLE AND IRREVERSIBLE

Reversible Attachment

The first step in biofilm formation is reversible attachment, a step wherein a bacterium first contacts a surface. To overcome surface repulsion, *P. aeruginosa* utilizes flagellar-mediated swimming motility (35). Biosynthesis of the flagellar machinery is tightly regulated in an intricate fashion and involves FleQ, a transcriptional regulator of flagellar expression, which binds to the upstream activation sequence of the *flhA* gene to start the cascade of flagellar gene expression (36). FleQ activity is regulated by at least two distinct methods: (i) sequestration of FleQ by FleN (36, 37) and (ii) conformational change of FleQ when bound to c-di-GMP (38). More specifically, it has been shown that both FleN and c-di-GMP can inhibit the ATPase activity of FleQ *in vitro*, which renders FleQ incapable of regulating downstream flagellar genes (39). And in an interesting twist, FleQ may also have an indirect role of modulating c-di-GMP levels since decreased intracellular levels of c-di-GMP were observed in a Δ*fleQ* mutant compared to wild type *P. aeruginosa* PAO1 (38). The redundancy in regulating flagellar biosynthesis, whether FleN- or c-di-GMP-mediated, not only underscores the importance of the flagellum and motility, but also provides flexibility in controlling flagellar gene expression.

The flagellar machinery includes the rotor, stator, flagellar hook, and filament proteins, along with a number of ring structures that anchor the complex in the membrane and peptidoglycan layers. Flagellar-mediated propulsion requires the stator complex, which in turn generates torque through proton motive force. Recent identification of a "bacterial clutch" protein (YcgR) implies the possibility of controlling flagellar function after this molecular machine is assembled (24, 40–44). YcgR contains a PilZ domain, a common feature among c-di-GMP-responsive proteins (21). While the specifics of this mechanism remain under investigation, YcgR is shown to interact with both the motor (FliG) and stator (MotA) proteins of the flagellar machine to reduce motility when c-di-GMP levels are high (40, 42, 44).

Despite the lack of an annotated *ycgR* gene in the *P. aeruginosa* genome database, sequence comparison of *Escherichia coli ycgR* with the *P. aeruginosa* PAO1 genome identified *flgZ* (encoded by PA3353) as a potential homolog (45). Like *E. coli* YcgR, FlgZ is a cytoplasmic protein with a predicted PilZ domain. And based on studies in other pseudomonads, FlgZ binds c-di-GMP and colocalizes to the polar flagellar machinery, similar to YcgR (45, 46). Expanding on these findings, follow-up functional analyses should help identify the role(s) of FlgZ in *P. aeruginosa* motility.

Thus, there appear to be two distinct mechanisms that control flagellar motility during early biofilm formation in response to increasing intracellular c-di-GMP: the FleQ-c-di-GMP complex represses transcription of the genes required for flagellar assembly, while YcgR forms a complex with c-di-GMP that modulates flagellar rotation. It is important to note that repression of flagellar function can occur prior to initial, reversible attachment, thereby blocking even the earliest stage in biofilm formation. Alternatively, suppression of flagellar function may occur after reversible attachment, which would stabilize bacterial-surface interactions and promote the transition to irreversible attachment.

Following reversible attachment, *P. aeruginosa* transitions into an irreversibly attached state and begins its progression toward a mature biofilm. Upon irreversible attachment, the cell encounters a number of changes that lay the groundwork for subsequent steps in biofilm formation. In the following section, we discuss how flagellar motility is modulated once bacteria have engaged a surface, as well as mechanisms that may control other types of surface motility, such as pili-mediated twitching motility.

Irreversible Attachment

Once bacteria have begun interacting with a surface, they have three possible fates. First, *P. aeruginosa* can move using twitching motility, powered by type IV pili (TFP), which extend and retract to tug bacterial cells across the surface. Alternatively, *P. aeruginosa* has a second form of motility known as swarming, which utilizes the flagellum as well as surfactants, to migrate on a substratum. Recent studies by Wong and colleagues have visualized such surface movements in real time (47–50), although the role of c-di-GMP and the control of these behaviors remain to be explored in detail. Third, bacteria can attach "irreversibly" via the long axis of the cell. This irreversible attachment is much more stable than the reversible attachment discussed above, is the first committed step in biofilm formation, and is typically associated with the so-called monolayer stage of biofilm formation (51).

In the following paragraphs, we will focus on the DGCs and PDEs of *P. aeruginosa* that may contribute to controlling irreversible attachment. As previously alluded to, the control of swarming motility, twitching motility, and irreversible attachment are tightly interlinked. For example, while these surface motility mechanisms likely allow *P. aeruginosa* to explore a newly colonized surface environment, suppression of these motility functions is required to allow this microbe to initiate and commit to the more stable, irreversible attachment. We will begin with a discussion of the control of swarming motility, followed by the control of twitching motility.

In the discussion of swimming motility above, we outlined the role of FleQ and YcgR in regulating flagellar-mediated swimming motility. Given the essential role of the flagellum in swarming motility, these same regulators may also impact flagellar-mediated surface motility as well. The role of FleQ and YcgR in modulating swarming has not been examined; however, enzymes involved in c-di-GMP metabolism that impact swarming have been identified. Based on the inverse relationship between c-di-GMP and motility, it is likely that DGC activity (producing c-di-GMP) will repress motility, whereas PDE activity (degrading c-di-GMP) will promote motility.

Consistent with this model, work from Merritt et al. and Kuchma et al. identified two proteins—a DGC and a PDE—called SadC and BifA, respectively, that inversely impact swarming and biofilm formation. *In vitro* activity assays and quantification of intracellular c-di-GMP levels indicated that the SadC and BifA proteins were indeed DGC and PDE enzymes, respectively (52, 53). The Δ*sadC* mutant is a hyper-swarming strain that is also defective in attachment, and conversely, the Δ*bifA* mutant has a hyper-biofilm phenotype (showing a ~5.5-fold increase in attachment) and is unable to swarm. Thus, SadC and BifA both appear to contribute to the control of surface motility and, thereby, are also required to establish irreversible attachment. When assayed for swimming motility, neither the Δ*sadC* nor Δ*bifA* mutant was statistically different from that of wild type *P. aeruginosa* PA14, and only the Δ*bifA* mutant showed a reduction in twitching motility (52, 53), which collectively suggest that these DGCs and PDEs preferentially impact swarming motility. Thus, SadC and BifA appear to regulate c-di-GMP levels predominantly for surface behaviors of *P. aeruginosa* PA14.

While identifying mutants with biofilm defects, O'Toole and Kolter discovered mutations in TFP (31). TFP are polar appendages that can serve as a bacteriophage receptor (54), and as previously mentioned, are required for a mode of surface motility known as twitching (54, 55). The complexity of TFP biosynthesis is well documented (55), and its expression is partly regulated by c-di-GMP. The FimX protein (encoded by PA4959) has a predicted DGC and a PDE domain. Based on *in vitro* studies and crystal structure, FimX's DGC domain was demonstrated to be degenerate (28, 56). Due to conflicting results, however, whether the FimX PDE domain is enzymatically active remains controversial. It is possible that the binding of c-di-GMP to the PDE domain may implicate FimX as an effector protein rather than an enzyme of c-di-GMP hydrolysis.

Using *P. aeruginosa* strain PA103, the intracellular levels of c-di-GMP in the Δ*fimX* mutant were shown to be on par with wild type. However, suppressor mutant studies using the Δ*fimX* mutant showed that an increase in c-di-GMP by overexpressing DGCs rescued twitching motility and phage sensitivity defects (27). Surprisingly, overexpression of the same DGCs in *P. aeruginosa* PAO1 Δ*fimX* mutants failed to restore twitching motility despite proper pili formation (27).

The differences in these findings may be due to differences in strain background or may reflect differential responses to the nonphysiological overexpression of DGCs. Regardless of these differences, FimX and its ability to regulate TFP biosynthesis and function remains consistent across *P. aeruginosa* strains.

Currently, the mechanism by which FimX impacts twitching motility is unclear. FimX may act as a canonical PDE and reduce c-di-GMP levels, or alternatively, this protein may act as part of an effector system, binding c-di-GMP and modulating twitching motility. FimX is a polar-localized protein in *P. aeruginosa* PAK (28) and thus may exert its effect on pili biogenesis and function via directly interacting with the pilus machinery. Subsequent studies identified a second c-di-GMP-binding protein, PilZ (encoded by PA2960), that was also defective in twitching motility and failed to export pilin subunits despite proper expression of the *pilA* gene (21, 57). Therefore, this protein may also regulate pilus assembly. Whether FimX and PilZ interact, colocalize, or compete for c-di-GMP molecules is unclear, but their converging role in regulating pilus assembly warrants further investigation. At the same time, the mechanism(s) by which c-di-GMP regulates pilus assembly and function deserves additional attention given the role of pili in irreversible attachment, as well as in the formation of the mushroom "caps" during biofilm maturation (58).

In addition to repressing surface motility, irreversible attachment also appears to require the production of exopolysaccharides (EPSs). Beyond the role of EPSs in irreversible attachment, these secreted polysaccharides, together with nucleic acids, proteins, and additional factors, comprise the matrix of the mature biofilm—a point discussed in more detail below. Here, we will focus on the role of EPSs in irreversible attachment.

Two polysaccharides, Pel and Psl, have been identified in *P. aeruginosa* PAO1, but only Pel is produced in *P. aeruginosa* strain PA14 (33, 34). These two EPSs are believed to be structurally different, with Pel thought to be mainly glucose-rich, although the structure of this EPS remains to be definitively established, while Psl is a mannose-rich polysaccharide (34). Colvin et al. demonstrated that *P. aeruginosa* PA14 Δ*pel* and *P. aeruginosa* PAO1 Δ*psl* mutants were both arrested at the monolayer stage of biofilm development, with concurrent reduction in accumulated biofilm biomass compared to their respective parental strains (59). These data suggest that either Pel or Psl can promote irreversible attachment. Wong and colleagues recently showed that *P. aeruginosa* PAO1 migrates across a substratum, leaving a trail of the Psl polysaccharide, which in turn enhances subsequent attachment, eventually driving microcolony formation (47), a key step in the production of the mature biofilm (51). In contrast, no specific role for the Pel EPS has been assigned.

Clinical isolates expressing both Pel and Psl have been identified (34, 60), indicating that the *P. aeruginosa* PA14 may be somewhat unusual in only having one of these EPS biosynthetic operons. Furthermore, no protein adhesin of *P. aeruginosa* has been identified as mediating irreversible attachment analogous to the cell-surface protein LapA of *Pseudomonas fluorescens* (61). All data considered, we speculate that perhaps at early stages of attachment, Pel and/or Psl serve as such an adhesin.

P. aeruginosa PAO1 produces Psl, a mannose-rich polysaccharide. Genetic analysis identified *pslACDEFGHIJKL* to be essential in Psl production and/or secretion, whereas mutations in *pslBMNO* genes failed to exhibit any change in surface-associated Psl and may be nonessential for the production of this polymer (22). The *psl* genes are constitutively expressed in PAO1 (62), indicating that this EPS may be regulated strictly via posttranscriptional mechanisms, at least in this strain and in the laboratory. It is possible that there are unknown environmental signals that drive regulation in the host or environment. To date, no PilZ domain protein has been identified as required for Psl biosynthesis, leaving open the question as to how the production of this EPS is regulated post-transcriptionally by c-di-GMP.

Pel polysaccharide production is dependent on a seven-gene operon in both *P. aeruginosa* PA14 and *P. aeruginosa* PAO1. The *pel* genes are transcriptionally regulated by FleQ, which binds at two sites on the *pelA* promoter region. Although FleQ acts as a repressor in low c-di-GMP environments, upon elevation of c-di-GMP, conformational change induces FleQ-dependent expression of the *pel* operon (38, 63). The PilZ domain protein PelD is crucial for proper Pel production (64). Similar to previously described PilZ domain proteins, PelD binds c-di-GMP (65, 66) and presumably requires activities of DGCs and PDEs as a means to ensure the regulated production of this EPS. To date, several studies have identified potential candidate DGCs and PDEs that may play such a role in EPS regulation.

In strain *P. aeruginosa* PA14, mutation of the *bifA* gene encoding a PDE was shown to boost Pel production, as well as biofilm formation (52). Merritt et al. also demonstrated a decrease in Pel-mediated EPS production when the DGC RoeA (encoded by PA1107) was mutated in *P. aeruginosa* PA14 (67). Mutating *roeA* reduced intracellular c-di-GMP levels but did not alter expression of the *pel* locus (52, 67), suggesting that RoeA likely contributes to the production of the c-di-GMP bound by PelD, although such a connection has not been formally demonstrated. Similarly, the boost in Pel production seen in a Δ*bifA* mutant was not associated with a change in *pel* transcript levels (52). In any event, loss of RoeA function reduced early biofilm formation and impacted the ability of *P. aeruginosa* PA14 to irreversibly attach to the surface (67).

An interesting observation emerged during analysis of the phenotypes of Δ*bifA*, Δ*roeA*, and Δ*sadC* mutants. While the Δ*bifA* mutant showed high levels of Pel production, the Δ*bifA*Δ*roeA* double mutant showed very little Pel EPS production, a phenotype similar to mutating RoeA or the Pel locus (67). In contrast, the Δ*bifA*Δ*sadC* double mutant showed levels of Pel production quite similar to the Δ*bifA* mutant. Surprisingly, both the Δ*bifA*Δ*roeA* and the Δ*bifA*Δ*sadC* double mutants showed similar levels of c-di-GMP, which were still ~4- to 5-fold higher than the measured levels of the wild type *P. aeruginosa* PA14 (67). These data indicated that quite distinct phenotypes could emerge from strains with similar global c-di-GMP pools, arguing strongly for localized pools of c-di-GMP as a mechanism to confer specific phenotypic outputs. Consistent with this hypothesis, Miller and colleagues, using FRET biosensors, argued for asymmetrical distribution of c-di-GMP pools (68). Localization studies of RoeA in the bacterium showed discrete patches, perhaps beginning to address a RoeA-specific impact on EPS production. However, the inability to make a functional fluorescent fusion protein to PelD did not allow a clear determination of whether RoeA and PelD were colocalized (67).

It is likely that other factors in addition to motility and EPS control irreversible attachment. For example, an in-frame deletion of a dual-domain protein MorA (encoded by PA4601) in strain PAO1 resulted in wild type swimming motility (69) but reduced biofilm formation (15, 69). The impact of this mutation on swarming or c-di-GMP levels was not examined, so it is difficult to make any conclusions as to its role in biofilm formation; however, it is likely that MorA contributes to irreversible attachment given its decreased biofilm formation in the microtiter dish biofilm assay, a model that identifies early defects in attachment. A Δ*morA* mutant in the small colony variant *P. aeruginosa* 20265 (a small-colony variant of strain *P. aeruginosa* PA14) increased swimming motility (70). The contribution of MorA may be strain specific (PAO1 vs. PA14), or the small-colony variant may uncover a new function of MorA not identified in *P. aeruginosa* PAO1.

Extracellular DNA (eDNA) is also important in the early stages of attachment in *P. aeruginosa*. Whitchurch and colleagues demonstrated that degradation of eDNA by DNAse I treatment prevented *P. aeruginosa* surface attachment, but this effect was marginal on more mature biofilms (71). On a cellular level, eDNA release facilitated cell-to-cell clumping even among planktonic cells (72). These data suggest that the early release of eDNA may facilitate cell-to-cell clumping and a stable surface attachment leading up to the irreversible attachment. Despite its contribution in biofilm formation, unlike other biofilm-relevant traits discussed thus far, eDNA release has yet to be correlated with c-di-GMP. We reiterate that there has been no single adhesin identified for *P. aeruginosa* to stabilize irreversible surface attachment, but the collective evidence we have provided thus far suggests a possibility wherein EPS, and perhaps eDNA, are regulated by c-di-GMP and contributes to irreversible attachment.

Taken together, it is likely that several c-di-GMP-regulated processes contribute to irreversible attachment and the commitment to a biofilm lifestyle in *P. aeruginosa*. Upon initial reversible attachment, *P. aeruginosa* can immediately transition to irreversible attachment but also has the capability to move across the substratum by TFP-mediated twitching motility or flagellar-mediated swarming motility. Thus, coregulation of these three surface behaviors—irreversible attachment, twitching, and swarming—are tightly interrelated. We have discussed how a single PDE, BifA, impacts biofilm formation, swarming, and twitching motility, but the DGCs SadC and RoeA impact swarming and EPS production, respectively, leading to irreversible attachment. Another example shows that twitching and swarming are linked through the actions of the PilY1 protein and the minor pilins, which are required for both TFP biogenesis and for regulation of swarming motility (73). Accordingly, to establish irreversible attachment and progress to the formation of a mature biofilm, *P. aeruginosa* must utilize various c-di-GMP-mediated pathways to repress both twitching and swarming motility subsequent to cell-to-surface contact.

BIOFILM MATURATION: MICROCOLONIES AND MACROCOLONIES

Mature biofilms are enclosed in an extracellular matrix composed of proteins, polysaccharides, nucleic acids, lipids, and other cellular components (1). This polymeric matrix forms a protective barrier for biofilm-associated cells from a plethora of threats including, but not limited to, bacteriophage, amoebae, host immune responses, and antibiotics (14). In addition to providing structural rigidity, polymeric substances that comprise the matrix of a biofilm are known to retard the diffusion of molecules including antibiotics (14), likely contributing to one mechanism of antibiotic tolerance associated with biofilms. In this section, we provide a picture of biofilm maturation, that is, the events following initial surface attachment and monolayer formation, and how c-di-GMP contributes to this overall process.

Microcolony Formation

The current model in the field postulates that the formation of microcolonies and macrocolonies, and thus mature biofilms, requires the production of the EPS component of the matrix. In general, *P. aeruginosa* relies on three different sugars as a central building block for its matrix: alginate, Pel, and Psl. This role in the formation of a mature biofilm for Pel and Psl is in addition to their contribution to irreversible attachment, as described above. We address how Pel and Psl contribute to biofilm maturation below. The third EPS, alginate, likely plays a critical role *in vivo*, particularly in the context of cystic fibrosis; however, there is little evidence that alginate contributes to biofilm formation in the typical *in vitro* models using glass or plastic as a substratum (74) despite the evidence that genes involved in alginate synthesis (*algC*, *algD*, and *algU*) are upregulated when *P. aeruginosa* attaches to a surface (75).

The regulation of alginate, as is the case for Pel and Psl, is linked to c-di-GMP. Studies using *P. aeruginosa* PDO300, a mucoid strain, identified the dual-domain GGDEF-EAL protein MucR (encoded by PA1727) as a regulator of the mucoidy phenotype (76). Loss of mucoidy observed in a Δ*mucR* mutant indicated that MucR might be a functional DGC. Perhaps not surprisingly, *P. aeruginosa* PAO1 and *P. aeruginosa* PA14 overexpressing MucR overproduced EPS (76), although one must be cautious in interpreting such nonphysiological, overexpression studies. Alg44, a PilZ domain protein, was subsequently identified (57, 77) and demonstrated to bind c-di-GMP *in vitro* (57), and a *alg44::Gm*^{R} insertional mutant failed to release alginate to the extracellular milieu (77), findings which are consistent with Alg44's expected role in stimulating alginate production. With its demonstrated affinity for c-di-GMP, it is possible that Alg44 responds to c-di-GMP produced by MucR.

How are Pel and Psl regulated to specifically contribute to the later stages of biofilm formation, in contrast to their potential role in irreversible attachment? One common characteristic of Pel and Psl production is their dependence on c-di-GMP. However, there appear to be several different c-di-GMP-metabolizing proteins that contribute to the temporal control of Pel and Psl production.

The WspR protein (encoded by PA3702) was identified as a regulator of EPS and contributes to biofilm microcolony formation and maturation (78). WspR is a component of a chemotaxis-like system that controls EPS production in a number of pseudomonads, including *P. aeruginosa*. Though the precise signal has yet to be determined, contact with a solid surface seems to be sufficient to trigger regulation through the Wsp system (79, 80). On an agar surface, the Asp70 residue of WspR is phosphorylated, which in turn is associated with subcellular clustering of this protein and induction of its WspR DGC activity (79). The resulting WspR-mediated accumulation of c-di-GMP in turn yields increased EPS production, which can readily be observed via a wrinkly colony morphology (78). This wrinkly colony phenotype requires a functional Pel/Psl system, and presumably the c-di-GMP produced by the Wsp system induces EPS production via increased expression of the *pel/psl* genes and activity of their products.

Irie et al. observed an unusual increase in intracellular c-di-GMP when *psl* genes were overexpressed. Subsequent supplementation of purified Psl *in trans* to a *P. aeruginosa* PAO1 culture led to the conclusion that Psl, and not the expression of *psl* genes, stimulated c-di-GMP production. Furthermore, this outcome is mediated by two DGCs: SiaD (encoded by PA0169) and SadC (81). Thus, c-di-GMP stimulates Psl production, and this EPS in turn stimulates c-di-GMP levels, establishing a positive feedback loop. How might this increased Psl impact biofilm formation? Using high-resolution microscopy and single-cell tracking studies, the stimulatory effect of Psl on biofilm formation appears to act by (i) promoting incorporation of planktonic cells into biofilms and/or their attachment to a surface and (ii) maintaining micro- and macrocolony structure.

In addition to changes in *psl* gene expression, an increase in c-di-GMP also affects *cdrA* expression. The relative levels of *cdrA* transcript in a Δ*wspF* mutant (high intracellular c-di-GMP) was 27-fold higher than that of wild type *P. aeruginosa* PAO1 (82). Originally a 220-kDa protein, it is processed and secreted as a 150-kDa molecule that can bind Psl in the extracellular milieu and mediate autoaggregation (82). This Psl-binding property is thought to assist biofilm formation, especially en route to a mature biofilm in *P. aeruginosa*. However, it is unlikely that CdrA acts as an early attachment adhesin since Δ*cdrA* and Δ*wspF*Δ*cdrA* mutants showed no noticeable defects in static microtiter biofilm assays after 20 hours (82). Nevertheless, the increase in *cdrA* expression in biofilm-grown cells (82) implies its significance in the later maturation of this surface-associated group behavior.

EPS appears to play roles in two discrete steps in biofilm formation: irreversible attachment and biofilm maturation. The DGCs/PDEs that control production of polysaccharide are also distinct, with RoeA acting early and WspR participating in later stages in biofilm formation. We suggest that multiple DGCs in *P. aeruginosa* control EPS production, but do so at different stages in this process. For example, perhaps RoeA is required for initial stable surface attachment, but to maintain this attachment, WspR and its surface-dependent activity must be induced to reinforce the attachment decision and maintain EPS production. Although temporal regulation of *cdrA* has not been explored, its dependence on c-di-GMP suggests a similar reliance on RoeA and/or WspR along with other relevant DGCs and PDEs for controlled expression. The observation that Psl can stimulate c-di-GMP levels also suggests the possibility of a self-reinforcing, and perhaps temporal, maintenance of the biofilm. Thus, we suggest that multiple, serial signals may be required before a bacterium commits to a life on the surface.

Macrocolony Formation

How microcolonies transition to macrocolonies is still poorly understood. The mechanism may be just the continued development of a microcolony over time, or *P. aeruginosa* may have evolved explicit mechanisms for inducing macrocolony formation at a particular phase. Alternatively, perhaps maturation of the biofilm is driven by a combination of genetic determinants required for microcolony formation intersecting with, and modified by, physiological factors and constraints.

Examining the localization of Psl during the maturation of a biofilm by *P. aeruginosa* PAO1 illustrates how microcolonies and macrocolonies may be distinct. Microcolonies of *P. aeruginosa* PAO1 probed with anti-Psl antibodies showed a uniform distribution of the polysaccharide across the biofilm (83). Based on the data above, initial production of Psl is likely further reinforced via stimulation of SiaD- and SadC-dependent c-di-GMP production. In the larger macrocolonies, however, the expression profile of Psl is much less uniform. In macrocolonies, Psl is primarily expressed on the periphery of the structure, whereas Psl expression in the central regions of the colonies is low or absent (83). Once expressed, Psl in the extracellular matrix is tethered together with the assistance of c-di-GMP-dependent CdrA (82). Moreover, CdrA is responsible for proper morphology of macrocolonies and stabilizing attachment to the substratum that enables the biofilm to withstand any changes in physical parameters of the environment (82). This differential expression of *psl* may be attributed to relatively higher metabolic activity observed in the periphery of the macrocolony and the lack of necessary energy to make Psl in the heart of the biofilm (84, 85), or Psl may be actively downregulated in the center of the macrocolony because it is only required on the macrocolony surface at this late stage of maturation. For example, the loss of Psl production together with other contributing factors, e.g., rhamnolipid production (86), may enable the formation and maintenance of channels between macrocolonies, which has been postulated as being necessary for transporting nutrients, metabolites, and waste (1, 86). Unfortunately, currently available data do not allow us to readily distinguish between these models.

BRINGING IT FULL CIRCLE: c-di-GMP-DEPENDENT DISPERSAL

Under appropriate conditions, dispersal enables a controlled release of *P. aeruginosa* cells from a mature biofilm structure back to a planktonic mode of growth. A number of signals have been identified that trigger dispersal of biofilm cells back to the planktonic state. Here, we focus on pathways that utilize c-di-GMP as a key second messenger in the dispersal process.

Yoon and colleagues postulated a role for nitrous oxide (NO) and nitrosative stress in biofilm biology (87). Based on these observations, Webb, Kjelleberg, and colleagues postulated a role for NO and related compounds in triggering release of bacteria from a biofilm. Subsequent studies showed that NO could indeed stimulate release of planktonic cells from an established biofilm (88). Sodium nitroprusside (SNP) is a known NO donor, and when added at a sublethal concentration of 500 nM to *P. aeruginosa* biofilms, this compound induced dispersion and reduced biofilm surface coverage by approximately 68% compared to wild type (89, 90). These findings supported the conclusion that NO can serve as a dispersal signal. This experiment was performed using an established, mature biofilm, so the underlying mechanism(s) is thought to mimic those operating in a natural dispersion. This dispersion event was subsequently linked to changes in PDE activity (90); that is, SNP promoted PDE activity in a dose-dependent manner (90). Intracellular c-di-GMP levels, in turn, decreased ~45 to 47% upon exposure to 5 uM of SNP (90). Additionally, c-di-GMP concentrations of dispersed cells closely resembled those of planktonic cells, which was approximately 10-fold lower than biofilm-associated cells (20). To demonstrate the causality between SNP/NO and biofilm dispersion, PTIO (NO scavenger; 2-phenyl-4,4,5,5-tetramethylimidazoline-1-oxyl-3-oxide) was added in conjunction with SNP, and the mixture of NO-generator and NO-sink effectively inhibited dispersion (90), although the impact of PTIO on PDE activity and c-di-GMP level was not explored.

Glutamate is a second molecule known to induce release of planktonic cells from the biofilm, and it does so in a process called nutrient-induced dispersion (91). A collection of strains carrying mutations in PDEs was screened to identify enzymes involved in glutamate-mediated dispersion, which led to the identification of two candidates: *dipA* (encoded by PA5017) and *rbdA* (encoded by PA0861). The *dipA*::Tn and *rbdA*::Tn mutants, unlike the wild type *P. aeruginosa* PA14 parent strain, maintained an intact biofilm even after treatment with glutamate. Interestingly, under SNP/NO-induced dispersion conditions, these mutants also resisted this dispersion inducer as well (20). Thus, the same mechanism may be responsible for biofilm dispersion under both glutamate- and NO-induced conditions.

Loss of the *dipA* gene, encoding a dual-domain GGDEF-EAL protein, resulted in increased intracellular c-di-GMP (20). The resulting change in c-di-GMP also increased EPS production, likely through posttranscriptional regulation since both *pelA* and *pslA* expression remained comparable to wild type *P. aeruginosa* PAO1 (20). The Δ*dipA* mutant showed an increase in intracellular c-di-GMP levels (~110 pmol/mg) compared to wild type *P. aeruginosa* PA14 (~80 pmol/mg). A recent report by Roy et al. utilized tetracycline inhibition of protein synthesis to show that expression of *dipA* occurs during the mature biofilm stage (4 days) (20). Effects of the Δ*dipA* mutation are strain-specific, because disruption of this gene in *P. aeruginosa* strain PA68 resulted in a reduction in swimming (92), implying DipA's role modulating motility and EPS production in this strain. Meanwhile, loss of the *rbdA* gene, which also encodes a dual-domain GGDEF-EAL protein, showed an increase in EPS production by *P. aeruginosa* PAO1 that was largely Pel-dependent (19). The transcriptional analysis of the *pelA* gene in the Δ*rbdA* mutant showed an ~1.5-fold increase in expression compared to wild type *P. aeruginosa* PAO1 (19), suggesting that EPS production was posttranscriptionally regulated. Unfortunately, the specific mechanism underlying DipA and RbdA's contribution to the control of EPS remains an open question. Also, reflecting on BifA's role in EPS production, comparative analysis of *bifA*, *rbdA*, and *dipA* expression levels during dispersal could be an interesting avenue to explore.

In a separate study investigating biofilm dispersion, the *bdlA* gene was identified.

Specifically, the Δ*bdlA* mutant biofilm failed to disperse upon NO exposure, but its intracellular c-di-GMP levels remained unchanged from wild type *P. aeruginosa* PAO1 after NO treatment (93). BdlA lacks domains associated with c-di-GMP metabolism (i.e., GGDEF, EAL, or HD-GYP domains) but has two putative PAS domains, as well as a TarH domain (94), both of which are common signal transduction domains. BdlA is also found in a larger, and smaller, proteolyzed variant. Functionally, both PAS domains are necessary to complement the dispersion phenotype in a Δ*bdlA* mutant biofilm, likely due to a ClpP-dependent cleavage of BdlA between the Met and Ala residues on position 130-131, which are located between the PASa and PASb motifs of BdlA (94). Additionally, phosphorylation on a tyrosine residue (Y238) is necessary to prime BdlA for cleavage; a Y→A substitution variant failed to complement the Δ*bdlA* mutant phenotype (94). This posttranslational modification of BdlA is consistent with previous reports (91, 93), but how does BdlA fit into the paradigm of c-di-GMP-regulated dispersion? It turns out that low c-di-GMP levels, particularly driven by DipA, impair proper cleavage of BdlA, while an increase in c-di-GMP mediated by the DGC encoded by PA4843 induced BdlA cleavage (94). These sets of phenotypes establish a convincing model wherein DipA, expressed during the late biofilm maturation stage, regulates intracellular c-di-GMP to promote dispersal via cleavage of BdlA. How the change in intracellular c-di-GMP levels stimulates ClpP-mediated cleavage of BdlA is unknown, as is the mechanism by which cleavage of BdlA stimulates release of bacteria from the biofilm.

CONCLUDING REMARKS

Since its initial discovery as an allosteric factor regulating cellulose biosynthesis in *Gluconacetobacter xylinus* (known as *Acetobacter xylinum* at the time) (95, 96), the list of functional outputs regulated by c-di-GMP has grown. We have focused this article on one of these c-di-GMP-regulated processes, namely, biofilm formation. The majority of DGCs and PDEs encoded in the *P. aeruginosa* genome still remain uncharacterized, and thus there is still a great deal to be learned about the link between c-di-GMP and biofilm formation in this microbe. In particular, while a number of c-di-GMP-metabolizing enzymes have been identified that participate in reversible and irreversible attachment, and biofilm maturation, there is still a significant knowledge gap regarding the c-di-GMP output systems in this organism. Even for the well-characterized Pel system, where c-di-GMP-mediated transcriptional regulation is now well documented, how binding of c-di-GMP by PelD stimulates Pel production is not understood in any detail. Similarly, c-di-GMP-mediated control of swimming, swarming, and twitching also remains to be elucidated. Thus, despite terrific advances in our understanding of *P. aeruginosa* biofilm formation and the role of c-di-GMP in this process since the last version of this book (indeed there was no chapter on c-di-GMP!) (97), there is still much to learn.

ACKNOWLEDGMENTS

This work was supported by NIH grant R37 AI 83256 to G.A.O and by the Rosaline Borison predoctoral fellowship to D.G.H.

Conflicts of interest: We disclose no conflicts.

CITATION

Ha D-G, O'Toole GA. 2015. c-di-GMP and its effects on biofilm formation and dispersion: a *pseudomonas aeruginosa* review. Microbiol Spectrum 3(2):MB-0003-2014.

REFERENCES

1. **Sutherland IW.** 2001. The biofilm matrix: an immobilized but dynamic microbial environment. *Trends Microbiol* **9:**222–227.

2. **Shanks RM, Caiazza NC, Hinsa SM, Toutain CM, O'Toole GA.** 2006. *Saccharomyces cerevisiae*-based molecular tool kit for manipulation of genes from Gram-negative bacteria. *Appl Environ Microbiol* **72:**5027–5036.
3. **Liberati NT, Urbach JM, Miyata S, Lee DG, Drenkard E, Wu G, Villanueva J, Wei T, Ausubel FM.** 2006. An ordered, nonredundant library of *Pseudomonas aeruginosa* strain PA14 transposon insertion mutants. *Proc Natl Acad Sci USA* **103:**2833–2838.
4. **Palmer KL, Aye LM, Whiteley M.** 2007. Nutritional cues control *Pseudomonas aeruginosa* multicellular behavior in cystic fibrosis sputum. *J Bacteriol* **189:**8079–8087.
5. **Palmer KL, Mashburn LM, Singh PK, Whiteley M.** 2005. Cystic fibrosis sputum supports growth and cues key aspects of *Pseudomonas aeruginosa* physiology. *J Bacteriol* **187:** 5267–5277.
6. **Filkins LM, Hampton TH, Gifford AH, Gross MJ, Hogan DA, Sogin ML, Morrison HG, Paster BJ, O'Toole GA.** 2012. Prevalence of streptococci and increased polymicrobial diversity associated with cystic fibrosis patient stability. *J Bacteriol* **194:**4709–4717.
7. **Dubern JF, Diggle SP.** 2008. Quorum sensing by 2-alkyl-4-quinolones in *Pseudomonas aeruginosa* and other bacterial species. *Mol Biosyst* **4:**882–888.
8. **Kirisits MJ, Parsek MR.** 2006. Does *Pseudomonas aeruginosa* use intercellular signalling to build biofilm communities? *Cell Microbiol* **8:** 1841–1849.
9. **Ng WL, Bassler BL.** 2009. Bacterial quorum-sensing network architectures. *Annu Rev Genet* **43:**197–222.
10. **Partridge JD, Harshey RM.** 2013. Swarming: flexible roaming plans. *J Bacteriol* **195:**909–918.
11. **Kearns DB.** 2010. A field guide to bacterial swarming motility. *Nat Rev Microbiol* **8:**634–644.
12. **Haussler S, Fuqua C.** 2013. Biofilms 2012: new discoveries and significant wrinkles in a dynamic field. *J Bacteriol* **195:**2947–2958.
13. **Berleman JE, Kirby JR.** 2009. Deciphering the hunting strategy of a bacterial wolfpack. *FEMS Microbiol Rev* **33:**942–957.
14. **Costerton JW, Stewart PS, Greenberg EP.** 1999. Bacterial biofilms: a common cause of persistent infections. *Science* **284:**1318–1322.
15. **Kulasakara H, Lee V, Brencic A, Liberati N, Urbach J, Miyata S, Lee DG, Neely AN, Hyodo M, Hayakawa Y, Ausubel FM, Lory S.** 2006. Analysis of *Pseudomonas aeruginosa* diguanylate cyclases and phosphodiesterases reveals a role for bis-(3′-5′)-cyclic-GMP in virulence. *Proc Natl Acad Sci USA* **103:**2839–2844.
16. **Ryan RP, Lucey J, O'Donovan K, McCarthy Y, Yang L, Tolker-Nielsen T, Dow JM.** 2009. HD-GYP domain proteins regulate biofilm formation and virulence in *Pseudomonas aeruginosa*. *Environ Microbiol* **11:**1126–1136.
17. **Galperin MY.** 2004. Bacterial signal transduction network in a genomic perspective. *Environ Microbiol* **6:**552–567.
18. **Galperin MY, Nikolskaya AN, Koonin EV.** 2001. Novel domains of the prokaryotic two-component signal transduction systems. *FEMS Microbiol Lett* **203:**11–21.
19. **An S, Wu J, Zhang LH.** 2010. Modulation of *Pseudomonas aeruginosa* biofilm dispersal by a cyclic-di-GMP phosphodiesterase with a putative hypoxia-sensing domain. *Appl Environ Microbiol* **76:**8160–8173.
20. **Roy AB, Petrova OE, Sauer K.** 2012. The phosphodiesterase DipA (PA5017) is essential for *Pseudomonas aeruginosa* biofilm dispersion. *J Bacteriol* **194:**2904–2915.
21. **Alm RA, Bodero AJ, Free PD, Mattick JS.** 1996. Identification of a novel gene, *pilZ*, essential for type 4 fimbrial biogenesis in *Pseudomonas aeruginosa*. *J Bacteriol* **178:**46–53.
22. **Byrd MS, Sadovskaya I, Vinogradov E, Lu H, Sprinkle AB, Richardson SH, Ma L, Ralston B, Parsek MR, Anderson EM, Lam JS, Wozniak DJ.** 2009. Genetic and biochemical analyses of the *Pseudomonas aeruginosa* Psl exopolysaccharide reveal overlapping roles for polysaccharide synthesis enzymes in Psl and LPS production. *Mol Microbiol* **73:**622–638.
23. **Dow JM, Fouhy Y, Lucey JF, Ryan RP.** 2006. The HD-GYP domain, cyclic di-GMP signaling, and bacterial virulence to plants. *Mol Plant Microbe Interact* **19:**1378–1384.
24. **Fang X, Gomelsky M.** 2010. A post-translational, c-di-GMP-dependent mechanism regulating flagellar motility. *Mol Microbiol* **76:**1295–1305.
25. **Huang B, Whitchurch CB, Mattick JS.** 2003. FimX, a multidomain protein connecting environmental signals to twitching motility in *Pseudomonas aeruginosa*. *J Bacteriol* **185:**7068–7076.
26. **Hurley JH.** 2003. GAF domains: cyclic nucleotides come full circle. *Sci STKE* **2003:**PE1.
27. **Jain R, Behrens AJ, Kaever V, Kazmierczak BI.** 2012. Type IV pilus assembly in *Pseudomonas aeruginosa* over a broad range of cyclic di-GMP concentrations. *J Bacteriol* **194:**4285–4294.
28. **Kazmierczak BI, Lebron MB, Murray TS.** 2006. Analysis of FimX, a phosphodiesterase that governs twitching motility in *Pseudomonas aeruginosa*. *Mol Microbiol* **60:**1026–1043.
29. **Hengge R.** 2009. Principles of c-di-GMP signalling in bacteria. *Nat Rev Microbiol* **7:**263–273.

30. **Caiazza NC, Merritt JH, Brothers KM, O'Toole GA.** 2007. Inverse regulation of biofilm formation and swarming motility by *Pseudomonas aeruginosa* PA14. *J Bacteriol* **189:**3603–3612.
31. **O'Toole GA, Kolter R.** 1998. Flagellar and twitching motility are necessary for *Pseudomonas aeruginosa* biofilm development. *Mol Microbiol* **30:**295–304.
32. **Toutain CM, Caizza NC, Zegans ME, O'Toole GA.** 2007. Roles for flagellar stators in biofilm formation by *Pseudomonas aeruginosa*. *Res Microbiol* **158:**471–477.
33. **Friedman L, Kolter R.** 2004. Genes involved in matrix formation in *Pseudomonas aeruginosa* PA14 biofilms. *Mol Microbiol* **51:**675–690.
34. **Friedman L, Kolter R.** 2004. Two genetic loci produce distinct carbohydrate-rich structural components of the *Pseudomonas aeruginosa* biofilm matrix. *J Bacteriol* **186:**4457–4465.
35. **Liang Y, Hilal N, Langston P, Starov V.** 2007. Interaction forces between colloidal particles in liquid: theory and experiment. *Adv Colloid Interface Sci* **134–135:**151–166.
36. **Dasgupta N, Ramphal R.** 2001. Interaction of the antiactivator FleN with the transcriptional activator FleQ regulates flagellar number in *Pseudomonas aeruginosa*. *J Bacteriol* **183:**6636–6644.
37. **Dasgupta N, Arora SK, Ramphal R.** 2000. *fleN*, a gene that regulates flagellar number in *Pseudomonas aeruginosa*. *J Bacteriol* **182:**357–364.
38. **Hickman JW, Harwood CS.** 2008. Identification of FleQ from *Pseudomonas aeruginosa* as a c-di-GMP-responsive transcription factor. *Mol Microbiol* **69:**376–389.
39. **Baraquet C, Harwood CS.** 2013. Cyclic diguanosine monophosphate represses bacterial flagella synthesis by interacting with the Walker A motif of the enhancer-binding protein FleQ. *Proc Natl Acad Sci USA* **110:**18478–18483.
40. **Ryjenkov DA, Simm R, Romling U, Gomelsky M.** 2006. The PilZ domain is a receptor for the second messenger c-di-GMP: the PilZ domain protein YcgR controls motility in enterobacteria. *J Biol Chem* **281:**30310–30314.
41. **Zorraquino V, Garcia B, Latasa C, Echeverz M, Toledo-Arana A, Valle J, Lasa I, Solano C.** 2013. Coordinated cyclic-di-GMP repression of *Salmonella* motility through YcgR and cellulose. *J Bacteriol* **195:**417–428.
42. **Boehm A, Kaiser M, Li H, Spangler C, Kasper CA, Ackermann M, Kaever V, Sourjik V, Roth V, Jenal U.** 2010. Second messenger-mediated adjustment of bacterial swimming velocity. *Cell* **141:**107–116.
43. **Ko M, Park C.** 2000. Two novel flagellar components and H-NS are involved in the motor function of *Escherichia coli*. *J Mol Biol* **303:**371–382.
44. **Paul K, Nieto V, Carlquist WC, Blair DF, Harshey RM.** 2010. The c-di-GMP binding protein YcgR controls flagellar motor direction and speed to affect chemotaxis by a "backstop brake" mechanism. *Mol Cell* **38:**128–139.
45. **Martinez-Granero F, Navazo A, Barahona E, Redondo-Nieto M, Gonzalez de Heredia E, Baena I, Martin-Martin I, Rivilla R, Martin M.** 2014. Identification of *flgZ* as a flagellar gene encoding a PilZ domain protein that regulates swimming motility and biofilm formation in *Pseudomonas*. *PLoS One* **9:**e87608. doi:10.1371/journal.pone.0087608.
46. **Ko J, Ryu KS, Kim H, Shin JS, Lee JO, Cheong C, Choi BS.** 2010. Structure of PP4397 reveals the molecular basis for different c-di-GMP binding modes by PilZ domain proteins. *J Mol Biol* **398:**97–110.
47. **Zhao K, Tseng BS, Beckerman B, Jin F, Gibiansky ML, Harrison JJ, Luijten E, Parsek MR, Wong GC.** 2013. Psl trails guide exploration and microcolony formation in *Pseudomonas aeruginosa* biofilms. *Nature* **497:**388–391.
48. **Jin F, Conrad JC, Gibiansky ML, Wong GC.** 2011. Bacteria use type-IV pili to slingshot on surfaces. *Proc Natl Acad Sci USA* **108:**12617–12622.
49. **Conrad JC, Gibiansky ML, Jin F, Gordon VD, Motto DA, Mathewson MA, Stopka WG, Zelasko DC, Shrout JD, Wong GC.** 2011. Flagella and pili-mediated near-surface single-cell motility mechanisms in *P. aeruginosa*. *Biophys J* **100:**1608–1616.
50. **Gibiansky ML, Conrad JC, Jin F, Gordon VD, Motto DA, Mathewson MA, Stopka WG, Zelasko DC, Shrout JD, Wong GC.** 2010. Bacteria use type IV pili to walk upright and detach from surfaces. *Science* **330:**197.
51. **Monds RD, O'Toole GA.** 2009. The developmental model of microbial biofilms: ten years of a paradigm up for review. *Trends Microbiol* **17:**73–87.
52. **Kuchma SL, Brothers KM, Merritt JH, Liberati NT, Ausubel FM, O'Toole GA.** 2007. BifA, a cyclic-di-GMP phosphodiesterase, inversely regulates biofilm formation and swarming motility by *Pseudomonas aeruginosa* PA14. *J Bacteriol* **189:**8165–8178.
53. **Merritt JH, Brothers KM, Kuchma SL, O'Toole GA.** 2007. SadC reciprocally influences biofilm formation and swarming motility via modulation of exopolysaccharide production and flagellar function. *J Bacteriol* **189:**8154–8164.

54. **Bradley DE.** 1980. A function of *Pseudomonas aeruginosa* PAO polar pili: twitching motility. *Can J Microbiol* **26:**146–154.
55. **Mattick JS.** 2002. Type IV pili and twitching motility. *Annu Rev Microbiol* **56:**289–314.
56. **Navarro MV, De N, Bae N, Wang Q, Sondermann H.** 2009. Structural analysis of the GGDEF-EAL domain-containing c-di-GMP receptor FimX. *Structure* **17:**1104–1116.
57. **Merighi M, Lee VT, Hyodo M, Hayakawa Y, Lory S.** 2007. The second messenger bis-(3′-5′)-cyclic-GMP and its PilZ domain-containing receptor Alg44 are required for alginate biosynthesis in *Pseudomonas aeruginosa. Mol Microbiol* **65:**876–895.
58. **Barken KB, Pamp SJ, Yang L, Gjermansen M, Bertrand JJ, Klausen M, Givskov M, Whitchurch CB, Engel JN, Tolker-Nielsen T.** 2008. Roles of type IV pili, flagellum-mediated motility and extracellular DNA in the formation of mature multicellular structures in *Pseudomonas aeruginosa* biofilms. *Environ Microbiol* **10:**2331–2343.
59. **Colvin KM, Irie Y, Tart CS, Urbano R, Whitney JC, Ryder C, Howell PL, Wozniak DJ, Parsek MR.** 2012. The Pel and Psl polysaccharides provide *Pseudomonas aeruginosa* structural redundancy within the biofilm matrix. *Environ Microbiol* **14:**1913–1928.
60. **Zegans ME, Wozniak D, Griffin E, Toutain-Kidd CM, Hammond JH, Garfoot A, Lam JS.** 2012. *Pseudomonas aeruginosa* exopolysaccharide Psl promotes resistance to the biofilm inhibitor polysorbate 80. *Antimicrob Agents Chemother* **56:**4112–4122.
61. **Boyd CD, O'Toole GA.** 2012. Second messenger regulation of biofilm formation: breakthroughs in understanding c-di-GMP effector systems. *Annu Rev Cell Dev Biol* **28:**439–462.
62. **Overhage J, Schemionek M, Webb JS, Rehm BH.** 2005. Expression of the *psl* operon in *Pseudomonas aeruginosa* PAO1 biofilms: PslA performs an essential function in biofilm formation. *Appl Environ Microbiol* **71:**4407–4413.
63. **Baraquet C, Murakami K, Parsek MR, Harwood CS.** 2012. The FleQ protein from *Pseudomonas aeruginosa* functions as both a repressor and an activator to control gene expression from the *pel* operon promoter in response to c-di-GMP. *Nucleic Acids Res* **40:**7207–7218.
64. **Whitney JC, Colvin KM, Marmont LS, Robinson H, Parsek MR, Howell PL.** 2012. Structure of the cytoplasmic region of PelD, a degenerate diguanylate cyclase receptor that regulates exopolysaccharide production in *Pseudomonas aeruginosa. J Biol Chem* **287:** 23582–23593.
65. **Lee VT, Matewish JM, Kessler JL, Hyodo M, Hayakawa Y, Lory S.** 2007. A cyclic-di-GMP receptor required for bacterial exopolysaccharide production. *Mol Microbiol* **65:**1474–1484.
66. **Li Z, Chen JH, Hao Y, Nair SK.** 2012. Structures of the PelD cyclic diguanylate effector involved in pellicle formation in *Pseudomonas aeruginosa* PAO1. *J Biol Chem* **287:**30191–30204.
67. **Merritt JH, Ha DG, Cowles KN, Lu W, Morales DK, Rabinowitz J, Gitai Z, O'Toole GA.** 2010. Specific control of *Pseudomonas aeruginosa* surface-associated behaviors by two c-di-GMP diguanylate cyclases. *MBio* **1:** e00183-10. doi:10.1128/mBio.00183-10.
68. **Christen M, Kulasekara HD, Christen B, Kulasekara BR, Hoffman LR, Miller SI.** 2010. Asymmetrical distribution of the second messenger c-di-GMP upon bacterial cell division. *Science* **328:**1295–1297.
69. **Choy WK, Zhou L, Syn CK, Zhang LH, Swarup S.** 2004. MorA defines a new class of regulators affecting flagellar development and biofilm formation in diverse *Pseudomonas* species. *J Bacteriol* **186:**7221–7228.
70. **Meissner A, Wild V, Simm R, Rohde M, Erck C, Bredenbruch F, Morr M, Romling U, Haussler S.** 2007. *Pseudomonas aeruginosa cupA*-encoded fimbriae expression is regulated by a GGDEF and EAL domain-dependent modulation of the intracellular level of cyclic diguanylate. *Environ Microbiol* **9:**2475–2485.
71. **Whitchurch CB, Tolker-Nielsen T, Ragas PC, Mattick JS.** 2002. Extracellular DNA required for bacterial biofilm formation. *Science* **295:** 1487.
72. **Allesen-Holm M, Barken KB, Yang L, Klausen M, Webb JS, Kjelleberg S, Molin S, Givskov M, Tolker-Nielsen T.** 2006. A characterization of DNA release in *Pseudomonas aeruginosa* cultures and biofilms. *Mol Microbiol* **59:**1114–1128.
73. **Kuchma SL, Ballok AE, Merritt JH, Hammond JH, Lu W, Rabinowitz JD, O'Toole GA.** 2010. Cyclic-di-GMP-mediated repression of swarming motility by *Pseudomonas aeruginosa*: the *pilY1* gene and its impact on surface-associated behaviors. *J Bacteriol* **192:**2950–2964.
74. **Wozniak DJ, Wyckoff TJ, Starkey M, Keyser R, Azadi P, O'Toole GA, Parsek MR.** 2003. Alginate is not a significant component of the extracellular polysaccharide matrix of PA14 and PAO1 *Pseudomonas aeruginosa* biofilms. *Proc Natl Acad Sci USA* **100:**7907–7912.

75. **Davies DG, Geesey GG.** 1995. Regulation of the alginate biosynthesis gene *algC* in *Pseudomonas aeruginosa* during biofilm development in continuous culture. *Appl Environ Microbiol* **61:**860–867.
76. **Hay ID, Remminghorst U, Rehm BH.** 2009. MucR, a novel membrane-associated regulator of alginate biosynthesis in *Pseudomonas aeruginosa. Appl Environ Microbiol* **75:**1110–1120.
77. **Oglesby LL, Jain S, Ohman DE.** 2008. Membrane topology and roles of *Pseudomonas aeruginosa* Alg8 and Alg44 in alginate polymerization. *Microbiology* **154:**1605–1615.
78. **Hickman JW, Tifrea DF, Harwood CS.** 2005. A chemosensory system that regulates biofilm formation through modulation of cyclic diguanylate levels. *Proc Natl Acad Sci USA* **102:**14422–14427.
79. **Huangyutitham V, Guvener ZT, Harwood CS.** 2013. Subcellular clustering of the phosphorylated WspR response regulator protein stimulates its diguanylate cyclase activity. *MBio* **4:** e00242-00213. doi:10.1128/mBio.00242-13.
80. **Guvener ZT, Harwood CS.** 2007. Subcellular location characteristics of the *Pseudomonas aeruginosa* GGDEF protein, WspR, indicate that it produces cyclic-di-GMP in response to growth on surfaces. *Mol Microbiol* **66:**1459–1473.
81. **Irie Y, Borlee BR, O'Connor JR, Hill PJ, Harwood CS, Wozniak DJ, Parsek MR.** 2012. Self-produced exopolysaccharide is a signal that stimulates biofilm formation in *Pseudomonas aeruginosa. Proc Natl Acad Sci USA* **109:**20632–20636.
82. **Borlee BR, Goldman AD, Murakami K, Samudrala R, Wozniak DJ, Parsek MR.** 2010. *Pseudomonas aeruginosa* uses a cyclic-di-GMP-regulated adhesin to reinforce the biofilm extracellular matrix. *Mol Microbiol* **75:**827–842.
83. **Ma L, Conover M, Lu H, Parsek MR, Bayles K, Wozniak DJ.** 2009. Assembly and development of the *Pseudomonas aeruginosa* biofilm matrix. *PLoS Pathog* **5:**e1000354. doi:10.1371/journal.ppat.1000354.
84. **Rani SA, Pitts B, Beyenal H, Veluchamy RA, Lewandowski Z, Davison WM, Buckingham-Meyer K, Stewart PS.** 2007. Spatial patterns of DNA replication, protein synthesis, and oxygen concentration within bacterial biofilms reveal diverse physiological states. *J Bacteriol* **189:**4223–4233.
85. **Werner E, Roe F, Bugnicourt A, Franklin MJ, Heydorn A, Molin S, Pitts B, Stewart PS.** 2004. Stratified growth in *Pseudomonas aeruginosa* biofilms. *Appl Environ Microbiol* **70:**6188–6196.
86. **Davey ME, Caiazza NC, O'Toole GA.** 2003. Rhamnolipid surfactant production affects biofilm architecture in *Pseudomonas aeruginosa* PAO1. *J Bacteriol* **185:**1027–1036.
87. **Yoon SS, Hennigan RF, Hilliard GM, Ochsner UA, Parvatiyar K, Kamani MC, Allen HL, DeKievit TR, Gardner PR, Schwab U, Rowe JJ, Iglewski BH, McDermott TR, Mason RP, Wozniak DJ, Hancock RE, Parsek MR, Noah TL, Boucher RC, Hassett DJ.** 2002. *Pseudomonas aeruginosa* anaerobic respiration in biofilms: relationships to cystic fibrosis pathogenesis. *Dev Cell* **3:**593–603.
88. **Webb JS, Thompson LS, James S, Charlton T, Tolker-Nielsen T, Koch B, Givskov M, Kjelleberg S.** 2003. Cell death in *Pseudomonas aeruginosa* biofilm development. *J Bacteriol* **185:**4585–4592.
89. **Barraud N, Hassett DJ, Hwang SH, Rice SA, Kjelleberg S, Webb JS.** 2006. Involvement of nitric oxide in biofilm dispersal of *Pseudomonas aeruginosa. J Bacteriol* **188:**7344–7353.
90. **Barraud N, Schleheck D, Klebensberger J, Webb JS, Hassett DJ, Rice SA, Kjelleberg S.** 2009. Nitric oxide signaling in *Pseudomonas aeruginosa* biofilms mediates phosphodiesterase activity, decreased cyclic di-GMP levels, and enhanced dispersal. *J Bacteriol* **191:**7333–7342.
91. **Sauer K, Cullen MC, Rickard AH, Zeef LA, Davies DG, Gilbert P.** 2004. Characterization of nutrient-induced dispersion in *Pseudomonas aeruginosa* PAO1 biofilm. *J Bacteriol* **186:** 7312–7326.
92. **Li Y, Xia H, Bai F, Xu H, Yang L, Yao H, Zhang L, Zhang X, Bai Y, Saris PE, Tolker-Nielsen T, Qiao M.** 2007. Identification of a new gene PA5017 involved in flagella-mediated motility, chemotaxis and biofilm formation in *Pseudomonas aeruginosa. FEMS Microbiol Lett* **272:**188–195.
93. **Morgan R, Kohn S, Hwang SH, Hassett DJ, Sauer K.** 2006. BdlA, a chemotaxis regulator essential for biofilm dispersion in *Pseudomonas aeruginosa. J Bacteriol* **188:**7335–7343.
94. **Petrova OE, Sauer K.** 2012. Dispersion by *Pseudomonas aeruginosa* requires an unusual posttranslational modification of BdlA. *Proc Natl Acad Sci USA* **109:**16690–16695.
95. **Ross P, Aloni Y, Weinhouse H, Michaeli D, Weinbergerohana P, Mayer R, Benziman M.** 1986. Control of cellulose synthesis in *Acetobacter xylinum:* a unique guanyl oligonucleotide is the immediate activator of the cellulose synthase. *Carbohyd Res* **149:**101–117.

96. **Ross P, Weinhouse H, Aloni Y, Michaeli D, Weinberger-Ohana P, Mayer R, Braun S, de Vroom E, van der Marel GA, van Boom JH, Benziman M.** 1987. Regulation of cellulose synthesis in *Acetobacter xylinum* by cyclic diguanylic acid. *Nature* **325**:279–281.
97. **Ghannoum MA, O'Toole GA.** 2004. *Microbial Biofilms*. ASM Press, Washington, DC.

Mechanisms of Competition in Biofilm Communities

16

OLAYA RENDUELES[1] and JEAN-MARC GHIGO[2]

> One general law, leading to the advancement of all organic beings, namely, multiply, vary, let the strongest live and the weakest die.
>
> CHARLES DARWIN, *The Origin of Species* (1)

INTRODUCTION

The rapid development of new sequencing technologies and the use of metagenomics revealed the great diversity of microbial life and enabled the emergence of a new perspective on population dynamics. Moreover, it has highlighted the central role of social interactions in ecological and evolutionary processes. Microbes living in multispecies communities are prevalent in nature and have been shown to extensively cooperate and compete. Both intra- and interspecies interactions are instrumental in major geochemical cycles and are important in human health and homeostasis (e.g., the human microbiome has been associated with several diseases) and in industrial and clinical settings (2). Few studies have addressed the role of individual species within mixed communities (3), and they generally focus on cooperative

[1]Institute for Integrative Biology, ETH Zürich, 8092 Zürich, Switzerland; [2]Institut Pasteur, Unité de Génétique des Biofilms, Département de Microbiologie, F-75015 Paris, France.

Microbial Biofilms 2nd Edition
Edited by Mahmoud Ghannoum, Matthew Parsek, Marvin Whiteley, and Pranab K. Mukherjee

doi:10.1128/microbiolspec.MB-0009-2014

interactions and increased benefits of community life (4, 5). However, recent work pointed out that most interactions are competitive rather than cooperative, suggesting that adaptation is more likely achieved by competitive success (6). A further degree of complexity in understanding multispecies interactions and dynamics is brought by increasing evidence suggesting that phenomena occurring in complex communities cannot be predicted by the observation of single-species communities (7).

How do biofilm characteristics contribute to shape the evolution of competitive social interactions? What are the different competitive strategies deployed by bacteria when forming biofilms? Here, we describe exploitative and interference competition strategies, with a special focus on the underlying molecular mechanisms involved; we offer some insights on the evolutionary origins and ecological consequences of competition and reflect on new venues for this exciting multidisciplinary field.

ECOLOGICAL AND EVOLUTIONARY PARAMETERS OPERATING WITHIN BIOFILMS

Our current understanding of microbial physiology is mostly based on studies performed in homogeneous batch cultures. This reductionist approach enabled an experimental simplicity that has led to major discoveries but has largely neglected the complexity of the microbial world. Indeed, extrapolating characteristics observed in liquid to traits potentially relevant to a community context could be misleading. In nature, bacteria display complex multicellular behaviors and influence each other, enabling them to perform a great diversity of tasks that they would not otherwise accomplish in liquid monocultures. The following is a review of some important parameters for the study of competition within biofilms (Fig. 1).

Consequences of Clustering: Group Effects

Life in groups or aggregates is a common trait found in all forms of life, from animals to bacteria. The Allee effect suggests that there is an inherent benefit to grouping, evidenced by a positive correlation between population size or density and mean individual fitness (8). This phenomenon has been extensively studied; for instance, it was demonstrated that grouping provides animals with enhanced resistance to predation (9), affects mating and reproduction efficiency in mobile organisms (10), and increases prokaryote resistance to desiccation (11). In bacteria, it has been shown that survival of *Streptococcus mutans* growing on the surface of teeth and experiencing acidic stress is strongly density-dependent (12). Other examples of the Allee effect include trade-offs between dispersal and survival in *Escherichia coli* populations (13) and an increased tolerance to several antibiotics when bacteria are at higher cellular densities. This tolerance to antibiotics is independent of biofilm formation, quorum sensing (QS), and antibiotic resistance pathways (14).

Although grouping is often beneficial, the physiological consequences of increased cell density in bacterial clusters can enhance competition. In the case of biofilms, bacteria are encased in a self-produced biofilm matrix that acts as a molecular sink or reservoir due to limited outward diffusion and/or retention of compounds by the matrix. Therefore, antagonist molecules released by bacteria are more concentrated in local areas and lead to increased efficiency against neighboring competitors (15). Finally, grouping enhances competitive interactions, the effects of which are strengthened when cellular densities are high and resources low, for example, via QS-dependent regulation and other strategies controlled by positive feedback loops (16).

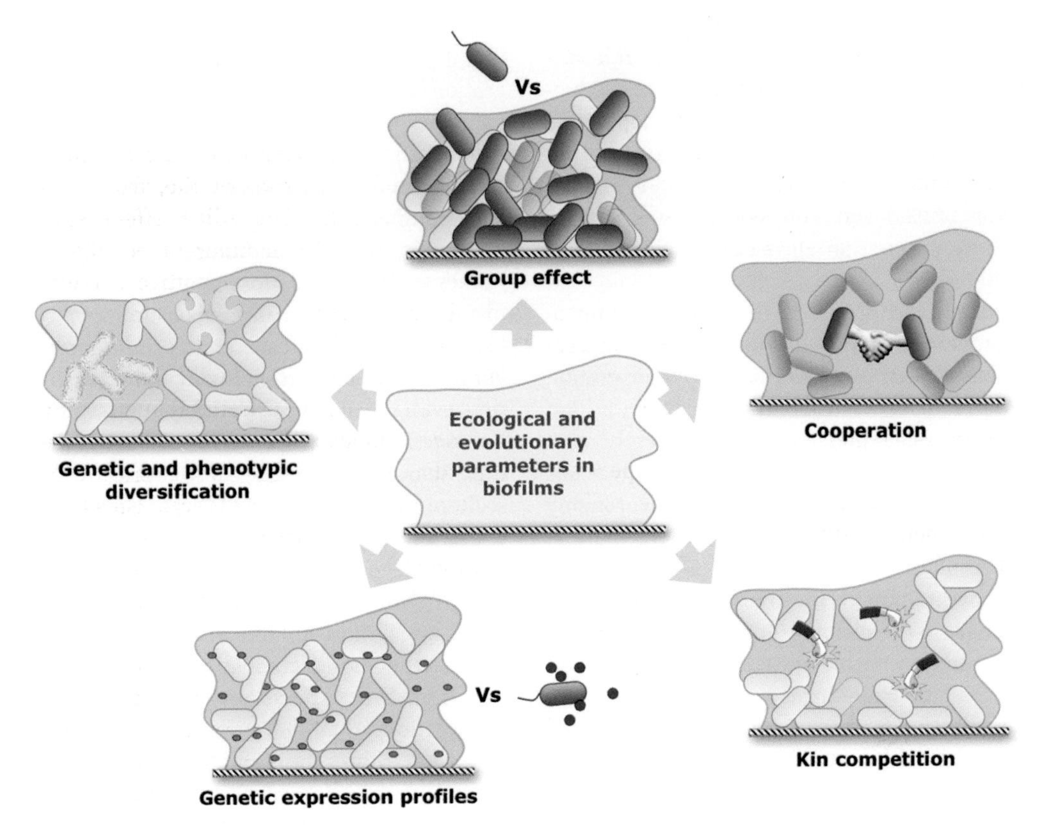

FIGURE 1 Ecological and evolutionary parameters operating within biofilm communities. Group effects: increase bacterial fitness compared to solitary life. Cooperation: biofilm bacteria can actively cooperate to increase their individual fitness. Kin competition: under high stress and low nutrient conditions, kin can become a source of competition and enhance spatial segregation. Genetic expression profiles: planktonic bacteria express different genes than those expressed by biofilm. Genotypic and phenotypic diversification: Due to competition, different variants can spontaneously appear within biofilm communities. doi:10.1128/microbiolspec.MB-0009-2014.f1

Cooperation in Multispecies Biofilms

In contrast to competitive forces, spatial proximity may also allow for cooperative interactions between microbes (17). Coaggregation, or recognition and adhesion between genetically distinct bacteria (18), can be advantageous and develop into mutualistic behaviors that impose few, if any, associated costs to the interacting partners. Examples of such interactions are (i) cross-feeding of two species (19, 20), (ii) more efficient degradation of herbicides by a three-species biofilm than comparable single-species or dual-species biofilms (21), and (iii) altruistic and/or synergistic degradation of toxic compounds, allowing growth of other sensitive species (22, 23).

However, more often than not, there are costs associated with cooperation, and understanding the evolution of such a social investment in Darwinian terms can be challenging, in particular, when the burden of cooperative behavior is costly enough to impact the individual reproductive fitness of the cooperators. During cooperative interactions, "cheaters," (individuals that benefit from the cooperation but do not pay the costs associated with it), can emerge and eventually invade the population due to higher individual fitness than cooperators. The dilemma of cooperation is also relevant during biofilm formation and matrix production, which strongly rely on the cooperative secretion of shared compounds, also referred to as "public goods," such as exopolysaccharides (EPSs) and iron siderophores, and to whose production all members should contribute (24).

Hamilton proposed that the evolution of cooperation can be explained by an indirect fitness effect, which occurs when cooperation is directed toward kin, individuals with high genetic relatedness (25). Consequently, perfect cooperation in biofilms should only occur in niches populated by one genotype or highly related genotypes. In this situation, there is no evolutionary competition, and genotypes behave optimally for the group (26). Additionally, several mechanisms have been described to control cheaters (27, 28) and to ensure the maintenance of cooperation (29). This includes cheater policing (30) or preferentially directing cooperation benefits toward kin (i.e., kin discrimination and green beard effect (mechanism by which natural selection favors altruistic behavior toward kin) [31]).

Interestingly, as predicted by theoretical models (32), biofilm characteristics such as proximity, limited dispersal, and spatial structure could enhance cooperation *per se* by increasing genetic relatedness between the interacting partners and avoiding the spread of cheating genotypes. This is supported by experimental work showing that cooperative bacteria have greater fitness than cheaters in biofilms (33). More particularly, thick biofilms limit the diffusion of public goods through the biofilm so that only producers and their immediate neighbors have access to it (15, 33).

Kin Competition in Biofilms

When resources are scarce, biofilm bacteria have to compete not only against other species but also against their own kin, therefore limiting the benefits of cooperative behaviors (34, 35). Reports of the persistence of cooperation found in nature, either by establishment of lifelong symbiosis or isolated collaborations, indicate that the biofilm lifestyle presents evolutionary advantages despite competition (kin or not) and/or cheater apparition (36). In addition to Hamilton's rule (see above) and aforementioned mechanisms, the evolution of cooperation can be favored by ecological factors influencing both benefits and strengths of the competition, even when kin competition is harsh (37). For instance, the spread of cheaters and their invasion is strongly determined by the scale of the competition (34). Moreover, cooperation is facilitated over competition if the public good increases group productivity (37). Taylor revisited Hamilton's rule and proposed a more integrative model for kin selection including competition, patch structure, and their effects on the evolution of cooperation (17). Such a model might more accurately represent biofilm communities.

A widespread weakness when studying kin competition is that only one social trait is usually taken into account (for instance, siderophore secretion). Reports that consider several cooperative traits (e.g., siderophore secretion, bacteriocin production, virulence factors, matrix production) at once and analyze how they are impacted by social interactions are scarce (38). Additionally, most studies investigating the emergence of cheaters are performed in single-species communities, without the additional level of complexity added when nonrelated species are introduced into the community (39, 40) (see below).

Genetic and Phenotypic Diversity

Biofilms are highly heterogeneous environments in which different microniches can coexist. They are characterized by a vertical heterogeneity where the deeper layers of the biofilm are thought to be under strong starvation conditions, reduced oxygen diffusion, and accumulation of waste products. There is also a heterogeneity corresponding to the presence of small patches of tightly packed cells separated throughout the biofilm by the extracellular polymeric matrix and water channels (41). Hence, biofilm bacteria might have different growth rates, suffer from different stresses, and accordingly, adopt different physiologies (2). The existence of small and diverse microniches within biofilms provides an excellent landscape for adaptive radiation and, consequently, genetic and phenotypic divergence. Several experimental studies indicate that structured habitats lead to increased diversity due to frequency-dependent selection and to adaptation of to unexploited microniches generated by spatial heterogeneity (42). Indeed, key innovations to exploit new niches in a biofilm rapidly and readily arise given the opportunity (43, 44) (see "Experimental Evolution in Biofilms," below).

Ecological competition plays a key role in maintaining diversity in biofilms, because it is unlikely that one genotype is well adapted to all environmental conditions of heterogeneous structures. Consequently, potential trade-offs across the different microniches of biofilms allow for the coexistence of different adapted genotypes (43). The resultant self-diversification ensures that in changing conditions a subset of the population will still be adapted and have an advantage (43). For example, *Pseudomonas aeruginosa* biofilm cells diversify into different genotypes that show enhanced dispersal or faster formation of biofilms (43).

Biofilms have two inherent characteristics that maintain diversity: limited dispersal of genotypes (clone-mates remain in the neighborhood) and the existence of defined spatial structure (i.e., heterogeneity). This results in the coexistence of different competing genotypes, whereas mixing and enhanced diffusion results in the establishment of the fittest genotype (45). Other mechanisms such as negative frequency-dependent selection also allow variants adapted to a microniche to be maintained (44). Finally, the existence of high genetic diversity also implies that the ability of one genotype to overtake an entire population is slowed down because it has to displace competitors with different beneficial mutations. Taken together, there is increased genetic diversity caused and maintained by the ecological competition and spatial structure found within biofilms.

Understanding the diversity within a community is a useful first step to elucidating its distribution patterns in a given environment and identifying mechanisms of competition within multispecies communities. Since the launch of the Human Microbiome Project, next-generation sequencing has made tremendous progress and considerably lowered the detection limits for underrepresented taxa in microbial populations. The use of new diversity indexes taking into account both high and low abundant taxa (46) will likely improve our understanding of ecological diversity in host (i.e., mouth, intestine, skin), and environmental biofilm niches across individuals or between populations.

Genetic Expression Profiles

Biofilm bacteria have very different gene expression profiles compared to planktonic lifestyles, with up to a 10% difference between transcriptomes (47, 48). Aside from genes involved in general metabolic function, these studies show that stress response genes, energy production, and envelope-associated genes are upregulated in biofilm bacteria (47, 49, 50). Differences between biofilm and planktonic cells may also be relevant during competition because many upregulated loci during biofilm development still have no known function (47, 50). Moreover, recent studies reported that biofilm gene expression profiles are highly dynamic and vary throughout the developmental stages of biofilm. For instance, dispersal of *Streptococcus pneumoniae* biofilm completely changes its gene expression profiles compared to planktonic biofilm and bacteria in early stages of biofilm formation (51). These changes in expression include genes involved in the synthesis of rhamnolipids and other molecules already known to have antibiofilm effects on other bacteria (52, 53). These novel biofilm gene expression patterns could lead to biofilm-specific functions associated with a potential role in ecological competition (54).

MECHANISMS OF BIOFILM COMPETITION

Ecological competition can be classified into two groups. The first type, known as "exploitative competition," refers to indirect interactions between organisms, by which one organism prevents access to and/or limits the use of resources by another organism (55, 56). Exploitative competition is common across the biological spectrum; however, it is particularly acute within biofilms, which are already nutrient-depleted environments (57). The second type of ecological competition is direct or "interference competition," which corresponds to specific mechanisms which damage competitors' survival or their access to an ecological niche or specific resources (Table 1). This can be illustrated by the production of antibiotics and other growth-inhibiting mechanisms that lower bacterial fitness. In the context of biofilms, numerous interference strategies that do not alter growth but do alter the ability to colonize a niche have been recently described (see reference 58 for review) (Table 1). Both exploitative and interference competition mechanisms strongly affect evolutionary outcomes and population dynamics. Interactions between both mechanisms remain largely unaddressed; nevertheless, theoretical studies predict that the combination of these two strategies should result in increased biodiversity or the coexistence of genotypes rather than exclusion or extinction (59).

Exploitative Competition

The current general view considers that most biofilm bacteria live under severe environmental conditions, characterized by low nutrient concentrations and low rates of gas renewal or exchanges. This situation is particularly common in the deep layers of a biofilm, where diffusion is difficult (41). Because growth is dependent on the concentration of limiting resources (iron, carbon, oxygen, etc.) in this biofilm environment, the availability of such resources drives competition fate.

TABLE 1 Interference competition. Summary of the interference strategies described in this chapter

Interference competition			
Growth-inhibition		Alteration of biofilm development	
Environment alteration		Inhibition of cell-to-cell communication	
Production of toxic metabolic byproducts		Inhibition of adhesion	Modification of adhesion by SACs
Small antimicrobial compounds	Colicins		Downregulation of adhesins
	Bacteriocins	Matrix degradation	
Contact-dependent mechanism	CDI	Induced dispersal	
	Type VI secretion system	Motility-based mechanisms	Surface-blanketing
Predation	Phagocytosis		Induction of motility
	Cell invasion		Penetrating the matrix
	Secretion of lytic diffusible factors	Resistance to colonization	

In air-liquid interface biofilms (also called pellicles), competition for oxygen can determine the success or failure of one species to cocolonize the surface. For instance, *Brevibacillus* is an obligate aerobe that can grow in pellicles in a multispecies consortium composed of four species including the facultative aerobe *Pseudoxanthomonas*. However, in pair-wise competition with *Pseudoxanthomonas* spp., *Brevibacillus* initially dominates, but its viability rapidly decreases due to severe competition for oxygen (60). These results show that competition between aerobes in the air-liquid interface is strongly driven by access to oxygen. Moreover, population dynamics can also be further altered by other ecological and biotic factors, such as the presence of other species in the consortium (60).

In addition to oxygen, exploitation of a variety of essential elements, including carbon, phosphorous, and especially iron have been reported (61–63). Iron is a key element in microbial physiology, controlling many bacterial functions including virulence and biofilm formation. Due to its low bioavailability, bacteria have evolved many iron-scavenging strategies, such as high-affinity iron-chelators or siderophores. Theoretical models have pointed out the advantages of producing such public goods (i.e., iron-sequestering molecules) during competition. Siderophore-producing genotypes increase their own growth rate while they deplete iron in their surroundings and potentially limit access to it for other genotypes (64). Besides iron limitation and competition for iron acquisition, iron sequestration can be used by bacteria to manipulate coexisting microorganisms. For instance, a *Bulkholderia* sp. is able to alter the physiology of neighboring *P. aeruginosa* bacteria through the production of the ornibactin siderophore. *P. aeruginosa* specifically responds to ornitobactin secreted by *Burkholderia* and induces genes usually expressed under genuinely low levels of iron (63). Whether ornitobactin affects the fitness of *P. aeruginosa* is still unclear. Taken together, these results suggest that iron might drive not only intra-species competition, but also interspecies interactions between *Pseudomonas* and *Burkholderia*.

Interference Competition

Different interference competition strategies, which are also relevant in biofilm conditions, have evolved; for instance, some target bacterial ability to form a biofilm, whereas others are less specific and kill or limit the growth of competing bacteria within mixed biofilms (Table 1). Many studies addressing interference competition usually start with the identification of antagonist molecules produced by the interacting strains in liquid monocultures. Although this might not be the optimal approach to understand the reality of ecological competition in natural settings, it has certainly been successful, for instance, leading to the identification of antibiotics and many other antibacterial toxins. However, this approach also potentially misses other competitive possibilities. As previously stated, changes in gene expression associated with the biofilm lifestyle (47, 48) can lead to production of biofilm-specific metabolites and polymers (54), some of which can display antagonist activities against other microorganisms in mixed species contexts (65–67). In addition to competitive interactions induced by the biofilm lifestyle, interference mechanisms might also be triggered in response to the presence of competition in the surroundings, also known as the competition-sensing hypothesis (57). Hence, bacterial defense mechanisms would only occur when bacteria are under stress due to the harmful action of other competing bacteria rather than being constitutively expressed. This hypothesis also differentiates harm imposed by other bacterial species (nonkin) from harm originating from cells with the same genotype (kin) for which there is no evolutionary competition because all bacteria seek the same interests (57). In the context of competition, kin could be potentially differentiated from nonkin by self-produced molecules such as QS signals. Furthermore, Cornforth and Foster propose that stress response regulatory networks could integrate the source of the competition (kin or nonkin) with the level of stress/damage experienced by the sensing bacteria and subsequently respond in an appropriate manner (57).

Here, we describe ecological interference mechanisms within biofilms (Table 1 and Fig. 2). Although most mechanisms are not biofilm-specific, selected examples gain special relevance in biofilms or were shown to play a role in mixed communities.

Interference mediated by growth inhibition

Environment alteration

Bacterial fitness, or survival, is environment-dependent. One mechanism by which bacteria can negatively impact competitors is by changing their local environment either directly or as a consequence of their secondary metabolism and physiological by-products. *Lactobacilli* spp. produce lactic acid that lowers environmental pH and hence limits the growth of other bacterial species. Indeed, growth of *Neisseria gonorrhoeae* in mixed cultures with vaginal *Lactobacilli* spp. was severely reduced due to the acidification of the environment. When a suitable pH is sensed, *N. gonorrhoeae* can resume growth (68). Other examples of molecules able to alter surrounding conditions are volatile compounds such as trimethylamine secreted by *E. coli* and other Gram-negative bacteria in intestinal or precursor-rich environments. These compounds increase environmental pH and affect biofilm formation and stress resistance (69). Although volatile compounds can potentially be secreted in liquid conditions, their production is strongly density-dependent, and therefore their effect in biofilms is magnified.

Toxic metabolic byproducts

Bacteria also use low-molecular-weight compounds derived from their own metabolism to kill and gain competitive advantage over surrounding bacteria. One of the best-described effects is hydrogen-peroxide-mediated killing. Hydrogen peroxide (H_2O_2) is a byproduct of aerobic metabolism from a reaction usually mediated by a pyruvate oxidase (70). The role of hydrogen peroxide in competition has been demonstrated in several environments. In the human nasopharynx, *S. pneumoniae* produces hydrogen peroxide against *Neisseria meningitidis* and

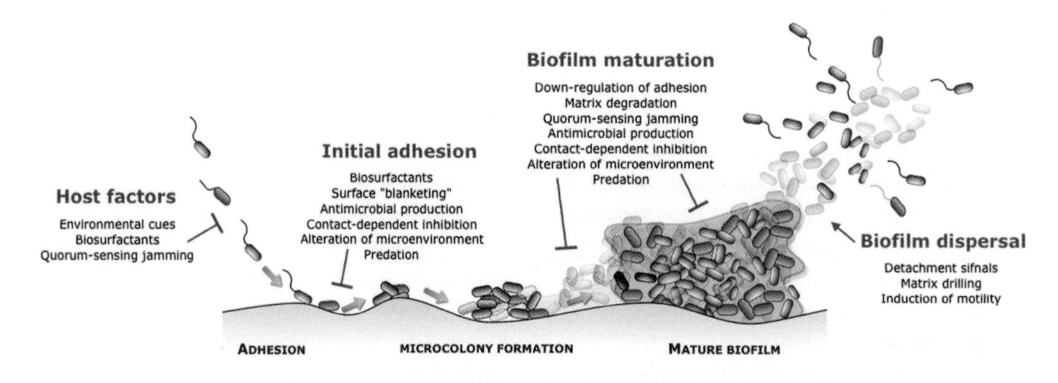

FIGURE 2 Mechanisms of competition within biofilms. Microbial interference can affect biofilm formation or dispersion through different mechanisms and strategies at different biofilm stages. These strategies include the secretion of growth or adhesion inhibitory molecules, jamming quorum sensing, altering biofilm regulation, and enhancing biofilm dispersal. (See text.) doi:10.1128/microbiolspec.MB-0009-2014.f2

Moraxella catarrhalis (70), whereas *Streptococcus sanguinis* and *Streptococcus gordonii* are able to inhibit *S. mutans* in dental plaque (71). However, this competitive advantage is strongly dependent on abiotic conditions. Levels of oxygen and glucose modulate the production of hydrogen peroxide by SpxB pyruvate oxidase in *S. sanguinis* and *S. gordonii* as well as the production of other growth-inhibiting factors, such as bacteriocins by *S. mutans* (71).

Small antimicrobial compounds

Colicins. A widespread mechanism of defense used by 50% of the *E. coli* biodiversity against closely related genotypes is the production of colicins and microcins (72). Colicins are typically encoded by small plasmids that enable horizontal transfer of colicin production. Colicinogenic bacteria are resistant to colicin biocidal action via a constitutively expressed immunity gene coded downstream of the colicin gene. For lethal activity, most colicins are released into the extracellular medium by cell lysis, presumably encoded by the *kil* gene carried in the plasmid. Colicins interact with target cells by direct binding to cell surface receptors and are subsequently internalized. Colicins display two distinct mechanisms of action: they can either form pores in the inner membrane (activity encoded in the C terminal of the colicin) or display an enzymatic activity that will degrade DNA or RNA, both of which will result in cell death (for a review on colicin biology, see reference 73). Both theoretical and experimental work showed that toxin producers always outcompete the nonproducer. In addition, colicin production also enables the invasion of pre-established microbial communities and resistance to other species invasions (74, 75).

Recently, a study described a new pore-forming colicin, colicin R, specifically produced within biofilms (65). Colicin expression is tightly regulated by the SOS response (76), which has been shown to be highly induced in mature biofilms (77). This stress response leads to the production of colicin R and subsequent local concentration of colicin within biofilms due to limited diffusion. The release of colicin R conferred a strong competitive advantage in mixed biofilms. Finally, colicin release by one strain within mixed communities induces the colicin production by a second strain and vice versa (78).

Other Gram-negative bacteria also produce small antimicrobial peptides similar to colicins, such as pesticins from *Yersinia pestis*, klebicins from *Klebsiella* spp., pyocins produced by *Pseudomonas* spp., or limicins of *Photorhabdus luminescens* (73).

Bacteriocins. A wide range of antimicrobial peptides are generically and misleadingly termed bacteriocins, but strictly speaking, bacteriocins are proteinaceous toxins produced by 3 to 23% of Gram-positive bacteria, notably, lactic acid bacteria (79). They are synthesized in the form of premature peptides with an N-terminal signal sequence, typically accompanied by an immunity protein to protect the producers. As observed for interference molecules, some bacteriocins can be produced in a biofilm-specific fashion, for instance, mutacin I of *S. mutans* (80) or bacitracin produced by *Bacillus licheniformis* (66). Bacitracin was only detected within biofilms, but addition of spent biofilm supernatant to planktonic cells of *B. licheniformis* triggered bacitracin production, suggesting the presence of some inducer in the biofilm supernatant (66).

Contact-dependent growth inhibition. Contact-dependent growth inhibition (CDI) is a mechanism of interbacterial competition originally described in uropathogenic *E. coli* strains (81) and now shown to exist in many species such as *Pseudomonas fluorescens* and *Bibersteinia trehalosi* (82). CDI functionality is encoded by three genes, *cdiBAI* (in *E. coli*), and mediated by a two-partner secretion system. The effector, CdiA, needs cell-to-cell contact to inhibit growth and be released into the cytoplasm of the targeted bacteria. Toxic activity, often due to a nuclease, is encoded in the terminal residues of the CdiA. CdIB aids the secretion across the outer membrane of CdiA, whereas CdiI encodes the immunity protein (81). One of the targets of CDI is the outer membrane protein, BamA. Further, *acrB* mutants, which fail to produce a protein that acts downstream of BamA, are also resistant to CDI (83). An analogous CDI system (BcpAIOB proteins) was recently reported in *Burkholderia thailandensis*. These proteins are involved in the biofilm formation process, suggesting a cooperative role. However, biofilm cocultures showed that BcpA also mediated intraspecific competition and was involved in community exclusion of nonkin (84).

Additionally, other CDI systems have been revealed by an experimental evolution setup using *E. coli* as a model. After serial passaging of independent *E. coli* populations, parallel mutations in the *glgC* gene emerged. This resulted in strong inhibition and subsequent domination of ancestral/competitive strains (85). This competition mechanism was induced during stationary phase (and termed SCDI, stationary contact-dependent inhibition), a physiological state that resembles biofilms, in which cells are not required to actively grow (86).

Another widespread contact-dependent system that recently received a lot of attention is the type 6 secretion system (T6SS), which is able to inject directly a toxic compound not only into eukaryotic cells, but also into other competitor bacteria (87, 88). Although it has been more studied for its role in virulence, T6SS fulfills several roles including bactericidal effect, kin recognition, and competitive growth (89). T6SS of *B. thailandensis* is required for colonization and persistence within biofilms. Moreover, in mixed flow cell biofilms, T6SS conferred an ecological advantage because it protected *B. thailandensis* from invasion by other competitor species, for example, *Pseudomonas putida* (90).

Predation. Several predation mechanisms have been described, including (i) phagocytosis by the unicellular amoeba *Dictyostellium discoideum*, (ii) cell invasion via penetration through the outer cell wall such as *Bdellovibrio bacteriovorus* mediated by type IV pili, (iii) predation by diffusible factors (secondary metabolites such as antibiotics) such as *Streptomyces* spp., and (iv) community-dependent predation displayed by the soil-dwelling bacterium *Myxococcus xanthus* (91). *M. xanthus*, often described as a "bacterial wolfpack" (91), hunts cooperatively by tightly coordinating community movement (rippling) and then releasing a myriad of secondary metabolites (92), antibiotics (93), polyketides, and degrading enzymes into the local environment. Despite these observations, the precise molecular mechanisms of predation remain to be elucidated. Myxobacterial ubiquity in soils world-wide and the ability to prey on a broad spectrum of microorganisms (bacteria and fungi) suggest that *M. xanthus* may have a great impact on population and evolutionary dynamics in the soil (94).

Interference mediated by alteration of biofilm development

The increase of fitness in biofilms relies on the ability of a given strain not only to adhere, settle, and develop as a biofilm, but also to inhibit others from doing so. Bacteria have evolved different strategies that prevent other bacteria to form or colonize existing biofilms. All steps of biofilm formation can be targeted: from inhibition of initial adhesion to matrix degradation to jamming of cell–cell communications, and induction of biofilm dispersion (58). Biofilm-inhibiting strategies have gained interest recently for their potential applications in the industrial and medical settings as an alternative to the use of antibiotics (for review see references 95–97) (Fig. 2).

Inhibition of cell-to-cell communication

Biofilm formation triggers quorum sensing (QS), a density- and dose-dependent communication system that coordinates gene expression at the community level (98, 99). Disruption of QS-regulated genes was shown to interfere with the ability to form a biofilm (100). Some bacteria have evolved mechanisms by which they degrade QS molecules; for instance, Gram-negative bacteria use acyl homoserine lactones that can be digested by several enzymes synthesized and released by competing bacteria (101–103). Spent bacterial supernatants containing phenolic groups and aliphatic amines inhibit biofilm formation of *P. aeruginosa* PAO1 (104), or production of AiiA, an AHL-lactonase, by *Bacillus cereus* inhibits *Vibrio cholerae* biofilm formation (105).

In mixed *P. aeruginosa* and *E. coli* biofilms, production of indole determines *E. coli* fitness, since mutant bacteria unable to synthesize indole are rapidly outcompeted by *P. aeruginosa*. It was shown that indole is able to block production of the pyocianin toxin and other QS-related phenotypes of *P. aeruginosa*, suggesting that indole affects QS signaling and alters population dynamics both in liquid cultures and in biofilms (106).

The oral environment provides several ecologically relevant examples of secreted degrading enzymes that inhibit communication signals of coexisting species. Colonization by *S. mutans*, the primary etiologic agent of human dental caries, relies on a competence-stimulating peptide (CSP), an essential QS molecule. However, competing streptococci, such as *S. gordonii*, *Streptococcus salivaris*, *S. sanguinis*, *Streptococcus mitis*, and *Streptococcus oralis*, are all capable of secreting degrading enzymes that inactivate CSP and inhibit *S. mutans* biofilm formation (107, 108). These CSP-degrading enzymes were shown to be relevant in mixed biofilms, because *S. salivarus* had a strong competitive advantage over *S. mutans*. Moreover, *S. mutans* resistance to bacteriocin is also regulated by CSP.

Although the prevalence of QS-inhibitory molecules has been documented in different marine and soil environments (109, 110), assessing the importance of QS-inhibitory strategies in nature remains difficult, because resistance can emerge readily (111, 112). Some authors suggest that bacteria could even escape from QS inhibition without undergoing genetic changes (113).

Inhibition of adhesion

The first stage of interaction between bacteria and surfaces (initial adhesion) are crucial for colonization and biofilm development. Mechanisms interfering with these first steps generally involve (i) secreting specific molecules that alter surface physicochemical properties of the surface itself, (ii) downregulation of key bacterial adhesion factors, or (iii) modulation of microbe-microbe interactions (Fig. 2).

Modification of adhesion by surface-active compounds. Surface-active compounds (SACs) produced by microorganisms or biosurfactants are amphipatic lipid-based molecules that lower interfacial tension (114, 115). SACs are diverse in their biochemical nature and include glycolipids (rhamnolipids, trehalolipids, sophorolipids), lipopeptides and lipoproteins, fatty acids and phospholipids, and other polymeric biosurfactants (115). Some of these surfactants display antimicrobial properties. However, in recent years, many SACs have also been described to inhibit biofilm formation without affecting growth (for review, see reference 116). These antibiofilm effects are likely due to alterations in the wettability and surface charges induced by SACs on treated surfaces (114). These modifications impact bacterial interactions and bacteria-surface interactions and weaken the bacteria's ability to adhere and form a biofilm. For instance, polysaccharides specifically produced within biofilms formed by *E. coli* natural isolates (Ec300 and Ec111) were shown to inhibit biofilm formation by Gram-positive bacteria (*Staphylococcus aureus*, *Streptococcus epidermidis*, *Enterococcus faecalis*) but not by Gram-negative bacteria (*E. coli*, *Enterobacter cloacae*, *P. aeruginosa*,

Klebsiella pneumoniae) (117). These polysaccharides were also shown to alter the properties of abiotic surfaces by lowering interfacial energy and increasing the hydrophilicity of treated glass surfaces, leading to inhibition of bacterial adhesion (117). The production of such bacterial surfactants protected *E. coli* biofilms from colonization by other biofilm-forming species (*S. aureus*), showing a competitive advantage for the production of otherwise costly polysaccharide (117). Other SAC polysaccharides, such as group 2 capsules produced in both planktonic and biofilm conditions by uropathogenic bacteria, affect abiotic surfaces by increasing their hydrophilicity and biotic surfaces by altering Lewis base properties (118, 119).

Downregulation of adhesion factors. Bacteria secrete different molecules into the extracellular medium including digestive enzymes and polysaccharides, both of which were shown to regulate gene expression of ecological competitors. For instance, released polysaccharides of *Lactobacillus acidophilus* are able to inhibit biofilm formation of a broad range of bacterial strains including enterohemorrhagic *E. coli* (EHEC), *Salmonella enteritidis, Salmonella enterica* serovar *Typhimurium, Yersinia enterocolitica, P. aeruginosa,* and *Listeria monocytogenes* and modify gene expression of *E. coli*. Exopolysaccharides downregulated genes involved in chemotaxis (*cheY*) and adhesion (curli-associated genes *crl, csgA,* and *csgB*) and resulted in a dramatic decrease of *E. coli* biofilms in mixed cultures with *L. acidophilus* (120).

Similarly, *Streptococcus intermedius* releases an arginine deaminase, which affects expression of different fimbriae of *Porphyromonas gingivalis,* a coexisting strain in the oral microbiota (121).

Matrix degradation

The extracellular matrix of a biofilm is an important element involved in cohesion and structure, resistance to both physical and chemical aggressions, and other functions (122). The biofilm matrix is composed of proteins, nucleic acids, polysaccharides, amyloid fibers, vesicles, and ions. Several compounds have been described as targeting the expression, assembly, or integrity of matrix components and can negatively impact biofilm fitness. Dissolution of the biofilm matrix results in reduced biomass, increased cell exposure to other growth-inhibiting molecules, bacterial dispersal, and ultimately, liberation of an ecological niche. Many broad-spectrum nucleases targeting DNA and RNA (123–125) as well as polysaccharide-degrading enzymes (126–128) have been described. More species-specific enzymes have also been studied; for instance, Esp secreted by *S. epidermidis* inhibits colonization and degrades matrix-associated proteins of *S. aureus* (129, 130). Similarly, *Streptococcus salivarius*, a commensal bacterium covering the oral epithelium, tongue, and throat produces an exo-beta-D-fructosidase or fructanase (FruA) shown to inhibit biofilm development of other oral coexisting bacteria, such as *S. mutans*, the primary etiologic agents of human dental caries (127).

Biofilm dispersal

Biofilm dispersal is a natural step in the biofilm lifecycle by which cells leave the biofilm and re-enter the planktonic lifestyle. Dispersal is triggered by different environmental cues such as nutrient and oxygen levels, physiological cues, and other cellular signals such as second messengers and QS (see reference 131 for a review). Dispersal often results in cell death, induction of bacterial motility systems, downregulation of adhesins, and secretion of matrix-degrading enzymes. Despite many published studies, dispersal remains one of the most complex and less understood steps of the biofilm cycle, which was mainly studied in single-species contexts. Hence, it is difficult to determine whether any of the identified dispersal molecules are primarily involved in biofilm self-control mechanisms or contribute to competitive strategy against other biofilm-forming bacteria.

Several diffusible signal factors have been described to trigger biofilm disassembly in different bacterial species. First described in the plant pathogen *Xanthomonas campestris*, small *cis* unsaturated fatty acids can induce biofilm dispersal of a broad range of bacteria and are also involved in cross-kingdom interactions (132). Nevertheless, these molecules also have a physiological impact on the strain producing them, and failure to produce diffusible signal factors can result in altered biofilm phenotype (133). Similarly, *P. aeruginosa* is also capable of producing other antibiofilm molecules such as cis-2-decenoic acid known to disperse *K. pneumoniae, E. coli, Bacillus subtilis*, and *S. aureus* (134) or rhamnolipids able to disperse *Bordetella bronchispetica* biofilms (135, 136).

Another well-studied dispersing signal is nitric oxide, which is released in the deeper layer of the biofilm, where oxygen is scarce. Nitric oxide upregulates a phosphodiesterase that degrades the second messenger c-di-GMP, resulting in low intracellular levels, which in turn trigger dispersal mechanisms of both monospecies and multispecies biofilms (137). Additionally, nitric oxide was reported to, among other things, downregulate the expression of adhesins in *P. aeruginosa* (138).

Recently, several other signals were reported to trigger biofilm disassembly in *B. subtilis*: D-amino acids and norspermidine (139, 140). These signals were shown to have biofilm-inhibitory activity against other competing strains (140, 141). However, whether D-amino acids and norspermidine are self-produced by *B. subtilis* and play a role in biologically relevant contexts is controversial (142, 143).

Motility-based interference

Bacterial motility *per se* or in combination with other mechanisms is a competitive strategy that is often neglected, and its impact on bacterial evolution has only recently received some attention (144, 145). In non-motile organisms, adaptation ultimately occurs through increases in the intrinsic average reproductive rate (growth rate), exploitation of novel resource niches (exploitative competition), or hindering direct competitors (interference competition) as shown in many experimental evolution studies. However, mobile organisms can additionally adapt by modifying the rate, pattern, or energetics of their movements, which could lead to a fitness increase. Moreover, motility allows microorganisms to escape from competition, or from an exhausted niche, and most remarkably, motility is instrumental in bacterial predation. Additionally, several elements implicated in motility are also instrumental in the early stages of biofilm formation, such as type IV pili (146).

Several mechanisms have been described to illustrate the importance of motility in competition within biofilms. First, An and colleagues described a strategy called "surface blanketing" (147). In this case, bacteria compete not for a limited resource but for access to an appropriate surface and subsequent colonization. In cocultures with *Agrobacterium tumefaciens*, *P. aeruginosa* displays an increased growth rate, which enables it to rapidly consume available resources and occupy the ecological niche. *P. aeruginosa* spreads through the surface via swarming and twitching motility, preventing *A. tumefaciens* adhesion. Consistently, a *P. aeruginosa flgK* motility-deficient mutant unable to spread quickly over a surface was no longer able to exclude *A. tumefaciens* (147). In addition to surface blanketing, in iron-depleted situations, *P. aeruginosa* is also able to secrete a compound(s) that inhibits and disperses *A. tumefaciens* independently of QS (148). Bacteria can also force out competitors by stimulating their motility. *E. coli* produces BdcA, a protein that binds cyclic diguanylate, c-di-GMP, a ubiquitous bacterial second messenger, regulating both dispersal and motility. Upon BdcA binding, intracellular levels of c-di-GMP are reduced and motility is stimulated, forcing bacteria out of the biofilm. In a multispecies biofilm composed of *E. coli, P. aeruginosa,* or *Rhizobium meliloti, E. coli* is able to transfer BdcA by conjugation, resulting in increased dispersal of the biofilm via enhanced motility (149).

Finally, motile bacteria can create holes or tunnels in the matrix of mature biofilms, increasing the susceptibility of those drilled biofilms to toxic molecules from the environment (150). Houry and colleagues screened motile bacteria for their ability to destabilize biofilms from other species, and in most cases, motile bacteria were able to penetrate and move through foreign biofilms (150). Such strategies, in combination with other competitive strategies, could lead to important ecological advantages. For instance, motile *Bacillus* spp. expressing antimicrobial compounds were able to tunnel through *S. aureus* biofilms and release the bioactive molecule directly into the inner core layers of the biofilm. This resulted in a complete clearing of the *S. aureus* biofilm and establishment of the *Bacillus* biofilm (150).

Resistance to colonization

One major fitness parameter in biofilm development is the ability to resist colonization of other genotypes that could invade the biofilm and eventually take over. Moreover, invasive genotypes can lower the relatedness of the interacting cooperators and threaten the maintenance of cooperation. Several of the previously discussed mechanisms can also be implicated in colonization resistance, for example, the aforementioned surfactant produced by *E. coli*, Ec300p, that not only inhibits initial adhesion of competing strains, but also provides a shield against invading bacteria (117). Similarly, T6SS in *B. thailandensis* protected the biofilm from competitors (90).

Due to its medical relevance, there has been increased interest in the ecology of host-associated microbiota and how colonization resistance plays a role in maintaining health and homeostasis. However, due to the complexity of multispecies communities, very few studies have shed light on the specific mechanisms. Recently, a transcriptomic study investigated how commensal populations of *E. coli* reacted when challenged with two pathogens, either invasive *E. coli* or *K. pneumoniae* (151). This approach revealed specific responses upon invasion and highlighted the role of several genes in limiting pathogen settling and spreading, including two genes, *yiaF* and *bssS*, implicated in *in vivo* resistance to colonization. However, the precise mechanisms by which these genes reduce colonization remain to be elucidated (151).

Similarly, in experimental oral biofilms, a microbial consortium of 10 species was able to resist colonization of *E. coli*. The consortium specifically sensed the presence of *E. coli* and responded by increasing H_2O_2 secretion that has a bactericidal activity and results in the killing of *E. coli* (152). In a later study, it was shown that three members of this consortium played very distinct and fundamental roles in this colonization resistance pathway. First, one species, named "the sensor," *Staphylococcus saprophyticus*, detected the presence of *E. coli* via cell-to-cell interactions, possibly through lipopolysaccharide detection. Then, "the mediator," *Streptococcus infantis*, responded to diffusible signals produced by the sensor *S. saprophyticus* and induced the production of H_2O_2 in a third bacterial species, "the killer," *Streptococcus sanguinis*. In this bacterial cascade, and in the absence of *E. coli, S. infantis* repressed H_2O_2 secretion of *S. sanguinis* (153). This example supports the competition-sensing hypothesis (57) and provides more evidence that some competitive outcomes observed in multispecies biofilms cannot be predicted by observing pair-wise competitions.

EVOLUTION OF COMPETITIVE INTERACTIONS

> Most of them [species] are doomed to rapid extinction, but a few may make evolutionary inventions, such as physiological, ecological, or behavioral innovations, that give these species improved competitive potential.
>
> Ernst Mayr, "Speciational Evolution or Punctuated Equilibria" (154)

Microorganisms are instrumental in the study of general evolutionary theory. Understanding the evolution of competitive interactions, the selective forces acting on multispecies communities, and the nature of adaptation mechanisms provides the opportunity to analyze (and predict) evolutionary fate and the distribution of possible outcomes. This is especially relevant for biofilms, which not only are a prevalent lifestyle in nature, but also play a central role in major natural processes as well as in human health and homeostasis.

Theoretical Modeling of Species' Competitive Interactions

Modeling of single-species communities was commonly used as a tool to gain insight into the evolution of bacterial interactions, namely, cooperation and competition (155–157). Few studies have directly taken into account the complex dynamics and heterogeneity of multispecies communities (158–160). Moreover, most models specifically addressing species interactions within biofilms remain to be experimentally tested (161, 162).

Xavier and Foster created a computational model to analyze competitive outcomes between two genotypes from a single species with various levels of exopolysaccharide (EPS) production within a biofilm. Counterintuitively, the model predicted that, although EPS producers have lower individual fitness due to associated costs of polymer production, they would prevail in the biofilm over the nonproducers by suffocating the latter cells in the bottom of the biofilm, where environmental conditions are harsher. EPS-producing strains were predicted to ascend through the biofilm and gain access to more oxygen-rich niches (163). Recent experimental work confirmed the model's predictions. Work carried out with *V. cholerae* showed that EPS-producing bacteria benefit from their clone-mates and gain competitive advantage against other nonproducers (164). However, the advantage in the biofilm environment came at a cost since, upon biofilm dispersal, the EPS producers were hindered in their dispersal capacity and subsequent colonization of new niches compared to nonproducers (164).

Another model predicts that when nutrient concentrations are low, cooperative bacteria secreting public goods, in this case a growth-promoting compound, would cluster together. This would result in the segregation of nonproducers, and because the benefits are preferentially directed toward other producers, the latter will have higher fitness than the nonproducers (32). By contrast, in nutrient-rich conditions, the different genotypes (producers and nonproducers) would more readily mix (32). However, this competitive outcome between genotypes of the same species can change when the complexity of the biofilm is increased (i.e., chimerism). In another study, Mitri and colleagues analyzed how the addition of a totally different species affects competition between producer and nonproducer genotypes from the same bacterial species (40). They showed that in low-nutrient environments, the new bacterial species could undermine the benefits of cooperation between the producer and nonproducer and eventually take over the biofilm (40). Moreover, competition imposed by the incoming bacterial species is more detrimental to the producers than it is to the nonproducers. This was shown in a study in which *S. aureus* was added to a

P. aeruginosa culture in which some cooperator cells secreted iron siderophores and others did not. *S. aureus* increased nonproducers' fitness over that of the producers (39).

Experimental Evolution in Biofilms

Experimental evolution is a powerful tool to decipher mechanisms of adaptation, including gain in competitiveness. This strategy is based on observing the effect of time on a given community. It is generally carried out by serial passaging of replicate populations for long periods of time under controlled conditions (165). Nevertheless, emergence of competitive traits might happen very fast (44). Experimental evolution approaches have been mostly used with single-species batch cultures. For example, the long-term experimental evolution set-up with *E. coli* has shown that key innovations result in the novel use of another carbon source (citrate) (166) and that fitness can increase for very long periods of time in an unchanged environment (167), among other important evolutionary findings. Recently, Poltak and Cooper described an experimental model that enables experimental evolution in biofilms (19). This method is based on the use of beads to which bacteria can adhere. Each day, a colonized bead is transferred into fresh media and a new bead is colonized, allowing cells to adhere, form mature biofilms, and disperse. This approach showed that self-diversified phenotypes became synergistic and significantly increased group productivity (19). Such new techniques should increase studies that experimentally address evolution in biofilm population dynamics and long-term consequences of biofilm selection.

Early work by Rainey and Travisano showed that when there is ecological opportunity in microcosms, exploitative competition arises readily and generates diversity in a population (44). After three days, the initial population of *P. fluorescens*, a generalist genotype, diverged into three stable specialist morphotypes that occupy different niches. This radiation was driven by competition for oxygen and achieved by several mutations. One morphotype evolved by mutations in the cellulose-like operon, resulting in the overexpression of the polymer and the formation of a self-sustainable biofilm in the air-liquid interface (168). The repeatability of the evolutionary outcome suggests very strong pressure for diversification. Such diversity did not evolve in homogeneous batch cultures with reduced ecological opportunities (44).

The study of more complex communities also revealed the rapid emergence of competitive interactions between different species (169), not only within species, as a result of self-diversification (44). In a dual-species evolution experiment, it was shown that spatial structure led to the emergence of exploitative interactions by a single mutation altering lipopolysaccharide biosynthesis of one of the interactants (169). More precisely, *P. putida* was initially grown with an *Acinetobacter* sp. in an environment in which *P. putida* was dependent on the *Acinetobacter* sp. After several days of coevolution, the dynamics and biofilm morphology changed; *P. putida* grew closer to and eventually overgrew *Acinetobacter* (169). The *P. putida* population developed a rough morphotype in response to the presence of *Acinetobacter* and as a result of developing within a biofilm, probably due to enhanced oxygen competition.

More recently, a study using *Burkholderia cenocepacia* as a model organism showed that there are several possible routes to adapt via interference competition within a population. This study showed that increased fitness by bacteria compared to their ancestor was achieved either by enhanced iron storage, enhanced metabolic efficiency, and/or alteration of polysaccharide and lipopolysaccharide structures (170). Moreover, this study highlighted the role of competition in maintaining genetic diversity within complex communities (170), in contrast to results obtained in batch cultures or more homogeneous environments in which competitive mutants quickly take over the population (170). This study also showed that for a given genotype to sweep the population, more time and more mutations are required, probably due to high cell density and absolute population numbers (171).

Finally, a recent study showed that the complexity of a community and the number of interacting partners strongly influences evolution in ways that were unpredicted by single-species experiments (7). More specifically, interspecies competition led to the use of alternative resources and shifted the population from generalists to specialists. This resulted in increased productivity of the whole community due to the complementarity in the use of resources. Interestingly, several species were able to feed on waste products of other members of the community. This and other examples suggest that some evolutionary pathways are only available when within multispecies communities and not in simpler settings (172, 173).

CONCLUDING REMARKS

The studies and experimental data presented in this article were carried out under lab-controlled conditions, so the question remains as to whether these processes are meaningful in the natural world. Indeed, given the inherent complexity of natural ecosystems, competition could be harsher and selective pressures stronger and applied through longer periods of time than the ones created or used in laboratory conditions.

Besides the ecological and evolutionary importance of understanding microbial competition, knowledge of corresponding underlying mechanisms could be particularly useful in dealing with specific applied problems, such as the ubiquity of antibiotic resistance in clinical settings (58, 174). Interference interactions have already inspired the design of alternatives to antibiotics in the war against pathogenic microorganisms (97). Additionally, well-defined antagonisms could be exploited to adequately use probiotics to prevent infection (175) or to restore and preserve the equilibrium of the flora to efficiently resist pathogen colonization (176).

Several studies highlighted that complex communities interact differently depending on their degree of chimerism and the diversity of their genotypes. Moreover, they have shown that several adaptive pathways and outcomes are only available with high levels of complexity and that they cannot be predicted by observing a simplification of the ecosystem in either single-species or dual-species studies (7, 153, 172). Recently, a new integrative approach, community systems (CoSy) biology, was proposed to further explore how social interactions in multispecies communities evolve (3, 177), and different alternative strategies based on multispecies cultures are emerging (178, 179). New multidisciplinary approaches combining population genetics, evolutionary and molecular biology, and bioinformatics are needed to decipher the genetic networks underlying community interactions and place them in their ecological and evolutionary context. Additionally, the spatial scale at which interactions occur is also an important parameter to take into account. Hence, the integration of physical and mathematical models of biological processes will advance our ability to predict evolutionary outcomes of bacterial competition in heterogeneous environments. Finally, genotypic diversity within naturally occurring biofilms should inspire future experimental work to progressively move away from simplified microbial systems toward the study of complex populations including more genotypes. This will lead to a broader understanding of bacterial biology and could expose new mechanisms of competition within multispecies communities, with potential impact on applied systems biology and control of colonization resistance in medically or industry-relevant biofilms.

ACKNOWLEDGMENTS

We are grateful to Elze Rackaityte and Christophe Beloin for critical reading of the manuscript. O.R. is the recipient of an EMBO long-term fellowship. J.-M.G. acknowledges support from the French government's Investissement d'Avenir Program, Laboratoire d'Excellence "Integrative Biology of Emerging Infectious Diseases" (grant ANR-10-LABX-62-IBEID), and the Fondation pour la Recherche Médicale grant "Equipe FRM DEQ20140329508."

Conflicts of interest: We disclose no conflicts.

CITATION

Rendueles O, Ghigo J-M. 2015. Mechanisms of competition in biofilm communities. Microbiol Spectrum 3(3):MB-0009-2014.

REFERENCES

1. **Darwin C.** 1859. *On the Origin of Species by Means of Natural Selection, or the Preservation of Favored Races in the Struggle for Life*. John Murray, London.
2. **Davey ME, O'Toole GA.** 2000. Microbial biofilms: from ecology to molecular genetics. *Microbiol Mol Biol Rev* **64:**847–867.
3. **Zengler K, Palsson BO.** 2012. A road map for the development of community systems (CoSy) biology. *Nat Rev Microbiol* **10:**366–372.
4. **Burmølle M, Ren D, Bjarnsholt T, Sørensen SJ.** 2014. Interactions in multispecies biofilms: do they actually matter? *Trends Microbiol* **22:**84–91.
5. **Elias S, Banin E.** 2012. Multi-species biofilms: living with friendly neighbors. *FEMS Microbiol Rev* **36:**990–1004.
6. **Foster KR, Bell T.** 2012. Competition, not cooperation, dominates interactions among culturable microbial species. *Curr Biol* **22:** 1845–1850.
7. **Lawrence D, Fiegna F, Behrends V, Bundy JG, Phillimore AB, Bell T, Barraclough TG.** 2012. Species interactions alter evolutionary responses to a novel environment. *PLoS Biol* **10:**e1001330. doi:10.1371/journal.pbio.1001330.
8. **Allee WC, Bowen E.** 1932. Studies in animal aggregations: mass protection against colloidal silver among goldfishes. *J Exp Zool* **61:**185–207.
9. **Ruxton GD, Sherratt TN.** 2006. Aggregation, defence and warning signals: the evolutionary relationship. *Proc Biol Sci* **273:**2417–2424.
10. **Gascoigne J, Berec L, Gregory S, Courchamp F.** 2009. Dangerously few liaisons: a review of mate-finding Allee effects. *Popul Ecol* **51:**355–372.
11. **Potts M.** 1994. Desiccation tolerance of prokaryotes. *Microbiol Rev* **58:**755–805.
12. **Li YH, Hanna MN, Svensater G, Ellen RP, Cvitkovitch DG.** 2001. Cell density modulates acid adaptation in *Streptococcus mutans*: implications for survival in biofilms. *J Bacteriol* **183:**6875–6884.
13. **Smith R, Tan CM, Srimani JK, Pai A, Riccione KA, Song H, You LC.** 2014. Programmed Allee effect in bacteria causes a tradeoff between population spread and survival. *Proc Natl Acad Sci USA* **111:**1969–1974.
14. **Butler MT, Wang Q, Harshey RM.** 2010. Cell density and mobility protect swarming bacteria against antibiotics. *Proc Natl Acad Sci USA* **107:**3776–3781.
15. **Julou T, Mora T, Guillon L, Croquette V, Schalk IJ, Bensimon D, Desprat N.** 2013. Cell-cell contacts confine public goods diffusion inside *Pseudomonas aeruginosa* clonal microcolonies. *Proc Natl Acad Sci USA* **110:** 12577–12582.
16. **Darch SE, West SA, Winzer K, Diggle SP.** 2012. Density-dependent fitness benefits in quorum-sensing bacterial populations. *Proc Natl Acad Sci USA* **109:**8259–8263.
17. **Taylor PD.** 1992. Altruism in viscous populations: an inclusive fitness model. *Evol Ecol* **6:** 352–356.
18. **Whittaker CJ, Klier CM, Kolenbrander PE.** 1996. Mechanisms of adhesion by oral bacteria. *Annu Rev Microbiol* **50:**513–552.
19. **Poltak SR, Cooper VS.** 2011. Ecological succession in long-term experimentally evolved biofilms produces synergistic communities. *ISME J* **5:**369–378.
20. **Ramsey MM, Rumbaugh KP, Whiteley M.** 2011. Metabolite cross-feeding enhances virulence in a model polymicrobial infection. *PLoS Pathog* **7:**e1002012. doi:10.1371/journal.ppat.1002012.
21. **Breugelmans P, Barken KB, Tolker-Nielsen T, Hofkens J, Dejonghe W, Springael D.** 2008. Architecture and spatial organization in a triple-species bacterial biofilm synergistically degrading the phenylurea herbicide linuron. *FEMS Microbiol Ecol* **64:**271–282.
22. **Burmolle M, Webb JS, Rao D, Hansen LH, Sorensen SJ, Kjelleberg S.** 2006. Enhanced biofilm formation and increased resistance to antimicrobial agents and bacterial invasion are

caused by synergistic interactions in multispecies biofilms. *Appl Environ Microbiol* **72:**3916–3923.

23. **Whiteley M, Ott JR, Weaver EA, McLean RJ.** 2001. Effects of community composition and growth rate on aquifer biofilm bacteria and their susceptibility to betadine disinfection. *Environ Microbiol* **3:**43–52.
24. **Sutherland I.** 2001. Biofilm exopolysaccharides: a strong and sticky framework. *Microbiology* **147:**3–9.
25. **Hamilton WD.** 1964. The genetical evolution of social behaviour. I & II. *J Theor Biol* **7:**1–52.
26. **Foster K.** 2005. Biomedicine. Hamiltonian medicine: why the social lives of pathogens matter. *Science* **308:**1269–1270.
27. **Nowak MA, May RM.** 1992. Evolutionary games and spatial chaos. *Nature* **359:**826–829.
28. **Doebeli M, Knowlton N.** 1998. The evolution of interspecific mutualisms. *Proc Natl Acad Sci USA* **95:**8676–8680.
29. **Travisano M, Velicer GJ.** 2004. Strategies of microbial cheater control. *Trends Microbiol* **12:**72–78.
30. **Manhes P, Velicer GJ.** 2011. Experimental evolution of selfish policing in social bacteria. *Proc Natl Acad Sci USA* **108:**8357–8362.
31. **Strassmann JE, Gilbert OM, Queller DC.** 2011. Kin discrimination and cooperation in microbes. *Annu Rev Microbiol* **65:**349–367.
32. **Nadell CD, Foster KR, Xavier JB.** 2010. Emergence of spatial structure in cell groups and the evolution of cooperation. *PLoS Comput Biol* **6:**e1000716. doi:10.1371/journal.pcbi.1000716.
33. **Drescher K, Nadell CD, Stone HA, Wingreen NS, Bassler BL.** 2014. Solutions to the public goods dilemma in bacterial biofilms. *Curr Biol* **24:**50–55.
34. **Griffin AS, West SA, Buckling A.** 2004. Cooperation and competition in pathogenic bacteria. *Nature* **430:**1024–1027.
35. **West SA, Pen I, Griffin AS.** 2002. Conflict and cooperation: cooperation and competition between relatives. *Science* **296:**72–75.
36. **Leigh EG.** 2010. The evolution of mutualism. *J Evol Biol* **23:**2507–2528.
37. **Platt TG, Bever JD.** 2009. Kin competition and the evolution of cooperation. *Trends Ecol Evol* **24:**370–377.
38. **Inglis RF, Brown SP, Buckling A.** 2012. Spite versus cheats: competition among social strategies shapes virulence in *Pseudomonas aeruginosa. Evolution* **66:**3472–3484.
39. **Harrison F, Paul J, Massey RC, Buckling A.** 2008. Interspecific competition and siderophore-mediated cooperation in *Pseudomonas aeruginosa. ISME J* **2:**49–55.
40. **Mitri S, Xavier JB, Foster KR.** 2011. Social evolution in multispecies biofilms. *Proc Natl Acad Sci USA* **108**(Suppl 2):10839–10846.
41. **Stewart PS.** 2003. Diffusion in biofilms. *J Bacteriol* **185:**1485–1491.
42. **Korona R, Nakatsu CH, Forney LJ, Lenski RE.** 1994. Evidence for multiple adaptive peaks from populations of bacteria evolving in a structured habitat. *Proc Natl Acad Sci USA* **91:**9037–9041.
43. **Boles BR, Thoendel M, Singh PK.** 2004. Self-generated diversity produces "insurance effects" in biofilm communities. *Proc Natl Acad Sci USA* **101:**16630–16635.
44. **Rainey PB, Travisano M.** Adaptive radiation in a heterogeneous environment. *Nature* **394:**69–72.
45. **Kerr B, Riley MA, Feldman MW, Bohannan BJ.** 2002. Local dispersal promotes biodiversity in a real-life game of rock-paper-scissors. *Nature* **418:**171–174.
46. **Li K, Bihan M, Yooseph S, Methe BA.** 2012. Analyses of the microbial diversity across the human microbiome. *PLoS One* **7:**e32118. doi:10.1371/journal.pone.0032118.
47. **Beloin C, Valle J, Latour-Lambert P, Faure P, Kzreminski M, Balestrino D, Haagensen JA, Molin S, Prensier G, Arbeille B, Ghigo JM.** 2004. Global impact of mature biofilm lifestyle on *Escherichia coli* K-12 gene expression. *Mol Microbiol* **51:**659–674.
48. **Whiteley M, Bangera MG, Bumgarner RE, Parsek MR, Teitzel GM, Lory S, Greenberg EP.** 2001. Gene expression in *Pseudomonas aeruginosa* biofilms. *Nature* **413:**860–864.
49. **Lazazzera BA.** 2005. Lessons from DNA microarray analysis: the gene expression profile of biofilms. *Curr Opin Microbiol* **8:**222–227.
50. **Schembri MA, Kjaergaard K, Klemm P.** 2003. Global gene expression in *Escherichia coli* biofilms. *Mol Microbiol* **48:**253–267.
51. **Marks LR, Davidson BA, Knight PR, Hakansson AP.** 2013. Interkingdom signaling induces *Streptococcus pneumoniae* biofilm dispersion and transition from asymptomatic colonization to disease. *MBio* **4:**e00438-13. doi:10.1128/mBio.00438-13.
52. **Campisano A, Overhage J, Rehm BH.** 2008. The polyhydroxyalkanoate biosynthesis genes are differentially regulated in planktonic- and biofilm-grown *Pseudomonas aeruginosa. J Biotechnol* **133:**442–452.
53. **Lequette Y, Greenberg EP.** 2005. Timing and localization of rhamnolipid synthesis gene expression in *Pseudomonas aeruginosa* biofilms. *J Bacteriol* **187:**37–44.

54. **Ghigo J-M.** 2003. Are there biofilm-specific physiological pathways beyond a reasonable doubt? *Res Microbiol* **154:**1–8.
55. **Case TJ, Gilpin ME.** 1974. Interference competition and niche theory. *Proc Natl Acad Sci USA* **71:**3073–3077.
56. **Vance RR.** 1984. Interference competition and the coexistence of two competitors on a single limiting resource. *Ecology* **65:**1349–1357.
57. **Cornforth DM, Foster KR.** 2013. Competition sensing: the social side of bacterial stress responses. *Nat Rev Microbiol* **11:**285–293.
58. **Rendueles O, Ghigo JM.** 2012. Multi-species biofilms: how to avoid unfriendly neighbors. *FEMS Microbiol Rev* **36:**972–989.
59. **Amarasekare P.** 2002. Interference competition and species coexistence. *Proc Biol Sci* **269:** 2541–2550.
60. **Yamamoto K, Haruta S, Kato S, Ishii M, Igarashi Y.** 2010. Determinative factors of competitive advantage between aerobic bacteria for niches at the air-liquid interface. *Microbes Environ* **25:**317–320.
61. **Bradshawa DJ, Marsha PD, Hodgson RJ, Visser JM.** 2002. Effects of glucose and fluoride on competition and metabolism within *in vitro* dental bacterial communities and biofilms. *Caries Res* **36:**81–86.
62. **Oehmen A, Lemos PC, Carvalho G, Yuan Z, Keller J, Blackall LL, Reis MA.** 2007. Advances in enhanced biological phosphorus removal: from micro to macro scale. *Water Res* **41:**2271–2300.
63. **Weaver VB, Kolter R.** 2004. *Burkholderia* spp. alter *Pseudomonas aeruginosa* physiology through iron sequestration. *J Bacteriol* **186:**2376–2384.
64. **Eberl HJ, Collinson S.** 2009. A modeling and simulation study of siderophore mediated antagonism in dual-species biofilms. *Theor Biol Med Model* **6:**30.
65. **Rendueles O, Beloin C, Latour-Lambert P, Ghigo JM.** 2014. A new biofilm-associated colicin with increased efficiency against biofilm bacteria. *ISME J* **8:**1275–1288.
66. **Yan L, Boyd KG, Adams DR, Burgess JG.** 2003. Biofilm-specific cross-species induction of antimicrobial compounds in bacilli. *Appl Environ Microbiol* **69:**3719–3727.
67. **Valle J, Da Re S, Schmid S, Skurnik D, D'Ari R, Ghigo JM.** 2008. The amino acid valine is secreted in continuous-flow bacterial biofilms. *J Bacteriol* **190:**264–274.
68. **Graver MA, Wade JJ.** 2011. The role of acidification in the inhibition of *Neisseria gonorrhoeae* by vaginal lactobacilli during anaerobic growth. *Ann Clin Microbiol Antimicrob* **10:**8.
69. **Létoffé S, Audrain B, Bernier SP, Delepierre M, Ghigo JM.** 2014. Aerial exposure to the bacterial volatile compound trimethylamine modifies antibiotic resistance of physically separated bacteria by raising culture medium pH. *MBio* **5:**e00944-13. doi:10.1128/mBio.00944-13.
70. **Pericone CD, Overweg K, Hermans PW, Weiser JN.** 2000. Inhibitory and bactericidal effects of hydrogen peroxide production by *Streptococcus pneumoniae* on other inhabitants of the upper respiratory tract. *Infect Immun* **68:**3990–3997.
71. **Kreth J, Zhang Y, Herzberg MC.** 2008. Streptococcal antagonism in oral biofilms: *Streptococcus sanguinis* and *Streptococcus gordonii* interference with *Streptococcus mutans*. *J Bacteriol* **190:**4632–4640.
72. **Gillor O, Kirkup BC, Riley MA.** 2004. Colicins and microcins: the next generation antimicrobials. *Adv Appl Microbiol* **54:**129–146.
73. **Cascales E, Buchanan SK, Duche D, Kleanthous C, Lloubes R, Postle K, Riley M, Slatin S, Cavard D.** 2007. Colicin biology. *Microbiol Mol Biol Rev* **71:**158–229.
74. **Gordon DM, Riley MA.** 1999. A theoretical and empirical investigation of the invasion dynamics of colicinogeny. *Microbiology* **145:**655–661.
75. **Riley MA, Gordon DM.** 1999. The ecological role of bacteriocins in bacterial competition. *Trends Microbiol* **7:**129–133.
76. **Gillor O, Vriezen JA, Riley MA.** 2008. The role of SOS boxes in enteric bacteriocin regulation. *Microbiology* **154:**1783–1792.
77. **Bernier SP, Lebeaux D, DeFrancesco AS, Valomon A, Soubigou G, Coppee JY, Ghigo JM, Beloin C.** 2013. Starvation, together with the SOS response, mediates high biofilm-specific tolerance to the fluoroquinolone ofloxacin. *PLoS Genet* **9:**e1003144. doi:10.1371/journal.pgen.1003144.
78. **Majeed H, Gillor O, Kerr B, Riley MA.** 2011. Competitive interactions in *Escherichia coli* populations: the role of bacteriocins. *ISME J* **5:**71–81.
79. **Gillor O, Etzion A, Riley MA.** 2008. The dual role of bacteriocins as anti- and probiotics. *Appl Microbiol Biotechnol* **81:**591–606.
80. **Qi F, Chen P, Caufield PW.** 2000. Purification and biochemical characterization of mutacin I from the group I strain of *Streptococcus mutans*, CH43, and genetic analysis of mutacin I biosynthesis genes. *Appl Environ Microbiol* **66:**3221–3229.
81. **Aoki SK, Pamma R, Hernday AD, Bickham JE, Braaten BA, Low DA.** 2005. Contact-dependent inhibition of growth in *Escherichia coli*. *Science* **309:**1245–1248.

82. **Aoki SK, Diner EJ, de Roodenbeke CT, Burgess BR, Poole SJ, Braaten BA, Jones AM, Webb JS, Hayes CS, Cotter PA, Low DA.** 2010. A widespread family of polymorphic contact-dependent toxin delivery systems in bacteria. *Nature* **468:**439–442.
83. **Aoki SK, Malinverni JC, Jacoby K, Thomas B, Pamma R, Trinh BN, Remers S, Webb J, Braaten BA, Silhavy TJ, Low DA.** 2008. Contact-dependent growth inhibition requires the essential outer membrane protein BamA (YaeT) as the receptor and the inner membrane transport protein AcrB. *Mol Microbiol* **70:**323–340.
84. **Anderson MS, Garcia EC, Cotter PA.** 2014. Kind discrimination and competitive exclusion mediated by contact-dependent growth inhibition systems shape biofilm community structure. *PLoS Pathog* **10:**e1004076. doi:10.1371/journal.ppat.1004076.
85. **Lemonnier M, Levin BR, Romeo T, Garner K, Baquero MR, Mercante J, Lemichez E, Baquero F, Blazquez J.** 2008. The evolution of contact-dependent inhibition in non-growing populations of *Escherichia coli*. *Proc Biol Sci* **275:**3–10.
86. **Waite RD, Papakonstantinopoulou A, Littler E, Curtis MA.** 2005. Transcriptome analysis of *Pseudomonas aeruginosa* growth: comparison of gene expression in planktonic cultures and developing and mature biofilms. *J Bacteriol* **187:**6571–6576.
87. **Kapitein N, Mogk A.** 2013. Deadly syringes: type VI secretion system activities in pathogenicity and interbacterial competition. *Curr Opin Microbiol* **16:**52–58.
88. **Pukatzki S, Ma AT, Sturtevant D, Krastins B, Sarracino D, Nelson WC, Heidelberg JF, Mekalanos JJ.** 2006. Identification of a conserved bacterial protein secretion system in *Vibrio cholerae* using the *Dictyostelium* host model system. *Proc Natl Acad Sci USA* **103:**1528–1533.
89. **Ho BT, Dong TG, Mekalanos JJ.** 2014. A view to a kill: the bacterial type VI secretion system. *Cell Host Microbe* **15:**9–21.
90. **Schwarz S, West TE, Boyer F, Chiang WC, Carl MA, Hood RD, Rohmer L, Tolker-Nielsen T, Skerrett SJ, Mougous JD.** 2010. *Burkholderia* type VI secretion systems have distinct roles in eukaryotic and bacterial cell interactions. *PLoS Pathog* **6:**e1001068. doi:10.1371/journal.ppat.1001068.
91. **Berleman JE, Kirby JR.** 2009. Deciphering the hunting strategy of a bacterial wolfpack. *FEMS Microbiol Rev* **33:**942–957.
92. **Krug D, Zurek G, Revermann O, Vos M, Velicer GJ, Muller R.** 2008. Discovering the hidden secondary metabolome of *Myxococcus xanthus*: a study of intraspecific diversity. *Appl Environ Microbiol* **74:**3058–3068.
93. **Xiao Y, Wei X, Ebright R, Wall D.** 2011. Antibiotic production by myxobacteria plays a role in predation. *J Bacteriol* **193:**4626–4633.
94. **Morgan AD, MacLean RC, Hillesland KL, Velicer GJ.** 2010. Comparative analysis of myxococcus predation on soil bacteria. *Appl Environ Microbiol* **76:**6920–6927.
95. **Cusumano CK, Hultgren SJ.** 2009. Bacterial adhesion: a source of alternate antibiotic targets. *IDrugs* **12:**699–705.
96. **LaSarre B, Federle MJ.** 2013. Exploiting quorum sensing to confuse bacterial pathogens. *Microbiol Mol Biol Rev* **77:**73–111.
97. **Rasko DA, Sperandio V.** 2010. Anti-virulence strategies to combat bacteria-mediated disease. *Nat Rev Drug Discov* **9:**117–128.
98. **Bassler BL, Losick R.** 2006. Bacterially speaking. *Cell* **125:**237–246.
99. **Fuqua C, Greenberg EP.** 2002. Listening in on bacteria: acyl-homoserine lactone signalling. *Nat Rev Mol Cell Biol* **3:**685–695.
100. **Davies DG, Parsek MR, Pearson JP, Iglewski BH, Costerton JW, Greenberg EP.** 1998. The involvement of cell-to-cell signals in the development of a bacterial biofilm. *Science* **280:**295–298.
101. **Bijtenhoorn P, Schipper C, Hornung C, Quitschau M, Grond S, Weiland N, Streit W.** 2011. BpiB05, a novel metagenome-derived hydrolase acting on N-acylhomoserine lactones. *J Biotechnol* **155:**86–94.
102. **Dong YH, Wang LY, Zhang LH.** 2007. Quorum-quenching microbial infections: mechanisms and implications. *Philos Trans R Soc Lond B Biol Sci* **362:**1201–1211.
103. **Shepherd RW, Lindow SE.** 2009. Two dissimilar N-acyl-homoserine lactone acylases of *Pseudomonas syringae* influence colony and biofilm morphology. *Appl Environ Microbiol* **75:**45–53.
104. **Musthafa KS, Saroja V, Pandian SK, Ravi AV.** 2011. Antipathogenic potential of marine *Bacillus* sp. SS4 on N-acyl-homoserine-lactone-mediated virulence factors production in *Pseudomonas aeruginosa* (PAO1). *J Biosci* **36:**55–67.
105. **Augustine N, Kumar P, Thomas S.** 2010. Inhibition of *Vibrio cholerae* biofilm by AiiA enzyme produced from *Bacillus* spp. *Arch Microbiol* **192:**1019–1022.
106. **Chu W, Zere TR, Weber MM, Wood TK, Whiteley M, Hidalgo-Romano B, Valenzuela E Jr, McLean RJ.** 2012. Indole production promotes *Escherichia coli* mixed-culture growth with *Pseudomonas aeruginosa* by

inhibiting quorum signaling. *Appl Environ Microbiol* **78:**411–419.

107. **Senadheera D, Cvitkovitch DG.** 2008. Quorum sensing and biofilm formation by *Streptococcus mutans. Adv Exp Med Biol* **631:**178–188.
108. **Tamura S, Yonezawa H, Motegi M, Nakao R, Yoneda S, Watanabe H, Yamazaki T, Senpuku H.** 2009. Inhibiting effects of *Streptococcus salivarius* on competence-stimulating peptide-dependent biofilm formation by *Streptococcus mutans. Oral Microbiol Immunol* **24:**152–161.
109. **Golberg K, Pavlov V, Marks RS, Kushmaro A.** 2013. Coral-associated bacteria, quorum sensing disrupters, and the regulation of biofouling. *Biofouling* **29:**669–682.
110. **Wang YJ, Leadbetter JR.** 2005. Rapid acylhomoserine lactone quorum signal biodegradation in diverse soils. *Appl Environ Microbiol* **71:**1291–1299.
111. **Garcia-Contreras R, Maeda T, Wood TK.** 2013. Resistance to quorum-quenching compounds. *Appl Environ Microbiol* **79:**6840–6846.
112. **Maeda T, Garcia-Contreras R, Pu M, Sheng L, Garcia LR, Tomas M, Wood TK.** 2012. Quorum quenching quandary: resistance to antivirulence compounds. *ISME J* **6:**493–501.
113. **Kalia VC, Wood TK, Kumar P.** 2013. Evolution of resistance to quorum-sensing inhibitors. *Microb Ecol.* [Epub ahead of print.] doi:10.1007/s00248-013-0316-y.
114. **Banat IM, Franzetti A, Gandolfi I, Bestetti G, Martinotti MG, Fracchia L, Smyth TJ, Marchant R.** 2010. Microbial biosurfactants production, applications and future potential. *Appl Microbiol Biotechnol* **87:**427–444.
115. **Desai JD, Banat IM.** 1997. Microbial production of surfactants and their commercial potential. *Microbiol Mol Biol Rev* **61:**47–64.
116. **Rendueles O, Kaplan JB, Ghigo JM.** 2013. Antibiofilm polysaccharides. *Environ Microbiol* **15:**334–346.
117. **Rendueles O, Travier L, Latour-Lambert P, Fontaine T, Magnus J, Denamur E, Ghigo JM.** 2011. Screening of *Escherichia coli* species biodiversity reveals new biofilm-associated antiadhesion polysaccharides. *MBio* **2:**e00043-11. doi:10.1128/mBio.00043-11.
118. **Travier L, Rendueles O, Ferrieres L, Herry JM, Ghigo JM.** 2013. *Escherichia coli* resistance to nonbiocidal antibiofilm polysaccharides is rare and mediated by multiple mutations leading to surface physicochemical modifications. *Antimicrob Agents Chemother* **57:**3960–3968.
119. **Valle J, Da Re S, Henry N, Fontaine T, Balestrino D, Latour-Lambert P, Ghigo JM.** 2006. Broad-spectrum biofilm inhibition by a secreted bacterial polysaccharide. *Proc Natl Acad Sci USA* **103:**12558–12563.
120. **Kim Y, Oh S, Kim SH.** 2009. Released exopolysaccharide (r-EPS) produced from probiotic bacteria reduce biofilm formation of enterohemorrhagic *Escherichia coli* O157:H7. *Biochem Biophys Res Commun* **379:**324–329.
121. **Christopher AB, Arndt A, Cugini C, Davey ME.** 2010. A streptococcal effector protein that inhibits *Porphyromonas gingivalis* biofilm development. *Microbiology* **156:**3469–3477.
122. **Flemming HC, Wingender J.** 2010. The biofilm matrix. *Nat Rev Microbiol* **8:**623–633.
123. **Lambert C, Sockett RE.** 2013. Nucleases in *Bdellovibrio bacteriovorus* contribute towards efficient self-biofilm formation and eradication of preformed prey biofilms. *FEMS Microbiol Lett* **340:**109–116.
124. **Tang J, Kang M, Chen H, Shi X, Zhou R, Chen J, Du Y.** 2011. The staphylococcal nuclease prevents biofilm formation in *Staphylococcus aureus* and other biofilm-forming bacteria. *Sci China Life Sci* **54:**863–869.
125. **Nijland R, Hall MJ, Burgess JG.** 2010. Dispersal of biofilms by secreted, matrix degrading, bacterial DNase. *PLoS One* **5:** e15668. doi:10.1371/journal.pone.0015668.
126. **Kaplan JB, Ragunath C, Velliyagounder K, Fine DH, Ramasubbu N.** 2004. Enzymatic detachment of *Staphylococcus epidermidis* biofilms. *Antimicrob Agents Chemother* **48:**2633–2636.
127. **Ogawa A, Furukawa S, Fujita S, Mitobe J, Kawarai T, Narisawa N, Sekizuka T, Kuroda M, Ochiai K, Ogihara H, Kosono S, Yoneda S, Watanabe H, Morinaga Y, Uematsu H, Senpuku H.** 2011. Inhibition of *Streptococcus mutans* biofilm formation by *Streptococcus salivarius* FruA. *Appl Environ Microbiol* **77:**1572–1580.
128. **Dusane DH, Damare SR, Nancharaiah YV, Ramaiah N, Venugopalan VP, Kumar AR, Zinjarde SS.** 2013. Disruption of microbial biofilms by an extracellular protein isolated from epibiotic tropical marine strain of *Bacillus licheniformis. PLoS One* **8:**e64501. doi:10.1371/journal.pone.0064501.
129. **Iwase T, Uehara Y, Shinji H, Tajima A, Seo H, Takada K, Agata T, Mizunoe Y.** 2010. *Staphylococcus epidermidis* Esp inhibits *Staphylococcus aureus* biofilm formation and nasal colonization. *Nature* **465:**346–349.
130. **Sugimoto S, Iwamoto T, Takada K, Okuda K, Tajima A, Iwase T, Mizunoe Y.** 2013. *Staphylococcus epidermidis* Esp degrades specific

proteins associated with *Staphylococcus aureus* biofilm formation and host-pathogen interaction. *J Bacteriol* **195:**1645–1655.

131. **McDougald D, Rice SA, Barraud N, Steinberg PD, Kjelleberg S.** 2012. Should we stay or should we go: mechanisms and ecological consequences for biofilm dispersal. *Nat Rev Microbiol* **10:**39–50.
132. **Wang LH, He Y, Gao Y, Wu JE, Dong YH, He C, Wang SX, Weng LX, Xu JL, Tay L, Fang RX, Zhang L.** 2004. A bacterial cell-cell communication signal with cross-kingdom structural analogues. *Mol Microbiol* **51:**903–912.
133. **Ryan RP, Dow JM.** 2011. Communication with a growing family: diffusible signal factor (DSF) signaling in bacteria. *Trends Microbiol* **19:**145–152.
134. **Davies DG, Marques CN.** 2009. A fatty acid messenger is responsible for inducing dispersion in microbial biofilms. *J Bacteriol* **191:** 1393–1403.
135. **Boles BR, Thoendel M, Singh PK.** 2005. Rhamnolipids mediate detachment of *Pseudomonas aeruginosa* from biofilms. *Mol Microbiol* **57:**1210–1223.
136. **Irie Y, O'Toole GA, Yuk MH.** 2005. *Pseudomonas aeruginosa* rhamnolipids disperse *Bordetella bronchiseptica* biofilms. *FEMS Microbiol Lett* **250:**237–243.
137. **Barraud N, Schleheck D, Klebensberger J, Webb JS, Hassett DJ, Rice SA, Kjelleberg S.** 2009. Nitric oxide signaling in *Pseudomonas aeruginosa* biofilms mediates phosphodiesterase activity, decreased cyclic di-GMP levels, and enhanced dispersal. *J Bacteriol* **191:**7333–7342.
138. **Firoved AM, Wood SR, Ornatowski W, Deretic V, Timmins GS.** 2004. Microarray analysis and functional characterization of the nitrosative stress response in nonmucoid and mucoid *Pseudomonas aeruginosa*. *J Bacteriol* **186:**4046–4050.
139. **Kolodkin-Gal I, Cao S, Chai L, Bottcher T, Kolter R, Clardy J, Losick R.** 2012. A self-produced trigger for biofilm disassembly that targets exopolysaccharide. *Cell* **149:**684–692.
140. **Kolodkin-Gal I, Romero D, Cao S, Clardy J, Kolter R, Losick R.** 2010. D-amino acids trigger biofilm disassembly. *Science* **328:**627–629.
141. **Hochbaum AI, Kolodkin-Gal I, Foulston L, Kolter R, Aizenberg J, Losick R.** 2011. Inhibitory effects of D-amino acids on *Staphylococcus aureus* biofilm development. *J Bacteriol* **193:**5616–5622.
142. **Hobley L, Kim SH, Maezato Y, Wyllie S, Fairlamb AH, Stanley-Wall NR, Michael AJ.** 2014. Norspermidine is not a self-produced trigger for biofilm disassembly. *Cell* **156:**844–854.
143. **Leiman SA, May JM, Lebar MD, Kahne D, Kolter R, Losick R.** 2013. D-amino acids indirectly inhibit biofilm formation in *Bacillus subtilis* by interfering with protein synthesis. *J Bacteriol* **195:**5391–5395.
144. **Taylor TB, Buckling A.** 2011. Selection experiments reveal trade-offs between swimming and twitching motilities in *Pseudomonas aeruginosa*. *Evolution* **65:**3060–3069.
145. **van Ditmarsch D, Boyle KE, Sakhtah H, Oyler JE, Nadell CD, Deziel E, Dietrich LE, Xavier JB.** 2013. Convergent evolution of hyperswarming leads to impaired biofilm formation in pathogenic bacteria. *Cell Rep* **4:**697–708.
146. **Jin F, Conrad JC, Gibiansky ML, Wong GCL.** 2011. Bacteria use type-IV pili to slingshot on surfaces. *Proc Natl Acad Sci USA* **108:**12617–12622.
147. **An D, Danhorn T, Fuqua C, Parsek MR.** 2006. Quorum sensing and motility mediate interactions between *Pseudomonas aeruginosa* and *Agrobacterium tumefaciens* in biofilm cocultures. *Proc Natl Acad Sci USA* **103:**3828–3833.
148. **Hibbing ME, Fuqua C.** 2012. Inhibition and dispersal of *Agrobacterium tumefaciens* biofilms by a small diffusible *Pseudomonas aeruginosa* exoproduct(s). *Arch Microbiol* **194:** 391–403.
149. **Ma Q, Zhang G, Wood TK.** 2011. *Escherichia coli* BdcA controls biofilm dispersal in *Pseudomonas aeruginosa* and *Rhizobium meliloti*. *BMC Res Notes* **4:**447.
150. **Houry A, Gohar M, Deschamps J, Tischenko E, Aymerich S, Gruss A, Briandet R.** 2012. Bacterial swimmers that infiltrate and take over the biofilm matrix. *Proc Natl Acad Sci USA* **109:**13088–13093.
151. **Da Re S, Valle J, Charbonnel N, Beloin C, Latour-Lambert P, Faure P, Turlin E, Le Bouguenec C, Renauld-Mongenie G, Forestier C, Ghigo JM.** 2013. Identification of commensal *Escherichia coli* genes involved in biofilm resistance to pathogen colonization. *PLoS One* **8:** e61628. doi:10.1371/journal.pone.0061628.
152. **He X, Tian Y, Guo L, Lux R, Zusman DR, Shi W.** 2010. Oral-derived bacterial flora defends its domain by recognizing and killing intruders: a molecular analysis using *Escherichia coli* as a model intestinal bacterium. *Microb Ecol* **60:**655–664.

153. **He X, McLean JS, Guo L, Lux R, Shi W.** 2014. The social structure of microbial community involved in colonization resistance. *ISME J* **8:**564–574.
154. **Mayr E.** 1989. Speciational evolution or punctuated equilibria. *J Soc Biol Struct* **12:**137–158.
155. **Rainey PB, Rainey K.** 2003. Evolution of cooperation and conflict in experimental bacterial populations. *Nature* **425:**72–74.
156. **Diggle SP, Griffin AS, Campbell GS, West SA.** 2007. Cooperation and conflict in quorum-sensing bacterial populations. *Nature* **450:**411–414.
157. **MacLean RC, Gudelj I.** 2006. Resource competition and social conflict in experimental populations of yeast. *Nature* **441:**498–501.
158. **Fagerlind MG, Webb JS, Barraud N, McDougald D, Jansson A, Nilsson P, Harlen M, Kjelleberg S, Rice SA.** 2012. Dynamic modelling of cell death during biofilm development. *J Theor Biol* **295:**23–36.
159. **Stolyar S, Van Dien S, Hillesland KL, Pinel N, Lie TJ, Leigh JA, Stahl DA.** 2007. Metabolic modeling of a mutualistic microbial community. *Mol Syst Biol* **3:**92.
160. **Zhuang K, Izallalen M, Mouser P, Richter H, Risso C, Mahadevan R, Lovley DR.** 2011. Genome-scale dynamic modeling of the competition between *Rhodoferax* and *Geobacter* in anoxic subsurface environments. *ISME J* **5:** 305–316.
161. **Wanner O, Gujer W.** 1986. A multispecies biofilm model. *Biotechnol Bioeng* **28:**314–328.
162. **Poplawski NJ, Shirinifard A, Swat M, Glazier JA.** 2008. Simulation of single-species bacterial-biofilm growth using the Glazier-Graner-Hogeweg model and the CompuCell3D modeling environment. *Math Biosci Eng* **5:**355–388.
163. **Xavier JB, Foster KR.** 2007. Cooperation and conflict in microbial biofilms. *Proc Natl Acad Sci USA* **104:**876–881.
164. **Nadell CD, Bassler BL.** 2011. A fitness trade-off between local competition and dispersal in *Vibrio cholerae* biofilms. *Proc Natl Acad Sci USA* **108:**14181–14185.
165. **Kussell E.** 2013. Evolution in microbes. *Annu Rev Biophys* **42:**493–514.
166. **Blount ZD, Borland CZ, Lenski RE.** 2008. Historical contingency and the evolution of a key innovation in an experimental population of *Escherichia coli. Proc Natl Acad Sci USA.* **105:**7899–7906.
167. **Wiser MJ, Ribeck N, Lenski RE.** 2013. Long-term dynamics of adaptation in asexual populations. *Science* **342:**1364–1367.
168. **Spiers AJ, Kahn SG, Bohannon J, Travisano M, Rainey PB.** 2002. Adaptive divergence in experimental populations of *Pseudomonas fluorescens*. I. Genetic and phenotypic bases of wrinkly spreader fitness. *Genetics* **161:**33–46.
169. **Hansen SK, Rainey PB, Haagensen JA, Molin S.** 2007. Evolution of species interactions in a biofilm community. *Nature* **445:**533–536.
170. **Traverse CC, Mayo-Smith LM, Poltak SR, Cooper VS.** 2013. Tangled bank of experimentally evolved *Burkholderia* biofilms reflects selection during chronic infections. *Proc Natl Acad Sci USA* **110:**E250–E259.
171. **Park SC, Krug J.** 2007. Clonal interference in large populations. *Proc Natl Acad Sci USA* **104:** 18135–18140.
172. **Lee KW, Periasamy S, Mukherjee M, Xie C, Kjelleberg S, Rice SA.** 2013. Biofilm development and enhanced stress resistance of a model, mixed-species community biofilm. *ISME J.* [Epub ahead of print.] doi:10.1038/ismej.2013.194.
173. **Turcotte MM, Corrin MSC, Johnson MTJ.** 2012. Adaptive evolution in ecological communities. *PLoS Biol* **10:**e1001332. doi:10.1371/journal.pbio.1001332.
174. **Boyle KE, Heilmann S, van Ditmarsch D, Xavier JB.** 2013. Exploiting social evolution in biofilms. *Curr Opin Microbiol* **16:**207–212.
175. **Reid G, Howard J, Gan BS.** 2001. Can bacterial interference prevent infection? *Trends Microbiol* **9:**424–428.
176. **Buffie CG, Pamer EG.** 2013. Microbiota-mediated colonization resistance against intestinal pathogens. *Nat Rev Immunol* **13:**790–801.
177. **Conrad D, Haynes M, Salamon P, Rainey PB, Youle M, Rohwer F.** 2013. Cystic fibrosis therapy: a community ecology perspective. *Am J Respir Cell Mol Biol* **48:**150–156.
178. **Ren D, Madsen JS, de la Cruz-Perera CI, Bergmark L, Sorensen SJ, Burmolle M.** 2013. High-throughput screening of multispecies biofilm formation and quantitative PCR-based assessment of individual species proportions, useful for exploring interspecific bacterial interactions. *Microb Ecol* [Epub ahead of print.] doi:10.1007/s00248-013-0315-z.
179. **Shank EA.** 2013. Using coculture to detect chemically mediated interspecies interactions. *J Vis Exp* **80:**e50863. doi:10.3791/50863.

17 Dispersal from Microbial Biofilms

NICOLAS BARRAUD,[1] STAFFAN KJELLEBERG,[1,2] and SCOTT A. RICE[1,2]

INTRODUCTION

For all organisms the ability to spread and colonize new habitats is crucial to ensure species continuity and prevent extinction (1). In sessile organisms this constraint has led to the evolution of a motile, dispersal phase in their life cycle, which in plants and corals involves the release of differentiated and often phenotypically diverse seeds or propagules. Similarly, sessile microbial biofilms have developed mechanisms to release differentiated, highly motile dispersal cells into the bulk liquid.

Dispersal has profound ecological consequences, allowing biofilm populations to spread and colonize new surfaces as well as to avoid overcrowding and regenerate bacteria in the biofilm core. These benefits represent strong evolutionary pressures for dispersal, propagation, and rejuvenation of biofilms. The ecological drivers of dispersal are complex, because dispersal is both costly (2) and risky when dispersal cells face an uncertain and uncontrolled environment (3). However, the inability to disperse may ultimately lead to the collapse of any sessile biological community through the accumulation of social cheaters, such as those that disturb metabolic cooperation (4), or

[1]Centre for Marine Bio-Innovation and School of Biotechnology and Biomolecular Sciences, The University of New South Wales, Sydney, NSW 2052, Australia; [2]Singapore Centre on Environmental Life Sciences Engineering, and the School of Biological Sciences, Nanyang Technological University, Singapore 639798.

Microbial Biofilms 2nd Edition
Edited by Mahmoud Ghannoum, Matthew Parsek, Marvin Whiteley, and Pranab K. Mukherjee

doi:10.1128/microbiolspec.MB-0015-2014

catastrophic events at one location resulting in either complete loss of the population leading to extinction or in genetic bottlenecks as a consequence of significant population die-off (5).

Sloughing is a passive process of cell loss from the biofilm, which may be a consequence of cell division at the biofilm–bulk liquid interface or shear forces that remove cells. In contrast, active or seeding dispersal is coordinated via regulatory systems and is energy dependent, and dispersal cells exhibit a distinct phenotype. Dispersal relies on a number of cues, both environmental and self-produced signals. Regulatory responses to changes in environmental conditions potentially enhance the chances of successful release of propagules and colonization of new surfaces. In addition, signals and cues from other biological organisms can also influence microbial dispersal. Cells that have dispersed from biofilms exhibit a specialized phenotype that is distinct from that of biofilm and planktonic cells. Moreover, the dispersal subpopulation often comprises a high level of heterogeneity, as a consequence of cell differentiation and phase variation as well as genetic diversification through mutations. The production of such variation in dispersal cells has been hypothesized to increase population fitness, e.g., enhanced colonization and survival due to complementarity and selection effects associated with the formation of such variants.

This article aims to describe features that are common to the dispersal process across species and presents specific examples to define those features. The underlying regulatory and signaling systems are examined in terms of both the extracellular cues and the internal signal transduction cascades that control active dispersal. Further, the consequences of dispersal for the microbial population are addressed. Finally, because dispersal offers unique opportunities to control biofilms, several strategies currently being developed to exploit dispersal pathways and design novel biofilm control measures are described.

THE MOTILE PHASE OF THE BIOFILM LIFE CYCLE

It is now well established that biofilm formation occurs through a series of stages that resemble developmental processes in multicellular eukaryotes. Recent studies have characterized specific regulators involved at each stage of development (6). Initial adhesion to a surface is thus controlled by nutrient availability, surface sensing, and quorum sensing (QS) (7). While flagella-mediated motility may play a role for planktonic bacteria to reach a surface and attach, the initiation of biofilm formation often leads to the loss of flagella (8). Biofilm bacteria are essentially sessile, embedded in a matrix of extracellular polymeric substances (EPSs), with the exception of a subpopulation of cells using surface motility that migrate within the biofilm to form the caps of three-dimensional (3D) microcolonies (9). In the model organism *Pseudomonas aeruginosa*, maturation of the biofilm involves production of specific matrix components and sequentially activated regulatory systems (10). These mechanisms include positive and negative feedback regulations, which are pivotal components of developmental biology. For example, the polysaccharide synthesis locus (Psl) is necessary for *P. aeruginosa* cells to attach onto a surface and subsequently serves to recruit incoming bacteria into the biofilm, establishing a feed-forward loop amplifying biofilm attachment and maturation (11). Over the course of the developmental process, bacteria adopt a succession of different phenotypes. While the biofilm phenotype appears adaptable to changes in environmental conditions, notably through stress responses, a core genetic program takes the biofilm through a series of checkpoints, eventually leading to maturation and dispersal of the biofilm.

During maturation, biofilms establish complex 3D structures comprised of differentiated bacteria, for instance exhibiting different expression of metabolic or EPS genes, and steep nutrient and oxygen (O_2) gradients, rendering the biofilm environment and the bacterial populations highly heterogeneous (12–14). The final stage of biofilm development involves the coordinated release of differentiated, motile, chemotactic cells known as dispersal cells (15). These specialized cells can colonize new surfaces and restart the biofilm life cycle. In several bacteria, biofilm dispersal correlates with the programmed death of a subpopulation of cells in mature microcolonies (16). Surviving cells are then able to escape the biofilm, leaving behind hollow structures in the biofilm. In many species, dispersal typically occurs from mature biofilm microcolonies and is preceded by localized death and lysis of cells in the center of these structures (16–22) (Fig. 1). Bacteria were observed to disperse from larger microcolonies, e.g., with a diameter greater than 40 μm and depth greater than 10 μm, but not from smaller ones (17, 23).

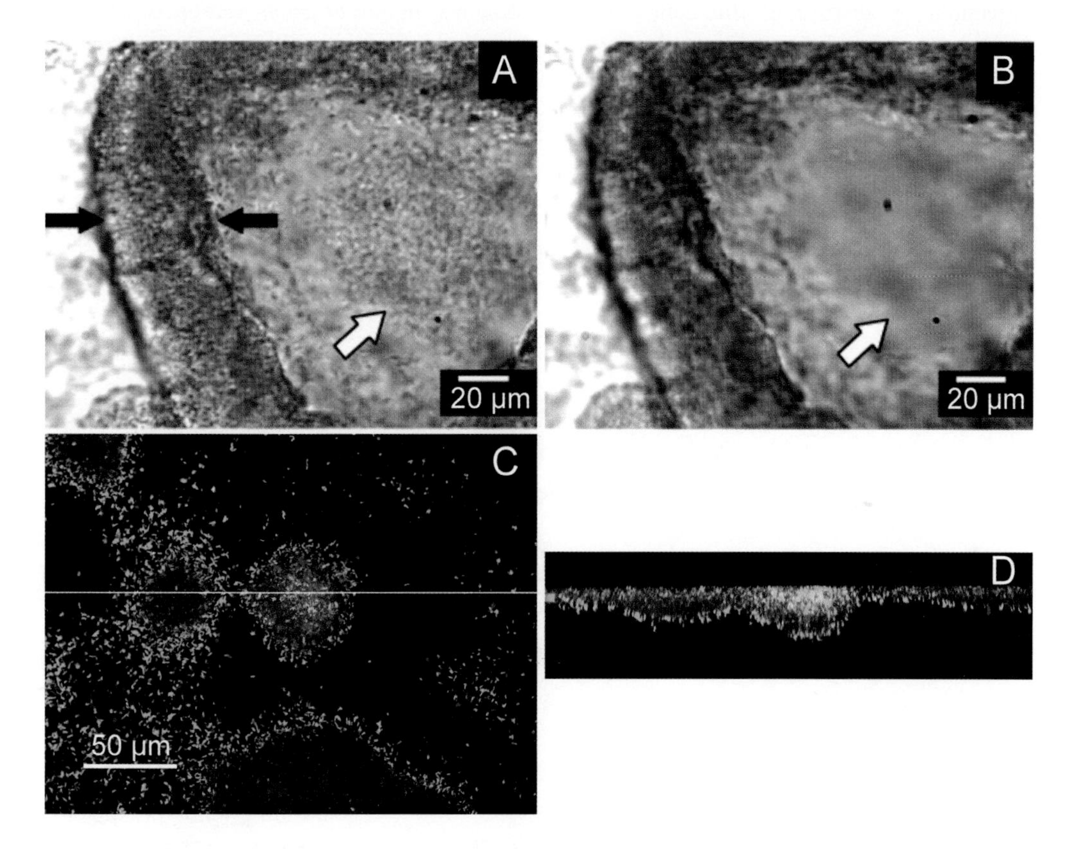

FIGURE 1 Microscopic images of biofilm microcolonies during seeding dispersal. (A-B) Motile cells appear in mature biofilm microcolonies. (A) Single frame of a mature microcolony. (B) Picture showing the average of 30 frames captured over a 1-second period. The highly motile cells "average" out and appear blurred in the center of the microcolony, demonstrating the extent of the motile region (white arrow in panels A and B). The sessile "wall" region is indicated by the black arrows in panel A (taken from reference 26, with permission from the publisher). (C) Live/dead staining of a 7-day-old biofilm reveals patterns of cell death inside biofilm structures that occur simultaneously with biofilm dispersal, as indicated by the formation of hollow biofilm structures. Live cells are green and dead cells are red (adapted from reference 56, copyright © American Society for Microbiology). (D) XZ cross-view of the biofilm in panel C (XY view) at the location indicated by the white line. doi:10.1128/microbiolspec.MB-0015-2014.f1

This suggests that a specific maturation stage, as defined by the microcolony size and most likely linked to the establishment of specific gradients, e.g., nutrients or O_2, is required to trigger dispersal events. In *P. aeruginosa*, cell lysis has been linked to the activation of a superinfective prophage (16), while in the marine bacterium *Pseudoalteromonas tunicata*, the expression and activity of an autolytic protein caused cell death in mature biofilm microcolonies (24).

It has been shown that dispersal can be prevented by disrupting the cells' ability to generate energy (25), and thus, the surviving cells may benefit from nutrients released from lysed cells, and this energy is used to activate dispersal responses. This is supported by microscopic observations of biofilms showing that mature microcolonies undergo a brief "seething" stage, where cells become highly motile within microcolonies, suggesting that biofilm EPSs have been fully solubilized and that cells have regained motility (17, 26) (Figs. 1 and 2). Dispersal cells then escape by coordinated evacuation from biofilm structures (26), resulting in the hollowing of biofilm microcolony structures leaving behind voids that are typically observed during the dispersal stage for many biofilms. Dispersal cells are characterized by downregulation of genes that encode for the sessile biofilm phenotype, such as exopolysaccharides and fimbriae, and upregulation of genes encoding factors important for the motile lifestyle, including flagella and chemotaxis (27–30).

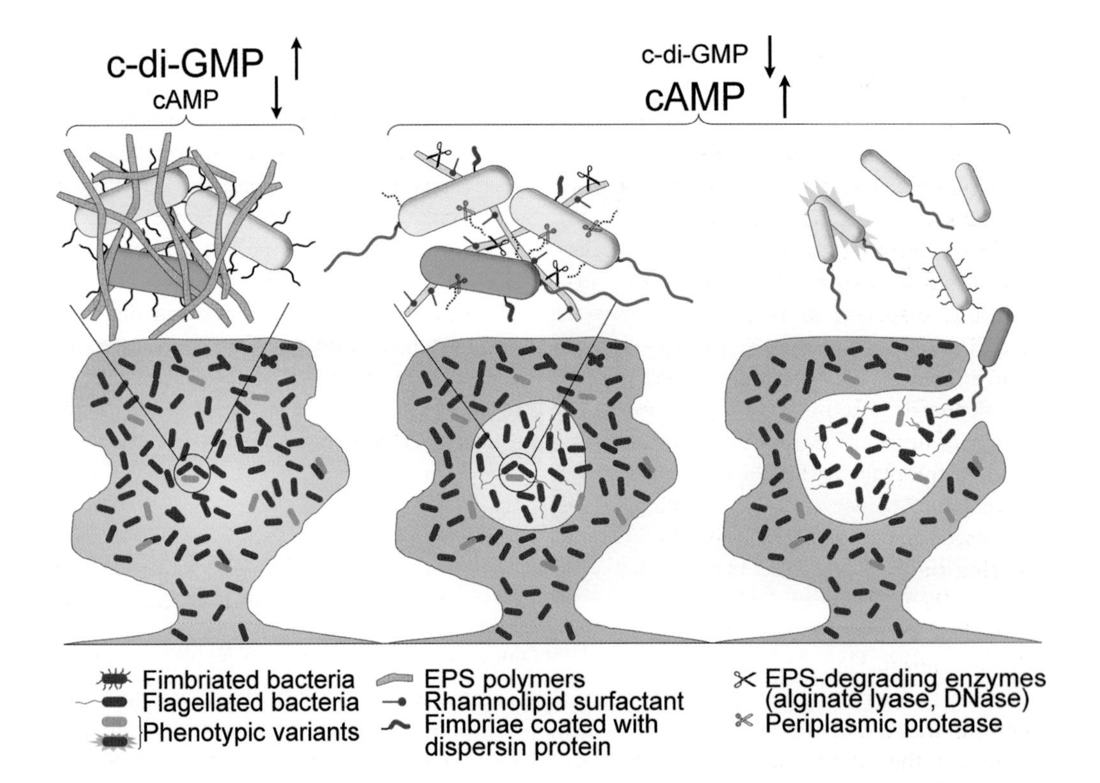

FIGURE 2 Effectors of biofilm dispersal. Bacteria within the center of microcolonies induce a number of mechanisms to degrade and solubilize the biofilm EPS matrix and extracellular appendages such as fimbriae that immobilize cells. When the interior of the microcolony becomes fluid, cells begin to show signs of motility, and a breach is made in the microcolony wall through which dispersal cells are released. doi:10.1128/microbiolspec.MB-0015-2014.f2

REGULATION AND COORDINATION IN SEEDING DISPERSAL

The release of dispersal cells from biofilms occurs in a precisely coordinated manner and in response to changes in environmental conditions. Growth in a biofilm can lead to stress; for instance, as the biofilm grows in size, bacteria in the lower layers will experience reduced access to nutrients and electron acceptors available from the bulk liquid interface or will accumulate waste products and toxins. Stress cues including nutrient starvation, either carbon (31, 32), phosphate (33), or iron sources (34), can trigger dispersal responses. Depletion of O_2 (35, 36) and addition of the metabolism inhibitor carbonylcyanide *m*-chlorophenylhydrazone (CCCP) (37) were also found to induce dispersal. In mature biofilms, dispersal may also be triggered by specific responses to the presence of soluble signals derived from protozoan predators, which are produced whether in the presence or absence of bacteria (unpublished observation). An increase in nutrient availability, such as carbon (38) or iron sources (39), can also trigger dispersal and lead to mass detachment events. Possibly, dispersal in response to nutrient upshift could represent a strategy to enhance chances of successful colonization by dispersal cells by ensuring an adequate supply of nutrient sources. Interestingly, biofilms of the plant pathogen *Xylella fastidiosa* were found to use cell-cell signaling to coordinate dispersal events in response to sensing the presence of a transmission vector, e.g., leafhopper insects (40). In this case it is clear that precise regulation and rapid signal transduction are necessary for the bacteria to trigger dispersal processes upon sensing the presence of the insect before it leaves the plant surface. Transmission vectors for dispersal and spreading of bacterial populations are common in nature. In a recent study, a nonmotile *Xanthomonas* sp. was found to associate with a motile *Paenibacillus vortex*, in a hitchhiking strategy, to spread and disperse on plant surfaces (41).

Cell-cell signaling is known to play a major role in regulating dispersal in various species (42). For example, QS via *N*-acyl homoserine lactones has been implicated in dispersal in *Vibrio* species (43, 44), *Serratia marcescens* (45), and *Rhodobacter sphaeroides* (46). In *Staphylococcus aureus*, the *agr*-mediated QS system was found to control biofilm dispersal (47). In *Vibrio cholerae*, QS regulates detachment (48), and add-back of cholera QS autoinducer-1 (CAI-1) and autoinducer-2 (AI-2) resuscitate dormant cells and induce dispersal of biofilm aggregates in environmental samples (49). In *Xanthomonas campestris*, cell-cell signaling via the diffusible signal cis-11-methyl-2-dodecenoic acid triggers dispersal responses (50). This class of signal molecule has more recently been shown to be conserved across a broader range of bacteria, where it induces dispersal (51), including cis-2-decenoic acid, a potent dispersal trigger in various bacterial species (23). Further, the addition of QS inhibitors, such as furanones, can induce detachment of biofilm bacteria and prevent biofilm formation (52, 53). The role of cell-cell signaling in regulating dispersal may allow biofilm bacteria to induce the switch at an appropriate stage of the developmental process, e.g., once a threshold density is reached, or to transmit a dispersal signal from one location to another within the biofilm population.

Several research groups have also focused their efforts on identifying signal molecules produced in mature cultures entering stationary phase that were thought to trigger dispersal events. The diffusible fatty acid cis-2-decenoic acid mentioned above was first identified after fractionation of *P. aeruginosa* spent medium that was active at inducing dispersal (23). In *Bacillus subtilis*, the re-addition of a conditioned medium from late biofilm cultures was also found to induce biofilm disassembly, and D-amino acids were tentatively identified as active compounds, suggesting a role in regulating dispersal (54). However, later studies from the same team revealed that the observed effects were due to a mutation in the experimental strain which caused addition of D-amino acids to reduce metabolism in these biofilms; thus, dispersal in these studies was found to rely entirely on reduced metabolism (55), which is similar to the starvation-induced dispersal for some species as described above.

In *P. aeruginosa*, studies of cell death and dispersal events led to the identification of nitric oxide (NO) as a major signal regulating dispersal (56). NO was found to be produced in mature biofilms and trigger detachment events. NO was also found to induce dispersal in a wide range of monospecies biofilms and mixed-species biofilms (reviewed in 57). Thus, biofilm dispersal appears to be controlled by a variety of cues and signals, some of which, such as *N*-acyl homoserine lactones, play dual roles in biofilm formation as well as dispersal (7).

One important system that is involved in integrating these signals is the regulatory network centered on the secondary messenger cyclic di-GMP, because it controls both attachment and dispersal (reviewed in 58, 59). Mainly found in Gram-negative bacteria, c-di-GMP is a central element of a signaling network that integrates one or multiple signals (input) sensed by a bacterium and activates cellular effectors (output), resulting in either biofilm attachment or dispersal. Intracellular levels of c-di-GMP are controlled through the opposing activities of diguanylate cyclases (DGCs) for the synthesis of c-di-GMP, and phosphodiesterases (PDEs) for its degradation. These are encoded by conserved GGDEF-domain- and EAL- or HDGYP-domain-containing genes, respectively. Many bacterial genomes encode multiple DGCs and PDEs often associated with other putative signaling domains, suggesting that their enzymatic activities may be responsive to a range of environmental cues, including NO, O_2, and nutrients. Recently, cAMP, which was previously known to control the stringent response, was found to play a role in the regulation of biofilm formation and dispersal (25, 60). The role that nucleotide secondary messengers such as c-di-GMP and cAMP play in signal transduction in bacteria is being increasingly recognized (61, 62; cf. online census http://www.ncbi.nlm.nih.gov/Complete_Genomes/SignalCensus.html), which is perhaps not surprising given the abundance of nucleotides in the intracellular milieu and their relationship to metabolic activity, a key dispersal trigger.

A link between NO and c-di-GMP in the regulation of biofilm dispersal was first established in *P. aeruginosa* (29) and was then also described in *Shewanella woodyi* (63) and *Legionella pneumophila* (64). In *P. aeruginosa* several c-di-GMP–specific PDEs have been identified that appear to be involved in NO-mediated dispersal, including DipA, RbdA, and NbdA (65, 66). In some bacteria, the sensor for NO has been identified. For example, *S. woodyi* encodes a heme NO/O_2–binding (HNOX) protein, which when complexed with NO, binds to and activates a PDE enzyme, resulting in dispersal. HNOX domains are conserved hemoproteins that are highly sensitive to NO, producing responses at femtomolar levels in *Clostridium botulinum* (67). They are found in several Gram-negative and Gram-positive bacterial genomes and are often associated with a DGC or PDE (68). However, many bacterial strains known to disperse in response to NO, including *P. aeruginosa* and *Escherichia coli*, do not have HNOX-domain-encoding proteins, suggesting that other systems can sense NO signals. In *E. coli*, a redox sensor associated with a c-di-GMP PDE, *Ec*DOS was shown to increase PDE activity upon binding of NO, O_2, and carbon monoxide (69). *Ec*DOS was later shown to be part of a two-gene operon for the control c-di-GMP in response to O_2 levels where sensing of O_2 activates both DGC and PDE activity. This suggests a feedback control mechanism, similar to a thermostat, to regulate biofilm formation and dispersal precisely, depending on O_2 levels (70).

In *P. aeruginosa* the methyl-accepting protein BdlA has been found to be involved in the dispersal response to both nutrient upshift (28) and NO (29) and interacts with a PDE, possibly by stabilizing the enzyme, resulting in lower intracellular levels of c-di-GMP and enhanced dispersal (65). Recently, an interesting mechanism of feedback regulation was uncovered when it was found that BdlA needed to be cleaved before activating dispersal. Further, cleavage of BdlA, involving the chaperone ClpD and protease ClpP, requires high levels of c-di-GMP (71). Thus, high levels of c-di-GMP, which are typical of biofilm cells, appear to be a prerequisite to induce dispersal events in mature biofilms in this strain.

EFFECTORS OF DISPERSAL

Upon sensing a dispersal cue, bacteria can activate a range of cellular effectors that allow them to break their bonds to the biofilm, including enzymes and surfactants that will degrade the biofilm EPS, resulting in dispersal (Fig. 2).

First, bacteria can produce and secrete enzymes that degrade the biofilm EPS matrix, the composition of which varies but is mainly comprised of polysaccharides, proteins, nucleic acids, and lipids. In the nonmotile species *Aggregatibacter actinomycetemcomitans*, a screen of transposon mutant biofilms deficient in the ability to release cells identified dispersinB, an endogenous β-*N*-acetylglucosaminidase capable of degrading the matrix polysaccharides, which was later found to induce biofilm dispersal in a range of bacterial species (72, 73). In *P. aeruginosa*, alginate is an essential component of the matrix, and early studies revealed that increased expression of alginate lyase leads to biofilm detachment (74). Furthermore, it was found that DNA is a major constituent of the EPS matrix, and biofilm dispersal could be induced upon treatment with DNase in *P. aeruginosa* and *S. aureus* (75, 76). DNase was also found to be endogenously produced and secreted in *Bacillus licheniformis* biofilms to induce dispersal (77). In *S. aureus*, extracellular protease activity mediated by the *agr* QS system was also found to trigger biofilm detachment (47). In *Pseudomonas fluorescens* (33), *Pseudomonas putida* (78), and *P. aeruginosa* (25), starvation-induced dispersal was linked to activation of a periplasmic cysteine protease, LapG. When activated, LapG has been shown to cleave the surface adhesion LapA to release cells from the surface (78). In *P. aeruginosa, in vitro* batch and continuous-flow biofilm assays revealed that a *lapG* knockout mutant strain was unable to disperse in response to nutrient starvation, O_2 depletion, or NO donor signals compared to a wild type strain (unpublished data). This suggests that dispersal events in response to these various cues all integrate via a c-di-GMP signaling cascade that relies on LapG activation to induce biofilm detachment.

Second, the connecting biopolymers that maintain bacteria in the biofilm may be modulated by the secretion of chemicals that specifically alter adhesion appendages. Studies of *E. coli* suggested that dispersal bacteria are coated with a protein, called dispersin, that counteracts fimbriae-mediated aggregative adherence (79). Further, dispersal of *E. coli* biofilms from human epithelial cells was associated with alterations of bundle-forming type IV pili structures from thin to much longer and thicker bundles that resulted in loss of adherence and aggregation (80). To reduce surface tension and allow cells to detach, biofilm bacteria also produce amphipathic molecules, such as rhamnolipids, to induce dispersal (81). In *P. putida*, endogenous biosurfactants called putisolvins produced upon entry into stationary growth phase were able to induce biofilm detachment (82). In this scenario, the dispersal signal is amplified within an entire subpopulation of cells, where neighboring cells may degrade the matrix surrounding cells that may not have yet switched to a dispersal phenotype.

Finally, the transition from a biofilm to a planktonic mode of growth involves the activation of flagella (swimming and swarming motility) or pili (twitching motility) in motile microorganisms. Thus, flagella synthesis genes were found to be expressed in dispersed cells of *E. coli* and *P. aeruginosa* biofilms (26, 27, 38, 83), while type IV pili were found to be activated in biofilm bacteria treated with the dispersal signal NO (29). Further, the analysis of microscopy movies of *P. aeruginosa* biofilms revealed that attached bacteria use type IV pili to adopt a vertical orientation and mediate surface detachment (84).

Overall, a variety of strategies have been observed for detachment of bacteria from biofilms between different species but also in the same species. It is likely that the mechanisms used will depend on the growth conditions, including the conditional composition of the matrix, as well as the age of the biofilm.

DISPERSAL CELLS PRESENT A SPECIALIZED PHENOTYPE

Biofilm populations often display a high level of phenotypic heterogeneity (13), and not surprisingly, the dispersal subpopulation released from mature biofilms also reveals a high level of phenotypic variation. The variable phenotypes of dispersal cells are the result of a combination of changes in gene expression (transient phenotypic differences) and altered genotypes (permanent phenotypic changes). This diversity in biofilms and dispersal cells has been linked to increased fitness and improved ability to colonize a range of habitats. Dispersal bacteria appear to be affected in key traits including attachment, metabolism, and antimicrobial resistance, as well as virulence and motility (30, 85–87).

Two studies of *P. aeruginosa* and *Streptococcus mutans* have compared stationary-phase planktonic cultures, resuspended biofilm cells, and bacteria collected from the biofilm effluent, and found differences in attachment properties and growth rates (30, 87). The entire dispersal population appeared to grow at rates similar to biofilm cells, both of which grew more slowly than planktonic cultures. While dispersal cells may have an overall slower metabolism, in *P. tunicata, P. aeruginosa,* and *S. mutans,* the dispersal populations have been found to utilize a wider range of nutrient sources (85–87). When variants isolated from biofilm effluents were analyzed separately, individual variants showed either decreased or increased metabolic rates compared to the parental strain (85–87). Further, dispersal cells in *P. aeruginosa* and *S. mutans* attached better and formed biofilms with greater biomass compared to both planktonic and resuspended biofilm cells (30, 87). Another study showed enhanced attachment and biofilm formation in dispersal cells of *Marinobacter hydrocarbonoclasticus* compared to planktonic cultures (88). This ability was linked to reshaping of the cell envelope and mobilization of storage reserves of cellular fatty acids and alcohol esters. Finally, increased resistance to antimicrobials and stress has been observed in biofilm-dispersed cells compared to planktonic cultures. Bacteria released from *S. mutans* biofilms had greater tolerance to chlorhexidine and acid treatments compared to stationary-phase planktonic cells (87). However, when a range of antibiotic treatments were tested by using a disc diffusion method, no difference between the dispersal and planktonic subpopulations was found in *P. aeruginosa* (30). When effluents from dual cultures of *P. aeruginosa* and *Burkholderia cepacia* biofilms were analyzed, the dispersal cells showed enhanced resistance to chlorine compared to planktonic chemostat cultures, and this resistance was proportional to the size of the detached clusters (89).

The physiological plasticity of dispersal bacteria is in part due to a high level of heterogeneity, resulting in the formation of phenotypic variants. Individual variants have been isolated from biofilm effluents with a frequency ranging from 5 to 60% of the dispersal subpopulation, which increased as the biofilm matured, in various strains including *P. aeruginosa,* especially from undomesticated clinical isolate biofilms (86, 90) and *S. marcescens* (91). In *P. tunicata,* a mutant unable to undergo cell lysis and dispersal released fewer variants in their effluent, with an overall reduced ability to form biofilms, reduced motility, and homogeneous growth rate. In contrast, wild type biofilms showed extensive variation and overall increased values in all three aspects (85). Interestingly, when all variants collected from the dispersal effluent were grown in mixed cultures and tested for resistance to predation, the assemblage of variants showed increased resistance compared to monocultures grown from individual variants or the parental strain (92). The effect of self-generated genetic variation on increasing resistance of the mono-species biofilm population is also observed when genetic diversity is increased at the species level in mixed-community biofilms (93).

Dispersal cells appear to have a specialized phenotype, different from biofilm bacteria and also clearly distinct from cells grown in planktonic cultures (Fig. 3). The dispersal phenotype appears to show various degrees of stability resulting from phase variation to more permanent genetic alterations. Variants that were isolated from biofilm effluents, e.g., in *P. aeruginosa* (86) or *S. marcescens* (91), were found to be stable and maintain a distinct phenotype after subculturing. The physiological characterizations of entire dispersal populations have been performed immediately after detachment events, within hours following induction or seeding dispersal, and it is not entirely clear whether the observed specialized traits, such as antibiotic resistance (e.g., persisters) or metabolic activity may be stable or transient and whether reculturing of these cells would lead to a phenotype closer to planktonic cells.

Genotypic diversity has been observed in dispersal cells; for example, in *S. marcescens* biofilms, variants isolated from the biofilm effluent were found to result from a single

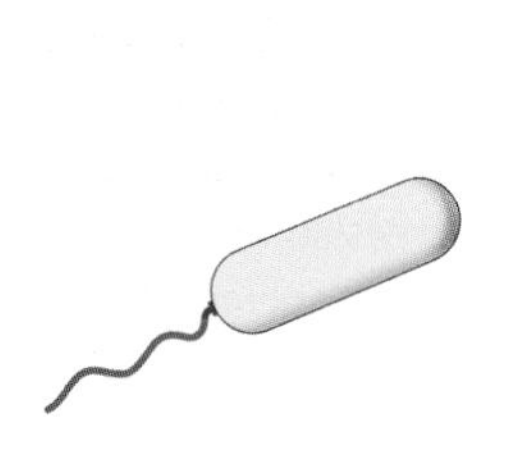

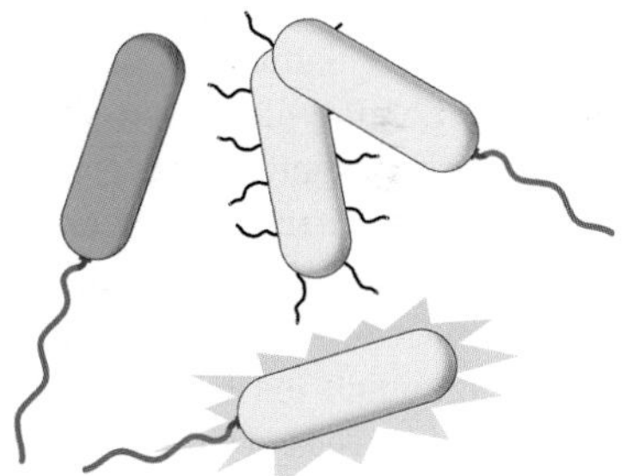

FIGURE 3 Physiological traits of planktonic, biofilm, and dispersal cells. Symbols are defined in Figure 2.
doi:10.1128/microbiolspec.MB-0015-2014.f3

mutation in a regulatory gene (92), which likely represents a favorable genetic switch as opposed to *de novo* mutation. Recently, whole-genome deep sequencing of biofilm dispersal populations revealed single nucleotide mutations in several hot spots, indicative of parallel evolution, but no extensive genetic variation could be observed (94). The latter observation is surprising given the high frequency of phenotypic variation observed.

Thus, it appears that the dispersal population harbors a high level of heterogeneity. However, there has been no investigation of whether this diversity may be directly linked to dispersal or its regulation or whether it was already found within the biofilm, although we can detect such variants within the biofilm biomass prior to dispersal (S.A. Rice et al., unpublished). The generation of phenotypic diversity in dispersal bacteria has been linked to a range of triggers including oxidative stress (22, 95), activation of a superinfective filamentous phage that is also linked to reactive oxygen and nitrogen species activity (96), as well as anaerobic conditions (97). In *P. aeruginosa*, the mismatch repair system was found to generate genetic mutants after repair of oxidative stress–induced DNA damage (95). The emergence of stable phenotypic variants was also associated with the activation of a superinfective prophage by nitrosative stress (56, 96). In *P. tunicata*, an autolysis protein, AlpP, was shown to generate production of hydrogen peroxide, which led to phenotypic diversification and increased fitness (85). Indeed, deletion of the filamentous phage, the AlpP protein, and RecA-mediated recombination resulted in the loss of variant formation and an increase in sensitivity of the biofilms formed by the mutants when exposed to stressors.

ECOLOGICAL AND EVOLUTIONARY ASPECTS OF BIOFILM DISPERSAL

From an evolutionary perspective, biofilm dispersal represents an important adaptive strategy with profound impact on the survival and fitness of bacteria. The nature of dispersal, whether it benefits the biofilm population as a whole or only individual dispersal cells, is still not entirely clear. Because biofilms display a high level of cell differentiation and show close similarities to multicellular organisms (98), biofilm development may share some similarities with the evolution and transition to multicellularity of higher organisms. In this context, dispersal has been considered an important step for the transition to multicellularity (99) and has been explained within the framework of each of the two main theories of multilevel (group) (100) and kin (individual) (101) selection. Group selection implies dispersal is an altruistic event that conserves the genetic inheritance of the population. In this scenario, diversity in the dispersal subpopulation should be restricted to reversible phenotypic switches and not derive from genetic mutations. In contrast, dispersal as an event selected at the individual level, kin selection, correlates with a high level of genetic variation within dispersal cells and thus a gradual change in genetic content or function relative to the parental population.

These questions of genetic evolution and dispersal in biofilms have important implications, for instance, in the context of the emergence of antibiotic resistance, a significant concern of our modern society. Biofilms play a major role in the resistance of bacteria toward antimicrobials and antibiotics, because bacteria in biofilms often display a high level of tolerance, via the generation of persister cells (102) or by accumulating resistant mutants via spontaneous mutations (103) as well as horizontal transfer of resistance genes (104). A direct link between increased genetic diversity specifically associated with a dispersal program rather than as a result of diversity that occurred within the biofilm has not been firmly established yet. In contrast, specific diversity in the dispersal population appears to result from stable cell differentiation or phase variation and thus derive from a program encoded within bacterial genomes to maximize chances of colonization and survival.

The lungs of cystic fibrosis (CF) patients chronically infected with bacterial pathogens, which are mostly clonal in nature due to the apparently infrequent transmission between patients, offer a rare opportunity to observe evolutionary adaptation of bacterial populations. Genomic analyses of *P. aeruginosa* infections in CF patients over 8 years showed strong positive selection signals (ratio of nonsynonymous to synonymous changes per site, dN/dS >1) (105). Later analyses of bacterial infections over 35 years, representing 200,000 generations, revealed that after a period of adaptation *P. aeruginosa* stabilizes its phenotype in CF lungs and then applies a negative selection for mutations (106, 107). Thus, it is clear that clonal populations in a restricted environment undergo genetic evolution, which is either positively or negatively selected. While bacterial infections in CF patients' lungs are often associated with the biofilm mode of growth, and thus chronic infections are likely to go through dispersal events, the role that biofilm dispersal plays in generating genetic mutations over time remains unclear. Another study specifically analyzed genetic changes in *B. cepacia* biofilms through successive dispersal events *in vitro*. Beneficial changes occurred in the population, leading to mixed communities of stable variants that were more productive (colony-forming units number) than any monoculture (108).

Further, the mixed-variant biofilms obtained in these experiments resembled that of CF infections, suggesting that parallel evolution occurs *in vitro* and in CF lungs (109). In particular, evolved *B. cepacia* populations harbored mutations in the *wsp* operon regulating c-di-GMP signaling, suggesting that the mutants were affected in their ability to form biofilms and disperse (109). This supports a close association between the evolution of strains in the CF lungs and biofilm dispersal events. Overall, it is likely that dispersal plays a critical role in evolutionary processes, because it allows spreading of any new phenotypic trait to colonize new habitats where competition with the parental population is limited. Interestingly, the importance of biofilm dispersal in evolutionary processes may be observed in the apparent evolutionary conservation of the regulatory pathways that mediate dispersal. For example, multiple components of the NO-mediated dispersal signaling pathways from biofilms, NO/c-di-GMP, are also conserved in higher multicellular organisms, where the NO/cyclic GMP system regulates diverse physiological functions in mammals, regulating vasodilation, platelet aggregation, and sensory systems, as well as in plants, regulating development and pathogen defense responses (110). These similarities suggest an intriguing relationship between biofilm dispersal signaling and the evolution of eukaryotic regulatory pathways (29).

Intriguingly, bacteriophages have been associated with phenotypic diversity in the dispersal population. In particular, in *P. aeruginosa* a filamentous phage, Pf4, which is present as a prophage, is highly expressed during biofilm development compared to planktonic cells (8). The activity of the Pf4 phage was linked to killing and lysis of a subpopulation of cells within biofilms (16) and the emergence of phenotypic variants (96). Addition of the superinfective phage to planktonic cultures resulted in the formation of morphotypic variants, which were not observed in uninfected control cultures (111). Further, the Pf4 phage was found to increase fitness and virulence in a murine infection environment (111). However, it is not yet understood whether the Pf4 phage superinfective switch is coregulated with dispersal pathways. In *P. aeruginosa*, modulating c-di-GMP levels had no effect on phage superinfection, and both the wild type and a prophage mutant dispersed equally upon NO addition, suggesting that the NO dispersal pathway is not dependent on the Pf4 phage (unpublished data). A lytic RNA-containing bacteriophage, PP7, has also been implicated in regulating phenotypic variation in *P. aeruginosa,* leading to diversification and the evolution of a small-rough colony phenotype (112).

Dispersal is thought to be beneficial to biofilms for several reasons and, in particular, allows microbial populations to spread and survive. Arguably, dispersal may be considered the most important phase of biofilm development. Mathematical models have suggested that dispersal mostly benefits a population in an open space, while it can lead to its collapse in restricted environments (5). Dispersal is beneficial when a population faces catastrophic events that may eradicate a colony at a particular location, while leaving the location habitable for new organisms.

The lack of detailed studies of biofilm dispersal in natural systems leaves a number of questions unanswered. Can biofilm dispersal be detrimental as the result of utilizing crucial resources while increasing competition in a local environment and thus ultimately damaging the parental biofilm population? Do biofilms adopt a strategy of mass dispersal in harsh environments, risking extinction by misallocating energy resources to dispersal cells that lose the inherent protection of the biofilm? In the context of infection, a number of reports revealed that dispersal is a critical strategy for pathogens to successfully colonize a host. A functioning dispersal response appears to be crucial to ensure transmission of disease in both animals (113, 114) and plants, for instance, transmission of the plant pathogen *X. campestris* on the Chinese radish (50). A recent study found that *P. aeruginosa* mutant strains impaired in their ability to disperse from biofilms had reduced virulence and reduced ability to cause persistent infections in both murine and plant *in vivo* infection models (115). These beneficial effects of dispersal in a context of infection may be due to the association of dispersal signaling pathways with activation of virulence in some pathogens, e.g., via a decrease in c-di-GMP levels leading to expression of an endo-β-1,4-mannanase in *X. campestris* (50) and activation of type III secretion in *P. aeruginosa* (116).

Further studies will be required to fully grasp the role of dispersal in complex ecological systems involving multispecies biofilms in highly heterogeneous habitats. Metagenomics and molecular analysis–based studies appear highly suitable to reveal complex patterns of diversity in microbial communities but require precise temporally and geographically linked sampling points to track and analyze biofilm dispersal processes in natural environments (117). While not biofilm specific, some elegant model studies have been performed to determine the impact of dispersal events on microbial community composition and changes, which showed that environmental filtering, functional plasticity, and competition are all important mechanisms influencing the fate of dispersed communities (118).

OPPORTUNITIES FOR BIOFILM CONTROL

Manipulation of the biofilm development program by inducing dispersal signals has emerged as a strategy for developing novel biofilm control measures in recent years (Fig. 4). One of the first strategies was to treat biofilms with EPS-degrading agents to disperse attached cells. These treatments were later combined with antibiotics to eradicate the dispersed cells, for instance dispersinB in combination with the broad-spectrum biocide triclosan (119), or alginate lyase together with the aminoglycoside tobramycin (120). Given the diversity in the components of biofilm EPS matrices (121), enzyme-based treatments are likely to be biofilm specific, which can be beneficial when designing narrow-spectrum treatments.

The recent discoveries of the central role of the secondary messenger c-di-GMP in regulating the transition between biofilm

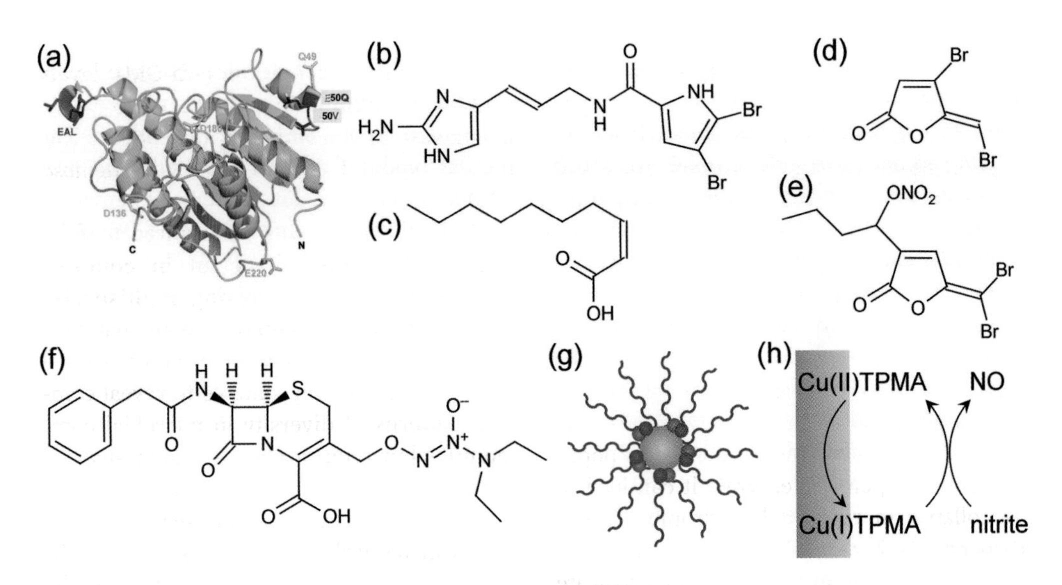

FIGURE 4 A range of strategies targeting dispersal have been developed to control biofilms and biofilm-related infections. (a) BdcA protein with enhanced c-di-GMP binding (124). (b) Oroidin and synthetic derivatives that were identified as potent dispersal inducers after screening chemical libraries (125). (c) Diffusible fatty acid signal cis-2-decenoic acid (23). (d) Furanone 30, a synthetic QS inhibitor derived from natural furanone compounds isolated from the red alga *Delisea pulchra* (131). (e) Fimbrolide-nitroester with dual action QS inhibition and NO release (134). (f) β-lactam-NO prodrugs for the targeted delivery of NO to infectious biofilms (132). (g) Controlled delivery of NO using nanoparticles (135). (h) Catalytic generation of NO from endogenous nitrite sources to disperse and prevent biofilm for long-term applications (138). doi:10.1128/microbiolspec.MB-0015-2014.f4

and free-swimming cells across many bacterial species attracted strong interest. Enzymes involved in c-di-GMP turnover appear very potent for controlling the switch, and the global signaling network has become a primary objective for developing novel antibiofilm measures. The first proof of concept studies of targeting c-di-GMP to treat infections showed that *in vivo* manipulation of c-di-GMP levels can effectively clear (by decreasing c-di-GMP) or prolong (by increasing c-di-GMP) *P. aeruginosa* infections in murine models (122, 123). Further, in an effort to develop a potential therapeutic drug, a BdcA protein was modified to enhance its c-di-GMP binding, thus reducing the intracellular c-di-GMP concentration. When added to biofilms *in vitro*, the modified BdcA protein caused nearly complete dispersal (124).

Improved high-throughput assays for biofilm dispersal allowed for the screening of large numbers of synthetic compounds and are particularly amenable to screening chemical libraries. Using this approach, 2-aminoimidazole derivatives targeting QS were designed and found to disperse established biofilms (125, 126). Studies of aerobic granules, a type of suspended biofilms used for wastewater treatment, revealed that the fatty acid 2-decenoic acid, which was previously identified to regulate dispersal in *P. aeruginosa* and a range of single-species biofilms, can also trigger dispersal in these mixed-species biofilms (127). In this context, strategies to inhibit 2-decenoic acid signaling may be useful to maintain granule integrity and sludge performance (127). QS signals and inhibitors have also been examined for their potential to enhance performance outputs in wastewater treatment biological reactors (128, 129), as well as for the treatment of biofilm-related infections and bacterial virulence (130, 131).

Of particular interest is NO, a simple and versatile dispersal signal that is highly conserved across biofilm species and activates c-di-GMP signaling to induce dispersal. Much progress has been made in recent years to design efficient NO delivery strategies, making it an outstanding candidate for novel therapeutic strategies. A range of NO-based technologies have been developed that offer a versatile range of solutions to control biofilms (reviewed in 57), which include a new class of β-lactam-NO prodrugs for the targeted delivery of NO to infectious biofilms (132, 133), dual quorum-sensing antagonists–NO releasing compounds (134), and novel polymers for sustained delivery of NO to prevent biofilm formation in a nontoxic fashion (135). NO-based treatments have been applied to the treatment of fouling on water filtration membranes (136, 137). A promising approach for long-term prevention and dispersal of biofilms in industrial and clinical settings is the use of copper-based catalytic technology to reduce endogenous nitrite ions to continuously produce NO at the surface (138). Recently, the first clinical trial was conducted to evaluate the use of low-dose inhaled NO gas combined with standard antibiotic therapy for the disruption of *P. aeruginosa* biofilms in patients with CF (139). The results demonstrated that patients who received NO gas at 5–10 ppm (~200 nM NO) showed significant reductions in the number of *Pseudomonas* biofilm aggregates compared to patients who received a placebo. These data suggest that using NO as an adjunctive therapy may be highly beneficial for the treatment of CF-related biofilm infections.

CONCLUDING REMARKS

Advances in biofilm and molecular biology studies have led to the elucidation of key signaling molecules and genes that govern biofilm development and dispersal processes. The existence of a developmental program appears clearer, but the challenge remains to address how these molecules and genes work together as control systems to generate patterns in space and time in biofilms. The biofilm mode of growth represents the predominant lifestyle for bacteria in environmental, industrial, and medical settings. A precise understanding of biofilm development and dispersal is crucial to better control microbial communities, with profound implications for global health, and to overcome the threat of antibiotic resistance as well as to develop innovative solutions for improved management of environmental microbes.

ACKNOWLEDGMENTS

Conflicts of interest: We disclose no conflicts.

CITATION

Barraud N, Kjelleberg S, Rice SA. 2015. Dispersal from microbial biofilms. Microbiol Spectrum 3(3):MB-0015-2014.

REFERENCES

1. **Ronce O.** 2007. How does it feel to be like a rolling stone? Ten questions about dispersal evolution. *Annu Rev Ecol Evol Syst* **38:**231–253.
2. **Nadell CD, Bassler BL.** 2011. A fitness trade-off between local competition and dispersal in *Vibrio cholerae* biofilms. *Proc Natl Acad Sci USA* **108:**14181–14185.
3. **Kisdi E.** 2002. Dispersal: risk spreading versus local adaptation. *Am Nat* **159:**579–596.
4. **Kerr B, Neuhauser C, Bohannan BJ, Dean AM.** 2006. Local migration promotes competitive restraint in a host-pathogen 'tragedy of the commons.' *Nature* **442:**75–78.
5. **Gyllenberg M, Parvinen K, Dieckmann U.** 2002. Evolutionary suicide and evolution of dispersal in structured metapopulations. *J Math Biol* **45:**79–105.
6. **Petrova OE, Sauer K.** 2009. A novel signaling network essential for regulating *Pseudomonas aeruginosa* biofilm development. *PLoS Pathog* **5:**e1000668. doi:10.1371/journal.ppat.1000668.
7. **Petrova OE, Sauer K.** 2012. Sticky situations: key components that control bacterial surface attachment. *J Bacteriol* **194:**2413–2425.

8. **Whiteley M, Bangera MG, Bumgarner RE, Parsek MR, Teitzel GM, Lory S, Greenberg EP.** 2001. Gene expression in *Pseudomonas aeruginosa* biofilms. *Nature* **413:**860–864.
9. **Klausen M, Aaes-Jorgensen A, Molin S, Tolker-Nielsen T.** 2003. Involvement of bacterial migration in the development of complex multicellular structures in *Pseudomonas aeruginosa* biofilms. *Mol Microbiol* **50:**61–68.
10. **Yang L, Hu Y, Liu Y, Zhang J, Ulstrup J, Molin S.** 2011. Distinct roles of extracellular polymeric substances in *Pseudomonas aeruginosa* biofilm development. *Environ Microbiol* **13:**1705–1717.
11. **Irie Y, Borlee BR, O'Connor JR, Hill PJ, Harwood CS, Wozniak DJ, Parsek MR.** 2012. Self-produced exopolysaccharide is a signal that stimulates biofilm formation in *Pseudomonas aeruginosa*. *Proc Natl Acad Sci USA* **109:** 20632–20636.
12. **Lenz AP, Williamson KS, Pitts B, Stewart PS, Franklin MJ.** 2008. Localized gene expression in *Pseudomonas aeruginosa* biofilms. *Appl Environ Microbiol* **74:**4463–4471.
13. **Stewart PS, Franklin MJ.** 2008. Physiological heterogeneity in biofilms. *Nat Rev Microbiol* **6:**199–210.
14. **Williamson KS, Richards LA, Perez-Osorio AC, Pitts B, McInnerney K, Stewart PS, Franklin MJ.** 2012. Heterogeneity in *Pseudomonas aeruginosa* biofilms includes expression of ribosome hibernation factors in the antibiotic-tolerant subpopulation and hypoxia-induced stress response in the metabolically active population. *J Bacteriol* **194:**2062–2073.
15. **McDougald D, Rice SA, Barraud N, Steinberg PD, Kjelleberg S.** 2012. Should we stay or should we go: mechanisms and ecological consequences for biofilm dispersal. *Nat Rev Microbiol* **10:**39–50.
16. **Webb JS, Thompson LS, James S, Charlton T, Tolker-Nielsen T, Koch B, Givskov M, Kjelleberg S.** 2003. Cell death in *Pseudomonas aeruginosa* biofilm development. *J Bacteriol* **185:**4585–4592.
17. **Tolker-Nielsen T, Brinch UC, Ragas PC, Andersen JB, Jacobsen CS, Molin S.** 2000. Development and dynamics of *Pseudomonas* sp. biofilms. *J Bacteriol* **182:**6482–6489.
18. **Entcheva-Dimitrov P, Spormann AM.** 2004. Dynamics and control of biofilms of the oligotrophic bacterium *Caulobacter crescentus*. *J Bacteriol* **186:**8254–8266.
19. **Lawrence JR, Chenier MR, Roy R, Beaumier D, Fortin N, Swerhone GD, Neu TR, Greer CW.** 2004. Microscale and molecular assessment of impacts of nickel, nutrients, and oxygen level on structure and function of river biofilm communities. *Appl Environ Microbiol* **70:**4326–4339.
20. **Manteca A, Fernandez M, Sanchez J.** 2005. A death round affecting a young compartmentalized mycelium precedes aerial mycelium dismantling in confluent surface cultures of *Streptomyces antibioticus*. *Microbiology* **151:** 3689–3697.
21. **Bayles KW.** 2007. The biological role of death and lysis in biofilm development. *Nat Rev Microbiol* **5:**721–726.
22. **Mai-Prochnow A, Lucas-Elio P, Egan S, Thomas T, Webb JS, Sanchez-Amat A, Kjelleberg S.** 2008. Hydrogen peroxide linked to lysine oxidase activity facilitates biofilm differentiation and dispersal in several Gram-negative bacteria. *J Bacteriol* **190:**5493–5501.
23. **Davies DG, Marques CN.** 2009. A fatty acid messenger is responsible for inducing dispersion in microbial biofilms. *J Bacteriol* **191:** 1393–1403.
24. **Mai-Prochnow A, Evans F, Dalisay-Saludes D, Stelzer S, Egan S, James S, Webb JS, Kjelleberg S.** 2004. Biofilm development and cell death in the marine bacterium *Pseudoalteromonas tunicata*. *Appl Environ Microbiol* **70:**3232–3238.
25. **Huynh TT, McDougald D, Klebensberger J, Al Qarni B, Barraud N, Rice SA, Kjelleberg S, Schleheck D.** 2012. Glucose starvation-induced dispersal of *Pseudomonas aeruginosa* biofilms is cAMP and energy dependent. *PLoS One* **7:**e42874. doi:10.1371/journal.pone.0042874.
26. **Purevdorj-Gage B, Costerton WJ, Stoodley P.** 2005. Phenotypic differentiation and seeding dispersal in non-mucoid and mucoid *Pseudomonas aeruginosa* biofilms. *Microbiology* **151:**1569–1576.
27. **Sauer K, Camper AK, Ehrlich GD, Costerton JW, Davies DG.** 2002. *Pseudomonas aeruginosa* displays multiple phenotypes during development as a biofilm. *J Bacteriol* **184:** 1140–1154.
28. **Morgan R, Kohn S, Hwang SH, Hassett DJ, Sauer K.** 2006. BdlA, a chemotaxis regulator essential for biofilm dispersion in *Pseudomonas aeruginosa*. *J Bacteriol* **188:**7335–7343.
29. **Barraud N, Schleheck D, Klebensberger J, Webb JS, Hassett DJ, Rice SA, Kjelleberg S.** 2009. Nitric oxide signaling in *Pseudomonas aeruginosa* biofilms mediates phosphodiesterase activity, decreased cyclic di-GMP levels, and enhanced dispersal. *J Bacteriol* **191:**7333–7342.
30. **Rollet C, Gal L, Guzzo J.** 2009. Biofilm-detached cells, a transition from a sessile to a planktonic phenotype: a comparative study of

adhesion and physiological characteristics in *Pseudomonas aeruginosa. FEMS Microbiol Lett* **290:**135–142.

31. **Gjermansen M, Ragas P, Sternberg C, Molin S, Tolker-Nielsen T.** 2005. Characterization of starvation-induced dispersion in *Pseudomonas putida* biofilms. *Environ Microbiol* **7:**894–906.
32. **Schleheck D, Barraud N, Klebensberger J, Webb JS, McDougald D, Rice SA, Kjelleberg S.** 2009. *Pseudomonas aeruginosa* PAO1 preferentially grows as aggregates in liquid batch cultures and disperses upon starvation. *PLoS One* **4:**e5513. doi:10.1371/journal.pone.0005513.
33. **Newell PD, Boyd CD, Sondermann H, O'Toole GA.** 2011. A c-di-GMP effector system controls cell adhesion by inside-out signaling and surface protein cleavage. *PLoS Biol* **9:** e1000587. doi:10.1371/journal.pbio.1000587.
34. **Banin E, Brady KM, Greenberg EP.** 2006. Chelator-induced dispersal and killing of *Pseudomonas aeruginosa* cells in a biofilm. *Appl Environ Microbiol* **72:**2064–2069.
35. **Thormann KM, Saville RM, Shukla S, Spormann AM.** 2005. Induction of rapid detachment in *Shewanella oneidensis* MR-1 biofilms. *J Bacteriol* **187:**1014–1021.
36. **An S, Wu J, Zhang LH.** 2010. Modulation of *Pseudomonas aeruginosa* biofilm dispersal by a cyclic-di-GMP phosphodiesterase with a putative hypoxia-sensing domain. *Appl Environ Microbiol* **76:**8160–8173.
37. **Saville RM, Rakshe S, Haagensen JA, Shukla S, Spormann AM.** 2011. Energy-dependent stability of *Shewanella oneidensis* MR-1 biofilms. *J Bacteriol* **193:**3257–3264.
38. **Sauer K, Cullen MC, Rickard AH, Zeef LA, Davies DG, Gilbert P.** 2004. Characterization of nutrient-induced dispersion in *Pseudomonas aeruginosa* PAO1 biofilm. *J Bacteriol* **186:** 7312–7326.
39. **Musk DJ, Banko DA, Hergenrother PJ.** 2005. Iron salts perturb biofilm formation and disrupt existing biofilms of *Pseudomonas aeruginosa. Chem Biol* **12:**789–796.
40. **Chatterjee S, Wistrom C, Lindow SE.** 2008. A cell-cell signaling sensor is required for virulence and insect transmission of *Xylella fastidiosa. Proc Natl Acad Sci USA* **105:**2670–2675.
41. **Hagai E, Dvora R, Havkin-Blank T, Zelinger E, Porat Z, Schulz S, Helman Y.** 2014. Surface-motility induction, attraction and hitchhiking between bacterial species promote dispersal on solid surfaces. *ISME J* **8:**1147–1151.
42. **Solano C, Echeverz M, Lasa I.** 2014. Biofilm dispersion and quorum sensing. *Curr Opin Microbiol* **18c:**96–104.
43. **Zhu J, Mekalanos JJ.** 2003. Quorum sensing-dependent biofilms enhance colonization in *Vibrio cholerae. Dev Cell* **5:**647–656.
44. **Kim SM, Park JH, Lee HS, Kim WB, Ryu JM, Han HJ, Choi SH.** 2013. LuxR homologue SmcR is essential for *Vibrio vulnificus* pathogenesis and biofilm detachment, and its expression is induced by host cells. *Infect Immun* **81:** 3721–3730.
45. **Rice SA, Koh KS, Queck SY, Labbate M, Lam KW, Kjelleberg S.** 2005. Biofilm formation and sloughing in *Serratia marcescens* are controlled by quorum sensing and nutrient cues. *J Bacteriol* **187:**3477–3485.
46. **Puskas A, Greenberg EP, Kaplan S, Schaefer AL.** 1997. A quorum-sensing system in the free-living photosynthetic bacterium *Rhodobacter sphaeroides. J Bacteriol* **179:**7530–7537.
47. **Boles BR, Horswill AR.** 2008. *agr*-Mediated dispersal of *Staphylococcus aureus* biofilms. *PLoS Pathog* **4:**e1000052. doi:10.1371/journal.ppat.1000052.
48. **Hammer BK, Bassler BL.** 2003. Quorum sensing controls biofilm formation in *Vibrio cholerae. Mol Microbiol* **50:**101–104.
49. **Bari SM, Roky MK, Mohiuddin M, Kamruzzaman M, Mekalanos JJ, Faruque SM.** 2013. Quorum-sensing autoinducers resuscitate dormant *Vibrio cholerae* in environmental water samples. *Proc Natl Acad Sci USA* **110:**9926–9931.
50. **Dow JM, Crossman L, Findlay K, He YQ, Feng JX, Tang JL.** 2003. Biofilm dispersal in *Xanthomonas campestris* is controlled by cell-cell signaling and is required for full virulence to plants. *Proc Natl Acad Sci USA* **100:**10995–11000.
51. **Deng Y, Wu J, Tao F, Zhang LH.** 2011. Listening to a new language: DSF-based quorum sensing in Gram-negative bacteria. *Chem Rev* **111:**160–173.
52. **Hentzer M, Riedel K, Rasmussen TB, Heydorn A, Andersen JB, Parsek MR, Rice SA, Eberl L, Molin S, Hoiby N, Kjelleberg S, Givskov M.** 2002. Inhibition of quorum sensing in *Pseudomonas aeruginosa* biofilm bacteria by a halogenated furanone compound. *Microbiology* **148:**87–102.
53. **Lonn-Stensrud J, Landin MA, Benneche T, Petersen FC, Scheie AA.** 2009. Furanones, potential agents for preventing *Staphylococcus epidermidis* biofilm infections? *J Antimicrob Chemother* **63:**309–316.
54. **Kolodkin-Gal I, Romero D, Cao S, Clardy J, Kolter R, Losick R.** 2010. D-Amino acids trigger biofilm disassembly. *Science* **328:**627–629.
55. **Leiman SA, May JM, Lebar MD, Kahne D, Kolter R, Losick R.** 2013. D-Amino acids

indirectly inhibit biofilm formation in *Bacillus subtilis* by interfering with protein synthesis. *J Bacteriol* **195:**5391–5395.

56. **Barraud N, Hassett DJ, Hwang SH, Rice SA, Kjelleberg S, Webb JS.** 2006. Involvement of nitric oxide in biofilm dispersal of *Pseudomonas aeruginosa. J Bacteriol* **188:**7344–7353.
57. **Barraud N, Kelso MJ, Rice SA, Kjelleberg S.** 2014. Nitric oxide: a key mediator of biofilm dispersal with applications in infectious diseases. *Curr Pharm Design* **21:**31–42.
58. **Sondermann H, Shikuma NJ, Yildiz FH.** 2012. You've come a long way: c-di-GMP signaling. *Curr Opin Microbiol* **15:**140–146.
59. **Römling U, Galperin MY, Gomelsky M.** 2013. Cyclic di-GMP: the first 25 years of a universal bacterial second messenger. *Microbiol Mol Biol Rev* **77:**1–52.
60. **Kalivoda EJ, Brothers KM, Stella NA, Schmitt MJ, Shanks RM.** 2013. Bacterial cyclic AMP-phosphodiesterase activity coordinates biofilm formation. *PLoS One* **8:**e71267. doi:10.1371/journal.pone.0071267.
61. **Galperin MY, Higdon R, Kolker E.** 2010. Interplay of heritage and habitat in the distribution of bacterial signal transduction systems. *Mol Biosyst* **6:**721–728.
62. **Gomelsky M.** 2011. cAMP, c-di-GMP, c-di-AMP and now cGMP: bacteria use them all! *Mol Microbiol* **79:**562–565.
63. **Liu N, Xu Y, Hossain S, Huang N, Coursolle D, Gralnick JA, Boon EM.** 2012. Nitric oxide regulation of cyclic di-GMP synthesis and hydrolysis in *Shewanella woodyi. Biochemistry* **51:**2087–2099.
64. **Carlson HK, Vance RE, Marletta MA.** 2010. H-NOX regulation of c-di-GMP metabolism and biofilm formation in *Legionella pneumophila. Mol Microbiol* **77:**930–942.
65. **Roy AB, Petrova OE, Sauer K.** 2012. The phosphodiesterase DipA (PA5017) is essential for *Pseudomonas aeruginosa* biofilm dispersion. *J Bacteriol* **194:**2904–2915.
66. **Li Y, Heine S, Entian M, Sauer K, Frankenberg-Dinkel N.** 2013. NO-induced biofilm dispersion in *Pseudomonas aeruginosa* is mediated by an MHYT domain-coupled phosphodiesterase. *J Bacteriol* **195:**3531–3542.
67. **Nioche P, Berka V, Vipond J, Minton N, Tsai AL, Raman CS.** 2004. Femtomolar sensitivity of a NO sensor from *Clostridium botulinum. Science* **306:**1550–1553.
68. **Plate L, Marletta MA.** 2013. Nitric oxide-sensing H-NOX proteins govern bacterial communal behavior. *Trends Biochem Sci* **38:**566–575.
69. **Tanaka A, Takahashi H, Shimizu T.** 2007. Critical role of the heme axial ligand, Met^{95}, in locking catalysis of the phosphodiesterase from *Escherichia coli* (*Ec* DOS) toward cyclic diGMP. *J Biol Chem* **282:**21301–21307.
70. **Tuckerman JR, Gonzalez G, Sousa EH, Wan X, Saito JA, Alam M, Gilles-Gonzalez MA.** 2009. An oxygen-sensing diguanylate cyclase and phosphodiesterase couple for c-di-GMP control. *Biochemistry* **48:**9764–9774.
71. **Petrova OE, Sauer K.** 2012. Dispersion by *Pseudomonas aeruginosa* requires an unusual posttranslational modification of BdlA. *Proc Natl Acad Sci USA* **109:**16690–16695.
72. **Kaplan JB, Meyenhofer MF, Fine DH.** 2003. Biofilm growth and detachment of *Actinobacillus actinomycetemcomitans. J Bacteriol* **185:**1399–1404.
73. **Kaplan JB, Ragunath C, Velliyagounder K, Fine DH, Ramasubbu N.** 2004. Enzymatic detachment of *Staphylococcus epidermidis* biofilms. *Antimicrob Agents Chemother* **48:**2633–2636.
74. **Boyd A, Chakrabarty AM.** 1994. Role of alginate lyase in cell detachment of *Pseudomonas aeruginosa. Appl Environ Microbiol* **60:**2355–2359.
75. **Whitchurch CB, Tolker-Nielsen T, Ragas PC, Mattick JS.** 2002. Extracellular DNA required for bacterial biofilm formation. *Science* **295:**1487.
76. **Mann EE, Rice KC, Boles BR, Endres JL, Ranjit D, Chandramohan L, Tsang LH, Smeltzer MS, Horswill AR, Bayles KW.** 2009. Modulation of eDNA release and degradation affects *Staphylococcus aureus* biofilm maturation. *PLoS One* **4:** e5822. doi:10.1371/journal.pone.0005822.
77. **Nijland R, Hall MJ, Burgess JG.** 2010. Dispersal of biofilms by secreted, matrix degrading, bacterial DNase. *PLoS One* **5:**e15668. doi:10.1371/journal.pone.0015668.
78. **Gjermansen M, Nilsson M, Yang L, Tolker-Nielsen T.** 2010. Characterization of starvation-induced dispersion in *Pseudomonas putida* biofilms: genetic elements and molecular mechanisms. *Mol Microbiol* **75:**815–826.
79. **Sheikh J, Czeczulin JR, Harrington S, Hicks S, Henderson IR, Le Bouguenec C, Gounon P, Phillips A, Nataro JP.** 2002. A novel dispersin protein in enteroaggregative *Escherichia coli. J Clin Invest* **110:**1329–1337.
80. **Knutton S, Shaw RK, Anantha RP, Donnenberg MS, Zorgani AA.** 1999. The type IV bundle-forming pilus of enteropathogenic *Escherichia coli* undergoes dramatic alterations in structure associated with bacterial adherence, aggregation and dispersal. *Mol Microbiol* **33:**499–509.
81. **Boles BR, Thoendel M, Singh PK.** 2005. Rhamnolipids mediate detachment of *Pseudomonas aeruginosa* from biofilms. *Mol Microbiol* **57:**1210–1223.

82. **Kuiper I, Lagendijk EL, Pickford R, Derrick JP, Lamers GEM, Thomas-Oates JE, Lugtenberg BJJ, Bloemberg GV.** 2004. Characterization of two *Pseudomonas putida* lipopeptide biosurfactants, putisolvin I and II, which inhibit biofilm formation and break down existing biofilms. *Mol Microbiol* **51:**97–113.
83. **Jackson DW, Suzuki K, Oakford L, Simecka JW, Hart ME, Romeo T.** 2002. Biofilm formation and dispersal under the influence of the global regulator CsrA of *Escherichia coli*. *J Bacteriol* **184:**290–301.
84. **Gibiansky ML, Conrad JC, Jin F, Gordon VD, Motto DA, Mathewson MA, Stopka WG, Zelasko DC, Shrout JD, Wong GC.** 2010. Bacteria use type IV pili to walk upright and detach from surfaces. *Science* **330:**197.
85. **Mai-Prochnow A, Ferrari BC, Webb JS, Kjelleberg S.** 2006. Ecological advantages of autolysis during the development and dispersal of *Pseudoalteromonas tunicata* biofilms. *Appl Environ Microbiol* **72:**5414–5420.
86. **Woo JK, Webb JS, Kirov SM, Kjelleberg S, Rice SA.** 2012. Biofilm dispersal cells of a cystic fibrosis *Pseudomonas aeruginosa* isolate exhibit variability in functional traits likely to contribute to persistent infection. *FEMS Immunol Med Microbiol* **66:**251–264.
87. **Liu J, Ling JQ, Zhang K, Wu CD.** 2013. Physiological properties of *Streptococcus mutans* UA159 biofilm-detached cells. *FEMS Microbiol Lett* **340:**11–18.
88. **Vaysse PJ, Sivadon P, Goulas P, Grimaud R.** 2011. Cells dispersed from *Marinobacter hydrocarbonoclasticus* SP17 biofilm exhibit a specific protein profile associated with a higher ability to reinitiate biofilm development at the hexadecane-water interface. *Environ Microbiol* **13:**737–746.
89. **Behnke S, Parker AE, Woodall D, Camper AK.** 2011. Comparing the chlorine disinfection of detached biofilm clusters with those of sessile biofilms and planktonic cells in single- and dual-species cultures. *Appl Environ Microbiol* **77:**7176–7184.
90. **Kirov SM, Webb JS, O'May CY, Reid DW, Woo JK, Rice SA, Kjelleberg S.** 2007. Biofilm differentiation and dispersal in mucoid *Pseudomonas aeruginosa* isolates from patients with cystic fibrosis. *Microbiology* **153:**3264–3274.
91. **Koh KS, Lam KW, Alhede M, Queck SY, Labbate M, Kjelleberg S, Rice SA.** 2007. Phenotypic diversification and adaptation of *Serratia marcescens* MG1 biofilm derived morphotypes. *J Bacteriol* **189:**119–130.
92. **Koh KS, Matz C, Tan CH, Le HL, Rice SA, Marshall DJ, Steinberg PD, Kjelleberg S.** 2012. Minimal increase in genetic diversity enhances predation resistance. *Mol Ecol* **21:**1741–1753.
93. **Lee KW, Periasamy S, Mukherjee M, Xie C, Kjelleberg S, Rice SA.** 2014. Biofilm development and enhanced stress resistance of a model, mixed-species community biofilm. *ISME J* **8:**894–907.
94. **McElroy KE, Hui JG, Woo JK, Luk AW, Webb JS, Kjelleberg S, Rice SA, Thomas T.** 2014. Strain-specific parallel evolution drives short-term diversification during *Pseudomonas aeruginosa* biofilm formation. *Proc Natl Acad Sci USA* **111:**E1419–E1427.
95. **Boles BR, Singh PK.** 2008. Endogenous oxidative stress produces diversity and adaptability in biofilm communities. *Proc Natl Acad Sci USA* **105:**12503–12508.
96. **Webb JS, Lau M, Kjelleberg S.** 2004. Bacteriophage and phenotypic variation in *Pseudomonas aeruginosa* biofilm development. *J Bacteriol* **186:**8066–8073.
97. **Fang H, Toyofuku M, Kiyokawa T, Ichihashi A, Tateda K, Nomura N.** 2013. The impact of anaerobiosis on strain-dependent phenotypic variations in *Pseudomonas aeruginosa*. *Biosci Biotechnol Biochem* **77:**1747–1752.
98. **Webb JS, Givskov M, Kjelleberg S.** 2003. Bacterial biofilms: prokaryotic adventures in multicellularity. *Curr Opin Microbiol* **6:**578–585.
99. **Hochberg ME, Rankin DJ, Taborsky M.** 2008. The coevolution of cooperation and dispersal in social groups and its implications for the emergence of multicellularity. *BMC Evol Biol* **8:**238.
100. **Boots M, Mealor M.** 2007. Local interactions select for lower pathogen infectivity. *Science* **315:**1284–1286.
101. **Wild G, Gardner A, West SA.** 2009. Adaptation and the evolution of parasite virulence in a connected world. *Nature* **459:**983–986.
102. **Barraud N, Buson A, Jarolimek W, Rice SA.** 2013. Mannitol enhances antibiotic sensitivity of persister bacteria in *Pseudomonas aeruginosa* biofilms. *PLoS One* **8:**e84220. doi:10.1371/journal.pone.0084220.
103. **Hoiby N, Bjarnsholt T, Givskov M, Molin S, Ciofu O.** 2010. Antibiotic resistance of bacterial biofilms. *Int J Antimicrob Agents* **35:**322–332.
104. **Roberts AP, Mullany P.** 2010. Oral biofilms: a reservoir of transferable, bacterial, antimicrobial resistance. *Expert Rev Anti-Infect Ther* **8:**1441–1450.

105. **Smith EE, Buckley DG, Wu Z, Saenphimmachak C, Hoffman LR, D'Argenio DA, Miller SI, Ramsey BW, Speert DP, Moskowitz SM, Burns JL, Kaul R, Olson MV.** 2006. Genetic adaptation by *Pseudomonas aeruginosa* to the airways of cystic fibrosis patients. *Proc Natl Acad Sci USA* **103:**8487–8492.
106. **Yang L, Jelsbak L, Marvig RL, Damkiaer S, Workman CT, Rau MH, Hansen SK, Folkesson A, Johansen HK, Ciofu O, Hoiby N, Sommer MO, Molin S.** 2011. Evolutionary dynamics of bacteria in a human host environment. *Proc Natl Acad Sci USA* **108:**7481–7486.
107. **Folkesson A, Jelsbak L, Yang L, Johansen HK, Ciofu O, Hoiby N, Molin S.** 2012. Adaptation of *Pseudomonas aeruginosa* to the cystic fibrosis airway: an evolutionary perspective. *Nat Rev Microbiol* **10:**841–851.
108. **Poltak SR, Cooper VS.** 2011. Ecological succession in long-term experimentally evolved biofilms produces synergistic communities. *ISME J* **5:**369–378.
109. **Traverse CC, Mayo-Smith LM, Poltak SR, Cooper VS.** 2013. Tangled bank of experimentally evolved *Burkholderia* biofilms reflects selection during chronic infections. *Proc Natl Acad Sci USA* **110:**E250–E259.
110. **Wendehenne D, Pugin A, Klessig DF, Durner J.** 2001. Nitric oxide: comparative synthesis and signaling in animal and plant cells. *Trends Plant Sci* **6:**177–183.
111. **Rice SA, Tan CH, Mikkelsen PJ, Kung V, Woo J, Tay M, Hauser A, McDougald D, Webb JS, Kjelleberg S.** 2009. The biofilm life cycle and virulence of *Pseudomonas aeruginosa* are dependent on a filamentous prophage. *ISME J* **3:**271–282.
112. **Brockhurst MA, Buckling A, Rainey PB.** 2005. The effect of a bacteriophage on diversification of the opportunistic bacterial pathogen, *Pseudomonas aeruginosa*. *Proc Biol Sci* **272:**1385–1391.
113. **Hall-Stoodley L, Stoodley P.** 2005. Biofilm formation and dispersal and the transmission of human pathogens. *Trends Microbiol* **13:**7–10.
114. **Chamot-Rooke J, Mikaty G, Malosse C, Soyer M, Dumont A, Gault J, Imhaus AF, Martin P, Trellet M, Clary G, Chafey P, Camoin L, Nilges M, Nassif X, Dumenil G.** 2011. Posttranslational modification of pili upon cell contact triggers *N. meningitidis* dissemination. *Science* **331:**778–782.
115. **Li Y, Petrova OE, Su S, Lau GW, Panmanee W, Na R, Hassett DJ, Davies DG, Sauer K.** 2014. BdlA, DipA and induced dispersion contribute to acute virulence and chronic persistence of *Pseudomonas aeruginosa*. *PLoS Pathog* **10:**e1004168. doi:10.1371/journal.ppat.1004168.
116. **Moscoso JA, Mikkelsen H, Heeb S, Williams P, Filloux A.** 2011. The *Pseudomonas aeruginosa* sensor RetS switches type III and type VI secretion via c-di-GMP signalling. *Environ Microbiol* **13:**3128–3138.
117. **Besemer K, Singer G, Hodl I, Battin TJ.** 2009. Bacterial community composition of stream biofilms in spatially variable-flow environments. *Appl Environ Microbiol* **75:**7189–7195.
118. **Székely AJ, Berga M, Langenheder S.** 2013. Mechanisms determining the fate of dispersed bacterial communities in new environments. *ISME J* **7:**61–71.
119. **Darouiche RO, Mansouri MD, Gawande PV, Madhyastha S.** 2009. Antimicrobial and antibiofilm efficacy of triclosan and dispersinB combination. *J Antimicrob Chemother* **64:**88–93.
120. **Lamppa JW, Griswold KE.** 2013. Alginate lyase exhibits catalysis-independent biofilm dispersion and antibiotic synergy. *Antimicrob Agents Chemother* **57:**137–145.
121. **Flemming HC, Wingender J.** 2010. The biofilm matrix. *Nat Rev Microbiol* **8:**623–633.
122. **Byrd MS, Pang B, Hong W, Waligora EA, Juneau RA, Armbruster CE, Weimer KE, Murrah K, Mann EE, Lu H, Sprinkle A, Parsek MR, Kock ND, Wozniak DJ, Swords WE.** 2011. Direct evaluation of *Pseudomonas aeruginosa* biofilm mediators in a chronic infection model. *Infect Immun* **79:**3087–3095.
123. **Christensen LD, van Gennip M, Rybtke MT, Wu H, Chiang WC, Alhede M, Hoiby N, Nielsen TE, Givskov M, Tolker-Nielsen T.** 2013. Clearance of *Pseudomonas aeruginosa* foreign-body biofilm infections through reduction of the cyclic di-GMP level in the bacteria. *Infect Immun* **81:**2705–2713.
124. **Ma Q, Yang Z, Pu M, Peti W, Wood TK.** 2011. Engineering a novel c-di-GMP-binding protein for biofilm dispersal. *Environ Microbiol* **13:**631–642.
125. **Rogers SA, Huigens RW 3rd, Cavanagh J, Melander C.** 2010. Synergistic effects between conventional antibiotics and 2-aminoimidazole-derived antibiofilm agents. *Antimicrob Agents Chemother* **54:**2112–2118.
126. **Frei R, Breitbach AS, Blackwell HE.** 2012. 2-Aminobenzimidazole derivatives strongly inhibit and disperse *Pseudomonas aeruginosa* biofilms. *Angew Chem-Int Edit* **51:**5226–5229.
127. **Cai PJ, Xiao X, He YR, Li WW, Yu L, Yu HQ.** 2013. Disintegration of aerobic granules induced by *trans*-2-decenoic acid. *Bioresour Technol* **128:**823–826.
128. **Yeon KM, Cheong WS, Oh HS, Lee WN, Hwang BK, Lee CH, Beyenal H, Lewandowski**

Z. 2009. Quorum sensing: a new biofouling control paradigm in a membrane bioreactor for advanced wastewater treatment. *Environ Sci Technol* **43:**380–385.

129. **Tan CH, Koh KS, Xie C, Tay M, Zhou Y, Williams R, Ng WJ, Rice SA, Kjelleberg S.** 2014. The role of quorum sensing signalling in EPS production and the assembly of a sludge community into aerobic granules. *ISME J.* [Epub ahead of print.] doi:10.1038/ismej.2013.240.
130. **Hentzer M, Givskov M.** 2003. Pharmacological inhibition of quorum sensing for the treatment of chronic bacterial infections. *J Clin Invest* **112:**1300–1307.
131. **Hentzer M, Wu H, Andersen JB, Riedel K, Rasmussen TB, Bagge N, Kumar N, Schembri MA, Song Z, Kristoffersen P, Manefield M, Costerton JW, Molin S, Eberl L, Steinberg P, Kjelleberg S, Hoiby N, Givskov M.** 2003. Attenuation of *Pseudomonas aeruginosa* virulence by quorum sensing inhibitors. *EMBO J* **22:**3803–3815.
132. **Barraud N, Kardak BG, Yepuri NR, Howlin RP, Webb JS, Faust SN, Kjelleberg S, Rice SA, Kelso MJ.** 2012. Cephalosporin-3′-diazeniumdiolates: targeted NO-donor prodrugs for dispersing bacterial biofilms. *Angew Chem-Int Edit* **51:**9057–9060.
133. **Yepuri NR, Barraud N, Shah Mohammadi N, Kardak BG, Kjelleberg S, Rice SA, Kelso MJ.** 2013. Synthesis of cephalosporin-3′-diazeniumdiolates: biofilm dispersing NO-donor prodrugs activated by β-lactamase. *Chem Commun* **49:**4791–4793.
134. **Kutty SK, Barraud N, Pham A, Iskander G, Rice SA, Black DS, Kumar N.** 2013. Design, synthesis, and evaluation of fimbrolide-nitric oxide donor hybrids as antimicrobial agents. *J Med Chem* **56:**9517–9529.
135. **Duong HT, Jung K, Kutty SK, Agustina S, Adnan NN, Basuki JS, Kumar N, Davis TP, Barraud N, Boyer C.** 2014. Nanoparticle (star polymer) delivery of nitric oxide effectively negates *Pseudomonas aeruginosa* biofilm formation. *Biomacromolecules.* [Epub ahead of print.] doi:10.1021/bm500422v.
136. **Barraud N, Storey MV, Moore ZP, Webb JS, Rice SA, Kjelleberg S.** 2009. Nitric oxide-mediated dispersal in single- and multi-species biofilms of clinically and industrially relevant microorganisms. *Microb Biotechnol* **2:**370–378.
137. **Barnes RJ, Bandi RR, Wong WS, Barraud N, McDougald D, Fane A, Kjelleberg S, Rice SA.** 2013. Optimal dosing regimen of nitric oxide donor compounds for the reduction of *Pseudomonas aeruginosa* biofilm and isolates from wastewater membranes. *Biofouling* **29:**203–212.
138. **Ren H, Wu J, Xi C, Lehnert N, Major T, Bartlett RH, Meyerhoff ME.** 2014. Electrochemically modulated nitric oxide (NO) releasing biomedical devices via copper(II)-tri (2-pyridylmethyl)amine mediated reduction of nitrite. *ACS Appl Mater Interfaces* **6:**3779–3783.
139. **Cathie K, Howlin RP, Barraud N, Carroll MP, Clarke SC, Connett GJ, Cornelius V, Daniels TW, Duignan C, Feelisch M, Fernandez B, Hall-Stoodley L, Jefferies JMC, Kelso MJ, Kjelleberg S, Legg JP, Pink S, Rice SA, Rogers G, Salib RJ, Smith C, Stoodley P, Sukhtankar P, Webb JS, Faust SN.** 2014. Low dose nitric oxide as adjunctive therapy to reduce antimicrobial tolerance of *Pseudomonas aeruginosa* biofilms in the treatment of patients with cystic fibrosis: report of a proof of concept clinical trial. *Am J Respir Crit Care Med* **189:**A2843.

Chemical Biology Strategies for Biofilm Control

18

LIANG YANG[1,2] and MICHAEL GIVSKOV[1,3]

SIGNALING PATHWAYS REGULATING BACTERIAL BIOFILM FORMATION

Signaling pathways are required for bacterial biofilm formation and antimicrobial resistance. Among them, quorum sensing (QS) and c-di-GMP signaling are the best characterized. QS is a widely distributed intercellular signaling mechanism by which microorganisms regulate gene expression in response to small diffusible signaling molecules (1). Bacteria have developed oligopeptides, *N*-acyl homoserine lactones (HSLs), and autoinducer-2 as signal molecules (1). When the QS signal molecules reach a local threshold concentration, they can interact with specific receptors and impact the expression of hundreds of genes. Many of the QS-regulated genes (motility, biosurfacant synthesis, EPS synthesis) are required for the biofilm formation and antibiotic resistance of various bacterial species (2).

Cyclic di-GMP (c-di-GMP) is a secondary messenger which mediates signal transduction in many Gram-negative species of bacteria (3). Based on widely distributed sequence similarities and known functionality in a few bacterial

[1]Singapore Centre on Environmental Life Sciences Engineering, Nanyang Technological University, Singapore; [2]School of Biological Sciences, Nanyang Technological University, Singapore 639798; [3]Costerton Biofilm Center, Department of International Health, Immunology, and Microbiology, University of Copenhagen, 2200 København N, Denmark.

Microbial Biofilms 2nd Edition
Edited by Mahmoud Ghannoum, Matthew Parsek, Marvin Whiteley, and Pranab K. Mukherjee

doi:10.1128/microbiolspec.MB-0019-2015

species, c-di-GMP is believed to play an essential role in determining the lifestyles of a wide range of bacteria. When the intracellular c-di-GMP content is high, bacteria cells tend to downregulate motility and increase synthesis of extracellular polymeric substances and thus facilitate biofilm formation. In contrast, lowering the intracellular c-di-GMP content enhances bacterial motility and causes biofilm dispersal. The intracellular c-di-GMP content is determined by a number of proteins equipped with diguanylate cyclase (DGC) activity. Such proteins can catalyze the synthesis of c-di-GMP, whereas proteins equipped with phosphodiesterase (PDE) activities degrade c-di-GMP. Many species contain multiple such proteins with DGC and PDE activities, which presumably offers bacteria a great degree of flexibility in how they can modulate their intracellular c-di-GMP content in response to different environmental stimuli (3). Recently, small noncoding RNAs (sRNAs) were found to participate in both QS and c-di-GMP signaling transduction and thus modulate biofilm formation posttranscriptionally. sRNAs often interact with regulatory proteins involved in QS and c-di-GMP pathways and can thus serve as potential targets for designing antibiofilm compounds (4).

CHEMICAL BIOLOGY STRATEGIES TARGETING QS

For decades, QS has been recognized as a key target for development of the next generation of antimicrobial agents. Chemical blockage of QS was shown to reduce production of virulence factors, attenuate bacteria in mouse models of infection, and reduce biofilm resistance by various bacterial pathogens (5–7). Efforts directed toward identifying functional QS inhibitors (QSIs) started in the 1990s (8, 9). Multiple strategies have been developed to interfere with bacterial QS including but not limited to (i) competitive binding of inhibitors to the QS receptors (10), (ii) inhibiting synthesis of QS signals (11), (iii) enzymatic degradation of QS signals (12), and (iv) posttranscriptional control of QS genes via sRNAs (13).

Among these strategies, preventing the QS receptors (also referred to as LuxR-type receptors) to perceive their native signals, seems to be the most successfully employed strategy to identify QSIs that would also show functionality in animal models of infection. Chemicals capable of blocking QS have been referred to as antipathogenic or antivirulence drugs (14). To find functional chemicals, Rasmussen and coworkers developed three QSI selector (QSIS) systems by fusing different report genes to QS-regulated promoters (Fig. 1) (10). Natural as well as synthetic compounds have been identified to specifically bind to the LuxR-type receptors (10). Besides the classic LuxR-type receptors,

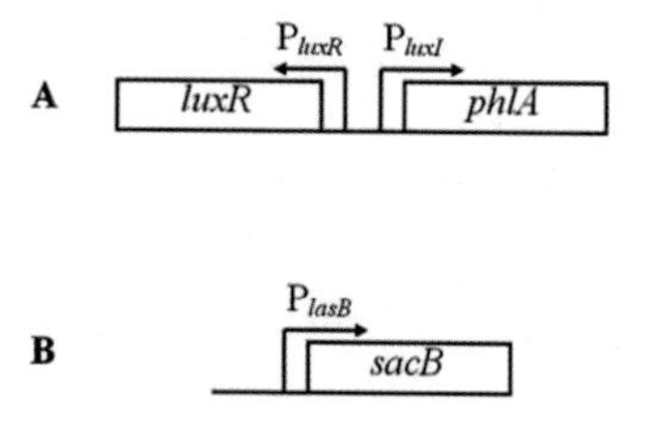

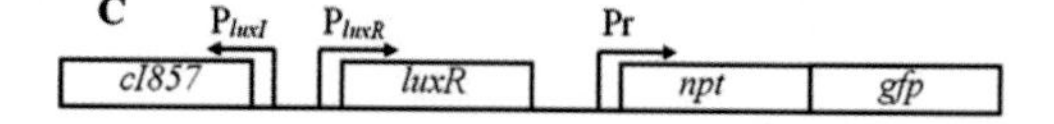

FIGURE 1 Example of three QSIS systems. (A) In QSIS1, an engineered vector expressing the *phlA* gene that encodes the toxic gene product under the control of LuxR was transformed to *E. coli*. (B) In QSIS2, the LasR-regulated *lasB* promoter controls the expression of the *sacB* gene, expression of which leads to cell death in the presence of sucrose. (C) The QSIS3 system is also based on LuxR regulation. The *npt* and *gfp* genes, conferring kanamycin resistance and green fluorescence, respectively, are controlled by the *cI* repressor, which in turn is regulated by QS through the *luxI* promoter. The system was established in *E. coli*. Figure adapted from Rasmussen et al. (10) with permission of the publisher. doi:10.1128/microbiolspec.MB-0019-2015.f1

QSIs targeting other types of receptors for QS signals have also been identified. *Pseudomonas aeruginosa* has an alkyl-quinolone (AQ)-dependent QS system in addition to its two *N*-acyl-HSL QS systems to regulate expression of a range of virulence genes (15). The *P. aeruginosa* AQ QS system has a LysR-type transcriptional regulator protein, PqsR (also known as MvfR), as the receptor of multiple 2-alkyl-4-quinolones including the *Pseudomonas* quinolone signal (2-heptyl-3-hydroxy-4[1H]-quinolone) and its precursor 2-heptyl-4-hydroxyquinoline (16).

Ilangovan and coworkers have obtained the crystal structures of the PqsR co-inducer binding domain and a complex with the native agonist 2-nonyl-4-hydroxyquinoline (17). Based on the structure analysis of PqsR, they employed a ligand-based design strategy and evaluated a series of 50 AQ and novel quinazolinone analogues for their impact on AQ QS. This study provides novel insights for development of antivirulence drugs by targeting the AQ receptor PqsR (17).

Some 20 years ago, screening for QSI compounds involved plate-based phenotypic assays, such as observing the change of QS-regulated motility and pigment production in the presence of potential QSIs (8, 18). The introduction of *gfp*-based QS reporter fusion vectors has made possible the application of microplate-based high-throughput screening of QSIs (19). Recent advances in QSI screening strategies include *in silico* virtual screening, cell-free screening, and *in vivo* screening (20). Virtual screening of QSI relied on high-resolution crystal structures of QS receptors. Recently, the crystal structure of the ligand-binding domain of the *P. aeruginosa* LasR protein bound to its native signal 3-oxo-C (12)-acyl-HSL lactone was solved and provided clues of atomic interactions between an acyl-HSL receptor and its cognate signal molecule. Molecular docking–based *in silico* virtual screening was performed based on the LasR crystal structures. Several natural compounds and recognized drugs with distinct structures to the 3-oxo-C (12)-acyl-HSL were identified as specific LasR inhibitors and were able to repress LasR-regulated virulence factors (21, 22). Because the synthesis of QS signals requires enzymes such as acyl-HSL synthases, a cell-free coupled enzyme assay based on the commercially available S-adenosyl homocysteine assay was developed by Christensen and coworkers (23) to screen inhibitors of acyl-HSL synthases. In this assay, the *Burkholderia mallei* acyl-HSL synthase, BmaI1, was chosen as the primary target for screening, and the fluorescent compound resorufin production is dependent on acyl-HSL synthesis. By using this screen, they identified and characterized several acyl-HSL synthase inhibitors.

Extracellular inactivation or interference of the function of QS signals is another strategy to reduce bacterial QS-mediated virulence. For example, *N*-acyl-HSL acylases are found to be produced by both prokaryotic and eukaryotic species and can degrade the *N*-acyl-HSLs produced by pathogens (12, 24). In addition to regulating production of virulence factors, the QS signals were reported to act directly as virulence factors. Evidence has shown that a low concentration of the *P. aeruginosa* 3-oxo-C (12)-acyl-HSL can repress NF-κB transcriptional activity and cause apoptosis (25, 26). To target the direct virulence effects of 3-oxo-C (12)-acyl-HSL, Valentine and coworkers developed a cell-based high-throughput screen to identify small molecules that restore NF-κB transcriptional activity of a fisher rat thyroid epithelial cell line in the presence of 3-oxo-C (12)-acyl-HSL. This cell line contains a NF-κB-dependent luciferase reporter, and luciferase expression can be efficiently induced after incubation with lipopolysaccharide derived from *P. aeruginosa*. The presence of 20 μM 3-oxo-C (12)-acyl-HSL lactone greatly repressed the lipopolysaccharide-induced luciferase expression. Screening of 25,600 synthetic, drug-like small molecules has lead to the identification of triazolo[4,3-a] quinolines that can restore NF-κB-induced responses in cell lines (27). This study shed light on the search for small molecules that could directly interfere with the virulence effects generated by the QS signals.

Several sRNAs have been reported to control QS at different levels. For example, in *P. aeruginosa*, the GacS/GacA two-component system activates expression of two sRNAs denoted RsmY and RsmZ, which are required to sequester a QS translational repressor, RsmA (28, 29). RsmA can prevent ribosome binding of its target mRNAs (including QS-regulated mRNAs) and thus represses their translation. High concentrations of RsmY and RsmZ are required to bind RsmA and abolish its function. Similarly, another sRNA, CrcZ, whose expression is under the control of the CbrA/CbrB two-component system (30), was required to sequester a *pqs* QS repressor, the catabolite repression control protein Crc (31). Another two sets of sRNAs, prrF1/prrF2 and PhrS, were shown to negatively and positively regulate expression of the *P. aeruginosa pqs* QS genes (32, 33). The abundance of sRNAs encoded by bacterial genomes has thus provided multiple potential targets for designing specific QS inhibitors. Active compounds that are able to interfere with the expression of sRNAs were already found to efficiently inhibit QS and virulence (13).

Unlike conventional antibiotics, QSIs do not directly inhibit the growth of bacteria and thus are believed to pose less selective pressure for the development-resistant strains (5). However, *P. aeruginosa* LasR mutants that cannot recognize the 3-oxo-C (12)-acyl-HSL were frequently observed in clinical isolates (34–36), which indicates that a functional QS system isn't required during long-term colonization. Mellbye and Schuster designed a proof-of-principle experiment to show that QSI resistance could not spread, because QS cheaters can affect nutrient uptake and delay growth of the entire population (37). In another study, a placebo-controlled, azithromycin (a recognized antibiotic with QSI properties) pneumonia-prevention trial was performed and showed that the *lasR* (QS)-mutants appeared less frequently among the population in the presence of QSI than in the absence of QSI and thus QSI selected for cooperation in *P. aeruginosa* populations during infection (38). It is therefore likely that QSI drugs may be functional during extended periods of clinical usage. More interestingly, recent work from Lee and coworkers reported that a new QS signal, IQS, can be employed by the *P. aeruginosa lasR* mutant to coordinate the expression of virulence genes that are normally regulated by the *las* QS system (39). The production of IQS can be activated by phosphate limitation in the absence of LasR. This study highlights the complexity of QS systems and suggests potential novel targets for designing QSIs.

CHEMICAL BIOLOGY STRATEGIES TARGETING c-di-GMP

It has been a while since researchers realized that bacterial c-di-GMP signaling could be a promising target for development of chemical biology–based biofilm control. Compounds targeting c-di-GMP can be used to prevent biofilm formation or disperse late-stage biofilms. Bacterial genomes often encode multiple copies of DGCs and PDEs, presumably to modulate c-di-GMP levels in response to diverse environmental cues. Understanding the working mechanisms of DGCs and PDEs can lead to identification of chemical compounds that inhibit DGCs or activate PDEs. C-di-GMP signal transduction requires PilZ domain proteins or riboswitches that can directly bind to c-di-GMP (3, 40). These c-di-GMP receptors can also be targets for designing compounds targeting c-di-GMP signaling.

Several fluorescence-based biosensors have been developed to monitor intracellular c-di-GMP levels. Binding of c-di-GMP to a receptor protein can often change its conformation. Based on this, several groups designed fluorescence resonance energy transfer (FRET)–based reporter vectors to monitor c-di-GMP levels within single bacterial cells. Christen and coworkers constructed a set of genetically encoded FRET-based biosensors by fusing PilZ proteins between cyan fluorescent proteins (CFPs) and yellow fluorescent proteins (YFPs) (41). They used the FRET/CFP emission ratio to reflect cellular c-di-GMP levels and found that there is an asymmetrical distribution of c-di-GMP in the progeny that correlates with the time of cell division and polarization for motile bacteria. The intracellular c-di-GMP levels were shown to be significantly lower in the flagellated cell than in the nonflagellated cells of *P. aeruginosa* (41). In another study, Ho and coworkers constructed two FRET-based c-di-GMP biosensors, cdg-S1 and cdg-S2, by fusing CFPs and YFPs to two PilZ proteins, MrkH and VCA0042, respectively (Fig. 2) (42). These two biosensors exhibit similar response dynamics intracellularly but with different dynamic ranges, so they can be used to test effects of different compounds to c-di-GMP, perturbation (Fig. 2) (42). Using these two biosensors, the authors evaluated the time-dependent changes in the FRET efficiencies of biosensor strains after exposure to compounds which can enhance or decrease c-di-GMP, and showed that *Escherichia coli* cells reduce intracellular c-di-GMP content shortly after being engulfed by macrophages (42). Rybtke and coworkers created a c-di-GMP reporter strain by transcriptionally fusing the c-di-GMP-responsive *P. aeruginosa cdrA* promoter to gfp (43). CdrA is an adhesive matrix protein that can bind the *P. aeruginosa* Psl exopolysaccharide and cause aggregation (44). The p_{cdrA}-*gfp* reporter construct can be transformed into different *P. aeruginosa* strains and reflects the intracellular c-di-GMP level in real time.

A range of synthetic and natural compounds have been discovered that can affect c-di-GMP signaling or competitively bind to DGCs via different approaches. Several groups synthesized analogs of c-di-GMPs and proved that these analogs could efficiently inhibit the enzymatic activity of DGCs (45, 46). High-throughput screens have also been employed to identify c-di-GMP-inhibiting compounds. Sambanthamoorthy and coworkers constructed a *Vibrio cholerae* c-di-GMP reporter strain to perform cell-based high-throughput screening of c-di-GMP interference compounds (47). Initially, they engineered a *V. cholerae* strain to constitutively express the DGC VC1216. Then they introduced a c-di-GMP reporter plasmid (which carries a transcriptionally fused c-di-GMP-inducible promoter of the VC1673 gene to the luciferase operon [P_{VC1673}-*lux*]) into the high c-di-GMP-level *V. cholerae* VC1216 background. The high intracellular c-di-GMP of this strain promotes the expression of P_{VC1673}-*lux* fusion with luciferase activity as an indicator of intracellular c-di-GMP levels. The researchers cultivated the *V. cholerae* (VC1216)/P_{VC1673}-*lux* strain in the presence of 66,000 compounds/natural-product extracts and identified compounds and natural-product extracts that reduced P_{VC1673}-*lux* fusion expression without negatively impacting *V. cholerae* growth. During this screening, they identified 358 compounds that exhibited greater than 50% inhibition of P_{VC1673}-*lux* fusion expression, and 184 identified compounds had IC_{50} of less than 10 μM (47).

In another study, Sambanthamoorthy and coworkers used an *in silico* pharmacophore-based screen to identify small-molecule inhibitors of DGCs (48). Two-dimensional pharmacophores were generated based on the PleD DGC of *Caulobacter crescentus*, and a focused library from the commercially available compound database was generated using queries derived from the two-dimensional phamacophores. *In silico* screening of this focused library was performed against the c-di-GMP binding site of the published crystal structure of PleD DGC (Pubmed: 15569936) by using Molecular Operating Environment (Chemical Computing Group, Quebec, Canada). The top-ranking compounds were then purchased and tested for their ability to inhibit DGC activity *in vitro*. Four small molecules were discovered in this

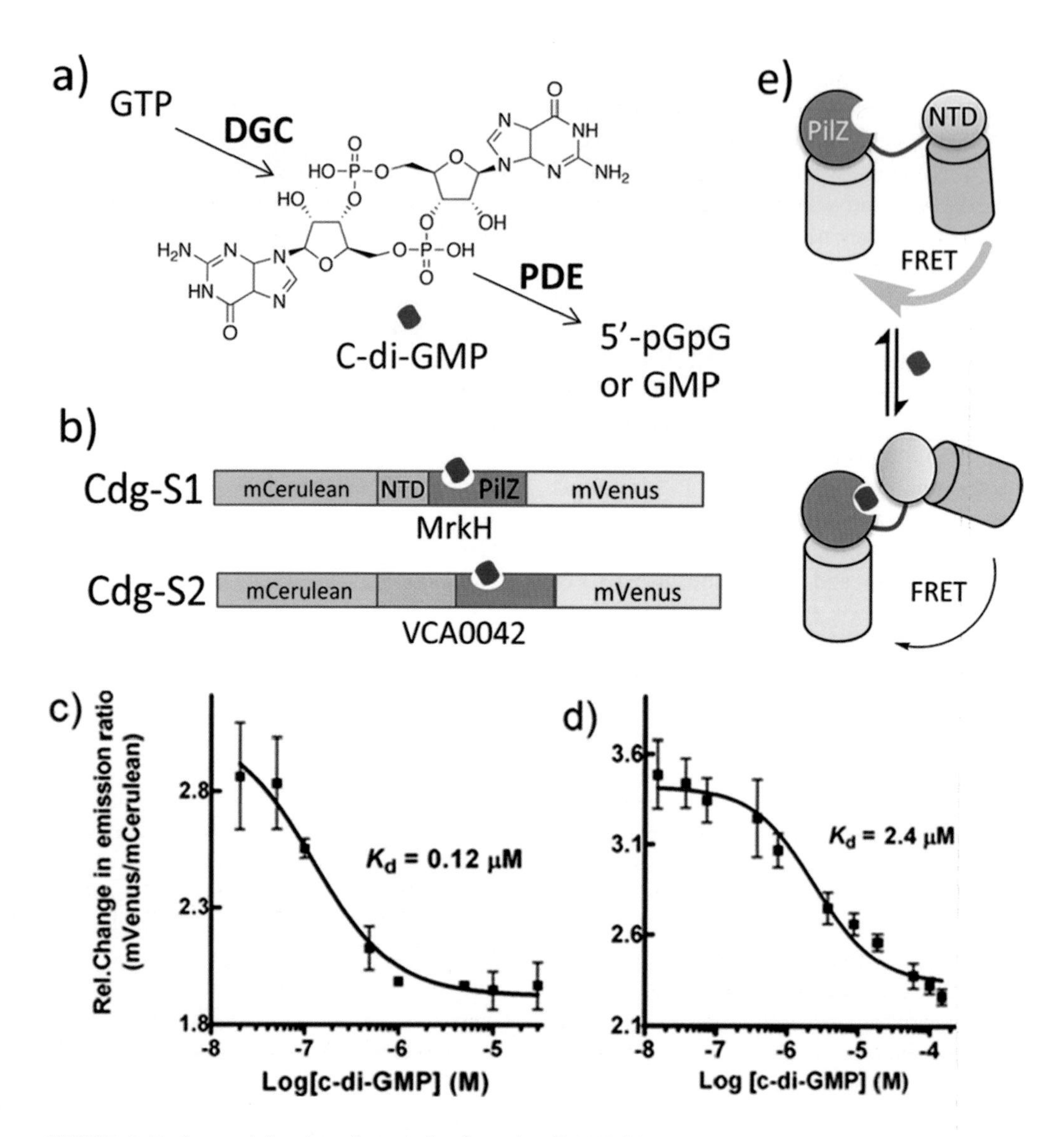

FIGURE 2 Design and *in vitro* characterization of c-di-GMP biosensors. (a) Principle of synthesis and degradation of c-di-GMP by DGCs and PDEs. (b) Construction of the genetically encoded FRET-based biosensors for c-di-GMP using MrkH and VCA0042. Both proteins contain a c-di-GMP binding PilZ domain and an N-terminal domain (NTD). (c, d) Fluorescence titration curves for cdg-S1 and cdg-S2. (e) Schematic illustration of the conformational change induced by binding c-di-GMP to cdg-S1 and cdg-S2. Figure adapted from Ho et al. (42) with permission of the publisher. doi:10.1128/microbiolspec.MB-0019-2015.f2

screening to antagonize DGC activities and inhibit biofilm formation by *P. aeruginosa* and *Acinetobacter baumannii* (48).

Several natural compounds and recognized drugs were found to inhibit c-di-GMP signaling of bacteria. Ohana and coworkers reported that the glycosylated triterpenoid saponin is a specific inhibitor of DGC from *Acetobacter xylinum* (49). An immunosuppressive drug, azathioprine, was found by Antoniani and coworkers to inhibit c-di-GMP synthesis via perturbation of bacterial intracellular nucleotide pools (50). Recently, nitric oxide (NO)–generating compounds were reported to be promising agents to control biofilms via c-di-GMP signaling interference (51, 52). Barraud and coworkers synthesized a chemically stable cephalosporin-3′-diazeniumdiolate NO-donor prodrug that can be activated by bacterial β-lactamases after exposure to biofilms and cause biofilm dispersal (53).

Similar to the case for QSI compounds, there is a profound lack of animal models to test identified c-di-GMP-signaling interference compounds. Christensen and coworkers carried out a proof-of-concept study to show that biofilm dispersal can be induced via reducing intracellular c-di-GMP content in a *P. aeruginosa* foreign-body mouse infection model (54). In this work, reduction of c-di-GMP in *P. aeruginosa* cells was achieved via a genetic approach to avoid pleotropic effects of chemical-based approaches. An *E. coli* PDE, YhjH, was fused to an arabinose-inducible p_{BAD} promoter and transformed into the *P. aeruginosa* wild-type PAO1 strain. Biofilms were formed by the PAO1/p_{BAD}-*yhjH* strain on the surface of silicone implants located in the peritoneal cavity of mice. *In vivo* dispersal of biofilms was observed after induction of the YhjH protein via arabinose (54). This study provides the first real-time observation of biofilm dispersal and can serve as a model system to test the impact of c-di-GMP interference compounds in the future.

SUMMARY

Bacterial biofilms are notorious in many industrial and hospital settings and cause a huge economic burden to our society. The fundamental biofilm mechanism studies in recent years have opened novel routes for development of biofilm-controlling approaches. Chemical biology approaches to combat biofilms are being widely investigated by researchers, including QS inhibitors, c-di-GMP-signaling interference compounds, anti-attachment compounds, and so on. A few of these compounds have been shown to be functional *in vivo* and to be promising antibiofilm agents for further development. These studies will further provide molecular targets and biomarkers for development of antibiofilm agents.

ACKNOWLEDGMENTS

Conflicts of interest: We disclose no conflicts.

CITATION

Yang L, Givskov M. 2015. Chemical biology strategies for biofilm control. Microbiol Spectrum 3(4):MB-0019-2015.

REFERENCES

1. **Miller MB, Bassler BL.** 2001. Quorum sensing in bacteria. *Annu Rev Microbiol* **55:**165–199.
2. **Parsek MR, Greenberg EP.** 2005. Sociomicrobiology: the connections between quorum sensing and biofilms. *Trends Microbiol* **13:**27–33.
3. **Romling U, Gomelsky M, Galperin MY.** 2005. C-di-GMP: the dawning of a novel bacterial signalling system. *Mol Microbiol* **57:**629–639.
4. **Chambers JR, Sauer K.** 2013. Small RNAs and their role in biofilm formation. *Trends Microbiol* **21:**39–49.
5. **Hentzer M, Wu H, Andersen JB, Riedel K, Rasmussen TB, Bagge N, Kumar N, Schembri MA, Song Z, Kristoffersen P, Manefield M, Costerton JW, Molin S, Eberl L, Steinberg P, Kjelleberg S, Hoiby N, Givskov M.** 2003. Attenuation of *Pseudomonas aeruginosa* virulence by quorum sensing inhibitors. *EMBO J* **22:**3803–3815.
6. **Ng WL, Perez L, Cong J, Semmelhack MF, Bassler BL.** 2012. Broad spectrum pro-quorum-sensing molecules as inhibitors of virulence in vibrios. *PLoS Pathog* **8:**e1002767. doi:10.1371/journal.ppat.1002767.
7. **Brackman G, Cos P, Maes L, Nelis HJ, Coenye T.** 2011. Quorum sensing inhibitors increase the susceptibility of bacterial biofilms to antibiotics *in vitro* and *in vivo*. *Antimicrob Agents Chemother* **55:**2655–2661.
8. **Givskov M, de Nys R, Manefield M, Gram L, Maximilien R, Eberl L, Molin S, Steinberg PD, Kjelleberg S.** 1996. Eukaryotic interfer-

ence with homoserine lactone-mediated prokaryotic signalling. *J Bacteriol* **178:**6618–6622.

9. **Schaefer AL, Hanzelka BL, Eberhard A, Greenberg EP.** 1996. Quorum sensing in *Vibrio fischeri*: probing autoinducer-LuxR interactions with autoinducer analogs. *J Bacteriol* **178:**2897–2901.
10. **Rasmussen TB, Bjarnsholt T, Skindersoe ME, Hentzer M, Kristoffersen P, Kote M, Nielsen J, Eberl L, Givskov M.** 2005. Screening for quorum-sensing inhibitors (QSI) by use of a novel genetic system, the QSI selector. *J Bacteriol* **187:**1799–1814.
11. **Chung J, Goo E, Yu S, Choi O, Lee J, Kim J, Kim H, Igarashi J, Suga H, Moon JS, Hwang I, Rhee S.** 2011. Small-molecule inhibitor binding to an N-acyl-homoserine lactone synthase. *Proc Natl Acad Sci USA* **108:**12089–12094.
12. **Dong YH, Wang LH, Xu JL, Zhang HB, Zhang XF, Zhang LH.** 2001. Quenching quorum-sensing-dependent bacterial infection by an N-acyl homoserine lactonase. *Nature* **411:**813–817.
13. **Perez-Martinez I, Haas D.** 2011. Azithromycin inhibits expression of the GacA-dependent small RNAs RsmY and RsmZ in *Pseudomonas aeruginosa*. *Antimicrob Agents Chemother* **55:**3399–3405.
14. **Hentzer M, Givskov M.** 2003. Pharmacological inhibition of quorum sensing for the treatment of chronic bacterial infections. *J Clin Invest* **112:**1300–1307.
15. **Gallagher LA, McKnight SL, Kuznetsova MS, Pesci EC, Manoil C.** 2002. Functions required for extracellular quinolone signaling by *Pseudomonas aeruginosa*. *J Bacteriol* **184:**6472–6480.
16. **Deziel E, Lepine F, Milot S, He J, Mindrinos MN, Tompkins RG, Rahme LG.** 2004. Analysis of *Pseudomonas aeruginosa* 4-hydroxy-2-alkylquinolines (HAQs) reveals a role for 4-hydroxy-2-heptylquinoline in cell-to-cell communication. *Proc Natl Acad Sci USA* **101:**1339–1344.
17. **Ilangovan A, Fletcher M, Rampioni G, Pustelny C, Rumbaugh K, Heeb S, Camara M, Truman A, Chhabra SR, Emsley J, Williams P.** 2013. Structural basis for native agonist and synthetic inhibitor recognition by the *Pseudomonas aeruginosa* quorum sensing regulator PqsR (MvfR). *PLoS Pathog* **9:** e1003508. doi:10.1371/journal.ppat.1003508.
18. **McClean KH, Winson MK, Fish L, Taylor A, Chhabra SR, Camara M, Daykin M, Lamb JH, Swift S, Bycroft BW, Stewart GS, Williams P.** 1997. Quorum sensing and *Chromobacterium violaceum*: exploitation of violacein production and inhibition for the detection of N-acylhomoserine lactones. *Microbiology* **143:**3703–3711.
19. **Muh U, Schuster M, Heim R, Singh A, Olson ER, Greenberg EP.** 2006. Novel *Pseudomonas aeruginosa* quorum-sensing inhibitors identified in an ultra-high-throughput screen. *Antimicrob Agents Chemother* **50:**3674–3679.
20. **Wu H, Song Z, Hentzer M, Andersen JB, Molin S, Givskov M, Hoiby N.** 2004. Synthetic furanones inhibit quorum-sensing and enhance bacterial clearance in *Pseudomonas aeruginosa* lung infection in mice. *J Antimicrob Chemother* **53:**1054–1061.
21. **Yang L, Rybtke MT, Jakobsen TH, Hentzer M, Bjarnsholt T, Givskov M, Tolker-Nielsen T.** 2009. Computer-aided identification of recognized drugs as *Pseudomonas aeruginosa* quorum-sensing inhibitors. *Antimicrob Agents Chemother* **53:**2432–2443.
22. **Tan SY, Chua SL, Chen Y, Rice SA, Kjelleberg S, Nielsen TE, Yang L, Givskov M.** 2013. Identification of five structurally unrelated quorum-sensing inhibitors of *Pseudomonas aeruginosa* from a natural-derivative database. *Antimicrob Agents Chemother* **57:**5629–5641.
23. **Christensen QH, Grove TL, Booker SJ, Greenberg EP.** 2013. A high-throughput screen for quorum-sensing inhibitors that target acyl-homoserine lactone synthases. *Proc Natl Acad Sci USA* **110:**13815–13820.
24. **Ozer EA, Pezzulo A, Shih DM, Chun C, Furlong C, Lusis AJ, Greenberg EP, Zabner J.** 2005. Human and murine paraoxonase 1 are host modulators of *Pseudomonas aeruginosa* quorum-sensing. *FEMS Microbiol Lett* **253:**29–37.
25. **Kravchenko VV, Kaufmann GF, Mathison JC, Scott DA, Katz AZ, Grauer DC, Lehmann M, Meijler MM, Janda KD, Ulevitch RJ.** 2008. Modulation of gene expression via disruption of NF-kappaB signaling by a bacterial small molecule. *Science* **321:**259–263.
26. **Tateda K, Ishii Y, Horikawa M, Matsumoto T, Miyairi S, Pechere JC, Standiford TJ, Ishiguro M, Yamaguchi K.** 2003. The *Pseudomonas aeruginosa* autoinducer N-3-oxododecanoyl homoserine lactone accelerates apoptosis in macrophages and neutrophils. *Infect Immun* **71:**5785–5793.
27. **Valentine CD, Zhang H, Phuan PW, Nguyen J, Verkman AS, Haggie PM.** 2013. Small molecule screen yields inhibitors of pseudomonas homoserine lactone-induced host responses. *Cell Microbiol.* [Epub ahead of print.] doi:10.1111/cmi.12176.
28. **Brencic A, McFarland KA, McManus HR, Castang S, Mogno I, Dove SL, Lory S.** 2009.

The GacS/GacA signal transduction system of *Pseudomonas aeruginosa* acts exclusively through its control over the transcription of the RsmY and RsmZ regulatory small RNAs. *Mol Microbiol* **73:**434–445.

29. **Kay E, Humair B, Denervaud V, Riedel K, Spahr S, Eberl L, Valverde C, Haas D.** 2006. Two GacA-dependent small RNAs modulate the quorum-sensing response in *Pseudomonas aeruginosa*. *J Bacteriol* **188:**6026–6033.
30. **Sonnleitner E, Abdou L, Haas D.** 2009. Small RNA as global regulator of carbon catabolite repression in *Pseudomonas aeruginosa*. *Proc Natl Acad Sci USA* **106:**21866–21871.
31. **Zhang L, Gao Q, Chen W, Qin H, Hengzhuang W, Chen Y, Yang L, Zhang G.** 2013. Regulation of pqs quorum sensing via catabolite repression control in *Pseudomonas aeruginosa*. *Microbiology* **159:**1931–1936.
32. **Oglesby AG, Farrow JM 3rd, Lee JH, Tomaras AP, Greenberg EP, Pesci EC, Vasil ML.** 2008. The influence of iron on *Pseudomonas aeruginosa* physiology: a regulatory link between iron and quorum sensing. *J Biol Chem* **283:**15558–15567.
33. **Sonnleitner E, Gonzalez N, Sorger-Domenigg T, Heeb S, Richter AS, Backofen R, Williams P, Huttenhofer A, Haas D, Blasi U.** 2011. The small RNA PhrS stimulates synthesis of the *Pseudomonas aeruginosa* quinolone signal. *Mol Microbiol* **80:**868–885.
34. **Bjarnsholt T, Jensen PO, Jakobsen TH, Phipps R, Nielsen AK, Rybtke MT, Tolker-Nielsen T, Givskov M, Hoiby N, Ciofu O.** 2010. Quorum sensing and virulence of *Pseudomonas aeruginosa* during lung infection of cystic fibrosis patients. *PLoS One* **5:**e10115. doi:10.1371/journal.pone.0010115.
35. **D'Argenio DA, Wu M, Hoffman LR, Kulasekara HD, Deziel E, Smith EE, Nguyen H, Ernst RK, Larson Freeman TJ, Spencer DH, Brittnacher M, Hayden HS, Selgrade S, Klausen M, Goodlett DR, Burns JL, Ramsey BW, Miller SI.** 2007. Growth phenotypes of *Pseudomonas aeruginosa* lasR mutants adapted to the airways of cystic fibrosis patients. *Mol Microbiol* **64:**512–533.
36. **Smith EE, Buckley DG, Wu Z, Saenphimmachak C, Hoffman LR, D'Argenio DA, Miller SI, Ramsey BW, Speert DP, Moskowitz SM, Burns JL, Kaul R, Olson MV.** 2006. Genetic adaptation by *Pseudomonas aeruginosa* to the airways of cystic fibrosis patients. *Proc Natl Acad Sci USA* **103:**8487–8492.
37. **Mellbye B, Schuster M.** 2011. The sociomicrobiology of antivirulence drug resistance: a proof of concept. *MBio* **2:**e00131-11. doi:10.1128/mBio.00131-11.
38. **Kohler T, Perron GG, Buckling A, van Delden C.** 2010. Quorum sensing inhibition selects for virulence and cooperation in *Pseudomonas aeruginosa*. *PLoS Pathog* **6:**e1000883. doi:10.1371/journal.ppat.1000883.
39. **Lee J, Wu J, Deng Y, Wang J, Wang C, Chang C, Dong Y, Williams P, Zhang LH.** 2013. A cell-cell communication signal integrates quorum sensing and stress response. *Nat Chem Biol* **9:**339–343.
40. **Sudarsan N, Lee ER, Weinberg Z, Moy RH, Kim JN, Link KH, Breaker RR.** 2008. Riboswitches in eubacteria sense the second messenger cyclic di-GMP. *Science* **321:**411–413.
41. **Christen M, Kulasekara HD, Christen B, Kulasekara BR, Hoffman LR, Miller SI.** 2010. Asymmetrical distribution of the second messenger c-di-GMP upon bacterial cell division. *Science* **328:**1295–1297.
42. **Ho CL, Chong KSJ, Oppong JA, Chuah MLC, Tan SM, Liang ZX.** 2013. Visualizing the perturbation of cellular cyclic di-GMP levels in bacterial cells. *J Am Chem Soc* **135:**566–569.
43. **Rybtke MT, Borlee BR, Murakami K, Irie Y, Hentzer M, Nielsen TE, Givskov M, Parsek MR, Tolker-Nielsen T.** 2012. Fluorescence-based reporter for gauging cyclic di-GMP levels in *Pseudomonas aeruginosa*. *Appl Environ Microbiol* **78:**5060–5069.
44. **Borlee BR, Goldman AD, Murakami K, Samudrala R, Wozniak DJ, Parsek MR.** 2010. *Pseudomonas aeruginosa* uses a cyclic-di-GMP-regulated adhesin to reinforce the biofilm extracellular matrix. *Mol Microbiol* **75:**827–842.
45. **Ching SM, Tan WJ, Chua KL, Lam Y.** 2010. Synthesis of cyclic di-nucleotidic acids as potential inhibitors targeting diguanylate cyclase. *Bioorg Med Chem* **18:**6657–6665.
46. **Zhou J, Watt S, Wang J, Nakayama S, Sayre DA, Lam YF, Lee VT, Sintim HO.** 2013. Potent suppression of c-di-GMP synthesis via I-site allosteric inhibition of diguanylate cyclases with 2'-F-c-di-GMP. *Bioorg Med Chem* **21:**4396–4404.
47. **Sambanthamoorthy K, Sloup RE, Parashar V, Smith JM, Kim EE, Semmelhack MF, Neiditch MB, Waters CM.** 2012. Identification of small molecules that antagonize diguanylate cyclase enzymes to inhibit biofilm formation. *Antimicrob Agents Chemother* **56:**5202–5211.
48. **Sambanthamoorthy K, Luo C, Pattabiraman N, Feng X, Koestler B, Waters CM, Palys TJ.** 2013. Identification of small molecules inhibiting diguanylate cyclases to control bacterial biofilm development. *Biofouling*. [Epub ahead of print.] doi:10.1080/08927014.2013.832224.

49. **Ohana P, Delmer DP, Volman G, Benziman M.** 1998. Glycosylated triterpenoid saponin: a specific inhibitor of diguanylate cyclase from *Acetobacter xylinum*. Biological activity and distribution. *Plant Cell Physiol* **39:**153–159.
50. **Antoniani D, Rossi E, Rinaldo S, Bocci P, Lolicato M, Paiardini A, Raffaelli N, Cutruzzola F, Landini P.** 2013. The immunosuppressive drug azathioprine inhibits biosynthesis of the bacterial signal molecule cyclic-di-GMP by interfering with intracellular nucleotide pool availability. *Appl Microbiol Biotechnol* **97:**7325–7336.
51. **Barraud N, Hassett DJ, Hwang SH, Rice SA, Kjelleberg S, Webb JS.** 2006. Involvement of nitric oxide in biofilm dispersal of *Pseudomonas aeruginosa*. *J Bacteriol* **188:**7344–7353.
52. **Barraud N, Schleheck D, Klebensberger J, Webb JS, Hassett DJ, Rice SA, Kjelleberg S.** 2009. Nitric oxide signaling in *Pseudomonas aeruginosa* biofilms mediates phosphodiesterase activity, decreased cyclic di-GMP levels, and enhanced dispersal. *J Bacteriol* **191:**7333–7342.
53. **Barraud N, Kardak BG, Yepuri NR, Howlin RP, Webb JS, Faust SN, Kjelleberg S, Rice SA, Kelso MJ.** 2012. Cephalosporin-3'-diazeniumdiolates: targeted NO-donor prodrugs for dispersing bacterial biofilms. *Angew Chem Int Ed Engl* **51:**9057–9060.
54. **Christensen LD, van Gennip M, Rybtke MT, Wu H, Chiang WC, Alhede M, Hoiby N, Nielsen TE, Givskov M, Tolker-Nielsen T.** 2013. Clearance of *Pseudomonas aeruginosa* foreign-body biofilm infections through reduction of the cyclic Di-GMP level in the bacteria. *Infect Immun* **81:**2705–2713.

From Biology to Drug Development: New Approaches to Combat the Threat of Fungal Biofilms

19

CHRISTOPHER G. PIERCE,[1,3] ANAND SRINIVASAN,[2,3]
ANAND K. RAMASUBRAMANIAN,[2,3] and JOSÉ L. LÓPEZ-RIBOT[1,3]

INTRODUCTION

Advances in modern medicine are prolonging the lives of severely ill individuals; however, at the same time they are creating an expanding population of compromised patients at increased risk of suffering from invasive fungal infections (1). These include surgical, transplant, cancer, intensive care unit, and HIV-infected patients, as well as neonates. The use of broad-spectrum antibiotics, parenteral nutrition, medical implant devices, and immune suppression, as well as disruption of mucosal barriers due to surgery, chemotherapy, and radiotherapy represent the most important predisposing factors for these infections. Unfortunately, the mortality rates associated with these fungal infections remain unacceptably high, which clearly points to the many limitations of current antifungal therapy, including the limited armamentarium of antifungal agents, their inherent toxicity, and the emergence of resistance (2–4). Fungi are eukaryotic organisms, and there is a paucity of selective targets which can be exploited for antifungal drug development, while at the same time this is also the main reason for

[1]Departments of Biology; [2]Biomedical Engineering; [3]South Texas Center for Emerging Infectious Diseases, The University of Texas at San Antonio, San Antonio, TX 78249.

Microbial Biofilms 2nd Edition
Edited by Mahmoud Ghannoum, Matthew Parsek, Marvin Whiteley, and Pranab K. Mukherjee

doi:10.1128/microbiolspec.MB-0007-2014

the elevated toxicity of some of the current agents (2, 3, 5). Only three classes of antifungal agents—azoles, polyenes, and echinocandins—constitute the mainstay of antifungal therapy for patients with life-threatening invasive fungal infections. Moreover, the antifungal drug pipeline is mostly dry and, with the exception of isavuconazole, no new agents are expected to reach the market any time soon (2).

The formation of biofilms by many pathogenic fungi further complicates treatment (4). Biofilms are attached and structured microbial communities surrounded by a protective exopolymeric matrix (6–8). Once formed, these biofilms can initiate or prolong infections by providing a safe haven from which cells can invade local tissue, seed new infection sites, resist eradication efforts, and also lead to failure of implanted medical devices (9, 10). For example, yeasts, and in particular *Candida* spp., are now the third most common microorganisms associated with catheter-related bloodstream infections (11). The net result is that fungal biofilm formation carries important negative clinical consequences, adversely impacting the health of an increasing number of at-risk patients, with soaring economic implications to our health care system. Clearly, there is an urgent and unmet need for the development of novel antifungal agents, in particular, those targeting fungal biofilm formation.

To efficiently treat fungal biofilms we need to understand their biology, physiology, and implications for pathogenesis. Below we briefly review our current understanding of fungal biofilm development and the mechanisms of antifungal drug resistance which are operative in biofilms. We will then discuss how this accumulated knowledge can be harnessed to develop new approaches, including biofilm-eradication strategies and development of new antibiofilm drugs, for the prevention and therapy of these difficult-to-treat infections.

FUNGAL BIOFILM DEVELOPMENT

Most information on fungal biofilm formation and on the structural characteristics of fungal biofilms comes from *in vitro* models developed during the last couple of decades (6, 7). Although early models of fungal biofilm formation were cumbersome, more recently, research on fungal biofilms has been greatly facilitated by the development of relatively simple methodologies, most notably the 96-well microtiter plate model for the formation of fungal biofilms (12, 13). *In vitro* models have the advantage that they are amenable to careful control and manipulation, because multiple parameters can be modified and examined at the same time. This has allowed for the implementation of powerful molecular techniques (i.e., genetics, genomics, and proteomics) for the dissection of pathways regulating fungal biofilm formation, along with state of the art microscopy techniques for the direct visualization and characterization of the overall structure of fungal biofilms.

It is now well established that fungal biofilm development is a complex process that occurs through multiple developmental stages (14, 15). These include initial adhesion to a substrate (which can be either a biologic or an abiotic surface), followed by cell division and proliferation, and biofilm maturation. In general, mature fungal biofilms exhibit a complex three-dimensional architecture with extensive spatial heterogeneity and are typically surrounded by a matrix of self-produced exopolymeric material that accumulates as the biofilms mature and also contributes to cohesion (16–18). This spatial arrangement likely represents an ideal adaptation that allows for the uptake of essential nutrients, the excretion of toxic products, and the communication between cells and the environment, and it is also responsible for the protection of sessile cells within the biofilms against immune defense mechanisms (8). Ultimately, cells can detach from biofilms, and this dispersal step represents the return to a planktonic mode of growth while subsequently allowing for the colonization of new surfaces at distal sites, thereby completing the biofilm lifecycle (19). Importantly, although most of this structural information on fungal biofilms has been gathered using *in vitro* models, different groups of investigators have reported similar architectural features for *in vivo*-formed biofilms, thereby corroborating the validity of these observations (20–25).

The development of fungal biofilms occurs in response to distinct environmental cues and is controlled, at the molecular level, by complex regulatory networks (6, 14, 26, 27). In the case of *Candida albicans*, the most frequent causative agent of fungal biofilm infections, its ability to form biofilms is intimately linked to filamentation, and this may also hold true for other filamentous fungi (7, 28). In addition, as in their bacterial counterparts, cell-cell communication (i.e., quorum sensing) plays a critical role during the biofilm mode of growth in fungi (29, 30).

MECHANISMS OF FUNGAL BIOFILM RESISTANCE

One of the most salient hallmarks of fungal biofilms is their extreme resistance against most conventional antifungal agents, which carries profound clinical implications. Very early studies already documented that sessile cells within these biofilms were up to 1,000 times more resistant to the fungistatic azoles (i.e., fluconazole) than their planktonic counterparts, and they also showed decreased susceptibility against the polyenes, such as amphotericin B and nystatin (16, 17). We note here that polyenes display a decent biofilm activity *in vitro*, but unfortunately, they do so at supraphysiological concentrations which are generally considered unsafe because of toxicity issues (31). Subsequent *in vitro* studies demonstrated excellent activity of therapeutic concentrations of liposomal formulations of amphotericin B and the echinocandins against *Candida* biofilms (31–33), although this new class of antifungal agents may not be universally active against biofilms formed by other fungal species, in particular, *Aspergillus fumigatus* (7, 34).

The resistance displayed by fungal biofilms is multifactorial in nature and may include molecular mechanisms of resistance that are relevant to planktonic cells but also those that are unique to the biofilm mode of growth. For three excellent recent and comprehensive reviews on this topic, including a detailed description of mechanisms of resistance, please refer to references 34–36. A potential list of contributory mechanisms to fungal biofilm resistance includes (i) the increased cellular density within the biofilm (i.e., "safety in numbers") (37), (ii) the existence of subpopulations of "persister" cells that develop tolerance to antifungal agents (38), (iii) the activation of efflux pumps, which may occur physiologically as a means to facilitate the removal of waste products but may concomitantly result in increased efflux of antifungal compounds (39), (iv) changes in the membrane sterols of sessile cells (40), (v) overall changes in the metabolic and physiological status of cells, including cellular stress responses mostly orchestrated by the Hsp90 molecular chaperone also with connections to the calcineurin pathway (41), and perhaps most notoriously, (vi) the physical blockage and overall protective effect exerted by the biofilm matrix, mostly due to drug binding by exopolymeric components and the presence of extracellular DNA (42, 43). Of note, the fact that fungal biofilm resistance is multifactorial poses a substantial challenge to the development of novel strategies: targeting just one of the multiple mechanisms may not be enough to circumvent resistance.

DEVELOPMENT OF NOVEL STRATEGIES AND THERAPEUTICS TO COMBAT ANTIFUNGAL BIOFILMS AND TO OVERCOME BIOFILM ANTIFUNGAL DRUG RESISTANCE

By now, the readers are probably appreciative of the intrinsic difficulties in treating fungal biofilm infections. The increasing understanding of fungal biofilms and the complexity of resistance mechanisms highlight many of the obstacles and the Herculean challenge ahead of us. Indeed, this is one of the urgent and unmet needs in medical mycology and arguably represents one of the top priorities in the field of antifungal drug research and development, which clearly indicates the dire medical, but also economical, need for new antifungal agents and strategies. The main question remains how we can harness these insights from fungal biofilm biology and pathogenesis, along with knowledge gained during the clinical management of patients, for the development of new strategies aimed at the control of fungal biofilms in clinical settings. When developing new drug candidates, it is imperative that the biofilm phenotype be taken into consideration, and the antifungal activity of the drug should be tested not only against planktonic cultures of a given organism but also against biofilms. Ideally, this will be accompanied by optimization of different parameters for the targeting of fungal cells within biofilms, such as mode of action, penetration, and pharmacodynamic/pharmacokinetic properties.

Obviously, prevention of biofilm formation should be considered the optimal strategy against a potential fungal biofilm infection, since once fully established these mature biofilms are much more difficult to treat and eradicate. Prophylactic regimens could represent important weapons, particularly against infections in implanted medical devices, including catheters (9, 10). These preventive approaches will target fungal cells while they remain planktonic and therefore still susceptible to therapy, or during the early stages of development by inhibiting their biofilm-forming ability. Nevertheless, some strategies can also be devised for the treatment of fully established (mature) fungal biofilms.

Below we present and discuss several of the main strategies that are the focus of current research on this topic and can be further developed, we hope in the near future, to accomplish these lofty goals.

Modified Surfaces and Antimicrobial Coatings

A wide range of biomaterials used for the fabrication of biomedical-assist devices have been shown to support adhesion, colonization, and subsequent biofilm formation by different pathogenic fungi (10). Thus, to prevent infections associated with biofilm formation on the surface of implanted devices used in clinical practice, there is an interest in the development of advanced biomaterials which do not support microbial (both fungal and bacterial) adhesion and colonization, as well as the development of antimicrobial coatings that can inhibit adhesion of fungal cells and also display direct inhibitory activity through the release of drug from the biomaterial. A recent study tested the effect of surface and serum on biofilm development by different *Candida* clinical isolates (44). The materials tested included polycarbonate, polystyrene, stainless steel, Teflon, polyvinyl chloride, and hydroxyapatite. Results indicated that a serum conditioning film increases the adherence to both metallic and nonmetallic materials and that roughness and hydrophobicity can modulate *C. albicans* biofilm formation. Another study reported that surface modifications of polyetherurethane, polycarbonateurethane, and poly(ethylene-terephthalate) influenced fungal biofilm formation *in vitro* (45).

Bachmann and colleagues described that impregnation with different concentrations of caspofungin displayed an inhibitory role on subsequent biofilm formation of *C. albicans* on polystyrene surfaces (18). More recently, a similar inhibitory effect was demonstrated for voriconazole, because surfaces previously coated with this new-generation azole (which is not active against preformed biofilms) showed reduced formation of biofilms by different *Candida* spp. (46). Thin-film coating of polymethylmethacrylate incorporating different antifungals, including nystatin, amphotericin B, and clorhexidine, effectively inhibited *C. albicans* biofilm formation as a potential preventive therapy for denture stomatitis; importantly, these novel formulations allowed for the slow release of drugs over time (47). In another study pre-coating the surface of acrylic disks with chlorhexidine or histatin 5 (a naturally occurring antimicrobial peptide in saliva), but not defensins, significantly inhibited *C. albicans* biofilm formation (48). Silver-coated catheters have been used in clinics to inhibit microbial growth, and Monteiro and colleagues recently demonstrated that silver nanoparticles exert activity against fungal biofilms (49, 50).

Although very few studies have focused exclusively on antifungal surfaces, a recent article provides a comprehensive review of biomaterial surfaces capable of resisting fungal attachment and biofilm formation, and in particular, surface coatings prepared with antifungal agents irreversibly immobilized (covalently bound to) the biomaterial surface. These agents may include cationic compounds, low molecular weight antiseptics, antimicrobial peptides, and polyenes. When tethered covalently onto a biomaterial surface, these agents are not released and do not become fouled, and in theory they should remain permanently active at high concentrations to continue to kill or inhibit microorganisms upon contact (51).

Catheter Lock Therapy

The widespread use of indwelling catheters in modern medicine, particularly central venous and hemodialysis catheters, has greatly contributed to the increasing incidence of fungal bloodstream infections, in particular candidiasis, which is the third most frequent intravascular catheter-related infection, with the overall highest crude mortality (9–11, 52). These infections may arise at any time during hospitalization, and most frequently, contamination may occur by (i) intraluminal colonization of the hub and lumen of the device, (ii) extraluminal colonization of the catheter, which typically originates from either the patient's skin microbiota or from medical personnel, (iii) hematogenous seeding from a distal site of infection, for example, from the gastrointestinal tract following cytotoxic therapy, and (iv) less commonly, from contaminated infusates (11, 52). Unfortunately, once a biofilm is formed on these catheters, removal of the medical device is often required to eliminate the infection; however, this is not always feasible due to clinical considerations (53, 54). Thus, the use of antifungal lock solutions is receiving increasing interest as a strategy to prevent and treat these infections (54, 55). Lock therapy utilizes prolonged instillation of a solution containing very high (suprapharmacological) concentrations of drugs, either antifungals or antiseptics or combinations of both, to try to sterilize the infected catheter.

Some of the most promising antifungal lock therapy strategies include the use of conventional antifungal agents (in particular, amphotericin B and the echinocandins because of their activity against preformed biofilms), antibacterial antibiotics which may display antifungal effects at high concentrations (such as minocycline, doxycycline, and rifampin), antimicrobial peptides, and antiseptics (usually ethanol) (54, 56, 57). Interestingly, *in vitro* and *in vivo* assays have demonstrated that the ethanol effect on *C. albicans* biofilm formation is also due to its biochemical inhibition of alcohol dehydrogenase (Adh1p), because the enzymatic activity of Adh1p contributes to the ability of *C. albicans* to form biofilms (58). Very often, antiocclusive agents such as heparin, citrate, or EDTA are also used in these solutions together with the antimicrobial agents (54). Despite the growing evidence for the efficacy of antifungal lock therapy from both *in vitro* and *in vivo* models of catheter-related fungal infections, to date, there is little comparative or clinical data between the different potential antifungal lock therapy strategies to permit specific recommendations for its use in the clinical management of patients (54).

High Throughput Phenotypic Screens for Identification of Small Molecule Compounds with Biofilm Inhibitory Activity

In high throughput screening (HTS), large libraries of small molecules are tested in *in vitro* assays to identify those molecules that affect a biological process of interest. These phenotypic screens can be used to investigate complex biological phenomena, such as biofilm formation by pathogenic fungi, while at the same time they yield promising leads for the development of new antifungal drugs (3). They differ significantly from molecularly targeted therapeutics in that they require no detailed structural information about the target molecule, but rather allow for the selection of bioactive compounds through sampling large numbers of small molecules present in chemical libraries or repositories. One main attribute of HTS techniques in the drug development arena is their "hunger for speed."

Recognizing the power of HTS, different groups during the last decade have used this approach for the identification of inhibitors of fungal biofilms. LaFleur and colleagues screened over 120,000 small molecule compounds present in the NIH Molecular Libraries Small Molecule Repository and identified compounds with activity against *C. albicans* biofilms, and also those which enhanced the antibiofilm activity of clotrimazole (an azole that does not display antibiofilm activity by itself). Several of the most potent clotrimazole potentiators contained a 1,3-benzothiazole scaffold (59). More recently, and using similar HTS techniques, this same group identified 2-adamantanamine (AC17) as a chemosensitizer and potentiator of another azole antifungal, miconazole (60). The Johnson group screened for inhibitors of the budded-to-hyphal-form transition in *C. albicans*; the identified inhibitors likely inhibit multiple signaling pathways involved in filamentation, many of which also regulate biofilm formation (61, 62). They subsequently reported that three of these inhibitors, buhytrinA, ETYA, and CGP-37157, were capable of inhibiting *C. albicans* biofilm formation, although the biofilm-inhibitory activity varied among different clinical isolates (62). A similar screen, but this time for inhibitors of *C. albicans* adhesion to plastic, resulted in the identification of "filastatin" (63). A more in-depth characterization of this compound indicated that besides adhesion, it also inhibits the yeast-to-hyphae transition and biofilm formation, both *in vitro* and *in vivo*, in a nematode model and in a mouse model of vulvovaginal candidiasis (63).

More recently, Wong and colleagues screened over 50,000 compounds in search for inhibitors of the *C. albicans* yeast-to-hyphae transition. Further examination of the antifungal activity of initial hits led to the discovery of SM21, which displays potent fungicidal activity and was more effective against *Candida* spp. biofilms than most current antifungals (64). *In vivo*, SM21 showed efficacy in the murine model of hematogenously disseminated candidiasis and in a mouse model of oral candidiasis. Pierce et al. performed cell-based phenotypic screens using three different chemical libraries from the National Cancer Institute's Open Chemical Repository collection and identified several compounds with inhibitory activity against *C. albicans* biofilm formation and/or filamentation (65). However, most of the identified inhibitors displayed unacceptable levels of toxicity, which may limit their potential as candidates for the development of new antifungal drugs.

Our group has recently developed a technique for the nanoscale culture of fungal biofilms on a chip format, which allows for ultra–high throughput drug screening (66, 67). The chip (termed *Ca*BChip, for *Candida albicans*biofilm chip) consists of a standard microscope glass slide containing over 1,200 individual *C. albicans* biofilms ("nano-biofilms"), each with a volume of approximately 30 nanoliters, encapsulated in a three-dimensional hydrogel matrix to simulate an environment which is close to the natural physiology (Fig. 1). Despite a close to 3,000-fold miniaturization over conventional biofilms formed on microtiter plates, these nano-biofilms display similar phenotypic characteristics, including structural features and antifungal drug resistance. This platform is robotically controlled and operates in nanoscale volume, thereby reducing manual labor, reagent costs, and assay duration. Although initially developed for *C. albicans* biofilms, these techniques can be adapted to other biofilm-forming fungal species, as well as to polymicrobial biofilms.

Drug Repurposing: in Search of Novel Antifungal and Antibiofilm Activity of "Old" Drugs

The development of an entirely new drug is a long, difficult, and extremely expensive process, estimated at 15 years of research and an investment of over US $1.5 billion. The overall success rate is extremely low and, furthermore, any new drug has to undergo a highly demanding approval process by the Food and Drug Administration (FDA). Thus, repositioning or reevaluating already FDA-approved drugs for new indications is gaining traction as an alternative path to accelerated drug development. Repurposing old drugs as antifungal agents may decrease the time and effort in bringing drugs with novel antifungal activity from the bench to the bedside (68).

An early report by the Douglas group already demonstrated the inhibitory effects

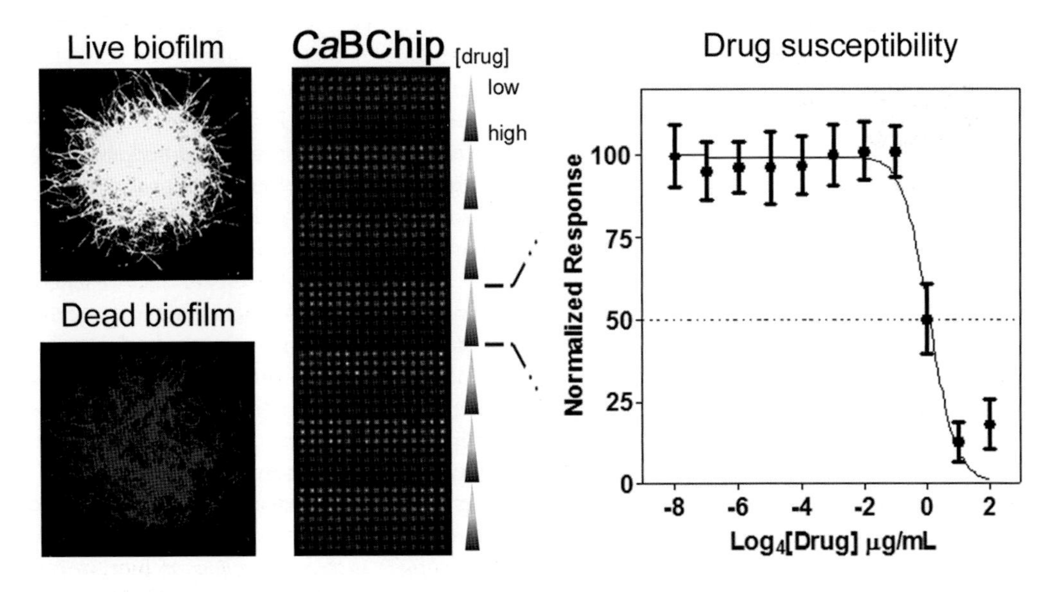

FIGURE 1 Images depicting the *Ca*BChip as a new technology platform for microbial culture at the nanoscale level. The technique allows for a variety of high-throughput applications, including ultra-high-throughput drug screening. doi:10.1128/microbiolspec.MB-0007-2014.f1

of acetylsalicylic acid (aspirin) and other nonsteroidal anti-inflammatory drugs (all cyclooxygenase inhibitors) on *C. albicans* biofilm formation (69). High aspirin concentrations were also inhibitory against biofilms produced by other non-*albicans Candida* spp. (70). More recently the nonsteroidal anti-inflammatory drug flufenamic acid, which has been specifically used for the treatment of rheumatoid arthritis, was reported to display inhibitory activity against *C. albicans* biofilms both alone and in combination with other antifungal agents (71). Another recent study reported on the screening of the Prestwick Library to identify inhibitors of *C. albicans* biofilm formation (72). This library consists of more than 1,200 FDA-approved, off-patent drugs with a wide range of functions and mechanisms of action and with well-characterized pharmacological and toxicological properties. Among the identified bioactive drugs there were several with no known or previously reported antifungal activity. Perhaps the most promising of these was auranofin, which is a gold-containing compound with anti-inflammatory properties that has been used to treat rheumatoid arthritis for over 25 years. Importantly, auranofin was also active against preformed *C. albicans* biofilms (72).

NATURAL PRODUCTS

Plants and other natural products have been used extensively in folk medicine and are generally considered good choices for a wide variety of drugs. Therefore, they constitute potential sources for new antifungal drugs. Following the development of easy and inexpensive microtiter plate-based models of fungal biofilm formation and antifungal susceptibility testing, there has been increasing interest in the examination of the antibiofilm activity of a variety of natural products. The compounds tested to date are too many to list individually, and readers are referred to a recent comprehensive review on this topic (73). An example illustrating the potential of natural products as antibiofilm agents was recently provided by Delattin and colleagues (74). This research group had previously identified a decapeptide, OSIP108, produced by the model plant *Arabidopsis thaliana* in response to infection by pathogenic fungi. They went on to demonstrate that OSIP108 inhibits *C. albicans* biofilm formation, enhances the antibiofilm activity of conventional antifungal agents, and displays no cytotoxicity against human cells (74). The mechanism of action of this decapeptide may be related to its ability to interfere with cell wall components and with the cell wall integrity pathway (74).

Another possibility is to use some of these natural products as potentiators of conventional antifungals; for example, You et al. reported that shearinines produced by *Penicillium* spp. synergistically enhanced the activity of amphotericin B against biofilms formed by *C. albicans* clinical isolates (75). Chitosan, a polymer of chitin isolated from crustacean exoskeletons, damaged *C. albicans* and *C. neoformans* biofilms, both *in vitro* and *in vivo* (76, 77).

MODULATION OF QUORUM SENSING

Fungal biofilms constitute a remarkable and highly regulated developmental program (15). When part of a multicellular consortium, cells shift their behavior when they sense that their population density has reached a threshold level, at which point they communicate though small signaling molecules (29). This process is referred to as "quorum sensing." Thus, another possibility for controlling fungal biofilms is by interrupting or modulating quorum sensing mechanisms.

In *C. albicans* farnesol and tyrosol constitute two main quorum sensing molecules that act inversely in the regulation of morphogenetic conversions which also play a very important role in biofilm development (30, 78–80). However, treatment with farnesol, which inhibits filamentation and biofilm formation *in vitro*, increased *C. albicans* virulence *in vivo* in the murine model of hematogenously disseminated candidiasis (81), perhaps due to its potential to promote dissemination to distal sites (19). Farnesol also acts as a modulator of efflux mediated by multidrug efflux pumps and synergizes with conventional antifungals (82). In addition to farnesol, culture supernatants of *C. albicans* contain a mixture of autoregulatory alcohols that include isoamyl alcohol, 2-phenylethanol, 1-dodecanol, and E-nerolidol. In *in vitro* tests these molecules, alone or in combination, are capable of inhibiting *C. albicans* morphogenetic transitions. Moreover, when injected into mice, a cocktail solution containing a mixture of these alcohols led to increased survival and reduced organ burdens (83, 84).

In nature fungi and bacteria coexist in a wide variety of environments, and polymicrobial (mixed bacterial-fungal) biofilms may be more frequent than single-species biofilms. Reciprocal influences between bacteria and fungi within these communities can have dramatic effects on survival, development, and pathogenesis. It has been demonstrated that some bacteria secrete quorum sensing and other antifungal molecules. For example, the 3-oxo-C12-acyl homoserine lactone-signaling molecule produced by *Pseudomonas aeruginosa* represses *C. albicans* morphologic transitions (85). Also, *P. aeruginosa*–produced phenazines inhibit filamentation, intercellular adherence, and biofilm development in *C. albicans* (86, 87). *Aspergillus*-bacterial biofilms are also associated with respiratory infections, particularly in the context of cystic fibrosis (7). Mixed bacterial–fungal communities are also commonly found in the oral environment, and mutanobactins secreted by *Streptococcus mutans* are capable of inhibiting or killing *C. albicans* biofilms (88, 89). A better understanding of these fungal-bacterial interactions may also serve as the basis for the development of cost-effective, low-risk strategies to control fungal infections through the use of probiotic organisms.

TARGETING THE BIOFILM MATRIX

A unique biofilm property is the production of an exopolymeric material (extracellular matrix) which encapsulates and protects fungal biofilm cells and contributes to the structural integrity of biofilms and their interaction with the surrounding environment. Two major components of this exopolymeric material, β-glucan and extracellular DNA, promote biofilm resistance to multiple antifungals. Therefore, some strategies could be devised to target the fungal biofilm matrix that could lead to increased penetration of antifungals and/or to disaggregation of biofilm cells. One such approach has been described recently involving the application of enzymes that degrade the biofilm matrix. Martins and colleagues described the presence of extracellular DNA as a component of the *C. albicans* biofilm exopolymeric material and demonstrated its role in biofilm maintenance and stability (41). Results of a subsequent series of experiments indicated improved *in vitro* efficacy of DNAse treatment of mature biofilms when used in combination with conventional antifungals. In the case of *A. fumigatus* biofilms, treatment with alginate lyase, an enzyme degrading the polysaccharides present within the biofilm matrix enhanced the activity of amphotericin B formulations, most likely by facilitating the drugs to reach the depths of the biofilm (90). Also, a series of elegant studies by the Andes and Mitchell groups have provided important insights into the pathway for matrix glucan delivery which is induced during the *C. albicans* biofilm mode of growth, including the identification of key regulators of matrix biosynthesis and the characterization of several glucan-modification enzymes, which could represent additional targets for the development of promising antibiofilm therapeutics (43, 91, 92).

ANTIBODIES

Antibodies against certain adhesins can inhibit the initial attachment of fungal cells and impede subsequent biofilm development (93). A polyclonal anti–*C. albicans* antibody produced in chicken egg yolk prevented the adherence and biofilm formation by multiple *Candida* species (94). Antibody blocking of the *C. albicans* complement receptor 3–related protein significantly reduced adhesion and subsequent biofilm formation (95). Martinez and colleagues studied the effect of combinations of conventional antifungals and specific monoclonal antibodies against *C. neoformans* biofilms; however, the antibodies displayed a mostly antagonistic effect in combination with the antifungal drugs (96, 97). Importantly, the identification of antibodies with protective effects against fungal biofilms may open new avenues for the development of adjunctive and vaccination strategies for biofilm-associated infections, an area of increasing interest.

Combination Therapy

The existence of different classes of antifungal drugs with different molecular targets opened new possibilities for combination therapy against fungal biofilms (4). However, an antagonistic trend has been described for combinations of azoles and echinocandins against biofilms, both when used concomitantly and in a sequential therapy regimen (azole, first followed by an echinocandin) (98, 99). Although biofilms formed by multiple clinical isolates of *C. albicans* displayed increased resistance to caspofungin after fluconazole pretreatment, the effect was not manifested by other non-*albicans Candida* spp. including *Candida glabrata, Candida dubliniensis*, and *Candida parapsilosis* (99).

Studies have implicated the molecular chaperone Hsp90 and the calcineurin pathway as key regulators of drug resistance in fungal biofilms (41, 100). Uppuluri et al. demonstrated that *C. albicans* cells in biofilms were resistant to individually delivered fluconazole or calcineurin inhibitors (cyclosporine, FK506-tacrolimus) but were exquisitely susceptible when the azole and the calcineurin inhibitor were used in combination (100). Moreover, combinatorial screening analysis demonstrated that different classes of antifungals that do not show activity against fungal biofilms can be made effective in synergistic combination with relatively small doses of FK506 (66). Genetic or pharmacologic (with geldanamycin) depletion of Hsp90 reduced *C. albicans* biofilm growth and maturation, reduced glucan levels in the biofilm matrix, and abrogated resistance of *C. albicans* biofilms to the azole antifungal agents both *in vitro* and *in vivo* (41). Moreover, treatment with the Hsp90 inhibitor geldanamycin enhanced the efficacy of azoles and echinocandins against *A. fumigatus* biofilms *in vitro* (41). Thus, it would seem that targeting Hsp90 may provide a potential strategy for the treatment of biofilm-associated infections caused by different fungal species.

PHOTODYNAMIC THERAPY

Photodynamic therapy (PDT) represents a relatively new therapeutic technique for the treatment of a variety of diseases, including infectious diseases. The cytotoxic effects of PDT are generally mediated by the action of reactive oxygen species generated by the activation of a photosensitizer (dye) by light. PDT has potential applications for the treatment of fungal biofilm infections, in particular, superficial mycoses such as denture stomatitis and oropharyngeal candidiasis (101–103). For example, when activated by a blue-light-emitting diode, the organic dye curcumin led to inactivation of both planktonic and sessile *C. albicans* cells (104, 105). Erythrosine activated by a green light reduced adherence to buccal epithelial cells and displayed significant antifungal effects against *C. albicans* biofilms formed *in vivo* in a murine model of oral candidiasis. Gold nanoparticle-enhanced photodynamic therapy of methylene blue demonstrated efficacy against otherwise recalcitrant biofilms formed by *C. albicans* (106).

CONCLUSIONS AND OUTLOOK

There is a desperate need for new antifungals, in particular those with antibiofilm activity. Fungal biofilms are now an area of intensive research, with hundreds of groups around the globe working in this field, including both clinical and molecular aspects of fungal biofilm infections. Insights provided by these studies should guide the development of novel strategies to combat the increasing threat of these infections and to overcome fungal biofilm resistance. This will likely require a multidisciplinary effort by clinicians and basic researchers, also with participation of governmental agencies as well as pharmaceutical and biotechnology companies. Successful development of these antibiofilm strategies should have a profound impact on the management of patients suffering from these devastating fungal infections, potentially saving the lives of thousands of patients and significantly reducing the expenses to our health care system.

ACKNOWLEDGMENTS

Biofilm work in the authors' laboratory is funded by grant number 1R01DE023510 from the National Institute of Dental & Craniofacial Research (to JLL-R), and by the Army Research Office of the Department of Defense under contract number W911NF-11-1-0136. Infectious diseases–related work in the AKR laboratory is funded by the National Institutes of Health (SC1HL112629). CGP and AS acknowledge the receipt of predoctoral fellowships from the American Heart Association, numbers 51PRE30004 and 13PRE17110093. The funders had no role in study design, data collection and analysis, decision to publish, or preparation of the manuscript, and the content is solely the responsibility of the authors.

Conflicts of interest: We disclose no conflicts.

CITATION

Pierce CG, Srinivasan A, Ramasubramanian AK, López-Ribot JL. 2015. From biology to drug development: new approaches to combat the threat of fungal biofilms. Microbiol Spectrum 3(3):MB-0007-2014.

REFERENCES

1. **Brown GD, Denning DW, Gow NA, Levitz SM, Netea MG, White TC.** 2012. Hidden killers: human fungal infections. *Sci Transl Med* **4:**165rv113.
2. **Ostrosky-Zeichner L, Casadevall A, Galgiani JN, Odds FC, Rex JH.** 2010. An insight into the antifungal pipeline: selected new molecules and beyond. *Nat Rev Drug Discov* **9:**719–727.
3. **Pierce CG, Lopez-Ribot JL.** 2013. Candidiasis drug discovery and development: new approaches targeting virulence for discovering and identifying new drugs. *Expert Opin Drug Discov* **8:**1117–1126.
4. **Pierce CG, Srinivasan A, Uppuluri P, Ramasubramanian AK, Lopez-Ribot JL.** 2013. Antifungal therapy with an emphasis on biofilms. *Curr Opin Pharmacol* **13:**726–730.
5. **Odds FC, Brown AJ, Gow NA.** 2003. Antifungal agents: mechanisms of action. *Trends Microbiol* **11:**272–279.
6. **Fanning S, Mitchell AP.** 2012. Fungal biofilms. *PLoS Pathog* **8:**e1002585. doi:10.1371/journal.ppat.1002585.
7. **Ramage G, Mowat E, Jones B, Williams C, Lopez-Ribot J.** 2009. Our current understanding of fungal biofilms. *Crit Rev Microbiol* **35:** 340–355.
8. **Ramage G, Saville SP, Thomas DP, Lopez-Ribot JL.** 2005. *Candida* biofilms: an update. *Eukaryot Cell* **4:**633–638.
9. **Kojic EM, Darouiche RO.** 2004. *Candida* infections of medical devices. *Clin Microbiol Rev* **17:**255–267.
10. **Ramage G, Martinez JP, Lopez-Ribot JL.** 2006. *Candida* biofilms on implanted biomaterials: a clinically significant problem. *FEMS Yeast Res* **6:**979–986.

11. **Crump JA, Collignon PJ.** 2000. Intravascular catheter-associated infections. *Eur J Clin Microbiol Infect Dis* **19:**1–8.
12. **Pierce CG, Uppuluri P, Tristan AR, Wormley FL Jr, Mowat E, Ramage G, Lopez-Ribot JL.** 2008. A simple and reproducible 96-well plate-based method for the formation of fungal biofilms and its application to antifungal susceptibility testing. *Nat Protoc* **3:**1494–1500.
13. **Ramage G, VandeWalle K, Wickes BL, Lopez-Ribot JL.** 2001. Standardized method for *in vitro* antifungal susceptibility testing of *Candida albicans* biofilms. *Antimicrob Agents Chemother* **45:**2475–2479.
14. **Blankenship JR, Mitchell AP.** 2006. How to build a biofilm: a fungal perspective. *Curr Opin Microbiol* **9:**588–594.
15. **Nett J, Andes D.** 2006. *Candida albicans* biofilm development, modeling a host-pathogen interaction. *Curr Opin Microbiol* **9:**340–345.
16. **Chandra J, Kuhn DM, Mukherjee PK, Hoyer LL, McCormick T, Ghannoum MA.** 2001. Biofilm formation by the fungal pathogen *Candida albicans*: development, architecture, and drug resistance. *J Bacteriol* **183:**5385–5394.
17. **Ramage G, VandeWalle K, Wickes BL, Lopez-Ribot JL.** 2001. Characteristics of biofilm formation by *Candida albicans*. *Rev Iberoam Micol* **18:**163–170.
18. **Bachmann SP, Patterson TF, Lopez-Ribot JL.** 2002. *In vitro* activity of caspofungin (MK-0991) against *Candida albicans* clinical isolates displaying different mechanisms of azole resistance. *J Clin Microbiol* **40:**2228–2230.
19. **Uppuluri P, Chaturvedi AK, Srinivasan A, Banerjee M, Ramasubramaniam AK, Kohler JR, Kadosh D, Lopez-Ribot JL.** 2010. Dispersion as an important step in the *Candida albicans* biofilm developmental cycle. *PLoS Pathog* **6:**e1000828.
20. **Andes D, Nett J, Oschel P, Albrecht R, Marchillo K, Pitula A.** 2004. Development and characterization of an *in vivo* central venous catheter *Candida albicans* biofilm model. *Infect Immun* **72:**6023–6031.
21. **Kucharikova S, Tournu H, Holtappels M, Van Dijck P, Lagrou K.** 2010. *In vivo* efficacy of anidulafungin against mature *Candida albicans* biofilms in a novel rat model of catheter-associated candidiasis. *Antimicrob Agents Chemother* **54:**4474–4475.
22. **Kucharikova S, Tournu H, Lagrou K, Van Dijck P, Bujdakova H.** 2011. Detailed comparison of *Candida albicans* and *Candida glabrata* biofilms under different conditions and their susceptibility to caspofungin and anidulafungin. *J Med Microbiol* **60:**1261–1269.
23. **Lazzell AL, Chaturvedi AK, Pierce CG, Prasad D, Uppuluri P, Lopez-Ribot JL.** 2009. Treatment and prevention of *Candida albicans* biofilms with caspofungin in a novel central venous catheter murine model of candidiasis. *J Antimicrob Chemother* **64:**567–570.
24. **Schinabeck MK, Long LA, Hossain MA, Chandra J, Mukherjee PK, Mohamed S, Ghannoum MA.** 2004. Rabbit model of *Candida albicans* biofilm infection: liposomal amphotericin B antifungal lock therapy. *Antimicrob Agents Chemother* **48:**1727–1732.
25. **Shuford JA, Rouse MS, Piper KE, Steckelberg JM, Patel R.** 2006. Evaluation of caspofungin and amphotericin B deoxycholate against *Candida albicans* biofilms in an experimental intravascular catheter infection model. *J Infect Dis* **194:**710–713.
26. **Nobile CJ, Fox EP, Nett JE, Sorrells TR, Mitrovich QM, Hernday AD, Tuch BB, Andes DR, Johnson AD.** 2012. A recently evolved transcriptional network controls biofilm development in *Candida albicans*. *Cell* **148:**126–138.
27. **Nobile CJ, Mitchell AP.** 2006. Genetics and genomics of *Candida albicans* biofilm formation. *Cell Microbiol* **8:**1382–1391.
28. **Lopez-Ribot JL.** 2005. *Candida albicans* biofilms: more than filamentation. *Curr Biol* **15:**R453–R455.
29. **Hogan DA.** 2006. Talking to themselves: autoregulation and quorum sensing in fungi. *Eukaryot Cell* **5:**613–619.
30. **Ramage G, Saville SP, Wickes BL, Lopez-Ribot JL.** 2002. Inhibition of *Candida albicans* biofilm formation by farnesol, a quorum-sensing molecule. *Appl Environ Microbiol* **68:**5459–5463.
31. **Ramage G, VandeWalle K, Bachmann SP, Wickes BL, Lopez-Ribot JL.** 2002. *In vitro* pharmacodynamic properties of three antifungal agents against preformed *Candida albicans* biofilms determined by time-kill studies. *Antimicrob Agents Chemother* **46:**3634–3636.
32. **Bachmann SP, VandeWalle K, Ramage G, Patterson TF, Wickes BL, Graybill JR, Lopez-Ribot JL.** 2002. *In vitro* activity of caspofungin against *Candida albicans* biofilms. *Antimicrob Agents Chemother* **46:**3591–3596.
33. **Kuhn DM, George T, Chandra J, Mukherjee PK, Ghannoum MA.** 2002. Antifungal susceptibility of *Candida* biofilms: unique efficacy of amphotericin B lipid formulations and echinocandins. *Antimicrob Agents Chemother* **46:**1773–1780.
34. **Ramage G, Rajendran R, Sherry L, Williams C.** 2012. Fungal biofilm resistance. *Int J Microbiol* **2012:**528521.

35. **Mathe L, Van Dijck P.** 2013. Recent insights into *Candida albicans* biofilm resistance mechanisms. *Curr Genet* **59:**251–264.
36. **Taff HT, Mitchell KF, Edward JA, Andes DR.** 2013. Mechanisms of *Candida* biofilm drug resistance. *Future Microbiol* **8:**1325–1337.
37. **Perumal P, Mekala S, Chaffin WL.** 2007. Role for cell density in antifungal drug resistance in *Candida albicans* biofilms. *Antimicrob Agents Chemother* **51:**2454–2463.
38. **LaFleur MD, Kumamoto CA, Lewis K.** 2006. *Candida albicans* biofilms produce antifungal-tolerant persister cells. *Antimicrob Agents Chemother* **50:**3839–3846.
39. **Ramage G, Bachmann S, Patterson TF, Wickes BL, Lopez-Ribot JL.** 2002. Investigation of multidrug efflux pumps in relation to fluconazole resistance in *Candida albicans* biofilms. *J Antimicrob Chemother* **49:**973–980.
40. **Mukherjee PK, Chandra J, Kuhn DM, Ghannoum MA.** 2003. Mechanism of fluconazole resistance in *Candida albicans* biofilms: phase-specific role of efflux pumps and membrane sterols. *Infect Immun* **71:**4333–4340.
41. **Robbins N, Uppuluri P, Nett J, Rajendran R, Ramage G, Lopez-Ribot JL, Andes D, Cowen LE.** 2011. Hsp90 governs dispersion and drug resistance of fungal biofilms. *PLoS Pathog* **7:** e1002257.
42. **Martins M, Uppuluri P, Thomas DP, Cleary IA, Henriques M, Lopez-Ribot JL, Oliveira R.** 2010. Presence of extracellular DNA in the *Candida albicans* biofilm matrix and its contribution to biofilms. *Mycopathologia* **169:**323–331.
43. **Nett JE, Sanchez H, Cain MT, Andes DR.** 2010. Genetic basis of *Candida* biofilm resistance due to drug-sequestering matrix glucan. *J Infect Dis* **202:**171–175.
44. **Frade JP, Arthington-Skaggs BA.** 2011. Effect of serum and surface characteristics on *Candida albicans* biofilm formation. *Mycoses* **54:** e154–e162.
45. **Chandra J, Patel JD, Li J, Zhou G, Mukherjee PK, McCormick TS, Anderson JM, Ghannoum MA.** 2005. Modification of surface properties of biomaterials influences the ability of *Candida albicans* to form biofilms. *Appl Environ Microbiol* **71:**8795–8801.
46. **Valentin A, Canton E, Peman J, Martinez JP.** 2012. Voriconazole inhibits biofilm formation in different species of the genus *Candida*. *J Antimicrob Chemother* **67:**2418–2423.
47. **Redding S, Bhatt B, Rawls HR, Siegel G, Scott K, Lopez-Ribot J.** 2009. Inhibition of *Candida albicans* biofilm formation on denture material. *Oral Surg Oral Med Oral Pathol Oral Radiol Endod* **107:**669–672.
48. **Pusateri CR, Monaco EA, Edgerton M.** 2009. Sensitivity of *Candida albicans* biofilm cells grown on denture acrylic to antifungal proteins and chlorhexidine. *Arch Oral Biol* **54:**588–594.
49. **Monteiro DR, Silva S, Negri M, Gorup LF, de Camargo ER, Oliveira R, Barbosa DB, Henriques M.** 2013. Antifungal activity of silver nanoparticles in combination with nystatin and chlorhexidine digluconate against *Candida albicans* and *Candida glabrata* biofilms. *Mycoses* **56:**672–680.
50. **Silva S, Pires P, Monteiro DR, Negri M, Gorup LF, Camargo ER, Barbosa DB, Oliveira R, Williams DW, Henriques M, Azeredo J.** 2013. The effect of silver nanoparticles and nystatin on mixed biofilms of *Candida glabrata* and *Candida albicans* on acrylic. *Med Mycol* **51:**178–184.
51. **Coad BR, Kidd SE, Ellis DH, Griesser HJ.** 2014. Biomaterials surfaces capable of resisting fungal attachment and biofilm formation. *Biotechnol Adv* **32:**296–307.
52. **Raad I.** 1998. Intravascular-catheter-related infections. *Lancet* **351:**893–898.
53. **Pappas PG, Kauffman CA, Andes D, Benjamin DK Jr, Calandra TF, Edwards JE Jr, Filler SG, Fisher JF, Kullberg BJ, Ostrosky-Zeichner L, Reboli AC, Rex JH, Walsh TJ, Sobel JD.** 2009. Clinical practice guidelines for the management of candidiasis: 2009 update by the Infectious Diseases Society of America. *Clin Infect Dis* **48:**503–535.
54. **Walraven CJ, Lee SA.** 2013. Antifungal lock therapy. *Antimicrob Agents Chemother* **57:**1–8.
55. **Cateau E, Rodier MH, Imbert C.** 2008. *In vitro* efficacies of caspofungin or micafungin catheter lock solutions on *Candida albicans* biofilm growth. *J Antimicrob Chemother* **62:**153–155.
56. **Sherertz RJ, Boger MS, Collins CA, Mason L, Raad II.** 2006. Comparative *in vitro* efficacies of various catheter lock solutions. *Antimicrob Agents Chemother* **50:**1865–1868.
57. **Sousa C, Henriques M, Oliveira R.** 2011. Mini-review: antimicrobial central venous catheters: recent advances and strategies. *Biofouling* **27:**609–620.
58. **Mukherjee PK, Mohamed S, Chandra J, Kuhn D, Liu S, Antar OS, Munyon R, Mitchell AP, Andes D, Chance MR, Rouabhia M, Ghannoum MA.** 2006. Alcohol dehydrogenase restricts the ability of the pathogen *Candida albicans* to form a biofilm on catheter surfaces through an ethanol-based mechanism. *Infect Immun* **74:**3804–3816.
59. **LaFleur MD, Lucumi E, Napper AD, Diamond SL, Lewis K.** 2011. Novel high-throughput

screen against *Candida albicans* identifies antifungal potentiators and agents effective against biofilms. *J Antimicrob Chemother* **66:**820–826.

60. **LaFleur MD, Sun L, Lister I, Keating J, Nantel A, Long L, Ghannoum M, North J, Lee RE, Coleman K, Dahl T, Lewis K.** 2013. Potentiation of azole antifungals by 2-adamantanamine. *Antimicrob Agents Chemother* **57:**3585–3592.
61. **Midkiff J, Borochoff-Porte N, White D, Johnson DI.** 2011. Small molecule inhibitors of the *Candida albicans* budded-to-hyphal transition act through multiple signaling pathways. *PLoS One* **6:**e25395. doi:10.1371/journal.pone.0025395.
62. **Grald A, Yargosz P, Case S, Shea K, Johnson DI.** 2012. Small-molecule inhibitors of biofilm formation in laboratory and clinical isolates of *Candida albicans. J Med Microbiol* **61:**109–114.
63. **Fazly A, Jain C, Dehner AC, Issi L, Lilly EA, Ali A, Cao H, Fidel PL Jr, Rao RP, Kaufman PD.** 2013. Chemical screening identifies filastatin, a small molecule inhibitor of *Candida albicans* adhesion, morphogenesis, and pathogenesis. *Proc Natl Acad Sci USA* **110:**13594–13599.
64. **Wong SS, Kao RY, Yuen KY, Wang Y, Yang D, Samaranayake LP, Seneviratne CJ.** 2014. *In vitro* and *in vivo* activity of a novel antifungal small molecule against *Candida* infections. *PLoS One* **9:**e85836. doi:10.1371/journal.pone.0085836.
65. **Pierce CG, Saville SP, Lopez-Ribot JL.** 2014. High content phenotypic screenings to identify inhibitors of *Candida albicans* biofilm formation and filamentation. *Pathog Dis* **70:**423–431.
66. **Srinivasan A, Leung KP, Lopez-Ribot JL, Ramasubramanian AK.** 2013. High-throughput nano-biofilm microarray for antifungal drug discovery. *mBio* **4:**e00331-13. doi:10.1128/mBio.00331-13.
67. **Srinivasan A, Uppuluri P, Lopez-Ribot J, Ramasubramanian AK.** 2011. Development of a high-throughput *Candida albicans* biofilm chip. *PLoS One* **6:**e19036. doi:10.1371/journal.pone.0019036.
68. **Tobinick EL.** 2009. The value of drug repositioning in the current pharmaceutical market. *Drug News Perspect* **22:**119–125.
69. **Alem MA, Douglas LJ.** 2004. Effects of aspirin and other nonsteroidal anti-inflammatory drugs on biofilms and planktonic cells of *Candida albicans. Antimicrob Agents Chemother* **48:**41–47.
70. **Stepanovic S, Vukovic D, Jesic M, Ranin L.** 2004. Influence of acetylsalicylic acid (aspirin) on biofilm production by *Candida* species. *J Chemother* **16:**134–138.
71. **Chavez-Dozal AA, Jahng M, Rane HS, Asare K, Kulkarny VV, Bernardo SM, Lee SA.** 2014. *In vitro* analysis of flufenamic acid activity against *Candida albicans* biofilms. *Int J Antimicrob Agents* **43:**86–91.
72. **Siles SA, Srinivasan A, Pierce CG, Lopez-Ribot JL, Ramasubramanian AK.** 2013. High-throughput screening of a collection of known pharmacologically active small compounds for the identification of *Candida albicans* biofilm inhibitors. *Antimicrob Agents Chemother* **57:**3681–3687.
73. **Sardi JC, Scorzoni L, Bernardi T, Fusco-Almeida AM, Mendes Giannini MJ.** 2013. *Candida* species: current epidemiology, pathogenicity, biofilm formation, natural antifungal products and new therapeutic options. *J Med Microbiol* **62:**10–24.
74. **Delattin N, De Brucker K, Craik DJ, Cheneval O, Frohlich M, Veber M, Girandon L, Davis TR, Weeks AE, Kumamoto CA, Cos P, Coenye T, De Coninck B, Cammue BP, Thevissen K.** 2014. The plant-derived decapeptide OSIP108 interferes with *Candida albicans* biofilm formation without affecting cell viability. *Antimicrob Agents Chemother* **58:**2647–2656.
75. **You J, Du L, King JB, Hall BE, Cichewicz RH.** 2013. Small-molecule suppressors of *Candida albicans* biofilm formation synergistically enhance the antifungal activity of amphotericin B against clinical *Candida* isolates. *ACS Chem Biol* **8:**840–848.
76. **Martinez LR, Mihu MR, Han G, Frases S, Cordero RJ, Casadevall A, Friedman AJ, Friedman JM, Nosanchuk JD.** 2010. The use of chitosan to damage *Cryptococcus* neoformans biofilms. *Biomaterials* **31:**669–679.
77. **Martinez LR, Mihu MR, Tar M, Cordero RJ, Han G, Friedman AJ, Friedman JM, Nosanchuk JD.** 2010. Demonstration of antibiofilm and antifungal efficacy of chitosan against candidal biofilms, using an *in vivo* central venous catheter model. *J Infect Dis* **201:**1436–1440.
78. **Alem MA, Oteef MD, Flowers TH, Douglas LJ.** 2006. Production of tyrosol by *Candida albicans* biofilms and its role in quorum sensing and biofilm development. *Eukaryot Cell* **5:**1770–1779.
79. **Chen H, Fujita M, Feng Q, Clardy J, Fink GR.** 2004. Tyrosol is a quorum-sensing molecule in *Candida albicans. Proc Natl Acad Sci USA* **101:**5048–5052.
80. **Hornby JM, Jensen EC, Lisec AD, Tasto JJ, Jahnke B, Shoemaker R, Dussault P, Nickerson KW.** 2001. Quorum sensing in

the dimorphic fungus *Candida albicans* is mediated by farnesol. *Appl Environ Microbiol* **67:**2982–2992.

81. **Navarathna DH, Hornby JM, Krishnan N, Parkhurst A, Duhamel GE, Nickerson KW.** 2007. Effect of farnesol on a mouse model of systemic candidiasis, determined by use of a DPP3 knockout mutant of *Candida albicans*. *Infect Immun* **75:**1609–1618.
82. **Sharma M, Prasad R.** 2011. The quorum-sensing molecule farnesol is a modulator of drug efflux mediated by ABC multidrug transporters and synergizes with drugs in *Candida albicans*. *Antimicrob Agents Chemother* **55:** 4834–4843.
83. **Martins M, Henriques M, Azeredo J, Rocha SM, Coimbra MA, Oliveira R.** 2007. Morphogenesis control in *Candida albicans* and *Candida dubliniensis* through signaling molecules produced by planktonic and biofilm cells. *Eukaryot Cell* **6:**2429–2436.
84. **Martins M, Lazzell AL, Lopez-Ribot JL, Henriques M, Oliveira R.** 2012. Effect of exogenous administration of *Candida albicans* autoregulatory alcohols in a murine model of hematogenously disseminated candidiasis. *J Basic Microbiol* **52:**487–491.
85. **Hogan DA, Vik A, Kolter R.** 2004. A *Pseudomonas aeruginosa* quorum-sensing molecule influences *Candida albicans* morphology. *Mol Microbiol* **54:**1212–1223.
86. **Gibson J, Sood A, Hogan DA.** 2009. *Pseudomonas aeruginosa-Candida albicans* interactions: localization and fungal toxicity of a phenazine derivative. *Appl Environ Microbiol* **75:**504–513.
87. **Morales DK, Jacobs NJ, Rajamani S, Krishnamurthy M, Cubillos-Ruiz JR, Hogan DA.** 2010. Antifungal mechanisms by which a novel *Pseudomonas aeruginosa* phenazine toxin kills *Candida albicans* in biofilms. *Mol Microbiol* **78:**1379–1392.
88. **Joyner PM, Liu J, Zhang Z, Merritt J, Qi F, Cichewicz RH.** 2010. Mutanobactin A from the human oral pathogen *Streptococcus mutans* is a cross-kingdom regulator of the yeast-mycelium transition. *Org Biomol Chem* **8:**5486–5489.
89. **Wang X, Du L, You J, King JB, Cichewicz RH.** 2012. Fungal biofilm inhibitors from a human oral microbiome-derived bacterium. *Org Biomol Chem* **10:**2044–2050.
90. **Bugli F, Posteraro B, Papi M, Torelli R, Maiorana A, Paroni Sterbini F, Posteraro P, Sanguinetti M, De Spirito M.** 2013. *In vitro* interaction between alginate lyase and amphotericin B against *Aspergillus fumigatus* biofilm determined by different methods. *Antimicrob Agents Chemother* **57:**1275–1282.
91. **Nobile CJ, Nett JE, Hernday AD, Homann OR, Deneault JS, Nantel A, Andes DR, Johnson AD, Mitchell AP.** 2009. Biofilm matrix regulation by *Candida albicans* Zap1. *PLoS Biol* **7:**e1000133. doi:10.1371/journal.pbio.1000133.
92. **Taff HT, Nett JE, Zarnowski R, Ross KM, Sanchez H, Cain MT, Hamaker J, Mitchell AP, Andes DR.** 2012. A *Candida* biofilm-induced pathway for matrix glucan delivery: implications for drug resistance. *PLoS Pathog* **8:**e1002848. doi:10.1371/journal.ppat.1002848.
93. **Lopez-Ribot JL, Casanova M, Murgui A, Martinez JP.** 2004. Antibody response to *Candida albicans* cell wall antigens. *FEMS Immunol Med Microbiol* **41:**187–196.
94. **Fujibayashi T, Nakamura M, Tominaga A, Satoh N, Kawarai T, Narisawa N, Shinozuka O, Watanabe H, Yamazaki T, Senpuku H.** 2009. Effects of IgY against *Candida albicans* and *Candida* spp. adherence and biofilm formation. *Jpn J Infect Dis* **62:**337–342.
95. **Bujdakova H, Paulovicova E, Paulovicova L, Simova Z.** 2010. Participation of the *Candida albicans* surface antigen in adhesion, the first phase of biofilm development. *FEMS Immunol Med Microbiol* **59:**485–492.
96. **Martinez LR, Casadevall A.** 2007. *Cryptococcus* neoformans biofilm formation depends on surface support and carbon source and reduces fungal cell susceptibility to heat, cold, and UV light. *Appl Environ Microbiol* **73:**4592–4601.
97. **Martinez LR, Christaki E, Casadevall A.** 2006. Specific antibody to *Cryptococcus neoformans* glucurunoxylomannan antagonizes antifungal drug action against cryptococcal biofilms *in vitro*. *J Infect Dis* **194:**261–266.
98. **Bachmann SP, Ramage G, VandeWalle K, Patterson TF, Wickes BL, Lopez-Ribot JL.** 2003. Antifungal combinations against *Candida albicans* biofilms *in vitro*. *Antimicrob Agents Chemother* **47:**3657–3659.
99. **Sarkar S, Uppuluri P, Pierce CG, Lopez-Ribot JL.** 2014. *In vitro* study of sequential fluconazole and caspofungin treatment against *Candida albicans* biofilms. *Antimicrob Agents Chemother* **58:**1183–1186.
100. **Uppuluri P, Nett J, Heitman J, Andes D.** 2008. Synergistic effect of calcineurin inhibitors and fluconazole against *Candida albicans* biofilms. *Antimicrob Agents Chemother* **52:** 1127–1132.

101. **Costa AC, Pereira CA, Freire F, Junqueira JC, Jorge AO.** 2013. Methods for obtaining reliable and reproducible results in studies of *Candida* biofilms formed *in vitro*. *Mycoses* **56:**614–622.

102. **Junqueira JC, Jorge AO, Barbosa JO, Rossoni RD, Vilela SF, Costa AC, Primo FL, Goncalves JM, Tedesco AC, Suleiman JM.** 2012. Photodynamic inactivation of biofilms formed by *Candida* spp., *Trichosporon mucoides,* and *Kodamaea ohmeri* by cationic nanoemulsion of zinc 2,9,16,23-tetrakis(phenylthio)-29H, 31H–phthalocyanine (ZnPc). *Lasers Med Sci* **27:**1205–1212.

103. **Pereira CA, Romeiro RL, Costa AC, Machado AK, Junqueira JC, Jorge AO.** 2011. Susceptibility of *Candida albicans, Staphylococcus aureus,* and *Streptococcus mutans* biofilms to photodynamic inactivation: an *in vitro* study. *Lasers Med Sci* **26:**341–348.

104. **Dovigo LN, Pavarina AC, Carmello JC, Machado AL, Brunetti IL, Bagnato VS.** 2011. Susceptibility of clinical isolates of *Candida* to photodynamic effects of curcumin. *Lasers Surg Med* **43:**927–934.

105. **Dovigo LN, Pavarina AC, Ribeiro AP, Brunetti IL, Costa CA, Jacomassi DP, Bagnato VS, Kurachi C.** 2011. Investigation of the photodynamic effects of curcumin against *Candida albicans*. *Photochem Photobiol* **87:**895–903.

106. **Khan S, Alam F, Azam A, Khan AU.** 2012. Gold nanoparticles enhance methylene blue-induced photodynamic therapy: a novel therapeutic approach to inhibit *Candida albicans* biofilm. *Int J Nanomed* **7:**3245–3257.

Index